AF595351

Structural and Functional Analysis of Extracts in Plants

Structural and Functional Analysis of Extracts in Plants

Editor

Stefania Lamponi

MDPI • Basel • Beijing • Wuhan • Barcelona • Belgrade • Manchester • Tokyo • Cluj • Tianjin

Editor
Stefania Lamponi
Departmentof Biotechnology,
Chemistry and Pharmacy
SienabioACTIVE
University of Siena
Siena
Italy

Editorial Office
MDPI
St. Alban-Anlage 66
4052 Basel, Switzerland

This is a reprint of articles from the Special Issue published online in the open access journal *Plants* (ISSN 2223-7747) (available at: www.mdpi.com/journal/plants/special_issues/structural_functional).

For citation purposes, cite each article independently as indicated on the article page online and as indicated below:

LastName, A.A.; LastName, B.B.; LastName, C.C. Article Title. *Journal Name* **Year**, *Volume Number*, Page Range.

ISBN 978-3-0365-1895-4 (Hbk)
ISBN 978-3-0365-1894-7 (PDF)

Contents

About the Editor

Stefania Lamponi

Stefania Lamponi is a Researcher and Assistant Professor of General and Inorganic Chemistry at the Department of Biotechnology, Chemistry and Pharmacy (Department of Excellence 2018-2022) of University of Siena (Italy). Moreover, she is a co-founder of the academic spin-off company SienabioACTIVE of the University of Siena. She graduated in Biological Sciences and has a PhD in Biomaterials.

Her main research interests concern the extraction, chemical characterization and evaluation of *in vitro* bioactivity, both in cellular and acellular systems, of bioactive substances from plants; the study of nanocarriers for the intracellular release of natural antioxidants and the *in vitro* analysis of cytotoxicity and genotoxicity of new drugs. Furthermore, a significant part of her research activity focuses on the *in vitro* evaluation of the biocompatibility of medical devices, new materials, modified surfaces, nanomaterials and hydrogels for biomedical applications.

Preface to "Structural and Functional Analysis of Extracts in Plants"

Today there is a growing interest in the study of classes of compounds obtained from plant species due to their specific activity, allowing their application for many selected purposes in humans. The role of their natural active principles is closely related to their chemical structure as the type and number of different classes of molecules in plant extracts influence their bioactivity and consequently their use. Moreover, agronomical and processing conditions, extraction techniques and solvents may influence the chemical profile of the herbal preparations and their pharmacological activities. For these reasons, chemical analysis should always be performed in order to correlate the type and amount of phytochemicals present with extract bioactivity so as to better select its specific applications.

Stefania Lamponi
Editor

 plants

Editorial

The importance of Structural and Functional Analysis of Extracts in Plants

Stefania Lamponi

Department of Biotechnologies, Chemistry and Pharmacy and SienabioACTIVE s.r.l., University of Siena, Via Aldo Moro 2, 53100 Siena, Italy; stefania.lamponi@unisi.it; Tel.: +39-0577-234386; Fax: +39-0577-234254

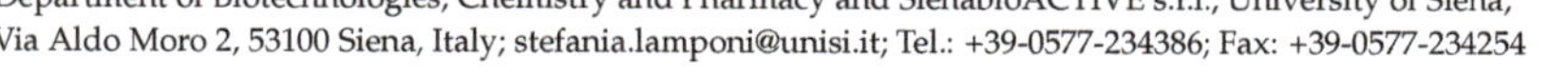

1. Introduction

Plants and their extracts have traditionally been used against various pathologies and in some regions are the only therapeutic source for the treatment and prevention of many chronic diseases [1]. Their numerous beneficial effects on human health are due to their content of natural bioactive compounds, molecules capable of modulating metabolic processes with positive properties such as antioxidant effects, inhibition of receptor activities and enzymes and induction of gene expression [2].

Many factors, such as agronomical and processing conditions used in plant cultivation, extraction methods and solvents, may influence the chemical profile of the herbal preparations and, consequently, their pharmacological activities [3–5]. For these reasons, chemical analysis should be always performed on each plant extract in order to correlate the type and the amount of phytochemicals present with its bioactivity so as to better select its specific applications [6].

Although numerous advances have been made in recent years regarding the structural and functional analysis of extracts in plants, this topic deserves more attention from the academic community; therefore, we have edited two Special Issues that include numerous research articles and reviews reflecting the advancements made thus far in this field.

 check for updates

Citation: Lamponi, S. The importance of Structural and Functional Analysis of Extracts in Plants. *Plants* **2021**, *10*, 1225. https://doi.org/10.3390/plants10061225

Received: 11 June 2021
Accepted: 15 June 2021
Published: 16 June 2021

2. Plant Extracts' Phytochemicals and Their Bioactivity

The health benefits of plant extracts mainly depend on their secondary metabolites, i.e., substances produced by plants that make them competitive in their own environment [7]. Secondary metabolites vary widely in chemical structure (types of functional groups, number and position with respect to the carbon skeleton, substitution in the aromatic ring, stereochemistry, side chain length, saturation, etc.) [8] and the most extensively studied are those with antioxidant properties that protect cellular systems from oxidative damage through a variety of mechanisms able to reduce the risk of chronic diseases such as cancer and cardiovascular disease [9].

The most important classes of secondary metabolites in plants and in their extracts are alkaloids, phytoestrogens, carotenoids, tocopherols, terpenes and phenolics.

Alkaloids are plant-derived compounds containing one or more nitrogen atoms, usually in a heterocyclic ring (amine functional group). They derive from amino acids as well as proteins, from which they differ in being alkaline [7]. Alkaloids demonstrate a large spectrum of activities and among them, there are compounds showing antibacterial, antiviral, anti-inflammatory and anticancer properties. For example, *Dicentra spectabilis*, *Corydalis lutea*, *Mahonia aquifolia*, *Fumaria officinalis*, *Meconopsis cambrica* and *Macleaya cordata* plant extracts are cytotoxic against human squamous carcinoma and adenocarcinoma cells and the extracts obtained from the stem bark of *Rutidea parviflora* against ovarian cancer [10]. The bisbenzylisoquinoline alkaloid, named curine, is able to modulate inflammatory effects in mice, by inhibiting macrophage activation, production of cytokines and neutrophil recruitment, and decreasing nitric oxide levels [11]. The antibacterial activity of alkaloids has been described for nigritanine, an alkaloid obtained from *Strychnos nigritana* belonging to the family of Loganiaceae, against *Staphylococcus aureus*, one of the most important pathogenic bacteria diffused worldwide [12]. Moreover, extracts from *Lepidium meyenii*, a plant in the Brassicaceae family rich in alkaloids, shows a strong antioxidant effect, higher than that of phenols [13]. Many alkaloids act on the nervous system. For example, poppy is narcotic, caffeine and nicotine are stimulants, while cocaine is an anesthetic, and scopolamine induces "twilight sleep". Codeine is frequently used in medical practice to suppress severe coughing [7].

Phytoestrogens are polyphenolic and non-steroid compounds that have a similar structure and biological activity as human estrogens. They are divided into two main subgroups, isoflavonoids and lignans. Isoflavones are divided into isoflavones and cumestanes, and the most representative compound of this second subgroup is coumestrol. Lignans include matairesinol, secoisolariciresinol, lariciresinol, pinoresinol and their metabolites, enterodiol, enterolactone and equol. Numerous studies reported in the literature have shown that phytoestrogens can have protective effects in estrogen-dependent diseases. This bioactivity is due to their structural and/or functional similarity to estradiol and to their capacity to bind the human estrogen receptors. Furthermore, the use of phytoestrogens can have a positive effect on insomnia and cognitive function in neuronal pathologies such as Alzheimer's disease. Phytoestrogens also exhibit antioxidant activity by acting as scavengers of free radicals or forming chelating complexes with metal ions [14,15].

Carotenoids are natural pigments and one of the main classes of phytochemicals in plants. They are derived from acyclic C_{40} isoprenoid lycopene, which can be classified as a tetraterpene [16]. Carotenoids are divided into carotenes (i.e., α-carotene, β-carotene, lycopene) and xanthophylls, which represent the oxygenated carotenoids fraction (lutein, zeaxanthin and β-cryptoxanthin). The importance of carotenoids is correlated, over their role as precursors of vitamin A (α-carotene, β-carotene and β-cryptoxanthin), with numerous bioactivities. In fact, it has been shown that carotenoids have antioxidant and antitumor activity, regulate gene function and gap-junction communication and modulate the immune response [17–19]. Carotenoids may protect light-exposed tissues from photooxidative damage which could be involved in the pathobiochemistry of several diseases affecting the skin and the eye. Lutein and zeaxanthin are the predominant carotenoids of the retina and are considered to act as photoprotectants, preventing retinal degeneration. Moreover, β-carotene is also used as an oral sun protectant for the prevention of sunburn and has been shown to be effective either alone or in combination with other carotenoids or antioxidant vitamins [20].

Tocopherols, together with carotenoids, are the most abundant group of lipid-soluble antioxidants in chloroplasts [21]. α-, β-, δ- and γ-tocopherols and α-, β-, δ- and γ-tocotrienols are different forms of vitamin E and, among them, α-tocopherol is the most predominant and active form in human tissues. Tocopherols are antioxidants, free-radical scavengers and membrane stabilizers, protect thylakoid components from oxidative damage, are involved in electron transport reactions, in the prevention of light-induced pathologies of skin and eyes and in photophosphorylation and also have hypocholestemic health benefits [22–24]. Other bioactivities of tocopherols, not correlated with their antioxidant effects, are inhibition of platelet aggregation and monocyte adhesion and anti-proliferative and neuroprotective effects [25].

Terpenes are hydrocarbons based on combinations of dimethylallyl pyrophosphate and isoprenyl diphosphate/pyrophosphate, while terpenoids (also known as isoprenoids) are terpenes with an oxygen moiety and additional structural rearrangements. Terpenoids are classified on the basis of the number of carbon atoms present in their structure in hemiterpenoids (C_5), monoterpenoids (C_{10}), homoterpenoids ($C_{11,16}$), sesquiterpenoids (C_{15}), diterpenoids (C_{20}), sesterpenoids (C_{25}), triterpenoids (C_{30}), tetraterpenoids (C_{40}) and polyterpenoids ($C_{>40}$, higher-order terpenoids) [26]. Recent studies have shown that many triterpenoids are effective and have pharmacological activities against cancer and other pathologies, such as cardiovascular diseases, diabetes and neurological disorders [27]. The pharmacological properties of triterpenoids in cancer prevention are attributed to multiple mechanisms, including antioxidant, anti-inflammatory and cell cycle regulatory properties, as well as epigenetic/epigenomic regulation.

Phenolic compounds are a class of molecules whose basic structural feature is an aromatic ring of hydroxyl groups. They include flavonoids (i.e., flavonols, flavones, flavan-3-ols, anthocyanidins, flavanones, isoflavones, condensed tannins) and nonflavonoids (i.e., phenolic acids, hydroxycinnamates, stilbenes, hydrolyzable tannins) depending on the number and arrangement of their carbon atoms. These compounds have high antioxidant activity, a protective effect against chronic pathologies such as cancer, inflammatory diseases and bacterial disorders and favorable effects in reducing the risks of coronary heart disease [28]. Much epidemiological evidence also reports their ability to reduce diabetes and human neurodegenerative pathologies, such as Parkinson's and Alzheimer's diseases. Moreover, anti-analgesic, anti-allergic, cardioprotective and anti-diabetic activities have also been documented for food phenolics [28].

3. Future Perspectives

All the studies reported in the literature demonstrated that natural plant products are an abundant source of biologically active compounds, many of which can be considered as the basis for the development of new lead chemicals for pharmaceuticals. Although numerous advances have been made in recent years in this field, further research is needed in order to translate experimental in vitro results into clinical applications. In particular, studying the in vivo mechanisms, identifying epigenetic regulatory switches, finding new analogs and increasing the bioavailability of plants metabolites, could help to identify more effective compounds to prevent and/or to treat chronic diseases. For the selection of natural compounds intended to represent the next generation of therapies based on natural formulations, and to enable them to compete with modern drugs, it is necessary to investigate different aspects of the processes necessary to obtain them, such as extraction techniques, and evaluation of the quality and of the bioactivity of crude extracts and their combinations. Moreover, new and advanced techniques for their purification and efficient animal studies along with appropriate clinical trials are required for justified use of these plant extracts with safety and efficacy.

Funding: This research received no external funding.

Acknowledgments: I would like to thank Abigail Yuan for the guidance and support throughout the entire process of these Special Issues. I also would like to thank the numerous reviewers and authors who contributed to this challenge with their science and expertise.

Conflicts of Interest: The author declares no conflict of interest.

References

1. Passero, L.F.; Laurenti, M.D.; Santos-Gomes, G.; Soares Campos, B.L.; Sartorelli, P.; Lago, J.H. Plants used in traditional medicine: Extracts and secondary metabolites exhibiting antileishmanial activity. *Curr. Clin. Pharmacol.* **2014**, *9*, 187–204. [CrossRef] [PubMed]
2. Carbonell-Capella, J.M.; Barba, F.J.; Esteve, M.J.; Frígola, A. Quality parameters, bioactive compounds and their correlation with antioxidant capacity of commercial fruit-based baby foods. *Food Sci. Technol. Int.* **2014**, *20*, 479–487. [CrossRef] [PubMed]
3. Ribeiro, A.; Caleja, C.; Barros, L.; Santos-Buelga, C.; Barreiro, M.F.; Ferreira, I.C.F.R. Rosemary extracts in functional foods: Extraction, chemical characterization and incorporation of free and microencapsulated forms in cottage cheese. *Food Funct.* **2016**, *7*, 2185–2196. [CrossRef] [PubMed]
4. Mulinacci, N.; Innocenti, M.; Bellumori, M.; Giaccherini, C.; Martini, V.; Michelozzi, M. Storage method, drying processes and extraction procedures strongly affect the phenolic fraction of rosemary leaves: An HPLC/DAD/MS study. *Talanta* **2011**, *85*, 167–176. [CrossRef] [PubMed]
5. Bicchi, C.; Binello, A.; Rubiolo, P. Determination of phenolic diterpene antioxidants in rosemary (*Rosmarinus officinalis* L.) with different methods of extraction and analysis. *Phytochem. Anal.* **2000**, *11*, 236–242. [CrossRef]
6. Lamponi, S.; Baratto, M.C.; Miraldi, E.; Baini, G.; Biagi, M. Chemical Profile, Antioxidant, Anti-Proliferative, Anticoagulant and Mutagenic Effects of a Hydroalcoholic Extract of Tuscan Rosmarinus officinalis. *Plants* **2021**, *10*, 97. [CrossRef] [PubMed]
7. Teoh, E.S. Secondary Metabolites of Plants. In *Medicinal Orchids of Asia*; Springer: Cham, Switzerland, 2016; pp. 59–73. [CrossRef]

8. Segneanu, A.E.; Velciov, S.M.; Olariu, S.; Cziple, F.; Damian, D.; Grozescu, I. Bioactive Molecules Profile from Natural Compounds. In *Amino Acid—New Insights and Roles in Plant and Animal*; Asao, T., Ed.; IntechOpen: London, UK, 2017; pp. 209–228.
9. Kris-Etherton, P.M.; Hecker, K.D.; Bonanome, A.; Coval, S.M.; Binkoski, A.E.; Hilpert, K.F.; Griel, A.E.; Etherton, T.D. Bioactive compounds in foods: Their role in the prevention of cardiovascular disease and cancer. *Am. J. Med.* **2002**, *113*, 71S–88S. [CrossRef]
10. Lelario, F.; De Maria, S.; Rivelli, A.R.; Russo, D.; Milella, L.; Bufo, S.A.; Scrano, L. A Complete Survey of Glycoalkaloids Using LC-FTICR-MS and IRMPD in a Commercial Variety and a Local Landrace of Eggplant (*Solanum melongena* L.) and their Anticholinesterase and Antioxidant Activities. *Toxins* **2019**, *11*, 230. [CrossRef]
11. Ribeiro-Filho, J.; Carvalho Leite, F.; Surrage Calheiros, A.; de Brito Carneiro, A.; Alves Azeredo, J.; Fernandes de Assis, E.; da Silva Dias, C.; Regina Piuvezam, M.T.; Bozza, P. Curine Inhibits Macrophage Activation and Neutrophil Recruitment in a Mouse Model of Lipopolysaccharide-Induced Inflammation. *Toxins* **2019**, *11*, 705. [CrossRef]
12. Casciaro, B.; Calcaterra, A.; Cappiello, F.; Mori, M.; Loffredo, M.R.; Ghirga, F.; Mangoni, M.L.; Botta, B.; Quaglio, D. Nigritanine as a New Potential Antimicrobial Alkaloid for the Treatment of Staphylococcus aureus-Induced Infections. *Toxins* **2019**, *11*, 511. [CrossRef] [PubMed]
13. Gan, J.; Feng, Y.; He, Z.; Li, X.; Zhang, H. Correlations between Antioxidant Activity and Alkaloids and Phenols of Maca (*Lepidium meyenii*). *J. Food Qual.* **2017**, *2017*, 1–10. [CrossRef]
14. Gajić, I.; Savic Gajic, I.M.; Tačić, A.; Savic, I.M. Classification and biological activity of phytoestrogens: A review. *Adv. Technol.* **2017**, *6*, 96–106.
15. Martins, S.; Cristóbal, N.; Aguilar, J.A.; Teixeira, S.I. Mussatto, Bioactive compounds (phytoestrogens) recovery from Larrea tridentata leaves by solvents extraction. *Sep. Purif. Technol.* **2012**, *88*, 163–167. [CrossRef]
16. Arvayo-Enríquez, H.; Mondaca-Fernández, I.; Gortárez-Moroyoqui, P.; López-Cervantes, J.; Rodríguez-Ramírez, R. Carotenoids extraction and quantification: A review. *Anal. Methods* **2013**, *5*, 2916–2924. [CrossRef]
17. Adadi, P.; Barakova, N.V.; Krivoshapkina, E.F. Selected Methods of Extracting Carotenoids, Characterization, and Health Concerns: A Review. *J. Agric. Food Chem.* **2018**, *66*, 5925–5947. [CrossRef]
18. Rao, A.V.; Rao, L.G. Carotenoids and human health. *Pharmacol. Res.* **2007**, *55*, 207–216. [CrossRef]
19. Paradiso, V.M.; Castellino, M.; Renna, M.; Santamaria, P.; Caponio, F. Setup of an Extraction Method for the Analysis of Carotenoids in Microgreens. *Foods* **2020**, *9*, 459. [CrossRef] [PubMed]
20. Stahl, W.; Sies, H. Bioactivity and protective effects of natural carotenoids. *Biochim. Biophys. Acta* **2005**, *1740*, 101–107. [CrossRef] [PubMed]
21. Della Penna, D.; Pogson, B.J. Vitamin synthesis in plants: Tocopherols and carotenoids. *Annu. Rev. Plant. Biol.* **2006**, *57*, 711–738. [CrossRef] [PubMed]
22. Fryer, M.J. The antioxidant effects of thylakoid vitamin E (α-tocopherol). *Plant. Cell Environ.* **1992**, *15*, 381–392. [CrossRef]
23. Giasuddin, A.S.M.; Diplock, A.T. The influence of vitamin E in membrane lipids of mouse fibroblasts in culture. *Arch. Biochem. Biophy.* **1981**, *210*, 348–362. [CrossRef]
24. Qureshi, A.A.; Bradlow, B.A.; Brace, L.; Manganello, J.; Peterson, D.M.; Pearce, B.C.; Wright, J.J.K.; Grapor, A.; Elson, C.E. Response of hypercholesterolemic subjects to administration of tocotrienols. *Lipids* **1995**, *30*, 1171–1177. [CrossRef] [PubMed]
25. Rimbach, G.; Minihane, A.M.; Majewicz, J.; Fischer, A.; Pallauf, J.; Virgli, F.; Weinberg, P.D. Regulation of cell signaling by vitamin E. *Proc. Nutr. Soc.* **2002**, *61*, 415–425. [CrossRef]
26. Boncan, D.A.T.; Tsang, S.S.K.; Li, C.; Lee, I.H.T.; Lam, H.-M.; Chan, T.F.; Hui, J.H.L. Terpenes and Terpenoids in Plants: Interactions with Environment and Insects. *Int. J. Mol. Sci.* **2020**, *21*, 7382. [CrossRef]
27. Li, S.; Kuo, H.D.; Yin, R.; Wu, R.; Liu, X.; Wang, L.; Hudlikar, R.; Peter, R.M.; Kong, A.N. Epigenetics/epigenomics of triterpenoids in cancer prevention and in health. *Biochem. Pharmacol.* **2020**, *1751*, 13890. [CrossRef]
28. Shahidi, F.; Yeo, J. Bioactivities of Phenolics by Focusing on Suppression of Chronic Diseases: A Review. *Int. J. Mol. Sci.* **2018**, *19*, 1573. [CrossRef] [PubMed]

Article

Chemical Profile, Antioxidant, Anti-Proliferative, Anticoagulant and Mutagenic Effects of a Hydroalcoholic Extract of Tuscan *Rosmarinus officinalis*

Stefania Lamponi [1,*], Maria Camilla Baratto [1], Elisabetta Miraldi [2], Giulia Baini [2] and Marco Biagi [2]

1 Department of Biotechnologies, Chemistry and Pharmacy, University of Siena, Via Aldo Moro 2, 53100 Siena, Italy; mariacamilla.baratto@unisi.it

2 Department of Physical Sciences, Earth and Environment, University of Siena, Strada Laterina 8, 53100 Siena, Italy; elisabetta.miraldi@unisi.it (E.M.); baini3@student.unisi.it (G.B.); marco.biagi@unisi.it (M.B.)

* Correspondence: lamponi@unisi.it; Tel.: +39-0577-234386; Fax: +39-0577-234254

Abstract: This study aimed to characterize the chemical profile of an ethanolic extract of Tuscan *Rosmarinus officinalis* (*Ro*ex) and to determine its in vitro bioactivity. The content of phenolic and flavonoid compounds, hydroxycinnamic acids and triterpenoids was determined, and high-performance liquid chromatography-diode array detection (HPLC-DAD) analysis revealed that rosmarinic acid and other hydroxycinnamic derivatives were the main constituents of the extract. *Ro*ex demonstrated to have both antioxidant activity and the capability to scavenge hydrogen peroxide in a concentration dependent manner. Moreover, NIH3T3 mouse fibroblasts and human breast adenocarcinoma cells MDA-MB-231 viability was influenced by the extract with an IC_{50} of 2.4×10^{-1} mg/mL and 4.8×10^{-1} mg/mL, respectively. The addition of *Ro*ex to the culture medium of both the above cell lines, resulted also in the reduction of cell death after H_2O_2 pretreatment. The Ames test demonstrated that *Ro*ex was not genotoxic towards both TA98 and TA100 strains, with and without S9 metabolic activation. The extract, by inactivating thrombin, showed to also have an anti-coagulating effect at low concentration values. All these biological activities exerted by *Ro*ex are tightly correlated to its phytochemical profile, rich in bioactive compounds.

Keywords: antioxidant activity; antiproliferative activity; mutagenicity; anticoagulant activity; *Rosmarinus officinalis*; hydroalcoholic extract

Citation: Lamponi, S.; Baratto, M.C.; Miraldi, E.; Baini, G.; Biagi, M. Chemical Profile, Antioxidant, Anti-Proliferative, Anticoagulant and Mutagenic Effects of a Hydroalcoholic Extract of Tuscan *Rosmarinus officinalis*. *Plants* **2021**, *10*, 97. https://doi.org/10.3390/plants10010097

Received: 14 December 2020
Accepted: 31 December 2020
Published: 6 January 2021

1. Introduction

Plants have been widely used all over the world for their numerous properties throughout the millennia, and today there is a growing interest in the study of classes of compounds obtained from plant species due to their specific activity, allowing their application for many selected purposes in humans [1]. The role of their natural active principles in the human organism is closely related to their chemical structure (functional group types, number and position related to carbon skeleton, substitution in aromatic ring, stereochemistry, side chain length, saturation, etc.) as the type and number of different classes of molecules in plant extracts influence their bioactivity and consequently their use [2]. For example, terpenoids show antimicrobial, antiviral, antibacterial, anticancer, antimalarial, anti-inflammatory effects; phenolics acids have anticarcinogenic and antimutagenic, anti-inflammation and anti-allergic activities; alkaloids have antispasmodic, antimalarial, analgesic and diuretic activities; flavonoids possess antioxidant, anti-inflammatory, antiviral, antibacterial, antifungal, activities, are cardiovascular and hepato-protective; saponins are antitumor, anti-inflammatory, immunostimulant, anti-hypoglycemic, antihepatotoxic and hepatoprotective, anticoagulant, neuroprotective, antioxidant; tannins are antioxidant, anti-carcinogenic, diuretics, hemostatic, anti-mutagenic, metal ion-chelators, antiseptic [2]

Rosmarinus officinalis L. is an aromatic, evergreen plant belonging to the family Lamiaceae, native of Mediterranean regions, where it grows wild, but now widely distributed all over the world. The dark green, needle-like leaves of the plant are usually used as spice for flavoring in food cooking, but rosemary is cultivated not only for its aromatic properties but mainly for its antioxidant activity [3–6]. Rosemary, in the form of extract derived from the leaves, contains several compounds which have been demonstrated to have antioxidant effects [7]. These compounds belong mainly to the classes of phenolic acids, flavonoids, diterpenes and triterpenes [8].

Thanks to their chemical composition, rosemary extracts are used in food and in cosmetic industries for preventing food deterioration and protecting skin from free radical damage, respectively [9,10].

Butylated hydroxy anisole (BHA) and butylated hydroxy toluene (BHT) are the commonly used synthetic antioxidants added to foods to preserve the lipid components from quality deterioration, and to cosmetics to inhibit oxidation or reactions promoted by oxygen, peroxides or free radicals. Their synthetic quality, however, will not produce the same health benefits as natural antioxidants [11].

R. officinalis is also known to be employed in traditional and complementary alternative medicine in many countries, thanks to its broad range of beneficial health properties such as anti-inflammatory [4], anti-proliferative [12], antibacterial [13], antithrombotic [14], anticancer [15,16], hepatoprotective [17], antidiabetic [18], hypocolesterolemic [19] and antihypertensive [20].

It is well known that the antioxidant and biological properties of rosemary extracts are mainly due to their phytochemical composition which includes phenolcarboxylic acids (rosmarinic acid, caffeic acid, vanillic acid, quinic acid, syringic acid) as the major chemical constituents [21]. Other compounds present are: flavonoids [21,22], diterpenes, [21,23], and triterpenes [21]. Considering the co-presence of all the above classes of chemical compounds in rosemary extracts [6,24], their metabolization [25–27] and the bioavailability of their metabolites [28], the bioactivity of rosemary cannot be attributed to a single class of compounds but rather to the synergistic contribution of the various bioactive components and their metabolites (that is to its phytocomplex).

Moreover, agronomical and processing conditions, used in plant cultivation and extraction respectively, may influence the chemical profile of the herbal preparations and, consequently, their pharmacological activities [29–31]. For these reasons, chemical analysis should be always performed in order to correlate the type and amount of phytochemicals present with extract bioactivity so as to better select its specific applications.

This study aimed to analyze the chemical profile of an ethanolic extract obtained from an Italian rosemary collected in Tuscany and to determine its in vitro antioxidant activity, both in non-cellular and cellular systems, anti-proliferative effect towards NIH3T3 mouse fibroblasts and human breast adenocarcinoma cell line MDA-MB231, mutagenicity and interference with human blood coagulation factor, in order to correlate chemical profile of the extract with its wide range of in vitro bioactivity and to prevent unwanted effects when used in humans.

2. Results

2.1. Chemical Analysis of Rosmarinus Officinalis Extract (Roex)

Extraction Yield and Chemical Characterization of Tuscan *R. officinalis* Extract

The solvent utilized for the production of *Ro*ex was a hydro-alcoholic mixture with 60% v/v ethanol. As reported in Table 1, the amount of dry extract obtained from each gram of dried rosemary subjected to extraction was 112 mg, and 1 mL of suspension contained 24.3 mg of dry material corresponding to a percentage yield of 11.2%.

In order to quantify total polyphenols and total flavonoids, the extract was analyzed by means of Folin–Ciocalteu colorimetric assay and direct spectrophotometry, respectively. Total sulfuric-vanillin assay-reactive triterpenoids were also quantified. Rosmarinic acid content was evaluated by means of high-performance liquid chromatography-diode array detection (HPLC-DAD) analysis, that was also used in order to identify other phenolic compounds in *Ro*ex.

Table 1. Extraction yield, total phenolic, flavonoids and triterpenoids content. Results are expressed as mean value ± standard deviation (SD) from three different preparations.

Dry Extract for mL of Suspension (mg/mL ± SD)	Extraction Yield (% ± SD)	Total Phenolic Content (mg GAE/g d.e. ± SD)	Total Flavonoids Content (mg Hyperoside/g d.e. ± SD)	Total Triterpenoids Content (mg β-sitosterol/g d.e. ± SD)
24.3 ± 1.2	11.2 ± 0.5	95.8 ± 1.1	6.5 ± 0.7	71.7 ± 7.2

The quantifications of total phenolic content, expressed as gallic acid equivalents, of total flavonoids expressed as hyperoside, and of total triterpenoids, expressed as β-sitosterol, were reported in Table 1. The amount of each class of compounds was expressed as mg per gram of dry extract (d.e.).

In order to identify its major components, the extract was analyzed by HPLC-DAD which recorded three main constituents at the following retention time (RT): 6.40 min, 6.95 min and 16.20 min. (Figure 1). By comparing RT and UV spectra of reference standards, it was possible to assign the peak at 16.20 min to rosmarinic acid (Figure 1), while the peak at 6.40 and 6.95 min were assigned to two other hydroxycinnamic derivatives, identified by characteristic UV spectra with λmax at 198–200, 288–298 and 322–332 nm (see Supplementary Materials). Other hydroxycinnamic derivatives identified in *Ro*ex were chlorogenic and caffeic acid, at RT = 7.80 and 10.00 min, respectively. The content of rosmarinic acid, chlorogenic acid, caffeic acid and total hydroxycinnamic derivatives (as the sum of rosmarinic, chlorogenic and caffeic acid and undefined hydroxycinnamic derivatives), expressed as rosmarinic acid, are reported in Table 2. Beside hydroxycinnamic acids, the chromatogram showed two main flavonoids at RT = 18.58 min and 20.55 min, identified by the characteristic UV profile of this class of metabolites with λmax at 194–200, the highest absorbance at 265–290 and a wide shoulder at 330–350 nm (see Supplementary Materials). These flavonoids could not be unambiguously identified, since RT and UV spectrum did not match any used reference standard. Nevertheless, by comparison with published literature on rosemary flavonoids [22,32] and polarity order of identified compounds, the main flavonoids in *Ro*ex could be likely related to luteolin glycosides.

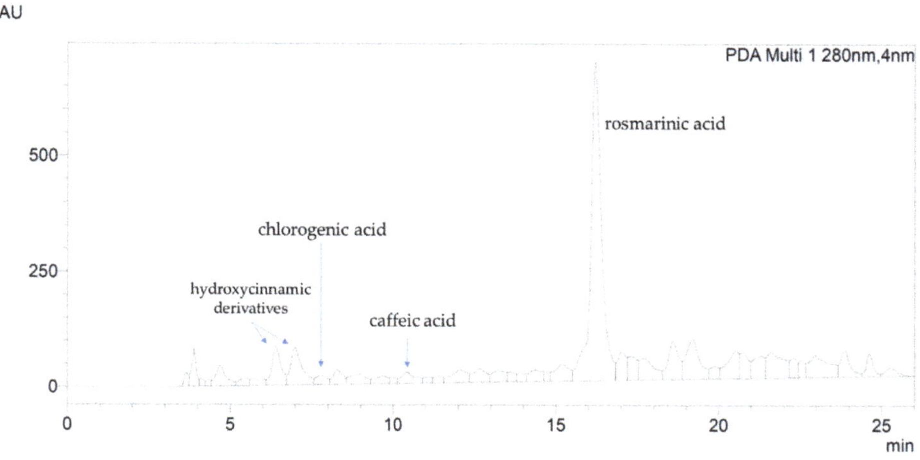

Figure 1. HPLC-DAD analysis of *Ro*ex recorded at 280 nm. Rosmarinic acid is the phenolic compound of the extract with the highest content; other hydroxycinnamic derivatives are present in high concentration.

Table 2. Content of rosmarinic, chlorogenic and caffeic acid, and total hydroxycinnamic derivatives, expressed as mg of rosmarinic acid/g of d.e.

Compound	mg/g d.e.
Rosmarinic acid	46.3 ± 5.0
Chlorogenic acid	2.0 ± 0.1
Caffeic acid	0.7 ± 0.1
Total hydroxycinnamic derivatives (expressed as rosmarinic acid)	69.4 ± 5.2

2.2. Antiradical Activity: DPPH Assay and EPR Analysis

The antioxidant behavior of the *Ro*ex was evaluated by its ability to scavenge the free DPPH radical by electron donation, by both spectrophotometric and EPR analysis. The results obtained by spectrophotometric measurements, demonstrated that the extract showed a great antiradical activity with the IC_{50} value around 50 µg/mL.

By EPR analysis, as shown in Figure 2, the addition of the antioxidant to the DPPH radical solution determined a reduction in intensity of the signal with a scavenger percentage equal to 99%. *Ro*ex showed a gallic acid-equivalent antioxidant activity towards DPPH radical equal to 0.777 ± 0.013 (standard deviation).

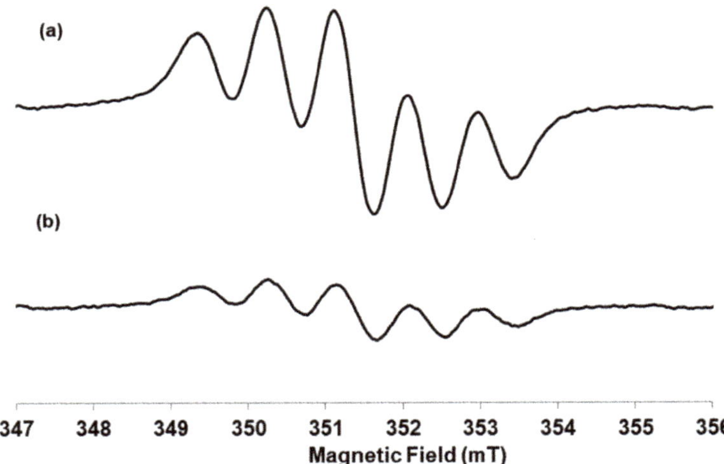

Figure 2. Room temperature X-band spectra of (**a**) DPPH radical alone, (**b**) DPPH radical after the addition of *Rosmarinus officinalis* extract. *Experimental conditions:* The spectra were recorded at microwave frequency ν = 9.86 GHz, microwave power 2 mW and 0.1 mT modulation amplitude.

2.3. Hydrogen Peroxide Scavenging Activity

As shown in Table 3, the extract had the capability to scavenge hydrogen peroxide in a concentration dependent manner (0.024–0.96 mg/mL) with an EC_{50} (half maximum effective concentration) of 0.48 mg/mL corresponding at a phenolic content of about 4.6×10^{-2} mg/mL, hydroxycinnamic acids content of 2.2×10^{-2} mg/mL, flavonoids 3.1×10^{-3} mg/mL and triterpenoids 3.4×10^{-2} mg/mL.

Table 3. Total phenolic content and percentage of scavenged hydrogen peroxide as a function of increasing concentrations of *Ro*ex.

Concentration of d.e. (mg/mL)	% Scavenged $H_2O_2 \pm$ SD	Total Phenolic Content (mg/mL)	Total Hydroxycinnamic Acids Content (mg/mL)	Total Flavonoid Content (mg/mL)	Total Triterpenoids (mg/mL)
0.024	15 ± 3	2.3×10^{-3}	1.1×10^{-3}	1.6×10^{-4}	1.7×10^{-3}
0.12	28 ± 4	1.2×10^{-2}	5.6×10^{-3}	7.8×10^{-4}	8.6×10^{-3}
0.24	41 ± 4	2.3×10^{-2}	1.1×10^{-2}	1.6×10^{-3}	1.7×10^{-2}
0.48	53 ± 6	4.6×10^{-2}	2.2×10^{-2}	3.1×10^{-3}	3.4×10^{-2}
0.72	66 ± 3	6.9×10^{-2}	3.3×10^{-2}	4.7×10^{-3}	5.2×10^{-2}
0.96	73 ± 5	9.2×10^{-2}	4.4×10^{-2}	6.2×10^{-3}	6.9×10^{-2}

2.4. In Vitro Anti-Proliferative Activity

Non-confluent adherent mouse fibroblasts NIH3T3 and human breast adenocarcinoma cells MDA-MB-231 were incubated with different concentrations of *Ro*ex diluted 1:5 with 60% ethanol. Cells were analyzed after 24 h of contact with the test samples and the results are reported in Figure 3. As shown, *Ro*ex exerted an anti-proliferative effect against both NIH3T3 and MDA-MB-231 cell lines in a concentration dependent manner, with lesser efficiency on breast adenocarcinoma cells compared to fibroblasts.

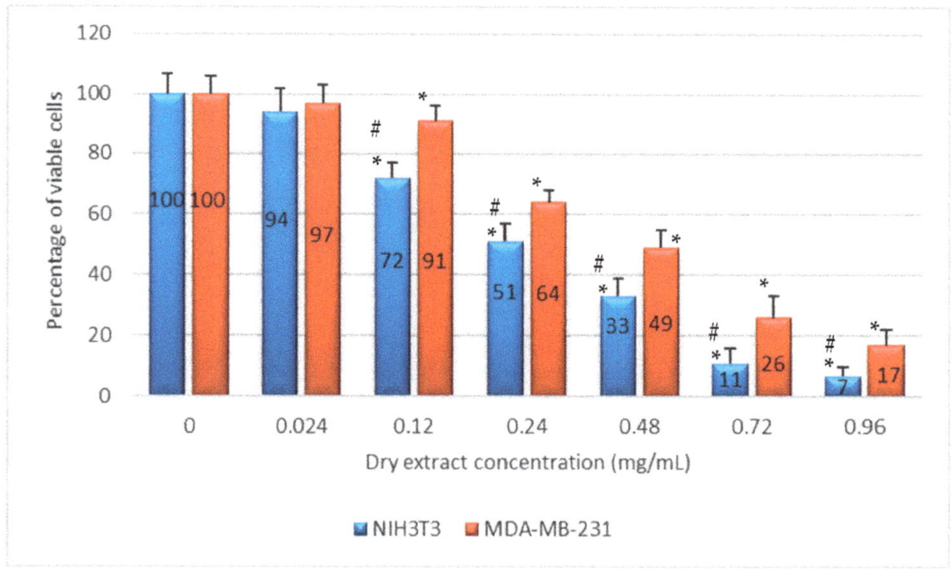

Figure 3. Percentage of viable NIH3T3 and MDA-MB-231 after 24 h of contact with increasing concentration of *Ro*ex dry extracts as determined by the Neutral Red Uptake. Date are mean ± SD of six replicates. * Values are statistically different versus negative control (complete medium), $p < 0.05$. # Values are statistically differently versus percentage of viable MDA-MB-231, $p < 0.05$.

2.5. Protective Effect against Hydrogen Peroxide Induced Oxidative Stress

Concentration-effect relationship for the cytotoxic action of H_2O_2 towards NIH3T3 and MDA-MB-231 cells were obtained by testing concentration values of the hydrogen peroxide ranging from 0.082 to 1.60 mM. Cytotoxicity was determined as decreasing of cell viability after H_2O_2 pre-treatment followed and not by contact with 7.20×10^{-3} mg/mL

*Ro*ex, the concentration value at which the extract showed the lowest degree of cytotoxicity for both cell lines.

As shown in Figure 4a, *Ro*ex was able to increase significantly NIH3T3 cell viability after pre-treatment with H_2O_2 concentrations ranging from 0.33 and 1.3 mM but not at highest values, i.e., 1.5 and 1.6 mM.

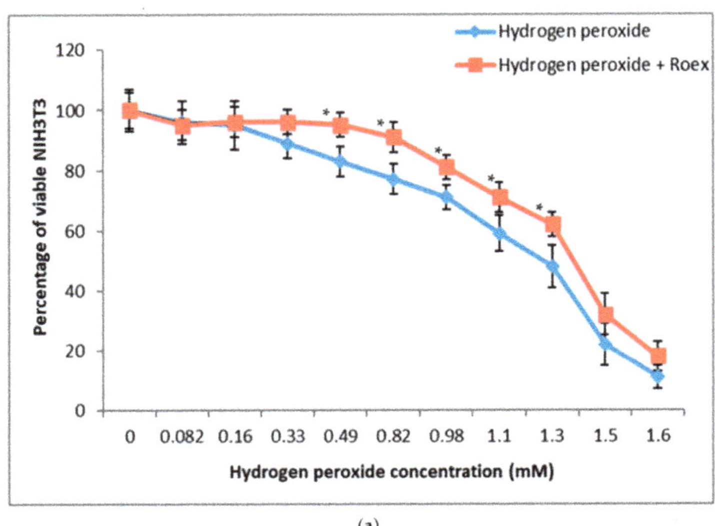

(**a**)

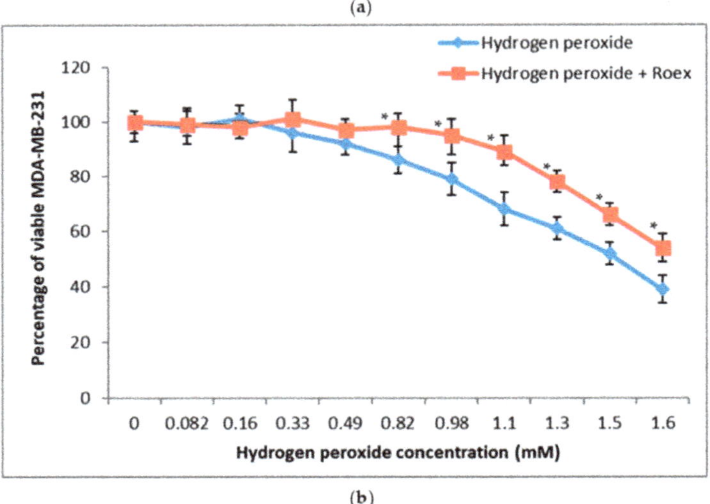

(**b**)

Figure 4. Influence of hydrogen peroxide pre-treatment on viability of (**a**) NIH3T3 and (**b**) MDA-MB-231. * Values are statistically different versus H_2O_2 treated cells, $p < 0.05$.

The addition of *Ro*ex increased the percentage of viable MDA-MB-231 after pre-treatment with hydrogen peroxide in a range of concentrations between 0.82 and 1.6 mM (Figure 4b). Human adenocarcinoma breast cells demonstrated to be more resistant to H_2O_2 pre-treatment in comparison to NIH3T3 fibroblasts and the rosmarinic extract improved cell viability neutralizing the toxic effect of H_2O_2 also at the highest concentrations, demonstrating to be able to protect cells and to reduce cell death.

2.6. Mutagenicity Assay: Ames Test

In *Salmonella* mutagenicity assay, six different concentrations of *Ro*ex were tested by Ames test on TA98, and TA100 strains with and without S9 metabolic activation. The results for the mutagenic effect of *Ro*ex reported in Table 4 demonstrated that all the concentrations tested were not genotoxic towards both TA98 and TA100 with and without S9 fraction. In fact, also at the highest concentration (24 mg/mL), the number of revertants was lower and statistically different in comparison to positive control ($p < 0.01$). The background level, as well as positive control values, were in all cases within the normal limit found in our laboratory and in accordance with literature data [33].

Table 4. Mean number of revertants in *S. typhimurium* strains 98 and 100, exposed to different concentrations of *Ro*ex with and without S9 fraction. The results are reported as the mean of revertants ± SD.

	***Salmonella typhimurium* TA 98**		***Salmonella typhimurium* TA 100**	
	Revertant (Mean ± SD)		**Revertant (Mean ± SD)**	
***Ro*ex d.e. Concentration (mg/mL)**	**−S9**	**+S9**	**−S9**	**+S9**
0	1.33 ± 0.52	1.67 ± 1.21	1.33 ± 1.37	1.33 ± 1.21
2.4×10^{-2}	1.67 ± 1.53	1.33 ± 0.58	1.33 ± 0.58	1.33 ± 1.53
1.2×10^{-1}	1.67 ± 1.15	2.33 ± 0.58	1.00 ± 0.00	2.00 ± 0.00
2.4×10^{-1}	2.00 ± 1.73	3.00 ± 1.00	1.33 ± 1.53	2.33 ± 2.08
4.8×10^{-1}	2.33 ± 1.53	333 ± 1.15	1.67 ± 2.08	1.67 ± 0.58
7.2×10^{-1}	2.67 ± 1.15	1.33 ± 1.15.	2.67 ± 1.15	2.33 ± 1.53
24	1.33 ± 1.53	2.33 ± 0.58	2.33 ± 1.15	2.33 ± 0.58
Positive Control	31.33 ± 3.06	34.00 ± 2.00	47.33 ± 1.15	47.67 ± 0.58
Baseline	1.85	2.88	2.54	2.70

2.7. Antithrombotic Activity: Thrombin Time (TT)

By looking at the TT values reported in Table 5, it was possible to note that all the five *Ro*ex tested concentrations significantly increased clotting time in comparison to the control, demonstrating the extract to have anti-coagulating effect also at the lowest concentration value.

Table 5. Thrombin Time of different concentration of *Ro*ex. Data are mean ± SD of six replicates. (*) Value statistical different versus control (human plasma diluted 1:1 with 60% *v*/*v* Et-OH solution).

	***Ro*ex d.e. Concentration (mg/mL)**						
	0	**2.4×10^{-2}**	**1.2×10^{-1}**	**2.4×10^{-1}**	**4.8×10^{-1}**	**7.2×10^{-1}**	**9.6×10^{-1}**
Thrombin Time (second ± SD)	18 ± 2	>120″ (*)	>120″ (*)	>120″ (*)	>120″ (*)	>120″ (*)	>120″ (*)

3. Discussion

Herbal extracts bioactivity depends on their chemical profile and the choice of the solvent is crucial in extraction processes as it directly affects the chemical composition of the final extracts and the mass extraction yield. Among the alternative solvents available for plants extraction, hydro-alcoholic mixtures are good candidates since they are rather few selective and let to extract a wide range of compounds [34–37]. Considering global extraction yields of *Rosmarinus officinalis*, a ratio ranging from 50% to 80% ethanol gives highest mass extraction yields and total content in target compounds in comparison to lower Et-OH values. For this reason, the solvent that we utilized for the production of *Ro*ex was hydro-alcoholic mixture with 60% ethanol which yielded 112 mg of dry extract from each gram of rosemary subjected to extraction, corresponding to a percentage yield of 11.2%. This lower percentage yield obtained in comparison to data reported by Jacotet-Navarro et al. [38] can be probably due to the environmental conditions as *Rosmarinus officinalis* is a sensitive plant to pedoclimatic variations that affect its chemical composition [30].

The *Ro*ex, showed a high total phenolic content, of about 95.8 mg/g dry extract, equivalent to 8.55 mg/g dry herbal material ca., in accordance with other extracts obtained from Mediterranean rosemary [21]. As previously reported for other rosemary [22,39], also in *Ro*ex the main phenolic subclass was identified in hydroxycinnamic acid derivatives, and rosmarinic acid was the main single compound, counting almost the half of total phenolic content. Still in accordance with other published papers, also in *Ro*ex total flavonoid content was low, about 6.5 mg/g. This analysis of the phenolic composition of the Italian rosemary extract, performed by means of HPLC-DAD technique, represents a good characterization of its phenolic fingerprint and demonstrated the range of molecules contributing to the definition of this matrix, and may assist in the study of its in vitro bioactive properties [8]. Surprisingly, we found also a high content in triterpenoids in *Ro*ex, 71.7 mg/g extract. Factors such as plant age, climate and stress conditions that inhibit or enhance the production of certain compounds might affect the chemical composition of the extracts [40]. These factors influence the metabolites qualitatively and quantitatively and may explain why certain compounds such as quercetin, rutin and other quercetin glycosides were not detectable in this study [40].

The good in vitro antioxidant activity in cell free assays showed by *Ro*ex, demonstrated both by its ability to scavenge the free DPPH radical via electron donation and by the capability to scavenge hydrogen peroxide in a concentration dependent manner (Figure 2), is very similar to others typical vegetal extracts with a good antiradical power [41], and is a consequence of its phytocompounds composition. Hydroxycinnamic acids, presents in high concentration in *Ro*ex, are potent antioxidants and, among them, rosmarinic acid and caffeic acid have a good dose-dependent DPPH scavenging activity and may contribute to the antiradical activity of the extract [42], together with rosmarinic acid. Moreover, also flavonoids and their metabolites are able to scavenge radicals and participate in antioxidant reactions as well as triterpenoids and total phenolic compounds [42].

A number of studies have demonstrated a good correlation of intrinsic antioxidant activity in a non-cellular assay with cytoprotection against an oxidant challenge in a cellular assay, demonstrating the ability of antioxidants to act intracellularly [43,44]. Cellular protection of *Ro*ex towards H_2O_2 pre-treatment, was evaluated against both cancer and non-cancer cell line, adenocarcinoma breast cancer cells MDA-MB-231 and mouse fibroblasts NIH3T3, respectively (Figure 4). According to data reported in literature [45], *Ro*ex was able to increase significantly NIH3T3 cell viability after pre-treatment with H_2O_2 concentrations ranging from 0.33 and 1.3 mM but not at highest values, i.e., 1.5 and 1.6 mM (Figure 4a). Human adenocarcinoma breast cells demonstrated to be more resistant to H_2O_2 pre-treatment in comparison to NIH3T3 fibroblasts and the rosmarinic extract improved cell viability neutralizing the toxic effect of H_2O_2 also at the highest concentrations (from 0.82 to 1.6 mM), demonstrating to be able to protect cells and to reduce cell death (Figure 4b). The importance of rosmarinic and hydrocynnamic derivative to protect cells in H_2O_2-induced cytotoxicity is well known. Rosmarinic acid exhibited substantial H_2O_2 scavenging activity and inhibited H_2O_2-induced intracellular ROS production [44]. Caffeic acid was found to scavenge intracellular reactive oxygen species, and 1,1-diphenyl-2-picrylhydrazyl radical, and thus prevented lipid peroxidation. Chlorogenic acid protect cells against H_2O_2-induced oxidative stress and apoptosis [45]. Intracellular dose-dependent uptake of flavonoids has been demonstrated in various cell types in vitro and is believed to be even higher in vivo than under normal culture conditions [16].

In vitro antiproliferative activity of *Ro*ex towards NIH3T3 and MDA-MB-231, noncancer and cancer cells, respectively, was evaluated by NRU test. The NRU results showed that *Ro*ex exerted a weak anti-proliferative effect against both NIH3T3 and MDA-MB-231 cell lines in a concentration dependent manner, with lesser efficiency on breast adenocarcinoma cells compared to fibroblasts (Figure 3). In particular, the concentration value of 2.4×10^{-1} mg/mL reduced NIH3T3 viability by 50%, while the same effect with MDA-MB-231 was obtained at a *Ro*ex concentration of 4.8×10^{-1} mg/mL, a higher value in comparison to data reported in literature where MDA-MB-231 cells proliferation is inhibited in a dose-dependent manner with an IC_{50} of about 2.04×10^{-2} mg/mL [46]. The antiproliferative effect of rosemary extracts towards cancer cells is influenced by their constituents and seems to be correlated with the presence of polyphenols, mainly caffeic and rosmarinic acid [47]. On the contrary, rosmarinic and caffeic acid demonstrated low cytotoxicity against non-cancer cells [48] and this could explain the weak antiproliferative activity of rosemary extract against non-cancer NIH3T3 cells.

A genotoxicity study is a key step for risk assessment during development of natural plant extracts for human applications because various genotoxic compounds can cause a DNA damage compromising human health [49].

In the bacterial reverse mutation assay (Ames test) performed on *Ro*ex, as expected, the positive control agents significantly induced genotoxicity but no genotoxic positive result was observed in *Ro*ex treated groups compared to control at any of the tested concentrations (Table 4). These results indicate that *Ro*ex and their phytochemical components, did not exhibit any genotoxic risk under the experimental conditions of this study. The preclinical evaluation of the antithrombotic potential of novel molecules requires the use of reliable reproducible experimental models. TT is one of the most commonly used tests to determine the efficacy of novel antithrombotic drugs. In our experiment, TT using pooled plasma from human healthy patients, was utilized to evaluate the anticoagulant effect of *Ro*ex. The results of TT assay (Table 5) showed that *Ro*ex prolonged coagulation times compared with the control sample, suggesting that the extract inhibited the activity of thrombin.

There is an evidence that coagulation and inflammation are related processes that may considerably affect each other [50]. On the basis of our current experimental studies, it can be hypothesized that inhibitory modulation of coagulation by extracts could give promising anti-inflammatory mediators. These anti-inflammatory and anticoagulant effects have been mostly attributed to the polyphenol and flavonoid compounds found in large quantities in these plants [51] as well as in our *Ro*ex. In the future, well-designed prospective studies are needed to prove this hypothesis.

4. Materials and Methods

4.1. Materials

Thrombin, Dulbecco's Modified Eagle's Medium (DMEM), trypsin solution, and all the solvents used for cell culture as well as Folin–Ciocalteu reactive, DPPH (2,2′-diphenyl-1-picrylhydrazyl), gallic acid and all the other reagents were of analytical grade and purchased from Sigma-Aldrich (Milan, Italy). Mouse immortalized fibroblasts NIH3T3 and human breast adenocarcinoma cells MDA-MB-231 were from American Type Culture Collection (Manassaa, VA, USA). Ames test kit was supplied from Xenometrix (Allschwil, Switzerland).

4.2. Rosmarinus officinalis Extract (Roex) Preparation

Rosmarinus officinalis L. (Tuscan blue cultivar, Linnean Herbarium number LINN 41.1, authenticated at the Botanical Museum and Garden of University of Siena) fresh leaves were collected in spring (April) in Val d'Orcia (Tuscany, Italy, 42°56′02″ N 11°38′17″ E). The fresh leaves were washed three times in distilled water and left to dry at room temperature for 24 h. Then, the leaves were chopped with a scalpel. The extract was obtained by putting 20 g of chopped leaves in 80 g of 60% (*v*/*v*) ethanol (EtOH) for 48 h at room temperature in a shaker incubator. At the end of the incubation, the suspension was filtered by a 0.45 µm Whatman membrane filter and dried using a rotary evaporator. The dry extract obtained was weighted and the percentage yield was expressed as air-dried weight of plant material. Samples then were stored at 2–4 °C until it was time to conduct further analysis.

4.3. Chemical Analysis

4.3.1. Determination of Total Phenolic and Flavonoid Content

Total polyphenols and flavonoids content of *Rosmarinus officinalis* ethanolic extract (*Ro*ex) was examined using spectrophotometric methods reported by Biagi et al., 2014 [52]. In detail, total polyphenols were determined by the colorimetric method of Folin-Ciocalteu. 0.01 mL of each extract were added to 2.990 mL of distilled water and 0.500 mL of Folin-Ciocalteu reagent 1:10 v/v in distilled water. After 30 s of shaking, 1.000 mL of Na_2CO_3 15% m/m in distilled water was added. After incubation at room temperature for 120 min, absorbance at 700 nm was read using a SAFAS UV-MC2 instrument (SAFAS, Monaco, Principality of Monaco). The polyphenols quantification was calculated by means of interpolation of calibration curve constructed using gallic acid.

Total flavonoids content of extracts was determined reading absorbance at 353 nm of 100 folds diluted extract according with Sosa et al. [53] and constructing calibration curve using hyperoside as standard.

4.3.2. Determination of Total Triterpenes

A total of 10 µL of the sample solution was added to 190 µL of glacial acetic acid, after which 300 µL of a solution of glacial acetic acid with 5% m/v vanillin and, after mixing for 30 s, 1 mL of perchloric acid.

The mixture was heated to 60 °C for 45 min and, after cooling, the volume was brought to 5 mL with glacial acetic acid [54].

The absorbance was read at 548 nm and the quantification of the total triterpenes in the extract calculated according to the calibration curve, constructed using β-sitosterol.

4.3.3. High-Performance Liquid Chromatography Analysis of Phenolics Compounds

With the aim of further investigating the polyphenolic fraction of the extract, a high-performance liquid chromatography-diode array detection (HPLC-DAD) analysis was carried out.

A Shimadzu Prominence LC 2030 3D instrument equipped with a Bondapak® C18 column, 10 µm, 125 Å, 3.9 mm × 300 mm column (Waters Corporation, Milford, MA, USA) was used.

Water + 0.1% v/v formic acid (A) and acetonitrile + 0.1% v/v formic acid (B) were used as mobile phase. The following program was applied: B from 10% at 0 min to 25% at 20 min, then B 50% at 26 min.; flow was set at 0.8 mL/min. Chromatograms were recorded at 280 nm.

Analyses were performed using 10 µL of extract; rosmarinic acid, caffeic acid, chlorogenic acid, quercetin, apigenin, luteolin, rutin and hyperoside were used as external standard. Calibration curves were established using reference standards ranging from 0.008 mg/mL to 0.500 mg/mL. The correlation coefficient (R^2) of each curve was > 0.99.

4.4. Antiradical Capacity: DPPH Assay

Antiradical capacity of *Ro*ex was evaluated both by spectrophotometric method using ascorbic acid as reference, and by EPR analysis using gallic acid as reference.

4.4.1. Spectrophotometric Assay

The antiradical capacity of *Ro*ex was tested by means of the validated DPPH (2,2-diphenyl-1-picrylhydrazyl) test. The DPPH solution was prepared in methanol at a concentration of 1×10^{-4} M. *Ro*ex was tested in seven 1:2 serial dilutions (ethanol 60% v/v) and ascorbic acid, used as reference.

All the samples were mixed with the DPPH solution (1:19), transferred into 1 cm path length cuvettes and incubated for 30 min at room temperature in the dark. Ethanol 60% v/v and DPPH (1:19) was used as positive control. The inhibition of DPPH was calculated according to the following Formula (1) where Absc is the absorbance of the positive control and Absx the absorbance of tested samples IC_{50} was calculated by constructing the curve of inhibition values for each tested concentration (in the linear range 10–75%) [55].

$$\% \text{ inibition} = (\text{Absc} - \text{Absx})/\text{Absc} \times 100 \quad (1)$$

4.4.2. Scavenger Activity by Electron Paramagnetic Resonance (EPR) Analysis

Continuous-wave X-band (CW, 9 GHz) EPR spectra were recorded using a Bruker E580 ELEXSYS Series spectrometer (Bruker, Rheinstetten, Germany), with the ER4122SHQE cavity. EPR measurements were performed filling 1.0 mm ID × 1.2 mm OD quartz capillaries placing them into a 3.0 mm ID × 4.0 mm OD suprasil tube.

A stock solution of DPPH was prepared (0.2 mM in ethanol) and the final concentration of the radical in each sample was 0.16 mM. The extract was diluted a hundred times in respect to the stock solution and it was added with one fourth volume in respect to that of the radical. The addition was incubated for 15 min.

The area of the EPR spectra was calculated by the double integral of the DPPH signal and the scavenger percentage was calculated using the Formula (2) where A_0 is the area of the DPPH signal without the addition of the extract, $A_{extract}$ is the area the DPPH signal after the addition of the extract.

The addition of the antioxidant to the DPPH radical solution determines a reduction in intensity of the EPR signal with a scavenger percentage equal to 99%.

$$\text{scavenger } \% = (A_0 - A_{extract})/A_0 \times 100 \quad (2)$$

4.4.3. Gallic Acid-Equivalent Antioxidant Activity through DPPH Assay

A stock solution of DPPH radical 1 mM in EtOH was freshly prepared and used within 5 h. A stock solution of gallic acid 0.2 mM in EtOH was prepared. The calibration curves were built using a standard solution of gallic acid. The gallic acid standard solution with different linear increasing volumes (10–100 μL) and a known volume (100 μL) of the extract was added to a fixed volume of the DPPH solution (100 μL). After 15 min of incubation in the dark at room temperature, the EPR spectra were recorded. The antioxidant activity was plotted through the decay percentage of the area of the DPPH signal versus increasing concentrations of gallic acid standard solution. The area of the EPR spectra was calculated through the double integral of the DPPH signal.

The decay percentage for the plotting refers to the Formula (3) where A_0 is the area of the DPPH signal without the addition of the antioxidant or extract, A_S is the area the DPPH signal after the addition of scavenger agent such as antioxidant gallic acid or the extract.

The decay area percentage expressed in gallic acid equivalent was obtained through the calibration curve built with the standard solution ($R^2 = 0.928$) and reporting the area of the DPPH EPR signal after the addition of the extract. The extract was diluted a hundred times in respect to the stock solution. The measurements were repeated in triplicate and the antioxidant activity of the extract was expressed as mole/g of acid gallic equivalent.

$$\text{decay}\% = (A_0 - A_S)/A_0 \times 100 \quad (3)$$

4.5. Hydrogen Peroxide Scavenging Assay

The ability of *Ro*ex to scavenge hydrogen peroxide was estimated according the method of Ruch et al. [56]. A solution of H_2O_2 (2 mM) was prepared in phosphate buffer (50 mM, pH = 7.4). Aliquots (0.05, 0.1, 0.2, 0.3 and 0.4 mL) of *Ro*ex at a concentration of 24 mg/mL, were transferred into test tubes and their volumes were made up to 0.4 mL with 50 mM phosphate buffer at pH = 7.4. After adding 0.6 mL of hydrogen peroxide solution, tubes were vortexed and the absorbance of H_2O_2 at 230 nm was determined after 10 min of incubation, against a blank containing phosphate buffer and Et-OH 60% without H_2O_2. The percentage of hydrogen peroxide scavenging was calculated as follows:

$$\% \text{ scavenged } H_2O_2 = [(A_i - A_t)/A_i] \times 100 \quad (4)$$

where A_i is the absorbance of control and A_t is the absorbance of test samples.

4.6. *Anti-Proliferative Assay and Protective Effect against Hydrogen Peroxide Induced Oxidative Stress*

4.6.1. Cell Cultures and Anti-Proliferative Test

In order to evaluate the in vitro anti-proliferative activity of new products, the direct contact test was used [57]. This test is suitable for sample with various shapes, sizes or physical status (i.e., liquid or solid). The evaluation of in vitro inhibition of cell growth does not depend on the final use for which the product is intended, and the document ISO 10995-5:2009 recommends many cell lines from American Type Collection. Among them, to test *Ro*ex cytotoxicity, NIH3T3 mouse fibroblasts were chosen [58]. Moreover, in order to evaluate the anti-proliferative activity of *Ro*ex towards tumoral cells, the same test was repeated by using human breast adenocarcinoma cells MDA-MB 231.

Both NIH3T3 and MDA-MB231 cells were propagated in DMEM supplemented with 10% fetal calf serum, 1% L-glutamine-penicillin-streptomycin solution, and 1% MEM non-essential amino acid solution, and incubated at 37 °C in a humidified atmosphere containing 5% CO_2. Once at confluence, the cells were washed with PBS 0.1M, separated with trypsin-EDTA solution and centrifuged at 1000 r.p.m. for 5 min. The pellet was re-suspended in complete medium (dilution 1:15). Cells (1.5×10^4) suspended in 1 mL of complete medium were seeded in each well of a 24 well round multidish and incubated at 37 °C in an atmosphere of 5% CO_2. Once reached the 50% of confluence (i.e., after 24 h of culture), the culture medium was discharged and the test compounds, properly diluted in completed medium, were added to each well. All samples were set up in six replicates. Complete medium was used as negative control. After 24 h of incubation, cell viability was evaluated by neutral red uptake (NRU) assay [59].

4.6.2. Protective Effect against Hydrogen Peroxide Induced Oxidative Stress

To determine the protective effect of alcoholic extract of *Ro*ex against oxidative stress, NIH3T3 and MDA-MB-231 cells were pre-incubated with different concentrations of hydrogen peroxide (0.1, 0.2, 0.3, 0.9, 1.0, 1.1, 1.3, 1.5, 1.6 μM) for 15 min, then washed and incubated for 24 h with 0.3% v/v *Ro*ex at a concentration of 7.2×10^{-3} mg/mL in complete culture medium. Cell viability was evaluated after 24 h of incubation at 37 °C in 5% CO_2 by NRU assay.

4.6.3. Evaluation of Cell Viability: NRU Assay

In order to determine the percentage of viable cells as follows, the following solutions were prepared:

1. Neutral Red (NR) stock solution: 0.33 g NR dye powder in 100 mL sterile H_2O
2. NR medium: 1.0 mL NR stock solution + 99.0 routine culture medium pre-warmed to 37 °C
3. NR desorb solution: 1% glacial acetic acid solution + 50% ethanol + 49% H_2O

At the end of incubation, the routine culture medium was removed from each plate and the cells were carefully rinsed with 1 mL pre-warmed D-PBS 0.1M. Plates were then gently blotted with paper towels. 1.0 mL NR medium was added to each dish and further incubated at 37 °C, 95% humidity, 5.0% CO_2 for 3 h. The cells were checked during incubation for NR crystal formation. After incubation, the NR medium was removed and the cells were carefully rinsed with 1 mL pre-warmed D-PBS 0.1M. PBS was decanted and blotted from the dishes and exactly 1 mL NR desorb solution was added to each sample. Plates were placed on a shaker for 20–45 min to extract NR from the cells and form a homogeneous solution. During this step the samples were covered to protect them from light. Five minutes after removal from the shaker, absorbance was read at 540 nm with a UV/visible spectrophotometer (Varian Cary 1E).

4.7. Mutagenicity Assay: Ames Test

The TA100 and TA98 strains of *Salmonella typhimurium* were utilized for mutagenicity assay in absence and presence of metabolic activation, i.e., with and without S9 liver fraction. The tester strains used were selected because they are sensitive and detect a large proportion of known bacterial mutagens and are most commonly used routinely within the pharmaceutical industry [60]. The following specific positive controls were used, respectively, with and without S9 fraction: 2-nitrofluorene (2-NF) 2 μg/mL + 4-nitroquinoline N-oxide (4-NQO) 0.1 μg/mL, and 2-aminoanthracene (2-AA) 5 μg/mL. The final concentration of S9 in the culture was 4.5%.

Approximately 10^7 bacteria were exposed to 6 concentrations of *Ro*ex, as well as to positive and negative controls, for 90 min in medium containing sufficient histidine to support approximately two cell divisions. After 90 min, the exposure cultures were diluted in pH indicator medium lacking histidine, and aliquoted into 48 wells of a 384-well plate. Within two days, cells which had undergone the reversion to His grew into colonies. Metabolism by the bacterial colonies reduced the pH of the medium, changing the color of that well. This color change can be detected visually. The number of wells containing revertant colonies were counted for each dose and compared to a zero-dose control. Each dose was tested in six replicates.

The material was regarded mutagenic if the number of histidine revertant colonies was twice or more than the spontaneous revertant colonies.

4.8. Antithrombotic Activity: Thrombin Time (TT)

The selected blood donors were normal, healthy men who had fasted for more than 8 h and had not received medication for at least 14 days. Blood samples were collected into 3.8% (m/V) tri-sodium citrate as anticoagulant at a volume ratio of 9 parts blood to 1-part citrate. The blood samples were then centrifuged at 3500 r.p.m. for 15 min to obtain PPP which was utilized to perform TT. In particular, 0.2 mL of each samples were added to 0.2 mL of a solution obtained by diluting 1:1 human plasma with a solution of 60% Et-OH (% v/v). TT was determined by incubating the aliquot (0.2 mL) of human plasma containing the sample at 37 °C for 2 min, after which 0.2 mL of thrombin (0.6 NIH) was added. The clotting time was revealed by an Automatic ElviDigiclot 2 Coagulimeter (from Logos SpA, Milan, Italy).

4.9. Statistical Analysis

All assays were carried out in six replicates and their results were expressed as mean ± standard deviation (SD). Multiple comparisons were performed by one-way ANOVA and individual differences tested by Fisher's test after the demonstration of significant inter-group differences by ANOVA. Differences with $p < 0.05$ were considered significant.

5. Conclusions

Taken together, the data obtained demonstrated that hydro-alcoholic extraction of Tuscan rosemary can be a good method for obtaining active compounds with high potential for application in many different fields, such as cosmetics, food or pharmaceutical research. The results of this study have demonstrated, indeed, that the *Ro*ex analyzed possessed good antioxidant and radical scavenging activities when tested in cellular and non-cellular assays, as well as a weak, anti-proliferative effects towards both cancer and noncancer cells, absence of genotoxic and ability to prolong thrombin time. All these bioactivities are tightly correlated to the *Ro*ex chemical profile in a dose dependent manner. In order to prevent unwanted effects when used in humans, it is essential that multiple aspects of the bioactivity of each *Rosmarinus officinalis* extract are tested.

Supplementary Materials: The following are available online at https://www.mdpi.com/2223-7747/10/1/97/s1.

Author Contributions: S.L. obtained the extract, performed biological experiments, and data/evidence collection; wrote the manuscript; M.C.B. performed the chemical characterization and data/evidence collection, revised the manuscript; E.M. performed the chemical characterization and data/evidence collection, revised the manuscript; G.B. performed antiradical capacity DPPH assay and spectrophotometric assay and data/evidence collection; M.B. performed biological experiments, and data/evidence collection, revised the manuscript. All authors have read and agreed to the published version of the manuscript.

Funding: This research received no external funding.

Institutional Review Board Statement: Ethical review and approval were waived for this study, due to the research involves no risk to subjects.

Informed Consent Statement: Patient consent was waived due to the research involves no risk to subjects.

Data Availability Statement: The data presented in this study are available in supplementary material.

Conflicts of Interest: The authors declare no conflict of interest.

References

1. Cavalcanti, R.N.; Forster-Carneiro, T.; Gomes, M.T.M.S.; Rostagno, M.A.; Prado, J.M.; Meireles, M.A.A. Uses and Applications of Extracts from Natural Sources. In *Natural Product Extraction: Principles and Applications, Green Chemistry Series*; Rostagno, M.A., Prado, J.M., Eds.; The Royal Society of Chemistry: London, UK, 2013; pp. 1–57.
2. Segneanu, A.E.; Velciov, S.M.; Olariu, S.; Cziple, F.; Damian, D.; Grozescu, I. Bioactive Molecules Profile from Natural Compounds. In *Amino Acid—New Insights and Roles in Plant and Animal*; Asao, T., Ed.; IntechOpen: London, UK, 2017; pp. 209–228. [CrossRef]
3. Gonzalez-Trujano, M.E.; Peña, E.I.; Martınez, A.L.; Moreno, J.; Guevara-Feferc, P.; Déciga-Camposd, M.; López-Muñoz, F.J. Evaluation of the antinociceptive effect of *Rosmarinus officinalis* L. using three different experimental models in rodents. *J. Ethnopharmacol.* **2007**, *111*, 476–482. [CrossRef]
4. Nogueira de Melo, G.A.; Grespan, J.; Fonseca, R.P.; Farinha, T.O.; Silva, E.L.; Romero, A.L.; Bersani-Amado, C.A.; Cuman, R.K.N. *Rosmarinus officinalis* L. essential oil inhibits in vivo and in vitro leukocyte migration. *J. Med. Food* **2011**, *14*, 944–949. [CrossRef]
5. Estévez, M.; Ramırez, R.; Ventanas, S.; Cava, R. Sage and rosemary essential oils versus BHT for the inhibition of lipid oxidative reactions in liver pâté. *LWT Food Sci. Technol.* **2007**, *40*, 58–65. [CrossRef]
6. Pérez-Fons, L.; Garzón, M.T.; Micol, V. Relationship between the antioxidant capacity and effect of rosemary (*Rosmarinus officinalis* L.) polyphenols on membrane phospholipid order. *J. Agric. Food Chem.* **2010**, *58*, 161–171. [CrossRef]
7. Brindisi, M.; Bouzidi, C.; Frattaruolo, L.; Loizzo, M.R.; Tundis, R.; Dugay, A.; Deguin, B.; Cappello, A.R.; Cappello, M.S. Chemical Profile, Antioxidant, Anti-Inflammatory, and Anti-Cancer Effects of Italian *Salvia rosmarinus* Spenn. Methanol Leaves Extracts. *Antioxidants* **2020**, *9*, 826. [CrossRef]
8. Mena, P.; Cirlini, M.; Tassotti, M.; Herrlinger, K.A.; Dall'Asta, C.; Del Rio, D. Phytochemical Profiling of Flavonoids, Phenolic Acids, Terpenoids, and Volatile Fraction of a Rosemary (*Rosmarinus officinalis* L.) Extract. *Molecules* **2016**, *21*, 1576. [CrossRef]
9. Huang, L.; Ding, B.; Zhang, H.; Kong, B.; Xiong, Y.L. Textural and sensorial quality protection in frozen dumplings through the inhibition of lipid and protein oxidation with clove and rosemary extracts. *J. Sci. Food Agric.* **2019**, *10*, 4739–4747. [CrossRef]
10. Nobile, V.; Michelotti, A.; Cestone, E.; Caturla, N.; Castillo, J.; Benavente-García, O.; Pérez-Sánchez, A.; Micol, V. Skin photoprotective and antiageing effects of a combination of rosemary (*Rosmarinus officinalis*) and grapefruit (*Citrus paradisi*) polyphenols. *Food Nutr. Res.* **2016**, *60*, 31871. [CrossRef]
11. Williams, G.M.; Iatropoulos, M.J.; Whysner, J. Safety assessment of butylated hydroxyanisole and butylated hydroxytoluene as antioxidant food additives. *J. Food Chem. Toxicol.* **1999**, *37*, 1027–1038. [CrossRef]
12. Tai, J.; Cheung, S.; Wu, M.; Hasman, D. Antiproliferation effect of rosemary (*Rosmarinus officinalis*) on human ovarian cancer cells in vitro. *Phytomedicine* **2012**, *19*, 436–443. [CrossRef]
13. Bozin, B.; Mimica-Dukic, N.; Samojlik, I.; Jovin, E. Antimicrobial and antioxidant properties of rosemary and sage (*Rosmarinus officinalis* L. and *Salvia officinalis* L., Lamiaceae) essential oils. *J. Agric. Food Chem.* **2007**, *55*, 7879–7885. [CrossRef]
14. Yamamoto, Y.; Yamada, K.; Naemura, A.; Yamashita, T.; Arai, R. Testing various herbs for antithrombotic effect. *Nutrition* **2005**, *21*, 580–587. [CrossRef]
15. Cheung, S.; Tai, J. Anti-proliferative and antioxidant properties of rosemary *Rosmarinus officinalis*. *Oncol. Rep.* **2007**, *17*, 1525–1531. [CrossRef] [PubMed]
16. Yesil-Celiktas, O.; Sevimli, C.; Bedir, E.; Vardar-Sukan, F. Inhibitory effects of rosemary extracts, carnosic acid and rosmarinic acid on the growth of various human cancer cell lines. *Plant Foods Hum. Nutr.* **2010**, *65*, 158–163. [CrossRef]
17. Harach, T.; Aprikian, O.; Monnard, I.; Moulin, J.; Membrez, M.; Béolor, J.C.; Raab, T.; Macè, K.; Darimont, C. Rosemary (*Rosmarinus officinalis* L.) Leaf extract limits weight gain and liver steatosis in mice fed a high-fat diet. *Planta Med.* **2010**, *76*, 566–571. [CrossRef]
18. Bakirel, T.; Bakirel, U.; Keles, O.U.; Ulgen, S.-G.; Yardibi, H. In vivo assessment of antidiabetic and antioxidant activities of rosemary (*Rosmarinus officinalis*) in alloxan-diabetic rabbits. *J. Ethnopharmacol.* **2008**, *116*, 64–73. [CrossRef]

19. Afonso, M.S.; De, O.; Silva, A.M.; Carvalho, E.B.; Rivelli, D.P.; Barros, S.B.; Rogero, M.M.; Lottenberg, A.M.; Torres, R.P.; Mancini-Filho, J. Phenolic compounds from Rosemary (*Rosmarinus officinalis* L.) attenuate oxidative stress and reduce blood cholesterol concentrations in diet-induced hypercholesterolemic rats. *Nutr. Metab.* **2013**, *10*, 19. [CrossRef]
20. Sasaki, K.; El Omri, A.; Kondo, S.; Han, J.; Isoda, H. *Rosmarinus officinalis* polyphenols produce anti-depressant like effect through monoaminergic and cholinergic functions modulation. *Behav. Brain. Res.* **2013**, *238*, 86–94. [CrossRef]
21. Borrás-Linares, I.; Stojanović, Z.; Quirantes-Piné, R.; Arráez-Román, D.; Švarc-Gajić, J.; Fernández-Gutiérrez, A.; Segura-Carretero, A. *Rosmarinus officinalis* leaves as a natural source of bioactive compounds. *Int. J. Mol. Sci.* **2014**, *15*, 585. [CrossRef]
22. Okamura, N.; Haraguchi, H.; Hashimoto, K.; Yagi, A. Flavonoids in *Rosmarinus officinalis* leaves. *Phytochemistry* **1994**, *37*, 1463–1466. [CrossRef]
23. Mahmoud, A.A.; Al-Shihry, S.S.; Son, B.W. Diterpenoid quinones from rosemary (*Rosmarinus officinalis* L.). *Phytochemistry* **2005**, *66*, 1685–1690. [CrossRef]
24. Kontogianni, V.G.; Tomic, G.; Nikolic, I.; Nerantzaki, A.A.; Sayyad, N.; Stosic-Grujicic, S.; Stojanovic, I.; Gerothanassis, I.P.; Tzakos, A.G. Phytochemical profile of *Rosmarinus officinalis* and *Salvia officinalis* extracts and correlation to their antioxidant and anti-proliferative activity. *Food Chem.* **2013**, *136*, 120–129. [CrossRef]
25. Romo-Vaquero, M.; García-Villalba, R.; Larrosa, M.; Yáñez-Gascón, M.J.; Fromentin, E.; Flanagan, J.; Roller, M.; Tomás-Barberán, F.A.; Espín, J.C.; García-Conesa, M.T. Bioavailability of the major bioactive diterpenoids in a rosemary extract: Metabolic profile in the intestine, liver, plasma, and brain of Zucker rats. *Mol. Nutr. Food. Res.* **2013**, *57*, 1834–1846. [CrossRef]
26. Del Rio, D.; Rodriguez-Mateos, A.; Spencer, J.P.E.; Tognolini, M.; Borges, G.; Crozier, A. Dietary (poly)phenolics in human health: Structures, bioavailability, and evidence of protective effects against chronic diseases. *Antioxid Redox Signal* **2013**, *18*, 1818–1892. [CrossRef]
27. Rodríguez-Mateos, A.; Vauzour, D.; Krueger, C.G.; Shanmuganayagam, D.; Reed, J.; Calani, L.; Mena, P.; Del Rio, D.; Crozier, A. Bioavailability, bioactivity and impact on health of dietary flavonoids and related compounds: An update. *Arch. Toxicol.* **2014**, *88*, 1803–1853. [CrossRef]
28. Fernández Ochoa, Á.; Borrás, I.; Pérez-Sánchez, A.; Barrajón-Catalán, E.; Gonzalez-Alvarez, I.; Arráez-Román, D.; Micol, V.; Segura, C. Phenolic compounds in rosemary as potential source of bioactive compounds against colorectal cancer: In situ absorption and metabolism study. *J. Funct. Food* **2017**, *33*, 202–210. [CrossRef]
29. Ribeiro, A.; Caleja, C.; Barros, L.; Santos-Buelga, C.; Barreiro, M.F.; Ferreira, I.C.F.R. Rosemary extracts in functional foods: Extraction, chemical characterization and incorporation of free and microencapsulated forms in cottage cheese. *Food Funct.* **2016**, *7*, 2185–2196. [CrossRef] [PubMed]
30. Mulinacci, N.; Innocenti, M.; Bellumori, M.; Giaccherini, C.; Martini, V.; Michelozzi, M. Storage method, drying processes and extraction procedures strongly affect the phenolic fraction of rosemary leaves: An HPLC/DAD/MS study. *Talanta* **2011**, *85*, 167–176. [CrossRef] [PubMed]
31. Bicchi, C.; Binello, A.; Rubiolo, P. Determination of phenolic diterpene antioxidants in rosemary (*Rosmarinus officinalis* L.) with different methods of extraction and analysis. *Phytochem. Anal.* **2000**, *11*, 236–242. [CrossRef]
32. Sharma, Y.; Velamuri, R.; Fagan, J.; Schaefer, J. Full-Spectrum Analysis of Bioactive Compounds in Rosemary (*Rosmarinus officinalis* L.) as Influenced by Different Extraction Methods. *Molecules* **2020**, *25*, 4599. [CrossRef] [PubMed]
33. Grillo, A.; Chemi, G.; Brogi, S.; Brindisi, M.; Relitti, N.; Fezza, F.; Fazio, D.; Castelletti, L.; Perdona, E.; Wong, A.; et al. Development of novel multipotent compounds modulating endocannabinoid and dopaminergic systems. *Eur. J. Med. Chem.* **2019**, *1*, 111674. [CrossRef]
34. Jacotet-Navarro, M.; Rombaut, N.; Fabiano-Tixier, A.S.; Danguien, M.; Bily, A.; Chemat, F. Ultrasound versus microwave as green processes for extraction of rosmarinic, carnosic and ursolic acids from rosemary. *Ultrason. Sonochem.* **2015**, *27*, 102–109. [CrossRef] [PubMed]
35. Ancillotti, C.; Ciofi, L.; Pucci, D.; Sagona, E.; Giordani, E.; Biricolti, S.; Gori, M.; Petrucci, W.A.; Giardi, F.; Bartoletti, R.; et al. Polyphenolic profiles and antioxidant and antiradical activity of Italian berries from *Vaccinium myrtillus* L. and *Vaccinium uliginosum* L. subsp. gaultherioides (Bigelow) S.B. *Young. Food Chem.* **2016**, *204*, 176–184. [CrossRef]
36. Liu, T.; Sui, X.; Zhang, R.; Yang, L.; Zu, Y.; Zhang, L.; Zhang, Y.; Zhang, Z. Application of ionic liquids based microwave-assisted simultaneous extraction of carnosic acid, rosmarinic acid and essential oil from *Rosmarinus officinalis*. *J. Chromatogr. A* **2011**, *1218*, 8480–8489. [CrossRef] [PubMed]
37. Biagi, M.; Noto, D.; Corsini, M.; Baini, G.; Cerretani, D.; Cappellucci, G.; Moretti, E. Antioxidant Effect of the Castanea sativa Mill. Leaf Extract on Oxidative Stress Induced upon Human Spermatozoa. *Oxid. Med. Cell Longev.* **2019**, *2019*, 8926075. [CrossRef] [PubMed]
38. Jacotet-Navarro, M.; Laguerre, M.; Fabiano-Tixier, A.S.; Tenon, M.; Feuillère, N.; Bily, A.; Chemat, F. What is the best ethanol-water ratio for the extraction of antioxidants from rosemary? Impact of the solvent on yield, composition, and activity of the extracts. *Electrophoresis* **2018**, *39*, 1946–1956. [CrossRef] [PubMed]
39. Megateli, S.; Krea, M. Enhancement of total phenolic and flavonoids extraction from *Rosmarinus officinalis* L using electromagnetic induction heating (EMIH) process. *Physiol. Mol. Biol. Plants* **2018**, *24*, 889–897. [CrossRef] [PubMed]
40. Zheng, W.; Wang, S.Y. Antioxidant activity and phenolic compounds in selected herbs. *J. Agric. Food Chem.* **2001**, *49*, 5165–5170. [CrossRef] [PubMed]

41. Chen, C.H.; Ho, C.-T. Antioxidant Activities of Caffeic Acid and Its Related Hydroxycinnamic Acid Compounds. *J. Agric. Food Chem.* **1997**, *45*, 2374–2378. [CrossRef]
42. Dueñas, M.; Surco-Laos, F.; González-Manzano, S.; González-Paramás, A.M.; Santos-Buelga, C. Antioxidant properties of major metabolites of quercetin. *Eur. Food Res. Technol.* **2011**, *232*, 103–111. [CrossRef]
43. Deiana, M.; Corona, G.; Incani, A.; Loru, D.; Rosa, A.; Atzeri, A.; Paola Melis, M.; Assunta Dessì, M. Protective effect of simple phenols from extravirgin olive oil against lipid peroxidation in intestinal Caco-2 cells. *Food Chem. Toxicol.* **2010**, *48*, 3008–3016. [CrossRef] [PubMed]
44. Hahn, H.J.; Kim, K.B.; An, I.S.; Ahn, K.J.; Han, H.J. Protective effects of rosmarinic acid against hydrogen peroxide-induced cellular senescence and the inflammatory response in normal human dermal fibroblasts. *Mol. Med. Rep.* **2017**, *16*, 9763–9769. [CrossRef]
45. Song, J.; Guo, D.; Bi, H. Chlorogenic acid attenuates hydrogen peroxide-induced oxidative stress in lens epithelial cells. *Int. J. Mol. Med.* **2018**, *41*, 765–772. [CrossRef] [PubMed]
46. Kook, D.; Wolf, A.H.; Yu, A.L.; Neubauer, A.S.; Priglinger, S.G.; Kampik, A.; Welge-Lussen, U.C. The Protective Effect of Quercetin against Oxidative Stress in the Human RPE In Vitro. *Investig. Ophthalmol. Vis. Sci.* **2008**, *49*, 1712–1720. [CrossRef]
47. Pelinson, L.P.; Assmann, C.E.; Palma, T.V.; Mânica da Cruz, I.B.; Pillat, M.M.; Mânica, A.; Stefanello, N.; CastagnaCezimbra Weis, G.; de Oliveira Alves, A.; Melazzo de Andrade, C.; et al. Antiproliferative and apoptotic effects of caffeic acid on SK-Mel-28 human melanoma cancer cells. *Mol. Biol. Rep.* **2019**, *46*, 2085–2092. [CrossRef] [PubMed]
48. Fujimoto, A.; Sakanashi, Y.; Matsui, H.; Oyama, T.; Nishimura, Y.; Masuda, T.; Oyama, Y. Cytometric Analysis of Cytotoxicity of Polyphenols and Related Phenolics to Rat Thymocytes: Potent Cytotoxicity of Resveratrol to Normal Cells. *Basic Clin. Pharmacol. Toxicol.* **2009**, *104*, 455–462. [CrossRef]
49. Makhafola, T.J.; Elgorashi, E.E.; McGaw, L.J.; Verschaeve, L.; Eloff, N.J. The correlation between antimutagenic activity and total phenolic content of extracts of 31 plant species with high antioxidant activity. *BMC Complementary Altern. Med.* **2016**, *16*, 490. [CrossRef]
50. TarikKhouya, T.; Ramchoun, M.; Hmidani, A.; Amrani, S.; Harnafi, H.; Benlyas, M.; Zegzouti, Y.F.; Alem, C. Anti-inflammatory, anticoagulant and antioxidant effects of aqueous extracts from Moroccan thyme varieties. *Asian Pac. J. Trop. Biomed.* **2015**, *5*, 636–644. [CrossRef]
51. Pawlaczyk, I.; Lewik-Tsirigotis, M.; Capek, P.; Matulová, M.; Sasinková, V.; Dąbrowski, P.; Witkiewicz, W.; Gancarz, R. Effects of extraction condition on structural features and anticoagulant activity of *F. vesca* L. conjugates. *Carbohydr. Pol.* **2013**, *92*, 741–750. [CrossRef]
52. Biagi, M.; Manca, D.; Barlozzini, B.; Miraldi, E.; Giachetti, D. Optimization of extraction: Of drugs containing polyphenols using an innovative technique. *Agro Food Ind. Hi Tech* **2014**, *25*, 60–65.
53. Sosa, S.; Bornancin, A.; Tubaro, A.; Loggia, R.D. Topical antiinflammatory activity of an innovative aqueous formulation of actichelated propolis vs two commercial propolis formulations. *Phytother. Res.* **2007**, *21*, 823–826. [CrossRef] [PubMed]
54. Governa, P.; Marchi, M.; Cocetta, V.; De Leo, B.; Saunders, P.T.K.; Catanzaro, D.; Miraldi, E.; Montopoli, M.; Biagi, M. Effects of Boswellia Serrata Roxb. and *Curcuma longa* L. in an In Vitro Intestinal Inflammation Model Using Immune Cells and Caco-2. *Pharmaceuticals* **2018**, *11*, 126. [CrossRef] [PubMed]
55. Biagi, M.; Collodel, G.; Corsini, M.; Pascarelli, N.A.; Moretti, E. Protective effect of Propolfenol® on induced oxidative stress in human spermatozoa. *Andrologia* **2018**, *50*. [CrossRef] [PubMed]
56. Ruch, R.J.; Cheng, S.J.; Klaunig, J.E. Prevention of cytotoxicity and inhibition of intercellular communication by antioxidant catechins isolated from Chinese green tea. *Carcinogenesis* **1989**, *10*, 1003–1008. [CrossRef]
57. ISO 10995-5:2009, Biological Evaluation of Medical Devices—Part 5: Tests for Cytotoxicity: In Vitro Methods. Available online: https://www.iso.org/standard/36406.html (accessed on 4 January 2021).
58. Lamponi, S.; Leone, G.; Consumi, M.; Greco, G.; Magnani, A. In vitro biocompatibility of new PVA-based hydrogels as vitreous body substitutes. *J. Biomater. Sci. Polym. Ed.* **2012**, *23*, 555–575. [CrossRef] [PubMed]
59. Lamponi, S.; Leone, G.; Consumi, M.; Greco, G.; Nelli, N.; Magnani, A. Porous multi-layered composite hydrogel as cell substrate for in vitro culture of chondrocytes. *Int. J. Polym. Mater.* **2020**. [CrossRef]
60. Purves, D.; Harvey, C.; Tweats, D.; Lumley, C.E. Genotoxicity testing: Current practices and strategies used by. the pharmaceutical industry. *Mutagenesis* **1995**, *10*, 297–312. [CrossRef]

MDPI

Article

Application of Deep Eutectic Solvents for the Extraction of Carnosic Acid and Carnosol from Sage (*Salvia officinalis* L.) with Response Surface Methodology Optimization

Martina Jakovljević [1], Stela Jokić [1], Maja Molnar [1] and Igor Jerković [2,*]

[1] Faculty of Food Technology Osijek, Josip Juraj Strossmayer University of Osijek, Franje Kuhača 18, 31000 Osijek, Croatia; mjakovljevic@ptfos.hr (M.J.); sjokic@ptfos.hr (S.J.); mmolnar@ptfos.hr (M.M.)
[2] Faculty of Chemistry and Technology, University of Split, Ruđera Boškovića 35, 21000 Split, Croatia
* Correspondence: igor@ktf-split.hr; Tel.: +385-21-329-434

Abstract: *Salvia officinalis* L. is a good source of antioxidant compounds such as phenolic diterpenes carnosic acid and carnosol. From 17 deep eutectic solvents (DESs) used, choline chloride: lactic acid (1:2 molar ratio) was found to be the most suitable for the extraction of targeted compounds. The influence of H_2O content, extraction time, and temperature (for stirring and heating and for ultrasound-assisted extraction (UAE)), H_2O content, extraction time, and vibration speed for mechanochemical extraction on the content of targeted compounds were investigated. Carnosic acid content obtained by the extraction assisted by stirring and heating was from 2.55 ± 0.04 to 14.43 ± 0.28 μg mg^{-1}, for UAE it was from 1.62 ± 0.29 to 14.00 ± 0.02 μg mg^{-1}, and for mechanochemical extraction the yield was from 1.80 ± 0.02 to 8.26 ± 0.45 μg mg^{-1}. Determined carnosol content was in the range 0.81 ± 0.01 to 4.83 ± 0.09 μg mg^{-1} for the extraction with stirring and for UAE it was from 0.56 ± 0.02 to 4.18 ± 0.05 μg mg^{-1}, and for mechanochemical extraction the yield was from 0.57 ± 0.11 to 2.01 ± 0.16 μg mg^{-1}. Optimal extraction conditions determined by response surface methodology (RSM) were in accordance with experimentally demonstrated values. In comparison with previously published or own results using conventional solvents or supercritical CO_2, used DES provided more efficient extraction of both targeted compounds.

Keywords: sage; optimization; stirring and heating extraction; ultrasound-assisted extraction; mechanochemical extraction

Citation: Jakovljević, M.; Jokić, S.; Molnar, M.; Jerković, I. Application of Deep Eutectic Solvents for the Extraction of Carnosic Acid and Carnosol from Sage (*Salvia officinalis* L.) with Response Surface Methodology Optimization. *Plants* **2021**, *10*, 80. https://doi.org/10.3390/plants10010080

Received: 4 November 2020
Accepted: 24 December 2020
Published: 2 January 2021

1. Introduction

The growth of the pharmaceutical industry and increased need for bioactive components has led to the increased development of new extraction and isolation methods [1]. The most important differences between these methods are better efficiency and shorter extraction time for modern techniques compared to the conventional ones. Furthermore, conventional solvents are very often flammable and toxic with their manufacture depending on fossil resources [2]. However, there are also certain issues associated with the modern techniques, such as poor selectivity and solubility of targeted components in the solvents used, such as H_2O, ethanol, or CO_2, as well as recovery of bioactive components and their chemical changes during the extraction period due to the reactions such as ionization, hydrolysis, and oxidation [3,4]. Over the past few years, deep eutectic solvents (DESs), first proposed by Abbott et al. [5,6], have been developed as analogues of ionic liquids (ILs), although they differ from them in the starting material and the method of preparation. DESs are mixtures of hydrogen bond acceptor (HBA) and hydrogen bond donor (HBD) with a lower melting point relative to the starting components. Their green character is also attributed to their low price, easy preparation, biodegradability, and low toxicity [7]. In addition to these properties, different studies reported that DES could dissolve several components better than organic solvents, due to dissolving lignocellulose which causes damage of plant cell wall and strengthens the mass transfer process [8]. DESs can be prepared from different combinations of the starting compounds, thus being tunable solvents with different functionality and solubility for various compounds. Therefore, a suitable combination of the starting solvents and their molar ratio can increase the solubility and the extraction efficiency of DESs for desired compounds [9]. However, their shortcomings should also be taken into account with an emphasis on viscosity and low vapor pressure which makes it difficult to isolate and purify desired components. In addition, high viscosity complicates industrial application due to the high energy consumption needed to ensure their liquid state [10].

One of the most commonly used HBA is choline chloride (ChCl), since it is an inexpensive, biodegradable, and non-toxic quaternary ammonium salt. ChCl can form DES with different nontoxic components as carboxylic acids, sugars, sugar alcohols, or amines which act as HBDs [5,6]. In the last few years, deep eutectic solvents are increasingly used for the extraction of phenolic compounds including phenolic acids, flavonoids, stilbenes, anthocyanins, and furanocoumarins [1,2,11–13]. In the research performed by Bi et al. [11] DESs were used to extract flavonoids such as myricetin and amentoflavone from *Chamaecyparis obtusa* (Siebold and Zucc.) Endl. leaves by alcohol-based DESs. Wang et al. [13] extracted polyphenols and furanocoumarins from fig (*Ficus carica* L.) with tailor-made DESs and showed that DESs were effective for the extraction of these components. Anthocyanins in the flower petals of *Catharanthus roseus* L. were extracted with natural deep eutectic solvents (NADESs) such as lactic acid—glucose and propane-1,2-diol—choline chloride, which provided higher stability for anthocyanins [12]. The extraction with DES can be improved by combination with ultrasound (UAE), microwaves and heating as well as by mechanochemical extraction (MCE). Therefore, Bosiljkov et al. [14] successfully extracted anthocyanins with UAE combined with DESs. Wang et al. [4] developed fast, efficient, and ecofriendly MCE for tanshinones as well as bioactive compounds from tea leaves [15]. Due to the desirability of DESs in the extraction of phenolic components, we decided to extract phenolic diterpenes, carnosic acid, and carnosol from sage (*Salvia officinalis* L.) applying DESs. These compounds have been suggested to account for over 90% of the sage antioxidant properties [16]. In addition to antioxidant activity, carnosic acid, and carnosol showed proapoptotic [17], antiproliferative [18], anti-angiogenic [19] and antitumor activity. So far, there are no available data on DESs extraction and optimization of the parameters for carnosic acid and carnosol from sage. There are few reports dealing with DESs extraction of sage by Bakirtzi et al. [2] and Georgantzi et al. [20] who investigated the influence of different DESs on the extraction of polyphenols from medicinal plants including sage. In paper by Bakirtzi et al. [2] lactic acid-based natural deep eutectic solvents in combination with ultrasound were used for the extraction of total polyphenols and flavonoids from sage, while Georgantzi et al. [20] investigated combination of lactic acid-based DES with cyclodextrin for UAE of total polyphenols and flavonoids.

Taking into account all the above mentioned, the objectives of this study were focused on (1) investigation on finding appropriate choline chloride based deep eutectic solvent for the extraction of carnosic acid and carnosol as well as (2) suitable extraction techniques (stirring with heating, UAE or MCE). Afterwards, the influence of various DES extraction parameters (H_2O content, time, and temperature for stirring and heating and UAE; H_2O content, extraction time and vibration speed for mechanochemical extraction) on the (3) content of carnosic acid and carnosol in sage extract analyzed by HPLC was investigated. In addition, (4) the optimal extraction conditions by RSM for desired antioxidant components (carnosic acid and carnosol) were determined.

2. Results and Discussion

2.1. Influence of DESs on the Obtained Amount of Carnosic Acid and Carnosol in the Extracts

Due to the different effects of viscosity, surface tension, polarity, and HBD interaction it is hard to estimate the suitability of a DES for the extraction of targeted compounds. Therefore, in order to select the best DES for the extraction of carnosic acid and carnosol from sage, the extraction was performed with different solvents and different H_2O addition at 30 °C (Figure A1). According to Dai et al. [12] and Bosiljkov et al. [14], H_2O addition in organic acid-based DESs causes decrease of the solvent polarity since these solvents are more polar than H_2O. Therefore, for targeted components it would be more suitable to lower H_2O addition (which is consistent with the results obtained). As can be seen from Figure A1, the solvents substantially differ in their ability to extract carnosic acid and carnosol. In addition to the influence of HBD, the amount of H_2O added also plays an important role for the extraction efficiency. For certain solvents, like choline chloride:malonic acid (1:1 molar ratio), the amount of carnosic acid was increased with increased H_2O content that may be due to the viscosity lowering effect. For DES choline chloride:citric acid (1:1 molar ratio), the highest amount of carnosic acid and carnosol was obtained with 30% H_2O addition that may be the consequence of a high viscosity of the solvent with 10% H_2O addition. H_2O amount was changed to reduce the viscosity which causes a slow mass transfer, thus affecting the extraction process. The viscosity of DESs can be reduced by the addition of a certain amount of H_2O as well as by increasing the temperature [12]. Several solvents such as choline chloride:glucose (1:1 molar ratio) were too viscous even with the addition of H_2O at 50% (v/v). Although the addition of H_2O can decrease the viscosity, an excessive concentration of H_2O can decrease the interactions between the components of DES as well the interactions between DES and desired components [11]. This is the reason why the focus was on H_2O addition in the range of 10–50% (v/v).

Carnosol can be extracted with all DESs applied, but choline chloride based DESs with butane-1,4-diol (1:2 molar ratio) (ChClB) and ethane-1,4-diol (1:2 molar ratio) (ChClE) were the most effective. The result of extraction with choline chloride:glucose (1:1 molar ratio) with 10% H_2O were not shown because of the high viscosity preventing further analysis. The same limitation was observed with choline chloride:fructose (1:1 molar ratio) (ChClF) and choline chloride:citric acid (1:1 molar ratio) (ChClC) at 10% of H_2O and at 30 °C. The higher amount of carnosol was observed at lower H_2O content (Figure A1). With choline chloride:urea (1:2) (ChClU) as the solvent, the highest amount of carnosol was extracted at 30 °C. According to the literature, higher content of carnosol is usually present in the extracts obtained at higher temperatures since it is one of the degradation products of carnosic acid [20].

On the other hand, not all applied solvents could extract carnosic acid. Basic solvents, such as ChCl:U, choline chloride:*N*-methlyurea (1:3 molar ratio) (ChClmU), and choline chloride:thiourea (1:2 molar ratio) (ChCltU), were not efficient for carnosic acid extraction. The highest amount of carnosic acid was extracted using acidic DESs, with emphasis on choline chloride:lactic acid (1:2 molar ratio) (ChClLa). Such solvents are significantly acidic (pH < 3) compared to the solvents where HBDs were sugars or alcohols (pH > 6) [7,12]. Additionally, the polarity of DESs should also be considered as an important criterion for the evaluation and selection of the solvents to achieve maximum extraction efficiency. Carnosic acid and carnosol are polar constituents, soluble in polar solvents (which is in agreement with the obtained results), such as DESs with organic acids as HBDs, which are more polar than the DESs with sugars as HBDs [12]. The highest amount of carnosic acid was obtained with the lowest amount of added H_2O (10%) in most solvents. Okamura et al. [21] have investigated the effect of temperature on the degradation of carnosic acid in acetone solution and reported that the increase of temperature affected the degradation of carnosic acid. In case of the extraction with DESs such as ChClLa and choline chloride:levulinic acid (1:2 molar ratio) (ChClL) the maximum amount of carnosic acid was extracted at 30 °C.

Since carnosol can be extracted with all DESs applied and due to the highest amount of extracted carnosic acid in the case of choline chloride:lactic acid (1:2 molar ratio) and the lactic acid properties as natural component, this solvent was selected for further optimization of the extraction with three extraction methods (stirring and heating, UAE, and MCE).

2.2. Comparison of the Used Extraction Methods

After selection of the appropriate solvent, the extractions performed by stirring and heating and UAE, applying the same temperature and H_2O content as well as MCE were compared. Both, stirring and heating and UAE increase mass transfer and speed up diffusion of the compounds. In the case of ultrasound, acoustic cavitation phenomenon leads to the disruption of cell walls and consequently improves the yield of extraction compared to maceration [22]. However, MCE can decrease the processing time and solvent consumption and reduce noise and radiation compared to UAE and to stirring and heating extraction.

Table 1 shows that slightly higher amounts of carnosol and carnosic acid were extracted by stirring and heating, compared to UAE. Such results can be explained by the positive influence of stirring on the mass transfer in such viscous solvent. For MCE, the utilization of the glass beads led to much better mixing of the plant material and solvent, thus extracting significant amounts of carnosic acid and carnosol in a shorter time compared to the other two extractions (Table 2).

Table 1. Experimental matrix and values (μg mg^{-1} of the plant material) of observed response for the extraction with choline chloride:lactic acid (1:2 molar ratio) obtained by stirring with heating and by ultrasound-assisted extraction (UAE). The results are expressed as mean value ± standard deviation (n = 3).

Run	H_2O (%)	Time (min)	Temperature (°C)	DES-MIXING		DES-UAE	
				Carnosic Acid (μg mg^{-1})	Carnosol (μg mg^{-1})	Carnosic Acid (μg mg^{-1})	Carnosol (μg mg^{-1})
1	30	60	50	3.86 ± 0.82	1.99 ± 0.05	7.37 ± 0.34	3.06 ± 0.21
2	10	90	50	14.43 ± 0.28	4.83 ± 0.09	9.94 ± 0.22	3.92 ± 0.09
3	10	60	70	10.16 ± 0.05	3.90 ± 0.37	14.00 ± 0.02	4.18 ± 0.05
4	30	60	50	4.53 ± 0.49	2.27 ± 0.09	4.18 ± 0.21	2.22 ± 0.14
5	30	60	50	5.16 ± 0.45	2.70 ± 0.17	5.04 ± 0.05	2.46 ± 0.29
6	10	30	50	8.71 ± 0.26	2.10 ± 0.06	8.48 ± 0.06	3.34 ± 0.10
7	30	30	70	5.64 ± 0.11	2.51 ± 0.22	1.62 ± 0.29	0.70 ± 0.02
8	50	90	50	3.16 ± 0.14	1.34 ± 0.21	2.09 ± 0.37	0.84 ± 0.29
9	30	90	70	7.41 ± 0.06	4.79 ± 0.38	8.06 ± 0.13	2.93 ± 0.02
10	10	60	30	10.68 ± 0.66	2.35 ± 0.03	2.66 ± 0.41	0.83 ± 0.38
11	50	60	70	3.87 ± 0.19	1.85 ± 0.15	3.69 ± 0.24	1.85 ± 0.11
12	50	30	50	2.68 ± 0.26	0.81 ± 0.01	2.00 ± 0.07	0.82 ± 0.13
13	30	60	50	5.65 ± 0.36	1.99 ± 0.09	4.37 ± 0.09	2.25 ± 0.17
14	50	60	30	2.55 ± 0.04	0.89 ± 0.03	1.94 ± 0.21	0.92 ± 0.09
15	30	60	50	5.44 ± 0.21	1.79 ± 0.11	2.46 ± 0.05	1.27 ± 0.16
16	30	90	30	5.27 ± 0.18	1.70 ± 0.08	2.80 ± 0.25	1.22 ± 0.12
17	30	30	30	5.37 ± 0.01	1.50 ± 0.19	2.18 ± 0.23	0.56 ± 0.02

For the constant H_2O content in all extraction methods, the extracted amounts of carnosic acid and carnosol obtained by different extractions were compared. The extracted amounts of selected compounds (8.26 and 7.92 μg mg^{-1} for carnosic acid and 1.87 and 2.02 μg mg^{-1} for carnosol) obtained by MCE at Run 12 and 17 can be compared to Run 6 and 9 obtained with stirring and heating extraction and UAE. The difference between the parameters of these extractions was the extraction time, so at 10% H_2O for MCE, 2 min were enough to obtain similar amounts of targeted compounds as for 30 min of stirring and heating extraction and UAE. In the case of 30% H_2O by MCE, 3 min were sufficient to obtain the amount of extracted components similar to the amount extracted for 90 min by stirring and heating extraction and UAE. It is important to note that MCE was carried out at room temperature (24–28 °C). In fact, by using mill, we wanted to show how much time was needed for the extraction at room temperature, and prolonging the extraction time would result in warming of the samples without the possibility of heating control.

Table 2. Experimental matrix and values (μg mg^{-1} of the plant material) of observed response for the extraction with choline chloride:lactic acid (1:2 molar ratio) obtained by mechanochemical extraction (MCE). The results are expressed as mean value ± standard deviation (n = 3).

Run	H_2O (%)	Time (min)	Vibration Speed (m s^{-1})	Carnosic Acid (μg mg^{-1})	Carnosol (μg mg^{-1})
1	30	3	1	3.50 ± 0.35	1.25 ± 0.02
2	30	2	3	3.38 ± 0.20	1.23 ± 0.06
3	50	3	3	2.51 ± 0.34	0.86 ± 0.02
4	50	1	3	2.01 ± 0.16	0.71 ± 0.04
5	50	2	5	2.98 ± 0.02	1.03 ± 0.13
6	10	3	3	4.29 ± 0.25	1.16 ± 0.04
7	50	2	1	1.80 ± 0.02	0.57 ± 0.11
8	10	2	1	3.37 ± 0.17	1.02 ± 0.20
9	10	1	3	4.06 ± 0.26	1.04 ± 0.37
10	30	1	5	4.69 ± 0.61	1.27 ± 0.11
11	30	2	3	3.31 ± 0.42	1.01 ± 0.09
12	10	2	5	8.26 ± 0.45	1.87 ± 0.33
13	30	2	3	3.63 ± 0.18	1.14 ± 0.01
14	30	2	3	3.45 ± 0.28	1.09 ± 0.12
15	30	1	1	2.45 ± 0.07	0.79 ± 21
16	30	2	3	3.41 ± 0.24	0.98 ± 0.03
17	30	3	5	7.92 ± 0.29	2.02 ± 0.14

2.3. *Influence of Various DES Extraction Parameters on the Content of Carnosol and Carnosic Acid*

The effect of H_2O addition, temperature or vibration speed and extraction time on carnosol and carnosic acid was investigated for three extraction techniques using DES choline chloride:lactic acid (1:2 molar ratio). In these experiments, the content of carnosic acid in sage extract obtained by stirring and heating was 2.55–14.43 μg mg^{-1}, depending on the applied extraction parameters. The lowest content of carnosic acid was obtained at 50% (v/v) H_2O added at 30 °C and 60 min, while the highest content was obtained at 10% (v/v) H_2O added at 50 °C and 90 min (Table 1). The content of carnosic acid obtained by UAE varied, depending on the parameters used, in the range 1.62–13.99 μg mg^{-1}. The lowest content of carnosic acid was obtained at 30% (v/v) H_2O added, 70 °C, and 30 min and the highest yield at 10% (v/v) H_2O added, 70 °C and 60 min (Table 1). The content of carnosic acid obtained by MCE varied depending on the parameters used in the range 1.80–8.26 μg mg^{-1}. The lowest content of carnosic acid was at 50% (v/v) H_2O added, vibration speed of 1 m/s and 2 min and the highest yield at 10% (v/v) H_2O added, 5 m/s and 2 min (Table 2). The content of carnosol obtained by mixing and heating was 0.81–4.83 μg mg^{-1} depending on the applied extraction parameters. The lowest content of carnosol was obtained at 50% (v/v) H_2O addition, 50 °C and 30 min, while the highest yield was obtained at 10% (v/v) of H_2O, 50 °C and 90 min. The content of carnosol, depending on the parameters used in UAE, was 0.56–4.18 μg mg^{-1} with the lowest content at 30% (v/v) H_2O addition, 30 °C and 30 min and the highest yield at 10% H_2O addition, 70 °C and 60 min (Table 1). The content of carnosol obtained by MCE was 0.57–2.02 μg mg^{-1} depending on the applied extraction parameters (Table 2). The lowest content of carnosol was obtained at 50% of H_2O (v/v), vibration speed of 1 m/s and time of 1 min, while the highest content was obtained at 30% of H_2O (v/v), 5 m/s and 3 min.

The addition of H_2O and extraction time (Figure 1 and Table A1) showed statistically significant influence on the content of carnosic acid ($p < 0.0001$; $p = 0.0202$) in the extracts obtained by stirring and mixing. The content of carnosic acid increased with prolonged extraction time and decreased with the increase of H_2O amount.

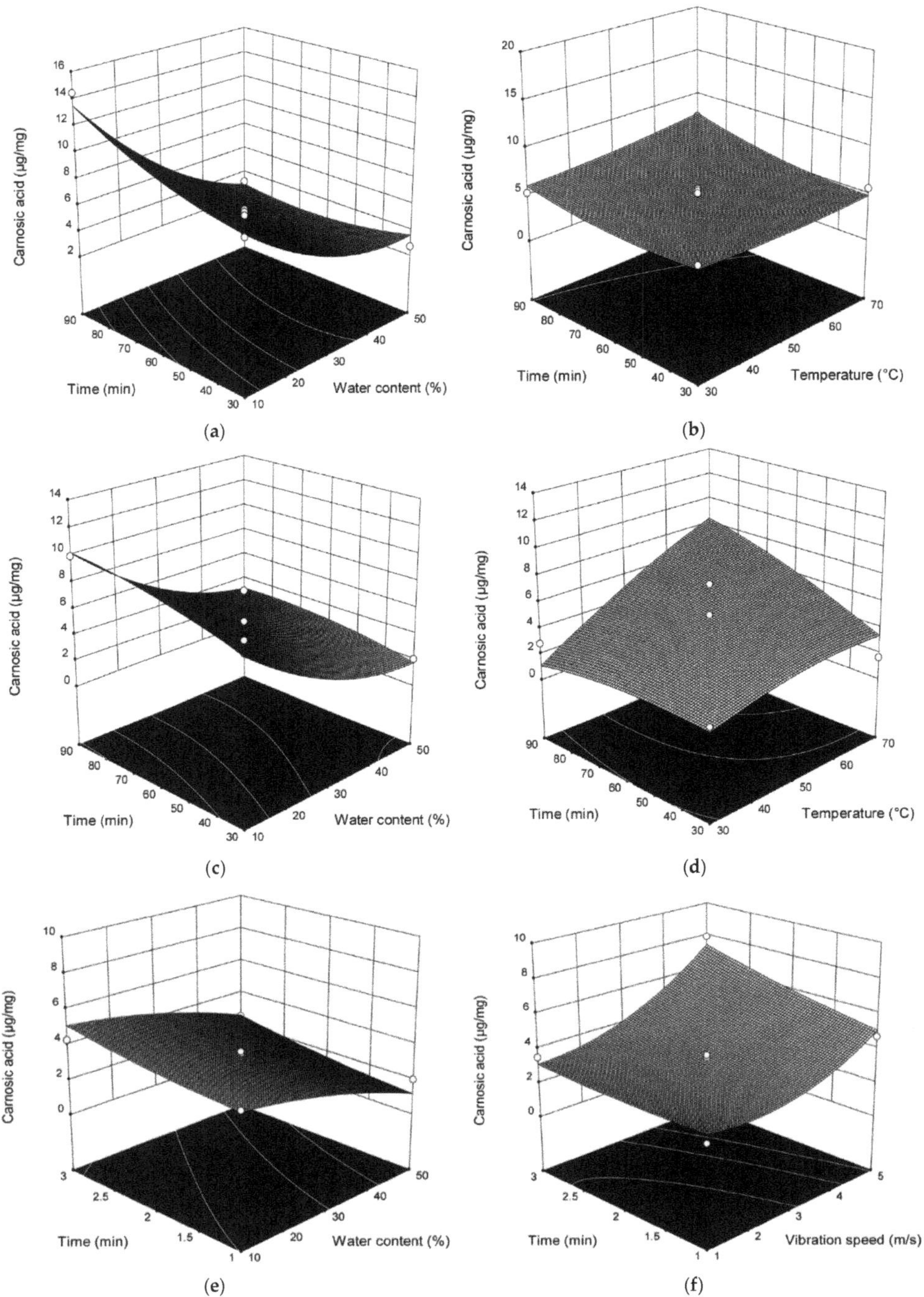

Figure 1. Three-dimensional plots for obtained content of carnosic acid as a function of the extraction time, temperature, and H_2O content for the extraction with mixing and heating (**a**,**b**), UAE (**c**,**d**) and for the extraction time, vibrational speed, and H_2O content for MCE (**e**,**f**).

The interactions between amount of H_2O added and extraction time ($p = 0.0259$) also showed a significant influence on the content of carnosic acid. In the extracts obtained by UAE, H_2O addition and temperature showed statistically significant influence on the content of carnosic acid ($p = 0.0025$; $p = 0.0144$). For this extraction technique, interactions between the amount of added H_2O and temperature ($p = 0.0433$) also showed a significant influence in terms of content of carnosic acid. The content of carnosic acid increased with increased extraction temperature and decreased with the increase of H_2O amount. In the extracts obtained by MCE, H_2O addition, time, and vibration speed ($p = 0.0006$; $p = 0.0266$; $p = 0.0002$) as well as the interactions between H_2O addition and vibration speed ($p = 0.0221$) showed statistically significant influence on the content of carnosic acid. The content of carnosic acid increased with prolonged extraction time and vibration speed and decreased with the increase of H_2O amount.

As can be seen from Figure 2 and Table A2, H_2O addition, extraction time, and temperature showed statistically significant influence on the content of carnosol ($p < 0.0001$; $p = 0.0008$; $p = 0.0003$) in the extracts obtained by stirring and mixing. The content of carnosol is increased with increased time and temperature of the extraction and with decreased H_2O amount.

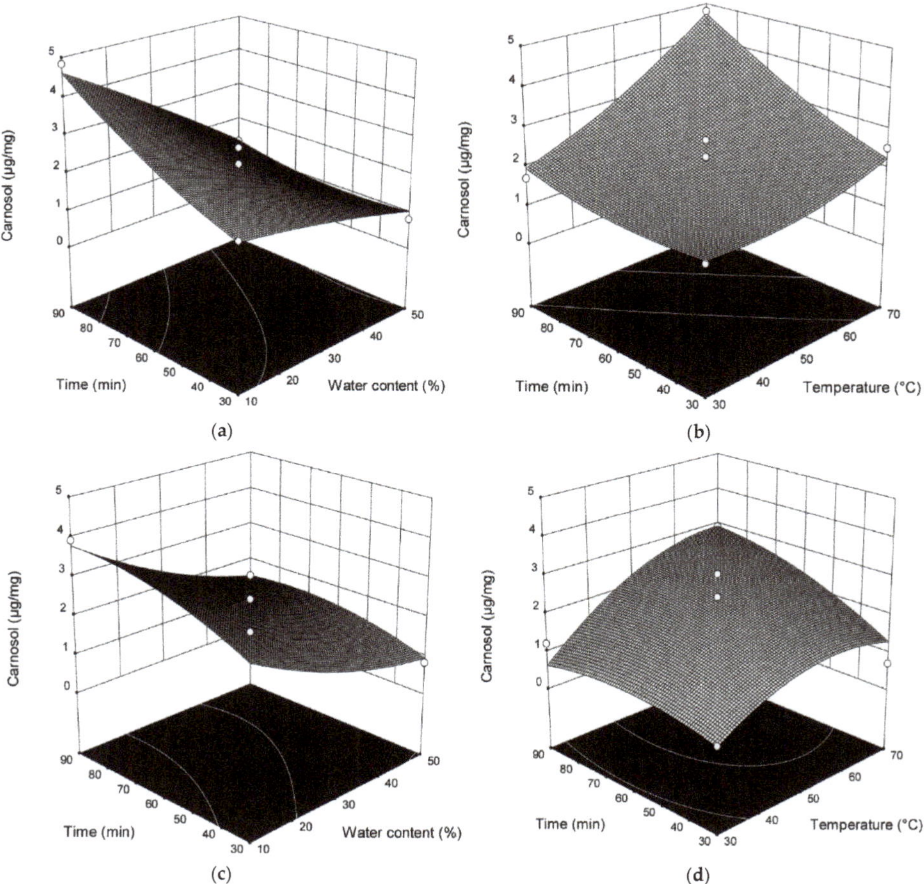

Figure 2. Three-dimensional plots for obtained content of carnosol as a function of the extraction time, temperature, and H_2O content in the extraction with mixing and heating (**a**,**b**) and MCE (**c**,**d**).

Interactions between amount of H_2O added and the extraction time and between the amount of H_2O added and temperature ($p = 0.0184$; $p = 0.0234$) also showed a significant influence for the content of carnosol. In the extracts obtained by MCE, H_2O addition, time, and vibration speed ($p = 0.0055$; $p = 0.0187$; $p = 0.0012$) showed statistically significant influence on the content of carnosol. The content of carnosol increased with prolonged extraction time and vibration speed and decreased with the increase of H_2O amount. Since model according to RSM is not significant for the extraction of carnosol with ultrasound ($p = 0.0708$), the results obtained for that extraction are not discussed. To optimize the extraction conditions of two different phenolic diterpenes 17 runs determined by BBD with three variables (percentage of H_2O added, time and temperature or vibration speed) at three levels were used to fit a second-order response surface. The amount of carnosic acid and carnosol were observed as the response (Tables A1 and A2).

The data describing the optimal conditions for the extraction of carnosic acid and carnosol from sage using DESs are not available in the literature, but there are few papers investigating the optimal conditions with other solvents. In paper by Fatma Ebru et al. [23] it was shown that 70% of ethanol was the most efficient solvent since it extracted 3.45 mg carnosol + carnosic acid per g of the extract. According to the optimization carried out, they showed that the amount of these bioactive components was in the function of extraction time. In addition, they also demonstrated that carnosol and carnosic acid degraded easily at higher temperatures over a longer period of time. Therefore, they have shown that the optimum conditions were temperatures of 40–50 °C, the extraction time 3–6 h, solvent-to-sage ratio 6:1 (v/w) and 70–80 wt.% ethanol for maceration. Similar results were also showed in paper by Durling et al. [24]. According to the optimization carried out, the amount of targeted components depended on several parameters such as particle size, temperature, time, and a solvent-to-sage ratio. The highest concentration of targeted components was obtained with the particle size 1 mm, 40 °C, the extraction time of 3 h, the solvent-to-sage ratio of 6:1 (v/w) and 55–75 wt.% ethanol. Under these conditions, the extract containing 10.6% carnosic compounds was obtained.

The optimization process of extraction is important for determining the most favorable conditions for achieving maximum yields of desired components in the extracts. Based on BBD, estimated coefficients of second order response models for carnosol and carnosic acid in sage extracts are given in Tables A1 and A2. R^2 for carnosic acid was 0.9630 and for carnosol was 0.9607 in the extracts obtained by stirring and heating, and for UAE R^2 for carnosic acid was 0.8660. In the case of MCE, R^2 for carnosic acid was 0.9442 and for carnosol R^2 was 0.9032. According to ANOVA, statistically significant models for carnosic acid (Table A3) and carnosol in the extraction by stirring and heating (Table A4) ($p \leq 0.05$) were obtained. Additionally, the obtained models showed non-significant lack of fit ($p = 0.2042$–0.4491), except in the case of MCE for carnosic acid ($p = 0.0008$).

According to RSM, optimum conditions are expressed as those at which it is possible to achieve the maximum amount of carnosic acid and carnosol. They are slightly different depending on the extraction technique used, so for the extraction with heating and mixing they were 10% H_2O addition, 90 min and 70 °C, while for UAE they were 11.05% of H_2O addition, 82.36 min and 69.84 °C. Under these optimal conditions, the content of carnosic acid and carnosol was calculated as 14.20 μg mg^{-1} and 6.47 μg mg^{-1} in case of stirring and heating and 14.72 μg mg^{-1} of carnosic acid for ultrasound extraction. The desirability for these optimizations was 0.990 and 1.0, respectively. The experimental results for the amount of carnosic acid and carnosol obtained at optimum conditions were 13.73 ± 0.26 and 6.15 ± 0.33 μg mg^{-1} for the extraction with stirring and heating, while for UAE this amount was 14.24 ± 0.21 μg mg^{-1}. Optimum conditions for MCE were 11.13% H_2O addition, time of extraction 2.90 min, and vibration speed 4.98 m s^{-1}. Under these optimal conditions, the content of carnosic acid and carnosol is calculated as 8.95 μg mg^{-1} and 2.02 μg mg^{-1} with the desirability 1.0 which was confirmed experimentally (8.90 ± 0.10; 2.03 ± 0.04 μg mg^{-1}).

2.4. Comparison with Other Extraction Methods

According to the literature, the most common solid–liquid extraction of sage has been performed with methanol. Due to the toxic effect of methanol, it is preferable to use ethanol which can be classified as bio-solvent and is much safer for the use [25,26]. In the paper by Abreu et al. [27] the content of carnosic acid and carnosol in methanolic extract of sage was 14.6 mg g^{-1} of dry weight and 0.4 mg g^{-1}, respectively. This is similar to our results for Run 2 (mixing and heating) and Run 3 (UAE), but with a significantly higher amount of carnosol in our case. Sage extraction with 80% methanol over 24 h at room temperature led to the extraction of carnosic acid only with the content of 273.8 mg 100 g^{-1} of the plant dry weight [28], much lower than our results. In other case, the extraction with 50% methanol during 60 min in ultrasound bath has brought carnosic acid content of 2.1 g kg^{-1} extract and carnosol content of 4.1 to 15.1 mg g^{-1} of plant dry weight [29].

According to Table 3, which shows our results obtained by the stirring with heating extraction of the same sage material with common solvents, it is observed that the most effective solvent is absolute ethanol, while H_2O is the least effective solvent for the extraction of carnosic acid and carnosol. The preparation of aqueous solutions of ethanol in the range of 30–70% (v/v) shows that the increase in the volume of ethanol (v/v) increased the amount of extracted components. In this case, methanol as the extraction solvent shows lower extraction efficiency compared to ethanol. In addition, the influence of extraction parameters such as extraction time and temperature can be observed in Table 3. However, when ethanol is used as the extraction solvent and with the most efficient extraction conditions applied (50 ° C and 90 min), lower amount of carnosic acid and carnosol was obtained compared to the selected DES (choline chloride:lactic acid 1:2).

Considering the adverse properties of organic solvents and in order to overcome their disadvantages, such as low selectivity for antioxidant compounds [30], safe or green solvents and processes have been used. Supercritical fluid extraction (SFE) has been used in the plant material extraction due to its ability to provide clean extracts without residual solvent [31]. In addition, SFE can be performed at low temperatures in short time, which is suitable for carnosic acid oxidation prevention during the extraction, also supported by the absence of air and light during the extraction process thus reducing its degradation [32]. In our previous work [33] we used the same herbal material for carnosic acid and carnosol extraction using SFE with CO_2 (SC-CO_2). Comparing the results, the highest amount of extracted carnosic acid using SC-CO_2 was 855.8 mg 100 g^{-1} of the plant material (30 MPa, 50 °C, 1 kg h^{-1} CO_2), while the extraction yield using DESs was 1443.22 mg 100 g^{-1} and 1399.22 mg 100 g^{-1}, depending on the extraction technique employed. In the case of carnosol, the highest amount was extracted under the same conditions of SC-CO_2 (446.35 mg 100 g^{-1}), and similar results were achieved using DESs (483.34 and 418.39 mg 100 g^{-1} of plant, depending on the extraction technique employed). However, certified reference material was not used and therefore minor changes in the composition of the plant material are possible with respect to the same sample used in our previously published data. In the paper published by Babovic et al. [34] the content of carnosic acid obtained by SC-CO_2 was 13.76 g per 100 g of the extract and carnosol content was 6.97 g per 100 g of the extract similar to our results (11.63 g carnosic acid per 100 g and 8.55 g carnosol per 100 g) [32]. Despite the fact that SC-CO_2 extraction conditions may reduce carnosic acid degradation, we still notice that more carnosic acid was extracted and preserved by DESs extraction even at higher extraction temperatures. On the other hand, preparing DESs is simple and inexpensive, i.e., the price is comparable to the cost of the conventional solvents. Moreover, this is sustainable process theoretically without generated waste [10] which makes this extraction process suitable for the extraction of bioactive components including carnosic acid and carnosol.

Table 3. The values (μg mg^{-1} of the plant material) of carnosic acid and carnosol for the extraction obtained by stirring with heating. The results are expressed as mean value ± standard deviation (n = 3).

Solvent	Time (min)	Temperature (°C)	Carnosic Acid (μg mg^{-1})	Carnosol (μg mg^{-1})
30% ethanol (v/v)	30	30	1.11 ± 0.03	2.63 ± 0.21
	60		1.33 ± 0.55	2.26 ± 0.02
	90		1.45 ± 0.53	2.18 ± 0.20
	30	50	2.32 ± 0.07	2.04 ± 0.05
	60		2.64 ± 0.04	1.99 ± 0.04
	90		2.26 ± 0.30	1.51 ± 0.00
	30	70	2.82 ± 0.03	2.92 ± 0.16
	60		2.13 ± 0.14	2.92 ± 0.311
	90		0.93 ± 0.17	2.27 ± 0.38
50% ethanol (v/v)	30	30	5.91 ± 0.19	4.65 ± 0.22
	60		3.07 ± 0.39	9.31 ± 0.29
	90		3.02 ± 0.15	9.73 ± 0.86
	30	50	7.17 ± 0.05	8.39 ± 0.47
	60		3.15 ± 0.02	9.06 ± 0.11
	90		2.11 ± 0.15	11.25 ± 0.35
	30	70	7.63 ± 0.44	6.73 ± 0.38
	60		4.43 ± 0.20	8.79 ± 0.79
	90		1.91 ± 0.00	11.23 ± 0.13
70% ethanol (v/v)	30	30	8.28 ± 0.53	3.04 ± 0.01
	60		7.40 ± 0.05	4.82 ± 0.29
	90		7.63 ± 0.0	5.93 ± 0.22
	30	50	7.73 ± 0.22	6.17 ± 0.59
	60		8.54 ± 0.28	7.21 ± 0.44
	90		8.71 ± 0.28	5.43 ± 0.51
	30	70	8.73 ± 0.14	4.37 ± 0.07
	60		7.53 ± 0.06	6.55 ± 0.22
	90		6.85 ± 0.32	6.89 ± 0.26
ethanol	30	30	11.21 ± 0.51	2.72 ± 0.27
	60		11.13 ± 0.13	3.57 ± 0.25
	90		12.77 ± 0.22	2.83 ± 0.06
	30	50	11.74 ± 0.09	3.57 ± 0.47
	60		12.80 ± 0.19	3.14 ± 0.06
	90		13.64 ± 0.10	3.47 ± 0.01
	30	70	13.36 ± 0.37	4.46 ± 0.11
	60		12.27 ± 0.11	3.29 ± 0.21
	90		11.27 ± 0.05	3.11 ± 0.09
methanol	30	30	8.71 ± 0.87	2.88 ± 0.48
	60		9.26 ± 0.06	4.44 ± 0.19
	90		10.50 ± 0.58	4.69 ± 0.58
	30	50	9.68 ± 0.25	5.03 ± 0.08
	60		11.85 ± 0.05	5.24 ± 0.07
	90		10.72 ± 0.30	4.85 ± 0.40
	30	70	9.69 ± 0.58	3.67 ± 0.24
	60		10.11 ± 0.43	4.41 ± 0.10
	90		10.41 ± 0.12	4.80 ± 0.03
H_2O	30	30	0.00	0.00
	60		0.00	0.24 ± 0.00
	90		0.00	0.24 ± 0.00
	30		0.00	0.29 ± 0.01
	60		0.00	0.27 ± 0.00
	90		0.75 ± 0.03	0.24 ± 0.00
	30		0.00	0.26 ± 0.03
	60		0.00	0.25 ± 0.01
	90		0.74 ± 0.00	0.42 ± 0.02

3. Materials and Methods

3.1. Chemicals

The standard compounds carnosic acid (≥95.0%) and carnosol (99.2%) (Sigma Chemical Co., St. Louis, MO, USA) were used for the chemical analyses. All solvents were of analytical grade and purchased from J.T. Baker (Avantor, Phillipsburg, NJ, USA).

3.2. Plant Material

Dried sage leaves (*Salvia officinalis* L.) were used for experiments. Prior to the extraction, the dried leaves were grounded and sieved using a vertical vibratory sieve shaker (LabortechnikGmbh, Ilmenau, Germany) as described in paper by Jokić et al. [35].

3.3. Preparation of DES

The choline chloride based DESs were prepared as described in our paper [36]. In this study, seventeen different choline chloride based DESs were prepared using inexpensive components as shown in Table 4.

Table 4. Preparation of deep eutectic solvents (DESs).

Name	Combination	Molar Ratio
ChClU	Choline chloride:urea	1:2
ChClmU	Choline chloride:N-methylurea	1:3
ChCltU	Choline chloride:thiourea	1:2
ChClG	Choline chloride:glucose	1:1
ChClF	Choline chloride:fructose	1:1
ChClX	Choline chloride:xylitol	1:1
ChClS	Choline chloride:sorbitol	1:1
ChClB	Choline chloride:butane-1,4-diol	1:2
ChClE	Choline chloride:ethane-1,2-diol	1:2
ChClGl	Choline chloride:glycerol	1:2
ChClA	Choline chloride:acetamide	1:2
ChClM	Choline chloride:malic acid	1:1
ChClC	Choline chloride:citric acid	1:1
ChClMa	Choline chloride:malonic acid	1:1
ChClO	Choline chloride:oxalic acid	1:1
ChClLa	Choline chloride:lactic acid	1:2
ChClL	Choline chloride:levulinic acid	1:1

3.4. Extraction of Carnosic Acid and Carnosol with DESs

Grounded *Salvia officinalis* L. dried leaves (50 mg) were mixed with 1 mL of the selected solvent, a pure DES or a mixture of DES and ultrapure H_2O (Millipore Simplicity 185, Darmstadt, Germany). Prepared samples were stirred at 1500 rpm in aluminum block (Stuart SHB) on a magnetic stirrer or ultrasound treated in temperature-controlled ultrasonic bath at specified temperature for the certain time (Table 1). The temperature-controlled ultrasonic bath (Elma P70 H, Singen, Germany) was set with frequency at 37 Hz and power at 50 W at the same temperature over the same time as in case of mixing in aluminum block (Table 1). Prepared samples (50 mg of plant + 1 g of glass beads with 1 mL of solvent) were also extracted on the BeadRuptor 12 ball mill (Omni International, Kennesaw, GA, USA) according to parameters in Table 2 at room temperature (24–28 °C). After the extraction, the mixture was centrifuged for 15 min and then decanted. The supernatant liquid was then diluted with methanol, filtered through a PTFE 0.45 µm filter, and subjected to HPLC analysis.

3.5. Extraction of Carnosic Acid and Carnosol with Conventional Solvents

Grounded *Salvia officinalis* L. dried leaves (50 mg) were mixed with 1 mL of selected solvent (Millipore Simplicity 185, Darmstadt, Germany). Prepared samples were stirred at 1500 rpm in aluminum block (Stuart SHB) on a magnetic stirrer at specified temperature for the certain time (Table 3).

3.6. Chemical Characterization of the Extracts

HPLC analyses of carnosic acid and carnosol was performed on an Agilent 1260 Infinity II (Agilent, Santa Clara, California, USA) with chromatographic separation obtained on a ZORBAX Eclipse Plus C18 (Agilent, Santa Clara, CA, USA) column (100 × 4.6 mm, 5 μm).

The separation of analyzed compounds was made with method described in our previous paper [31], but since analysis was performed on different device, linearity of the calibration curve, LOQ and LOD was confirmed. Standard stock solutions for carnosic acid and carnosol were prepared in a methanol and calibration was obtained at eight concentrations (concentration range 10.0, 20.0, 30.0, 50.0, 75.0, 100.0, 150.0, and 200.0 mg L^{-1}). Due to R^2 = 0.99789 for carnosic acid and R^2 = 0.99968 for carnosol, calibration curve was confirmed. Limit of detection were 0.795 mg L^{-1} and 0.971 mg L^{-1} for carnosic acid and carnosol, respectively. Limit of quantification were 2.648 mg L^{-1} and 7.416 mg L^{-1} for carnosic acid and carnosol, respectively. Retention time for carnosic acid was 7.416 min, while for carnosol was 4.217 min. The chromatograms of the standard and real sample are shown in the Appendix A (Figure A2). For the validation of the HPLC method for the determination of carnosic acid and carnosol, in addition to linearity, retention time comparison and absorption spectrum comparison with standards, repeatability of measurements and solution preparation as well as accuracy were performed, which is also shown in the Appendix A (Table A5).

3.7. Statistical Experimental Design

BBD explained in detail by Bas and Boyaci [37] was used for determination of optimal DES (stirring and heating), UAE-DES and MCE-DES extraction conditions in terms of getting higher amount of carnosic acid and carnosol in the *S. officinalis* extracts. Independent variables in design were H_2O content (X_1), time (X_2) and temperature (X_3) and vibration speed (X_3) and tested levels were reported in Table 5. Design-Expert® Commercial Software (ver. 9, Stat-Ease Inc., Minneapolis, MN, USA) was used for data analysis. The analysis of variance (ANOVA) was also used to evaluate the quality of the fitted model, and the test of statistical difference was based on the total error criteria with a confidence level of 95.0%.

Table 5. Coded and real levels of independent variables for the designed experiment.

Type of Extraction	Independent Variable	Symbol	Level		
			Low (−1)	Middle (0)	High (+1)
Stirring and heating	H_2O (%)	X_1	10	30	50
	Time (min)	X_2	30	60	90
	Temperature (°C)	X_3	30	50	70
Ultrasound assisted extraction	H_2O (%)	X_1	10	30	50
	Time (min)	X_2	30	60	90
	Temperature (°C)	X_3	30	50	70
Mechanochemical extraction	H_2O (%)	X_1	10	30	50
	Time (min)	X_2	1	2	3
	Vibration speed (ms^{-1})	X_3	1	3	5

4. Conclusions

In present study, determination of suitable deep eutectic solvent and optimization of the extraction of carnosol and carnosic acid from sage were performed. Among 17 different solvents, choline chloride:lactic acid (1:2 molar ratio) was selected for the extraction by heating and mixing, as well as for ultrasound and mechanochemical extraction. The content of carnosic acid and carnosol was slightly higher in the extracts obtained by stirring and heating and mechanochemical extraction. The influence of H_2O content, extraction time and temperature (for stirring and heating and for ultrasound-assisted extraction (UAE)), H_2O content, extraction time and vibration speed for mechanochemical extraction on the content of targeted compounds were investigated. Optimal extraction conditions determined by response surface methodology (RSM) were in accordance with experimentally demonstrated values.

Compared to SC-CO_2 extraction, we observed that more carnosic acid is extracted using DESs, with emphasis on ChClLa, while the amount of carnosol detected in the extract obtained by ChClLa is similar to that obtained by SC-CO_2. In addition, the comparison with the solvents such as ethanol, H_2O, aqueous solutions of ethanol (30–70% (v/v)) and methanol under the same extraction conditions, showed that choline chloride:lactic acid (1:2 molar ratio) was more efficient for the extraction of carnosic acid and carnosol compared to used conventional solvents.

Given the amounts of carnosic acid achieved at high temperatures in DES in further research it would be useful to examine the stability of the component over the certain period of time.

Author Contributions: Conceptualization, M.M. and M.J.; methodology, M.J., S.J., M.M., and I.J.; formal analysis, M.J. and S.J.; investigation, M.J. and M.M.; resources, I.J.; data curation, M.M.; writing—original draft preparation, M.J.; writing—review and editing, S.J., M.M., and I.J.; visualization, S.J., M.M., and I.J.; supervision, M.M., S.J., and I.J.; funding acquisition, I.J. All authors have read and agreed to the published version of the manuscript.

Funding: This research was funded by Croatian Government and the European Union (European Regional Development Fund—the Competitiveness and Cohesion Operational Programme—KK.01.2.2.03) for funding this research through project CEKOM 3LJ (KK.01.2.2.03.0017), granted to Institution for research and knowledge development of nutrition and health CEKOM 3LJ.

Institutional Review Board Statement: Not applicable.

Informed Consent Statement: Not applicable.

Data Availability Statement: The data presented in this study are available from the authors.

Acknowledgments: We would like to thank Croatian Government and the European Union (European Regional Development Fund—the Competitiveness and Cohesion Operational Programme—KK.01.2.2.03.0017) for funding this research through project CEKOM 3LJ (KK.01.2.2.03.0017), granted to Institution for research and knowledge development of nutrition and health CEKOM 3LJ.

Conflicts of Interest: The authors declare no conflict of interest.

Appendix A

Table A1. Regression coefficient of polynomial function of all response surfaces for carnosic acid.

Terms	Coefficients	Standard Error	*F*-Value	*p*-Value
Carnosic acid (stirring)				
Intercept	4.93	0.42		
X_1	−3.97	0.33	145.19	0.0001
X_2	0.98	0.33	8.95	0.0202
X_3	0.40	0.33	1.49	0.2619
X_1^2	1.60	0.45	12.51	0.0095
X_2^2	0.71	0.45	2.46	0.1608
X_3^2	0.28	0.45	0.39	0.5530
X_1X_2	−1.31	0.47	7.94	0.0259
X_1X_3	0.46	0.47	0.98	0.3562
X_2X_3	0.47	0.47	1.00	0.3495
R^2 = 0.9630				
Carnosic acid (UAE)				
Intercept	4.76	0.87		
X_1	−3.17	0.69	21.24	0.0025
X_2	1.08	0.69	2.45	0.1617
X_3	2.22	0.69	10.45	0.0144
X_1^2	1.39	0.95	2.14	0.1871
X_2^2	−0.52	0.95	0.30	0.6019
X_3^2	−0.57	0.95	0.37	0.5645
X_1X_2	−0.34	0.97	0.12	0.7349
X_1X_3	−2.39	0.97	6.06	0.0433
X_2X_3	1.46	0.97	2.24	0.1780
R^2 = 0.8660				

Table A1. *Cont.*

Terms	Coefficients	Standard Error	*F*-Value	*p*-Value
Carnosic acid (MCE)				
Intercept	3.44	0.28		
X_1	−1.34	0.22	35.66	0.006
X_2	0.63	0.22	7.82	0.0266
X_3	1.59	0.22	50.46	0.0002
X_1^2	−0.38	0.31	1.50	0.2608
X_2^2	0.16	0.31	0.27	0.6209
X_3^2	1.05	0.31	11.49	0.0116
X_1X_2	0.066	0.32	0.043	0.8409
X_1X_3	−0.93	0.32	8.56	0.0221
X_2X_3	0.54	0.32	2.95	0.1294
$R^2 = 0.9442$				

Table A2. Regression coefficient of polynomial function of all response surfaces for carnosol.

Terms	Coefficients	Standard Error	*F*-Value	*p*-Value
Carnosol (stirring)				
Intercept	2.15	0.16		
X_1	−1.07	0.13	70.76	0.0001
X_2	0.72	0.13	31.74	0.0008
X_3	0.86	0.13	45.54	0.0003
X_1^2	−0.093	0.18	0.28	0.6146
X_2^2	0.22	0.18	1.52	0.2580
X_3^2	0.26	0.18	2.26	0.1768
X_1X_2	−0.55	0.18	9.34	0.0184
X_1X_3	−0.22	0.18	1.48	0.2630
X_2X_3	0.52	0.18	8.34	0.0234
$R^2 = 0.9607$				
Carnosol (MCE)				
Intercept	1.09	0.077		
X_1	−0.24	0.061	15.65	0.0055
X_2	0.19	0.061	9.28	0.0187
X_3	0.32	0.061	27.85	0.0012
X_1^2	−0.18	0.084	4.61	0.0690
X_2^2	0.029	0.084	0.12	0.7436
X_3^2	0.21	0.084	6.33	0.0401
X_1X_2	6.381×10^{-3}	0.086	5.511×10^{-3}	0.9429
X_1X_3	−0.096	0.086	1.26	0.2988
X_2X_3	0.074	0.086	0.74	0.4189
$R^2 = 0.9032$				

Table A3. Analysis of variance (ANOVA) of the modelled responses for carnosic acid.

Source	Sum of Squares	Degree of Freedom	Mean Square	*F*-Value	*p*-Value *
Carnosic acid (mixing)					
The recovery					
Model	157.64	9	17.52	20.22	0.0003
Residual	6.06	7	0.87		
Lack of fit	3.92	3	1.31	2.44	0.2042
Pure error	2.14	4	0.54		
Total	163.70	16			

Table A3. *Cont.*

Source	Sum of Squares	Degree of Freedom	Mean Square	*F*-Value	*p*-Value *
Carnosic acid (UAE)					
The recovery					
Model	171.12	9	19.01	5.03	0.0224
Residual	26.48	7	3.78		
Lack of fit	13.99	3	4.66	1.49	0.3444
Pure error	12.49	4	3.12		
Total	197.60	16			
Carnosic acid (MCE)					
The recovery					
Model	47.49	9	5.28	13.17	0.0013
Residual	2.81	7	0.40		
Lack of fit	2.75	3	0.92	64.44	0.008
Pure error	0.057	4	0.014		
Total	50.30	16			

* $p < 0.01$ highly significant; $0.01 \leq p < 0.05$ significant; $p \geq 0.05$ not significant.

Table A4. Analysis of variance (ANOVA) of the modelled responses for carnosol.

Source	Sum of Squares	Degree of Freedom	Mean Square	*F*-Value	*p*-Value *
Carnosol (mixing)					
The recovery					
Model	22.28	9	2.48	19.04	0.0004
Residual	0.91	7	0.13		
Lack of fit	0.41	3	0.14	1.09	0.4491
Pure error	0.50	4	0.13		
Total	23.19	16			
Carnosol (UAE)					
The recovery					
Model	18.29	9	2.03	3.18	0.0708
Residual	4.48	7	0.64		
Lack of fit	2.81	3	0.94	2.24	0.2256
Pure error	1.67	4	0.42		
Total	22.77	16			
Carnosol (MCE)					
The recovery					
Model	1.93	9	0.21	7.26	0.0080
Residual	0.21	7	0.030		
Lack of fit	0.17	3	0.055		
Pure error	0.042	4	0.010		
Total	2.14	16			

* $p < 0.01$ highly significant; $0.01 \leq p < 0.05$ significant; $p \geq 0.05$ not significant.

Table A5. HPLC method validation parameters.

	Carnosic Acid ($\mu g\ mg^{-1}$)	Mean Value ($\mu g\ mg^{-1}$)	SD	RSD (%)	Carnosol ($\mu g\ mg^{-1}$)	Mean Value ($\mu g\ mg^{-1}$)	SD	RSD (%)	
	100.406				99.999				
	100.042				100.004				
Repeatability of measurements	100.913	100.39	0.43	0.43%	100.100				
	100.842				99.947				
	100.305				99.977				
	99.831				99.987	100.00	0.05	0.05%	
	104.65				102.231				
	108.45				101.996				
Repeatability of solution preparation	109.22	107.46	2.40	2.24%	103.260				
	109.09				102.161				
	109.24				101.677				
	104.13				101.996	102.22	0.54	0.53%	
		Added amount of carnosic acid ($\mu g\ mg^{-1}$)	Expected value of carnosic acid ($\mu g\ mg^{-1}$)	Obtained value of carnosic acid ($\mu g\ mg^{-1}$)	Recovery (%)	Added amount of carnosol ($\mu g\ mg^{-1}$)	Expected value of carnosol ($\mu g\ mg^{-1}$)	Obtained value of carnosol ($\mu g\ mg^{-1}$)	Recovery (%)
	Sample			10.36				6.50	
Accuracy	Sample + std 1 (1:1)	22.570	16.461	16.426	99%	21.329	13.915	13.213	95%
	Sample + std 2 (1:1)	100.406	55.385	55.797	99%	99.947	53.223	52.999	99%
	Sample + std 3 (1:1)	212.859	111.608	107.049	96%	209.221	107.860	106.995	99%

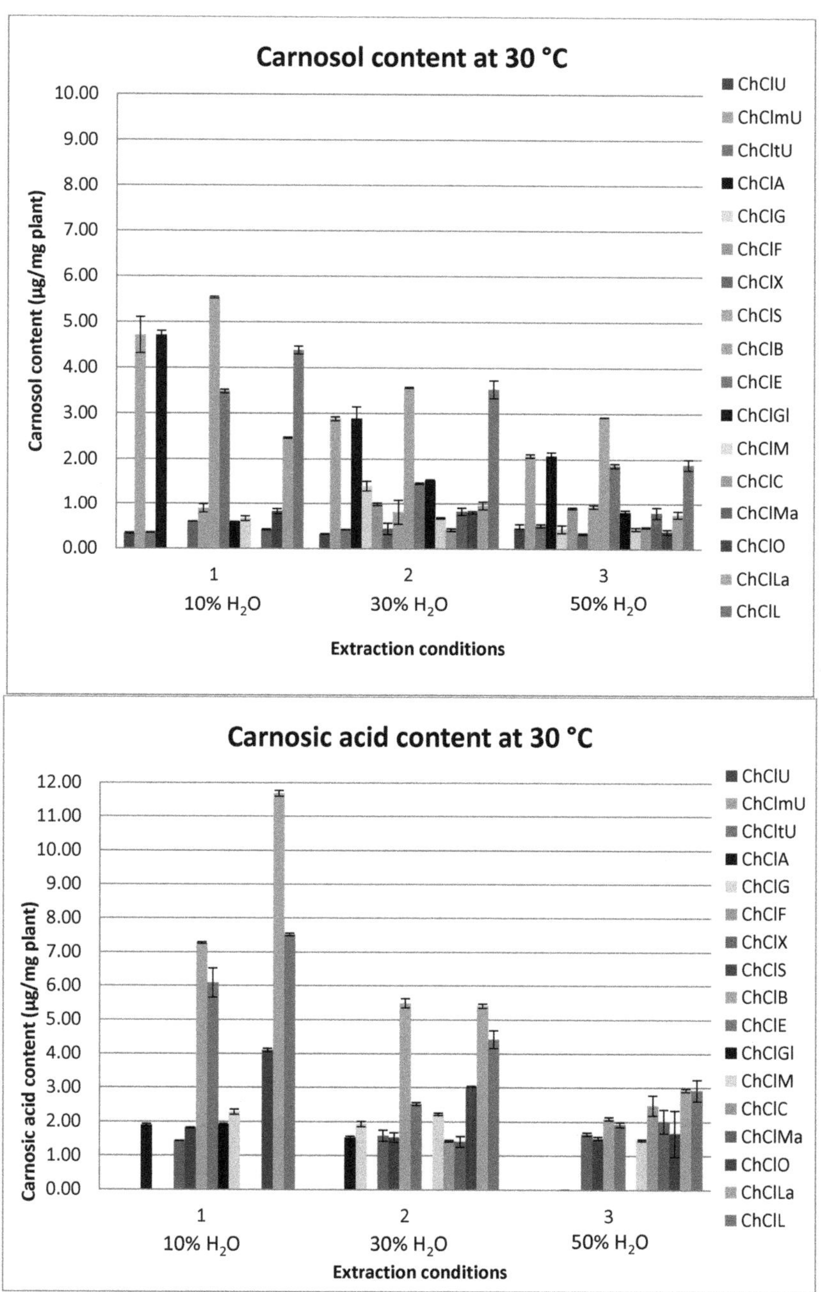

Figure A1. Carnosol and carnosic acid content obtained by stirring and heating with DESs depending on the temperature and H_2O content (n = 3). The columns represent determined amount of carnosol and carnosic acid in the samples, and the color of the column indicates DES used according to the abbreviations recorded in the legend to the right and Table 4.

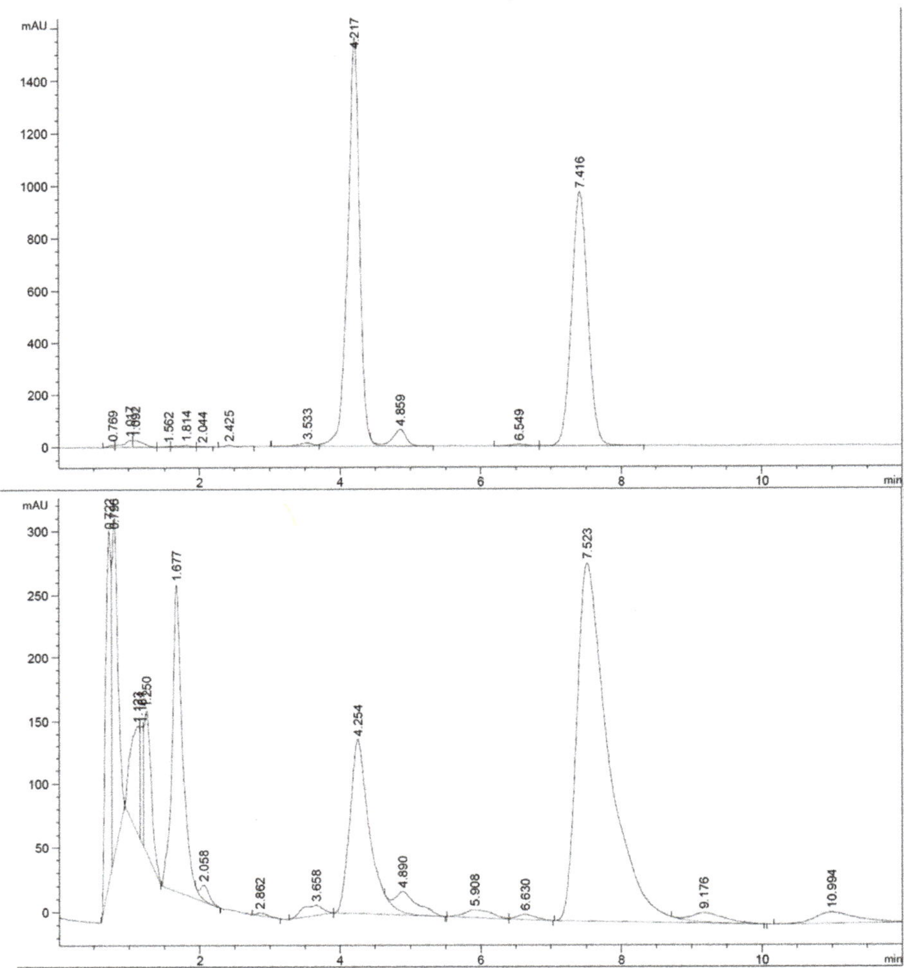

Figure A2. Obtained chromatogram of carnosol and carnosic acid standards (picture **above**) and chromatogram of real sample (picture **below**).

References

1. Zainal-Abidin, M.H.; Hayyan, M.; Hayyan, A.; Jayakumar, N.S. New horizons in the extraction of bioactive compounds using deep eutectic solvents: A review. *Anal. Chim. Acta* **2017**, *979*, 1–23. [CrossRef]
2. Bakirtzi, C.; Triantafyllidou, K.; Makris, D.P. Novel lactic acid-based natural deep eutectic solvents: Efficiency in the ultrasound-assisted extraction of antioxidant polyphenols from common native Greek medicinal plants. *J. Appl. Res. Med. Aromat. Plants* **2016**, *3*, 120–127. [CrossRef]
3. Wang, L.; Weller, C.L. Recent advances in extraction of nutraceuticals from plants. *Trends Food Sci. Technol.* **2006**, *17*, 300–312. [CrossRef]
4. Wang, M.; Wang, J.; Zhang, Y.; Xia, Q.; Bi, W.; Yang, X.; Chen, D.D.Y. Fast environment-friendly ball mill-assisted deep eutectic solvent-based extraction of natural products. *J. Chromatogr. A* **2016**, *1443*, 262–266. [CrossRef] [PubMed]
5. Abbott, A.P.; Boothby, D.; Capper, G.; Davies, D.L.; Rasheed, R.K. Deep eutectic solvents formed between choline chloride and carboxylic acids: Versatile alternatives to ionic liquids. *J. Am. Chem. Soc.* **2004**, *126*, 9142–9147. [CrossRef]
6. Abbott, A.P.; Capper, G.; Davies, D.L.; Rasheed, R.K.; Tambyrajah, V. Novel solvent properties of choline chloride/urea mixtures. *Chem. Commun.* **2003**, *9*, 70–71. [CrossRef]

7. Radošević, K.; Ćurko, N.; Gaurina Srček, V.; Cvjetko Bubalo, M.; Tomašević, M.; Kovačević Ganić, K.; Radojčić Redovniković, I. Natural deep eutectic solvents as beneficial extractants for enhancement of plant extracts bioactivity. *LWT Food Sci. Technol.* **2016**, *73*, 45–51. [CrossRef]
8. Zhang, C.W.; Xia, S.Q.; Ma, P.S. Facile pretreatment of lignocellulosic biomass using deep eutectic solvents. *Bioresour. Technol.* **2016**, *219*, 1–5. [CrossRef]
9. Min, W.N.; Jing, Z.; Min, S.L.; Ji, H.J.; Lee, J. Enhanced extraction of bioactive natural products using tailor-made deep eutectic solvents: Application to flavonoid extraction from *Flossophorae*. *Green Chem.* **2015**, *17*, 1718–1727.
10. Cvjetko Bubalo, M.; Vidović, S.; Radojčić Redovniković, I.; Jokić, S. New perspective in extraction of plant biologically active compounds by green solvents. *Food Bioprod. Process.* **2018**, *109*, 52–73. [CrossRef]
11. Bi, W.; Tian, M.; Row, K.H. Evaluation of alcohol based deep eutectic solvent in extraction and determination of flavonoids with response surface methodology optimization. *J. Chromatogr. A* **2013**, *1285*, 22–30. [CrossRef] [PubMed]
12. Dai, Y.; Rozema, E.; Verpoorte, R.; Hae Choi, Y. Application of natural deep eutectic solvents to the extraction of anthocyanins from *Catharanthus roseus* with high extractability and stability replacing conventional organic solvents. *J. Chromatogr. A* **2016**, *1434*, 50–56. [CrossRef] [PubMed]
13. Wang, T.; Jiao, J.; Gai, Q.Y.; Wang, P.; Guo, N.; Niu, L.L.; Fu, Y.J. Enhanced and green extraction polyphenols and furanocoumarins from Fig (*Ficus carica* L.) leaves using deep eutectic solvents. *J. Pharm. Biomed. Anal.* **2017**, *145*, 339–345. [CrossRef] [PubMed]
14. Bosiljkov, T.; Dujmić, F.; Cvjetko Bubalo, M.; Hribar, J.; Vidrih, R.; Brnčić, M.; Zlatić, E.; Radojčić Redovniković, I.; Jokić, S. Natural deep eutectic solvents and ultrasound-assisted extraction: Green approaches for extraction of wine lees anthocyanins. *Food Bioprod. Process.* **2017**, *102*, 195–203. [CrossRef]
15. Wang, M.; Wang, J.; Zhou, Y.; Mingyue, Z.; Xia, Q.; Bi, W.; Chen, D.D.Y. Ecofriendly Mechanochemical Extraction of Bioactive Compounds from Plants with Deep Eutectic Solvents. *ACS Sustain. Chem. Eng.* **2017**, *5*, 6297–6303. [CrossRef]
16. Aruoma, O.I.; Halliwell, B.; Aeschbach, R.; Loligers, J. Antioxidant and prooxidant properties of active rosemary constituents: Carnosol and carnosic acid. *Xenobiotica* **1992**, *22*, 257–268. [CrossRef]
17. Tsai, C.W.; Lin, C.Y.; Wang, Y.J. Carnosic acid induces the NAD(P)H: Quinone oxidoreductase 1 expression in rat clone 9 cells through the p38/nuclear factor erythroid-2 related factor 2 pathway. *J. Nutr.* **2011**, *141*, 2119–2125. [CrossRef]
18. Kontogianni, V.G.; Tomic, G.; Nikolic, I.; Nerantzaki, A.A.; Sayyad, N.; Stosic-Grujicic, S.; Stojanovic, I.; Gerothanassis, I.P.; Tzakos, A.G. Phytochemical profile of *Rosmarinus officinalis* and *Salvia officinalis* extracts and correlation to their antioxidant and anti-proliferative activity. *Food Chem.* **2013**, *136*, 120–129. [CrossRef]
19. Lopez-Jimenez, A.; Garcia-Caballero, M.; Medina, M.A.; Quesada, A.R. Antiangiogenic properties of carnosol and carnosic acid, two major dietary compounds from rosemary. *Eur. J. Nutr.* **2013**, *52*, 85–95. [CrossRef]
20. Georgantzi, C.; Lioliou, A.E.; Paterakis, N.; Makris Dimitris, P. Combination of Lactic Acid-Based Deep Eutectic Solvents (DES) with β-Cyclodextrin: Performance Screening Using Ultrasound-Assisted Extraction of Polyphenols from Selected Native Greek Medicinal Plants. *Agronomy* **2017**, *7*, 54. [CrossRef]
21. Okamura, N.; Fujimoto, Y.; Kuwabara, S.; Yagi, A. High-performance liquid chromatographic determination of carnosic acid and carnosol in *Rosmarinus oficinalis* and *Salvia oficinalis*. *J. Chromatogr. A* **1994**, *679*, 381–386. [CrossRef]
22. Vinatoru, M. An overview of the ultrasonically assisted extraction of bioactive principles from herbs. *Ultrason. Sonochem.* **2001**, *8*, 303–313. [CrossRef]
23. Fatma Ebru, K.; Ayse, A.; Caglar, K. Extraction and HPLC Analysis of Sage (*Salvia officinalis*) Plant. *Nat. Prod. Res.* **2018**, *5*, 298.
24. Durling, N.E.; Catchpole, O.J.; Grey, J.B.; Webby, R.F.; Mitchell, K.A.; Foo, L.Y.; Perry, N.B. Extraction of phenolics and essential oil from dried sage (*Salvia officinalis*) using ethanol–water mixtures. *Food Chem.* **2007**, *101*, 1417–1424. [CrossRef]
25. Chemat, F.; Vian, M.A.; Cravotto, G. Green extraction of natural products: Concept and principles. *Int. J. Mol. Sci.* **2012**, *13*, 8615–8627. [CrossRef] [PubMed]
26. Castañeda-Ovando, A.; Pacheco-Hernández, M.; Páez-Hernández, M.E.; Rodríguez, J.A.; Galán-Vidal, C.A. Chemical studies of anthocyanins: A review. *Food Chem.* **2009**, *113*, 859–871. [CrossRef]
27. Abreu, M.E.; Müller, M.; Alegre, L.; Munné-Bosch, S. Phenolic diterpene and α-tocopherol contents in leaf extracts of 60 Salvia species. *J. Sci. Food Agric.* **2008**, *88*, 2648–2653. [CrossRef]
28. Shan, B.; Cai, Y.Z.; Sun, M.; Corke, H. Antioxidant Capacity of 26 Spice Extracts and Characterization of Their Phenolic Constituents. *J. Sci. Food Agric.* **2005**, *53*, 7749–7759. [CrossRef]
29. Lamien-Meda, A.; Nell, M.; Lohwasser, U.; Börner, A.; Franz, C.; Novak, J. Investigation of Antioxidant and Rosmarinic Acid Variation in the Sage Collection of the Genebank in Gatersleben. *J. Agric. Food Chem.* **2010**, *58*, 3813–3819. [CrossRef]
30. Juntachote, T.; Berghofer, E.; Bauer, F.; Siebenhand, S. The application of response surface methodology to the production of phenolic extracts of lemon grass, galangal, holy basil and rosemary. *Int. J. Food Sci. Technol.* **2006**, *41*, 121–133. [CrossRef]
31. Tena, M.T.; Valcarcel, M.; Hidalgo, P.J.; Ubera, J.L. Supercritical Fluid Extraction of Natural Antioxidants from Rosemary: Comparison with Liquid Solvent Sonication. *Anal. Chem.* **1997**, *69*, 521–526. [CrossRef] [PubMed]
32. Cháfer, A.; Fornari, T.; Berna, A.; Ibañez, E.; Reglero, G. Solubility of solid carnosic acid in supercritical CO_2 with ethanol as a co-solvent. *J. Supercrit. Fluid* **2005**, *34*, 323–329. [CrossRef]
33. Pavić, V.; Jakovljević, M.; Molnar, M.; Jokić, S. Extraction of Carnosic Acid and Carnosol from Sage (*Salvia officinalis* L.) Leaves by Supercritical Fluid Extraction and Their Antioxidant and Antibacterial Activity. *Plants* **2019**, *8*, 16. [CrossRef] [PubMed]

34. Babovic, N.; Djilas, S.; Jadranin, M.; Vajs, V.; Ivanovic, J.; Petrovic, S.; Zizovic, I. Supercritical carbon dioxide extraction of antioxidant fractions from selected *Lamiaceae* herbs and their antioxidant capacity. *Innov. Food Sci. Emerg. Technol.* **2010**, *11*, 98–107. [CrossRef]
35. Jokić, S.; Molnar, M.; Jakovljević, M.; Aladić, K.; Jerković, I. Optimization of supercritical CO_2 extraction of *Salvia officinalis* L. leaves targeted on Oxygenated monoterpenes, α-humulene, viridiflorol and manool. *J. Supercrit. Fluid* **2018**, *133*, 253–262. [CrossRef]
36. Molnar, M.; Jakovljević, M.; Jokić, S. Optimization of the process conditions for the extraction of rutin from *Ruta graveolens* L. by choline chloride based deep eutectic solvents. *Solvent Extr. Res. Dev.* **2018**, *25*, 109–116. [CrossRef]
37. Bas, D.; Boyaci, I.H. Modeling and optimization I: Usability of response surface methodology. *J. Food Eng.* **2007**, *78*, 836–845. [CrossRef]

Article

The Impact of the Method Extraction and Different Carrot Variety on the Carotenoid Profile, Total Phenolic Content and Antioxidant Properties of Juices

Aleksandra Purkiewicz [1], Joanna Ciborska [1], Małgorzata Tańska [2], Agnieszka Narwojsz [1], Małgorzata Starowicz [3], Katarzyna E. Przybyłowicz [1] and Tomasz Sawicki [1,*]

1 Department of Human Nutrition, Faculty of Food Sciences, University of Warmia and Mazury in Olsztyn, Słoneczna 45F, 10-719 Olsztyn, Poland; aleksandra.purkiewicz@student.uwm.edu.pl (A.P.); joanna.ciborska@uwm.edu.pl (J.C.); agnieszka.narwojsz@uwm.edu.pl (A.N.); katarzyna.przybylowicz@uwm.edu.pl (K.E.P.)

2 Chair of Plant Raw Materials Chemistry and Processing, Faculty of Food Sciences, University of Warmia and Mazury in Olsztyn, Pl. Cieszyński 1, 10-726 Olsztyn, Poland; m.tanska@uwm.edu.pl

3 Institute of Animal Reproduction and Food Research, Polish Academy of Science, Tuwima 10, 10-748 Olsztyn, Poland; m.starowicz@pan.olsztyn.pl

* Correspondence: tomasz.sawicki@uwm.edu.pl

Received: 29 November 2020; Accepted: 9 December 2020; Published: 11 December 2020

Abstract: The study assesses the antioxidant activity (AA), carotenoid profile and total phenolic content (TPC) of carrot juices obtained from three different varieties (black, orange and yellow) and prepared using high- (HSJ) and low-speed juicer (LSJ). The AA assessment was carried out using four assays (DPPH, ABTS, PCL ACW and PCL ACL). The content of carotenoids was conducted by high performance liquid chromatography equipped with a diode array detector (HPLC-DAD) method, while the total phenolic content by the spectrophotometric method. It was shown that orange carrot juices contain more carotenoids than yellow and black carrot juices, approximately ten and three times more, respectively. The total carotenoid content in orange carrot juice made by the HSJ was higher (by over 11%) compared to juice prepared by the LSJ. The highest total phenolic content was noticed in black carrot juices, while the lowest in orange carrot juices. In black carrot juices, a higher range of TPC was found in juices made by HSJ, while in the case of the orange and yellow carrots, the highest content of TPC was detected in juices prepared by the LSJ. AA of the juices was highly dependent on the carrot variety, juice extraction method. The most assays confirmed the highest AA values in black carrot juices. Furthermore, it was shown that the HSJ method is more preferred to obtain orange and yellow carrot juices with higher antioxidant properties, while the LSJ method is more suitable for black carrot juice extraction.

Keywords: carrot juices; carotenoids; polyphenols; antioxidant activity; high-speed juicer; low-speed juicer; food processing

1. Introduction

Carrot is a root vegetable [1] that contains many bioactive compounds, for example, carotenoids (α-carotene, β-carotene lutein, zeaxanthin and lycopene) [2], phenolic acids (chlorogenic, ferulic, p-coumaric, caffeic) [3] and anthocyanins (cyanidin-3-O-xylosyl (sinapoylglucosyl) galactoside, cyanidin-3-O-xylosyl (feruolylglucosyl) galactoside, cyanidin-3-O-xylosyl (coumaroylglucosyl) galactoside) [1]. In addition, carrots are great sources of vitamins (ascorbic acid, thiamine, riboflavin, niacin, pyridoxine, folic acid, vitamin K and A) as well as minerals (calcium, iron, magnesium, phosphorus, potassium, sodium and zinc) [2]. Thanks to the presence of the mentioned bioactive compounds, a carrot has significant health-promoting properties. It was demonstrated that carrots display powerful antioxidant and radical scavenging activities. Moreover, the consumption of carrots has been linked with a lower risk of diseases such as atherosclerosis, cataract, diabetes and cancer [1].

Carrots are commonly classified by the color of roots into white, black, orange, yellow, purple and red [4]. The most common carrot variety is the orange ones which is a genetic crossword between purple, white and yellow carrots [5]. The color of the root has a significant impact on the presence of bioactive compounds. The orange carrot root contains high amounts of α-carotene and is the richest source of β-carotene (precursor of vitamin A). The black carrot is an excellent source of anthocyanins, red carrot root is rich in lycopene, the yellow ones, in turn, was demonstrated to accumulate lutein [6,7]. In addition to the root's color, the growing and season conditions, the ripeness of carrots as well as part of the root also influence the presence of bioactive compounds [8].

It should be emphasized that the content of bioactive compounds and their biological activity is also influenced by technological processes. However, effect of technological parameters and carrot variety interactions on juice nutrition value is still not fully understood Determination of the composition and content of biologically active substances in both fresh and processed vegetables is crucial for the food industry. Such information is also essential for consumers with an increasingly developing nutritional awareness.

The available literature data show that to date, only a high-speed juicer and traditional blenders were used to examine the impact of the extraction method on the juices' antioxidant activity. No studies have evaluated the impact of using a high-speed juicer or a low-speed juicer on the antioxidant activity of carrot juices [9,10]. In addition, in the previously published papers, the juices' antioxidant activity was tested only by DPPH and ABTS assays [9,11,12]. Therefore we intended to examine the influence of the extraction methods on the concentration of main bioactive compounds in the juices obtained from roots of different carrot varieties. To meet this goal we applied low- and high-speed juicer and roots of yellow, orange and black carrots. In addition, to examine the effects of the extraction methods on the antioxidant activity of carrot juices four different methods (DPPH, ABTS, PCL ACW and PCL ACL) were used.

2. Results

2.1. Content of Carotenoids in Carrot Juices

The results of the total and individual compound content of carotenoids in the tested carrot juices are presented in Table 1. The carotenoid profile in tested carrot juices were clearly more dependent on the carrot variety and this differentiation is visible in Figure 1.

Table 1. The content (mg/100 mL) of carotenoids in black, orange and yellow carrot juices.

Compounds	Black Carrot		Orange Carrot		Yellow Carrot	
	HSJ	LSJ	HSJ	LSJ	HSJ	LSJ
lutein	1.15 ± 0.49 ^{a}	1.98 ± 0.08 ^{A}	0.61 ± 0.03 ^{c*}	0.14 ± 0.05 ^{C}	0.81 ± 0.18 ^{b}	0.77 ± 0.02 ^{B}
zeaxanthin	0.31 ± 0.06 ^{b}	0.14 ± 0.03 ^{A}	0.61 ±0.00 ^{a*}	0.12 ± 0.03 ^{A}	0.06 ± 0.08 ^{c}	0.14 ± 0.02 ^{A}
α-carotene	0.14 ± 0.05 ^{b}	0.16 ± 0.04 ^{B}	2.90 ±0.34 ^{a}	2.70 ± 0.66 ^{A}	ND	ND
13-*cis*-β-carotene	0.24 ± 0.14 ^{b*}	0.35 ± 0.03 ^{B}	8.30 ± 1.33 ^{a*}	7.88 ± 0.08 ^{A}	ND	ND
β-carotene	ND	0.68 ± 0.06 ^{B}	15.08 ± 2.84 ^{a*}	13.85 ± 0.01 ^{A}	0.08 ± 0.01 ^{b}	0.12 ± 0.02 ^{C}
Total	1.84 ± 0.47 ^{b*}	3.31 ± 0.79 ^{B}	27.50 ± 0.71 ^{a*}	24.69 ± 0.51 ^{A}	0.95 ± 0.06 ^{c}	1.03 ± 0.06 ^{C}

Abbreviations: Rt—retention time; HSJ—high-speed juicer; LSJ—low-speed juicer; ND—non-detected. Data are expressed as means ± SD (n = 3), means in the line with different letters (abc/ABC) are significantly different ($p < 0.05$). Statistically significant differences ($p < 0.05$) between method of extraction each carrot variety are marked by *.

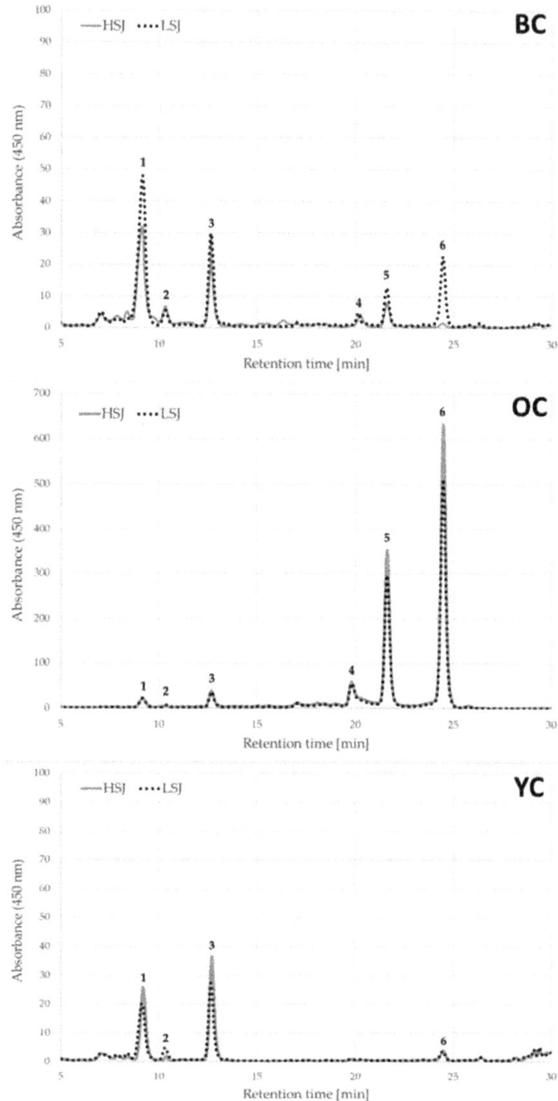

Figure 1. The high performance liquid chromatography (HPLC) chromatograms of carotenoid compounds identified in the carrot juices. Identified peaks are as follows: 1—lutein, 2—zeaxantin, 3—β-apo-8′carotenal (used as an internal standard), 4—13-*cis*-β-carotene, 5—α-carotene and 6—β-carotene. Abbreviations: BC—black carrot, OC—orange carrot, YC—yellow carrot, HSJ—high-speed juicer, LSJ—low-speed juicer.

The highest carotenoid content was found in orange carrot juices, while the lowest in yellow ones. The content of t in orange carrot juices was significantly different ($p < 0.05$) from the content in black and yellow carrot juices.

Orange carrot juices contained almost thirty and twenty four times more carotenoids in juices obtained by the use of high-speed juicer and a low-speed juicer than yellow ones and fifteen times (for high speed juicer (HSJ) method) and seven times (for low speed juicer (LSJ) method) more than black carrot juices. The significant difference ($p < 0.05$) was also detected in the carotenoid content of black and yellow carrot juices. Black carrot juices were characterized by two times and over three times more carotenoids content in juices prepared by HSJ and LSJ, respectively, than yellow ones.

In the examined carrot juices five different carotenoids were detected (Table 1). Three compounds belong to the carotenes group (α-carotene, β-carotene and 13-*cis*-β-carotene) and two compounds belong to the xanthophylls group (lutein and zeaxanthin). The richest profile of carotenoids was found in the orange carrot juices. Five compounds (α-carotene, β-carotene, 13-*cis*-β-carotene, lutein and zeaxanthin) were found in the carrot juice obtained by high-speed juicer (HSJ) and by low-speed juicer (LSJ).

The dominant carotenoid in orange carrot juices was β-carotene, constituting 55% and 56% of the total content of carotenoids in juices prepared by the HSJ and LSJ methods, respectively. The second dominant compound, in juices obtained from orange carrots, was 13-*cis*-β-carotene, which was 30% of the total content of carotenoids for juice obtained by the use of HSJ and 32% obtained by the use of LSJ. Juices obtained from yellow and black carrots contained three (lutein, zeaxanthin and β-carotene) and five compounds (lutein, zeaxanthin, β-carotene and 13-*cis*-β-carotene), respectively. In the yellow carrot juices, the dominant carotenoid was lutein which constituted 85% of the total carotenoids content in juices prepared by the use of high-speed juicer and 75% by low-speed juicer. The dominant compounds in the black carrot juices were lutein (63% for HSJ method and 60% for LSJ method) and zeaxanthin (17% for HSJ method and 4% for LSJ method) and 13-*cis*-β-carotene (13% for HSJ method and 11% for LSJ method). Significantly higher amounts ($p < 0.05$) of β-carotene and 13-*cis*-β-carotene were determined in the orange carrot juices compared to the black carrot ones. In turn, the orange carrot juices obtained by HSJ method contained thirty five times more 13-*cis*-β-carotene, whereas juices made by LSJ method contained twenty three times and twenty times more of 13-*cis*-β-carotene and β-carotene, respectively. The highest amounts of lutein were determined in the black carrot juices. The content of this compound was two and fourteen times higher in the black carrot juices than in orange carrot juices made by HSJ and LSJ, respectively. It should be emphasized that, in comparison to other carrot varieties, orange carrot juice contained the lowest percentage of lutein ($p < 0.05$).

It was showed that method of juice extraction significantly affected the content of lutein, zeaxanthin, 13-*cis*-β-carotene and β-carotene in orange carrot juices ($p < 0.05$). Juice prepared by the high-speed juicer contained 77%, 80%, 5% and 8% more these compounds compared to juice obtained using the low-speed juicer (Table 1). Furthermore, the total content of carotenoids in the juice obtained from orange carrots prepared by the HSJ method was significantly higher (by over 111%) than in juice made by the LSJ method ($p < 0.05$).

2.2. *Total Phenolic Content in Carrot Juices*

The total phenolic content (TPC) in the tested juices differed among carrot varieties (Figure 2). The significantly highest TPC was demonstrated in the juices obtained from black carrots ($p < 0.05$), while the lowest content of these compounds was detected in the orange carrot juices ($p < 0.05$). The black carrot juice prepared with the use of LSJ method contained 7% more total phenolic content than the juice prepared with the use of HSJ ($p < 0.05$). In comparison, the juice obtained from orange carrot using HSJ method was characterized by a higher TPC and contained 10% more these compounds than the juice obtained by LSJ. The TPC demonstrated in the juice obtained from yellow carrot did not differ between the extraction method. The TPC demonstrated in black carrot juices prepared with the HSJ and LSJ methods was eight and nine times higher than TPC demonstrated in orange carrot juices prepared with the same methods, respectively. and six times higher than in yellow carrot juices.

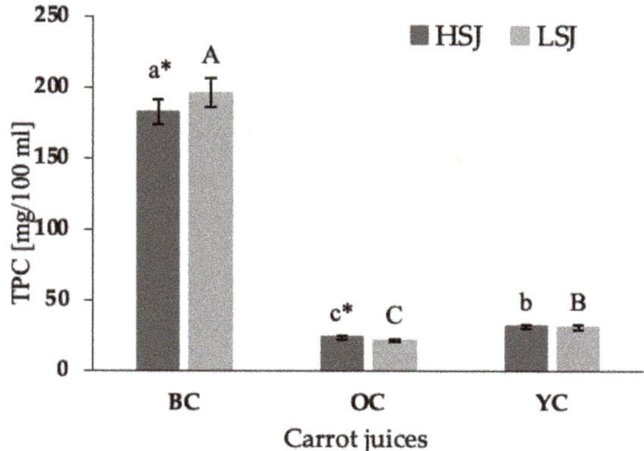

Figure 2. Results of total phenolic content (TPC) in black, orange and yellow carrot juices. Abbreviations: BC—black carrot, OC—orange carrot, YC—yellow carrot, HSJ—high-speed juicer, LSJ—low-speed juicer. Each bar corresponds to the mean of three independent replicates with error bars indicating the standard deviations. Different letters (abc/ABC) indicate significant differences among samples ($p < 0.05$). Statistically significant differences ($p < 0.05$) between method of extraction each carrot variety are marked by *.

2.3. Antioxidant Capacity of Carrot Juices

Four methods (ABTS, DPPH, PCL ACW and PCL ACL) were used to determine the antioxidant activity of the carrot juices (Table 2). The results of the PCL methods were presented both separately and as a sum of ACL and ACE measurements.

Table 2. The antioxidant activity of black, orange and yellow carrot juices determined by different assays.

Antioxidant Activity Assays [μmol Trolox/mL]	Black Carrot		Orange Carrot		Yellow Carrot	
	HSJ	LSJ	HSJ	LSJ	HSJ	LSJ
DPPH	28.36 ± 0.14 ^{c*}	27.66 ± 0.27 ^{C}	28.82 ± 0,27 ^{b}	28.54 ± 0.35 ^{B}	29.08 ± 0.21 ^{a}	29.12 ± 0.59 ^{A}
ABTS	4.16 ± 0.03 ^{a*}	4.10 ± 0.00 ^{A}	3.47 ± 0.04 ^{b*}	3.28 ± 0.01 ^{B}	3.28 ± 0.05 ^{c*}	3.15 ± 0.01 ^{C}
PCL ACW	9.88 ± 0.17 ^{a*}	32.18 ± 0.32 ^{A}	4.02 ± 0.03 ^{c*}	3.87 ± 0.12 ^{C}	6.38 ± 0.32 ^{b}	6.28 ± 0.32 ^{B}
PCL ACL	122.13 ± 1.73 ^{a*}	140.63 ± 5.98 ^{A}	85.43 ± 0.46 ^{b*}	77.28 ± 0.11 ^{C}	86.83 ± 2.16 ^{b}	83.35 ± 3.75 ^{B}
PCL	132.00 ± 1.56 ^{a*}	172.80 ± 5.66 ^{A}	89.45 ± 0.43 ^{c*}	81.14 ± 0.01 ^{C}	93.20 ± 2.47 ^{b}	89.63 ± 3.43 ^{B}

Abbreviations: HSJ—high-speed juicer; LSJ—low-speed juicer. Data are expressed as means ± SD ($n = 3$), means in the line with different letters (abc/ABC) are significantly different ($p < 0.05$). Statistically significant differences ($p < 0.05$) between method of extraction each carrot variety are marked by *.

The obtained juices were characterized by different antioxidant activity depending on the carrot variety and extraction method. The results of DPPH assay showed that yellow carrot juices demonstrated the highest antioxidant activity among examined varieties ($p < 0.05$) and the juice made by the LSJ method was characterized by slightly higher values of antioxidant activity than those made by the HSJ. In comparison, black and orange carrot juices obtained by the HSJ method were characterized by higher values of antioxidant activity than the juices obtained by the LSJ. A significant relationship ($p < 0.05$) between the extraction method and the antioxidant activity was demonstrated only for black carrot juices (Table 2). The antioxidant activity demonstrated in juice obtained from orange and yellow carrots did not differ between the extraction method ($p > 0.05$).

The results of ABTS assay demonstrated that the highest antioxidant activity was shown for the black carrot juices and the lowest activity was demonstrated for the yellow ones. The antioxidant activity of black carrot juice obtained by HSJ method was 17% and 21% higher than the antioxidant activity of orange and yellow carrot juices, respectively. A similar tendency was observed for the black carrot juices obtained with the use of LSJ. Moreover, the significant relationship between the extraction method and the antioxidant activity estimated by the ABTS test of juices obtained from each carrot variety was noted. The antioxidant activity demonstrated in juices obtained by HSJ method was higher than those shown in juice obtained with the use of LSJ (1.5%-black, 5.5%-orange and 4%-yellow carrot juice) ($p < 0.05$).

The highest antioxidant activity were demonstrated for the black carrot juices, while the lowest values were demonstrated for the orange carrot juices ($p < 0.05$) in all applied assays (PLC ACW, PLC ACL, PLC). The results of PCL ACW assay demonstrated that differences in juices' antioxidant activity values depends on the carrot variety. The antioxidant activity of black carrot juices method was higher in comparison to the antioxidant activity of orange and yellow carrot juices made (regardless of the extraction method) ($p < 0.05$). It should be noticed that, in case of black and orange carrot, the extraction method had a significant effect on the antioxidant activity determined by all applied assays. The results of all assays showed that the black carrot juice obtained by LSJ had higher antioxidant activity than the juice obtained by HSJ ($p < 0.05$). The opposite situation was observed in case of orange and yellow carrot juices, where the higher antioxidant values were measured in juices obtained by HSJ (Table 2).

2.4. *Association Between Obtained Data*

2.4.1. Linear Pearson's Correlation Coefficients

The values of antioxidant activity of carrot juices analyzed by the ABTS assay was strongly positively correlated with the values obtained by the PCL ACW ($r = 0.89$), PCL ACL ($r = 0.96$) assays and the sum of PCL assay ($r = 0.96$). While the results of the AC measured by the DPPH assay were only slightly correlated with other assays; negatively with all PCL assays and positively with ABTS assay (Table 3).

Table 3. Correlation coefficients (r) calculated for the relationships between carotenoids, total phenolic content and antioxidant activity assays (r marked by * are statistically significant at $p < 0.05$).

Correlation	DPPH	ABTS	PCL ACW	PCL ACL	Sum of PCL
zeaxanthin	0.05	0.18	−0.24	−0.05	−0.11
lutein	−0.91 *	0.91 *	0.89 *	0.96 *	0.96 *
α-carotene	0.09	−0.32	−0.42	−0.51	−0.49
13-*cis*-β-carotene	0.10	−0.31	−0.44	−0.51	−0.50
β-carotene	0.11	−0.33	−0.43	−0.51	−0.50
total carotenoids	0.07	−0.29	−0.40	−0.48	−0.47
TPC	−0.83 *	0.96 *	0.79	0.98 *	0.95 *
DPPH	0.05	0.18	−0.24	−0.05	−0.11
ABTS	−0.91 *	0.91 *	0.89 *	0.96 *	0.96 *
PCL ACW	0.09	−0.32	−0.42	−0.51	−0.49

Relationships between the content of carotenoids and antioxidant activity were also estimated (Table 3). Statistically significant ($p < 0.05$) and positive correlation coefficients were shown between the content of lutein and the results of ABTS ($r = 0.91$), PCL ACW ($r = 0.89$) and PCL ACL assays ($r = 0.96$) and sum of PCL ($r = 0.96$), while the negative correlation coefficient between the content of lutein and results of the DPPH assay ($r = -0.91$). In turn, average negative correlation coefficients (but not significant at $p < 0.05$) for relationships between content of total carotenoids, α-, β- and 13-*cis*-β-carotene and results of the ABTS, PCL ACW, PCL ACL and PCL assays were noted (r values ranging from −0.29 to −0.51). It was also shown that TPC was highly positively correlated with most applied antioxidant assays; r values in range of 0.79–0.98 for ABTS, PCL ACW, PCL ACL and total PCL assays. However, a strong negative correlation between TPC and DPPH ($r = -0.83$) was found.

2.4.2. Principal Component Analysis (PCA)

Principal component analysis (PCA) was performed on all samples and variables (individual carotenoid concentration, total phenolic content, antioxidant capacity (ABTS, DPPH, PCL ACW, PCL

ACL assays and the sum of PCL) to investigate the structure and regularity in the relationships between variables and cases. The first two principal components (PC) explained 90.61% of total data variance. The correlations between the original variables and the obtained principal components are shown in Figure 3a. Each of the variables is represented by a vector. The direction and lengths of the vectors indicate to what extent the given variables affect the principal components. In our study, most input variables are located near the circle, which means that the information in these variables is transferred by principal components. PCA analysis showed a strong positive correlation between α-carotene, β-carotene, 13-*cis*-β-carotene and between lutein and TPC, ABTS, PCL ACW, PCL ACL and the sum of PCL. However, the opposite variables are negatively correlated. The strong negative correlation between lutein and zeaxanthin and the DPPH assay and between ABTS, PCL ACW, PCL ACL, the sum of PCL and DPPH were noted. Moreover, the graph shows that TPC was also negatively correlated with the antioxidant capacity determined by the DPPH assay and positively correlated with the ABTS, PCL ACW, PCL ACL assays and sum of the PCL.

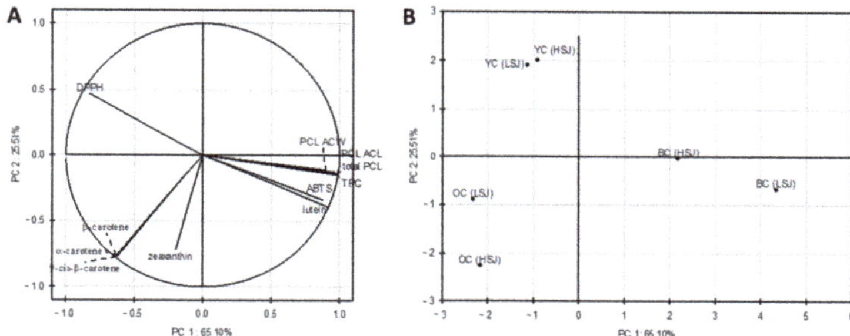

Figure 3. Principal components plot, variations in the parameters (ABTS, DPPH, PCL ACW, PCLACL assays and sum of the PCL) of the analyzed carrot juices (**A**) score plot of the obtained juices (**B**). Abbreviations: BC—black carrot, OC—orange carrot, YC—yellow carrot, HSJ—high-speed juicer, LSJ—low-speed juicer.

Figure 3b presents the score plot on the plane of principal components, which shows the similarity between the types of carrot juices tested. The analyzed cases' position concerning each other proves different antioxidant properties of juices obtained from black, orange and yellow carrots. Moreover, based on the analysis, it was observed that the juices from black carrots were the most diverse (LSJ and HSJ methods), while the juices from yellow carrots were the most similar. There was no similarity between the juices from different carrot cultivars.

3. Discussion

The carotenoid content in carrot juices varies and depends on many factors, for example, type of raw material and/or storage time [13]. In this study, the highest content of carotenoids was demonstrated in juices made of orange carrot 27.50 mg/100 mL in the juice prepared by the use of HSJ and 24.69 mg/100 mL in the juice obtained by the use of LSJ. In comparison, Amal et al. [13] shown that the total content of carotenoids in the juice from orange carrot was 6.6 mg/100 mL. The difference in the results may arise from the variety of carrot use in the research, growing conditioning and cultivation practice [14]. The lowest carotenoid content was identified in juices from the yellow carrot. Similarly, Sun et al. [15] presented that yellow (as well as white) carrot varieties contain the lowest amount of carotenoid pigments. Therefore, the carrot root color used to juice preparation has an essential contribution to the content of selected compounds from the carotenoid group.

In the conducted research the amount of β-carotene in juices made from orange carrot represented more than 50% of total carotenoids content. In juices made from black carrot the β-carotene content was 20% and in juices from yellow carrot the amount of β-carotene represented 8% (HSJ) and 12 % (LSJ).

Juices made of black and yellow carrot contained a higher amount of xanthophylls than carotenes. The group of xanthophylls includes lutein. The available literature data confirms that lutein is the main compound of the yellow carrot and the yellow carrot juices can contain from 0.1 to 0.5 mg/100 mL of lutein [5]. In our experiment, the lutein content in yellow carrot juice was from 0.77 mg/100 mL (LSJ method) to 0.81 mg/100 mL (HSJ method). Black carrot juice possesses 1.5 (from HSJ method) and 2.5 (from LSJ method) times higher content of lutein in comparison to yellow carrot juice. Black carrot juices are considered poor source carotenes but a better source of xanthophylls [5]. More than 79% of carotenoids in juices made of black carrots are zeaxanthin and lutein, which are assigned to the xanthophylls group. Higher content of β-carotene and 13-*cis*-β-carotene was demonstrated in orange juices prepared by the HSJ method and in yellow and black carrot juices prepared by the LSJ method. Similarly to the result obtained in the present study, the result of other researchers demonstrated that the juice preparation method did not significantly influence carotenoid content in juices [11]. On the other hand, the study conducted by Ma et al. [16] showed that the peeling method, blanching and enzyme liquefaction treatment had an impact on β-carotene, α-carotene and lutein contents in the carrot juice. In the case of peeling and blanching, the authors showed a decrease in carotenoid compounds content. On the other hand, use of enzymes in the carrot juice production may significantly increase the content of carotenoids [16].In the present study, the total polyphenols' content fluctuates in the wide range and depends on carrot variety and juices extraction methods. The highest amount of TPC was observed in black carrot juices. The high content of total phenolic (TPC) in black carrot juices is result of their high concentration in raw material [17]. Total phenolic content in carrot juices determined in the present study was similar to those demonstrated previously by the other authors [18].

The study showed that the extraction method affected phenolics' content in black and orange carrot juices. In the black carrot juice more phenolics were found in juices prepared with the use of LSJ, while in the orange carrot juices more phenolics contained juices obtained by HSJ. The studies of the other authors also confirm the relationship between the method of juice extraction and the content of polyphenols. For example, the studies of Pyo et al. [9] and Mphahlele et al. [19] demonstrated that juices prepared with the use of blender contain a significantly higher total phenolic content in compared to squeeze juice. Juices made with a blender contain more flesh which is rich in polyphenols, whereas squeezed juices are devoid of pomace to which most polyphenolic compounds pass [9,20].

In presented study, the strong correlations between the content of phenolics and antioxidant capacity measured by ABTS, PCL ACW, PCL ACL and the sum of PCL assays were found. It showed that phenolics play a significant role as antioxidants in carrot juices. The literature data also provided some information on the antioxidant capacity of carrot juices. Kim et al. [12] indicated the higher antioxidant activity of juices from the low-speed juicer. MacDonald-Wicks et al. [21], noticed that antioxidant capacity is the sum of the antioxidant activity of all types of antioxidant compounds present in the product that trap free radicals. In carrots, these compounds include polyphenols, carotenoids and antioxidant vitamins—C and E. The number of values of antioxidant activity of carrot juices for the methods used was as follows: sum of PCL > PCL, ACL > DPPH > PCL, ACW > ABTS. The highest antioxidant activity was established in ABTS, PCL ACW, PCL ACL and the sum of PCL for black carrot juices. According to the literature, methanolic extracts from black carrots possess the highest antioxidant activity determined with the use of DPPH assay [22]. Similarly to the results of present study, the results demonstrated by the other authors showed that, in comparison to juice obtained from yellow and orange carrots, black carrot juices exhibited the highest antioxidants activity [15,22–24].

The ratio of antioxidant potential of the lipophilic fraction (ACL) and hydrophilic fraction (ACW) was established at 12.4 and 4.4 in black carrot juices, 21.3 and 20.0 in orange carrot juices and 13.6 and 13.3 in orange carrot juices, obtained by the HSJ and LSJ methods, respectively. These proportions indicated that in a carrot juice, in which hydrophobic antioxidants are predominated, a species-diverse antioxidant profile is presented. The juices' antiradical activity for fat-soluble compounds (ACL) was significantly higher than the antiradical activity of water-soluble compounds (ACW). The PCL ACL assay indicated antioxidant activity in lipids, in which the soluble carotenoids contained in carrots were found; therefore, a strong positive correlation with the ABTS test was demonstrated. Due to the presence of water-soluble antioxidants in carrot juices (polyphenols and vitamin C) the PCL ACL assay was negatively correlated with the DPPH assay. The strong positive correlation between TPC and the ABTS, PCL ACL, PCL ACW assays and the sum of PCL and the strong negative correlation between TPC and the DPPH assay indicate that polyphenolic compounds that contributed to the increase in the activity of scavenging free oxygen radicals were characterized by different hydrophilicity [22]. In study, the antioxidant activity of carrot juices, measured with the ABTS, PCL ACL, PCL ACW assays and sum of PCL, is positively correlated mostly with the content of lutein, which indicates an important role of this carotenoid in formation of the antioxidant potential of carrot juices.

The juice preparation method had a significant impact on the antioxidant activity of black carrot juices. In the DPPH and ABTS assays, juices obtained with the use of HSJ were characterized by higher antioxidant activity than those obtained with the use of LSJ. In the PCL ACW, PCL ACL and total PCL assays, significantly higher antioxidant activity was confirmed in juices obtained with the use of LSJ method. The highest differences in antioxidant activity in the black carrot juices were demonstrated in the PCL ACW assay and LSJ-obtained juices showed three times higher antioxidant activity than HSJ-obtained juices. In orange carrot juices, in all the performed assays (except DPPH assay), a significant correlation was found between the method of juice extraction and antioxidant activity. In contrast to black carrot juices, in orange carrot juices in each of the assays, higher antioxidant activity was found for juices prepared by the use of HSJ. In yellow carrot juices, only in the ABTS test, there was a significant correlation between the method of juice extraction and the level of antioxidant activity. Higher values of antioxidant activity were obtained in juices prepared with the use of HSJ. In the studies on the influence of the juice extraction method on juices' antioxidant activity, various relationships were noted—the higher activity of juices prepared with LSJ method [25] or no significant influence of the extraction method on the antioxidant activity of juices [11].

4. Materials and Methods

4.1. Chemicals and Reagents

2,2′-azinobis(3-ethylbenzothiazoline-6-sulphonic-acid) diammonium salt (ABTS), 2,2-diphenyl-1-picrylhydrazyl (DPPH) and 6-hydrozy-2,5,7,8-tetramethylchroman-2-carboxylic acid (Trolox) were purchased from Sigma Chemical Co. (St. Louis, MO, USA). ACW (hydrophilic condition) and ACL (lipophilic condition) kits for the photochemiluminescence (PCL) assay were received from Analytik Jena AG (Jena, Germany). Hexane, acetone, ethanol, methanol, toluene, sodium thiosulfate, isopropanol, Folin-Ciocalteu reagent, gallic acid, methanol, methylene chloride were purchased from Sigma Chemical Co. (St. Louis, MO, USA). B-apo-8′-carotenal was received from Sigma-Aldrich (St. Louis, MO, USA).

4.2. Plant Material and Juices Extraction

The carrot (*Daucus carota* L.) varieties (Bangor—orange, Gele peen—yellow, Peen zwart—black) were obtained from a local market in Warsaw (Poland). The obtained carrots (5 kg for each variety) were thoroughly cleaned and washed. Carrot extracts were obtained using a low-speed (Kenwood Pure Juice JMP600WH, Warsaw, Poland) and high-speed (Waring Commercial Juice Extractor WJX50, China) juicers. The obtained juices were collected for analysis in appropriately labelled tubes. Carrot extracts were stored at −20 °C until the analysis (up to 2 days).

4.3. Determination of Carotenoids Content

The method described by Marx et al. [26] was used to determine the content of carotenoids. Carrot juice sample (5 g) was extracted three times in an amber glass separatory funnel with a 30 mL mixture of acetone and hexane (1:1, *v/v*). The emulsion formed was removed by adding 50 mL sodium chloride solution (10%, *w/v*). After separation, the hexane layer was washed three times with water (50 mL) to remove acetone. Butylated hydroxytoluene (BHT) was added as an antioxidant (0.1%) and the extract was dried with sodium sulfate (2 g). The separated hexane phase was evaporated to dryness under a nitrogen stream at 40 °C and dissolved in 10 mL of methanol-methylene chloride (45:55, *v/v*). The solution was filtered through a membrane filter (ReZist® syringe filter 30 mm, pore size 0.2 µm) and was analyzed by high performance liquid chromatography (HPLC) according to the procedure described by Czaplicki et al. [27]. For analysis it was used an Agilent Technologies 1200 series apparatus (Palo Alto, CA, USA) equipped with a diode array detector (DAD) from the same manufacturer. The separation was carried out on a YMC-C30 150 × 4.6 mm, 3 µm column (YMC-Europe GmbH, Dinslaken, Germany) at 30 °C. The mobile phases consisted of methanol (solvent A) and methyl tert-butyl ether (MTBE) (solvent B) were used. The solvent gradient was as follows: 0–5 min, 95% A, 1 mL/min; 25 min, 72% A, 1.25 mL/min; 33 min, 5% A, 1.25 mL/min; 40–60 min, 95% A, 1 mL/min. The absorbance was measured at the wavelength of 450 nm. Carotenoids were identified based on the retention times of available standards and by comparing the UV-Visible absorption spectra. The content of carotenoids (mg/100 mL) was calculated based on the concentration of the internal standard and expressed in mg/100 mL juice.

4.4. Determination of Total Phenolic Content (TPC)

The TPC assay was adapted from the method described by Singleton and Rossi [28]. 0.1 mL of juice sample was mixed with 0.5 mL of Folin-Ciocalteu reagent, diluted 1: 2 with water. Then 1.5 mL of a 14% sodium carbonate solution was added and mixed. The prepared solution was kept in the dark at room temperature for 2 h. Absorbance was measured at 765 nm using an Optizen Pop UV/VIS spectrophotometer (Metasys Co. Ltd., Daejeon, Korea) against a blank sample. The total phenolics content was calculated based on the gallic acid calibration curve (concentrations in the ranges of 10–500 mg/L) and expressed in mg/100 mL juice.

4.5. Antioxidant Activity Assays

4.5.1. DPPH Assay

Determination of antioxidant properties by the DPPH method was conducted according to the procedure developed by Brand-Williams et al. [29] and modified by Thaipong et al. [30]. The DPPH radical solution was prepared by dissolving 10 mg of DPPH in 250 mL of 80% methanol. To perform the spectrophotometric test, 300 µL of DPPH solution and 20 µL of juice sample or Trolox solution were mixed. The resulting mixture was left for 30 min at room temperature in the dark. Decreasing absorbance of the resulting solution was monitored at 517 nm using an Infinite M1000 plate reader (Tecan Group AG, Switzerland). The standard curve was determined based on the lag phase's length compared to Trolox concentrations in the range of 0.01–0.75 mM. The antioxidant activity was expressed as µmol Trolox/mL juice.

4.5.2. ABTS Assay

The ABTS test described by Re et al. [31] was used to determine carrot extract's antioxidant activity. The measurement required dilution of the ABTS solution using a methanol/water mixture (80:20, *v/v*) to achieve an absorbance level of 0.70 ± 0.02 at 734 nm. For the spectrophotometric test, 290 µL of ABTS solution and 10 µL of the Trolox or juice sample were mixed and absorbance was measured directly after 6 min The standard curve was determined based on the lag phase's length compared to Trolox concentrations in the range of 0.01–0.75 mM. Measurements were carried out using the Infinite M1000 PRO plate reader (Tecan Group AG, Männedorf, Switzerland). The antioxidant activity was expressed as µmol Trolox/mL juice.

4.5.3. Photochemiluminescence (PCL) Assays

The PCL method with the Photochem apparatus (Analytik Jena, Leipzig, Germany) was used to measure antioxidants' effectiveness against superoxide anion radicals. Antioxidant activity was analyzed using the ACW (antioxidative capacities of water-soluble) and ACL (antioxidative capacities of lipid-soluble) kits. The assay was conducted as previously described by Sawicki et al. [32]. For ACW and ACL tests, the luminal reagent and Trolox working solution were prepared according to the protocol. Juice solution concentration added that the generated luminescence was in the range limits of the standard curve. Therefore, the lag time (180 s) for the ACW test was used as a free radical scavenging activity. The antioxidant activity was calculated by comparing it with the Trolox standard curve (0.5–3 nmol) and expressed as μmol Trolox/mL juice. In the ACL test, the kinetic light emission curve, showing no lag phase, was monitored for 180 s and expressed in μmol Trolox/mL. The antioxidant test was performed in triplicate for each sample.

4.6. Statistical Analysis

The values were expressed as mean ± standard deviation (SD). The results were subjected to a one-way analysis of variation (ANOVA) using Duncan's test. A linear Pearson's correlation coefficients were calculated to show relationship between bioactive compounds and antioxidant activity and $p < 0.05$ was considered significant. Principal Component Analysis (PCA) was also carried out to show differences between juice samples. The statistical analysis was performed using Statistica 13.1 (Statsoft Inc., Tulsa, OH, USA).

5. Conclusions

The carrot juices obtained in the current study were a rich source of carotenoids and phenolic compounds and the content of these bioactive depended on carrot variety and juice extraction method. It was shown that the orange carrot juices were most abundant in the carotenoids, while black carrot juice was characterized by the highest TPC. The orange carrot juices obtained with the use of HSJ contained a higher amount of carotenoids and phenolic compounds in comparison to the juice obtained using LSJ. Contrary, black carrot juices prepared with the use of HSJ contained significantly lower content of phenolic compounds compared to the juice obtained with the use of LSJ. Moreover, the extraction method had a significant impact on the antioxidant activity of the obtained juices. Among the carrot varieties, the highest antioxidant activity exhibited black carrot juice. The experiments performed with the use of PCL antioxidant activity assays demonstrated that black carrot juice obtained with the use of LSJ was characterized by a higher antioxidant activity in comparison to that obtained with HSJ. Contrary, for orange and yellow carrots juices, a higher antioxidant activity was demonstrated for juices obtained with the use of HSJ. It may be concluded that juices prepared with the use of a low-speed juicer were not always characterized by the higher content of bioactive compounds and antioxidant potential.

Author Contributions: Conceptualization, J.C., M.T. and T.S.; methodology, J.C., M.T., A.N., M.S. and T.S.; formal analysis, A.P., M.T., A.N., M.S. and T.S.; investigation, A.P., M.T., A.N., M.S. and T.S.; writing—original draft preparation, A.P., J.C., M.T., A.N. and T.S.; writing—review and editing, M.S. and T.S.; visualization, A.P. and T.S.; supervision, J.C., M.T. and K.E.P.; funding acquisition, K.E.P. All authors have read and agreed to the published version of the manuscript.

Funding: This research was supported by Minister of Science and Higher Education in the range of the program entitled "Regional Initiative of Excellence" for the years 2019–2022, Project No. 010/RID/2018/19, amount of funding 12.000.000 PLN.

Conflicts of Interest: The authors declare no conflict of interest.

References

1. Sharma, K.; Karki, S.; Thakur, N.S.; Attri, S. Chemical composition, functional properties and processing of carrot—A review. *J. Food Sci. Technol.* **2012**, *49*, 22–32. [CrossRef] [PubMed]
2. Bystrická, J.; Kavalcová, P.; Musilová, J.; Vollmannová, A.; Tóth, T.; Lenková, M. Carrot (*Daucus carota* L. ssp. sativus (Hoffm.) Arcang.) as source of antioxidants. *Acta Agric. Slov.* **2015**, *105*, 303–311. [CrossRef]
3. Ahmad, T.; Cawood, M.; Iqbal, Q.; Ariño, A.; Batool, A.; Tariq, R.M.S.; Azam, M.; Akhtar, S. Phytochemicals in Daucus carota and Their Health Benefits—Review Article. *Foods* **2019**, *8*, 424. [CrossRef] [PubMed]
4. Que, F.; Hou, X.-L.; Wang, G.-L.; Xu, Z.-S.; Tan, G.-F.; Li, T.; Wang, Y.-H.; Khadr, A.; Xiong, A.-S. Advances in research on the carrot, an important root vegetable in the Apiaceae family. *Hortic. Res.* **2019**, *6*, 1–15. [CrossRef] [PubMed]
5. Arscott, S.A.; Tanumihardjo, S.A. Carrots of Many Colors Provide Basic Nutrition and Bioavailable Phytochemicals Acting as a Functional Food. *Compr. Rev. Food Sci. Food Saf.* **2010**, *9*, 223–239. [CrossRef]
6. Stange, C.; Rodriguez-Concepcion, M. Carotenoids in Carrot. In *Pigments in Fruits and Vegetables*; Springer Science and Business Media LLC: Berlin/Heidelberg, Germany, 2015; pp. 217–228.
7. Molldrem, K.L.; Li, J.; Simon, P.W.; Tanumihardjo, S.A. Lutein and β-carotene from lutein-containing yellow carrots are bioavailable in humans. *Am. J. Clin. Nutr.* **2004**, *80*, 131–136. [CrossRef]
8. Alasalvar, C.; Grigor, J.M.; Zhang, D.; Quantick, P.C.; Shahidi, F. Comparison of Volatiles, Phenolics, Sugars, Antioxidant Vitamins, and Sensory Quality of Different Colored Carrot Varieties. *J. Agric. Food Chem.* **2001**, *49*, 1410–1416. [CrossRef]
9. Pyo, Y.-H.; Jin, Y.-J.; Hwang, J.-Y. Comparison of the Effects of Blending and Juicing on the Phytochemicals Contents and Antioxidant Capacity of Typical Korean Kernel Fruit Juices. *Prev. Nutr. Food Sci.* **2014**, *19*, 108–114. [CrossRef]
10. Park, S.Y.; Kang, T.M.; Kim, M.J.; Kim, M.J. Enzymatic browning reaction of apple juices prepared using blender and a low speed masticating hausehold juicer. *Biosci. Biotech. Biochem.* **2018**, *82*, 2000–2006. [CrossRef]
11. Khaksar, G.; Assatarakul, K.; Sirikantaramas, S. Effect of cold-pressed and normal centrifugal juicing on quality attributes of fresh juices: Do cold-pressed juices harbor a superior nutritional quality and antioxidant capacity? *Heliyon* **2019**, *5*, e01917. [CrossRef]
12. Kim, M.-J.; Jun, J.-G.; Park, S.-Y.; Choi, M.-J.; Park, E.; Kim, J.-I.; Kim, M.-J. Antioxidant activities of fresh grape juices prepared using various household processing methods. *Food Sci. Biotechnol.* **2017**, *26*, 861–869. [CrossRef] [PubMed]
13. Matter, A.A.M.; Mahmoud, S.M.E.; Salem, R.H. A New Pasteurized Nectars Prepared from Husk Tomato and Carrot Juices. *J. Food Dairy Sci.* **2010**, *1*, 531–541. [CrossRef]
14. Yoon, G.-A.; Yeum, K.-J.; Cho, Y.-S.; Chen, C.-Y.O.; Tang, G.; Blumberg, J.B.; Russell, R.M.; Yoon, S.; Lee-Kim, Y.C. Carotenoids and total phenolic contents in plant foods commonly consumed in Korea. *Nutr. Res. Pr.* **2012**, *6*, 481–490. [CrossRef] [PubMed]
15. Sun, T.; Simon, P.W.; Tanumihardjo, S.A. Antioxidant Phytochemicals and Antioxidant Capacity of Biofortified Carrots (*Daucus carota* L.) of Various Colors. *J. Agric. Food Chem.* **2009**, *57*, 4142–4147. [CrossRef] [PubMed]
16. Ma, T.; Tian, C.; Luo, J.; Sun, X.; Quan, M.; Zheng, C.; Zhan, J. Influence of technical processing units on the α-carotene, β-carotene and lutein contents of carrot (*Daucus carrot* L.) juice. *J. Funct. Foods* **2015**, *16*, 104–113. [CrossRef]
17. Akhtar, S.; Rauf, A.; Imran, M.; Qamar, M.; Riaz, M.; Mubarak, M.S. Black carrot (*Daucus carota* L.), dietary and health promoting perspectives of its polyphenols: A review. *Trends Food Sci. Technol.* **2017**, *66*, 36–47. [CrossRef]
18. Leja, M.; Kamińska, I.; Kramer, M.; Maksylewicz-Kaul, A.; Kammerer, D.; Carle, R.; Baranski, R. The Content of Phenolic Compounds and Radical Scavenging Activity Varies with Carrot Origin and Root Color. *Plant Foods Hum. Nutr.* **2013**, *68*, 163–170. [CrossRef]
19. Mphahlele, R.; Fawole, O.; Mokwena, L.; Opara, U.L. Effect of extraction method on chemical, volatile composition and antioxidant properties of pomegranate juice. *S. Afr. J. Bot.* **2016**, *103*, 135–144. [CrossRef]
20. Rajasekar, D.; Akoh, C.C.; Martino, K.G.; MacLean, D. Physico-chemical characteristics of juice extracted by blender and mechanical press from pomegranate cultivars grown in Georgia. *Food Chem.* **2012**, *133*, 1383–1393. [CrossRef]

21. MacDonald-Wicks, L.K.; Wood, L.G.; Garg, M.L. Methodology for the determination of biological antioxidant capacity in vitro: A review. *J. Sci. Food Agric.* **2006**, *86*, 2046–2056. [CrossRef]
22. Algarra, M.; Fernandes, A.; Mateus, N.; De Freitas, V.; Da Silva, J.C.E.; Casado, J. Anthocyanin profile and antioxidant capacity of black carrots (*Daucus carota* L. ssp. sativus var. atrorubens Alef.) from Cuevas Bajas, Spain. *J. Food Compos. Anal.* **2014**, *33*, 71–76. [CrossRef]
23. Scarano, A.; Gerardi, C.; D'Amico, L.; Accogli, R.; Santino, A. Phytochemical Analysis and Antioxidant Properties in Colored Tiggiano Carrots. *Agriculture* **2018**, *8*, 102. [CrossRef]
24. Gajewski, M.; Szymczak, P.; Elkner, K.; Dąbrowska, A.; Kret, A.; Danilčenko, H. Some Aspects of Nutritive and Biological Value of Carrot Cultivars with Orange, Yellow and Purple-Coloured Roots. *Veg. Crop. Res. Bull.* **2007**, *67*, 149–161. [CrossRef]
25. Kim, M.-J.; Kim, J.-I.; Kang, M.-J.; Kwon, B.; Jun, J.-G.; Choi, J.-H.; Kim, M.-J. Quality evaluation of fresh tomato juices prepared using high-speed centrifugal and low-speed masticating household juicers. *Food Sci. Biotechnol.* **2015**, *24*, 61–66. [CrossRef]
26. Chen, B.H.; Peng, H.Y.; Chen, H.E. Changes of Carotenoids, Color, and Vitamin A Contents during Processing of Carrot Juice. *J. Agric. Food Chem.* **1995**, *43*, 1912–1918. [CrossRef]
27. Czaplicki, S.; Tańska, M.; Konopka, I. Sea-buckthorn oil in vegetable oils stabilisation. *Ital. J. Food Sci.* **2016**, *28*, 412–425.
28. Singleton, V.L.; Rossi, J.A. Colorimetry of Total Phenolics with Phosphomolybdic-Phosphotungstic Acid Reagents. *Am. J. Enol. Vitic.* **1965**, *16*, 144–158.
29. Brand-Williams, W.; Cuvelier, M.; Berset, C. Use of a free radical method to evaluate antioxidant activity. *LWT* **1995**, *28*, 25–30. [CrossRef]
30. Thaipong, K.; Boonprakob, U.; Crosby, K.; Cisneros-Zevallos, L.; Byrne, D.H. Comparison of ABTS, DPPH, FRAP, and ORAC assays for estimating antioxidant activity from guava fruit extracts. *J. Food Compos. Anal.* **2006**, *19*, 669–675. [CrossRef]
31. Re, R.; Pellegrini, N.; Proteggente, A.; Pannala, A.; Yang, M.; Rice-Evans, C. Antioxidant activity applying an improved ABTS radical cation decolorization assay. *Free. Radic. Biol. Med.* **1999**, *26*, 1231–1237. [CrossRef]
32. Sawicki, T.; Wiczkowski, W.; Hrynkiewicz, M.; Bączek, N.; Hornowski, A.; Honke, J.; Topolska, J. Characterization of the phenolic acid profile and in vitro bioactive properties of white beetroot products. *Int. J. Food Sci. Technol.* **2020**. [CrossRef]

Article

The Effect of Different Extraction Protocols on *Brassica oleracea* var. *acephala* Antioxidant Activity, Bioactive Compounds, and Sugar Profile

Nikola Major [1,*], Bernard Prekalj [1,2], Josipa Perković [1], Dean Ban [1,2], Zoran Užila [1,2] and Smiljana Goreta Ban [1,2,*]

[1] Institute of Agriculture and Tourism, K. Huguesa 8, 52440 Poreč, Croatia; bernard@iptpo.hr (B.P.); josipa@iptpo.hr (J.P.); dean@iptpo.hr (D.B.); zoran@iptpo.hr (Z.U.)

[2] Centre of Excellence for Biodiversity and Molecular Plant Breeding, Svetošimunska 25, 10000 Zagreb, Croatia

* Correspondence: nikola@iptpo.hr (N.M.); smilja@iptpo.hr (S.G.B.)

Received: 26 November 2020; Accepted: 16 December 2020; Published: 17 December 2020

Abstract: The extraction of glucosinolates in boiling aqueous methanol from freeze dried leaf tissues is the most common method for myrosinase inactivation but can be hazardous because of methanol toxicity. Although freeze drying is the best dehydration method in terms of nutritional quality preservation, the main drawbacks are a limited sample quantity that can be processed simultaneously, a long processing time, and high energy consumption. Therefore, the aim of this study is to evaluate the effects of applying high temperature for myrosinase inactivation via hot air drying prior to the extraction step, as well as the effects of cold aqueous methanol extraction on total antioxidant activity, total glucosinolates, total phenolic content, and sugar profile in 36 landraces of kale. The results from our study indicate that cold aqueous methanol can be used instead of boiling aqueous methanol with no adverse effects on total glucosinolate content. Our results also show that hot air drying, compared to freeze drying, followed by cold extraction has an adverse effect on antioxidant activity measured by DPPH radical scavenging, total glucosinolate content, as well as on the content of all investigated sugars.

Keywords: antioxidant activity; extraction; glucosinolates; kale (*Brassica oleracea* var. *acephala*); phenolics

1. Introduction

Vegetables from the Brassicaceae family are known for their excellent nutritional value and are abundant in carbohydrates, vitamins (ascorbic acid, folic acid, β-carotene, α-tocopherol), macro and micro elements (iron, calcium, selenium, copper, manganese, zinc), as well as secondary metabolites, including glucosinolates, phenolics (tannins, phenolic acids, anthocyanidins, flavonols, coumarins, flavones), and other bioactive molecules, such as phytosterols and terpenoids [1,2]. They are also known for their antimicrobial and anticancerogenic activity [3,4]. One of the well-known members of the Brassicaceae family is kale (*Brassica oleracea* var. *acephala*). It originates and is traditionally used in Mediterranean countries, but has gained special attention and popularity in the U.S., and later worldwide over the last decade [5]. It is characterized by leaves, which do not form a head, unlike other leafy vegetables from the Brassicaceae family as white cabbage, savoy cabbage, Brussels sprouts, and Chinese cabbage [6].

The most abundant sulfur containing compounds in plants from the Brassicaceae family are glucosinolates and S-methylcysteine sulfoxide [7]. Cruciferous vegetables are extensively researched regarding their beneficial compounds especially ones active in cancer prevention. Isothiocyanates are degradation products from glucosinolates are known for their antioxidant, immunostimulatory, anti-inflammatory, antiviral and antibacterial properties [8–10]. Isothiocyanates are converted from glucosinolates in a reaction catalyzed by the enzyme myrosinase which activates and releases upon plant tissue injury. The myrosinase enzyme was first discovered in 1840 from the mustard seed after which the Brassica studies were focused on clinical studies and prospective benefits of cruciferous vegetables [1]. Myrosinase catalyzes glucosinolate hydrolysis after the plant tissue is injured or in any way disrupted. Therefore, accurate glucosinolate analysis depends on myrosinase inactivation [11]. Differences in glucosinolate content regarding Brassica species and environmental factors can be an obstacle when comparing different studies regarding nutritional and bioactive quality of cruciferous vegetables. Moreover, presence of the enzyme myrosinase, activated in plant handling and extraction processes that precede glucosinolate determinations, is also an aggravating circumstance, which can lead to a decrease in total glucosinolate concentration [12]. Most published glucosinolate analysis methods employ dehydration by freeze drying for myrosinase inactivation and denaturation by moderately high temperature during the glucosinolate extraction step [12,13]. This process is outlined in the ISO9167:2019 standard, as well as in the work by Grosser and van Dame [14]. Freeze drying is a dehydration process based on water sublimation under vacuum, which protects the primary structure and shape of the sample [15]. Although freeze drying is considered the best form of dehydration, the process is time consuming and relatively expensive [16], especially for high throughput methods. On the other hand, hot air drying is simple, fast, and can handle large quantities of samples.

Most published methods for glucosinolate analysis employ boiling aqueous methanol (70/30, methanol/water, *v*/*v*) in the glucosinolate extraction step [17–21]. Methanol is a common organic solvent and reagent in organic synthetic procedures [22]. Poisoning can occur via methanol ingestion, skin absorption, or inhalation [23]. Acute methanol toxicity evolves in a well-understood pattern and results in metabolic acidosis via formic acid formation and superimposed toxicity to the visual system [24]. The use of boiling methanol in the glucosinolates extraction step increases the risk of poisoning and researchers offered several less hazardous alternatives to the standard method, including the use of hot water or the use of cold (ambient temperature) aqueous methanol [12]. The 2019 revision of the ISO9167 method for the determination of glucosinolates in rapeseed and rapeseed meals by HPLC replaces aqueous methanol (70/30, *v*/*v*) by aqueous ethanol (50/50, *v*/*v*) for lower toxicity [25]. Although the ISO9167:2019 official method supports the use of ethanol instead of methanol for glucosinolate extraction in rapeseed and rapeseed meals, aqueous methanol is the recommended extraction solvent due to the structural and biological diversity of seed and plant matrices, but can be replaced by other solvents if appropriate validation procedures are applied [26].

The effect of hot air drying versus freeze drying on phenolic compounds and antioxidant activity was a subject of numerous studies [27–30] but there is not enough data on the possible effects on glucosinolate and sugar content. While there are several studies concerning the comparison of cold versus hot methanol extraction [12] on glucosinolate levels, the number of samples involved could have been higher, and the studies did not include other bioactive compounds or antioxidant activity in the comparison.

Therefore, the aim of this work is to evaluate the effect of applying high temperature for myrosinase inactivation via hot air (oven) drying prior to the extraction step, as well as the effect of cold (ambient temperature) aqueous methanol extraction on total glucosinolates content, total phenolic content, total antioxidant activity, and sugar content in 36 landraces of kale (*Brassica oleracea* var. *acephala*).

2. Results

2.1. The Effect of Different Extraction Protocols on Antioxidant Activity, Bioactive Compounds, and Sugar Content in Brassica

The extraction of glucosinolates by the hot methanol extraction step from freeze dried kale leaves yielded significantly lower amounts of total glucosinolates (30.3 ± 0.6 mg sinigrin equivalents (SEQ)/g dry weight (DW), compared to the cold methanol extraction step (34.3 ± 0.9 mg SEQ/g DW) (Figure 1). Significantly higher content of total glucosinolates was found in extracts obtained from freeze dried kale leaves (34.3 ± 0.9 mg SEQ/g DW) compared to oven dried leaves followed by cold extraction (31.9 ± 0.7 mg SEQ/g DW) (Figure 1).

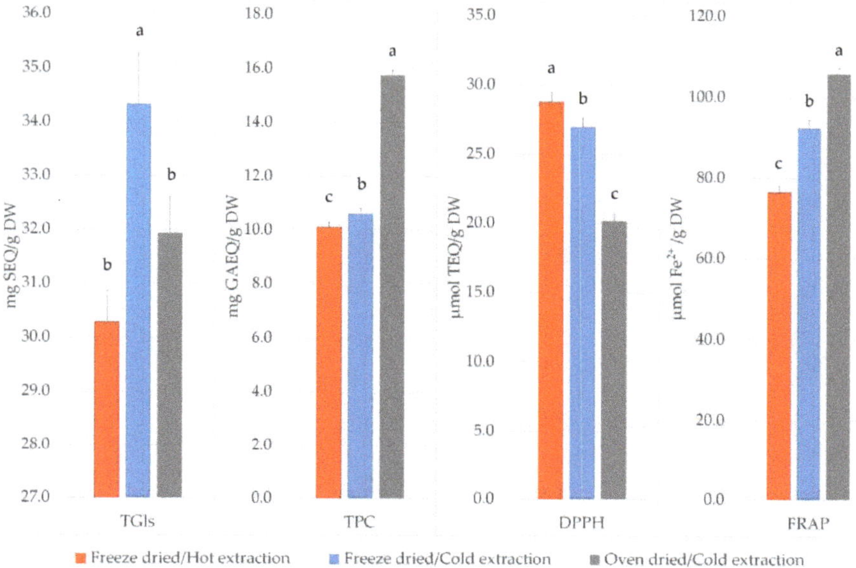

Figure 1. The effect of different extraction protocols on total glucosinolates content, total phenolic content, DPPH radical scavenging activity and ferric ion reducing antioxidant power in *B. oleracea* var. acephala leaf extracts. Values are expressed as mean ± SE (N = 324). The different letters above bars denote significant difference by Tukey's Unequal N Honestly Significant Difference (HSD) test, $p < 0.05$. TGls—total glucosinolates; TPC—total phenolic content; DPPH—DPPH radical scavenging activity; FRAP—Ferric ion Reducing Antioxidant Power.

The same effect was observed for total phenolic content, where significantly lower amounts were extracted with the hot methanol method (10.1 ± 0.2 mg gallic acid equivalents (GAEQ)/g DW) compared to the cold methanol extraction (10.6 ± 0.2 mg GAEQ/g DW) in freeze dried leaves (Figure 1). On the other hand, total phenolic content was found to be higher in oven dried leaves (15.7 ± 0.2 mg GAEQ/g DW) compared to freeze dried B. oleracea var. acephala leaves followed by cold extraction (10.6 ± 0.2 mg GAEQ/g DW) (Figure 1).

As for the antioxidant activity we found significantly higher Ferric Reducing Antioxidant Power (FRAP) values from samples extracted from freeze dried leaves with cold methanol (92.5 ± 1.8 µmol Fe^{2+}/g DW) compared to hot methanol extraction (76.5 ± 1.7 µmol Fe^{2+}/g DW) (Figure 1). The hot methanol extraction method yielded higher antioxidant power measured by 2,2-diphenyl-1-picrylhydrazyl (DPPH) radical scavenging activity (28.8 ± 0.7 µmol Trolox equivalents (TEQ)/g DW) compared to the cold methanol extraction method (27.0 ± 0.7 µmol TEQ/g DW) from freeze dried B. oleracea var. acephala leaves (Figure 1). DPPH radical scavenging was found to be significantly higher in freeze dried B. oleracea var. acephala leaves (27.0 ± 0.7 µmol TEQ/g DW) compared to oven dried samples (20.2 ± 0.6 µmol TEQ/g DW) followed by cold extraction (Figure 1). FRAP values were significantly higher in oven dried B. oleracea var. acephala leaves (105.7 ± 1.5 µmol Fe^{2+}/g DW) compared to freeze dried leaves (92.5 ± 1.8 µmol Fe^{2+}/g DW) followed by cold extraction (Figure 1).

Both sucrose and fructose content was significantly lower in the freeze dried extracts obtained by the hot methanol extraction method (7.8 ± 0.3 g/100g DW and 4.4 ± 0.2 g/100g DW for sucrose and fructose, respectively) compared to the cold methanol extraction method (10.0 ± 0.3 g/100g DW and 9.4 ± 0.2 g/100g DW for sucrose and fructose, respectively) (Figure 2). Glucose content was found to be no different in freeze dried sample extracts obtained by hot or cold methanol extraction (8.0 ± 0.3 g/100g DW and 8.0 ± 0.2 g/100g DW, respectively) (Figure 2). Extracts obtained by cold extraction from oven dried tissues exhibited significantly lower sucrose, glucose and fructose levels (5.2 ± 0.3 g/100g DW, 2.9 ± 0.1 g/100g DW and 4.6 ± 0.1 g/100g DW, respectively) compared to those obtained from freeze dried leaf samples (10.0 ± 0.3 g/100g DW, 8.0 ± 0.2 g/100 g DW and 9.4 ± 0.2 g/100g DW, respectively) (Figure 2).

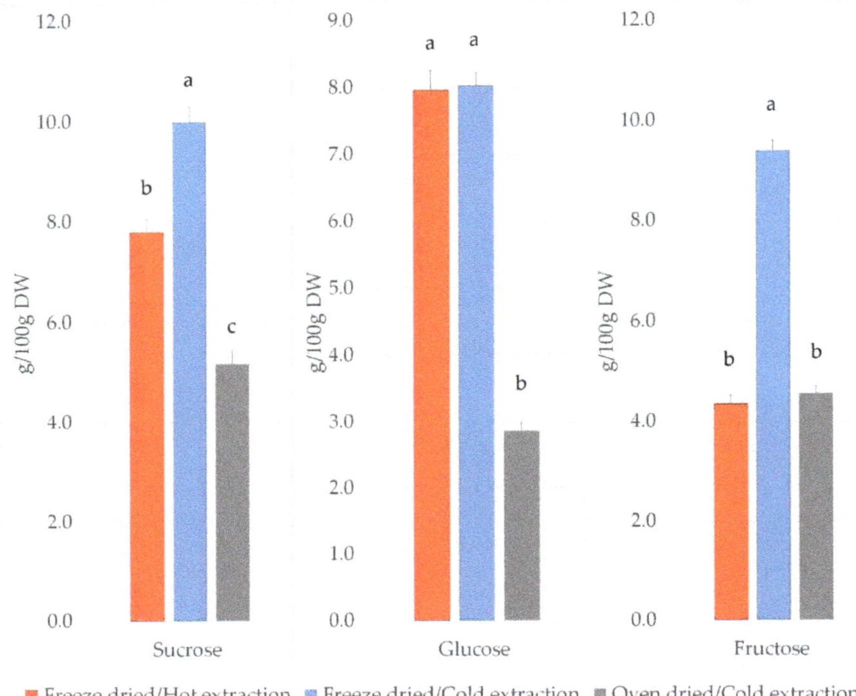

Figure 2. The effect of different extraction protocols on sucrose, glucose and fructose content in *B. oleracea* var. acephala extracts. Values are expressed as mean ± SE (N = 324). The different letters above bars denote significant difference by Tukey's Unequal N HSD test, $p < 0.05$.

2.2. Multivariate Analysis of the Data Obtained by Different Extraction Protocols

The obtained data was processed with Partial Least Squares–Discriminant Analysis (PLS-DA) to further clarify the relations between different extraction protocols. The PLS-DA model was employed due to the supervised nature of the learning algorithm.

Overall, the extracts which included oven drying leaf tissue following cold methanol extraction were separated on the primary axis from the extracts obtained from freeze dried leaf tissue followed by cold or hot methanol extraction (Figure 3). The highest contribution to the sample placement on the primary axis had total phenolic content, DPPH radical scavenging activity, glucose, and sucrose content (Figure 3). On the second axis, the extracts, which were obtained from freeze dried leaf tissues following hot or cold methanol extraction, were separated according to fructose, total glucosinolates, and FRAP values (Figure 3).

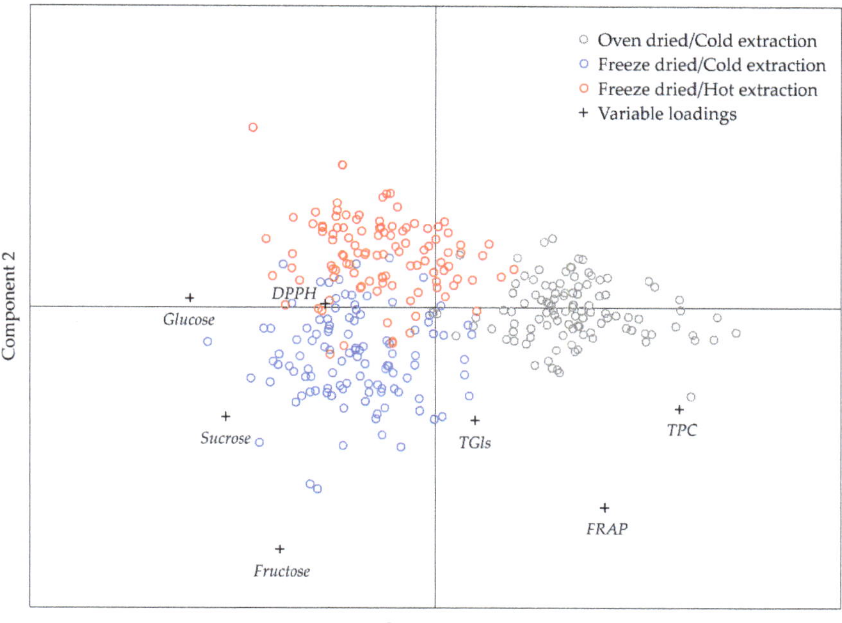

Figure 3. Partial Least Squares–Discriminant Analysis (PLS-DA) analysis of data obtained by different extraction protocols (N = 324); TGls—total glucosinolates; TPC—total phenolic content; DPPH—DPPH radical scavenging activity; FRAP—Ferric ion Reducing Antioxidant Power.

The model showed that the largest difference between extraction protocols was in fructose content where the freeze drying and cold methanol combination yielded the highest amount of the compound. Total phenolic content was the second most important variable in the differentiation between the observed extraction protocols where the oven dried and cold methanol combination exhibited the highest yield of phenolic compounds. The third most important variable in differentiating among extraction protocols was glucose where similar amounts were extracted with either hot or cold methanol from freeze dried leaf tissue, whereas several-fold lower amounts were extracted from oven dried tissue with cold methanol. The next most important variable was FRAP where the highest values were observed in extracts obtained from oven dried tissues by cold methanol. Sucrose was the fifth most important variable where the oven dried leaf tissues had lower yields of the compound compared to extracts from freeze dried tissues. The sixth variable, according to the discriminating power, was DPPH radical scavenging, where again the extracts obtained from oven dried tissues exhibited lower antioxidant activity compared to extracts obtained from freeze dried leaf tissues. The variable with the lowest discriminating power according to the obtained model was total glucosinolates where cold methanol extraction from freeze dried Brassica tissue exhibited the highest yield compared to the other two extraction protocols.

2.3. Correlations between Bioactive Compounds, Antioxidant Activity, and Sugar Content in Brassica Extracts

Statistically significant correlations ($p \geq 0.05$) were observed between bioactive compounds content, antioxidant activity, as well as between sugar content (Table 1). Positive correlations were observed between total phenolic content, total glucosinolates, and FRAP values (Table 1). Negative correlations were observed between total phenolic content and sucrose, glucose, and fructose content, as well as total glucosinolates and glucose content (Table 1). DPPH scavenging activity correlated positively with sugar content but FRAP correlated negatively with sucrose and glucose content (Table 1). Positive correlations were observed between sucrose, glucose, and fructose content (Table 1).

Table 1. Pearson's correlations between bioactive compounds content, antioxidant activity and sugar content in B. oleracea var. acephala extracts (N = 324).

	TGls	TPC	DPPH	FRAP	Sucrose	Glucose	Fructose
TGls	1.00						
TPC	0.14 [1]	1.00					
DPPH	0.06	−0.09	1.00				
FRAP	0.31	0.78	0.09	1.00			
Sucrose	−0.08	−0.44	0.27	−0.23	1.00		
Glucose	−0.13	−0.65	0.26	−0.41	0.47	1.00	
Fructose	−0.01	−0.23	0.12	0.09	0.55	0.43	1.00

[1] Numbers highlighted in red are statistically significant ($p \leq 0.05$); TGls—total glucosinolates; TPC—total phenolic content; DPPH—DPPH radical scavenging activity; FRAP—Ferric ion Reducing Antioxidant Power.

3. Discussion

Biological activity of kale containing phytochemicals are associated with antioxidant, anti-cancerogenic activity and protection of gastrointestinal and cardiovascular system [4,31–33]. Sikora et al. found that kale had much higher antioxidant activity than broccoli, Brussels sprouts, and cauliflower [34]. Šamec et al. on the other hand found that white cabbage and kale sprouts had significantly higher antioxidant capacity, polyphenol and glucosinolate content compared to arugula, broccoli, and Chinese cabbage sprouts [20]. Kale is also abundant in organic acids, including citric, malic, pyruvic, shikimic fumaric, and aconitic acids [35].

The freeze drying process is regarded as the best method for sample dehydration because it can prolong shelf life without compromising nutritional quality [36]. The major drawback of this method is limited sample quantity that can be processed simultaneously, long processing time (up to several days) and high energy consumption [37]. On the other hand, oven drying is readily accessible to most laboratories, sample quantity is seldom an issue and the drying time can be as low as one hour but rarely exceeds 24 h.

In our study, freeze dried leaf tissues followed by cold extraction yielded significantly higher total glucosinolate content compared to either oven dried leaf tissue followed by cold extraction or freeze dried leaf tissue followed by hot extraction. The results are in accordance with Rutnakornpituk et al., who investigated the effect of freeze and oven drying on phenolic and glucosinolate content, as well as, antioxidant activity in five cruciferous vegetables where the authors concluded that freeze dried leaf tissues yielded higher glucosinolate content in all investigated species [38]. On the other hand, Tetteh et al. studied the effect of different various drying methods on post-harvest glucosinolate content in *Moringa oleifera* leaf tissues where no differences were observed between oven dried and freeze dried leaf tissue samples [39]. Doheny-Adams et al. investigated the effect of hot and cold methanol extraction on glucosinolate content in five Brassicaceae species [12]. The results from their study showed that the hot extraction step can be replaced with cold extraction with no losses in glucosinolate content, which is in line with the findings presented in our study [12].

Oven drying, compared to freeze drying followed by cold extraction yielded significantly higher total phenolic content and FRAP values and lower total glucosinolate content and DPPH scavenging activity in *B. oleracea* var. *acephala* leaf extracts. Similarly, Managa et al. showed that the majority of investigated phenolic compounds in Chinese cabbage significantly increased in content in oven dried compared to freeze dried samples [40]. Although the phenolic compound content in Chinese cabbage was higher, the antioxidant activity measured by FRAP, DPPH, and ABTS, was significantly lower in the oven dried compared to freeze dried samples [40]. On the other hand, Korus found that freeze dried kale had significantly higher total phenolic content and antioxidant activity compared to air dried kale, albeit at a lower temperature (55 °C) compared to this study (105 °C) [41]. Papoutsis et al.

found higher total phenolic content and DPPH scavenging activity in hot air dried compared to freeze dried lemon [28]. Que et al. also found higher total phenolic content and antioxidant activity in hot air dried compared to freeze dried pumpkin flour [27]. Das et al. studied the antioxidant properties of freeze dried and oven dried wheatgrass where higher phenolic content but lower FRAP and DPPH radical scavenging activity was found in hot air dried compared to freeze dried samples [42]. Hot air drying also exhibited elevated total phenolic content and antioxidant activity compared to freeze drying in olive leaves as shown by Ahmad-Qasem et al. [43]. Hot air dried leaf tissue extracts tend to exhibit higher phenolic compounds content, as found in this study, but the link with antioxidant capacity is not clear. In our study we found FRAP having significant correlation with total glucosinolate content as well as total phenolic content, but no significant correlation with DPPH scavenging activity. Interestingly, we found at the same time the highest DPPH radical scavenging activity and lowest FRAP values in freeze dried tissues followed by hot extraction, while exactly the opposite was determined in hot air dried samples followed by cold extraction. The obtained results are contrary to previously published [42,44,45] data where FRAP and DPPH radical scavenging assays have high correlations, especially since it is known that both depend mainly on the electron transfer mechanism which measures the antioxidant's reducing ability [46,47]. FRAP values correlated highly with total phenolic content and negatively with sucrose and fructose content while the data obtained by the DPPH radical scavenging assay correlated positively with the samples' sugar content possibly indicating interferences with the meta-products of Maillard reactions occurring during the hot air drying process [48].

Hot air drying *B. oleracea* var. *acephala* leaf tissues followed by cold extraction induced significantly lower content of all investigated sugars compared to freeze drying followed by cold extraction. If freeze dried tissues were subjected to hot extraction sucrose and fructose levels were significantly lower compared to the cold extraction process but higher than oven dried leaf tissues except for fructose. Fante and Zapata Norena studied the quality of hot air dried and freeze dried garlic and found that inulin content decreased with hot air drying while glucose and fructose content increased compared to freeze dried garlic [49]. Iombor et al. studied changes in soursop flour composition as affected by oven and freeze drying and found 40% decrease in carbohydrate content in hot air dried samples [50]. Zhang et al. studied the changes in chestnut starch properties during different drying methods and found lower starch but higher reducing sugar content in oven dried compared to freeze dried samples [51]. Karaman et al. found lower fructose and glucose content in oven dried compared to freeze dried persimmon powders [52]. On the other hand, Gao et al. showed that there was no difference in sucrose, glucose and fructose content in oven and freeze dried jujube samples [53]. The observed decrease in sugar content in our study could be attributed to Maillard reactions occurring at elevated temperatures during hot air drying [54] or possibly the caramelization of sugars [55]. Michalska et al. determined that increased drying temperature of different plum cultivars resulted in decreased total sugar content while early and intermediate Maillard reaction products increased in content [48]. Similarly, Li et al. observed a decrease in reducing sugar content during hot air drying opposed to freeze drying of instant *Tremella fuciformis,* which the authors attributed to an intense Maillard reaction occurring between the carbonyl group of the reducing sugars and amino acids at elevated temperatures [56]. Previously published data on oligosaccharide and simple sugars extraction in various solvent mixtures and extraction temperatures showed that, with increased temperature, especially at the solvent boiling point, sugar content yields were higher [57]. We, on the other hand, observed lower yields of sucrose and fructose in freeze dried leaves hot extracts, compared to the cold extraction step while glucose levels remained unaffected.

4. Materials and Methods

4.1. Plant Material

Thirty-six *B. oleracea* var *acephala* ecotypes (accessions IPT379, IPT390, IPT386, IPT391, IPT381, IPT392, IPT393, IPT419, IPT394, IPT395, IPT383, IPT396, IPT384, IPT385, IPT397, IPT398, IPT420,

IPT399, IPT400, IPT387, IPT401, IPT 402, IPT403, IPT404, IPT405, IPT406, IPT174, IPT202, IPT206, IPT407, IPT421, IPT408, IPT409, IPT410, IPT380, IPT411) used in this study were grown under the same agro-climatic conditions on the experimental farm of the Institute of Agriculture and Tourism, Poreč, Croatia (N 45°13′20.30″, E 13°36′6.49″) and are part of the National Program for the Conservation and Sustainable Use of Plant Genetic Resources for Food and Agriculture (accession IPT399 and IPT403 are shown in Figure 4). The leaves used in this study were harvested when the plants reached technological maturity. Harvested leaves were fully developed without any signs of physiological, pest or diseases injury. Fresh plant samples (three biological replicates per sample), immediately after harvesting, were either kept at −80 °C until the freeze drying process or placed in an oven (Memmert UF160, Schwabach, Germany) at 105 °C overnight. Frozen plant samples were placed in a freeze dryer (Labogene Coolsafe 95-15 Pro, Allerød, Denmark) and lyophilized over a period of 48 h. Lyophilized or oven-dried samples were ground to powder (0.2 mm) using an ultra-centrifugal mill (Retsch ZM200, Haan, Germany).

Figure 4. *Brassica oleracea* var. *acephala* accession (**a**) IPT399 and (**b**) IPT403.

4.2. Hot Methanol Extraction

Freeze dried plant material (30 mg) was preheated to 75 °C for 3 min in a heating/cooling dry block (Biosan CH100, Riga, Latvia) and 1.5 mL of preheated 70: 30 methanol:water (*v*/*v*) at 75 °C was added. The samples were incubated for 10 min at 75 °C and manually shaken every 2 min. Afterwards the samples were centrifuged at 15,000 G for 5 min (Domel Centric 350, Železniki, Slovenia) and the supernatant was filtered through a 0.22 µm nylon filter and transferred to a clean tube. The samples were stored at −80 °C until further analysis.

4.3. Cold Methanol Extraction

Freeze dried or oven dried plant material (30 mg) was extracted with 1.5 mL of 80:20 methanol: water (*v*/*v*) at 20 °C over a period of 30 min in an ultrasonic bath (MRC 250H, Holon, Israel). The samples were centrifuged at 15,000 G for 5 min (Domel Centric 350, Železniki, Slovenia) and the supernatant was transferred to a clean tube. The samples were stored at −80 °C until further analysis.

4.4. Determination of Total Antioxidant Activity

Total antioxidant activity was evaluated using the FRAP assay [58] ant the DPPH radical scavenging activity assay [59]. Briefly, 100 µL of the sample was mixed with 200 µL of either freshly prepared FRAP reagent or 0.02M DPPH radical for the FRAP or DPPH assays, respectively. The antioxidant activity using the FRAP assay was evaluated after 10 min of reaction time at 25 °C by reading the absorbance at 593 nm while the DPPH radical scavenging ability was evaluated after 30 min of reaction time at 25 °C by reading the absorbance at 517 nm (Tecan Infinite 200 Pro M Nano+, Männedorf, Switzerland). FRAP values were calculated against a Fe^{2+} calibration curve (y = 0.0168x − 0.002; serial dilutions of Fe^{2+}—20, 40, 80, 120, 160, 200, 250 µM; coefficient of determination, R^2 = 0.9999, recovery: 101.8 ± 1.6 %) and expressed as µmol Fe^{2+}/g DW. DPPH radical scavenging ability values were calculated against a standard curve of Trolox (y = −0.0128x + 0.0125; serial dilutions of Trolox—2, 5, 10, 25, 50, 75, 100 µM; coefficient of determination, R^2 = 0.9995, recovery: 103.7 ± 1.2 %) and expressed as µmol TEQ/g DW, respectively.

4.5. Determination of Total Glucosinolates and Total Phenolic Content

Total glucosinolates were determined according to Ishida et al. [60] with some modifications. Briefly, 10 µL of plant extract was mixed with 300 µL of 2 mM Palladium (II) chloride and after 30 min of reaction time at 25 °C absorbance was read at 425 nm (Tecan Infinite 200 Pro M Nano+, Männedorf, Switzerland). The results were calculated against a standard curve of sinigrin (y = 3.8021x + 0.1682; serial dilutions of sinigrin—0.3, 0.6, 1.2, 1.8, 2.4, 3.0 mg/L; coefficient of determination, R^2 = 0.9995, recovery: 99.7 ± 2.9%) and expressed as SEQ/g DW. Total phenolic content was determined according to Singleton and Rossi [61] with some modifications. The methanolic extracts (20 µL) were mixed with 140 µL of freshly prepared 0.2M Folin–Ciocalteu reagent. After 1 min, 140 µL of 6% solution of Calcium carbonate was added to the mixture. The absorbance was read at 750 nm (Tecan Infinite 200 Pro M Nano+, Männedorf, Switzerland) after 60 min of reaction time at 25 °C. The results were calculated against a standard curve of gallic acid (y = 3.7867x − 0.2144; serial dilutions of gallic acid—12.5, 25, 50, 75, 100, 150, 250 mg/L; coefficient of determination, R^2 = 0.9999, recovery: 102.0 ± 2.9 %) and expressed as mg GAEQ/g DW.

4.6. Sugar Analysis by HPLC

The analysis of sucrose, fructose and sucrose content was carried out using a HPLC system consisting of a solvent delivery unit (Varian 210, Palo Alto, CA, USA), an autosampler (Varian 410, Palo Alto, CA, USA), column oven (Varian CM500, Palo Alto, CA, USA) and a refractive index detector (Varian 350, Palo Alto, CA, USA). Chromatographic separation was achieved by injecting 10 µL of the sample on a 300 × 8 mm, 9 µm particle size, calcium cation exchange column (Dr. Maisch ReproGel Ca, Ammerbuch, Germany) held at 80 °C using deionized water as the mobile phase (1 mL/min, isocratic elution). Retention times and peak areas of the investigated sugars were compared to analytical standards for identification and quantification, respectively. Linear calibration curves were obtained with serial dilutions of 0.25, 0.50, 1.00, 2.50, 5.00, 7.50, 10.00 g/L of sucrose (y = 2265.73x + 29.97, coefficient of determination, R^2 = 0.9998, recovery: 99.9 ± 2.3 %), glucose (y = 2224.75x + 28.33, coefficient of determination, R^2 = 0.9999, recovery: 99.8 ± 1.8 %) and fructose (y = 2233.53x + 47.37, coefficient of determination, R^2 = 0.9998, recovery: 100.0 ± 0.6 %).

4.7. Statistical Analysis

To determine the effect of the investigated extraction protocols on total antioxidant activity, total glucosinolate, total phenolic and sugar content in *B. oleracea* var. *acephala* leaves the results were processed by analysis of variance (ANOVA) and reported as mean ± SE. Pearson's correlations were calculated to evaluate the connection between bioactive compounds content, antioxidant activity, and sugar content. Further investigation on the impact of different extraction protocols on the studied compounds was carried out by employing PLS-DA as a supervised multivariate method. All statistical analyses were performed using Statistica 13.4.0.14. (Tibco Inc., Palo Alto, CA, USA). Significant differences were determined at $p \leq 0.05$ and homogenous group means were compared by Tukey–Kramer Unequal N HSD test.

5. Conclusions

The results from our study confirm the results published by Doheny-Adams et al. [12] that cold (ambient temperature) aqueous methanol can be used instead of boiling aqueous methanol with no adverse effects on total glucosinolate content. In addition to higher total glucosinolate content we observed an increase in antioxidant activity measured by FRAP, total phenolic content, sucrose content, and fructose content when the cold extraction step was applied on freeze dried *B. oleracea* var. *acephala* plant tissues while glucose levels remained unaffected. Our results also show that hot air drying, compared to freeze drying, followed by cold extraction has an adverse effect on antioxidant activity measured by DPPH radical scavenging, total glucosinolate content, as well as, on the content of all studied sugars in *B. oleracea* var. *acephala* leaves, indicating it would be inferior to freeze drying. On the other hand, if phenolics are the compounds of interest in *B. oleracea* var. *acephala* leaves, hot air drying may be a viable alternative to freeze drying.

Author Contributions: Conceptualization, N.M. and S.G.B.; methodology, N.M., B.P., S.G.B., Z.U., D.B.; investigation, N.M., B.P., Z.U.; data curation, N.M.; writing—original draft preparation, N.M. and J.P.; writing—review and editing, S.G.B. and D.B.; visualization, N.M.; funding acquisition, S.G.B. and D.B. All authors have read and agreed to the published version of the manuscript.

Funding: This research was supported by the project KK.01.1.1.01.0005 Biodiversity and Molecular Plant Breeding, Centre of Excellence for Biodiversity and Molecular Plant Breeding (CoE CroP-BioDiv), Zagreb, Croatia.

Conflicts of Interest: The authors declare no conflict of interest.

References

1. Manchali, S.; Chidambara Murthy, K.N.; Patil, B.S. Crucial facts about health benefits of popular cruciferous vegetables. *J. Funct. Foods* **2012**, *4*, 94–106. [CrossRef]
2. Singh, J.; Upadhyay, A.K.; Prasad, K.; Bahadur, A.; Rai, M. Variability of carotenes, vitamin C, E and phenolics in Brassica vegetables. *J. Food Compos. Anal.* **2007**, *20*, 106–112. [CrossRef]
3. Talalay, P.; Fahey, J.W. Phytochemicals from cruciferous plants protect against cancer by modulating carcinogen metabolism. *J. Nutr.* **2001**, *131*, 3027S–3033S. [CrossRef]
4. Ayaz, F.A.; Hayirlioglu-Ayaz, S.; Alpay-Karaoglu, S.; Grúz, J.; Valentová, K.; Ulrichová, J.; Strnad, M. Phenolic acid contents of kale (*Brassica oleraceae* L. var. acephala DC.) extracts and their antioxidant and antibacterial activities. *Food Chem.* **2008**, *107*, 19–25. [CrossRef]
5. Šamec, D.; Urlić, B.; Salopek-Sondi, B. Kale (*Brassica oleracea* var. acephala) as a superfood: Review of the scientific evidence behind the statement. *Crit. Rev. Food Sci. Nutr.* **2019**, *59*, 2411–2422. [CrossRef]
6. Lešić, R.; Borošić, J.; Buturac, I.; Herak Ćustić, M.; Poljak, M.; Romić, D. *Povrćarstvo (Vegetable Crops)*, 3rd ed.; Zrnski d.d.: Čakovec, Croatia, 2016.
7. Capriotti, A.L.; Cavaliere, C.; La Barbera, G.; Montone, C.M.; Piovesana, S.; Zenezini Chiozzi, R.; Laganà, A. Chromatographic column evaluation for the untargeted profiling of glucosinolates in cauliflower by means of ultra-high performance liquid chromatography coupled to high resolution mass spectrometry. *Talanta* **2018**, *179*, 792–802. [CrossRef]
8. Novío, S.; Cartea, M.E.; Soengas, P.; Freire-Garabal, M.; Núñez-Iglesias, M.J. Effects of Brassicaceae isothiocyanates on prostate cancer. *Molecules* **2016**, *21*, 626. [CrossRef]
9. Higdon, J.V.; Delage, B.; Williams, D.E.; Dashwood, R.H. Cruciferous vegetables and human cancer risk: Epidemiologic evidence and mechanistic basis. *Pharmacol. Res.* **2007**, *55*, 224–236. [CrossRef]
10. Herr, I.; Büchler, M.W. Dietary constituents of broccoli and other cruciferous vegetables: Implications for prevention and therapy of cancer. *Cancer Treat. Rev.* **2010**, *36*, 377–383. [CrossRef]

11. Mellon, F.A.; Bennett, R.N.; Holst, B.; Williamson, G. Intact glucosinolate analysis in plant extracts by programmed cone voltage electrospray LC/MS: Performance and comparison with LC/MS/MS methods. *Anal. Biochem.* **2002**, *306*, 83–91. [CrossRef]
12. Doheny-Adams, T.; Redeker, K.; Kittipol, V.; Bancroft, I.; Hartley, S.E. Development of an efficient glucosinolate extraction method. *Plant Methods* **2017**, *13*, 17. [CrossRef]
13. Ares, A.M.; Nozal, M.J.; Bernal, J.L.; Bernal, J. Effect of Temperature and Light Exposure on the Detection of Total Intact Glucosinolate Content by LC-ESI-MS in Broccoli Leaves. *Food Anal. Methods* **2014**, *7*, 1687–1692. [CrossRef]
14. Grosser, K.; van Dam, N.M. A straightforward method for glucosinolate extraction and analysis with high-pressure liquid chromatography (HPLC). *J. Vis. Exp.* **2017**, *2017*, 55425. [CrossRef]
15. Ratti, C. Hot air and freeze-drying of high-value foods: A review. *J. Food Eng.* **2001**, *49*, 311–319. [CrossRef]
16. Patel, S.M.; Doen, T.; Pikal, M.J. Determination of end point of primary drying in freeze-drying process control. *AAPS PharmSciTech* **2010**, *11*, 73–84. [CrossRef]
17. Song, L.; Thornalley, P.J. Effect of storage, processing and cooking on glucosinolate content of Brassica vegetables. *Food Chem. Toxicol.* **2007**, *45*, 216–224. [CrossRef]
18. Volden, J.; Borge, G.I.A.; Bengtsson, G.B.; Hansen, M.; Thygesen, I.E.; Wicklund, T. Effect of thermal treatment on glucosinolates and antioxidant-related parameters in red cabbage (*Brassica oleracea* L. ssp. capitata f. rubra). *Food Chem.* **2008**, *109*, 595–605. [CrossRef]
19. Gratacós-Cubarsí, M.; Ribas-Agustí, A.; García-Regueiro, J.A.; Castellari, M. Simultaneous evaluation of intact glucosinolates and phenolic compounds by UPLC-DAD-MS/MS in *Brassica oleracea* L. var. botrytis. *Food Chem.* **2010**, *121*, 257–263. [CrossRef]
20. Šamec, D.; Pavlović, I.; Radojčić Redovniković, I.; Salopek-Sondi, B. Comparative analysis of phytochemicals and activity of endogenous enzymes associated with their stability, bioavailability and food quality in five Brassicaceae sprouts. *Food Chem.* **2018**, *269*, 96–102. [CrossRef]
21. Bhandari, S.R.; Rhee, J.; Choi, C.S.; Jo, J.S.; Shin, Y.K.; Lee, J.G. Profiling of Individual Desulfo-Glucosinolate Content in Cabbage Head (*Brassica oleracea* var. capitata) Germplasm. *Molecules* **2020**, *25*, 1860. [CrossRef]
22. Tephly, T.R. The toxicity of methanol. *Life Sci.* **1991**, *48*, 1031–1041. [CrossRef]
23. Ashurst, J.V.; Nappe, T.M. *Methanol Toxicity*; StatPearls Publishing: Treasure Island, FL, USA, 2020.
24. Kavet, R.; Nauss, K.M. The toxicity of inhaled methanol vapors. *Crit. Rev. Toxicol.* **1990**, *21*, 21–50. [CrossRef]
25. ISO-ISO 9167:2019-Rapeseed and Rapeseed Meals—Determination of Glucosinolates Content—Method Using High-Performance Liquid Chromatography. Available online: https://www.iso.org/standard/72207.html (accessed on 17 November 2020).
26. Clarke, D.B. Glucosinolates, structures and analysis in food. *Anal. Methods* **2010**, *2*, 310–325. [CrossRef]
27. Que, F.; Mao, L.; Fang, X.; Wu, T. Comparison of hot air-drying and freeze-drying on the physicochemical properties and antioxidant activities of pumpkin (*Cucurbita moschata* Duch.) flours. *Int. J. Food Sci. Technol.* **2008**, *43*, 1195–1201. [CrossRef]
28. Papoutsis, K.; Pristijono, P.; Golding, J.B.; Stathopoulos, C.E.; Bowyer, M.C.; Scarlett, C.J.; Vuong, Q.V. Effect of vacuum-drying, hot air-drying and freeze-drying on polyphenols and antioxidant capacity of lemon (*Citrus limon*) pomace aqueous extracts. *Int. J. Food Sci. Technol.* **2017**, *52*, 880–887. [CrossRef]
29. Mao, L.C.; Pan, X.; Que, F.; Fang, X.H. Antioxidant properties of water and ethanol extracts from hot air-dried and freeze-dried daylily flowers. *Eur. Food Res. Technol.* **2006**, *222*, 236–241. [CrossRef]
30. Valadez-Carmona, L.; Plazola-Jacinto, C.P.; Hernández-Ortega, M.; Hernández-Navarro, M.D.; Villarreal, F.; Necoechea-Mondragón, H.; Ortiz-Moreno, A.; Ceballos-Reyes, G. Effects of microwaves, hot air and freeze-drying on the phenolic compounds, antioxidant capacity, enzyme activity and microstructure of cacao pod husks (*Theobroma cacao* L.). *Innov. Food Sci. Emerg. Technol.* **2017**, *41*, 378–386. [CrossRef]
31. Michalak, M.; Szwajgier, D.; Paduch, R.; Kukula-Koch, W.; Waśko, A.; Polak-Berecka, M. Fermented curly kale as a new source of gentisic and salicylic acids with antitumor potential. *J. Funct. Foods* **2020**, *67*, 103866. [CrossRef]
32. Olsen, H.; Grimmer, S.; Aaby, K.; Saha, S.; Borge, G.I.A. Antiproliferative effects of fresh and thermal processed green and red cultivars of curly kale (*Brassica oleracea* L. convar. acephala var. sabellica). *J. Agric. Food Chem.* **2012**, *60*, 7375–7383. [CrossRef]
33. Kim, S.Y.; Yoon, S.; Kwon, S.M.; Park, K.S.; Lee-Kim, Y.C. Kale Juice improves coronary artery disease risk factors in hypercholesterolemic men. *Biomed. Environ. Sci.* **2008**, *21*, 91–97. [CrossRef]

34. Sikora, E.; Cieślik, E.; Leszczyńska, T.; Filipiak-Florkiewicz, A.; Pisulewski, P.M. The antioxidant activity of selected cruciferous vegetables subjected to aquathermal processing. *Food Chem.* **2008**, *107*, 55–59. [CrossRef]
35. Sousa, C.; Taveira, M.; Valentão, P.; Fernandes, F.; Pereira, J.A.; Estevinho, L.; Bento, A.; Ferreres, F.; Seabra, R.M.; Andrade, P.B. Inflorescences of Brassicacea species as source of bioactive compounds: A comparative study. *Food Chem.* **2008**, *110*, 953–961. [CrossRef]
36. Silva-Espinoza, M.A.; Ayed, C.; Foster, T.; Camacho, M.d.M.; Martínez-Navarrete, N. The Impact of Freeze-Drying Conditions on the Physico-Chemical Properties and Bioactive Compounds of a Freeze-Dried Orange Puree. *Foods* **2019**, *9*, 32. [CrossRef]
37. Krokida, M.K.; Karathanos, V.T.; Maroulis, Z.B. Effect of freeze-drying conditions on shrinkage and porosity of dehydrated agricultural products. *J. Food Eng.* **1998**, *35*, 369–380. [CrossRef]
38. Rutnakornpituk, B.; Boonthip, C.; Sanguankul, W.; Sawangsup, P.; Rutnakornpituk, M. Study in Total Phenolic Contents, Antioxidant Activity and Analysis of Glucosinolate Compounds in Cruciferous Vegetables. *Naresuan Univ. J. Sci. Technol.* **2018**, *26*, 27–37.
39. Tetteh, O.N.A.; Ulrichs, C.; Huyskens-Keil, S.; Mewis, I.; Amaglo, N.K.; Oduro, I.N.; Adarkwah, C.; Obeng-Ofori, D.; Förster, N. Effects of harvest techniques and drying methods on the stability of glucosinolates in Moringa oleifera leaves during post-harvest. *Sci. Hortic.* **2019**, *246*, 998–1004. [CrossRef]
40. Managa, M.G.; Sultanbawa, Y.; Sivakumar, D. Effects of Different Drying Methods on Untargeted Phenolic Metabolites, and Antioxidant Activity in Chinese Cabbage (*Brassica rapa* L. subsp. chinensis) and Nightshade (*Solanum retroflexum* Dun.). *Molecules* **2020**, *25*, 1326. [CrossRef]
41. Korus, A. Effect of preliminary processing, method of drying and storage temperature on the level of antioxidants in kale (*Brassica oleracea* L. var. acephala) leaves. *LWT-Food Sci. Technol.* **2011**, *44*, 1711–1716. [CrossRef]
42. Das, A.; Raychaudhuri, U.; Chakraborty, R. Effect of freeze drying and oven drying on antioxidant properties of fresh wheatgrass. *Int. J. Food Sci. Nutr.* **2012**, *63*, 718–721. [CrossRef]
43. Ahmad-Qasem, M.H.; Barrajón-Catalán, E.; Micol, V.; Mulet, A.; García-Pérez, J.V. Influence of freezing and dehydration of olive leaves (var. Serrana) on extract composition and antioxidant potential. *Food Res. Int.* **2013**, *50*, 189–196. [CrossRef]
44. Clarke, G.; Ting, K.; Wiart, C.; Fry, J. High Correlation of 2,2-diphenyl-1-picrylhydrazyl (DPPH) Radical Scavenging, Ferric Reducing Activity Potential and Total Phenolics Content Indicates Redundancy in Use of All Three Assays to Screen for Antioxidant Activity of Extracts of Plants from the Malaysian Rainforest. *Antioxidants* **2013**, *2*, 1–10. [CrossRef]
45. Sulaiman, S.F.; Yusoff, N.A.M.; Eldeen, I.M.; Seow, E.M.; Sajak, A.A.B.; Supriatno; Ooi, K.L. Correlation between total phenolic and mineral contents with antioxidant activity of eight Malaysian bananas (*Musa* sp.). *J. Food Compos. Anal.* **2011**, *24*, 1–10. [CrossRef]
46. Prior, R.L.; Wu, X.; Schaich, K. Standardized methods for the determination of antioxidant capacity and phenolics in foods and dietary supplements. *J. Agric. Food Chem.* **2005**, *53*, 4290–4302. [CrossRef]
47. Huang, D.; Boxin, O.U.; Prior, R.L. The chemistry behind antioxidant capacity assays. *J. Agric. Food Chem.* **2005**, *53*, 1841–1856. [CrossRef]
48. Michalska, A.; Honke, J.; Łysiak, G.; Andlauer, W. Effect of drying parameters on the formation of early and intermediate stage products of the Maillard reaction in different plum (*Prunus domestica* L.) cultivars. *LWT-Food Sci. Technol.* **2016**, *65*, 932–938. [CrossRef]
49. Fante, L.; Noreña, C.P.Z. Quality of hot air dried and freeze-dried of garlic (*Allium sativum* L.). *J. Food Sci. Technol.* **2015**, *52*, 211–220. [CrossRef]
50. Iombor, T.T.; Olaitan, I.N.; Ede, R.A. Proximate composition, antinutrient content and functional properties of soursop flour as influenced by oven and freeze drying methods. *Curr. Res. Nutr. Food Sci.* **2014**, *2*, 106–110. [CrossRef]
51. Zhang, M.; Chen, H.; Zhang, Y. Physicochemical, thermal, and pasting properties of Chinese chestnut (*Castanea mollissima* Bl.) starches as affected by different drying methods. *Starch-Stärke* **2011**, *63*, 260–267. [CrossRef]
52. Karaman, S.; Toker, O.S.; Çam, M.; Hayta, M.; Doğan, M.; Kayacier, A. Bioactive and Physicochemical Properties of Persimmon as Affected by Drying Methods. *Dry. Technol.* **2014**, *32*, 258–267. [CrossRef]

53. Gao, Q.H.; Wu, C.S.; Wang, M.; Xu, B.N.; Du, L.J. Effect of drying of jujubes (*Ziziphus jujuba* Mill.) on the contents of sugars, organic acids, α-tocopherol, β-carotene, and phenolic compounds. *J. Agric. Food Chem.* **2012**, *60*, 9642–9648. [CrossRef]
54. Hu, Q.-g.; Zhang, M.; Mujumdar, A.S.; Du, W.-h.; Sun, J.-c. Effects of Different Drying Methods on the Quality Changes of Granular Edamame. *Dry. Technol.* **2006**, *24*, 1025–1032. [CrossRef]
55. Woo, K.S.; Kim, H.Y.; Hwang, I.G.; Lee, S.H.; Jeong, H.S. Characteristics of the thermal degradation of glucose and maltose solutions. *Prev. Nutr. Food Sci.* **2015**, *20*, 102–109. [CrossRef]
56. Li, Y.; Chen, J.; Lai, P.; Tang, B.; Wu, L. Influence of drying methods on the physicochemical properties and nutritional composition of instant *Tremella fuciformis*. *Food Sci. Technol.* **2020**, *40*, 741–748. [CrossRef]
57. Johansen, H.N.; Glitsø, V.; Bach Knudsen, K.E. Influence of extraction solvent and temperature on the quantitative determination of oligosaccharides from plant materials by high-performance liquid chromatography. *J. Agric. Food Chem.* **1996**, *44*, 1470–1474. [CrossRef]
58. Benzie, I.F.F.; Strain, J.J. The Ferric Reducing Ability of Plasma (FRAP) as a Measure of "Antioxidant Power": The FRAP Assay. *Anal. Biochem.* **1996**, *239*, 70–76. [CrossRef]
59. Brand-Williams, W.; Cuvelier, M.E.; Berset, C. Use of a free radical method to evaluate antioxidant activity. *LWT-Food Sci. Technol.* **1995**, *28*, 25–30. [CrossRef]
60. Ishida, M.; Nagata, M.; Ohara, T.; Kakizaki, T.; Hatakeyama, K.; Nishio, T. Small variation of glucosinolate composition in Japanese cultivars of radish (*Raphanus sativus* L.) requires simple quantitative analysis for breeding of glucosinolate component. *Breed. Sci.* **2012**, *62*, 63–70. [CrossRef]
61. Singleton, V.L.; Rossi, J.A. Colorimetry of Total Phenolics with Phosphomolybdic-Phosphotungstic Acid Reagents. *Am. J. Enol. Vitic.* **1965**, *16*, 144–158.

Publisher's Note: MDPI stays neutral with regard to jurisdictional claims in published maps and institutional affiliations.

Article

Optimization of a Green Ultrasound-Assisted Extraction of Different Polyphenols from *Pistacia lentiscus* L. Leaves Using a Response Surface Methodology

Cassandra Detti [1], Luana Beatriz dos Santos Nascimento [1,*], Cecilia Brunetti [1,2], Francesco Ferrini [1,2] and Antonella Gori [1,2,*]

1 Department of Agriculture, Food, Environment and Forestry (DAGRI), University of Florence, Sesto Fiorentino, 50019 Florence, Italy; cassandra.detti@unifi.it (C.D.); cecilia.brunetti@ipsp.cnr.it (C.B.); francesco.ferrini@unifi.it (F.F.)

2 National Research Council of Italy, Institute for Sustainable Plant Protection (IPSP), Sesto Fiorentino, 50019 Florence, Italy

* Correspondence: luanabeatriz.dossantosnascimento@unifi.it (L.B.d.S.N.); antonella.gori@unifi.it (A.G.)

Received: 28 September 2020; Accepted: 30 October 2020; Published: 3 November 2020

Abstract: *Pistacia lentiscus* leaves are used in several applications, thanks to their polyphenolic abundance. Thiswork aimed to characterize the polyphenols and to optimize the extraction conditions to shorten the time, decrease the consumption of solvent, and to maximize the yield of different classes of phenolics, which have diverse industrial applications. The variables were optimized by applying a Box–Behnken design. Galloyl and myricetin derivatives were the most abundant compounds, and two new tetragalloyl derivatives were identified by LC-MS/MS. According to the models, the maximum yields of polyphenols (51.3 ± 1.8 mg g^{-1} DW) and tannins (40.2 ± 1.4 mg g^{-1} DW) were obtained using 0.12 L g^{-1} of 40% ethanol at 50 °C. The highest content of flavonoids (10.2 ± 0.8 mg g^{-1} DW) was obtained using 0.13 L g^{-1} of 50% ethanol at 50 °C, while 0.1 L g^{-1} of 30% ethanol at 30 °C resulted in higher amounts of myricitrin (2.6 ± 0.19 mg g^{-1} DW). Our optimized extraction decreased the ethanolic fraction by 25% and halved the time compared to other methods. These conditions can be applied differently to obtain *P. lentiscus* extracts richer in tannins or flavonoids, which might be employed for various purposes.

Keywords: Anacardiaceae; design of experiments (DOEs); flavonoids; green extraction; HPLC-DAD; LC-MS/MS; tannins; ultrasound assisted-extraction (UAE)

---

1. Introduction

Pistacia lentiscus L. (Anacardiaceae), known as mastic orlentisk, is an evergreen shrub, widespread over many areas in the Mediterranean basin [1]. This species is largely distributed in dry ecosystems, characterized by nutrient and water scarcity due to the long periods of drought, high irradiation, and temperatures [2,3].

Several studies have demonstrated that *P. lentiscus* leaves are rich in polyphenolic compounds [4,5] including gallotannins and flavonoids (mainly quercetin and myricetin derivatives) [6–9]. These two main classes of compounds have different industrial and commercial applications. Flavonoids are intensively used in the food industry as preservatives and flavoring agents [10], the cosmetic industry as skin protectors [11], and in agricultureas an anti-infective agents [12]. Tannins, otherwise, are widely applied in the leather industry, as well as beverages additives, corrosion inhibitors of metals in shipbuilding, wood adhesives, and foams [13].

From a pharmacological point of view, both classes of phenolics have long been suggested to have high antioxidant capacities and other several biological activities [14]. *Pistacia lentiscus* leaves have traditionally been used in folk medicine for the treatment of various diseases such as hypertension, stomach aches, and kidney stones [15–18]. Moreover, anti-ulcer, anti-inflammatory, cytoprotective, acetylcholinesterase inhibition, and anticancer activities have been already described for its leaf extracts [15,19–21].

Therefore, leaves of *P. lentiscus* represent a reliable source of polyphenols to be exploited by several industries [22]. Thus, obtaining extracts enriched in different classes of these compounds is of high interest.

The quality and the content of polyphenols in plant leaf extracts depend on several factors such as the harvest moment and seasonality, the plant phenological stage, the leaf age, and the applied extraction process [6,23–25]. Well-established conventional extraction methodologies have been associated with significant economic and environmental impacts such as high solvent consumption and prolonged extraction times [23,26]. Nowadays, with the development of the concept of "green extraction", environmentally friendly techniques should be developed, avoiding hazardous reagents and optimizing extraction parameters such as time, temperature, and solvent type [27,28]. These green techniques, include ultrasound-assisted extraction (UAE), enabling the maximum yield of active compounds with low energy and less time consumed [29–31]. Ultrasound-assisted extraction represents one of the best and cheapest technologies with limited instrumental requirements [32,33], being efficiently used for extracting phenolic compounds from several plant materials [23,34,35].

The increasing interest inthe improvement of extraction processes from plants has triggered the application of mathematical models for the optimization of extraction conditions. In this sense, response surface methodology (RSM), widely applied for industrial purposes, has become the most preferable approach for optimizing extraction procedures that apply multiple variables at the same time [36,37]. In this sense, the Box–Behnken design (BBD) is one of the most used RSMs. This design requires a small number of runs and, therefore, avoids time-consuming experiments and has been largely applied for optimizing extractions of single or classes of molecules from different plant materials [38–40].

Response surface methodologieshave already been applied for optimizing the ultrasound-assisted extraction of polyphenols from *P. lentiscus* leaves [41]. In this study, the authors tested the UAE by using a central composite design with high solvent volume and leaf material [41]. In addition, the authors quantified the total phenolic content using the Folin–Ciocalteu reagent, not considered specific for phenolics, since this reagent can be reduced by other compounds that might cause interferences in the results [42]. As such, a detailed optimization of different classes of polyphenols in *P. lentiscus* leaf extracts has not yet been conducted.

In this context, this study aimed to:

- Evaluate the effect of different variables (solvent ratio, temperature, extraction time, and ethanol volume) on the UAE of *P. lentiscus* leaves using a first-step screening design;
- Optimize the extraction process, using a Box–Behnken design, in order to obtain extracts with higher amounts of different classes of polyphenols (quantified by high performance liquid chromatography coupled to diode array detection, HPLC-DAD) and applying a greener method than those conventionally used for the extraction of leaves of the species;
- Characterize the major compounds present in the extract with the highest content in polyphenols using liquid chromatography-mass spectrometry (LC-MS/MS).

2. Results and Discussions

2.1. Screening Design and Determination of the Important Factors

Several factors can influence the efficiency of an extraction such as solvent type, time, particle size, and temperature [43]. Consequently, it is important to verify how different variables affect the extraction of target compounds [23,44]. The UAE is one of the most appropriate extraction processes due to the fact of its efficacy, cleanliness, facility of use, and speed [45,46]. Among the different factors that should be considered in ultrasound-assisted extraction, the polarity of the solvent and the solvent ratio is very important [45,47]. Moreover, time and temperature also affect the yield of the compounds and the costs of the whole process. Indeed, it is desirable to develop methods that lead to a higher extraction of target compounds using lower temperatures, shorter time, and lower concentration of organic solvent than possible [43,47].

The calculated coefficients for the different answers (i.e., total tannins content (TTC), total flavonoids content (TFC), and total polyphenolic content (TPC)) are shown in Table 1. The solvent ratio (x_3) showed to be the most important factor affecting positively all the responses as can be inferred by the positive and significant value of the b_3 coefficient. As such, an increase in the solvent ratio from 0.06 to 0.1 L g^{-1} led to higher amounts of polyphenols. Therefore, even higher solvent ratios (solvent volumes) were chosen for the further optimization steps.

Table 1. Screening fractional factorial design matrix (FFD-2^{4-1}) of trials conducted (trials 1 to 9, this last the central point), with the independent variables (x_1 to x_4) and answers (y).

Experimental Trials	Independent Variables (x)							
	x_1 (Temperature, in °C)		x_2 (Time, in min)		x_3 (Solvent Ratio, in L g^{-1})		x_4 (Ethanol Fraction, in % *v*/*v*)	
1	5	(−)	15	(−)	0.06	(−)	50	(−)
2	25	(+)	15	(−)	0.06	(−)	75	(+)
3	5	(−)	30	(+)	0.06	(−)	75	(+)
4	25	(+)	30	(+)	0.06	(−)	50	(−)
5	5	(−)	15	(−)	0.1	(+)	75	(+)
6	25	(+)	15	(−)	0.1	(+)	50	(−)
7	5	(−)	30	(+)	0.1	(+)	50	(−)
8	25	(+)	30	(+)	0.1	(+)	75	(+)
9	15	(0)	22.5	(0)	0.08	(0)	62.5	(0)
Answers (y)	**Calculated coefficients**							
TTC	b_1	0.338	b_2	−0.104	b_3	0.571 *	b_4	−0.604 *
TFC	b_1	−0.092	b_2	−0.058	b_3	0.258 *	b_4	0.008
TPC	b_1	0.375	b_2	−0.108	b_3	1.267 *	b_4	−0.700 *

The trials were conducted in triplicate. The coefficients (b_1 to b_4), correspondent to each variable (x_1 to x_4), were calculated by regression. The asterisks (*) indicate significant coefficients ($p \leq 0.05$).Total tannins content (TTC); total flavonoids content (TFC); total polyphenolic content (TPC).

The fraction of ethanol (%) (x_4) showed to be a determinant for the extraction of tannins (TTC) and total polyphenols (TPC), negatively affecting both. The b_4 coefficient was significant and negative, indicating that the extractions conducted using a smallerpercentage of ethanol resulted in higher yields of these compounds. This phenomenon could be explained by the fact that with an increase in ethanol concentration, the solvent polarity may decrease as well as the molecular movements, reducing the solubility of the polar compounds [48]. In addition, by raising the surface tension of the solvent, an increase in the molecular interactions is induced, consequently raising the extraction [32], while the addition of water to the organic solvent may help break the hydrogen bonding and facilitate the extraction of polyphenols [49]. Based on these results, lower percentages of ethanol were studied in the optimization step in order to conduct a greener extraction [32,50]. In fact, ethanol and water are solvents widely used by food and pharmaceutical industries due to the fact of their safer handling [51].

For the temperature (x_1), the calculated coefficients (b_1) were also high for TTC and TPC. For both, this factor showed a positive effect. According to this, higher temperatures (30, 40, and 50 °C) were evaluated during the optimization, which should also be considered when higher amounts of water are applied. The use of higher temperatures during the UAE can increase the efficiency of the extraction process due to the increase in the number of cavitation bubbles formed [45,52]. Moreover, temperature influences the mass transfer process by improving the solvent penetration in plant cells due to the reduction in its viscosity. In addition, higher temperatures increase the degradation of the plant matrix, and weaken the interactions of the polyphenols with other cell constituents, making their extraction easier [49,53].

The coefficient b_2 showed to be low for all the answers, indicating that the variable x_2 (time) has no effect on the answers. Therefore, this factor was kept constant in the optimization step, and the shorter extraction time tested (15 min) was chosen. For further industrial purposes, this is desirable, since less time can reduce the energy consumed [54].

2.2. Optimization Design: Models and Response Surfaces Analysis

For an efficient extraction process, not only the method used (e.g., UAE, microwave assisted or conventional extractions), but also the variables applied are of a great importance as well as their linear, quadratic, and interactive effects. The multi-factorial study of them, such as applying experimental designs and RSM, allows the maximization of responses with minimal energy loss and solvent consumption [47].

The results of tannins (TTC), flavonoids (TFC), myricitrin (MYC), and total polyphenols (TPC) (calculated as the sum of individual phenolics, Supplementary Materials Tables S1 and S2) obtained for the experimental trials conducted are presented in Table 2 and were used to obtain regression equations (models, Supplementary Materials Table S3). Trials 9 to 11 were shownto have the best conditions for achieving higher amounts of TPC and TTC (Table 2), all of them, interestingly using 40% ethanol as solvent. For the flavonoids content (TFC and MYC) instead, the best trials were from trials 5 to 8, all using 0.15 L g^{-1} of solvent ratio (Table 2). It is interesting to note that for all the responses, an extraction conducted using a solvent ratio of 0.2 L g^{-1} at 40 °C resulted in the lowest amounts of polyphenols (Table 2).

Table 2. Box–Behnken design (BBD) matrix with natural and coded values for the independent variables and responses. Fifteen experimental trials were conducted with triplicates of the central point (trial 13).

Trials	Independent Variables (Factors)						Dependent Variables (Responses)			
	x_1 (Ethanol Fraction, in % v/v)		x_2 (Solvent Ratio, in L g^{-1})		x_3 (Temperature, in °C)		TTC	TFC	MYC	TPC
1	30	(−)	0.1	(−)	40	(0)	30.5	8.6	2.0	42.6
2	50	(+)	0.1	(−)	40	(0)	30.1	8.9	1.4	39.9
3	30	(−)	0.2	(+)	40	(0)	24.3	6.5	1.6	33.2
4	50	(+)	0.2	(+)	40	(0)	26.5	7.1	1.6	35.6
5	30	(−)	0.15	(0)	30	(−)	31.9	9.5	2.2	44.0
6	50	(+)	0.15	(0)	30	(−)	33.6	9.6	2.1	46.0
7	30	(−)	0.15	(0)	50	(+)	35.7	8.5	1.9	45.9
8	50	(+)	0.15	(0)	50	(+)	36.6	10.1	2.2	49.4
9	40	(0)	0.1	(−)	30	(−)	37.6	8.0	2.0	49.0
10	40	(0)	0.2	(+)	30	(-)	37.8	7.5	1.6	48.6
11	40	(0)	0.1	(−)	50	(+)	37.7	6.9	1.6	48.6
12	40	(0)	0.2	(+)	50	(+)	34.5	7.1	1.9	44.6
13	40	(0)	0.15	(0)	40	(0)	34.9 ± 1.3	5.8 ± 0.60	1.4 ± 0.15	44.3 ± 1.7

For the central point (trial 13), values of the mean ± SD are presented. All the values of the responses are expressed in mg g^{-1} DW. Total tannins content (TTC); total flavonoids content (TFC); myricitrin content (MYC), total polyphenolic content (TPC).

According to the results, second-order polynomial regression models based on the coded coefficient values were obtained for each response (Supplementary Materials Table S3). To verify the fitting of the mathematical models, the data were statistically analyzed. The quadratic model applied is usually assumed to fit the data sufficiently well to indicate the more suitable and the better regions of work. Statistically, the quality of a model is evaluated by the significance of the regression according to the ANOVA test; the lack of fit (LOF), used to measure the adequacy of these models [55]; the multiple determination coefficient (R^2), which represents the variation of the response explained by the model; and the adjusted multiple determination coefficient (R^2_{adj}), which indicates the capacity of the model to be predictive [55–57]. The analysis of the models obtained for the different responses (TTC, TFC, MYC, and TPC) are presented in Table 3.

Table 3. Statistical parameters after data analysis and fit of the models obtained for the different responses.

Responses	Analysis of the Model				Lack of Fit (LOF)	
	R^2	R^2_{adj}	F-Value of Model	*p*-Value of Model	F-Value of Lack of Fit	*p*-Value of Lack of Fit
TTC	0.95	0.74	5.32	0.040 *	3.56	0.23
TFC	0.90	0.73	5.13	0.043 *	1.52	0.42
MYC	0.89	0.71	4.79	0.049 *	1.25	0.47
TPC	0.90	0.71	4.80	0.049 *	3.07	0.26

* Significant values ($p \leq 0.05$).

For a good fit, R^2 should be at least 80% [58]. In all our analysis, the R^2 values were higher than 0.89 (89%, Table 3), suggesting that the models described well the behavior of these responses. Moreover, for all of them, the R^2_{adj} were higher than 0.71, indicating a good predictive power, since in a good statistical model R^2_{adj} should be comparable and similar to R^2, with differences less than 0.2–0.3. Furthermore, the values for the LOF were not significant to an extent with the pure error (p>0.05, for all). A model will fit the experimental data when a significant regression and a non-significant LOF are found [59]. Therefore, considering these results, as well as the *p*-value (all p≤0.05) (Table 3), the models showed to be suitable and appropriate to well describe the relationship betweenthe responses (TTC, TFC, MYC, and TPC) and the independent variables (x_1 to x_3).

The significance of the coefficients was also determined (Supplementary Materials Table S3). For the response of total polyphenols (TPC) and total tannin (TTC) contents, the coefficient b_2 (x_2–solvent ratio) showed to be the highest, compared with the coefficients b_1 (x_1–ethanol fraction, %) and b_3 (x_3–temperature). Besides, b_2 was negative, suggesting that the use of less solvent is better for the extraction of TPC and TTC. Similarly, for the total content of flavonoids (TFC), the coefficient b_2 showed higher values compared to b_1 and b_3, also negatively affecting the response. It means that less solvent should be better for the extraction of flavonoids (Supplementary Materials Table S3). The analysis of variance (ANOVA) for the polynomial models indicated that the quadratic terms of the variable x_1 (ethanol%, b_{11}) and x_3 (temperature, b_{33}) were the most important, significantly influencing the responses ($p < 0.05$) (Supplementary Materials Table S3). In fact, temperature and type of solvent are important factors to be considered in UAE [32].

To provide a better visualization of the effects of the factors in the responses, contour response surface plots were generated from the models (Figure 1—TTC and TPC and Figure 2—MYC and TFC), by plotting the responses with regard to ethanol concentration (x_1) and solvent ratio (x_2) at each temperature 30, 40, and 50 °C (x_3). These response surfaces can be used for the prediction of the responses (polyphenols contents) in the investigated experimental domain.

For the total polyphenolic content (TPC, Figure 1a–c), 30 °C and 50 °C predicted maximum amounts (>50.0 mg g^{-1} DW, Figure 1a,c), with 50 °C being better, since a more extended and stable optimal region of extraction wasobtained (Figure 1c). At this temperature, 35% to 45% of ethanol in a solvent ratio of 0.1 to 0.15 L g^{-1} should be used (Figure 1c). At 40 °C, a good region was also found, however, resulting in lower amounts (~44.0 mg g^{-1} DW). In fact, the optimal conditions proposed by the model were 40% ethanol in a ratio of 0.12 L g^{-1} at 50 °C, resulting in 51.3 ± 1.8 mg g^{-1} DW of TPC.

A similar percentage of ethanol was also proposed by a previous study focused on the optimization of the phenolic extraction of *P. lentiscus* leaves using a microwave-assisted method [22]. The authors showed that percentages of ethanol around 30% to 40% significantly raised the total phenolic content, spectrophotometrically quantified by the Folin–Ciocalteu reagent [22]. In addition, a similar effect of ethanol percentage was also reported for the extraction of phenolic compounds from other plant sources such as green tea [60]. Considering temperatures analogous to our findings, 45–50 °C was shown to maximize the extraction of polyphenols in *Pistacia atlantica* leaves [36].

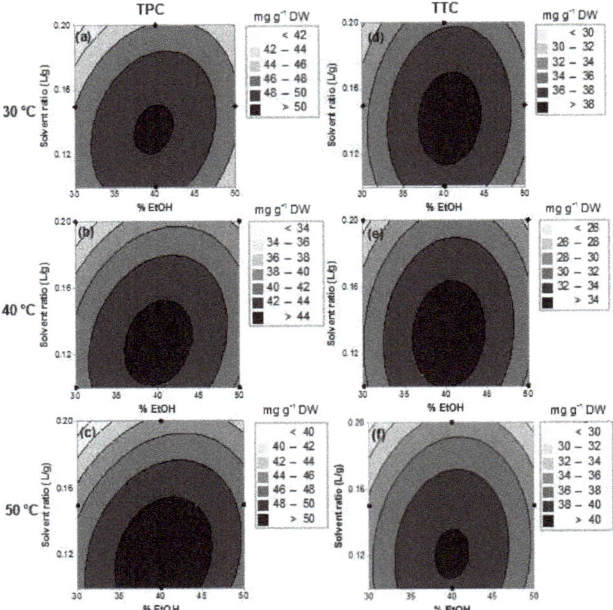

Figure 1. Response surface (contour plots) for predicting the TPC (**a–c**) and TTC (**d–f**) in *Pistacia lentiscus* leaf extracts with regard to the ethanol fraction (%, x_1) and solvent ratio (L g^{-1}, x_2), at each temperature (x_3, 30, 40, and 50 °C). The regions with the darkest-gray color represent the domains of working conditions assuring the maximum values for the evaluated compounds (total polyphenols and tannins).

The total tannin content (TTC, Figure 1d–f) showed very similar response surfaces and optimal conditions to TPC. Indeed, *P. lentiscus* leaves are rich in tannins, which represent around 70% of the TPC [6]. Therefore, a similar behavior should be expected. For TTC, temperatures of 30 °C and 50 °C should be also used to reach higher amounts of tannins (~40.0 mg g^{-1} DW). However, different from total polyphenols, a narrow optimal region was observed at 50 °C (Figure 1f). At any temperature, an extraction using around 0.13 L g^{-1} of 40% ethanol provided better results (Figure 1d–f). Indeed, the optimal conditions for maximization of the content of tannins were the same for TPC (0.12 L g^{-1}, 40% ethanol, 50 °C), yielding 40.2 ± 1.4 mg g^{-1} DW.

For TFC and MYC (Figure 2), at any temperature, a decrease in the content of these compounds was observed around the medium percentages of ethanol with the optimal regions being obtained when extreme values of ethanol fractions (30% or 50%) and temperatures (30 or 50 °C) are chosen (Figure 2). Therefore, higher contents of flavonoids (~10 mg g^{-1} DW, Figure 2a,c) and myricitrin (>2.5 mg g^{-1} DW, Figure 2d,f) are predicted under these conditions. As such, extractions conducted at 50 °C, using 50% ethanol in 0.1 to 0.17 L g^{-1} result in greater amounts of flavonoids (Figure 2c). Decreasing the temperature to 40 °C, lesser amounts of flavonoids are obtained (~8.5 mg g^{-1} DW) (Figure 2b). The increase in the temperature can cause higher solubility and diffusion coefficients of polyphenols, such as flavonoids, which result in a higher extraction rate [61]. The optimal conditions predicted by the model for maximization of the flavonoid content in *P. lentiscus* leaf extracts are 50% ethanol in 0.13 L g^{-1} at 50 °C, resulting in 10.2 ± 0.8 mg g^{-1} DW.

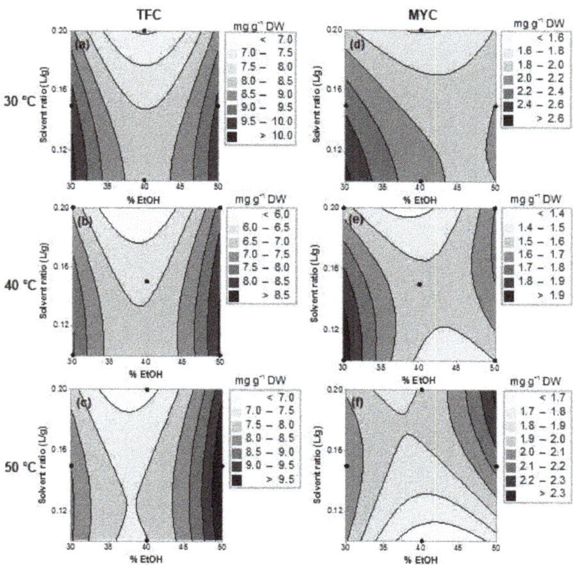

Figure 2. Response surface (contour plots) for predicting the TFC (**a–c**) and MYC (**d–f**) in *P. lentiscus* leaf extracts with regard to the ethanol fraction (%, x_1) and solvent ratio (x_2) used at each temperature (x_3, 30, 40, and 50 °C). The regions with the darkest-gray or black color represent the domains of working conditions assuring the maximum values for the evaluated compounds (total flavonoids and myricitrin).

Myricitrin content (Figure 2d–f) showed similar response surfaces to the TFC (Figure 2a–c). The extraction conducted at 30 °C with 30% ethanol in 0.1 L g^{-1} should result in maximal contents of this flavonoid (2.6 ± 0.19 mg g^{-1} DW). However, at 50 °C very similar amounts were also obtained, but 50% ethanol in slightly higher volumes should be used (Figure 2f). As can be noticed, the TFC and MYC have similar behaviors. This could be explained because myricitrin is the most abundant flavonoid detected in *P. lentiscus* leaves (Figure 3), also justifying our choice in maximizing its content. This compound has also been described as a major compound in lentisk leaf extracts in previous studies [6,7,62].

To validate the adequacy of the mathematical models, verification experiments were carried out in triplicate under the optimal conditions. Mean values of 39.8 ± 4.1 mg g^{-1} DW for TTC, 50.9 ± 4.9 mg g^{-1} DW for TPC, and 9.9 ± 1.4 mg g^{-1} DW for TFC were obtained from the real experiments and demonstrated the validation of the models for these three responses (p_{TTC} = 0.88; p_{TPC} = 0.90; p_{TFC} = 0.76).

We observed that among the variables tested, the same solvent ratio (~0.13 L g^{-1}) and temperature (50 °C) should be used during the UAE process to obtain the maximal yields of tannins and flavonoids. However, the ethanol percentage showed to differ between both classes of compounds. While for tannins (TTC) 40% ethanol should be used, 50% is preferable for the extraction of flavonoids (TFC). This difference can be explained by the distinct solubility of these compounds [47]. The polarity of the ethanol–water mixture decreases with the addition of ethanol, stimulating the extraction of less polar compounds from plant cells. Flavonols, such as myricetin derivatives, show higher solubility with increasing concentration of alcohol, consequently reaching greater extraction yields when less polar solvents are used [23]. Tannins with low molecular weight (galloyl derivatives) occurring in *P. lentiscus* leaf extracts (Figure 3, Table 4) are more polar than the flavonoids detected. Therefore, it is reasonable that the extraction of these types of tannins (and consequently the overall polyphenolic content) is stimulated by the utilization of more polar solvents (i.e., 40% ethanol). Indeed, higher concentrations of ethanol and methanol are more beneficial for the extraction of flavonoids than for tannins, which generally need higher amounts of water [47]. In agreement with our results, Barbouchi et al. [63] obtained higher phenolic contents in *P. lentiscus* leaf extracts when more polar extraction solvents were used. In addition, ethanol was considered the most suitable solvent for the recovery of flavonoids from this species [2].

The extraction conditions optimized here are suitable for experimental and further industrial applications, since they apply a green solvent (ethanol:water) in low quantity (~0.13 L g^{-1}) for a short time (15 min), using moderate temperatures (50 °C). Considering that more toxic solvents, such as methanol, chloroform, and ethyl acetate, are intensively used to extract *P. lentiscus* leaves [19,64–66] and that the typical extraction methods apply higher percentages of ethanol for longer times [2,6,7], our optimization led to a greener extraction procedure if compared to conventional extraction methods, by decreasing the ethanolic fraction by at least 25% and halved the time used.

A green extraction is defined as a procedure able to reduce energy consumption and use of organic solvents, saving the quality of the process [29]. In particular, three major points should be considered: the improvement and the optimization of the existing methods; the use of simple equipment; and the innovation in the use of alternative solvents [67]. In this sense, the optimization of a standard extraction procedure, especially employing the minimum amount of organic solvent, could be considered green, even if moderate temperatures are applied.

2.3. Polyphenolic Composition of the Richest P. lentiscus Extract

Figure 3 shows the polyphenolic profile of *P. lentiscus* leaf extract obtained using the conditions of trial 9 (BBD, Table 2), corresponding to the extract with the highest content of total polyphenols (TPC). The LC-MS/MS analysis was performed to provide a more comprehensive characterization of the polyphenols present in the leaves of the species as well as to confirm previous characterizations reported in the literature [7,9,62,64].

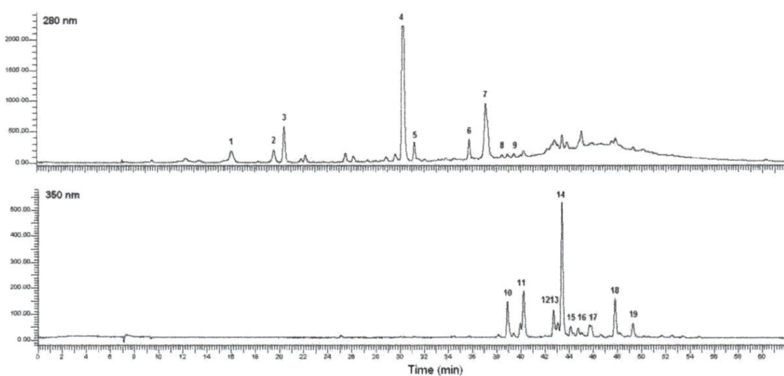

Figure 3. Chromatograms of *P. lentiscus* leaf extracts obtained using the extraction conditions of trial 9 (BBD, Table 2) acquired at 280 nm (above) and 350 nm (below).

The UV-Vis and MS/MS spectra allowed us to identify 19 compounds (Table 4), classified into three main classes: gallic acid derivatives (peaks 1 and 2), gallotannins (peaks 3–9), and flavonoids (peaks 10–19). For flavonoids, three peaks were identified as myricetin derivatives (peaks 10, 11, 14), six as quercetin derivatives (12, 13, 15–18), and one as a kaempferol derivative (19).

Among the four main peaks detected, three of them (peaks 3, 4, and 7) showed the fragmentation correspondent to mono-, di-, and trigalloylquinic acids, respectively (Table 4). These metabolites have already been described in the literature using different kinds of detectors such as triple quadrupole (QQQ) [7,9,64] and quadrupole time-of-flight (Q-TOF) mass spectrometers [62]. Indeed, according to these previous reports, the monogalloylquinic acid (peak 3) was defined by the fragments *m/z* 343 $[M - H]^-$ and 191; this lastresulted from the loss of the galloyl moiety [M152-H]. In addition, the digalloylquinic acid and its isomer (peaks 4 and 5) were characterized by the fragments *m/z* 495 $[M - H]^-$, 343, 191, and 169, that are consistent with the successive loss of two galloyl units, and correspondent to the gallic acid itself (*m/z* 169). Finally, the trigalloylquinic acid (peak 7) and its isomer (peak 6) showed the fragments m/z 647 $[M - H]^-$, 495, 343, 191, and 169, consistent with a trigalloyl substitution. Two minor peaks with the UV spectra and the mass fragmentation typical of quinic acid derivatives were also detected, with precursor ions of *m/z* 799 and fragments 495, 343, 191, and 169, correspondent to four consecutive losses of galloyl moieties (Supplementary Materials Figure S1). As such, these peaks (peaks 8 and 9) were tentatively identified as tetragalloylquinic acid (and its isomer), here, firstly reported in *P. lentiscus* leaf extracts.

Table 4. LC–DAD-MS/MS characterization of the main polyphenols present in extracts of *P. lentiscus* leaves. Compounds numbers correspond to those indicated in Figure 3 (sh, shoulder).

Peak	t_R, (min)	λ_{max}, (nm)	Collision Energy, (V)	$[M-H]^-$,(m/z)	MS^2, (m/z)	Peak Assignment
1	16.23	234,270	10	331	169,151,125	Monogalloyl glucose
2	19.71	234,272	10	169	125	Gallic acid
3	20.53	236,272	15	343	191	Monogalloyl quinic acid
4	30.24	236,276	15	495	343,191,169	Digalloyl quinic acid (isomer 1)
5	31.22	236,276	15	495	343,191,169	Digalloyl quinic acid (isomer 2)
6	35.72	256,356	20	647	495,343,191,169	Trigalloyl quinic acid (isomer 1)
7	37.09	256,356	20	647	343,191,169	Trigalloyl quinic acid (isomer 2)
8	38.47	265,355	20	799	495,343,191,169	Tetragalloyl quinic acid (isomer 1)
9	38.59	265,355	20	799	495,191,169	Tetragalloyl quinic acid (isomer 2)
10	38.83	264,314,346 sh	10	479	317,316	Myricetin-3-*O*-galactoside
11	40.15	268,314,348sh	15	625	479,316,317	Myricetin-3-*O*-rutinoside
12	42.62	256,350	10	493	301	Quercetin derivative
13	42.94	256,350	10	463	381,300,301	Quercetin-*O*-hexoside 1
14	43.29	260,358,346 sh	10	463	316,271,179	Myricitrin (Myricetin-3-*O*-rhamnoside)
15	44.23	270,350,300 sh	10	463	381,300,301	Quercetin-*O*-hexoside 2
16	44.97	256,350	15	585	525,301,179	Quercetin-*O*-galloyl-pentoside
17	45.57	256,350,300 sh	10	433	300,301	Quercetin-3-*O*-arabinoside
18	47.69	266,350,300 sh	15	447	300,301	Quercitrin (Quercetin-3-*O*-rhamnoside)
19	49.19	265,348	15	447	415,365,285	Kaempferol-*O*-hexoside

Among the gallic acid derivatives (peaks 1 and 2), the peak 1 was assigned as monogalloyl glucose (glucogallin), based on the literature [7] and according to its ion fragments *m/z* 331 $[M - H]^-$, 169 (resulted from the loss of the glucose, *m/z* 162) and 125 (derived from the decarboxylation of galloyl). The peak 2, instead, was identified as gallic acid, based on its characteristic mass spectra, with the precursor ion at *m/z* 169 $[M - H]^-$ and the fragment *m/z* 125 (decarboxylation of the galloyl). The identification of this compound was also confirmed based on the comparison with the specific external standard.

Flavonoids (from peak 12 to 18) were identified based on the mass fragment of their corresponding aglycon units, namely, myricetin (*m/z* 317), quercetin (*m/z* 301), and kaempferol (*m/z* 285). This was confirmed by the injection of the external standards myricitrin, rutin, and kampferol-3-*O*-rutinoside. The sugar moieties were characterized based on the neutral losses of 132 (presence of pentosides: xylose or arabinose), 162 (hexosides: galactose or glucose), and 146 (deoxyhexoside: rhamnose). Thus, in agreement with the fragmentation patterns described in the literature [62,64], the following flavonoids were tentatively identified as: myricetin-3-*O*-galactoside (peak 10), myricetin-3-*O*-rutinoside (peak 11), quercetin-*O*-hexosides 1 and 2 (peaks 13 and 15), quercetin-*O*-galloyl-pentoside (peak 16), quercetin-3-*O*-arabinoside (peak 17), quercetin-3-*O*-rhaminoside (peak 18), and kaempferol-*O*-hexoside (peak 19). The identification of the major flavonoidic peak (14), myricetin-3-*O*-rhamnoside (myricitrin), was obtained by comparison with the specific analytical standard. The remaining peak (12) was tentatively identified as a quercetin derivative based on the UV–Vis spectra, in the absence of conclusive mass-spectrometric data and reference in the literature.

High contents of gallotannins (galloylquinic acid derivatives) and myricetin derivatives were previously described in *P. letiscus* leaves [6,7,9,17,68]. These compounds represent approximately 90% of the polyphenolic composition of the leaf extracts [6] and are possibly the main responsible for their biological properties such as: anti-inflammatory, cytoprotective, hepatoprotective, enzymatic-inhibitory, antitumor, and anti-diabetes [7,62,69–71]. It is noteworthy that the two molecules here tentatively identified as tetragalloylquinic acid derivatives have not been described yet in lentisk leaves. These compounds have shown to possess high activity against bronchial hyperreactivity and allergic reactions [72].

All these reports show the importance of developing new methodologies in order to increase the content of active compounds in *P. lentiscus* extracts. This could lead to a wider application of the extracts as nutraceuticals, medicines, or as sources of substances for different commercial and industrial applications.

In fact, the compounds detected here showed important biological activities. In particular, galloyl derivatives of quinic acid have been shown to have effective inhibition of $Fe_2{}^+$-induced lipid peroxidation in cells [73], anti-HIV, anti-allergic [74], and high antioxidant activities [68]. This class of molecules is among the most pharmacologically active natural products detected in several plant species [75]. In addition, gallotannins are applied as wood adhesives in the leather manufacturing as well as in the construction sector [13].

Flavonoids, especially with quercetin and myricetin skeleton, are considered powerful antioxidants distributed in several plant species with proven anti-inflammatory and anti-cancer actions [76–79]. Besides their application as medicinal compounds, they are specially utilized in cosmetic and nutraceutical products [11].

Moreover, the most abundant flavonoid detected in all extracts, the myricetin-3-*O*-rhamnoside, showed a noticeable lipid peroxidation inhibitionin in vitro tests, with very low IC_{50} (inhibitory concentration at 50%) [80], being even more effective as antioxidant than vitamin C [81]. In addition, this molecule has demonstrated positive effects against the oxidative stress induced by hyperglycemia in C2C12 cells [82] and has shown significant inhibition in peroxynitrite-mediated DNA damage [83].

3. Materials and Methods

3.1. Plant Material

Fully expanded leaves from branches at the top of the canopy were randomly collected from adult plants of *P. lentiscus* growing in the coastal dunes of Southern Tuscany, Italy (42°46′ N, 10°53′ E). Harvesting was conducted in July 2019, around midday in order to ensure the high polyphenolic composition of the leaves [6].

After collection, the leaves were cleaned (to remove damaged parts, dust, and other contaminants from the natural habitat), immediately frozen in liquid nitrogen, freeze-dried, ground into a fine powder, and kept at −80 °C until the moment of extraction.

3.2. Ultrasound-Assisted Extraction (UAE) Procedure

Freeze-dried ground leaves (0.15 gweighted on a digital analytical balance Precisa® 125A) were extracted using ethanol in different percentages and volumes according to the design matrixes (Tables 1 and 2). The UAE was conducted in an ultrasonic bath (BioClass® CP104) using a constant frequency of 39 kHz and an input power of 100 W. The different temperatures and times, according to each trial (Tables 1 and 2), were monitored with a thermometer (Weber® 6750, Springfield, Illinois, USA) and a timer (Fisher Scientific®, Los Angeles, CA, USA). After the extraction, the samples were centrifuged (5 min, 9000 rpm, 5 °C-ALC® 4239R, Milan, Italy) and the supernatants were partitioned with 3 × 5 mL *n*-hexane, in order to remove lipophilic compounds that could interfere with the analysis. The hydroethanolic phase was reduced to dryness using a rotavapor (BUCHI® P12, Cornaredo, Italy; coupled to a vacuum controller V-855), and the residue was resuspended with 1.0 mL of MeOH:

Milli-Q_{H2O} solution (1:1 *v/v*, pH 2.5 adjusted with HCOOH). These samples were used to conduct the HPLC-DAD analysis for the quantification of the different classes of polyphenols to construct the model. In addition, the extract with the highest polyphenolic content was chosen for the LC-MS/MS analysis in order to furnish a detailed characterization of its chemical composition.

3.3. HPLC-DAD Quantification and LC-MS/MS Characterization of the Extracts

High performance liquid chromatography coupled to diode array detection (HPLC-DAD) was used for quantification of the different polyphenolic classes of the extractsobtained at the different conditions tested.

The samples (5 µL) were injected into a Perkin® Elmer Flexar liquid chromatography equipped with a quaternary 200Q/410 pump and an LC 200 diode array detector (DAD) (all from Perkin Elmer®, Branford, Connecticut, USA). The stationary phase consisted in a Zorbax® C-18 column (250 mm × 4.6 mm, 5 µm particle size) and the eluents were (A) acidified water (0.1% HCOOH) and (B) acetonitrile (0.1% HCOOH).The following gradient was applied: 1 min (3% B), 1–55 min (3–40% B), 55–60 min (40% B), 60–61 min (3% B), with 62 min of total analysis time, in a flow rate of 0.6 mL min^{-1}. Ten minutes of conditioning step were used to return to the initial conditions of the method.

The identification of the polyphenols with HPLC-DAD was carried out based on the retention time, UV-Vis spectral characteristics, comparison with those of the authentic standards acquired at 280 and 350 nm, as well as on the subsequent LC-MS/MS analysis. Quantifications were made by HPLC-DAD. The standards (gallic acid, myricitrin, rutin, and kaempferol-3-*O*-rutinoside from Sigma–Aldrich®–Merck®KGaA, Darmstadt, Germany) were used to obtain five-point calibration curves. If a commercial standard was not available, quantification was performed using the calibration curve of standards from the same phenolic group. The linearity of these calibration curves was determined by the coefficient of determination (R^2), being higher than 0.999 for all the three standards. The limit of detection (LOD) and quantification (LOQ), both expressed as µg/mL, were calculated using signal-to-noise ratio of 3 and 10, respectively [84]. The following limits of detection and quantification were found for the standards: $LOD_{gallic\ acid} = 0.3$ and $LOQ_{gallic\ acid} = 0.85$; $LOD_{rutin} = 0.28$ and $LOQ_{rutin} = 0.6$; $LOD_{myricitrin} = 0.12$ and $LOQ_{myricitrin} = 0.38$; $LOD_{kaempferol\text{-}3\text{-}O\text{-}rutinoside} = 0.21$ and $LOQ_{kaempferol\text{-}3\text{-}O\text{-}rutinoside} = 0.49$).

All the extracts were analyzed in triplicate. The quantitative results of the polyphenols (reported as mg per g of dry weight, DW) were expressed as: myricitrin (the most abundant flavonoid detected in the *P. lentiscus* leaf extracts), total tannin, total flavonoid, and total polyphenols contents, represented as the sum of individual tannins (TTC), flavonoids (TFC) and polyphenols (TPC) detected by HPLC-DAD analysis in each extract (Supplementary Materials Tables S1 and S2).

The characterization of polyphenols was conducted utilizing a LC–DAD-MS/MS system consisted of a Shimadzu® LCMS-8030 triple quadrupole mass spectrometer (Kyoto, Japan) operated in the electrospray ionization (ESI) negative mode and a Shimadzu®Nexera HPLC system (Kyoto, Japan), coupledto a diode array detector (DAD). A reversed-phase Waters® Nova-Pak C18 column (4.9 × 250 mm, 4 µm; Waters®, Milford, MA, USA) was used. The mobile phase consisted of water (1% HCOOH, solvent A) and acetonitrile (1% HCOOH, solvent B) and the separation was conducted using the following gradient: 2% B isocratic (10 min), from 2% to 98% B (30 min), 98% B isocratic (7 min) in a flow rate of 0.6 mL min^{-1} and 10 µL of injection volume. The conditions for MS analysis were nitrogen as nebulizing and drying gas (at flow rates of 3.0 and 15.0 L min^{-1}, respectively); interface voltage of –3.5 kV; desolvation line temperature of 250 °C; heat block temperature of 400 °C. The spectrometer operated in product ion scan mode using analyte-specific precursor ions; and argon was used as collision-induced dissociation (CID) gas (at 230 kPa). Identification of individual phenolics was carried out by comparison with retention times, UV-Vis, MS and MS/MS spectra, bibliographic data, and available external standards injected in the same conditions (gallic acid, myricitrin, rutin, and kampferol-3-*O*-rutinoside, all from SigmaAldrich®–Merck®KGaA, Darmstadt, Germany).

3.4. Experimental Designs: Optimization Procedure and Data Analysis

The factors affecting the ultrasound-assisted extraction (Section 3.2.) were firstly screened using a fractional factorial design (FFD) (2^{4-1}) in order to select the variables and levels to be applied during the optimization step. Based on the results, a Box–Behnken design was conducted to determine the best combination of the important variables selected [56].

3.4.1. Screening Fractional Factorial Design: Selection of the Important Variables for the Extraction Optimization

Four factors: temperature (x_1; 5 °C or 30 °C), time (x_2; 15 or 30 min), solvent ratio (x_3; 0.06 or 0.1 L g^{-1}) and ethanol fraction (x_4; 50% or 75% *v*/*v*) were chosen as independent variables and analyzed in two levels (+1, −1; FFD 2^{4-1}; Table 1). The variables and their levels were initially chosen based on their importance for the UAE of plant materials [47]. Nine trials were conducted (8 trials + central point), in different combinations of the variables (x_1 to x_4 inTable 1), all in triplicate. The details of the UAE process conducted are described in the Section 3.2.

The main effects of each factor (x_1 to x_4) in the following responses: total tannins (TTC), total flavonoids (TFC), and total polyphenolic contents (TPC) were estimated by the calculation of the coefficients of each variable (b_1, b_2, b_3, and b_4) using the statistical software Minitab® 18 (LCC, Pennsylvania, USA). The factors that were significant in the regression analysis ($p \leq 0.05$) were considered to have an impact on the responses and selected for the optimization step.

3.4.2. Box–Behnken Design for Optimization of the Extraction Conditions

After the determination of the most important factors, these variables were optimized using a Box–Behnken design, a simple and more efficient three-level factorial design in comparison to other 3^3 designs [39]. Three independent variables (factors) were analyzed in three levels: temperature (x_1; 30, 40, and 50 °C), solvent ratio (x_2; 0.1, 0.15, and 0.2 L g^{-1}), and ethanol fraction (x_3; 30, 40 and 50%, *v*/*v*).

Fifteen experimental trials, resulted from the combination of the three levels (−1, 0, 1) of each variable and three replicates of the central point were thus conducted in triplicate, following the BBD matrix (Table 2). The response variables (i.e., TTC, TFC, MYC, and TPC) were fitted to a second-order polynomial model equation (Equation(1)) that was used to predict the optimum conditions of extraction process and to construct the response surfaces (RSM).

$$Y = \beta 0 + \sum_{i=1}^{k} \beta_i X_i + \sum_{i=1}^{k} \beta_{ii} X_i^2 + \sum_{i=1}^{k} \sum_{j=i+1}^{k-1} \beta_{ij} Xij \quad (1)$$

where Y represents the response variables, Xi and Xj are the independent variables affecting the response, $\beta 0$, βi, βii, and βij are the regression coefficients of the model (intercept, linear, quadratic, and interaction terms, respectively), and k is the number of variables ($k = 3$). The variables and their levels, with both coded (−1, 0, 1) and uncoded (real values) are given in Table 2.

The Minitab®18 software (LCC, State College, PA, USA) was used for the RSM data analysis. To test the significance of the models, an ANOVA with 95% confidence level was carried out for each response. Furthermore, a lack of fit (LOF) test was performed to check the variability of the residues of the proposed models. The estimated coefficients of multiple determination (R^2) of the quadratic models and the adjusted coefficients of multiple determination (R^2_{Adj}) were also calculated. These coefficients reflect the fraction of the total variability in the response that is explained by the model.

In order to verify and validate the predicted optimal UAE conditions, experimental extractions were under the conditions selected as optimal for TTC, TPC (0.12 L g^{-1} of 40% ethanol, at 50 °C), and TFC (0.13 L g^{-1} of 50% ethanol, at 50 °C). The predicted and experimental responses were compared by a *t*-test and the model validation was confirmed if $p > 0.05$.

4. Conclusions

In conclusion, this study was able to define the optimal UAE conditions to obtain higher amounts of different polyphenolic classes from *P. lentiscus* leaves in a greener way when compared to conventional extraction methods, using a low percentage of organic solvent and less time consumed.

According to our findings, among the variables tested (i.e., temperature, ethanol fraction, and volume), the optimal conditions were slightly different only in terms of ethanol percentage: 40% for tannins and 50% for flavonoids but similar in solvent ratio (~0.13 L g^{-1}), temperature (50 °C), and time (15 min). A good agreement between the experimental and the predicted values at these optimal conditions showed the adequacy of the models obtained. Furthermore, this work brings novelty in the characterization of *P. lentiscus* leaf extracts, putatively identifying for the first time the presence of two tetragalloyllquinic acid derivatives.

These results are important considering the wide commercial applications of the different polyphenolic classes of this species, as well as the new trend in the green chemistry. Moreover, they may constitute the basis for future UAE processes applied in larger scale conditions.

Supplementary Materials: The following are available online at http://www.mdpi.com/2223-7747/9/11/1482/s1, Table S1:Polyphenolic content (in mg g^{-1} DW) of individual tannins and gallic acid derivatives obtained in each trial of the optimization BBD design; Table S2: Flavonoidic content (in mg g^{-1} DW) of individual compounds obtained in each trial of the optimization BBD design; Table S3: Regression coefficients (intercept, linear, quadratic, and interaction) of the models obtained for each response (total tannin—TTC, total flavonoids—TFC, myricitrin—MYC, and total polyphenols—TPC contents); Figure S1. MS/MS spectra of tetragalloyl quinic acids (isomer 1, A and isomer 2, B) corresponding to peak 8 and 9 of Table 4, respectively

Author Contributions: Conceptualization, A.G., C.B. and L.B.d.S.N.; methodology, A.G., C.B. and L.B.d.S.N.; software, L.B.d.S.N. and C.D.; formal analysis, L.B.d.S.N. and C.D.; investigation, L.B.d.S.N., C.D., A.G.; data curation, L.B.d.S.N.; writing—original draft preparation, C.D. and L.B.d.S.N.; writing—review and editing, C.B., A.G., L.B.d.S.N. and F.F.; supervision, A.G., C.B., F.F. and L.B.d.S.N. All authors have read and agreed to the published version of the manuscript.

Funding: This research was funded by "Progetto FOE ECONOMIA CIRCOLARE, CNR" project: FOE-2019 DBA.AD003.139.

Acknowledgments: The authors extend their gratitude to "Analytical Food, Firenze" and to Cristian Marinelli for the technical support during the LC-MS/MS analyses.

Conflicts of Interest: The authors declare no conflict of interest.

References

1. Zrira, S.; Elamrani, A.; Benjilali, B. Chemical composition of the essential oil of *Pistacia lentiscus* L. from Morocco-a seasonal variation. *Flavour. Fragr. J.* **2003**, *18*, 475–480. [CrossRef]
2. Bampouli, A.; Kyriakopoulou, K.; Papaefstathiou, G.; Louli, V.; Aligiannis, N.; Magoulas, K.; Krokida, M. Evaluation of total antioxidant potential of *Pistacia lentiscus* var. *chia* leaves extracts using UHPLC–HRMS. *J. Food. Eng.* **2015**, *167*, 25–31.
3. Margaris, N.S. Adaptive strategies in plants dominating Mediterranean-type ecosystems. In *Ecosystems of the World*; FAO: Roma, Italy, 1981.
4. Marzano, M.C.; Gori, A.; Baratto, M.C.; Lamponi, S.; Tattini, M.; Brunetti, C.; Ferrini, F. Antioxidant capacity and cytotoxicity of different polyphenolic extracts of *Pistacia lentiscus*. *Planta Med.* **2016**, *82*, P364. [CrossRef]
5. Nahida, A.; Siddiqui, A.N. *Pistacia lentiscus*: A review on phytochemistry and pharmacological properties. *Int. J. Pharm.* **2012**, *4*, 16–20.
6. Gori, A.; Nascimento, L.B.; Ferrini, F.; Centritto, M.; Brunetti, C. Seasonal and Diurnal Variation in Leaf Phenolics of Three Medicinal Mediterranean Wild Species: What Is the Best Harvesting Moment to Obtain the Richest and the Most Antioxidant Extracts? *Molecules* **2020**, *25*, 956. [CrossRef]
7. Remila, S.; Atmani-Kilani, D.; Delemasure, S.; Connat, J.L.; Azib, L.; Richard, T.; Atmani, D. Antioxidant, cytoprotective, anti-inflammatory and anticancer activities of *Pistacia lentiscus* (Anacardiaceae) leaf and fruit extracts. *Eur. J. Integr. Med.* **2015**, *7*, 274–286. [CrossRef]

8. Azaizeh, H.; Halahleh, F.; Abbas, N.; Markovics, A.; Muklada, H.; Ungar, E.D.; Landau, S.Y. Polyphenols from *Pistacia lentiscus* and *Phillyrea latifolia* impair the exsheathment of gastro-intestinal nematode larvae. *Vet. Parasitol.* **2013**, *191*, 44–50. [CrossRef]
9. Romani, A.; Pinelli, P.; Galardi, C.; Mulinacci, N.; Tattini, M. Identification and quantification of galloyl derivatives, flavonoid glycosides and anthocyanins in leaves of *Pistacia lentiscus* L. *Phytochem. Anal.* **2002**, *13*, 79–86. [CrossRef]
10. Ruiz-Cruz, S.; Chaparro-Hernández, S.; Hernández-Ruiz, K.L.; Cira-Chávez, L.A.; Estrada-Alvarado, M.I.; Gassos Ortega, L.E.; Lopez Mata, M.A. Flavonoids: Important biocompounds in food. In *Flavonoids: From Biosynthesis to Human Health*; Goncalo, C.J.J., Ed.; IntechOpen: London, UK, 2017; pp. 353–370.
11. Brodowska, K.M. Natural flavonoids: Classification, potential role, and application of flavonoid analogues. *Eur. J. Biol. Res.* **2017**, *7*, 108–123.
12. Pretorius, J.C. Flavonoids: A review of its commercial application potential as anti-infective agents. *Curr. Med. Chem.* **2003**, *2*, 335–353. [CrossRef]
13. Pizzi, A. Tannins: Prospectives and actual industrial applications. *Biomolecules* **2019**, *9*, 344. [CrossRef]
14. Gonçalves, S.; Moreira, E.; Grosso, C.; Andrade, P.B.; Valentao, P.; Romano, A. Phenolic profile, antioxidant activity and enzyme inhibitory activities of extracts from aromatic plants used in Mediterranean diet. *J. Food. Sci. Technol.* **2017**, *54*, 219–227. [CrossRef] [PubMed]
15. Bozorgi, M.; Memariani, Z.; Mobli, M.; Salehi Surmaghi, M.H.; Shams-Ardekani, M.R.; Rahimi, R. Five Pistacia species (*P. vera*, *P. atlantica*, *P. terebinthus*, *P. khinjuk*, and *P. lentiscus*): A review of their traditional uses, phytochemistry, and pharmacology. *Sci. World. J.* **2013**, *2013*. [CrossRef] [PubMed]
16. Ljubuncic, P.; Song, H.; Cogan, U.; Azaizeh, H.; Bomzon, A. The effects of aqueous extracts prepared from the leaves of *Pistacia lentiscus* in experimental liver disease. *J. Ethno. Pharmacol.* **2005**, *100*, 198–204. [CrossRef]
17. Landau, S.; Muklada, H.; Markovics, A.; Azaizeh, H. Traditional uses of *Pistacia lentiscus* in veterinary and human medicine. In *Medicinal and Aromatic Plants of the Middle-East; Medicinal and Aromatic Plants of the World*; Zohara, Y., Nativ, D., Eds.; Springer: Dordrecht, The Netherlands, 2014; Volume 2, pp. 163–180.
18. Moeini, R.; Memariani, Z.; Asadi, F.; Bozorgi, M.; Gorji, N. *Pistacia* Genus as a Potential Source of Neuroprotective Natural Products. *Planta Med.* **2019**, *85*, 1326–1350. [CrossRef]
19. El Bishbishy, M.H.; Gad, H.A.; Aborehab, N.M. Chemometric discrimination of three Pistacia species via their metabolic profiling and their possible in vitro effects on memory functions. *J. Pharm. Biomed. Anal.* **2020**, *177*, 112840. [CrossRef] [PubMed]
20. Dellai, A.; Souissi, H.; Borgi, W.; Bouraoui, A.; Chouchane, N. Antiinflammatory and antiulcerogenic activities of *Pistacia lentiscus* L. leaves extracts. *Ind. Crops. Prod.* **2013**, *49*, 879–882. [CrossRef]
21. Gacem, M.A.; El Hadj-Khelil, A.O.; Boudjemaa, B.; Gacem, H. Phytochemistry, Toxicity and Pharmacology of *Pistacia lentiscus*, *Artemisia herba-alba* and *Citrullus colocynthis*. In *Sustainable Agriculture Reviews 39*; Lichtfouse, E., Ed.; Springer: Cham, Switzerland, 2020; Volume 39, pp. 57–93.
22. Dahmoune, F.; Spigno, G.; Moussi, K.; Remini, H.; Cherbal, A.; Madani, K. *Pistacia lentiscus* leaves as a source of phenolic compounds: Microwave-assisted extraction optimized and compared with ultrasound-assisted and conventional solvent extraction. *Ind. Crop. Prod.* **2014**, *61*, 31–40. [CrossRef]
23. Oreopoulou, A.; Tsimogiannis, D.; Oreopoulou, V. Extraction of polyphenols from aromatic and medicinal plants: An overview of the methods and the effect of extraction parameters. In *Polyphenols in Plants*, 2nd ed.; Watson, R.R., Ed.; Academic Press: London, UK, 2019; pp. 243–259.
24. Nkhili, E.; Tomao, V.; El Hajji, H.; El Boustani, E.S.; Chemat, F.; Dangles, O. Microwave-assisted water extraction of green tea polyphenols. *Phytochem. Anal.* **2009**, *20*, 408–415. [CrossRef]
25. Farhat, M.B.; Jordán, M.J.; Chaouch-Hamada, R.; Landoulsi, A.; Sotomayor, J.A. Changes in phenolic profiling and antioxidant capacity of *Salvia aegyptiaca* L. by-products during three phenological stages. *LWT* **2015**, *63*, 791–797. [CrossRef]
26. Spigno, G.; Tramelli, L.; De Faveri, D.M. Effects of extraction time, temperature and solvent on concentration and antioxidant activity of grape marc phenolics. *J. Food. Eng.* **2007**, *81*, 200–208. [CrossRef]
27. Ameer, K.; Shahbaz, H.M.; Kwon, J.H. Green extraction methods for polyphenols from plant matrices and their by-products: A review. *Compr. Rev. Food. Sci.* **2017**, *16*, 295–315. [CrossRef]
28. Sasidharan, S.; Chen, Y.; Saravanan, D.; Sundram, K.M.; Latha, L.Y. Extraction, isolation and characterization of bioactive compounds from plants' extracts. *Afr. J. Tradit. Complement. Altern. Med.* **2011**, *8*, 1–10. [CrossRef]

29. Chemat, F.; Vian, M.A.; Fabiano-Tixier, A.S.; Nutrizio, M.; Jambrak, A.R.; Munekata, P.E.; Cravotto, G. A review of sustainable and intensified techniques for extraction of food and natural products. *Curr. Green. Chem.* **2020**, *22*, 2325–2353. [CrossRef]
30. Panja, P. Green extraction methods of food polyphenols from vegetable materials. *J. Food. Sci.* **2018**, *23*, 173–182. [CrossRef]
31. Sarker, S.D.; Nahar, L. Hyphenated techniques and their applications in natural products analysis. In *Natural Products Isolation Methods and Protocols*, 3rd ed.; Sarker, S.D., Nahar, L., Eds.; Humana Press: Mumbai, India; Springer: Mumbai, India, 2009; Volume 864, pp. 75–88.
32. Chemat, F.; Rombaut, N.; Meullemiestre, A.; Turk, M.; Perino, S.; Fabiano-Tixier, A.S.; Abert-Vian, M. Review of green food processing techniques. Preservation, transformation, and extraction. *Innov. Food Sci. Emerg. Technol.* **2017**, *41*, 357–377. [CrossRef]
33. Rostagno, M.A.; Palma, M.; Barroso, C.G. Ultrasound-assisted extraction of soy isoflavones. *J. Chromatogr. A* **2003**, *A1012*, 119–128. [CrossRef]
34. Alcántara, C.; Žugčić, T.; Abdelkebir, R.; García-Pérez, J.V.; Jambrak, A.R.; Lorenzo, J.M.; Barba, F.J. Effects of Ultrasound-Assisted Extraction and Solvent on the Phenolic Profile, Bacterial Growth, and Anti-Inflammatory/Antioxidant Activities of Mediterranean Olive and Fig Leaves Extracts. *Molecules* **2020**, *25*, 1718. [CrossRef]
35. Azmir, J.; Zaidul, I.S.M.; Rahman, M.M.; Sharif, K.M.; Mohamed, A.; Sahena, F.; Omar, A.K.M. Techniques for extraction of bioactive compounds from plant materials: A review. *J. Food. Eng.* **2003**, *117*, 426–436. [CrossRef]
36. Ben Ahmed, Z.; Yousfi, M.; Viaene, J.; Dejaegher, B.; Demeyer, K.; Mangelings, D.; Vander Heyden, Y. Seasonal, gender and regional variations in total phenolic, flavonoid, and condensed tannins contents and in antioxidant properties from *Pistacia atlantica* ssp. leaves. *Pharm. Biol.* **2017**, *55*, 1185–1194. [CrossRef]
37. Teng, H.; Choi, Y.H. Optimization of ultrasonic-assisted extraction of bioactive alkaloid compounds from rhizomacoptidis (*Coptis chinensis* Franch.) using response surface methodology. *Food. Chem.* **2014**, *142*, 299–305. [CrossRef]
38. Abd El-Salam, E.A.; Morsy, N.F. Optimization of the extraction of polyphenols and antioxidant activity from *Malva parviflora* L. leaves using Box–Behnken design. *Prep. Biochem. Biotech.* **2019**, *49*, 876–883. [CrossRef]
39. Montgomery, D.C. *Design and Analysis of Experiments*; John Wiley & Sons: Hoboken, NJ, USA, 2017.
40. Khuri, A.I.; Cornell, J.A. *Response Surfaces: Designs and Analyses*; CRC Press: Boca Raton, FL, USA, 2018.
41. Dahmoune, F.; Remini, H.; Dairi, S.; Aoun, O.; Moussi, K.; Bouaoudia-Madi, N.; Mouni, L. Ultrasound assisted extraction of phenolic compounds from *P. lentiscus* L. leaves: Comparative study of artificial neural network (ANN) versus degree of experiment for prediction ability of phenolic compounds recovery. *Ind. Crop. Prod.* **2015**, *77*, 251–261. [CrossRef]
42. Alessandri, S.; Ieri, F.; Romani, A. Minor polar compounds in extra virgin olive oil: Correlation between HPLC-DAD-MS and the Folin-Ciocalteu spectrophotometric method. *J. Agric. Food. Chem.* **2014**, *62*, 826–835. [CrossRef]
43. Braga, M.E.M.; Seabra, I.J.; Dias, A.M.A.; De Sousa, H.C. Recent trends and perspectives for the extraction of natural products. In *Natural Product Extraction: Principles and Applications*; Rostagno, M.A., Prado, J.M., Eds.; RSC Publishing: Cambridge, UK, 2013; pp. 231–275.
44. Nascimento, L.B.S.N.; de Aguiar, P.F.; Leal-Costa, M.V.; Coutinho, M.A.S.; Borsodi, M.P.G.; Rossi-Bergmann, B.; Costa, S.S. Optimization of aqueous extraction from *Kalanchoe pinnata* leaves to obtain the highest content of an anti-inflammatory flavonoid using a response surface model. *Phytochem. Anal.* **2018**, *29*, 308–315. [CrossRef]
45. Chemat, F.; Rombaut, N.; Sicaire, A.G.; Meullemiestre, A.; Fabiano-Tixier, A.S.; Abert-Vian, M. Ultrasound assisted extraction of food and natural products. Mechanisms, techniques, combinations, protocols and applications. A review. *Ultrason. Sonochem.* **2017**, *34*, 540–560. [CrossRef]
46. Rivera-Mondragón, A.; Broeckx, G.; Bijttebier, S.; Naessens, T.; Fransen, E.; Kiekens, F.; Foubert, K. Ultrasound-assisted extraction optimization and validation of an HPLC-DAD method for the quantification of polyphenols in leaf extracts of *Cecropia* species. *Sci. Rep.* **2019**, *9*, 1–16. [CrossRef]
47. Belwal, T.; Ezzat, S.M.; Rastrelli, L.; Bhatt, I.D.; Daglia, M.; Baldi, A.; Anandharamakrishnan, C. A critical analysis of extraction techniques used for botanicals: Trends, priorities, industrial uses and optimization strategies. *Trends Anal. Chem.* **2018**, *100*, 82–102. [CrossRef]

48. Yang, L.; Jiang, J.G.; Li, W.F.; Chen, J.; Wang, D.Y.; Zhu, L. Optimum extraction process of polyphenols from the bark of *Phyllanthus emblica* L. based on the response surface methodology. *J. Sep. Sci.* **2009**, *32*, 1437–1444. [CrossRef]
49. Jovanović, A.A.; Đorđević, V.B.; Zdunić, G.M.; Pljevljakušić, D.S.; Šavikin, K.P.; Gođevac, D.M.; Bugarski, B.M. Optimization of the extraction process of polyphenols from *Thymus serpyllum* L. herb using maceration, heat-and ultrasound-assisted techniques. *Sep. Purif. Technol.* **2017**, *179*, 369–380. [CrossRef]
50. Zhang, Z.S.; Li, D.; Wang, L.J.; Ozkan, N.; Chen, X.D.; Mao, Z.H.; Yang, H.Z. Optimization of ethanol–water extraction of lignans from flaxseed. *Sep. Purif. Technol.* **2007**, *57*, 17–24. [CrossRef]
51. Gam, D.H.; Yi Kim, S.; Kim, J.W. Optimization of Ultrasound-Assisted Extraction Condition for Phenolic Compounds, Antioxidant Activity, and Epigallocatechin Gallate in Lipid-Extracted Microalgae. *Molecules* **2020**, *25*, 454. [CrossRef]
52. Lavilla, I.; Bendicho, C. Fundamentals of ultrasound-assisted extraction. In *Water Extraction of Bioactive Compounds, From Plant to Drug Development*; Gonzàlez, H.D., Muñoz, M.J.G., Eds.; Elsevier: Amsterdam, The Netherlands, 2017; pp. 291–316.
53. Arruda, H.S.; Pereira, G.A.; Pastore, G.M. Optimization of extraction parameters of total phenolics from *Annona crassiflora* Mart. (Araticum) fruits using response surface methodology. *Food. Anal. Methods* **2017**, *10*, 100–110. [CrossRef]
54. Zhao, S.; Baik, O.D.; Choi, Y.J.; Kim, S.M. Pretreatments for the efficient extraction of bioactive compounds from plant-based biomaterials. *Crit. Rev. Food. Sci. Nutr.* **2014**, *54*, 1283–1297. [CrossRef]
55. Rusu, M.E.; Fizeșan, I.; Pop, A.; Gheldiu, A.M.; Mocan, A.; Crișan, G.; Tomuta, I. Enhanced recovery of antioxidant compounds from hazelnut (*Corylus avellana* L.) involucres based on extraction optimization: Phytochemical profile and biological activities. *Antioxidants* **2019**, *8*, 460. [CrossRef]
56. Bezerra, M.A.; Santelli, R.E.; Oliveira, E.P.; Villar, L.S.; Escaleira, L.A. Response surface methodology (RSM) as a tool for optimization in analytical chemistry. *Talanta* **2008**, *76*, 965–977. [CrossRef]
57. Emeko, H.A.; Olugbogi, A.O.; Betiku, E. Appraisal of artificial neural network and response surface methodology in modeling and process variable optimization of oxalic acid production from cashew apple juice: A case of surface fermentation. *Bioresources* **2015**, *10*, 2067–2082. [CrossRef]
58. Joglekar, A.M.; May, A.T.; Graf, E.; Saguy, I. Product excellence through experimental design. *Cereal Food World* **1987**, *32*, 857–868.
59. Wu, J.; Yu, D.; Sun, H.; Zhang, Y.; Zhang, W.; Meng, F.; Du, X. Optimizing the extraction of anti-tumor alkaloids from the stem of *Berberis amurensis* by response surface methodology. *Ind. Crops. Prod.* **2015**, *69*, 68–75. [CrossRef]
60. Pan, X.; Niu, G.; Liu, H. Microwave-assisted extraction of tea polyphenols and tea caffeine from green tea leaves. *Chem. Eng. Process.* **2003**, *42*, 129–133. [CrossRef]
61. Silva, E.M.; Rogez, H.; Larondelle, Y. Optimization of extraction of phenolics from *Inga edulis* leaves using response surface methodology. *Sep. Purif. Technol.* **2007**, *55*, 381–387. [CrossRef]
62. Rodríguez-Pérez, C.; Quirantes-Piné, R.; Amessis-Ouchemoukh, N.; Madani, K.; Segura-Carretero, A.; Fernández-Gutierrez, A. A metabolite-profiling approach allows the identification of new compounds from *Pistacia lentiscus* leaves. *J. Pharm. Biomed. Anal.* **2013**, *77*, 167–174. [CrossRef]
63. Barbouchi, M.; Elamrani, K.; El Idrissi, M. A comparative study on phytochemical screening, quantification of phenolic contents and antioxidant properties of different solvent extracts from various parts of *Pistacia lentiscus* L. *J. King Saud Univ. Sci.* **2020**, *32*, 302–306. [CrossRef]
64. Pacifico, S.; Piccolella, S.; Marciano, S.; Galasso, S.; Nocera, P.; Piscopo, V.; Monaco, P. LC-MS/MS profiling of a mastic leaf phenol enriched extract and its effects on H_2O_2 and Aβ (25–35) oxidative injury in SK-B-NE (C)-2 cells. *J. Agric. Food. Chem.* **2014**, *62*, 11957–11966. [CrossRef]
65. Saliha, D.; Seddik, K.; Djamila, A.; Abdrrahmane, B.; Lekhmici, A.; Noureddine, C. Antioxidant proprieties of *Pistacia lentiscus* L. leaves extracts. *Pharmacogn. Commun.* **2013**, *3*, 28. [CrossRef]
66. Cherbal, A.; Kebieche, M.; Madani, K.; El-Adawi, H. Extraction and valorization of phenolic compounds of leaves of Algerian *Pistacia lentiscus*. *Asian. J. Plant Sci.* **2012**, *11*, 131–136. [CrossRef]
67. Chemat, F.; Vian, M.A.; Cravotto, G. Green extraction of natural products: Concept and principles. *Int. J. Mol. Sci.* **2012**, *13*, 8615–8627. [CrossRef]
68. Baratto, M.C.; Tattini, M.; Galardi, C.; Pinelli, P.; Romani, A.; Visioli, F.; Pogni, R. Antioxidant activity of galloyl quinic derivatives isolated from *P. lentiscus* leaves. *Free Radic. Res.* **2003**, *37*, 405–412. [CrossRef]

69. Beghlal, D.; El Bairi, K.; Marmouzi, I.; Haddar, L.; Mohamed, B. Phytochemical, organoleptic and ferric reducing properties of essential oil and ethanolic extract from *Pistacia lentiscus* (L.). *Asian. Pac. J. Trop. Dis.* **2016**, *6*, 305–310. [CrossRef]
70. Mehenni, C.; Atmani-Kilani, D.; Dumarçay, S.; Perrin, D.; Gérardin, P.; Atmani, D. Hepatoprotective and antidiabetic effects of *Pistacia lentiscus* leaf and fruit extracts. *J. Food Drug. Anal.* **2016**, *24*, 653–669. [CrossRef]
71. Foddai, M.; Kasabri, V.; Afifi, F.U.; Azara, E.; Petretto, G.L.; Pintore, G. In vitro inhibitory effects of Sardinian *Pistacia lentiscus* L. and *Pistacia terebinthus* L. on metabolic enzymes: Pancreatic lipase, α-amylase, and α-glucosidase. *Starch-Stärke* **2015**, *67*, 204–212. [CrossRef]
72. Neszmelyi, A.; Kreher, B.; Müller, A.; Dorsch, W.; Wagner, H. Tetragalloylquinic acid, the major antiasthmatic principle of Galphimia glauca. *Planta Med.* **1993**, *59*, 164–167. [CrossRef]
73. Bouchet, N.; Barrier, L.; Fauconneau, B. Radical scavenging activity and antioxidant properties of tannins from *Guiera senegalensis* (Combretaceae). *Phytother. Res.* **1998**, *12*, 159–162. [CrossRef]
74. Karas, D.; Ulrichová, J.; Valentová, K. Galloylation of polyphenols alters their biological activity. *Food Chem. Toxicol.* **2017**, *105*, 223–240. [CrossRef]
75. Hussain, S.; Schönbichler, S.A.; Güzel, Y.; Sonderegger, H.; Abel, G.; Rainer, M.; Bonn, G.K. Solid-phase extraction of galloyl-and caffeoylquinic acids from natural sources (*Galphimia glauca* and *Arnicae flos*) using pure zirconium silicate and bismuth citrate powders as sorbents inside micro spin columns. *J. Pharm. Biomed. Anal.* **2013**, *84*, 148–158. [CrossRef]
76. Lin, S.; Zhu, Q.; Wen, L.; Yang, B.; Jiang, G.; Gao, H.; Chen, F.; Jiang, Y. Production of quercetin, kaempferol and their glycosidic derivatives from the aqueous organic extracted residue of litchi pericarp with *Aspergillus awamori*. *Food Chem.* **2014**, *145*, 220–227. [CrossRef] [PubMed]
77. Lin, Y.; Shi, R.; Wang, X.; Shen, H.M. Luteolin a flavonoid with potential for cancer prevention and therapy. *Curr. Cancer. Drug. Targets* **2008**, *8*, 634–646. [CrossRef]
78. Mazaki, T.; Kitajima, T.; Shiozaki, Y.; Sato, M.; Mino, M.; Yoshida, A.; Nakamura, M.; Yoshida, Y.; Tanaka, M.; Ozaki, T.; et al. In vitro and in vivo enhanced osteogenesis by kaempferol found by a high-through put assay using human mesenchymal stromal cells. *J. Funct. Foods* **2014**, *6*, 241–247. [CrossRef]
79. Miean, K.H.; Mohamed, S. Flavonoid (myricetin, quercetin, kaempferol, luteolin, and apigenin) content of edible tropical plants. *J. Agric. Food Chem.* **2001**, *49*, 3106–3112. [CrossRef] [PubMed]
80. Azib, L.; Debbache-Benaida, N.; Da Costa, G.; Atmani-Kilani, D.; Saidene, N.; Ayouni, K.; Atmani, D. *Pistacia lentiscus* L. leaves extract and its major phenolic compounds reverse aluminium-induced neurotoxicity in mice. *Ind. Crop. Prod.* **2019**, *137*, 576–584. [CrossRef]
81. Adebayo, H.A.; Abolaji, A.O.; Kela, R.; Ayepola, O.O.; Olorunfemi, T.B.; Taiwo, O.S. Antioxidant activities of the leaves of *Chrysophyllum Albidum* G. *Pak. J. Pharm. Sci.* **2011**, *24*, 545–551. [PubMed]
82. Ahangarpour, A.; Oroojan, A.A.; Khorsandi, L.; Kouchak, M.; Badavi, M. Antioxidant effect of myricitrin on hyperglycemia-induced oxidative stress in C2C12 cell. *Cell Stress Chaperones* **2018**, *23*, 773–781. [CrossRef]
83. Chen, W.; Zhuang, J.; Li, Y.; Shen, Y.; Zheng, X. Myricitrin protects against peroxynitrite-mediated DNA damage and cytotoxicity in astrocytes. *Food Chem.* **2013**, *141*, 927–933. [CrossRef] [PubMed]
84. Miller, J.N. Basic statistical methods for analytical chemistry. Part 2. Calibration and regression methods. A review. *Analyst* **1991**, *116*, 3–14. [CrossRef]

Article

Optimized Extraction of Polysaccharides from *Bergenia emeiensis* Rhizome, Their Antioxidant Ability and Protection of Cells from Acrylamide-induced Cell Death

Chen Zeng and Shiling Feng *

College of Life Science, Sichuan Agricultural University, Ya'an 625014, China; zc@stu.sicau.edu.cn
* Correspondence: fsl@sicau.edu.cn

Received: 21 July 2020; Accepted: 30 July 2020; Published: 31 July 2020

Abstract: *Bergeniaemeiensis* is a traditional herb in Chinese folk medicine. Most related studies are focused on the bioactivity of bergenin, neglecting other bioactive compounds. In our previous work, polysaccharides were identified in *B. emeiensis* rhizome. To evaluate the extraction process and the antioxidant ability of these polysaccharides, a response surface method and antioxidant assays were applied. The results showed that the yield of polysaccharides was highly affected by extraction time, followed by temperature and solvent-to-sample ratio. Under the optimal conditions (43 °C, 30 min and 21 mL/g), the yield was 158.34 ± 0.98 mg/g. After removing other impurities, the purity of the polysaccharides from *B. emeiensis* (PBE) was 95.97 ± 0.92%. The infrared spectrum showed that PBE had a typical polysaccharide structure. Further investigations exhibited the PBE could scavenge well DPPH and ABTS free radicals and chelate Fe^{2+}, showing an excellent antioxidant capacity. In addition, PBE also enhanced the cell viability of HEK 239T and Hep G2 cells under acrylamide-exposure conditions, exhibiting great protection against the damage induced by acrylamide. Therefore, PBE can be considered a potential natural antioxidant candidate for use in the pharmaceutical industry as a health product.

Keywords: *Bergenia emeiensis*; polysaccharides; optimization; RSM; acrylamide-induced damage

---

1. Introduction

Bergenia emeiensis (*B. emeiensis*), belonging to the Saxifragaceae family, is a special species in China. In southwest China, *B. emeiensis* is found distributed at altitudes of 1300–1500 m. The plants of the genus *Bergenia* were well known in folk medicine for its beneficial effect on treating many diseases. In India, *Bergenia* was called Paashaanbhed and used for its hepatoprotective, diuretic and antipyretic properties. The aqueous extract of *Bergenia ligulata* rhizome could inhibit in vitro growth of CaC_2O_4 and calcium hydrogen phosphate dihydrate crystals [1]. Moreover, *Bergenia* plants were widely applied to treat diarrhea, vomiting, fever, cough, pulmonary infections, menorrhagia, excessive uterine hemorrhage, kidney stones and ulcer of large intestines [2–4]. Bergenin and arbutin were found to be the main polyphenol components in *Bergenia* [5]. Quercetin, kaemipferol and other flavonoids were also isolated from *Bergenia* [6]. Meanwhile, other flavonoids, anthraquinones and phenolic constituents were identified in some species of *Bergenia* [5–8]. However, analyses of the content and pharmacological activity of polysaccharides from *Bergenia* were rarely reported.

Polysaccharides from natural plants have been proven to possess a good biological activity. The polysaccharides from *Cyclocarya paliurus* showed great anti-hyperlipidemic and anti-diabetic ability [9]. Since the anticancer effect on polysaccharides has been proved, the antitumor ability of polysaccharides has gradually become a research hotspot [10]. In addition, natural polysaccharides could also effectively scavenge free radicals and increase the activity of antioxidant enzymes, thus exerting an excellent antioxidant capacity in vitro and in vivo [11,12].

Polysaccharides are condensed monosaccharide residues, containing a large number of hydrophilic groups with a high polarity. Thus, polysaccharides have good thermal stability and are easily soluble in water and insoluble in organic reagents such as petroleum ether, acetone, and ethanol. Therefore, the hot water extraction method is widely used to extract polysaccharides from natural plants with its advantages of mild extraction conditions, low equipment requirements and simple operation, but the method exhibits disadvantages such as high energy consumption, long time consumption and low efficiency. In addition to the extraction method, the extraction yield was also affected individually and interactively by extraction parameters such as extraction temperature, extraction time, solution pH and solvent to sample ratio. Therefore, a mathematical statistical technology—the response surface method (RSM)—was established to optimize the extraction process. Bb building a mathematical model based on a small number of experimental results, RSM has been widely applied to predict the optimal extraction process for its accurate estimation of the interactions between extraction parameters [13,14].

B. emeiensis was used as a folk medicine to treat respiratory diseases on the area of Mount Emei (Sichuan, China). However, the definite identity of the active compounds remains unknown. In this study, polysaccharides were considered as the index to investigate the content in *B. emeiensis* rhizomes. RSM was applied to optimize the extraction process. Moreover, the antioxidant ability of polysaccharides of *B. emeiensis* was evaluated by determined the scavenging ability on DPPH and ABTS free radicals and also the chelating capacity on Fe^{2+}. The results may provide a theoretical basis for the potential application of the plant in the food, pharmaceuticals and cosmetics industries.

2. Results

2.1. Single-factor Experiment Analysis

To investigate the effect of a single extraction method on the yield of polysaccharides, single-factor experiments were carried out. When the liquid-solid ratio was changed from 10 to 20 mL/g, the yield of polysaccharides also increased. When the ratio was over 20 mL/g, the output decreased gradually (Figure 1A). Thereby, 20 mL/g was chosen as the center point for the RSM design. As shown in Figure 1B, the yield of polysaccharides reached the peak when the extraction temperature was 40 °C. The content tended to be stable as the temperature was increased over 40 °C. Hence, 40 °C was picked as the central point for the RSM design. Appropriately prolonging the extraction time could help the leaching of polysaccharides. As shown in Figure 1C, as the extraction time was extended from 30 to 60 min, the yield of polysaccharides was also promoted, but, over 60 min, the extraction rate declined and although the yield went up again at 150 min, it was still lower than that at 60 min. Therefore, 60 min was selected as the central point for the subsequent RSM design.

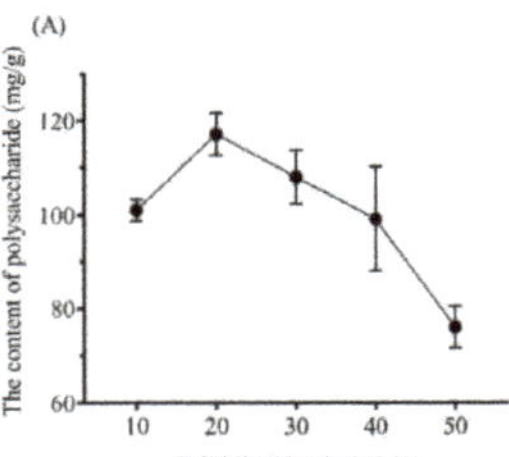

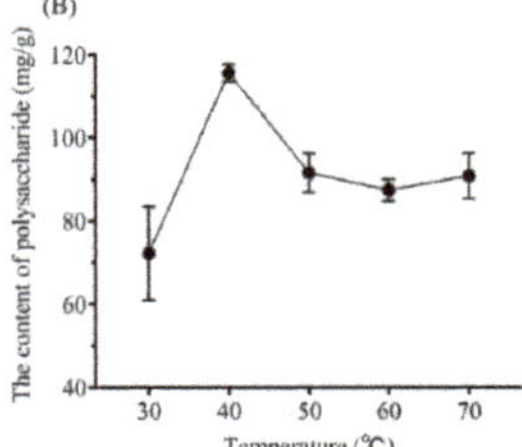

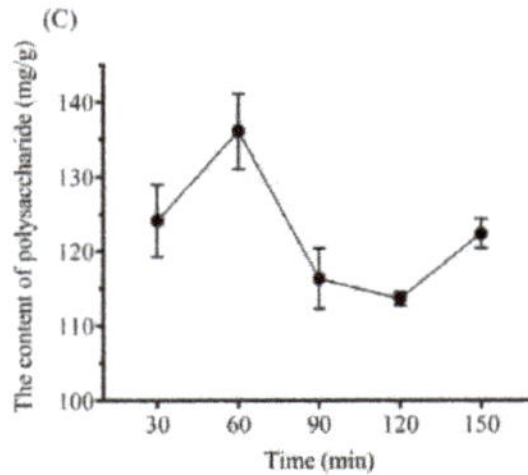

Figure 1. Effects of different extraction parameters on the yield of polysaccharides. (**A**) liquid-solid ratio, (**B**) extraction temperature and (**C**) extraction time.

2.2. Analysis of the Response Surface

Based on the outcomes of the single-factor assays, suitable extraction conditions were selected and used to design the RSM (Table 1). According to the experimental results and the multiple regression analysis, the yield of polysaccharides could be expressed using the following formula:

$$Y = 156.06 - 1.52A - 3.21B - 0.8744C - 3.48AB - 2.32AC - 2.05BC - 2.9A^2 - 0.3684B^2 - 3.35C^2 \quad (1)$$

where Y was the yield of polysaccharides; A was the extraction temperature; B was the extraction time and C was liquid-sold ratio.

Table 1. Box-Behnken experimental design and the results of these experiments.

Run	A-Temperature (°C)	B-Time (min)	C-Liquid-Solid Ratio (mL/g)	Y-Yield of Polysaccharides (mg/g)
1	40(0)	30(−1)	10(−1)	154.03
2	40(0)	90(1)	30(1)	146.55
3	40(0)	90(1)	10(−1)	151.91
4	40(0)	60(0)	20(0)	157.90
5	30(−1)	60(0)	10(−1)	150.05
6	30(−1)	60(0)	30(1)	152.44
7	40(0)	30(−1)	30(1)	156.89
8	50(1)	60(0)	10(−1)	151.82
9	50(1)	30(−1)	20(0)	157.98
10	30(−1)	30(−1)	20(0)	154.24
11	40(0)	60(0)	20(0)	155.55
12	40(0)	60(0)	20(0)	155.01
13	30(−1)	90(1)	20(0)	154.58
14	40(0)	60(0)	20(0)	155.20
15	50(1)	90(1)	20(0)	144.39
16	50(1)	60(0)	30(1)	144.95
17	40(0)	60(0)	20(0)	156.65
Experim.	43.11	30	20.69	159.25
Actual	43	30	21	158.34 ± 0.98

The ANOVA analysis for the fitted quadratic model is shown in Table 2. The low p value of the model means the regression model was highly significant and a high p value of lack of fit indicated the model equation was suitable for predicting the yield of polysaccharides from *B. emeiensis*. A lower value of the coefficient of variation (0.6278%) meant the results were highly reliable. Moreover, the R^2 was 0.9777 and adj R^2 was 0.9491, both of which was close to 1, thus meaning the model exhibited a good accuracy and a high fitting degree for efficiently predicting the yield under different extraction conditions. In addition, as shown in Table 2, the p values of linear coefficients (A, B and C) were lower than 0.05. Interactions (AB, AC and BC) and quadratic terms (A^2 and C^2) were also significant ($p < 0.05$). Hence the relationship between three factors and the yield of polysaccharides was not simple linear relationship.

A 3D surface map can intuitively reflect the interaction effects of factors in a binary regression model. As shown in Figure 2A, by fixing the liquid-solid ratio, appropriately increasing the extraction temperature could enhance the yield. When the temperature was fixed, the extraction rate increased as the extraction time decreased (Figure 2C). As shown in Figure 2B, fixing the extraction time, the yield was enhanced with the increase of the liquid-solid ratio and extraction temperature while the yield was decreased when the liquid-solid ratio and temperature were higher than the central point. Therefore, the interaction effects between two factors significantly affected the yield of polysaccharides.

Table 2. The variance analysis of regression model.

Source	Sum of Squares	Mean Square	F-Value	*p*-Value
Model	283.37	31.49	34.15	<0.0001 ***
A	18.56	18.56	20.13	0.0028 **
B	82.57	82.57	89.55	<0.0001 ***
C	6.12	6.12	6.63	0.0367 *
AB	48.48	48.48	52.57	0.0002 **
AC	21.47	21.47	23.28	0.0019 **
BC	16.87	16.87	18.3	0.0037 **
A^2	35.34	35.34	38.32	0.0004 **
B^2	0.5713	0.5713	0.6196	0.457
C^2	47.26	47.26	51.25	0.0002 **
Residual	6.45	0.9221		
Lack of Fit	0.6367	0.2122	0.1459	0.9271
R^2	0.9777			
Adj R^2	0.9491			
C.V. %	0.6278			

Notes: * means $p < 0.05$; ** means $p < 0.01$ and *** means $p < 0.0001$.

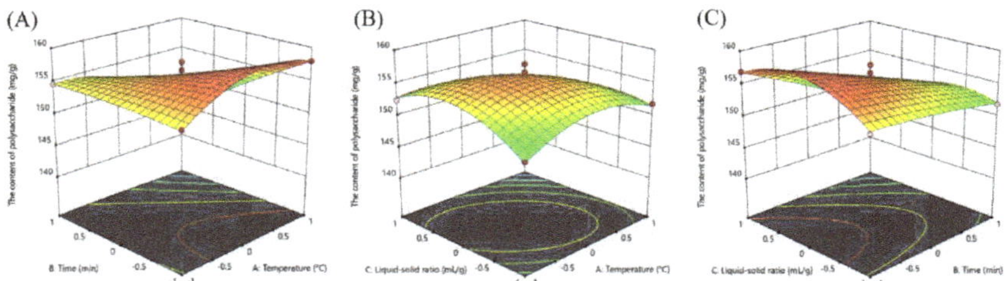

Figure 2. The response surface plots expressing the effect of extraction parameters on the yield of polysaccharides. (**A**) extraction temperature and extraction time; (**B**) extraction temperature and liquid-solid ratio; (**C**) extraction time and liquid-solid ratio.

Based on the RSM results the optimal extraction conditions were obtained: 43.11 °C, 30 min and 20.69 mL/g, which gave the highest yield of 159.25 mg/g. Considering the practical experiments, the extraction conditions were modified to 43 °C, 30 min and 21 mL/g. Under the modified conditions, the yield of polysaccharides was 158.34 ± 0.98 mg/g, which was close to the predicted value. Hence, RSM was successfully applied in this paper to optimize the extraction process.

2.3. Characteristics and Chemical Composition of Polysaccharides

To explore the primary structure of polysaccharides, the chemical composition and FT-IR scanning spectrum were determined. After purification, the content of polysaccharides from *B. emeiensis* (PBE) reached 95.97 ± 0.92% (Table 3). The protein that might be combined with PBE accounted for only 0.83 ± 0.025%. Therefore, we can state that most of the impurities were fully removed from the PBE. In addition, the molecular weight was 1.42×10^5 Da according to a standard curve made using dextrans. As shown in Figure 3, a broad peak appeared at 3200–3400 cm^{-1} (3396 cm^{-1}), that is attributed to the bending vibration absorption peak of the OH bonds caused by the stretching vibrations of the hydroxyl OH bonds related with the intermolecular hydrogen bonding [15]. A characteristic absorption peak of the polysaccharide substance ranging from 3000 to 2800 cm^{-1} (2933 cm^{-1}) was the stretching vibration of the C-H bond, thus confirming that PBE was a mixture of polysaccharides. A strong absorption at 2300–2400 cm^{-1} indicated that the polysaccharides might contain CO_2. In addition, the peaks between the range of 1300–1450 cm^{-1} showed the presence of -CH_3 and -CH_2 bending. The absorption peak at 1432 cm^{-1} was associated with the stretching and bending vibration of C-H bonds in the polysaccharides. Peaks at 1622 cm^{-1} and 1733 cm^{-1} were derived from carboxylate ion stretching bands and ester carbonyl groups, respectively [16]. Besides, the absorption peaks at 1000–1200 cm^{-1} (1078/1047 cm^{-1}) corresponded to C-O-C stretching vibrations and proved that PBE had pyranose rings [17]. The absorption peak at 1542 cm^{-1} indicated that PBE contained other characteristic absorption peaks of polysaccharides (C-H and C-H_2 stretching, C-OH bending vibrations) [18]. Many faint peaks in the region of 530–1000 cm^{-1} indicated the existence of β-glycosidic bonds with a pyranose ring (896.46 cm^{-1}) [19].

Table 3. The chemical composition of PBE.

Components	Content
Polysaccharides	95.97 ± 0.92%
Protein	0.83 ± 0.025%
Molecular weight	1.42 × 10^5 Da

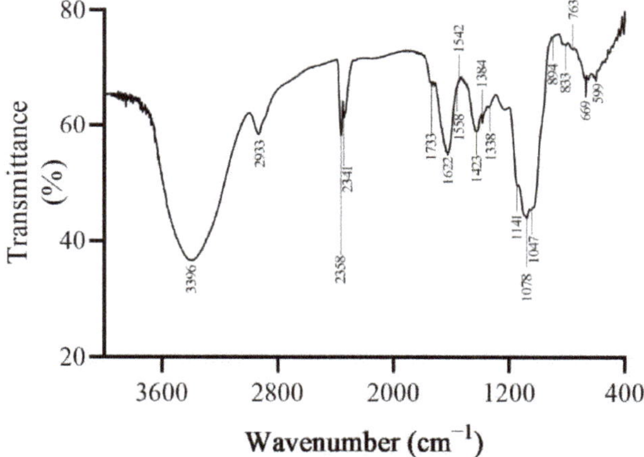

Figure 3. The FT-IR scanning spectrum of PBE.

2.4. Antioxidant Ability in vitro

Polysaccharides from plants were already proved to possess a strong antioxidant ability [20]. Polysaccharides from the fruit of *Rosa laevigata* had a noticeable effect on the radical scavenging of ABTS and DPPH [21]. In this study, scavenging on DPPH and ABTS radicals and total reducing power assays were conducted to investigate the antioxidant activity of PBE. As shown in Figure 4A, PBE could well clear DPPH at 1.0 mg/mL with a maximum clearance of 83.66%. Ascorbic acid (Vc) used as reference could scavenge 90.32% of DPPH radicals. When the concentration was higher than 0.6 mg/mL, the DPPH radical scavenging rate tended to be stable and close to that of Vc, with an EC_{50} of 0.31 mg/mL. As the concentration of PBE was increased, the ABTS radical clearance rate was also promoted, with an EC_{50} of 0.36 mg/mL (Figure 4B). Those two assays demonstrated that PBE possessed a strong free radical scavenging ability. Moreover, PBE also could chelate well Fe^{2+} (Figure 4C). The total reducing power of PBE also increased as the concentration increased (Figure 4D).

Polysaccharides from thirteen Boletus mushrooms were also proved to display a greater chelating activity with EC_{50} values ranging from 250.5 to 1413.8 μg/mL [22], and the EC_{50} of PBE was 0.48 mg/mL, which was higher than some natural polysaccharides, indicating PBE could become a potential antioxidant candidate.

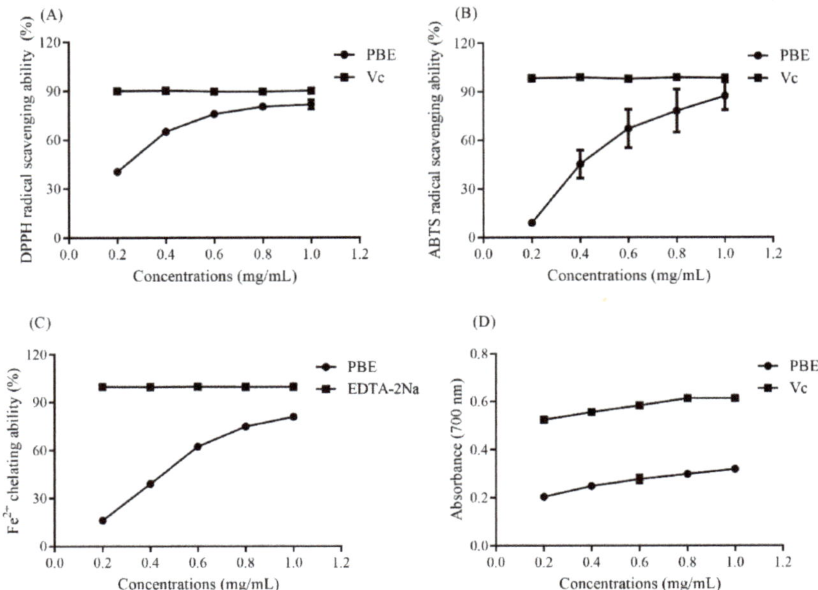

Figure 4. The antioxidant ability of PBE on (**A**) DPPH radical scavenging (**B**) ABTS radical clearance (**C**) Fe^{2+} chelating capacity and (**D**) the total reducing power.

2.5. Protection Against Acrylamide

Acrylamide (AC) may cause potential harm to organs. Since *B. emeiensis* was used as a raw material for treating some injuries such as pulmonary infections and ulcers of the large intestine [2,3], to further explore the protective effects of PBE at a cellular level against the damage caused by AC, different concentrations of PBE were used to treat HEK 293T and Hep G2 cells. As shown in Figure 5A, none of the trial concentrations of PBE had a negative effect on the growth of Hep G2 cells, while the proliferation of HEK 293T cells was inhibited with 400 µg/mL of PBE (Figure 4B). When both kinds of cells were exposed to AC, the cell viability was reduced compared with the control group, but when pretreated with PBE for 24 h, the survival of Hep G2 was significantly increased (Figure 4C). Moreover, 0–200 µg/mL PBE also could enhance the cell viability of HEK 293T cell under AC exposure conditions. However, 400 µg/mL PBE suppressed the growth of HEK 293T, which may be related to a toxicity of high concentrations of drug (as 400 µg/mL PBE negatively affected HEK 293T) plus the toxicity of AC. Therefore, PBE showed a strong protective capacity for cells against the toxicity of AC, making PBE a promising alternative health product.

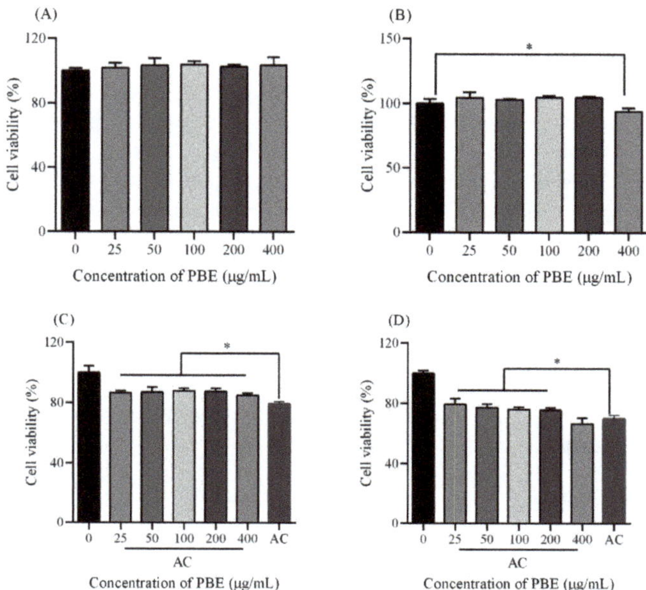

Figure 5. The toxicity of PBE on (**A**) Hep G2 cells and (**B**) HEK 293T cells. After treated with PBE for 24 h and exposed to AC for 10 h, the cell viability of (**C**) Hep G2 cells and (**D**) HEK 293T cells. Note: * means $p < 0.05$.

3. Discussion

Usually, a high temperature and a long extraction time would enhance the yield of polysaccharides obtained from many plants. Polysaccharides from the fruit of *Nitraria tangutorum* Bobr. was extracted at 60 °C for 7 h and the yield was 14.01% [23]. Polysaccharide from the wild mushroom *Paxillus involutus* was also extracted at 79 °C for 3 h to give the highest yield [24]. However, the extraction time for polysaccharides from flowers of *Dendrobium devonianum* was only 53.10 min and the polysaccharide exhibited an excellent antioxidant ability [25]. In this study, 40 °C and 30 min could make the output reach its highest value, indicating the polysaccharides from *B. emeiensis* could be distinguished from other polysaccharides and could possess a different bioactivity. The steep slope of the 3D graphs reflected the response sensitivity of the yield to the factors. A steeper 3D graph indicates a larger response value [26]. As shown in Figure 2, the three response surface plots were all steep and the contour maps were elliptic rather than a regular round shape meaning the interactions significantly influenced the yield. The results were exactly confirmed by the ANOVA analysis that the cross coefficients (AB, AC and BC) were significant ($p < 0.05$) (Table 3). Thereby, the RSM model were successfully established in this study to optimize the extraction process.

Reactive oxygen species (ROS) play an important role in maintaining the balance of homeostasis, signal transduction, and regulation of growth and development [27]. However, high concentrations of ROS could cause damage to proteins, lipids and nucleic acids, eventually leading to chronic diseases such as diabetes, atherosclerosis, cancer, and neurodegeneration [28]. Polysaccharides can not only directly bind to protons (H^+) ionized by groups such as -OH and -COOH, but also chelate metal ions to prevent the generation of free radicals. Polysaccharides from purple sweet potato exhibited moderate DPPH radical scavenging activity and reducing power [29]. Polysaccharides from *Astragalus cicer* L. also could scavenge well ABTS and DPPH free radicals [30]. In this study, PBE could well clear DPPH and ABTS free radicals with the IC_{50} of 0.31 mg/mL and 0.36 mg/mL, respectively, exerting a higher antioxidant ability than other natural polysaccharides [31]. Maybe the extraction time and extraction temperature were lower than in other extraction conditions, which might prevent the -OH and -COOH from oxidation and even decomposition providing more functional groups meaning a great antioxidant ability [32]. Like other natural polysaccharides, PEB showed a strong free radical scavenging capacity, but there are few polysaccharides that could chelate metal ions. PBE also could chelate Fe^{2+}, which was proved to a metal ion that can catalyze the Fenton reaction. Hence, the side outcomes such as lipid peroxidation would be prevented.

Acrylamide (AC), listed as a category 2A carcinogen, was proved to cause damage to the human body [33,34]. Now, some natural plant extracts were found to relieve the toxicity caused by AC. Resveratrol could ameliorate the oxidative damage in rats under AC exposure conditions by activating the antioxidant system [35]. Curcumin also could lower the high ROS levels induced by AC in Hep G2 cells, and reduce DNA fragment formation, thus decreasing the toxicity of AC [36]. Moreover, AC causes damage to cells by increasing the ROS level then inducing oxidative injuries to cells and even causing death [33,35,36]. PBE showed a strong antioxidant ability to scavenge free radicals (Figure 3), hence, the protection of PBE on HEK 293T and Hep G2 cells against AC may be associated with the significantly free radical clearance since AC might cause oxidative stress to cells. However, more specific investigations shall be carried out to find out the mechanism.

4. Materials and Methods

4.1. Materials and Chemicals

Bergenia emeiensis was collected in the area of the Mount Emei (Sichuan, China) at an altitude of 1300–1500 m. The rhizomes were washed with distilled water and naturally air-dried. The rhizomes were ground into a fine powder, and kept at −20 °C for the following experiments.

2,2-Diphenyl-1-picrylhydrazyl (DPPH), was purchased from Sigma Chemical Co. (St. Louis, MO, USA). Ethanol, phenol, sulfuric acid, chloroform, butanol, trichloroacetic acid and glucose were purchased from the Chengdu Kelong Chemical Factory (Chengdu, China). FBS was obtained from Gibco (New York, NY, USA). Dulbecco's modified eagle medium (DMEM), PBS and digestive enzyme were purchased from Hyclone (Logan, UT, USA). All chemicals were analytical grade.

4.2. Determination of Polysaccharides

The content of polysaccharides was determined by the phenol-sulfuric acid method [37]. Briefly, 100 μL phenol was mixed with 200 μL sample solution and 500 μL sulfuric acid. The reaction solution was standing for 30 min after shaking. Then the absorbance was measured at 490 nm. The standard curve was made with glucose.

4.3. Extraction Polysaccharides from *Bergenia emeiensis*

The powder of *B. emeiensis* was mixed with distilled water to different liquid-solid ratios (10, 20, 30, 40 and 50 mL/g) under various temperature (30, 40, 50, 60, and 70 °C) for 30, 60, 90, 120 and 150 min. Then the extract solutions were centrifuged for 10 min at a speed of 8000 r/min. The supernatant was collected to determine the content of polysaccharides.

4.4. Response Surface Design

Based on the outcomes of single factors assays, a three-level-three-factor Box-Behnken design with RSM was employed to optimize the extraction process. The yield of polysaccharides was taken as the response. The coded and actual levels of the three factors were presented in Table 4. Analysis of variance (ANOVA) was performed to detect the individual linear, quadratic and interaction regression coefficients using Design Expert 11 trial version (Stat-Ease, Inc., Minneapolis, MN, USA). The significance of the dependent variables was statistically analyzed by computing the p value.

Table 4. Box-behnken design table of factors and levels.

Factors	Coding Level		
Extraction time	−1(30 min)	0(60 min)	1(90 min)
Extraction temperature	−1(30 °C)	0(40 °C)	1(50 °C)
Liquid-solid ratio	−1(10 mL/g)	0(20 mL/g)	1(30 mL/g)

4.5. *Purification of Polysaccharides*

The extract was filtered and concentrated using a rotary evaporator. An equivalent volume of trichloroacetic acid (20%) was added to the concentrate and the mixture was centrifuged at a speed of 5000 rpm for 5 min to remove the brown solids. Then ethanol was added to a final concentration of 85%. After standing at 4 °C for 24 h, the precipitate was separated and dissolved in distilled water. The solution was centrifuged again to remove impurities. Then supernatant was mixed 15 times with Savage reagent (chloroform:butanol = 4:1) to deproteinize it. The resulting aqueous solution was dialyzed for 3 days after removing the Savage reagent by using a rotary evaporator, and lyophilized to obtain polysaccharides (PBE).

4.6. *Characterization of Purified Polysaccharides*

4.6.1. FT-IR Assay

5 mg of dry polysaccharides was milled with 500 mg KBr and then pressed onto a disc. The FT-IR spectrum of polysaccharides was identified using an infrared spectrometer (FT-IR-8400S, Shimadzu, Kyoto, Japan) in the range of 4000–400 cm^{-1}.

4.6.2. Determination of Protein

Protein content was measured by the Coomassie blue staining method [38]. Polysaccharides was dissolved in distilled water to 1.0 mg/mL. Coomassie brilliant blue G-250 (100 mg) was mixed with 50 mL 95% ethanol and 100 mL 85% phosphoric acid then diluted to 1 L with distilled water. Sample solution (0.1 mL) was mixed with 0.5 mL of Coomassie blue staining solution and 0.9 mL distilled water. The mixture was placed at 37 °C for 15 min in dark before reading the absorbance at 595 nm. The content of protein was calculated according to a standard curve made with bovine serum albumin.

4.6.3. Determination of Molecular Weight

The molecular weight of polysaccharides was measured by gel permeation chromatography (Sephadex G100, 1 × 30 cm). Purified polysaccharides were resolved in distilled water to 10 mg/mL and loaded onto the chromatography column at 0.5 mL/min. The eluent was collected and determined the content of polysaccharides by phenol-sulfuric acid method to definite retention time. The standard curve was made with standard dextrans (T-10, T-40, T-70, T-100, T-500, and T-2000 series).

4.7. *Determination of Antioxidant Ability*

4.7.1. Clearance Ability on DPPH Radicals

The scavenging ability on DPPH free radicals was conducted according to a previous method [39]. Briefly, 20 μL of different concentration of sample solution were mixed with 100 μL water and 80 μL DPPH ethanol solution (0.8 mM). Ascorbic acid (Vc) was used as the positive control. The reaction solution was placed at dark for 10 min before read the absorbance at 517 nm. The clearance rate was calculated according to the following formula:

$$\text{Clearance rate (\%)} = 1 - A1/A0 \quad (2)$$

where A1 was the absorbance of sample group and A0 was the absorbance of the group with distilled water instead.

4.7.2. Clearance Ability on ABTS Radical

The ABTS radical scavenging ability was tested according to a previous report with some modifications [40]. ABTS solution (7.4 mM) was equally mixed with potassium persulfate (2.6 mM) in the dark for 16 h. Then the mixture was diluted to an absorbance of 0.7. Sample solution (100 μL) was added to 100 μL of ABTS working fluid for 10 min before the absorbance of the reaction misture was read at 734 nm. The ABTS radical scavenging rate was determined by the following formula:

$$\text{Clearance rate } (\%) = 1 - \text{A1}/\text{A0} \tag{3}$$

where A1 was the absorbance of the sample group and A0 was the absorbance of the group with distilled water instead.

4.7.3. Fe^{2+} Chelating Ability

The Fe^{2+} chelating ability assay was conducted according to a previous method with some modifications [22]. Briefly, 50 μL of sample solution was mixed with 100 μL of ferrous sulfate (0.00625 mM) and 50 μL of ferrozine (1 mM). The mixture was placed in the dark for 30 min before the absorbance was read at 562 nm with EDTA-2Na as the positive control. The chelating ability was calculated according to the following formula:

$$\text{Chelating rate } (\%) = 1 - \text{A1}/\text{A0} \tag{4}$$

where A1 was the absorbance of sample group and A0 was the absorbance of the group with distilled water instead.

4.7.4. Total Reducing Power

The total reducing power was measured by q previous method [41]. Sample solution (0.2 mL) was mixed with 0.5 mL $K_3[Fe(CN_6)]$ (1% *w/v*) and 0.5 mL phosphate buffer (0.2 M). The mixture was placed at 50 °C for 20 min then cooled at 0 °C after added 0.5 mL TCA (10% *w/v*). Subsequently, 0.1 mL aliquot of supernatant fluid was transferred into 1 mL distilled water and 0.1 mL $FeCl_3$ (0.1% *w/v*). The absorbance of the reaction solution was read at 700 nm. An increasing absorbance meant an enhanced total reducing power.

4.8. The Protection Against Acrylamide

4.8.1. The Cell Culture

HEK 293T and Hep G2 cells were kindly provided by the Biotechnology Center of Sichuan University (Sichuan, China). Cells were maintained in DMEM medium containing 10% FBS, at 37 °C in a 5% CO_2 atmosphere. When the cell density was about 80% of the flask, the cells were digested from the bottle for following assay.

4.8.2. The Toxicity of PBE on Cells

PBE was dissolved in PBS to different concentrations. Cell suspension (10^5 cells/mL) was added into 96-well plate for 6. Then sample solution was transferred into the plate for 24. Subsequently, the solution was replaced with fresh medium containing 10% CCK-8 and placed at 37 °C for 30 min before the absorbance was read at 450 nm. The cell viability was calculated according to the kit instructions.

4.8.3. The Remission on Acrylamide-induced Damage

Cell suspension was adjusted to 10^5 cells/mL, then 90 μL solution was seeded into the plate for 6 h before 10 μL sample solution was transferred. After 24 h treatment, the cells were exposed to AC (6 mM) for 10 h, then the medium was replaced with new culture and CCK-8 and the plate was incubated at 37 °C for 30 min. The absorbance of each well was read at 450 nm then the cell viability was calculated.

4.9. Statistical Analyses

The data obtained in this study were analyzed statistically by ANOVA (GraphPad Prism 6, (GraphPad Software, Inc., La Jolla, CA, USA). Design-Expert 11 software (trial version, State-Ease Inc., Minneapolis, MN, USA) was applied to analyze the experimental results of the response surface design. All experimental results were expressed as mean ± standard deviation (SD). A value of $p < 0.05$ was considered to be significantly different.

5. Conclusions

In this work, a high content of polysaccharides as found in *Bergenia emeiensis* rhizome. The response surface method was successfully applied to optimize the extraction process. The optimal extraction temperature was 43 °C, the extraction time was 30 min and the solvent to sample ratio was 21 mL/g. Under these conditions, the highest yield was 158.34 ± 0.98 mg/g. After purification, the polysaccharides in PBE was 95.97 ± 0.92%, with a molecular weight of 1.42×10^5 Da and containing only 0.83 ± 0.025% of protein. The FT-IR spectrum showed PBE contained a typical polysaccharide structure and also had pyranose rings. Moreover, PBE could well clear DPPH and ABTS free radicals and also chelate Fe^{2+} ion. Therefore, PBE possessed a strong antioxidant ability. Besides, PBE could also ameliorate and protect HEK 239T and Hep G2 cells from the damage induced by AC. More structural characterization of PBE should be conducted and also the antioxidant ability in cell level and in vivo need further investigation.

Author Contributions: Investigation and writing-original draft preparation, C.Z.; writing-review and editing, S.F. All authors have read and agreed to the published version of the manuscript.

Funding: This work was supported by the Innovation Training Program of Sichuan Agricultural University, Grant Number: 201910626047.

Acknowledgments: The authors are grateful to Siyuan Luo for great help.

Conflicts of Interest: The authors declare no conflict of interest.

Sample Availability: Samples of *B. emeiensis* are available from the authors.

References

1. Joshi, V.S.; Parekh, B.B.; Joshi, M.J.; Vaidya, A.D.B. Inhibition of the growth of urinary calcium hydrogen phosphate dihydrate crystals with aqueous extracts of Tribulus terrestris and Bergenia ligulata. *Urol. Res.* **2005**, *33*, 80–86. [CrossRef] [PubMed]
2. Ahmed, E.; Arshad, M.; Ahmad, M.; Saeed, M.; Ishaque, M. Ethnopharmacological survey of some medicinally important plants of Galliyat areas of NWFP, Pakistan. *Asian J. Plant Sci.* **2004**, *3*, 410–415. [CrossRef]
3. Sinha, S.; Murugesan, T.; Maiti, K.; Gayen, J.R.; Pal, B.; Pal, M.; Saha, B.P. Antibacterial activity of Bergenia ciliate rhizomes. *Fitoterpia* **2001**, *72*, 550–552. [CrossRef]
4. Uniyal, S.K.; Singh, K.N.; Jamwal, P.; Lal, B. Traditional use of medicinal plants among the tribal communities of Chhota Bhangal, Western Himalaya. *J. Ethnobiol. Ethnomed.* **2006**, *2*, 14–21. [CrossRef] [PubMed]
5. Chen, X.; Yoshida, T.; Hatano, T.; Fukushima, M.; Okuda, T. Galloylarbutin and other polyphenols from Bergenia purpurascens. *Phytochemistry* **1987**, *26*, 515–517. [CrossRef]
6. Zhao, J.; Liu, J.; Zhang, X.; Liu, Z.; Tsering, T.; Zhong, Y.; Nan, P. Chemical composition of the volatiles of three wild Bergenia species from western China. *Flavour Frag. J.* **2006**, *21*, 431–434. [CrossRef]
7. Bohm, B.A.; Donevan, L.S.; Bhat, U.G. Flavonoids of some species of Bergenia, francoa, parnassia and lepuropetalon. *Biochem. Syst. Ecol.* **1986**, *14*, 75–77. [CrossRef]

8. Yuldashev, M.P.; Batirov, È.K.; Malikov, V.M. Anthraquinones of Bergenia hissarica. *Chem. Nat. Compd.* **1993**, *29*, 543–544. [CrossRef]
9. Mohib, U.K.; Muhammad, N.; Muhammad, S.; Shicong, Z.; Madiha, R.; Sundas, F.; Robina, M.; Yulin, D.; Rongji, D. A review on structure, extraction, and biological activities of polysaccharides isolated from Cyclocarya paliurus (Batalin) Iljinskaja. *Int. J. Biol. Macromol.* **2020**, *156*, 420–429. [CrossRef]
10. Chen, L.; Huang, G. Antitumor activity of polysaccharides: An Overview. *Curr. Drug Targets* **2018**, *19*, 89–96. [CrossRef]
11. Wu, G.H.; Hu, T.; Li, Z.Y.; Huang, Z.L.; Jiang, J.G. In vitro antioxidant activities of the polysaccharides from Pleurotus tuber-regium (fr.) sing. *Food Chem.* **2014**, *148*, 351–356. [CrossRef] [PubMed]
12. Liu, Q.; Zhu, M.; Geng, X.; Wang, H.; Tb, T.B. Characterization of polysaccharides with antioxidant and hepatoprotective activities from the edible mushroom oudemansiella radicata. *Molecules* **2017**, *22*, 234. [CrossRef] [PubMed]
13. Deniz, B.; Ismail, H.B. Modeling and optimization I: Usability of response surface methodology. *J. Food Eng.* **2007**, *78*, 836–846. [CrossRef]
14. Said, K.A.M.; Amin, M.A.M. Overview on the response surface methodology (RSM) in extraction processes. *J. Appl. Sci. Process. Eng.* **2015**, *2*, 1. [CrossRef]
15. Shi, M.J.; Wei, X.; Xu, J.; Chen, B.J.; Zhao, D.Y.; Cui, S.; Zhou, T. Carboxymethylated degraded polysaccharides from Enteromorpha prolifera: Preparation and in vitro antioxidant activity. *Food Chem.* **2017**, *215*, 76–83. [CrossRef]
16. Guo, Q.; Cui, S.W.; Kang, J.; Ding, H.; Wang, Q.; Wang, C. Non-starch polysaccharides from American ginseng: Physicochemical investigation and structural characterization. *Food Hydrocolloid.* **2015**, *44*, 320–327. [CrossRef]
17. Chen, Y.; Xue, Y. Purification, chemical characterization and antioxidant activities of a novel polysaccharide from Auricularia polytricha. *Int. J. Biol. Macromol.* **2018**, *120*, 1087–1092. [CrossRef]
18. Su, Y.; Li, L. Structural Characterization and antioxidant activity of polysaccharide from four Auriculariales. *Carbohyd. Polym.* **2019**, *229*, 115407. [CrossRef]
19. Lin, Y.; Zeng, H.; Wang, K.; Lin, H.; Li, P.; Huang, Y.; Zhou, S.; Zhang, W.; Chen, T.; Fan, H. Microwave-assisted aqueous two-phase extraction of diverse polysaccharides from Lentinus edodes: Process optimization, structure characterization and antioxidant activity. *Int. J. Biol. Macromol.* **2019**, *136*, 305–315. [CrossRef]
20. Kardošová, A.; Machová, E. Antioxidant activity of medicinal plant polysaccharides. *Fitoterapia* **2006**, *77*, 367–373. [CrossRef] [PubMed]
21. Liu, X.; Gao, Y.; Li, D.; Liu, C.; Jin, M.; Bian, J.; Lv, M.; Sun, Y.; Zhang, L.; Gao, P. The neuroprotective and antioxidant profiles of selenium containing polysaccharides from the fruit of Rosa laevigata. *Food Funct.* **2018**, *9*, 1800–1808. [CrossRef] [PubMed]
22. Zhang, L.; Hu, Y.; Duan, X.; Tang, T.; Shen, Y.; Hu, B.; Liu, A.; Chen, H.; Li, C.; Liu, Y. Characterization and antioxidant activities of polysaccharides from thirteen boletus mushrooms. *Int. J. Biol. Macromol.* **2018**, *113*, 1–7. [CrossRef] [PubMed]
23. Zhao, B.; Liu, J.; Chen, X.; Zhang, J.; Wang, J. Purification, structure and anti-oxidation of polysaccharides from the fruit of Nitraria tangutorum Bobr. *RSC Adv.* **2018**, *8*, 11731–11743. [CrossRef]
24. Liu, Y.; Zhou, Y.; Liu, M.; Wang, Q.; Li, Y. Extraction optimization, characterization, antioxidant and immunomodulatory activities of a novel polysaccharide from the wild mushroom Paxillus involutus. *Int. J. Biol. Macromol.* **2018**, *112*, 326–332. [CrossRef]
25. Wang, D.; Fan, B.; Wang, Y.; Zhang, L.; Wang, F. Optimum extraction, characterization, and antioxidant activities of polysaccharides from flowers of Dendrobium devonianum. *Int. J. Anal. Chem.* **2018**, *2018*, 3013497. [CrossRef]
26. Majeed, M.; Hussain, A.I.; Chatha, S.A.S.; Khosa, M.K.K.; Kamal, G.M.; Kamal, M.A.; Zhang, X.; Liu, M. Optimization protocol for the extraction of antioxidant components from Origanum vulgare leaves using response surface methodology. *Saudi J. Biol. Sci.* **2016**, *23*, 389–396. [CrossRef]
27. Sauer, H.; Wartenberg, M.; Hescheler, J. Reactive oxygen species as intracellular messengers during cell growth and differentiation. *Cell. Physiol. Biochem.* **2001**, *11*, 173–186. [CrossRef]
28. Vincent, A.M.; Russell, J.W.; Low, P.; Feldman, E.L. Oxidative stress in the pathogenesis of diabetic neuropathy. *Endocr. Rev.* **2004**, *25*, 612–628. [CrossRef]

29. Sun, J.; Zhou, B.; Tang, C.; Gou, Y.; Chen, H.; Wang, Y.; Jin, C.; Liu, J.; Niu, F.; Kan, J.; et al. Characterization, antioxidant activity and hepatoprotective effect of purple sweetpotato polysaccharides. *Int. J. Biol. Macromol.* **2018**, *115*, 69–76. [CrossRef]
30. Shang, H.; Wang, M.; Li, R.; Duan, M.; Wu, H.; Zhou, H. Extraction condition optimization and effects of drying methods on physicochemical properties and antioxidant activities of polysaccharides from Astragalus Cicer L. *Sci. Rep.* **2018**, *8*, 3359. [CrossRef]
31. Rjeibi, I.; Feriani, A.; Hentati, F.; Hfaiedh, N.; Michaud, P.; Pierre, G. Structural characterization of water-soluble polysaccharides from Nitraria retusa fruits and their antioxidant and hypolipidemic activities. *Int. J. Biol. Macromol.* **2019**, *129*, 422–432. [CrossRef]
32. Wang, J.; Hu, S.; Nie, S.; Yu, Q.; Xie, M. Reviews on mechanisms of in vitro antioxidant activity of polysaccharides. *Oxid. Med. Cell Longev.* **2016**, *2016*, e5692852. [CrossRef]
33. Yousef, M.I.; El-Demerdash, F.M. Acrylamide-induced oxidative stress and biochemical perturbations in rats. *Toxicology* **2006**, *219*, 133–141. [CrossRef]
34. Barber, D.S.; Hunt, J.R.; Ehrich, M.F.; Lehning, E.J.; LoPachin, R.M. Metabolism, toxicokinetics and hemoglobin adduct formation in rats following subacute and subchronic acrylamide dosing. *NeuroToxicology* **2001**, *22*, 341–353. [CrossRef]
35. Alturfan, A.A.; Tozan-Beceren, A.; Şehirli, A.Ö.; Demiralp, E.; Şener, G.; Omurtag, G.Z. Resveratrol ameliorates oxidative DNA damage and protects against acrylamide-induced oxidative stress in rats. *Mol. Biol. Rep.* **2011**, *39*, 4589–4596. [CrossRef] [PubMed]
36. Cao, J.; Liu, Y.; Jia, L.; Jiang, L.P.; Geng, C.Y.; Yao, X.F.; Kong, Y.; Jiang, B.N.; Zhong, L.F. Curcumin attenuates acrylamide-induced cytotoxicity and genotoxicity in HepG2 cells by ROS scavenging. *J. Agr. Food Chem.* **2008**, *56*, 12059–12063. [CrossRef] [PubMed]
37. Dubois, M.; Gilles, K.; Hamilton, J.K.; Rebers, P.A.; Smith, F. A colorimetric method for the determination of sugars. *Nature* **1951**, *168*, 167. [CrossRef] [PubMed]
38. Smith, P.K.; Krohn, R.I.; Hermanson, G.T. Measurement of protein using bicinchoninic acid. *Anal. Biochem.* **1985**, *163*, 76–85. [CrossRef]
39. Brand, W.W.M.; Cuvelier, M.E.; Berset, C. Use of a free radical method to evaluate antioxidant activity. *LWT Food Sci. Technol.* **1995**, *28*, 25–30. [CrossRef]
40. Xu, G.Y.; Liao, A.M.; Huang, J.H.; Zhang, J.G.; Thakur, K.; Wei, Z.J. Evaluation of structural, functional, and anti-oxidant potential of differentially extracted polysaccharides from potatoes peels. *Int. J. Biol. Macromol.* **2019**, *129*, 778–785. [CrossRef]
41. Jia, Z.; Tang, M.; Wu, J. The determination of flavonoid contents in mulberry and their scavenging effects on superoxide radicals. *Food Chem.* **1999**, *64*, 555–559. [CrossRef]

Article

Optimized Extraction of Total Triterpenoids from Jujube (*Ziziphus jujuba* Mill.) and Comprehensive Analysis of Triterpenic Acids in Different Cultivars

Lijun Song [1], Li Zhang [2], Long Xu [1], Yunjian Ma [1], Weishuai Lian [1], Yongguo Liu [3] and Yonghua Wang [1,*]

1 School of Food Science and Engineering, South China University of Technology, Guangzhou, Guangdong 510641, China; 201710104471@mail.scut.edu.cn (L.S.); 201610104342@mail.scut.edu.cn (L.X.); femayj@mail.scut.edu.cn (Y.M.); 201410104818@mail.scut.edu.cn (W.L.)
2 College of Life Science, Tarim University, Alar, Xinjiang 843300, China; 120100039@taru.edu.cn
3 Beijing Advanced Innovation Centre for Food Nutrition and Human Health, Beijing Key Laboratory of Flavor Chemistry, Beijing Technology and Business University, Beijing 100048, China; liuyg@th.btbu.edu.cn
* Correspondence: yonghw@scut.edu.cn; Tel.: +86-20-87113848

Received: 19 February 2020; Accepted: 23 March 2020; Published: 27 March 2020

Abstract: Triterpenoid compounds are one of the main functional components in jujube fruit. In this study, the optimal process for ultrasound-assisted extraction (UAE) of total triterpenoids from jujube fruit was determined using response surface methodology (RSM). The optimal conditions were as follows: temperature of 55.14 °C, ethanol concentration of 86.57%, time of 34.41 min, and liquid-to-solid ratio of 39.33 mL/g. The triterpenoid yield was 19.21 ± 0.25 mg/g under optimal conditions. The triterpenoid profiles and antioxidant activity were further analyzed. Betulinic acid, alphitolic acid, maslinic acid, oleanolic acid, and ursolic acid were the dominant triterpenoid acids in jujube fruits. Correlation analysis revealed a significant positive correlation between the major triterpenic acids and antioxidant activities. The variations of triterpenoid profiles and antioxidant activity within the jujube fruits and the degree of variation were evaluated by hierarchical cluster analysis (HCA) and principal component analysis (PCA), respectively. The results provide important guidance for the quality evaluation and industrial application of jujube fruit.

Keywords: hierarchical cluster analysis; principal component analysis; ultrasound-assisted extraction; triterpenic acid; *Ziziphus jujuba*

---

1. Introduction

Jujube (*Ziziphus jujuba* Mill.), belonging to the Rhamnaceae family, is widespread in Asia, Europe, and America [1]. In China, jujube has been cultivated for 4000 years and there are more than 700 cultivars of the fruits [2]. More than four million tons of jujube fruits are harvested in China per year, which represents 90% of the total yield globally [3]. The fruit has been commonly used in Traditional Chinese Medicine (TMC) for its various pharmacological activities, such as its anticancer, antiepileptic, anti-inflammatory, anti-insomnia, and neuroprotective effects [4,5]. In general, the beneficial effects of health are derived from a variety of bioactive compounds, such as triterpenes, alkaloids, flavonoids, and polysaccharides [6].

Triterpenes, belonging to the Phytosterol family, are naturally occurring bioactive components that are commonly found in cereals and vegetables [7]. Modern studies have shown that triterpenes and triterpenic acids, derivatives of pentacyclic triterpenes, have a variety of biological effects, such as antioxidative, anti-inflammatory, anticancer, hepatoprotective, and anti-microbial activities, combined with low toxicity [8–10]. Triterpenic acids in jujube fruit have been demonstrated to be a group of major bioactive compounds [11–13]. For example, triterpenic acids have been reported to be the most active part in jujuba for the inhibitory effects on inflammatory cells [14]. Alphitolic acid and 3-O-trans-coumaroyl alphitolic acid in jujube can significantly reduce nitric oxide (NO) release and the inducible nitric oxide synthase (iNOS) expression in macrophages [15]. Furthermore, betulinic acid isolated from jujube can cause apoptosis of human breast cancer cell line MCF-7 cells through the mitochondria transduction pathway [16].

The discovery of new natural and safe health products in the form of plant extracts represents a real challenge today. Thus, efficient extraction and further utilization of bioactive triterpenes of jujube and its products have been attracting attention in recent years. [17,18]. To obtain the highest recovery of triterpenoids, it is vital to select the best extraction method and optimize the parameters [19,20]. Compared to conventional extraction methods, such as maceration and Soxhlet extraction, ultrasonic-assisted extraction (UAE) is a green and efficient technology used for its short extraction time, reduced consumption of solvents and energy, and higher extraction yield of bioactive compounds [19–21]. This technique has been successfully used to extract triterpene acids from olive pomace [20], pomegranate flowers [22], and *Rosmarinus officinalis* leaves [12]. However, to the best of our knowledge, the application of UAE processes for extracting triterpene compounds from jujube has not been reported before.

Previous research has proved that many factors, such as solvent concentration, extraction temperature, time, and liquid/solid ratio, can affect the extraction efficiency from plant materials. Considering all of these factors and their levels, it is a tedious task to optimize extraction conditions, during which not only does the number of experimental run increase but also the interactive effect cannot be determined [23]. Response surface methodology (RSM) is a statistical method that uses multifactorial modeling to optimize complex processes. It gives a free space wherein the experimental terms can be defined based on the response value, and the levels of factors can be adjusted according to the requirement of the experiment [23,24]. Therefore, this method may be an ideal strategy for the optimization of triterpenoid extraction from jujube.

In addition, the differences in contents of the triterpenes in the materials also affects the composition of the extracts. The compositional profile of bioactive compounds presented in jujube has been found to be influenced by factors, such as cultivar, geographical environment, processing conditions, and storage conditions [23–25]. However, because of the difference between the chemical compositions of different cultivars, there are some difficulties in the breeding and planting of jujube varieties, as well as in the quality evaluation and standardization of the developed products. Therefore, it is of great significance for customers and the industry to explore the profiles of triterpenic acids of different jujube cultivars without regional disparity.

The aims of this study were: (1) to optimize the UAE conditions for triterpenoids from jujube fruit using RSM. The effects of extraction temperature, ethanol concentration, time, and the solvent-to-solid ratio on the total triterpenoid yield were studied. (2) To analyze the antioxidant activities and major triterpenic acids profiles in the extracts of different jujube samples. (3) To study the differences in the contents of triterpenic acids and antioxidant activities among different cultivars using principal component analysis (PCA) and hierarchical cluster analysis (HCA). This study provides a comprehensive triterpenoid acid profile of different jujube cultivars, irrespective of the origin differences, and the results provide substantial information on the understanding and utilization of the phytochemical properties of these jujube cultivars for further research.

2. Results and Discussion

2.1. UAE Process Optimization

2.1.1. Model Fitting

The merits of RSM include the use of a lower number of experimental measurements, the provision of a statistical interpretation of the data, and also the identification of the interaction amongst variables [23,24]. In this study, the Box–Behnken design (BBD) was employed to determine the interactions among X_1 (temperature), X_2 (ethanol concentration), X_3 (time), and X_4 (liquid-to-solid ratio), as well as to optimize the UAE conditions. Table 1 shows the experimental results, and Table 2 summarizes the results of the analysis of variance (ANOVA).

Table 1. The experimental results.

Run	X_1: Temperature (°C)	X_2: Ethanol Concentration (%)	X_3: Time (min)	X_4: Liquid-to-Solid Ratio (mL/g)	Y: Total Triterpenoid Yield (mg/g)
1	40	90	35	35	16.60
2	50	80	40	35	18.68
3	40	85	35	25	14.13
4	60	85	40	35	17.72
5	50	85	30	25	16.82
6	40	85	35	45	16.40
7	50	85	35	35	19.25
8	40	85	30	35	16.13
9	50	80	35	25	17.18
10	60	80	35	35	17.96
11	50	85	35	35	19.18
12	50	90	40	35	18.58
13	60	90	35	35	18.08
14	50	90	35	45	18.93
15	50	85	35	35	19.12
16	50	85	40	25	16.78
17	50	80	35	45	18.81
18	50	90	30	35	18.84
19	40	80	35	35	16.15
20	60	85	35	45	18.01
21	50	80	30	35	18.64
22	50	90	35	25	17.04
23	50	85	30	45	18.73
24	50	85	35	35	19.05
25	60	85	30	35	17.95
26	40	85	40	35	16.42
27	50	85	35	35	19.08
28	50	85	40	45	18.69
29	60	85	35	25	16.38

Table 2. Analysis of variance (ANOVA) for the response surface quadratic model.

Source	Sum of Squares	df	Mean Square	F Value	p-value
Model	45.29	14	3.24	229.64	<0.0001
X_1: Temperature	8.79	1	8.79	623.9	<0.0001
X_2: Ethanol concentration	0.035	1	0.035	2.5	0.1362
X_3: Time	0.0048	1	0.0048	0.34	0.5687
X_4: Liquid-to-solid ratio	10.53	1	10.53	747.33	<0.0001
$X_1 X_2$	0.027	1	0.027	1.93	0.1862
$X_1 X_3$	0.068	1	0.068	4.8	0.0459
$X_1 X_4$	0.1	1	0.1	7.27	0.0174
$X_2 X_3$	0.023	1	0.023	1.6	0.2269
$X_2 X_4$	0.017	1	0.017	1.2	0.2919
$X_3 X_4$	0	1	0	0	1
X_1^2	21.3	1	21.3	1512.04	<0.0001
X_2^2	0.089	1	0.089	6.32	0.0248

Table 2. *Cont.*

Source	Sum of Squares	df	Mean Square	*F* Value	*p*-value
X_3^2	0.61	1	0.61	43.09	<0.0001
X_4^2	7.37	1	7.37	523.14	<0.0001
Residual	0.2	14	0.014		
Lack of Fit	0.17	10	0.017	2.67	0.1785
Pure Error	0.026	4	0.00643		
Cor Total	45.49	28			
Adeq Precision	56.589				
R^2 = 0.9957; Adj R^2 = 0.9913; Pred R^2 = 0.9774					

ANOVA can fully reflect the significance and reliability of the response surface quadratic regression model [23,24]; as indicated in Table 2, the model was highly significant (F = 229.64, p < 0.0001). The p-value for the lack of fit was not significant (F = 2.67, p = 0.1785), which indicates the adequate predictive relevance of the model to explain the associations of independent variables with dependent variables. The linear coefficients (X_1, X_4), quadratic coefficients (X_1^2, X_2^2, X_3^2, and X_4^2), and interaction coefficients (X_1 X_3, X_1 X_4) were significant (p < 0.05). The R^2 value of 0.9957 indicates a reasonable fit of the model to the experimental data. An R^2 value (multiple correlation coefficient) closer to one denotes better correlation between the observed and predicted values. In this study, the values of R^2 (0.9957), Pred R^2 (0.9774), and Adj R^2 (0.9913) indicate a good correlation between the experimental and predicted values, which shows that the model was significant. In addition, "Adeq Precision" (a measure of the signal-to-noise ratio) of 56.589 indicates an adequate signal. It can be concluded that the model was statistically credible and reliable.

By using multiple regression analysis, the correlation between the response and the tested independent variables was established by Equation (1), and the X_i demonstrates the coded variables in the formula.

$$Y = 19.14 + 0.86\ X_1 + 0.054\ X_2 - 0.020\ X_3 + 0.94\ X_4 - 0.083\ X_1\ X_2 - 0.13\ X_1\ X_3 - 0.16\ X_1\ X_4 - 0.075\ X_2\ X_3 + 0.065\ X_2\ X_4 - 1.81\ X_1^2 - 0.12\ X_2^2 - 0.31\ X_3^2 - 1.07\ X_4^2 \quad (1)$$

2.1.2. Model Validation

The contour plot and the three-dimensional (3D) surface plot response surfaces described by the regression model are represented in Figure 1, and the maximum yield total triterpenoid was recorded under follow conditions: temperature of 55.14 °C, ethanol concentration of 86.57%, time of 34.41 min, and liquid-to-solid ratio of 39.33 mL/g. In order to check if the model was valid, extraction was carried out in triplicate under the optimal conditions. Further, the measured values (19.21 ± 0.25 mg/g) were in the range of the 95% confidence interval (95% CI) of the predicted value (19.44 mg/g), which verifies the predictability of the proposed model.

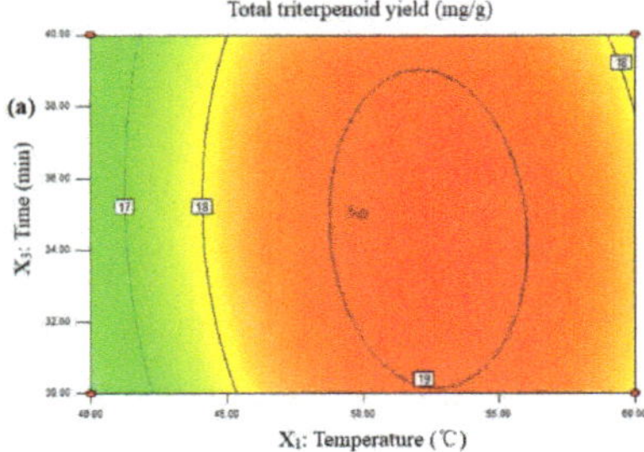

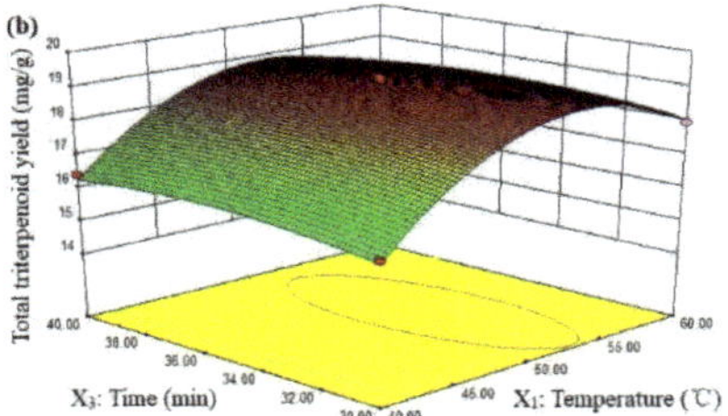

Figure 1. *Cont.*

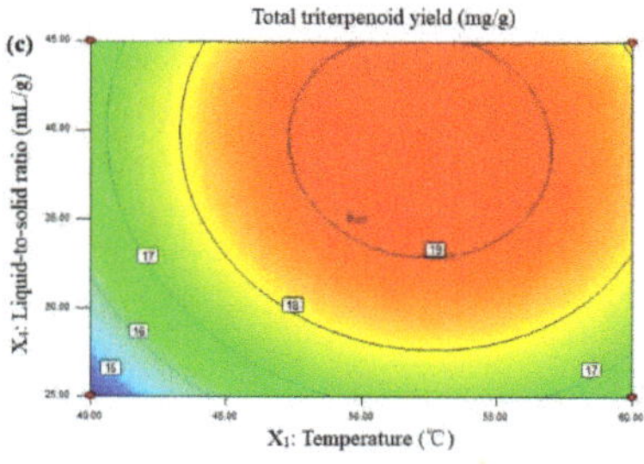

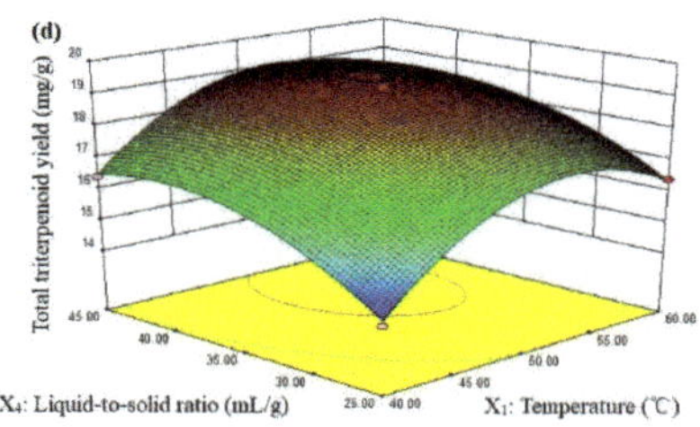

Figure 1. Contour plot (**a**,**c**) and three-dimensional (3D) surface plot (**b**,**d**) showing the interaction effects of the process variables on the total triterpenoid yield. (**a**,**b**): the interaction between temperature (X_1) and time (X_3) on total triterpenoid yield (Y); (**c**,**d**): the interaction between temperature (X_1) and liquid-to-solid ratio (X_4) on total triterpenoid yield (Y).

2.2. Triterpenic Acid Contents in the 99 Jujube Samples

The triterpenic acids extracted at optimal conditions from 99 cultivars of jujube samples were analyzed by ultra-performance liquid chromatography-mass spectrometry (UPLC–MS). The typical chromatograms of the 99 jujube samples are shown in Figure 2.

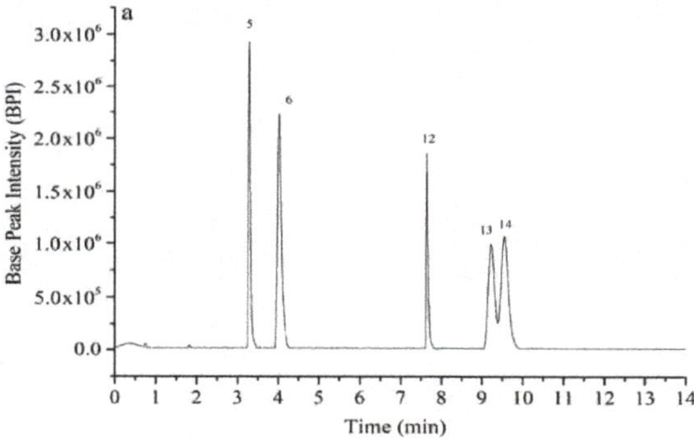

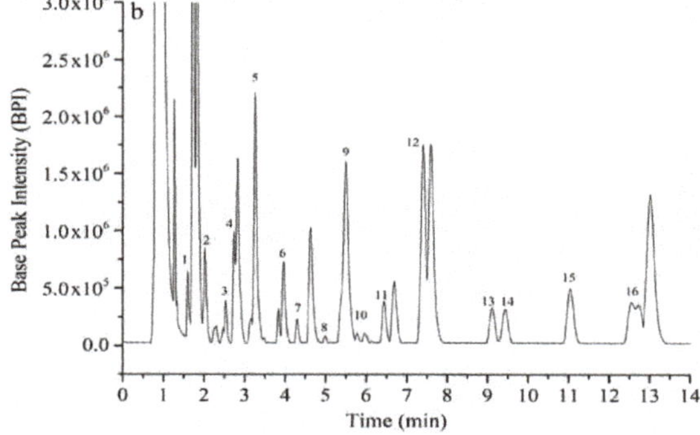

Figure 2. Ultra-performance liquid chromatography (UPLC) chromatograms of mixed standards (**a**) and sample (**b**). 1: Maslinic acid isomer-1 (Ma1); 2: Maslinic acid isomer-2 (Ma2); 3: Maslinic acid isomer-3 (Ma3); 4: Maslinic acid isomer-4 (Ma4); 5: Alphitolic acid (Aa); 6: Maslinic acid (Ma); 7: 2α-hydroxy ursolic acid (2αHa); 8: Maslinic acid isomer-5 (Ma5); 9: Oleanolic acid isomer-1 (Oa1); 10: Maslinic acid isomer-6 (Ma6); 11: Maslinic acid isomer-7 (Ma7); 12: Betulinic acid (Ba); 13: Oleanolic acid (Oa); 14: Ursolic acid (Ua); 15: Betulonic acid (Ba′); 16: Oleanonic acid + Ursonic acid (Oa′ + Ua′).

Sixteen peaks were observed on the chromatogram, including alphitolic acid, maslinic acid, 2α-hydroxy ursolic acid, betulinic acid, oleanolic acid, ursolic acid, betulonic acid, oleanonic acid, and ursonic acid, which were then identified and quantified. The other peaks were preliminarily identified as maslinic acid isomers and one ursolic acid isomer (Table 3). The quantitative results are shown in Figure 3 (detailed data are shown in Table S2).

Table 3. The retention time, mass spectrum (MS) parameters, and regression equations of standards and isomers.

Peak No.	Retention Time (min)	Compound	$[M + H]^-$ (m/z)	Regression Equation	R^2
1	1.72	Maslinic acid isomer-1	471.34	—	—
2	2.16	Maslinic acid isomer-2	471.34	—	—
3	2.56	Maslinic acid isomer-3	471.34	—	—
4	2.76	Maslinic acid isomer-4	471.34	—	—
5	3.29	Alphitolic acid	471.34	y = 138799x + 1595.7	0.9994
6	3.99	Maslinic acid	471.34	y = 140553x + 1065.1	0.9996
7	4.34	2α-hydroxy ursolic acid	471.34	—	—
8	5.04	Maslinic acid isomer-5	471.34	—	—
9	5.41	Oleanolic acid isomer-1	455.35	—	—
10	5.72	Maslinic acid isomer-6	471.34	—	—
11	6.3	Maslinic acid isomer-7	471.34	—	—
12	7.64	Betulinic acid	455.35	y = 125572x + 1121.5	0.9991
13	9.17	Oleanolic acid	455.35	y = 90033x - 1164.7	0.9992
14	9.49	Ursolic acid	455.35	y = 113372x + 1035.8	0.9997
15	11.09	Betulonic acid	455.35	—	—
16	12.59	Oleanonic acid + Ursonic acid	455.35	—	—

The triterpenes, secondary metabolites of plants, are distributed in several peels, leaves, stems and barks of plants, such as birch bark, olive leaves, mistletoe sprouts, clove flower, apple pomace, *Camellia sinensis*, etc. [26]. However, jujube is one of the few fruits with a high content of triterpenes. In this study, a significant difference in the total triterpenic acid content was observed among the jujube cultivars. The total triterpenic acid content ranged from 1082.775 to 7915.451 μg/g dry weight (DW), with a mean value of 3730.970 μg/g DW. Meanwhile, cultivar Jing39 (C41) had the highest content and Wanshuyuanling (C9) the lowest content of triterpenic acid. These results are consistent with previous results (166–6126 μg/g DW) [15].

Until now, more than 15 triterpenoid acids were found in the fruit of jujube [18]. A previous study reported that the identified triterpenic acids, including alphitolic acid, ceanothic acid, maslinic acid, 2a-hydroxyursolic acid, betulinic acid, ursolic acid, betulonic acid, oleanonic acid, and ursonic acid, showed large variations at different stages of growth [15]. In the present study, betulinic acid (516.409–4097.962 μg/g DW), alphitolic acid (198.195–3282.203 μg/g DW), maslinic acid (13.905–751.855 μg/g DW), oleanolic acid (36.696–837.463 μg/g DW), and ursolic acid (5.267–685.325 μg/g DW) were the dominant triterpenoid acids in jujube. Other triterpenoid acids, such as betulonic acid (9.417–304.731 μg/g DW), 2α-hydroxy ursolic acid (0.005–438.165 μg/g DW), and oleanonic acid + ursonic acid (9.834–244.797 μg/g DW), were relatively low. Notably, preliminary findings revealed seven isomers of maslinic acid; however, the detailed structure and chemical formula need to be investigated.

Some by-products of plant raw materials are good sources of triterpenoid acids. For example, olive pomace, a valuable by-product, contains maslinic acid and oleanolic acid in the proportions of 381.200 mg/g and 29.800 mg/g, respectively [20]. The outer bark of birch contains 11.600 mg/g of triterpenoid acids (betulinic + oleanolic) [27]. The extraction of ursolic, oleanolic, and rosmarinic acids in rosemary leaves reached a maximum of 15.800, 12.200, and 15.400 mg/g, respectively [12]. The ursolic acid content in other plant raw materials are as follows: *Calendula officinalis* flowers (20.530 mg/g DW), *Lamii albi flos* (110.400 mg/g DW), *Malus domestica* fruit peel (14.300 mg/g DW), and *Silphium* sp. flowers (17.950–22.050 mg/g DW) [19].

The 2,2-azinobis (3-ethylbenzothiazoline-6-sulphonic acid) (ABTS) and ferric reducing antioxidant power kit (FRAP) assays were carried out to differentiate the antioxidant properties of extracts from different jujube (Figure 4). The ABTS assay is based on hydrogen-donating antioxidants against nitrogen radicals, while the FRAP assay reflects the ferric ion-reducing antioxidant power of antioxidant. These methods have been widely used to evaluate the antioxidant capacity of food extracts [28]. In the present study, the ABTS and FRAP values ranged from 0.753 to 5.421 mM TE/100 g (C24, *Junzao*) and 0.968 to 5.529 mM FE/100 g (C86, *Huizaobianzhongyihao*), respectively.

In order to explore the effect of the antioxidant capacity in jujube, correlations among the triterpenic acids and the antioxidant activities were also analyzed (Table 4). As documented in Table 4, a significant positive correlation was observed between $ABTS^+$ radical scavenging activity and the contents of alphitolic acid, maslinic acid, betulinic acid, ursolic acid, betulonic acid, and total triterpenic acids ($p < 0.05$). Meanwhile, alphitolic acid, maslinic acid, betulinic acid, oleanolic acid, ursolic acid, betulonic acid, and total triterpenic acids also showed a positive correlation with the FRAP value ($p < 0.05$).

Obviously, the major triterpenic acids that widely exist in jujube are one of the main antioxidants with various important physiological and pharmacological properties. Previous researchers have proved that pentacyclic triterpenes, such as maslinic acid, alphitolic acid, maslinic acid, oleanolic acid, ursolic acid, glycyrrhetinic acid, betulinic acid, and lupeol, contribute various important physiological and pharmacological properties [29]. For example, ursolic acid and its isomer, oleanolic acid, have been reported have many beneficial effects, such as antioxidative, antimicrobial, anti-inflammatory, anticancer, anti-hyperlipidemic, analgesic, hepatoprotectory, gastroprotective, anti-ulcer, anti-HIV, cardiovascular, antiatherosclerotic, and immunomodulatory effects [19]. Betulinic acid has been reported to have anti-inflammatory, anti-cancer, anti-leukemia, anti-viral, and antihelmintic activities [30]. Due to its selective cytotoxicity against tumor cells and favorable therapeutic index, betulinic acid is considered a promising chemotherapeutic agent against HIV infection and cancers [31]. Maslinic acid has been shown to have antioxidant, anti-inflammatory, antimalarial, and antiprotozoal activities [29]. Therefore, the results of this study provide important guidance for health product development based on jujube.

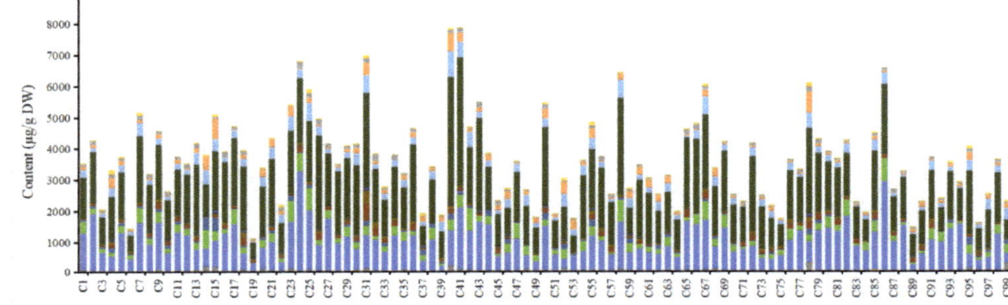

Figure 3. Contents (μg/g dry weight (DW)) of triterpenic acids in different jujube samples.

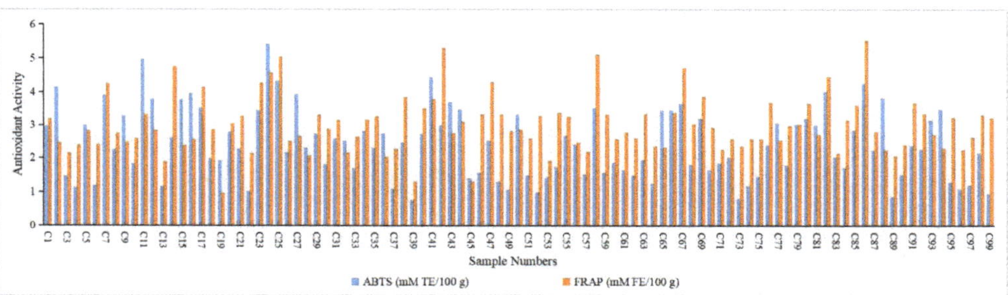

Figure 4. The antioxidant activities of the extracts of different jujube samples. ABTS = 2,2-azinobis (3-ethylbenzothiazoline-6-sulphonic acid) FRAP = ferric reducing antioxidant power kit.

Table 4. Correlation coefficients (r) of the studied triterpenic acids and the antioxidant activity of jujube cultivars.

	Ma1	Ma 2	Ma3	Ma4	Aa	Ma	2αHa	Ma5	Oa1	Ma6	Ma7	Ba	Oa	Ua	Ba′	Oa′ + Ua′	Total	ABTS	DPPH
ABTS	0.1274	0.1943	−0.0398	0.2576 [a]	0.9285 [b]	0.5205 [b]	0.0146	0.1462	−0.0792	−0.0141	−0.051	0.4838 [b]	0.1736	0.3378 [b]	0.464 [b]	−0.009	0.694 [b]	-	0.5205 [b]
FRAP	0.1174	0.0713	0.2629	0.1938	0.5549 [b]	0.9475 [b]	0.1955	−0.1852	0.0911	−0.013	−0.0387	0.3508 [b]	0.6123 [b]	0.2048 [a]	0.0111	0.1888	0.583 [b]	0.4993 [b]	-

[a] Significant at $p < 0.05$. [b] Significant at $p < 0.01$.

2.3. HCA and PCA

HCA and PCA are effective tools for multivariate analysis, which can be used to explore the existing differences among groups. HCA indicates the similarity among different cultivars, while PCA indicates the significant differences among the cultivars, thus reducing the dimensionality and increasing the interpretability of large datasets [24]. In this study, HCA and PCA were carried out based on the triterpenic acid content and antioxidant activities.

As shown in Figure 5, the 99 cultivars were divided into five clusters. The mean values of the detected compounds and antioxidant activities in each cluster are listed in Table 5.

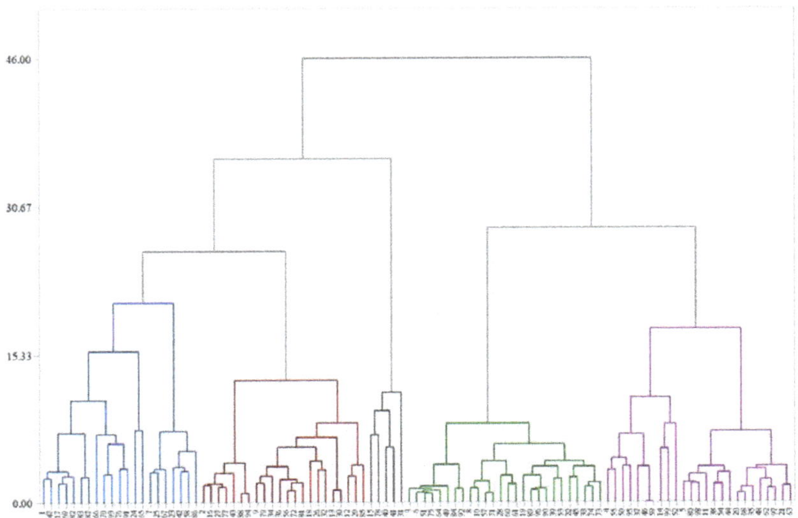

Figure 5. Hierarchical cluster analysis (HCA) of 99 cultivars of jujube samples. Cultivar lines with the same color are in the same cluster.

Table 5. The mean values of the detected compounds in different clusters.

Variables	Cluster 1	Cluster 2	Cluster 3	Cluster 4	Cluster 5
Maslinic acid isomer-1	**0.046** [b]	0.022	0.040	**0.011** [a]	0.020
Maslinic acid isomer-2	0.355	1.706	**7.917** [b]	0.332	**0.266** [a]
Maslinic acid isomer-3	25.352	17.883	**161.294** [b]	**6.081** [a]	65.097
Maslinic acid isomer-4	**11.305** [b]	2.839	2.802	0.655	**0.316** [a]
Alphitolic acid	**1535.713** [b]	1245.100	1189.970	**555.637** [a]	826.015
Maslinic acid	**449.873** [b]	203.267	328.307	**136.641** [a]	287.112
2α-hydroxy ursolic acid	40.014	29.682	**216.845** [b]	**29.278** [a]	127.938
Maslinic acid isomer-5	82.301	149.580	**216.391** [b]	79.385	**49.260** [a]
Oleanolic acid isomer-1	**12.814** [b]	8.929	7.839	**7.487** [a]	8.074
Maslinic acid isomer-6	49.880	74.380	**160.215** [b]	51.110	38.100 [a]
Maslinic acid isomer-7	**2.833** [a]	9.938	**70.101** [b]	3.801	16.411
Betulinic acid	1827.349	1909.479	**3158.536** [b]	**1043.670** [a]	1411.514
Oleanolic acid	317.869	217.850	**568.121** [b]	**159.657** [a]	308.816
Ursolic acid	**92.502** [a]	**92.296** [a]	**530.525** [b]	112.549	216.542
Betulonic acid	78.186	**109.488** [b]	107.236	48.038	**42.063** [a]
Oleanonic acid + Ursonic acid	40.246	33.499	**86.389** [b]	**26.385** [a]	50.759
Total	4572.323	4104.121	**6814.528** [b]	**2258.410** [a]	3448.962
ABTS	3.346	2.938	**3.066** [b]	**1.488** [a]	2.132
FRAP	**3.931** [b]	2.645	3.172	**2.336** [a]	3.208

Extreme values are in bold; [a] the element with the lowest mean value among the five clusters; [b] the highest mean value.

Cluster 1 contained 21 samples, which were generally clustered together according to the higher values of alphitolic acid (mean of 1535.713 μg/g DW), maslinic acid (mean of 449.873 μg/g DW), FRAP (mean of 3.931 mM TE/100 g), and lower content of ursolic acid (mean of 92.502 μg/g DW). Cluster 2 consisted of 22 samples, which were the cultivars with the highest content of betulonic acid (mean of 109.488 μg/g DW) and relatively low levels of ursolic acid (mean of 92.296 μg/g DW), respectively. It is noteworthy that cluster 3 consisted of five samples (C15, C31, C40, C41, and C78), in which the mean contents of 2α-hydroxy, betulinic acid, oleanolic acid, ursolic acid, and total triterpenic acids were the highest among those five clusters. The mean values were 216.845, 3158.536, 568.121, 530.525, and 6814.528 μg/g DW, respectively. Accordingly, these samples also had relatively high level of ABTS and FRAP, with mean values of 3.066 and 3.172 mM TE/100 g, respectively. On the contrary, cluster 3 contained 26 samples, which had lower concentrations of most of the major triterpenic acids. For example, the mean levels of alphitolic acid, maslinic acid, 2α-hydroxy ursolic acid, betulinic acid, oleanolic acid, and total triterpenic acids, as well as the antioxidant activities (ABTS and FRAP assay), were the lowest in the five clusters. Meanwhile, the 25 samples in cluster 5 also had relatively lower contents of most compounds.

According to the above results, all of the studied variables might contribute to sample classification. Typically, clusters 3 represented the groups with higher contents of triterpenic acids and higher antioxidant activity, while cluster 4 was indicative of the groups with lower levels.

PCA was carried out to analyze the differences among the 99 cultivars of jujube. The extraction sums of squared loadings are listed in Table 6.

Table 6. Total variance explained by principal component analysis (PCA).

Component	Eigenvalue	Percentage of Variance (%)	Cumulative (%)
PC1	6.04	31.79	31.79
PC2	3.89	20.46	52.25
PC3	2.70	14.23	66.49
PC4	1.34	7.04	73.53

The ellipses of the constant distance of the PCA method were calculated with a 95% confidence interval. Four principal components (PCs) were extracted, and the accumulative contribution rate of the four principal components was 73.53%. PC1, PC2, and PC3 explained 31.79%, 20.46%, and 14.23% of the total variance, together accounting for 66.49% of the total variance (Table 6). From the component matrix of the four principal components (Table S3), it can be inferred that the weight occupied by different triterpenic acids showed significant differences among the different main components. Overall, almost all of the compounds may contribute to the classification of the samples. A reduction in date dimension was successfully achieved. These four principal components, to a large extent, are indicative of the original 18 variables.

In addition, the scatter plot produced by PCA is very important and powerful, since it displays all samples in two- or three-dimensional graphs, and comparisons can be carried out among samples on the basis of the response variables applied in the study [28].

The interrelationship between different cultivars is clearly shown in the 3D score plot of PCA (Figure 6a). A significance of differences between groups can be observed in Figure 6b, which shows the profiles of the 95% confidence ellipse for different groups. If there is no intersection between two ellipses, it means that those two groups have significant differences [24]. In this case, since groups 1, 2, and 5 had an intersection between them, they could not be completely distinguished from each other. In other words, because all of the samples were collected from the same origin with similar cultivation conditions, the variability between the samples was not sufficient enough to classify all of the groups accurately. However, groups 3 and 4 were completely separated from each other, which indicates that there is a significant difference between these two groups. This result is consistent with

HCA, which revealed that groups 3 and 4 were the cultivars with highest and lowest triterpenic acids contents, respectively.

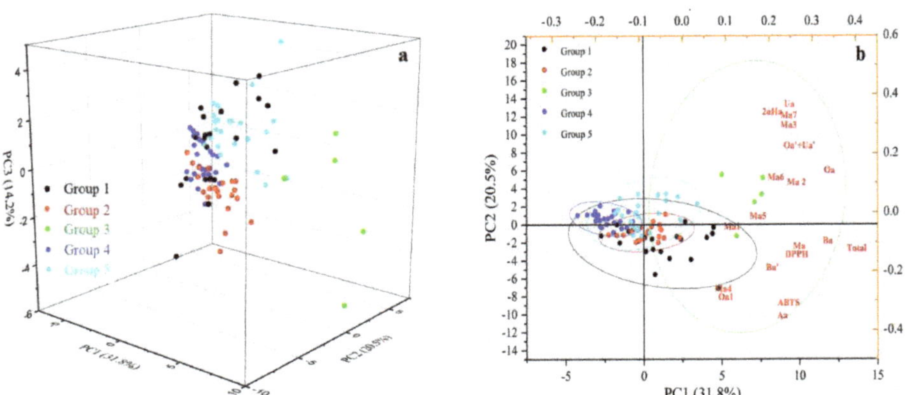

Figure 6. The 3D plots (**a**) and biplots (**b**) of PCA. Groups 1, 2, 3, 4, and 5 are the classified clusters of jujube by HCA. The ellipses with different colors represent the 95% confidence ellipse for different clusters.

Similar results have also been reported by previous researchers when PCA was used to distinguish between different clusters of jujube [24], jujube leaves [32], and finger millets [28]. The results provide important support for cultivation and breeding, quality evaluation, and product development of jujube. Furthermore, research on the main mechanism behind these differences of the different cultivars is urgently needed by molecular biological techniques, such as genomics and enzymology.

3. Materials and Methods

3.1. Plant Materials

Jujube of 99 cultivars (red maturity stage) (Table S1) were picked from the Germplasm Resources Base of Tarim University at Alaer City of Xinjiang Province, China, by the end of October 2019. Fruits without disease and mechanical injury and uniformly shaped were collected randomly from each side of the trees. After harvesting, all samples were lyophilized and then ground to fine powders and stored below −18 °C before analysis.

3.2. Chemicals

The following standards were purchased from ANPEL Co., Ltd (Shanghai, China): alphitolic acid (SPR01052), maslinic acid (Lot 67050010), betulinic acid (B330270), oleanolic acid (Lot Y4430050), and ursolic acid (Lot 40920050). Chemicals, such as acetonitrile, methanol, and ammonium formate, were all of HPLC grade and were purchased from Merck (Darmstadt, Germany). A 2,2-azinobis (3-ethylbenzothiazoline-6-sulphonic acid) (ABTS) kit (Art. No. A015-2) and a ferric reducing antioxidant power kit (FRAP, Art. No. A015-3) were provided by Jiancheng Biology Engineering Institute (Nanjing, Jiangsu, China). Other reagent solutions were of analytical grade (Solarbio Life Sciences Co., Ltd., Beijing, China).

3.3. Determination of Total Triterpenoid Content (TTC)

The TTC was measured using the vanillin–perchloric acid assay method [5]. The results were expressed as oleanolic acid equivalents (OAE, mg/g DW) through the standard calibration curve ($y = 16.005x - 0.0256$, $R^2 = 0.9987$). The total triterpene yield was measured using the following equation:

$$\text{Total triterpene yield (mg/g)} = \frac{\text{the mass of extracted triterpenes(mg)}}{\text{the mass of dried sample (g)}}. \quad (2)$$

3.4. Analysis of Triterpenic Acids by UPLC–MS

The extracts (extracted at optimum conditions) from the 99 cultivars of jujube were analyzed using a Waters ACQUITY UPLC H-CLASS system coupled with a Waters Xevo G2-XS QTof (Waters, Milford, MA) [15]. A Waters BEH C18 column (100 × 2.1 mm, 1.7 μm) operated at 30 °C was used. The injection volume was 2.0 μL, and the flow rate was 0.3 mL/min. The mobile phase was composed of A (3 mmol of ammonium formate) and B (methanol mixed with equal volume of acetonitrile) with a gradient elution of 0–2 min, 24–23% A; 2–18 min, 23% A. The parameters for the MS were set as follows: capillary voltage of 3.0 kV, source temperature of 110 °C, desolvation temperature of 450 °C, cone gas flow rate of 50 L/h, and desolvation gas flow rate of 800 L/h. Mass spectra in negative ion modes were recorded within the range of 100–1000 *m/z*. Concentrations of the compounds were calculated using the peak areas of the sample and the corresponding standards.

3.5. Analysis of Antioxidant Activities

The antioxidant activities were analyzed according to the methods reported by our lab [33]. Briefly, the samples were extracted according to the optimized extraction process. After centrifugation and lyophilization, the extract was then diluted to 10 mL for further antioxidant activity analysis. The antioxidant activities were evaluated using an ABTS kit and a FRAP kit, and the operational steps were performed in compliance with the instructions of the kits. The ABTS radical scavenging activity was expressed as millimoles of Trolox equivalent per 100 g (mM TE/100 g) of dry sample. The FRAP results were expressed as millimoles of ferrous sulfate equivalent per 100 g (mM FE/100 g) of dry sample.

3.6. UAE Procedures

Two grams (2.0 g) of dried jujube powder (C24, Junzao) was placed in an ultrasonic extractor (XY–2008; Xiyu Instruments Co., Ltd., Shanghai, China) at different influencing factors, including temperature (30–70 °C), ethanol concentration (55–95%), time (20–40 min), and liquid-to-solid ratio (15:1–55:1 mL/min). After extraction, the extracts were centrifuged at the speed of 5000 rpm for 5 min (Centrifuge 5804; Eppendorf AG, Germany). Then, the supernatants were evaporated, lyophilized, and stored below −18 °C until further analysis. Based on the results of single-factor experiments, the factors that have a major influence and the levels of those influences were determined and applied in the RSM design.

3.7. RSM Experimental Design

A four-factor three-level experimental RSM was employed to determine the optimal conditions. The yield of total triterpenoids (y, mg/g) was regarded as a dependent variable. Furthermore, the four-factor ranges were determined according to the previous single-factor experiments (data not shown). Table 7 shows the experimental design.

Table 7. Independent variable codes and levels in experimental design.

Code	X_1: Temperature (°C)	X_2: Ethanol Concentration (%)	X_3: Time (min)	X_4: Liquid-to-solid Ratio (mL/g)
−1	40	80	30	25:1
0	50	85	35	35:1
+1	60	90	40	45:1

3.8. Statistical Analysis

Design Expert 8.0.5 (Stat-Ease Inc., Minneapolis, MN, USA) was used to analyze RSM optimization and regression. The values of the dependent parameters obtained from the experiments were fitted to the second-order polynomial model as shown below:

$$Y = \beta_0 + \sum_{i=1}^{k} \beta_i X_i + \sum_{i=1}^{k} \beta_{ii} X_{\mathrm{i}}^2 + \sum_{i=1}^{k-1} \sum_{j>1}^{k} \beta_{ij} X_i X_j. \quad (3)$$

Y stands for the estimated response, X_j and X_i represent the independent variables, while k suggests variable number. Meanwhile, β_i, β_0, β_{ij}, and β_{ii} represent the regression coefficients of the linear, intercept, interaction, and quadratic terms, respectively.

All data were collected in triplicate and expressed as mean ± SD. The one-way ANOVA analysis, HCA, and PCA were carried out by SPSS 22.0 software (SPSS, Chicago, IL, USA).

4. Conclusions

In this study, the ultrasound-assisted extraction of total triterpenoids from jujube was optimized by RSM. The optimal conditions obtained were as follows: temperature of 55.14 °C, ethanol concentration of 86.57%, time of 34.41 min, and liquid-to-solid ratio of 39.33 mL/g. The triterpenoid yield was 19.21 ± 0.25 mg/g under the optimal conditions.

The triterpenoid acid profile of the extracts obtained from 99 cultivars of jujube were further analyzed by UPLC–MS. Betulinic acid (mean of 1602.008 μg/g DW), alphitolic acid (mean of 1017.060 μg/g DW), maslinic acid (mean of 265.568 μg/g DW), oleanolic acid (mean of 264.445 μg/g DW), ursolic acid (mean of 151.166 μg/g DW), betulonic acid (mean of 69.570 μg/g DW), and 2α-hydroxy ursolic acid (mean of 66.032 μg/g DW) were found to be the main triterpenoid acids in jujube of different cultivars. According to HCA and PCA, the 99 cultivars were categorized into five clusters, among which cluster 3 had relatively higher contents of most triterpenoid acids.

These results indicate that jujube is a potential natural source of triterpenic acids for the development of functional foods, and the differences in the compositional profile of cultivars may lead to their different applications. UAE is an efficient method to extract triterpenoids from jujube, and RSM is a useful method to optimize the UAE parameters of triterpenoid compounds from jujube. This study would be further contributable for the deep processing and utilization of jujube.

Supplementary Materials: The following are available online at http://www.mdpi.com/2223-7747/9/4/412/s1. Table S1. Summary of the information of jujube samples; Table S2. Contents (μg/g DW) of triterpenic acids in jujube fruit samples; Table S3. Component matrix of principal component analysis.

Author Contributions: Conceptualization, L.S. and L.Z.; methodology, Y.W.; software, L.X.; validation, Y.M.; formal analysis, W.L.; investigation, L.S.; resources, L.S. and L.Z.; data curation, Y.L. and Y.W.; writing—original draft preparation, L.S.; writing—review and editing, L.Z.; supervision, Y.W. All authors read and agreed to the published version of the manuscript.

Funding: This work was funded by the National Natural Science Foundation of China (31560462), the Beijing Advanced Innovation Center for Food Nutrition and Human Health, Beijing Technology and Business University (BTBU) (20171049), and the Projects of Innovation and Development Pillar Program for Key Industries in Southern Xinjiang of Xinjiang Production and Construction Corps (2018DB002).

Acknowledgments: The authors sincerely thank Minjuan Lin for her assistance during the sample collection.

Conflicts of Interest: The authors declare no conflicts of interest.

References

1. Wang, R.; Ding, S.; Zhao, D.; Wang, Z.; Wu, J.; Hu, X. Effect of dehydration methods on antioxidant activities, phenolic contents, cyclic nucleotides, and volatiles of jujube fruits. *Food Sci. Biotechnol.* **2016**, *25*, 137–143. [CrossRef] [PubMed]
2. Wang, Y.K.; Li, D.K.; Sui, C.L.; Zhao, A.L.; Du, X.M. Conservation, characterization, evaluation and utilization of Chinese jujube germplasm resources. *Int. Jujube Symp.* **2008**, *840*, 235–240. [CrossRef]
3. Wang, J.; Nakano, K.; Ohashi, S.; Kubota, Y.; Takizawa, K.; Sasaki, Y. Detection of external insect infestations in jujube fruit using hyperspectral reflectance imaging. *Biosyst. Eng.* **2011**, *108*, 345–351. [CrossRef]
4. Choi, S.H.; Ahn, J.B.; Kim, H.J.; Im, N.K.; Kozukue, N.; Levin, C.E.; Friedman, M. Changes in free amino acid, protein, and flavonoid content in jujube (*Ziziphus* jujube) fruit during eight stages of growth and antioxidative and cancer cell inhibitory effects by extracts. *J. Agric. Food Chem.* **2012**, *60*, 10245–10255. [CrossRef]
5. Kou, X.H.; Chen, Q.; Li, X.H.; Li, M.F.; Kan, C.; Chen, B.R.; Zhang, Y.; Xue, Z.H. Quantitative assessment of bioactive compounds and the antioxidant activity of 15 jujube cultivars. *Food Chem.* **2015**, *173*, 1037–1044. [CrossRef]
6. San, B.; Yildirim, A.N. Phenolic, alpha-tocopherol, beta-carotene and fatty acid composition of four promising jujube (*Ziziphus jujuba* Miller) selections. *J. Food Compos. Anal.* **2010**, *23*, 706–710. [CrossRef]
7. Siddique, H.R.; Saleem, M. Beneficial health effects of lupeol triterpene: A review of preclinical studies. *Life Sci.* **2011**, *88*, 285–293. [CrossRef]
8. Fujiwara, Y.; Hayashida, A.; Tsurushima, K.; Nagai, R.; Yoshitomi, M.; Daiguji, N.; Sakashita, N.; Takeya, M.; Tsukamoto, S.; Ikeda, T. Triterpenoids isolated from *Zizyphus* jujuba inhibit foam cell formation in macrophages. *J. Agric. Food Chem.* **2011**, *59*, 4544–4552. [CrossRef]
9. Miao, L.J.; Liu, M.J.; Liu, X.G.; Geng, J.N.; Wang, J.; Ning, Q. Study on the extraction of triterpenoids from jujube. *J. Agric. Univ. Hebei* **2008**, *31*, 68–75.
10. Bernatoniene, J.; Cizauskaite, U.; Ivanauskas, L.; Jakstas, V.; Kalveniene, Z.; Kopustinskiene, D.M. Novel approaches to optimize extraction processes of ursolic, oleanolic and rosmarinic acids from *Rosmarinus officinalis* leaves. *Ind. Crop. Prod.* **2016**, *84*, 72–79. [CrossRef]
11. Guo, S.; Duan, J.A.; Tang, Y.P.; Yang, N.Y.; Qian, D.W.; Su, S.L.; Shang, E.X. Characterization of triterpenic acids in fruits of *ziziphus* species by HPLC-ELSD-MS. *J. Agric. Food Chem.* **2010**, *58*, 6285–6289. [CrossRef] [PubMed]
12. Junhai, L.; Hongguang, G.E.; Zhizhou, L.; Feng, N. Methods for Extracting Ursolic Acid from Red Jujube. China Patent CN 102,321,144, 18 January 2012.
13. Guo, S.; Duan, J.A.; Qian, D.W.; Tang, Y.P.; Wu, D.W.; Su, S.L.; Wang, H.Q.; Zhao, Y.N. Content variations of triterpenic acid, nucleoside, nucleobase, and sugar in jujube (*Ziziphus* jujuba) fruit during ripening. *Food Chem.* **2015**, *167*, 468–474. [CrossRef] [PubMed]
14. Gao, Q.H.; Wu, C.S.; Wang, M. The jujube (*Ziziphus jujuba* mill.) fruit: A review of current knowledge of fruit composition and health benefits. *J. Agric. Food Chem.* **2013**, *61*, 3351–3363. [CrossRef] [PubMed]
15. Masullo, M.; Montoro, P.; Autore, G.; Marzocco, S.; Pizza, C.; Piacente, S. Quali-quantitative determination of triterpenic acids of *Ziziphus* jujuba fruits and evaluation of their capability to interfere in macrophages activation inhibiting NO release and iNOS expression. *Food Res. Int.* **2015**, *77*, 109–117. [CrossRef]
16. Sun, Y.F.; Song, C.K.; Viernstein, H.; Unger, F.; Liang, Z.S. Apoptosis of human breast cancer cells induced by microencapsulated betulinic acid from sour jujube fruits through the mitochondria transduction pathway. *Food Chem.* **2013**, *138*, 1998–2007. [CrossRef]
17. Cao, Y.P.; Yang, X.L.; Xue, C.H. Study on extraction technology of oleanolic acid in *Zizyphus* jujuba date. *Food Sci.* **2007**, *10*, 163–167.
18. Zhang, H.Q.; Liu, P.; Duan, J.A.; Dong, L.; Shang, E.X.; Qian, D.W.; Xiao, P.; Zhao, M.; Li, W.W. Hierarchical extraction and simultaneous determination of flavones and triterpenes in different parts of *Trichosanthes kirilowii Maxim*. By ultra-high-performance liquid chromatography coupled with tandem mass spectrometry. *J. Pharm. Biomed.* **2019**, *167*, 114–122. [CrossRef]

19. López-Hortas, L.; Pérez-Larrán, P.; González-Muñoz, M.J.; Falqué, E.; Domínguez, H. Recent developments on the extraction and application of ursolic acid. A review. *Food Res. Int.* **2018**, *103*, 130–149. [CrossRef]
20. Xie, P.J.; Huang, L.X.; Zhang, C.H.; Deng, Y.J.; Wang, X.J.; Cheng, J. Enhanced extraction of hydroxytyrosol, maslinic acid and oleanolic acid from olive pomace: Process parameters, kinetics and thermodynamics, and greenness assessment. *Food Chem.* **2019**, *276*, 662–674. [CrossRef]
21. Wen, C.T.; Zhang, J.X.; Zhang, H.H.; Dzah, C.S.; Zandile, M.; Duan, Y.Q.; Ma, H.L.; Luo, X.P. Advances in ultrasound assisted extraction of bioactive compounds from cash crops—A review. *Ultrason. Sonochem.* **2018**, *48*, 538–549. [CrossRef]
22. Fu, Q.; Zhang, L.; Cheng, N.; Jia, M.; Zhang, Y. Extraction optimization of oleanolic and ursolic acids from pomegranate (*Punica granatum* L.) flowers. *Food Bioprod. Process.* **2014**, *92*, 321–327. [CrossRef]
23. Li, J.W.; Fan, L.P.; Ding, S.D.; Ding, X.L. Nutritional composition of five cultivars of chinese jujube. *Food Chem.* **2007**, *103*, 454–460. [CrossRef]
24. Wang, L.N.; Fu, H.Y.; Wang, W.Z.; Wang, Y.Q.; Zheng, F.P.; Ni, H.; Chen, F. Analysis of reducing sugars, organic acids and minerals in 15 cultivars of jujube (*Ziziphus jujuba* mill.) fruits in China. *J. Food Compos. Anal.* **2018**, *73*, 10–16. [CrossRef]
25. Gao, Q.H.; Wu, P.T.; Liu, J.R.; Wu, C.S.; Parry, J.W.; Wang, M. Physicochemical properties and antioxidant capacity of different jujube (*Ziziphus jujuba* Mill.) cultivars grown in loess plateau of China. *Sci. Hortic.* **2011**, *130*, 67–72. [CrossRef]
26. Jäger, S.; Trojan, H.; Kopp, T.; Laszczyk, M.; Scheffler, A. Pentacyclic triterpene distribution in various plants-rich sources for a new group of multi-potent plant extracts. *Molecules* **2009**, *14*, 2016–2031. [CrossRef]
27. Popov, S.A.; Sheremet, O.P.; Kornaukhova, L.M.; Grazhdannikov, A.E.; Shults, E.E. An approach to effective green extraction of triterpenoids from outer birch bark using ethyl acetate with extractant recycle. *Ind. Crop. Prod.* **2017**, *102*, 122–132. [CrossRef]
28. Xiang, J.L.; Li, W.H.; Ndolo, V.U.; Beta, T. A comparative study of the phenolic compounds and in vitro antioxidant capacity of finger millets from different growing regions in *Malawi*. *J. Cereal Sci.* **2019**, *87*, 143–149. [CrossRef]
29. Sharma, H.; Kumar, P.; Deshmukh, R.R.; Bishayee, A.; Kumar, S. Pentacyclic triterpenes: New tools to fight metabolic syndrome. *Phytomedicine* **2018**, *15*, 166–177. [CrossRef]
30. Chen, Q.H.; Liu, J.; Zhang, H.F.; He, G.Q.; Fu, M.L. The betulinic acid production from betulin through biotransformation by fungi. *Enzyme Microb. Technol.* **2009**, *45*, 175–180. [CrossRef]
31. Cichewicz, R.H.; Kouzi, S.A. Chemistry, biological activity, and chemotherapeutic potential of betulinic acid for the prevention and treatment of cancer and HIV infection. *Med. Res. Rev.* **2004**, *24*, 90–114. [CrossRef]
32. Song, L.J.; Zheng, J.; Zhang, L.; Yan, S.J.; Huang, W.J.; He, J.; Liu, P.Z. Phytochemical profiling and eingerprint analysis of Chinese iujube (*Ziziphus jujuba* Mill.) leaves of 66 cultivars from Xinjiang province. *Molecules* **2019**, *24*, 4528. [CrossRef] [PubMed]
33. Song, L.J.; Liu, P.Z.; Yan, Y.Z.; Huang, Y.; Bai, B.Y.; Hou, X.J.; Zhang, L. Supercritical CO_2 fluid extraction of flavonoid compounds from Xinjiang jujube (*Ziziphus jujuba* Mill.) leaves and associated biological activities and flavonoid compositions. *Ind. Crop. Prod.* **2019**, *139*, 111508. [CrossRef]

Article

Underutilized Mexican Plants: Screening of Antioxidant and Antiproliferative Properties of Mexican Cactus Fruit Juices

Elda M. Melchor Martínez, Luisaldo Sandate-Flores, José Rodríguez-Rodríguez, Magdalena Rostro-Alanis, Lizeth Parra-Arroyo, Marilena Antunes-Ricardo, Sergio O. Serna-Saldívar, Hafiz M. N. Iqbal * and Roberto Parra-Saldívar *

Tecnologico de Monterrey, School of Engineering and Sciences, Monterrey 64849, Mexico; elda.melchor@tec.mx (E.M.M.M.); a00812589@itesm.mx (L.S.-F.); jrr@tec.mx (J.R.-R.); magda.rostro@tec.mx (M.R.-A.); A00812589@exatec.tec.mx (L.P.-A.); marilena.antunes@tec.mx (M.A.-R.); sserna@tec.mx (S.O.S.-S.)

* Correspondence: hafiz.iqbal@tec.mx (H.M.N.I.); r.parra@tec.mx (R.P.-S.)

Abstract: Cacti fruits are known to possess antioxidant and antiproliferative activities among other health benefits. The following paper evaluated the antioxidant capacity and bioactivity of five clarified juices from different cacti fruits (*Stenocereus* spp., *Opuntia* spp. and *M. geomettizans*) on four cancer cell lines as well as one normal cell line. Their antioxidant compositions were measured by three different protocols. Their phenolic compositions were quantified through high performance liquid chromatography and the percentages of cell proliferation of fibroblasts as well as breast, prostate, colorectal, and liver cancer cell lines were evaluated though in vitro assays. The results were further processed by principal component analysis. The clarified juice from *M. geomettizans* fruit showed the highest concentration of total phenolic compounds and induced cell death in liver and colorectal cancer cells lines as well as fibroblasts. The clarified juice extracted from yellow *Opuntia ficus-indica* fruit displayed antioxidant activity as well as a selective cytotoxic effect on a liver cancer cell line with no toxic effect on fibroblasts. In conclusion, the work supplies evidence on the antioxidant and antiproliferative activities that cacti juices possess, presenting potential as cancer cell proliferation preventing agents.

Keywords: Underutilized Mexican plants; Cactus fruits; Antioxidant activities; Antiproliferative properties

Citation: Martínez, E.M.M.; Sandate-Flores, L.; Rodríguez-Rodríguez, J.; Rostro-Alanis, M.; Parra-Arroyo, L.; Antunes-Ricardo, M.; Serna-Saldívar, S.O.; Iqbal, H.M.N.; Parra-Saldívar, R. Underutilized Mexican Plants: Screening of Antioxidant and Antiproliferative Properties of Mexican Cactus Fruit Juices. *Plants* **2021**, *10*, 368. https://doi.org/10.3390/plants10020368

Academic Editor: Stefania Lamponi

Received: 23 January 2021
Accepted: 10 February 2021
Published: 14 February 2021

1. Introduction

According to a recent report by the World Health Organization (WHO), cancer is a collection of diseases that is the second leading cause of mortality worldwide. The condition starts in any organ/tissue of the body and spreads beyond its initial boundaries to invade adjoining tissues until eventually it reaches other organs and tissues. The uncontrolled stage of this collection of illnesses is called metastasis and is a major cause of death. Cancer disease cause of death globally accounted for an estimated 9.6 million deaths in 2018, which could have increased by 50% to 15 million by 2020 [1]. The tissues most commonly affected were lung, breast, colorectal, prostate, skin, and stomach. Liver cancer alone caused 782,000 deaths [1]. In order to decrease the mortality rate, efforts should focus on preventing or treating cancer at its early stages. Mexican plants have been proven to act as natural antioxidant products as well as providing anticancer activities [2]. Cactaceae plants are a group of uncommon species worldwide that grow in arid areas. Mexico is home of 518 species of which 47.7% are endemic [3]. Cacti fruits are highly valued in the region for their chemical composition which grants them their distinctive organoleptic properties such as color and taste [4]. Few studies detail the chemical characterization of cacti fruits, and their bioactive properties.

Fruits from the *Stenocereus* genus (pitayas) can be harvested semi-annually from April to May and September to October. They are ovoid in shape, pigmented, and possess small thorns [5] (Figure 1). Partial characterizations of pitaya fruits have shown that they contain betalains, flavonoids, and phenolic compounds (caffeic, ferulic, and *p*-coumaric) [3,6]. Their uses have been reported against insect bites, rheumatism, hemorrhages, and gastrointestinal issues [3,7].

Figure 1. Cactus fruits (**a**) *Stenocereus pruinosus* yellow fruit (SY), (**b**) *Stenocereus pruinosus* red fruit (SR), (**c**) *Opuntia ficus-indica* yellow fruit (OPY), (**d**) *Opuntia ficus-indica* red fruit (OPR), (**e**) *Myrtillocactusgeomettizans* fruit (MG).

Prickly pears are oval shaped fruits commonly referred to as "tunas" (Figure 1) and are extensively distributed throughout Mexico [8]. They have been known to contain flavonoids, phenolic acids (caffeic, coumaric, vanillic, among others), betalains, ascorbic acid, fatty acids, lignans, and sterols [3,6]. Additionally, they possess a variety of health benefits, yet not restricted to a high antioxidant capacity [9], including cytotoxic activity [10,11], and an antiproliferative effect on cancer colon (HT29/Caco-2) and prostate cancer cell lines (PC-3) in vitro [12,13].

Myrtillocactus geomettizan produces cacti fruits called garambullo, also known as the berry cactus, which ripens from May to July. Garambullo fruits are 1.5 cm in length, globular, and purple (Figure 1) [14]. The berry cactus fruits contain flavonoids, betalains, ascorbic acid, and many phenolic acids (caffeic, gallic, vanillic) [3,6] with associated health properties such as the improvement of renal functions in rats as well as decreased glucose and cholesterol levels in blood [14].

There are only a few characterization studies about the Cactaceae family and their fruits. The exploration of nature as a source of novel compounds to treat chronic diseases, such as cancer, is growing. Considering the above properties of underutilized Mexican plants, herein, we report the chemical characterization and compositional analyses of five clarified juices extracted from different cacti fruits (*Stenocereus* sp., *Opuntia* sp. and *M. geomettizans*). The scientific rational behind this study was to test and report the correlation between antioxidant capacity and antiproliferative properties of Mexican cactus fruit juices. Furthermore, we studied the in vitro bioactive potential of in-house extracted juices from different cacti fruits against cancer cell lines and a normal cell line, with the aim of proposing their medicinal valorization.

2. Results and Discussion

2.1. Total Soluble Solids, Betacyanin, Betaxanthin Content, and Antioxidant Activity

There are few reports quantifying the chemical content of the cactus fruits here mentioned in order to establish a comparison. The clarified juices obtained had low values around 0.2 °Brix, of soluble solid content (Table S1). The juice of Pitaya (*Stenocereus pruinosus*) reported 9.8 °Brix [15], a higher value of the total solids than in SY (0.2 °Brix) and SR (0.2 °Brix). One possible reason for the discrepancy between the results from this paper and those in the literature is a difference in the procedure used to obtain the clarified juices, as showed in the Supplementary material.

The content of betalains is shown in Table 1. The OPR clarified juice had the highest concentration of betacyanins of the clarified juices (403.56 ± 1.41 µg/g of FS-fresh sample). The analysis of the different *Opuntia* fruits showed similar betacyanin compositions to that of *Opuntia robusta*, that goes by the common name of Tapón, reported in the literature (300.5 ± 8.8 µg/g FS) [12]. Regarding the betaxanthin concentrations, the SR clarified juice had the highest value (404.59 ± 2.33 µg/g of FS) followed by the OPR juice (263.24 ± 36.36 µg/g of FS), and the *Opuntia rastrera* juice (86.2 ± 22.3 µg/g FS) [12]. Betalains are water soluble which is why a higher content was observed in the clarified juices than in the pulps. A similar phenomenon was observed in freeze-dried cherry [16]. The clarified juice of *Myrtillocactus geomettizan* had less betacyanins (103.50 ± 0.01 µg/g FS) and betaxanthins (45.76 ± 0.42 µg/g FS), and even betacyanins in the fruit pulp (36.9 ± 3.7 µg/g FS) [17]. This may be due to the stage of maturity and geographical origin of the cactus berries [18].

Table 1. Betacyanin and betaxanthin content and antioxidant activity of clarified juices.

Parameter	SY	SR	OPY	OPR	MG
Betacyanins (µg/g FS)	47.99 ± 0.18 ^{d}	334.06 ± 2.08 ^{b}	13.29 ± 0.06 ^{e}	403.56 ± 1.41 ^{a}	103.50 ± 0.01 ^{c}
Betaxanthins (µg/g FS)	240.52 ± 0.88 ^{b}	404.59 ± 2.33 ^{a}	59.28 ± 0.41 ^{c}	263.24 ± 36.36 ^{b}	45.76 ± 0.42 ^{c}
Total phenolic compounds (mg GA/100 mL FS)	57.16 ± 1.52 ^{e}	108.75 ± 0.35 ^{c}	79.73 ± 1.04 ^{d}	111.72 ± 0.35 ^{b}	138.38 ± 0.14 ^{a}
ABTS (µmol TE/100 g FS)	542.62 ± 7.20 ^{d}	994.40 ± 32.28 ^{c}	370.41 ± 10.69 ^{e}	1097.35 ± 20.20 ^{b}	3123.77 ± 26.15 ^{a}
DPPH (µmol TE/100 mL FS)	80.71 ± 6.65	854.60 ± 17.60 ^{b}	379.87 ± 70.86 ^{c}	1115.25 ± 86.46 ^{a}	329.24 ± 9.48 ^{c}
FRAP (µmol TE/100 mL FS)	865.20 ± 10.24 ^{c}	2744.48 ± 42.16 ^{a}	480.20 ± 7.80 ^{d}	2532.05 ± 48.63 ^{b}	ND

Values represented as mean ± standard deviation (n = 3), different lowercase letters (**a–d**) indicate statistical significance difference ($p < 0.05$), FS = fresh sample, GA gallic acid equivalents, TE Trolox equivalents. ND = not determined *Stenocereus pruinosus* yellow fruit (SY), *Stenocereus pruinosus* red fruit (SR), *Opuntia ficus-indica* yellow fruit (OPY), *Opuntia ficus-indica* red fruit (OPR), and *Myrtillocactus geomettizans* fruit (MG). 2,2′-Azino-bis-(3-ethylbenzothiazoline-6-sulfonic acid) (ABTS), α-α-Diphenyl-β-picrylhydrazyl (DPPH), Ferric Reducing Antioxidant Power (FRAP.

Using the Folin–Ciocalteu method, the MG clarified juice (138.38 ± 0.14 mg GA/100 mL of FS) had the highest antioxidant content followed by the OPR clarified juice (111.72 ± 0.36 mg GA/100 mL of FS). A similar behavior to that of the MG clarified juice has been reported in red fruits such as raspberry *Rubusidaeus L.* (148.9 ± 0.45 mg GA/100 mL FS) [19]. The above results could be explained by the resulting effect of a wide range of structures such as monophenols, catechols, pyrogallols, phloroglucinols, resorcinols, p-hydroquinols, naphthols, anthracenes, flavonoid aglycones, glycosides, hydroxycoumarins, aminophenols [20], and vitamins as dehydroascorbic acid [21] that are detected by this technology.

The ABTS method corroborated the antioxidant results from the Folin–Ciocalteu method; the MG clarified juice (3123.77 ± 26.15 µmol TE/100 g FS) had the highest antioxidant content followed by the OPR clarified juice (1097.35 ± 20.20 µmol TE/100 g FS). Red fruits reported higher antioxidant activity through the DPPH procedure; OPR clarified juice being the highest (1115.25 ± 56.46 µmol TE/100 g of F) followed by SR juice (854.601 ± 17.60 µmol TE/100 g of FS). A similar trend was observed in the FRAP results of red fruits. However, the SR clarified juice had the highest antioxidant effect (2744.48 ± 42.16 µmol TE/100 g of FS). Antioxidants assays are classified in electron transfer (ET) methods such as ABTS, FRAP, and DPPH and hydrogen atom transfer (HAT) like methods such as ORAC [22]. The cacti fruits complex matrix contains a variety of chemical compounds that cause their antioxidant properties. In order to obtain an accurate understanding of the antioxidant capacity of cacti fruits and to be able to compare results across different labs with varying conditions, multiple assays with distinct mechanisms must be run. Nevertheless, juices from passion fruit (176.42 ± 23.40 µmol TE/100 mL FS) and lemon juice (56.75 ± 26.63 µmol TE/100 mL FS) [23], demonstrated a lower antioxidant capacity than all clarified juices analyzed here.

Recent studies have found that the flowers also have phenolic compounds, such as *Artocarpus lakoocha Roxb* [24]. For this reason, it will be interesting to study the extract of flowers of *Stenocereus pruinosus*, *Opuntia ficus-indica*, and *Myrtillocactus geomettizans* in future works.

2.2. Phenolic Composition Analysis

Due to their availability and cytotoxic activity, MG and OPY were analyzed by HPLC in order to detect the content of *p*-coumaric acid, gallic acid, caffeic acid, and vanillic acid in each clarified juice, using each phenolic acid's respective standard. Figures S2–S5 displays the HPLC profile of the MG and OPY clarified juices. The results indicated a high content of *p*-coumaric acid (60.60 ± 0.25 mg/L of FS) in MG clarified juice. The OPY clarified juice had a high amount of gallic acid (21.75 ± 0.75 mg/L FS) as well as *p*-coumaric acid (16.85 ± 1.02 mg/L FS) (Table 2).

Table 2. Phenolic acids' composition of *Myrtillocactus geomettizans* fruit (MG) and *Opuntia ficus-indica* yellow fruit (OPY) detected by HPLC.

Parameter	MG Fruit mg/L FS	OPY Fruit mg/L FS
p-Coumaric acid	60.60 ± 0.25 [a]	16.85 ± 1.02 [b]
Gallic acid	14.95 ± 0.01 [a]	21.75 ± 0.75 [a]
Caffeic acid	1.90 ± 0.01 [b]	4.60 ± 0.01 [a]
Vanillic acid	10.05 ± 0.04 [a]	13.00 ± 0.45 [a]

Values represented as mean ± standard deviation (n = 3), different lowercase letters (**a**,**b**) indicate statistical significance difference ($p < 0.05$). FS = fresh sample.

The HPLC analysis of the phenolic acids previously mentioned in *Opuntia joconostle* reported significant amounts of protocatechuic and caffeic acid. Vanillic acid was not detected in the whole fruit [25]. Additionally, *p*-coumaric and caffeic acid derivatives were detected in *Opuntia ficus-indica* by chromatography coupled to high resolution time of flight mass spectrometry (UPLC-QTOF-MS) [26], as the present work demonstrated. In the literature, *Myrtillocactus* was reported to have a higher amount of gallic acid than caffeic acid [18], in agreement with the present work.

2.3. Cell Viability Assay

A fast screening of the cytotoxic assay was performed, using a single concentration of the clarified juices of 2% (v/v) in order to identify the species with the highest cytotoxicity on the cancer cell lines while affecting the normal cell line as minimally as possible. The results showed, HepG2 cells were more sensitive to the clarified juices of OPY, OPR, SR, and MG with cell viability percentages of (49.02 ± 1.32), (63.82 ± 1.08), (64.35.78 ± 2.84), and (69.75 ± 3.70) respectively, compared to the Caco-2 cell line which was only affected by the MG clarified juice with cell viability percentage of (57.50 ± 4.58). All clarified juices except the SR one had a similar effect on the PC3 cell line. The cell viability percentage of the SR clarified juice was not detected in the PC3 cell line, the blank had high values of absorbance, leading to negative cytotoxicity percentages. The antiproliferative effect of the five clarified fruit juices on MCF7 could not be demonstrated as the cell viability percentage was more than 100 percent. The NIH/3T3, normal cell line was used as a control and the SY, MG, and SR clarified juices diminished the cell line proliferation with percentages of (43.15 ± 3.27), (55.68 ± 2.09), and (58.59 ± 4.56) respectively (Figure 2). A future investigation may be performed in order to evaluate the dose and time-dependence of the cactus juices with the potential on cytotoxicity.

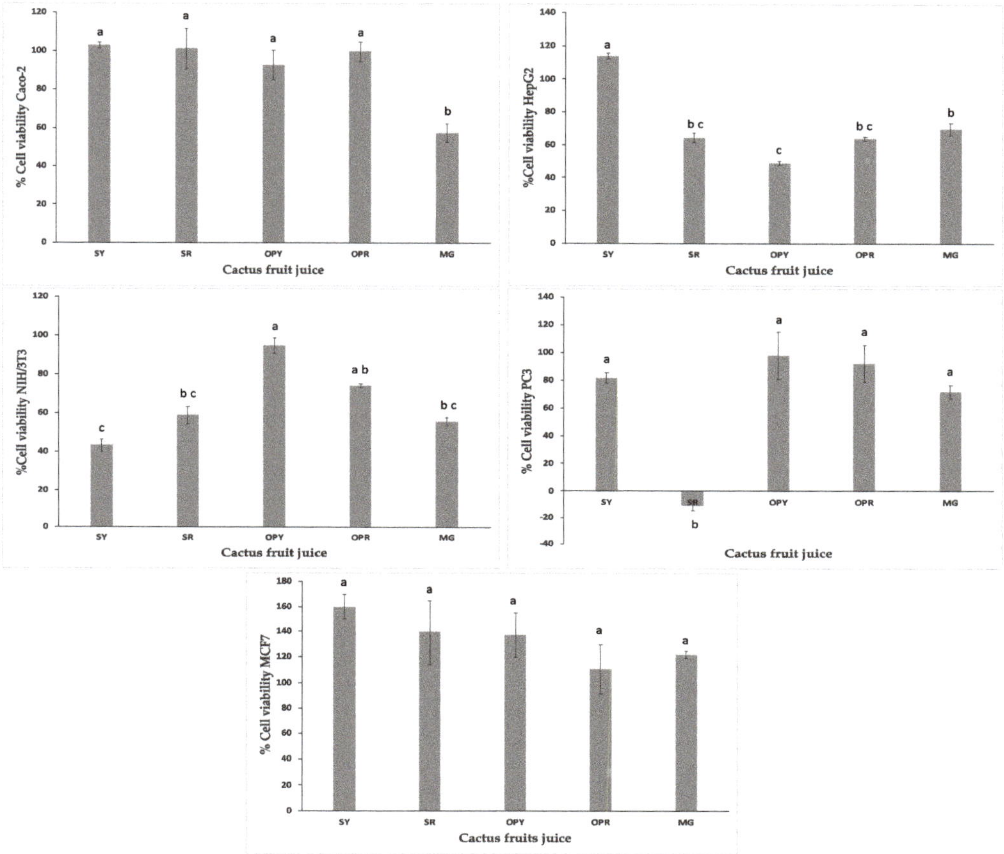

Figure 2. Cell viability of Caco-2, HepG2, NIH/3T3, PC3, and MCF7 cell-lines treated with cactus clarified juices at 2% *v*/*v*. Cell viability was expressed in terms of percentage of living cells relative to the non-treated control. Results were expressed as means of triplicate experiments and error was expressed as Standard error of mean (SEM). Different lowercase letters (a–c) indicate a statistical significance difference ($p < 0.05$). *Stenocereus pruinosus* yellow fruit (SY), *Stenocereus pruinosus* red fruit (SR), *Opuntia ficus-indica* yellow fruit (OPY), *Opuntia ficus-indica* red fruit (OPR), and *Myrtillocactus geomettizans* fruit (MG).

In vitro studies of juices extracted from fruits of *Opuntia ficus-indica* showed antiproliferative effects on hepatic cancer cells while no effect on normal fibroblast viability [12], corresponding with the findings of the present work. In order to understand if phenolic acids were responsible for the potential antiproliferative effect of clarified juices, the individual phenolic acids measured by HPLC analysis were calculated in µg/100 µL in the clarified juice at 2% in the cell proliferation assay. This calculation showed that the concentration of the phenolic acids previously mentioned was too low to have significantly contributed to the antiproliferative effect observed (Table S3). Further investigation to detect specific compounds such as quercetin, isorhamnetin, kaempferol, and betalains by HPLC is necessary to prove that the antiproliferative effect above described was due to the phenolic compounds. Evidence of these compounds has been reported in cactus juices. The fruit juices of *Stenocereus pruinosus* and *Stenocereus stellatus* were evaluated by HPLC-DAD-ESI/MS, to quantify the amounts of quercetin 3-*O*- rutinoside, kaempferol hexoside, isorhamnetin hexoside, isorhamnetin 3-*O*- glucoside, nine betacyanins, and two betaxanthins [6,15]. Isorhamnetin, quercetin, conjugated phenolic acids, indicaxanthin and coumarins were observed using UPLC-QTOF-MS in three *Opuntia ficus-indica* juices [26]. Quercetin was found in the cactus berry (*Myrtillocactus geometrizans*) fruit at different maturity stages before and after storage by HPLC-DAD [18]. The antiproliferative activities of flavonoids and betalains have been reported in extracts of *Opuntia robusta* and *Opuntia ficus-indica* fruit juices as they diminished human colorectal cancer cell line HT29 proliferation. The antiproliferative compounds identified were betacyanins, ferulic acid, and isorhamnetin derivatives [13]. In order to evaluate the therapeutic potential of the clarified juices for cancer, the molecular mechanism should be investigated and elucidated on normal and cancer cell lines of the same tissue. Previous studies have demonstrated the effect of plant extracts on proliferation, morphology, and cell death in MCF-7 breast adenocarcinoma and non-carcinogenic MCF-12A cell lines, where MCF-7 cell line was more susceptible to plant extracts exposure [27].

2.4. Principal Component Analysis (PCA)

In this study, PCA was used to correlate nine experimental variables: content of betacyanins and betaxanthins, antioxidant activity by both ABTS and DPPH methods, total phenolic composition by Folin–Ciocalteu assay according to cytotoxic activity of each clarified juice on HepG2, Caco-2, PC3, and NIH/3T3. The MCF7 cell line was not considered in the PCA analysis due to its high cell viability percentage and non-significant difference in all clarified juices. FRAP activity was not included in the PCA plot due to missing experimental results from MG fruit because it was unavailable during the analysis, due to the time at which it was collected. In Figure 3 we identified two components, principal component 1 (cell viability of cancer cell lines HepG2, Caco-2 which have a correlation with total phenolic composition by Folin–Ciocalteu assay and antioxidant activity by ABTS) and principal component 2 (betacyanins and betaxanthins which strongly correlated with antioxidant activity by DPPH and prostate cytotoxic activity on the PC3 cell line with red cactus fruits juices, SR and OPR). The variance of the data for the Principal Component 1 (PC1) was 36.81% and for the Principal Component 2 (PC2) was 31.89% of the variance in the data. The two principal components contributing to 68.7% of the total variance of the results. Based on PCA, the cell viability of the NIH/3T3 cell line was not correlated in either of the two groups. It is a normal cell line that was not observably influenced by the chemical content of the clarified juices in the grouping. Compared to cancer cell lines, PC3 was influenced by the chemical content of betalains and antioxidant compounds measured by DPPH; whilst Caco-2 and HepG2 were influenced by chemical content measured by ABTS and Folin–Ciocalteu methods. OPY, MG, and SY were excluded from principal component 2, due to their lower content of betalains. The PCA analysis evidenced that red cactus fruits showed a higher content of betalains which positively correlated with antioxidant activity by the DPPH method as shown in Table 1. The data is presented by two or three principal components defined as a linear combination and correlation between each other, therefore, it reveals clusters of the observed variables in terms of their similarities and dissimilarities [28]. However, additional investigation is needed in order to demonstrate SR antiproliferative activity on the PC3 cancer cell line. Previous investigations have used PCA to explain attributes of the sample and generate a global analysis of results; for example, to correlate the total phenolic content or flavonoids with antioxidant activity by hydrogen peroxide (H_2O_2), hydroxyl ($\cdot OH$), peroxyl ($ROO\cdot$) and ABTS radicals from *Opuntia elata* (Arumbeva) fruit extract [29]. Clear correlations were evidenced between total phenolics, fatty acids, phenolic compounds, and antioxidant activity of *Opuntia ficus-barbarica* A. Berger fruit pulp and seed oil harvested at different times [30].

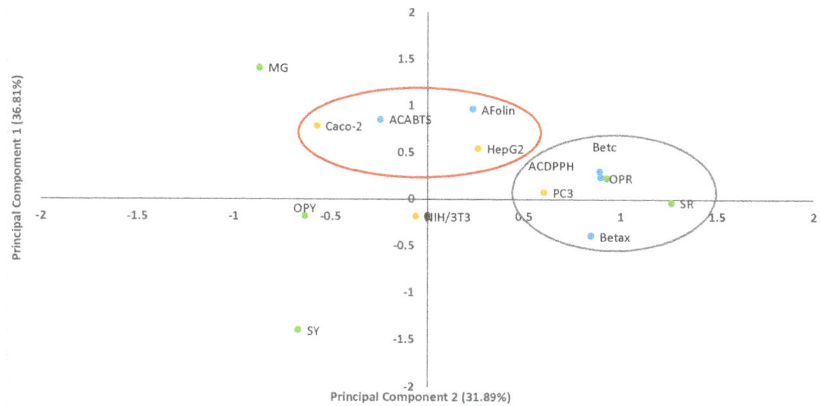

Figure 3. Distribution of five clarified juices along principal components 1 and 2 using nine variables. Betacyanins (Betac), betaxanthins (Betx),antioxidant activity by ABTS method (ACABTS), antioxidant activity by DPPH method (ACDPPH), total phenolic composition (AFolin), fibroblast cell line (NIH/3T3), colon cancer cell line (Caco-2), hepatic cancer cell line (HepG2), prostate cancer cell line (PC3) *Stenocereus pruinosus* yellow fruit clarified juice(SY), *Stenocereus pruinosus* red fruit clarified juice (SR), *Opuntia ficus-indica* yellow fruit clarified juice (OPY), *Opuntia ficus-indica* red fruit (OPR) clarified juice, and *Myrtillocactus geomettizans* fruit (MG) clarified juice. The oranges circles are cancer cell lines, the blue circles are the analytical methods and green circles are the clarified juices.

3. Materials and Methods

3.1. Chemical and Reagents

2,2-Diphenyl-L-picryl-hydrazyl and sodium carbonate were purchased from Sigma Aldrich (Steinheim, Germany). 2,2′-Azino-bis (3-ethylbenzothiazoline-6-sulfonic acid) (ABTS), Folin & Ciocalteu's Phenol Reagent, potassium persulfate, *p*-coumaric acid (concentration $\geq$ 98%), caffeic acid (concentration $\geq$ 98%), vanillic acid (concentration $\geq$ 97%) were purchased from Sigma-Aldrich (St. Louis, MO). Disodium phosphate and potassium chloride were acquired from Productos Químicos Monterrey S.A. de C.V. (Nuevo Leon, Mexico). was purchased from Productos Quimicos Monterrey S.A. de C.V. (Nuevo Leon, Mexico). Iron (lll) chloride hexahydrate and gallic acid monohydrate were obtained from Sigma Aldrich (Shanghai, China), Glacial acetic acid and methyl alcohol were purchased from Tedia High Purity Solvents (Fairfield, OH), hydrochloric acid was purchased from CTR Scientific (Nuevo Leon, Mexico) and 2,4,6-tris(2-pyridyl)-s-triazine (TPTZ) was from Sigma Aldrich (Buchs, Switzerland). The milli-Q water purification system was used to filter the water that was used to perform the procedures (Q-POD, Darmstadt, Germany). Potassium phosphate monobasic was acquired from Sigma-Aldrich (Tokyo, Japan). Sodium chloride was bought from Desarrollo de Especialidades Químicas, S.A. de C.V. (Nuevo Leon, Mexico).

3.2. Production of Clarified Juice

3.2.1. Preparation of Pitaya, Garambullo, and Tuna Pulps

Around, 3 kg of *Myrtillocactus geometrizan* fruit (MG) and 5 kg of yellow prickly pear *Opuntia ficus-indica* (OPY) from Ahualulco (San Luis Potosí, México); 5 kg of yellow pitaya *Stenocereus pruinosus* fruit (SY) and 5 kg of red pitaya *Stenocereus pruinosus* from Ahuatlán (Puebla, México); and 5 kg of red prickly pear *Opuntia ficus-indica* fruits (OR) from San Nicolás (Nuevo León, México) were refrigerated, while ensuring not handling for longer than 48 hours after being gathered. The fruits were washed with tap water and Extran MA05 (Merck, Item 1400001403, Lot Mx1400005004, Estado de Mexico, Mexico). Afterwards, the spines and peels were detached manually. Ultimately, the pulp, and seeds were separated using a juice extractor (Model TU05, Turmix ML, Estado de México). Seedless pulp was obtained from this procedure and its moisture was measured [31].

3.2.2. Production of Clarified Juice

The following procedure was carried out in the dark. The previously acquired seedless pulp (30g) was centrifuged (4000 g, 4 °C, 10 min, Model SL 40R, Thermo Fisher Scientific, Langenselbold, Germany) in 50 mL polypropylene conical tubes (Corning®, Tewksbury, MA, USA). 30 g of pulp of each fruit (SY, SR, OPY, OPR, and MG) was prepared to obtain clarified juice as shown in Figure S1. The supernatant was strained though 150 mm of Whatman paper grade 4 (item 1009150, GE Healthcare Life Sciences, Little Chalfont, UK), the supernatant of this second filtration was considered the clarified juice. Water was not added to the clarified juices.

3.3. *Total Soluble Solids*

The total soluble solids (°Brix) were determined in the clarified juices. One mL of each clarified juice was placed in the refractometer HI96811 (HANNA, Smithfield, RI, USA). Three samples of each clarified juice were used in this procedure.

3.4. *Betacyanin and Betaxanthin Content and Antioxidant Activity Assay*

3.4.1. Quantification of Betacyanin and Betaxanthin

In order to determine the pulp's betacyanin and betaxanthin content (µg/g, fresh weight) the spectrophotometric method described in [32–34] was carried out on clarified juices using a Model DR 500 spectrophotometer (Hach Lange GmbH, Düsseldorf, Germany). The clarified juices samples were diluted as shown in Table S2 in 5 mL volumetric flasks using Milli-Q water. The extinction coefficients used were E1% = 60,000 L mol^{-1} cm^{-1}, λ = 540 nm for betacyanin, and E1% = 48,000 L mol^{-1} cm^{-1}, λ = 480 for betaxanthin.

3.4.2. Antioxidant Activity by 2,2′-Azino-bis-(3-ethylbenzothiazoline-6-sulfonic acid) Diammonium Salt Capacity (ABTS)

The ABTS, a single electron transfer (ET) reaction-based assay, was carried out following the method proposed by Re et al. [35]. The Phosphate buffered saline (PBS) used was created with 0.8 g of NaCl, 0.02 g of KH_2PO_4, 0.115 g of Na_2HPO_4, 0.02 g of KCl, and 0.02 g of NaN_3. The volume was filled up to 100 mL, the difference being Milli-Q water. To create the ABTS reagent the following were used: 38.4 mg of ABTS 1 mM, 6.62 mg of potassium persulfate 2.45 mM, and 10 mL of the solution of PBS. The solutions were left mixing in the dark for 16 h. The absorbance was measured at 734 nm, with a spectrophotometer (Model DR 500, Hach Lange GmbH, Düsseldorf, Germany) and the dilution of the initial reagent that read 0.7 absorbance units was used (40 mL of PBS with 3 mL of ABTS solution). 20 µL of diluted clarified juice of each fruit and 2 mL of ABTS solution (with the absorbance of 0.7) were placed in a water bath at 30 °C for six minutes. Thereafter, the absorbance was read using Trolox as a standard in concentrations ranging from 5 to 200 ppm. The blank was created using Milli-Q water and the procedure was triplicated.

3.4.3. Antioxidant Activity by α-α-Diphenyl-β-picrylhydrazyl (DPPH)

Based on Brand-Williams et al., protocol [36] 0.0148 g of DPPH were added to a 25 mL volumetric flask and filled to the mark with methanol (mother solution). One mL of this mother solution was added to a new 25 mL volumetric flask and filled to the mark with methanol, creating a diluted solution. The solutions were placed in 4.0 mL cuvettes (75 µL of clarified juices dilutions and 3 mL DPPH solution), and left to react for 16 min before being read by the spectrophotometer (Model DR 500, Hach Lange GmbH, Düsseldorf, Germany) at 515 nm [37]. All measurements were made in triplicates with a calibration curve of Trolox at varying concentrations from 5 to 200 ppm.

3.4.4. Ferric Reducing Antioxidant Power (FRAP)

The antioxidant capacity of each clarified juice was determined through a modified method [38]. The FRAP reagent was prepared with acetate buffer 300 mM pH 3.6, which is a mixture of sodium acetate trihydrate, glacial acetic acid, and distilled water, a solution of iron triplicidyl triazine (TPTZ), concentrated HCl, and distilled water and finally a solution with $FeCl_3{\cdot}6H_2O$ and Milli-Q water. The solutions were mixed (10:1:1) respectively and incubated at 30 °C for 30 min in darkness. Then, 100 μL of each clarified juice was added to 3 mL of FRAP reagent. Concentrations of Trolox ranging from 10 to 200 ppm were used as standards in a calibration curve. Lastly, the absorbance was measured in a spectrophotometer (Model DR 500, Hach Lange GmbH, Düsseldorf, Germany) at 593 nm. All measurements were made in triplicate.

3.5. *Total Phenolic Composition*

In order to determine the overall phenolic composition, the Folin–Ciocalteu colorimetric procedure was used [39]. Twenty μL of the diluted clarified juices were added to a 96 well plate. Subsequently, 100 μL of 10% Folin reagent was added to each well and after a five minutes incubation an additional 80 μL 7.5% w/v of sodium carbonate was placed per well. The plates were incubated for 1.5 h in the absence of light at 37 °C. Once the incubation period had elapsed, microplates were read at 765 nm and 25 °C. In order to create the calibration curve, solutions of 50 to 200 mg/L of gallic acid were made in Milli-Q water. The blank was created with the same solvent and the procedure was replicated twice (in triplicate).

3.6. *HPLC-DAD Analysis*

After filtering the clarified juices through a 0.2 μm nylon filter (Waters, Milford, MA, USA), their chromatographic profile was analyzed through equipment from Altus Perkin Elmer with an autosampler, photodiode array detector (PDA) and a Zorbax Eclipse XDB C18 column (5 μm, 150 × 4.6 mm). In order to analyze the phenolic composition, a gradient method was achieved with Solvent A, consisting of a mixture of water and acetic acid (pH 2.5), and Solvent B, methanol. The mobile phase composition started at 100% solvent A for 3 min, followed by an increase of solvent B up to 30% from minutes 3 to 8 min, 50% from minutes 8 to 15 min, 30% from 15 to 20 min, and then returning the mobile phase composition back to 100% solvent A for the end of the run. Around, 20 μL of the clarified juices were injected at a flow rate of 0.8 mL/min at column temperature of 25 °C. The UV absorption spectra were documented of clarified juices and standards at 270 nm. The phenolic acids, caffeic, gallic, p-coumaric, and vanillic acids were used as standards to compare retention time and identified compounds in the clarified juices. The standards were dissolved in milli-Q water to prepare the calibration curves.

3.7. *Cell Viability Assay*

3.7.1. Cell Culture

A normal fibroblast cell line (NIH/ 3T3) and four different mammalian cancer cell lines: mammary (MCF-7), prostate (PC3), colon (Caco-2), and hepatic (HepG2), were cultivated in DMEM-F12 medium containing 10% FBS (Fetal Bovine Serum) (Gibco, Grand Island, NY, USA) and kept in a 5% CO_2 atmosphere at 37 °C and 80% humidity.

3.7.2. Cell Proliferation Assay

In order to determine viability, a cytotoxicity assay was carried out in 96-well plates with 100 μL of 5×10^5 cells/mL per well. Cancer cell lines (MCF-7, PC3, Caco-2 and HepG2) and NIH/3T3 were seeded as a control and incubated for 12 h to reach confluence. All the clarified juices were evaluated at a final concentration, 2% *v*/*v*. The plates were left in the incubator at 37 °C, with less than 5% CO_2 for 48 h. Subsequently, 20 μL of Cell Titer 96®AQueous One Solution Cell Proliferation Assay (Promega, Madison, WI) was added to each well, and the 96-well-plate was incubated at 37 °C, with 5% CO_2 for 1 h, then the absorbance was read at 490 nm using a microplate reader (Synergy HT, Bio-Tek, Winooski, VM). The viability was determined through the calculation of average absorbance units per well and conveyed as a percentage of the untreated cell wells. The experiment was done in triplicate, culture medium without cells was used as a blank, and the cells that only had medium were a positive viability control [12].

3.8. Principal Component Analysis (PCA)

PCA is a tool used to highlight the relationships among a group of experimental variables based on multivariate statistical analysis, where a map is generated to show how variables are distributed. The correlation of antioxidant results and antiproliferative effect of the clarified juices of the cactus fruits was determined by inspection of the principal component analysis (PCA). The variables analyzed were percentage of cell viability on Caco-2, HepG2, PC3 and NIH/3T3 cell lines; antioxidant activity regarding DPPH (μmol TE/100 g FS), ABTS methods (μmol TE/100 g FS), total phenolic content obtained by the Folin–Ciocalteu method (μg GA/g of fresh sample) and betacyanin and betaxanthin (μg/g FS).

3.9. Statistical Analysis

All experiments were performed in triplicate. Results were analyzed by ANOVA and different means were compared using the Tukey test with a level of significance of $p > 0.05$. The computer software used was MINITAB 18. The multivariate analysis of Principal component analysis (PCA) above described was done by SPSS (Version 19, IBM Corp, Chicago, IL, USA.

4. Conclusions

The results presented in this paper prove the antioxidant properties of five Mexican native cactus fruits by ABTS, DPPH, FRAP and Folin–Ciocalteu methods, as well as the in vitro cell cytotoxicity on cell lines. Our findings exemplified the antiproliferative effect of *Myrtillocactus geomettizans* clarified juices through the diminished cell viability of Caco-2, HepG2, and PC3 as well as, on the normal cell line NIH/3T3. The red fruit *Stenocereus pruinosus* clarified juices affected the cancer cell line HepG2 as well as the NIH/3T3 cell line. *Opuntia ficus-indica* yellow fruit may potentially be used as a cancer preventing agent due to its selective cytotoxicity on only cancer cells, with demonstrated activity against HepG2 and no effect on the normal cell NIH/3T3. A future investigation may be performed in order to evaluate the time and dose-dependence of the cactus juices on cytotoxicity. The total phenolic compounds could be the main contributors of the bioactivity, although a synergistic effect between phenolic acids, flavonoids, coumarins, alkaloids, and vitamins could explain the cactus fruit juices' cancer potential therapeutic use. More research is needed in order to identify the bioactive compounds and mechanisms that give the juices these properties.

Supplementary Materials: The following are available online at https://www.mdpi.com/2223-7747/10/2/368/s1, Table S1: Total soluble solids (°Brix) in clarified juices, Table S2: Dilutions of the clarified juices in the different techniques, Table S3: Calculated concentration of phenolic acids composition of *Myrtillocactus geomettizans* fruit (MG) and *Opuntia ficus-indica* yellow fruit (OPY) detected by HPLC. Figure S1: General procedure to obtain clarified juices from cactus fruits Figure S2: HPLC detection of phenolic compounds (vanillic acid and coumaric acid) from *Myrtillocactus geomettizans* fruit (MG), Figure S3: HPLC detection of phenolic compounds (gallic acid and caffeic acid) from *Myrtillocactus geomettizans* fruit (MG), Figure S4: HPLC detection of phenolic compounds (vanillic acid and coumaric acid) from *Opuntia ficus-indica* yellow fruit (OPY), Figure S5: HPLC detection of phenolic compounds (gallic acid and caffeic acid) from *Opuntia ficus-indica* yellow fruit (OPY).

Author Contributions: Data curation, J.R.-R.; Formal analysis, E.M.M.M. and L.S.-F.; Funding acquisition, R.P.-S.; Investigation, M.A.-R. and S.O.S.-S.; Methodology, E.M.M.M. and L.S.-F.; Project administration, M.R.-A.; Software, L.S.-F. and J.R.-R.; Supervision, J.R.-R., M.R.-A. and R.P.-S.; Validation, L.S.-F.; Writing—original draft, E.M.M.M., M.R.-A., L.P.-A., S.O.S.-S., H.M.N.I. and R.P.-S.; Writing—review & editing, E.M.M.M., H.M.N.I. and R.P.-S. All authors have read and agreed to the published version of the manuscript.

Funding: This research was supported by Consejo Nacional de Ciencia y Tecnología (CONACYT) Doctoral Fellowship No. 492030 awarded to author Luisaldo Sandate-Flores.

Institutional Review Board Statement: Not applicable.

Informed Consent Statement: Not applicable.

Data Availability Statement: All data, belongs to this work, is given and presented herein the manuscript. Additional data can be found in the supplementary material.

Acknowledgments: The authors appreciate the support from the FEMSA-Biotechnology Center and Latin America and Caribbean Water Center of Tecnologico de Monterrey, México. The authors would thank to J. Villela-Castrejón for supporting in cell viability experiments. The authors acknowledge to Opciones de vida.

Conflicts of Interest: The authors declare no conflict of interest.

References

1. WHO. Cancer. 2020. Available online: https://www.who.int/health-topics/cancer#tab=tab_1 (accessed on 22 January 2021).
2. Govea-Salas, M.; Morlett-Chávez, J.; Rodriguez-Herrera, R.; Ascacio-Valdés, J. Some Mexican Plants Used in Traditional Medicine. *Aromat. Med. Plants Back Nat. Intech.* **2017**, *1*, 191–200. [CrossRef]
3. Ramírez-Rodríguez, Y.; Martínez-Huélamo, M.; Pedraza-Chaverri, J.; Ramírez, V.; Martínez-Tagüeña, N.; Trujillo, J. Ethnobotanical, Nutritional and Medicinal Properties of Mexican Drylands Cactaceae Fruits: Recent Findings and Research Opportunities. *Food Chem.* **2020**, *312*, 126073. [CrossRef]
4. Chuck-Hernández, C.; Parra-Saldívar, R.; Sandate-Flores, L. Pitaya (*Stenocereus* Spp.). *Encycl. Food Health* **2015**, 385–391. [CrossRef]
5. Quiroz-González, B.; Rodriguez-Martinez, V.; Welti-Chanes, J.; García-Mateos, M.d.R.; Corrales-García, J.; Ybarra-Moncada, M.C.; Leyva-Ruelas, G.; Torres, J.A. Refrigerated storage of high hydristatic pressure (HHP) treated pitaya (*Stenocereus pruinosus*) juice. *Rev. Mex. Ing. Quim.* **2020**, *19*, 387–399. [CrossRef]
6. García-Cruz, L.; Dueñas, M.; Santos-Buelgas, C.; Valle-Guadarrama, S.; Salinas-Moreno, Y. Betalains and Phenolic Compounds Profiling and Antioxidant Capacity of Pitaya (*Stenocereus* Spp.) Fruit from Two Species (*S. Pruinosus* and *S. Stellatus*). *Food Chem.* **2017**, *234*, 111–118. [CrossRef] [PubMed]
7. Corzo-Rios, L.J.; Bautista-Ramírez, M.E.; Gómez y Gómez, Y.D.L.M.; Torres-Bustillos, L.G. Frutas de cactáceas: Compuestos bioactivos y sus propiedades nutracéuticas. In *Propiedades Funcionales de Hoy*; OmniaScience: Barcelona, España, 2016; pp. 35–65. [CrossRef]
8. Cota-Sanchez, J.H. Nutritional composition of the prickly pear (*Opuntia ficus-indica*) fruit. In *Nutritional Composition of Fruit Cultiva*, 1st ed.; Saskatoon, Ed.; Academic Press: London, UK, 2016; pp. 691–712. [CrossRef]
9. Albano, C.; Negro, C.; Tommasi, N.; Gerardi, C.; Mita, G.; Miceli, A.; de Bellis, L.; Blando, F. Betalains, Phenols and Antioxidant Capacity in Cactus Pear [*Opuntia Ficus-Indica* (L.) Mill.] Fruits from Apulia (South Italy) Genotypes. *Antioxidants* **2015**, *4*, 269–280. [CrossRef] [PubMed]
10. Abd El-Moaty, H.I. Structural Elucidation of Phenolic Compounds Isolated from Opuntia Littoralis and Their Antidiabetic, Antimicrobial and Cytotoxic Activity. *S. Afr. J. Bot.* **2020**, *131*, 320–327. [CrossRef]
11. Allegra, M.; D'anneo, A.; Frazzitta, A.; Restivo, I.; Livrea, M.A.; Attanzio, A.; Tesoriere, L. The Phytochemical Indicaxanthin Synergistically Enhances Cisplatin-Induced Apoptosis in Hela Cells via Oxidative Stress-Dependent P53/P21waf1 Axis. *Biomolecules* **2020**, *10*, 1–16. [CrossRef] [PubMed]
12. Chavez-Santoscoy, R.A.; Gutierrez-Uribe, J.A.; Serna-Saldívar, S.O. Phenolic Composition, Antioxidant Capacity and In Vitro Cancer Cell Cytotoxicity of Nine Prickly Pear (*Opuntia* Spp.) Juices. *Plant Foods Hum. Nutr.* **2009**, *64*, 146–152. [CrossRef]
13. Serra, A.T.; Poejo, J.; Matias, A.A.; Bronze, M.R.; Duarte, C.M.M. Evaluation of Opuntia Spp. Derived Products as Antiproliferative Agents in Human Colon Cancer Cell Line (HT29). *Food Res. Int.* **2013**, *54*, 892–901. [CrossRef]
14. Reynoso-Camacho, R.; Martinez-Samayoa, P.; Ramos-Gomez, M.; Guzmán, H.; Salgado, L.M. Antidiabetic and Renal Protective Properties of Berrycactus Fruit (*Myrtillocactus geometrizans*). *J. Med. Food* **2015**, *18*, 565–571. [CrossRef] [PubMed]
15. García-Cruz, L.; Valle-Guadarrama, S.; Salinas-Moreno, Y.; Joaquín-Cruz, E. Physical, Chemical, and Antioxidant Activity Characterization of Pitaya (*Stenocereus pruinosus*) Fruits. *Plant Foods Hum. Nutr.* **2013**, *68*, 403–410. [CrossRef]
16. Tatar, B.C.; Sumnu, G.; Oztop, M.; Ayaz, E. Effects of centrifugation, encapsulation method and different coating materials on the total antioxidant activity of the microcapsules of powdered cherry laurels. *Int. J. Nut. Food Engin.* **2017**, *10*, 901–904.

17. Guzmán-Maldonado, S.H.; Herrera-Hernández, G.; Hernández-López, D.; Reynoso-Camacho, R.; Guzmán-Tovar, A.; Vaillant, F.; Brat, P. Physicochemical, nutritional and functional characteristics of two underutilised fruit cactus species (*Myrtillocactus*) produced in central Mexico. *Food Chem.* **2010**, *121*, 381–386. [CrossRef]
18. Herrera-Hernández, M.G.; Guevara-Lara, F.; Reynoso-Camacho, R.; Guzmán-Maldonado, S.H. Effects of Maturity Stage and Storage on Cactus Berry (*Myrtillocactus geometrizans*) Phenolics, Vitamin C, Betalains and Their Antioxidant Properties. *Food Chem.* **2011**, *129*, 1744–1750. [CrossRef]
19. Vázquez-Araújo, L.; Chambers, E.; Adhikari, K.; Carbonell-Barrachina, Á.A. Sensory and physicochemical characterization of juices made with pomegranate and blueberries, blackberries, or raspberries. *J. Food Sci.* **2010**, *75*, 398–404. [CrossRef]
20. Singleton, V.L.; Orthofer, R.; Lamuela-Raventós, R.M. Analysis of total phenols and other oxidation substrates and antioxidants by means of folin-ciocalteu reagent. In *Methods in Enzymology*; Academic Press: Boston, MA, USA, 1999; Volume 299, pp. 152–178.
21. Sánchez-Rangel, J.C.; Benavides, J.; Heredia, J.B.; Cisneros-Zevallos, L.; Jacobo-Velázquez, D.A. The Folin-Ciocalteu assay revisited: Improvement of its specificity for total phenolic content determination. *Anal. Methods* **2013**, *5*, 5990–5999. [CrossRef]
22. Zeghad, N.; Ahmed, E.; Belkhiri, A.; Heyden, Y.V.; Demeyer, K. Antioxidant Activity of Vitis Vinifera, Punica Granatum, Citrus Aurantium and Opuntia Ficus Indica Fruits Cultivated in Algeria. *Heliyon* **2019**, *5*, e01575. [CrossRef] [PubMed]
23. Konan, A.; Konan, Y.; Kone, M. Polyphenols Content and Antioxidant Capacity of Traditional Juices Consumed in Côte d'Ivoire. *J. Appl. Biosci.* **2015**, *87*, 8015. [CrossRef]
24. Gupta, A.K.; Rather, M.A.; Jha, A.K.; Shashank, A.; Singhal, S.; Sharma, M.; Pathak, U.; Sharma, D.; Mastinu, A. Artocarpus lakoocha roxb. And artocarpus heterophyllus lam. flowers: New sources of bioactive compounds. *Plants* **2020**, *9*, 1329. [CrossRef]
25. Osorio-Esquivel, O.; Álvarez, V.B.; Dorantes-Álvarez, L.; Giusti, M.M. Phenolics, Betacyanins and Antioxidant Activity in Opuntia Joconostle Fruits. *Food Res. Int.* **2011**, *44*, 2160–2168. [CrossRef]
26. Farag, M.A.; Sallam, I.E.; Fekry, M.I.; Zaghloul, S.S.; El-Dine, R.S. Metabolite Profiling of Three Opuntia Ficus-Indica Fruit Cultivars Using UPLC-QTOF-MS in Relation to Their Antioxidant Potential. *Food Biosci.* **2020**, *36*, 100673. [CrossRef]
27. Vorster, C.; Stander, A.; Joubert, A. Differential signaling involved in Sutherlandia frutescens-induced cell death in MCF-7 and MCF-12A cells. *J. ethnopharmacol.* **2012**, *140*, 123–130. [CrossRef] [PubMed]
28. Kumirska, J.; Migowska, N.; Caban, M.; Plenis, A.; Stepnowski, P. Chemometric Analysis for Optimizing Derivatization in Gas Chromatography-Based Procedures. *J. Chemom.* **2011**, *25*, 636–643. [CrossRef]
29. Rockett, F.C.; de Oliveira Schmidt, H.; Schmidt, L.; Rodrigues, E.; Tischer, B.; de Oliveira, V.R.; Rios, A. Phenolic compounds and antioxidant activity in vitro and in vivo of Butia and Opuntia fruits. *Food. Res. Int.* **2020**, *137*, 109740. [CrossRef] [PubMed]
30. Al Juhaimi, F.; Ghafoor, K.; Uslu, N.; Ahmed, I.A.M.; Babiker, E.E.; Özcan, M.M.; Fadimu, G.J. The effect of harvest times on bioactive properties and fatty acid compositions of prickly pear (*Opuntia ficus-barbarica* A. *Berger*) fruits. *Food Chem.* **2020**, *303*, 125387. [CrossRef]
31. Association of Official Analytical Chemistry (AOAC). *Official Method of Analysis of the Association of Official Analytical Chemists*, 15th ed.; Association of Official Analytical Chemistry (AOAC): Arlington, VA, USA, 1990; pp. 141–144.
32. Sandate-Flores, L.; Rodríguez-Rodríguez, J.; Calvo-Segura, S.; Mayorga-Martínez, A.; Parra-Saldivar, R.; Chuck-Hernández, C. Evaluation of different methods for betanin quantification in pitaya (*Stenocereus* spp). *Agro. Food. Ind. Hi Tech.* **2016**, *27*, 20–25.
33. Sandate-Flores, L.; Romero-Esquivel, E.; Rodríguez-Rodríguez, J.; Rostro-Alanis, M.; Melchor-Martínez, E.M.; Castillo-Zacarías, C.; Parra-Saldívar, R. Functional Attributes and Anticancer Potentialities of Chico (*Pachycereus Weberi*) and Jiotilla (*Escontria Chiotilla*) Fruits Extract. *Plants* **2020**, *9*, 1623. [CrossRef]
34. Nilsson, T. Studies into the pigments in beetroot (*Beta vulgaris L.* ssp. *Vulgaris var.* rubra L.). *Lantbrukshögskolans Ann.* **1970**, *36*, 179–219.
35. Re, R.; Pellegrini, N.; Proteggenete, A.; Pannala, A.; Yang, M.; Rice-Evans, C. Antioxidant activity applying an improved ABTS radical cation decolorization assay. *Free Radic. Biol.Med.* **1999**, *26*, 1231–1237. [CrossRef]
36. Brand-Williams, W.; Cuvelier, M.E.; Berset, C. Use of a free radical method to evaluate antioxidant activity. *LWT Food Sci. Technol.* **1995**, *28*, 25–30. [CrossRef]
37. Rostro-Alanis, M.D.J.; Báez-González, J.; Torres-Alvarez, C.; Parra-Saldívar, R.; Rodriguez-Rodriguez, J.; Castillo, S. Chemical composition and biological activities of oregano essential oil and its fractions obtained by vacuum distillation. *Molecules* **2019**, *24*, 1904. [CrossRef] [PubMed]
38. Benzie, I.F.; Strain, J.J. The ferric reducing ability of plasma (FRAP) as a measure of "antioxidant power": The FRAP assay. *Anal. Biochem.* **1996**, *239*, 70–76. [CrossRef] [PubMed]
39. Singleton, V.L.; Rossi, J.A. Colorimetry of total phenolics with phosphomolybdic-phosphotungstic acid reagents. *Am. J. Enol. Vitic.* **1965**, *16*, 144–158.

Article

Accumulation of Anthocyanins and Other Phytochemicals in American Elderberry Cultivars during Fruit Ripening and its Impact on Color Expression

Yucheng Zhou [1], Yu Gary Gao [2,3,*] and M. Monica Giusti [1,*]

1 Department of Food Science and Technology, The Ohio State University, 2015 Fyffe Rd, Columbus, OH 43210, USA; zhou.1140@buckeyemail.osu.edu
2 Department of Extension, The Ohio State University, 2120 Fyffe Rd, Columbus, OH 43210, USA
3 OSU South Centers, The Ohio State University, 1864 Shyville Rd, Piketon, OH 45661, USA
* Correspondence: Gao.2@osu.edu (Y.G.G.); Giusti.6@osu.edu (M.M.G.)

Received: 18 November 2020; Accepted: 3 December 2020; Published: 7 December 2020

Abstract: American elderberry (*Sambucus canadensis*) is a plant native to North America with anthocyanin-rich fruits. Our objective was to investigate the effects of cultivar and ripeness on the phytochemical characteristics of its fruits and the corresponding color performance. Cultivars 'Adams', 'Johns', 'Nova', 'Wyldewood', and 'York' were examined for their °Brix, pH, anthocyanin (pH-differential method), and phenolic content (Folin-Ciocalteau method). Extract composition were analyzed by uHPLC-PDA-MS/MS. Color and spectra were determined using a plate reader. All characteristics evaluated were significantly affected by ripeness and cultivar, except for °Brix and total phenolic content, which did not vary significantly among cultivars. Most anthocyanins (63–72%) were acylated with *p*-coumaric acid, with cyanidin-3-(trans)-coumaroylsambubioside-5-glucoside the most predominant. The proportion of acylated anthocyanins was the only characteristic evaluated that decreased during ripening (from 80 to 70%). Extract from fully-ripened fruits exhibited red ($l_{vis\text{-}max}$ ~520 nm) and blue hues ($l_{vis\text{-}max}$ ~600 nm) at acidic and alkaline pH, respectively. Extracts from half-ripe fruit rendered yellowish tones and overall dull color. C-18 semi-purified extracts displayed higher color saturation (smaller L* and larger C^*_{ab}) than crude extracts. The vibrant and broad color expression of fully-ripened fruit extract, especially after C-18 purification, suggests this North American native plant as a promising natural colorant source.

Keywords: polyphenols; fruit development; natural food colorant; *Sambucus canadensis*

1. Introduction

Elderberry (*Sambucus* spp. L.) is a large perennial shrub or small deciduous tree in the Adoxaceae family with purplish-black small fruits that mature in late August to September in different parts of the USA [1]. It is distributed mostly in the temperate and subtropical regions of the Northern Hemisphere and includes 9 to 40 species depending on the taxonomy [1–3]. The most economically important species is European elderberry (syn. *S. nigra* L. subsp. *nigra*), which has been widely cultivated in the world from Europe to North America and East Asia [2]. Its fruits have been used as natural food colorants, and used to produce jam, jellies, yogurt, and wine [4]. In the USA, American elderberry (syn. *S. nigra* L. subsp. *canadensis* [L.] Bolli) is the most important domesticated species [3]. The first American elderberry cultivar 'Adams' was released in the 1920s, with few additional cultivars developed since then [4], mostly from the New York and Nova Scotia Agricultural Experiment Station breeding programs [1], and the Missouri State University and the University of Missouri Elderberry Improving Program [5].

American elderberry cultivars have shown to have higher yield in the midwestern states in the USA than their European counterparts [6]. A life cycle assessment of these American elderberry cultivars could reveal lower labor and pesticide input due to greater adaptability by following a similar assessment method on wine production [7]. This is because American elderberry can be completely pruned to the ground level in March for a more concentrated harvest in late summer to early autumn, thereby leading to a higher harvest efficiency and less carbon dioxide emission from tractors used for spraying and harvest. Complete removal of old shoots in American elderberry also help control elder shoot borer [8].

Elderberry has been traditionally used as a medicinal crop in many indigenous cultures [4]. The therapeutic effects of elderberry extracts, including flu symptoms alleviation, blood pressure regulation, diabetes and obesity control, and immune system enhancement, have been confirmed in a number of in vitro and in vivo studies [9,10]. Researchers attribute its medicinal and nutritional benefits to its abundant polyphenol content, composed mostly of anthocyanins, flavonols, phenolic acids, and proanthocyanidins [9]. Those polyphenols are known to be high in antioxidant capacity and possess antiviral, antibacterial, and antifungal activities [9]. The demand for valuable natural sources of antioxidant compounds has been going for decades, and this market is expected to gain an annual growth of ~6% globally during the period of 2019–2029 [11], forecasting a considerable growth potential of attention for elderberry in the coming years.

Anthocyanins contribute to both the high antioxidant capacity and the dark-violet color of elderberry fruits. These water-soluble natural pigments have been widely applied as food colorants since they are viewed as safer than synthetic colorants [12]. According to the FDA CFR, fruit (21CFR73.250) and vegetable (21CFR73.260) juice concentrates are approved for use as natural food colorants, but their practical application is restricted by their stability and color shades. Acylated anthocyanins is expected to have enhanced stability to pH, heat treatment, and light exposure than non-acylated anthocyanins through intramolecular copigmentation and self-association [13]. Vegetable sources, such as red cabbage and radish, are usually more abundant in acylated anthocyanins, but they may impart undesirable aromas or flavors to food products [13]. Anthocyanin profiles of European and American elderberry were previously compared, and acylated anthocyanins were only found in American elderberry in considerable quantity [1].

European elderberries have been evaluated for their anthocyanin and phenolic composition [1,14]; and the stability of their anthocyanin extract to heat [15,16], light [15,16], and processing [17]; as well as coloring strength [18]. Due to its high pigment content, European elderberry concentrate has been commercialized as a food coloring agent in the E.U. Despite its unique anthocyanin profile, American elderberry has not been as extensively studied as its European counterpart [18]. Our knowledge regarding American elderberry anthocyanins and phenolics is primarily related to their composition, with little research on their practical applications. Considering its high acylated anthocyanin content and milder flavor, American elderberry can potentially be a promising source of natural color. Its phytochemicals, mostly anthocyanins, may vary among cultivars and during ripening, and this variation could ultimately affect the colorimetric and spectrophotometric properties, therefore merit a more systematic examination.

Our objectives were to investigate the impact of cultivar and ripeness on the accumulation of anthocyanins and other phytochemicals in American elderberry, as well as the color performance of their anthocyanin extracts under a wide pH range. Our research aimed to reveal valuable information to help shape the potential application of American elderberry extract as a natural colorant for food industry.

2. Results

2.1. Phytochemical Attributes of Different Cultivars

Significant differences in pH, monomeric anthocyanin content, polymeric color, and anthocyanin/phenolic ratio were observed among five cultivars ($p < 0.05$) (Table 1). American elderberry had a pH between 4.5 and 4.9, with 'Wyldewood' having significantly lower pH than that of others. The monomeric anthocyanin content varied between 354 and 581 mg C3GE/100 g FW, and the anthocyanin/phenolic ratio was between 0.61 and 0.84, with 'Johns' and 'York' obtaining the highest and the lowest on both attributes, respectively. However, significantly higher percentage of polymeric color was detected in 'York' (10.8%) than in 'Johns' (5%), partially explained by their different monomeric anthocyanin content. The °Brix ranged from 12.0 (in 'Nova' and 'Wylderwood') to 13.1 (in 'York') and the total phenolic content was between 582 (in 'York') and 707 mg GAE/100 g FW (in 'Johns') with no significant differences among them.

Table 1. Fruits' phytochemical attributes comparison among different American elderberry cultivars. Results expressed as mean ± SD ($n = 3$). Cultivars with different superscripts were significantly different ($p < 0.05$).

Cultivar	°Brix	pH	Mono ACN [1]	Poly Color [2]	TP [3]	ACN/TP [4]
Adams	12.9 ± 1.1 ^{a}	4.7 ± 0.11 ^{a,b}	499 ± 91 ^{a,b}	7.5 ± 0.8 ^{a,b}	684 ± 88 ^{a}	73 ± 4 ^{a}
Johns	12.0 ± 1.1 ^{a}	4.8 ± 0.05 ^{b}	595 ± 26 ^{b}	5.0 ± 0.5 ^{a}	706 ± 22 ^{a}	84 ± 4 ^{b}
Nova	12.4 ± 0.3 ^{a}	4.8 ± 0.06 ^{b}	456 ± 73 ^{a,b}	9.1 ± 1.5 ^{a,b}	700 ± 77 ^{a}	65 ± 4 ^{a}
Wyldewood	12.0 ± 1.1 ^{a}	4.5 ± 0.08 ^{a}	471 ± 36 ^{a,b}	7.3 ± 0.4 ^{a,b}	637 ± 51 ^{a}	74 ± 2 ^{a,b}
York	13.1 ± 0.4 ^{a}	4.9 ± 0.12 ^{b}	354 ± 59 ^{a}	10.8 ± 1.1 ^{b}	582 ± 52 ^{a}	61 ± 3 ^{a}

[1] Monomeric anthocyanin (mg cyanidin-3-glucosides equivalent/100 g fresh weight); [2] Polymeric color (%); [3] Total phenolics (mg gallic acid equivalent/100 g fresh weight); [4] Anthocyanin/Total phenolic (%).

2.2. Color Development and Phytochemicals Accumulation during Ripening

Elderberry fruits reached a visibly darker surface color when fully ripe, as denoted by the significantly smaller L* (lightness) value of the fully-ripened fruits (L* = 19.1 ± 0.5) compared to the half-ripe counterpart (L* = 23.5 ± 0.5). The mean a* and b* values of the fully-ripe fruits were determined to be 2.4 ± 0.2 and 0.3 ± 0.2, respectively, similar to those of the half-ripe fruits (a* = 2.4 ± 0.2; b* = −0.2 ± 0.1).

°Brix, pH, anthocyanin, and phenolic content all significantly increased during fruit ripening except the pH of 'Wyldewood' (Figure 1). °Brix increased ~4% in both 'Wyldewood' and 'Nova' during ripening. Half-ripe berries were low in monomeric anthocyanin content, with means of 39 and 72 mg C3GE/100 g FW of 'Nova' and 'Wyldewood', respectively, consisting with the less intense redness of the fruits, but later greatly increased to 593 and 619 C3GE/100 g FW when fully ripe. Although the total phenolic content multiplied almost threefold during ripening, its rate of increase was not as high as that of the anthocyanin content. For this reason, the contribution of anthocyanins to total phenolic content was modest at the half-ripe stage, but accounted for almost 80% when ripe. Polymeric color proportion was the only monitored attribute that decreased during ripening.

2.3. Major Anthocyanins in American Elderberry

Major anthocyanins in the extract from fully-ripe fruits (FRFE) (listed in order of elution, Figure 2) were identified as cyanidin-3,5-diglucoside (peak 1), cyanidin-3-sambubioside-5-glucoside (peak 2), cyanidin-3-(*cis*)-coumaroylsambubioside-5-glucoside (peak 3), cyanidin-3-(*trans*)-coumaroyl-glucoside (peak 4), cyanidin-3-(*trans*)-coumaroylsambubioside-5-glucoside (peak 5) according to their visible spectra, MS data, and retention times. The extract from half-ripe fruits (HRFE) presented the same anthocyanins with the exception of peak 4. All anthocyanins identified were cyanidin-derivatives, with cyanidin-3-(*trans*)-coumaroylsambubioside-5-glucoside being the most predominant, representing 65–70%

of the total anthocyanin content. The *cis* isomers of cyanidin-3-coumaroylsambubioside-5-glucoside (peak 3) were present at both ripening stages, accounted for 7.6% (in HRFE) and 1.8% (in FRFE) of the peak area. The *cis* isomers eluted earlier than its *trans* counterpart showed lower absorption at 310–360 nm, and displayed a larger $l_{vis\text{-}max}$ (~525 nm). *p*-Coumaric acid was the only acylation found in American elderberry anthocyanins. Acylated anthocyanins constituted ~80% of the total pigments at the half-ripe stage, and ~70% when fully ripened. Nevertheless, ripened berries were much more abundant in both acylated and non-acylated anthocyanins due to their higher total anthocyanin content.

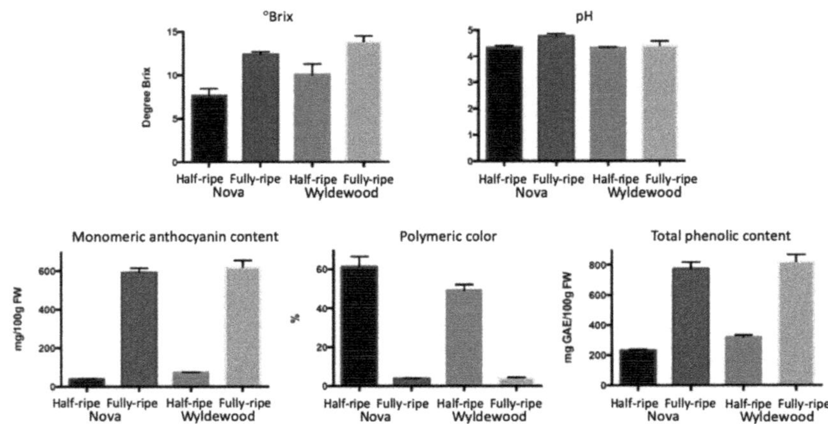

Figure 1. Comparison of 'Nova' and 'Wyldewood' fruits' phytochemical attributes (°Brix, pH, monomeric anthocyanins, polymeric color, total phenolic content) at the half and fully-ripe stages. Results expressed as mean ± SD (n = 3).

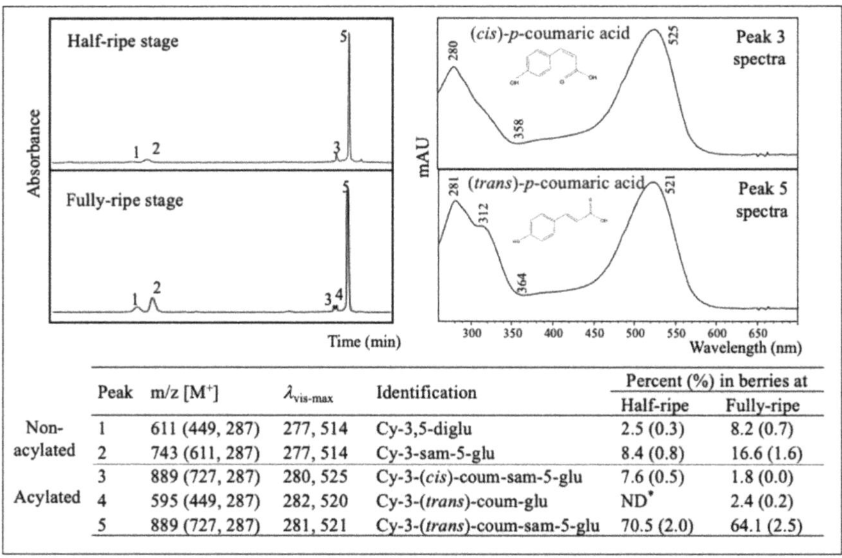

	Peak	m/z [M⁺]	$\lambda_{vis\text{-}max}$	Identification	Percent (%) in berries at	
					Half-ripe	Fully-ripe
Non-acylated	1	611 (449, 287)	277, 514	Cy-3,5-diglu	2.5 (0.3)	8.2 (0.7)
	2	743 (611, 287)	277, 514	Cy-3-sam-5-glu	8.4 (0.8)	16.6 (1.6)
Acylated	3	889 (727, 287)	280, 525	Cy-3-(*cis*)-coum-sam-5-glu	7.6 (0.5)	1.8 (0.0)
	4	595 (449, 287)	282, 520	Cy-3-(*trans*)-coum-glu	ND*	2.4 (0.2)
	5	889 (727, 287)	281, 521	Cy-3-(*trans*)-coum-sam-5-glu	70.5 (2.0)	64.1 (2.5)

Figure 2. HPLC chromatograms of half and full-ripe 'Nova' extracts detected at 520 nm, their peak identifications, and quantifications, and the UV-Vis spectra of peaks 3 and 5. Cy: Cyanidin; glu: glucoside; sam: sambubioside; coum: coumaroyl. *ND: Not detected.

2.4. Anthocyanin Profile of Different Cultivars

Different cultivars had similar anthocyanin profiles with minor differences in the proportion of individual peak areas. Among all cultivars tested, 'Wyldewood' was overall higher in non-acylated pigments but contained the lowest proportion of the *trans*-isomers ($p < 0.05$) (Table 2). 'Adam' and 'York' had the highest levels cyanidin-3-(*cis*)-coumaroylsambubioside-5-glucoside, while 'York' contained a significantly lower percentage of cyanidin-3,5-diglucoside ($p < 0.05$). Nevertheless, a higher percentage of a certain compound in one cultivar does not necessarily represent higher amount since cultivars also varied on total anthocyanin content.

Table 2. Percentage of individual anthocyanin in different American elderberry cultivars. Anthocyanins are listed in the order of elution. Results expressed as mean ± SD ($n = 3$). Cultivars with different superscripts were significantly different ($p < 0.05$).

Peak		Adam	Johns	Nova	Wyldewood	York	Overall
Non-acylated	1	7.2 ± 0.7^{b}	7.9 ± 0.7^{b}	8.2 ± 0.7^{b}	8.5 ± 0.5^{b}	5.2 ± 0.2^{a}	7.4 ± 1.3
	2	16.2 ± 2.3^{a}	15.4 ± 1.1^{a}	16.6 ± 1.6^{a}	18.6 ± 0.5^{a}	16.1 ± 0.7^{a}	16.6 ± 1.7
	Sum	23.3 ± 3^{a}	23.3 ± 1.8^{a}	24.8 ± 2.2^{a}	27.2 ± 1.0^{b}	21.3 ± 0.9^{a}	24.0 ± 2.6
Acylated	3	2.3 ± 0.2^{b}	1.7 ± 0.1^{a}	1.8 ± 0.0^{a}	1.9 ± 0.1^{a}	2.3 ± 0.2^{b}	2.0 ± 0.3
	4	2.4 ± 0.2^{b}	2.1 ± 0.1^{b}	2.4 ± 0.2^{b}	1.5 ± 0.1^{a}	2.1 ± 0.3^{b}	2.1 ± 0.4
	5	67.1 ± 2.7^{b}	67.0 ± 2.3^{b}	$64.1 \pm 2.5^{a,b}$	59.8 ± 0.6^{a}	67.0 ± 1.8^{b}	65.0 ± 3.5
	Sum	71.9 ± 2.4^{b}	70.8 ± 2.2^{b}	$68.3 \pm 2.6^{a,b}$	63.2 ± 0.6^{a}	71.4 ± 1.6^{b}	69.1 ± 3.7

2.5. Spectral Properties of American Elderberry Extract under Various pH

All extracts showed similar $l_{vis\text{-}max}$ ~520 nm at pH 2, but some differences on $l_{vis\text{-}max}$ were observed at higher pH. Crude HRFE did not show a clear $l_{vis\text{-}max}$ at pH 4–6, while the $l_{vis\text{-}max}$ of its fully-ripe counterpart increased from 523 to 549 nm when the pH increased from 4 to 7, although its visible absorbance was weak in this pH range (Figure 3, Table 3). When the pH increased to alkaline region, the $l_{vis\text{-}max}$ of both HRFE and FRFE shifted to 580–600 nm region. Interestingly, FRFE always obtained larger $l_{vis\text{-}max}$ than HRFE. The HRFE generally showed higher absorbance at 400–490 nm, rendering orange-yellowish undertones at all pH.

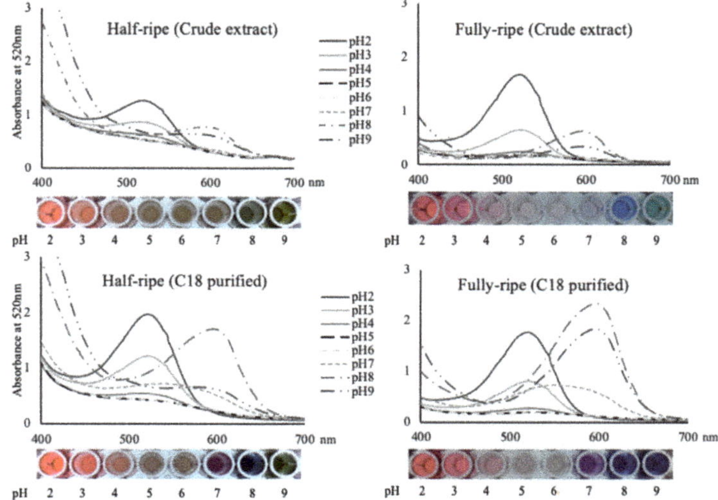

Figure 3. Spectral characteristics of American elderberry 'Nova' extracts at two ripening stages before and after C-18 semi-purification in pH 2–9 buffers. Data was collected 1 h after mixing.

Table 3. Spectrophotometric ($l_{vis\text{-}max}$, nm) and colorimetric (CIE-L*, C*, h*) data of half and fully-ripe 'Nova' extracts before and after C-18 semi-purification in pH 2–9 buffers ($n = 3$). Data was collected 1 h after mixing.

Extracts		pH2	pH3	pH4	pH5	pH6	pH7	pH8	pH9
				Lambda Max (l_{max}, nm)					
Half-ripe	CE [1]	520 ± 1	515 ± 1	ND [2]	ND	ND	ND	593 ± 2	580 ± 0
	C18	519 ± 3	522 ± 1	516 ± 3	ND	ND	540 ± 3	596 ± 1	577 ± 5
Fully-ripe	CE	519 ± 2	524 ± 3	523 ± 1	529 ± 9	528 ± 2	549 ± 1	600 ± 0	600 ± 0
	C18	520 ± 0	522 ± 0	524 ± 3	527 ± 1	523 ± 6	551 ± 2	597 ± 1	597 ± 1
				L* (Lightness)					
Half-ripe	CE	54.8 ± 0.3	59.8 ± 0.1	64.2 ± 1.5	66.9 ± 0.1	68.0 ± 0.1	66.0 ± 0.3	54.6 ± 0.6	54.6 ± 0.4
	C18	55.5 ± 0.1	60.0 ± 0.5	70.5 ± 0.4	73.1 ± 0.3	74.1 ± 2.3	56.3 ± 0.4	35.5 ± 0.4	54.9 ± 0.5
Fully-ripe	CE	64.0 ± 0.7	75.6 ± 0.4	85.4 ± 0.3	88.1 ± 0.3	88.3 ± 0.2	83.0 ± 0.3	73.8 ± 0.3	73.4 ± 0.7
	C18	63.7 ± 0.1	73.0 ± 0.2	85.1 ± 0.1	87.3 ± 0.5	87.8 ± 0.7	58.3 ± 0.9	39.4 ± 0.1	45.3 ± 0.4
				C* (Chroma)					
Half-ripe	CE	50.8 ± 0.5	40.5 ± 0.2	33.7 ± 0.3	32.5 ± 0.1	31.4 ± 0.3	30.3 ± 0.3	27.0 ± 0.5	48.9 ± 0.4
	C18	67.5 ± 0.1	52.6 ± 0.2	30.7 ± 0.1	26.3 ± 0.2	25.5 ± 2.1	22.4 ± 0.1	7.4 ± 0.4	46.5 ± 0.5
Fully-ripe	CE	52.6 ± 0.3	28.6 ± 0.3	8.1 ± 0.2	3.2 ± 0.2	2.5 ± 0.2	2.7 ± 0.3	17.5 ± 0.2	14.3 ± 0.6
	C18	66.0 ± 0.1	45.7 ± 0.5	15.4 ± 0.2	7.7 ± 0.2	7.1 ± 0.2	34.4 ± 0.7	41.5 ± 0.4	30.9 ± 0.2
				h* (Hue Angle)					
Half-ripe	CE	31.9 ± 0	42.5 ± 0.1	58.0 ± 0.3	67.4 ± 0.2	70.8 ± 0.2	72.2 ± 0.2	103.1 ± 0.1	88.3 ± 0.4
	C18	23.9 ± 0.1	23.2 ± 0.3	44.6 ± 0.3	60.5 ± 0.3	64.7 ± 1.2	32.4 ± 0.4	168.0 ± 4.6	85.5 ± 0.5
Fully-ripe	CE	1.1 ± 0.6	356.9 ± 0.2	7.5 ± 0.7	36.2 ± 1.4	51.6 ± 0.8	341.8 ± 4.0	249.2 ± 0.3	153.4 ± 1.0
	C18	9.9 ± 0.1	1.4 ± 0.1	10.9 ± 0.6	28.9 ± 0.6	36.7 ± 1.6	316.2 ± 0.5	270.4 ± 0.4	238.1 ± 0.2

[1] Crude extract. [2] Not detected.

Both HRFE and FRFE exhibited sharper peaks and higher absorbance at their $l_{vis\text{-}max}$ after C-18 semi-purification, particularly at low and alkaline pH, concurrent with a more vibrant color expression (Figure 3). The shape of the absorption spectra and the $l_{vis\text{-}max}$ of the HRFE more closely resembled those of FRFE after semi-purification. In spite of this, semi-purified HRFE still showed higher absorbance at 400–490 nm, and lower absorbance at $l_{vis\text{-}max}$ than FRFE at most pH. Semi-purification of FRFE had a negligible impact on $l_{vis\text{-}max}$. However, it reduced the absorption between 400–440 nm and increased the absorbance at $l_{vis\text{-}max}$ notably, particularly under alkaline pH, showing bolder color expression.

2.6. Colorimetric Properties of American Elderberry Extract under Various pH

The FRFE expressed colors from red to colorless to purple and blue when the pH increased from 2 to 9 (Figure 3). Under pH 2–3, the FRFE displayed intense red hues with h^*_{ab} between 0.4° and 7.3° and C^*_{ab} between 28.6–52.6 (Table 3). As the pH increased to mildly acidic (4–6), the L* increased by over 20 units and C^*_{ab} were generally small (≤ 10) due to the deprotonation of flavylium cations, with extracts appearing almost colorless. When the pH further increased into the alkaline region, a purple-bluish color appeared due to the formation of quinonoidal bases, exhibiting h^*_{ab} between 249.2–341.8°, and C^*_{ab} increased back to 17.5 at pH 8.

The h^*_{ab} of HRFE was between 31.9° (red)–103.1° (yellow) under the tested pH range, as predicted from our spectral data (Figure 3, Table 3). The HRFE generally had smaller L* and larger C^*_{ab} than its fully-ripe counterpart, especially under mildly acidic pH.

The color properties (L*, C^*_{ab}, h^*_{ab}) of FRFE before and after semi-purification were similar at acidic pH, while they largely differed at alkaline pH (Table 3). At pH 7–9, the semi-purification resulted in a decrease in the L* of 24.7–34.4 units and an increase in C^*_{ab} of 16–31.7 units. Semi-purification also resulted in the shifts of h^*_{ab} from 341.8° to 316.2° (purple hues) at pH 7 and from 249.2° to 270.4° (blue hues) at pH 8. The C-18 semi-purification greatly enhanced color intensity and saturation of the extract, rendering bluer color hues under neutral to alkaline pH.

At acidic pH, extracts from all cultivars except 'York' showed high similarities on their color properties ($\Delta E < 5$ in a pairwise comparison) (Table 4). At pH 3, the ΔE between the extracts from 'York' and other cultivars fell between 7.28 and 8.25, mainly resulting from their different L* (lightness) value as the L* of York extracts (76.13) were significantly lower than others (82.56–83.93, data not shown). Larger ΔE (up to 14.4) was observed when comparing the color of these extracts under alkaline pH. At pH 8, most pairwise ΔE were larger than 5. Different from the findings at acidic pH where the variance mostly came from L* value, the relatively large ΔE at alkaline pH was mainly contributed by Δb* (blue-yellow color) value. At pH 8, the a* values were similar among cultivars, ranging between −5.98 ('York') and −7.35 ('Johns'), while the b* value varied between −7.69 ('York') and −21.52 ('Adam'), leading to an overall large ΔE among cultivars.

Table 4. Mean (n = 3) color differences (ΔE) of the extracts from different cultivars of fully-ripe fruits at pH 2–9.

	Johns	Nova	Wyldewood	York
		Adam		
pH2	1.25	4.61	0.91	3.31
pH3	0.81	4.34	1.60	8.05
pH4	1.09	0.86	0.17	3.52
pH7	1.18	2.18	1.61	4.38
pH8	6.06	6.11	8.78	14.44
		Johns		
pH2		5.34	1.81	4.21
pH3		4.79	2.16	8.21
pH4		1.75	1.05	3.60
pH7		2.71	12.13	4.62
pH8		1.49	3.00	9.21
		Nova		
pH2			3.96	3.69
pH3			2.92	8.25
pH4			0.78	3.30
pH7			2.48	2.40
pH8			2.75	8.59
		Wyldewood		
pH2				2.50
pH3				7.28
pH4				3.41
pH7				3.97
pH8				6.73

3. Discussion

American elderberry fully-ripe fruits had a °Brix of 12.0–13.1 and a pH of 4.5–4.9, similar to the previously reported data [1,19]. The °Brix of American elderberry is comparable to those of European elderberry (8.9–14.6) [1,20] and blueberry (10.3–13.9) [21], but higher than those of blackberry (4.9–8.0) [22] and raspberry (9.4–11.5) [23]. The pH of most berries ranges between 2.8–4.2 [20–23]. High °Brix and low acidity are usually associated with a mild, pleasant taste. Nevertheless, American elderberry is seldom consumed as fresh fruit and used most extensively as processed food and beverages [2]. This would be ascribed to its rather small fruit size [24], and tart, tangy or bitter sensory attributes brought by its abundant polyphenols contained [25]. Most anthocyanins have negligible color expression at pH ~4.5 as they transited into the colorless hemiketal form, especially for 3,5-glycosides derivatives [26]. American elderberry was abundant in 3,5-diglycosides (~90%) with a fruit pH ~4.5 (Table 1, Figure 2). Its intense dark purplish-black coloration could be explained by the lower pH of the vacuoles, where anthocyanins are usually localized together with considerable organic acids [27]. The pH of grape vacuole has been reported to be ~1 unit lower than the pH of grape pulp [27]. Similarly, the pH of American elderberry vacuoles is expected to be lower than the fruit pH, therefore protects the integrity of anthocyanins.

Relatively large variability in anthocyanin and phenolic content within the cultivar was observed in both this and previous studies, where berries were sorted by their surface color ahead of phytochemical analysis [20]. Surface color is usually used as a maturity indicator of fruits; nevertheless, a slight variation in pH might cause considerable changes on the intensity of fruit color [27]. Moreover, fruits may exhibit the same color but have varying anthocyanin content, as the quantification of anthocyanin content is related to fruit size and water content, as well as anthocyanin distribution.

Reports on the effect of cultivar on phytochemical content vary among different studies. For example, Mathieu et al. reported higher anthocyanin and phenolic content of 'Nova' than 'York' during a two-year observation [20], consistent with our findings, while an opposite observation was reported by Perkins-Veazie et al. [19]. Conflicting results may also occur within the same study, as Lee and Finn observed significantly higher anthocyanins content of 'Adams' than 'Johns' and 'York' in 2005, but not in 2004 [1]. Such variability suggests an interplay involving both genetics and environmental factors. Geographic, climatic, edaphic conditions, and even the position of the sampled fruits on the mother-plants can all be origins of the variance displayed [27]. American elderberry appeared to be highly responsive to these environmental factors; thus, cultivar selection with interested biochemicals should be in accordance with specific environmental conditions and cultivation approaches.

American elderberry fruits went through significant phytochemical changes during ripening, which was reflected by increased °Brix, pH, phenolic (including anthocyanins) content, and decreased percentage of polymeric color. During red fruit ripening, an increase in °Brix and pH is the most common, along with an accumulation of anthocyanins. The increased °Brix and pH are attributed to the acids in the fruit converting to sugars during the ripening process [2,27]. Although both increased significantly, the total phenolic content did not increase as much as the anthocyanin content did. This phenomenon was also observed during the development of other berries, like grape and raspberry [28,29]. The total phenolic content is an equilibrium between biosynthesis and metabolism, thus can ascend, decline, or stay flat during maturation depending on enzyme activities and precursor availability [27]. On the other hand, anthocyanins are continuously synthesized during fruit development, thereby the accumulation of anthocyanin and phenolic is not expected to be correlated.

Acylated anthocyanins are a group of more stable anthocyanins and more commonly found in vegetable sources like red radish, black carrot, and red cabbage [13]. Conventional fruit sources, such as cranberry, blackberry, blueberry, or European elderberry usually only contain non-acylated anthocyanins [13]. Red grape may contain about 30% anthocyanins acylated with aromatic or aliphatic acids [30]. Black goji berry was reported to be abundant in anthocyanins acylated with *p*-coumaric acids [12], but it is mainly distributed in central and east Asia with limited availability. The high acylated anthocyanin content as well as its accessibility in North America make American elderberry an excellent candidate as a natural colorant with many attractive traits.

Both exogenous and endogenous factors can initiate anthocyanin composition evolution. For example, in blueberry, cyanidin derivatives were more abundant in ripe fruits, whereas malvidin derivatives were predominant in overripe fruits [27]; the stink bug infestation decreased malvidin-derivatives and increased other aglycone derivatives [31]. Our study revealed a higher ratio of acylated anthocyanins in American elderberry at the half-ripe stage (Figure 2). Similar fluctuation has been reported in some vegetables, such as colored waxy corn [32] or red cabbage [33], both featuring a higher acylated anthocyanin ratio at an earlier ripening stage. Fruit sources, like blueberry, usually express variation on anthocyanin aglycone.

Cyanidin-3-(*cis*)-coumaroylsambubioside-5-glucoside made up 7.6% and 1.8% of the total pigments in HRFE and FRFE, respectively (Figure 2). This isomer differed from its *trans* counterpart only on the spatial configuration of the acyl group but occurs much more rarely in nature with different UV-Vis spectral characteristics. An absorption band is generally observed in the 310–360 nm range for anthocyanins acylated with aromatic acids, giving a higher ratio of $A_{310–360}/A_{vis\text{-}max}$ than for aliphatic acylated or non-acylated anthocyanins [13]. However, our UV-Vis spectra of *cis* isomers revealed a much lower absorption in that range, being approximately half of $A_{310}/A_{vis\text{-}max}$ ratio of *trans* isomers. Apart from that, the *cis* isomers displayed a larger $l_{vis\text{-}max}$ (525 nm) than their *trans* counterparts (521 nm) in the same solvent. The coexistence of *cis* and *trans* isomers are seldom found in edible sources, and current studies about the impact of *cis-trans* configuration on the color of these pigments are few and limited only to petunidin or delphinidin derivatives [34]. Therefore, our research provided novel information about the impact of acyl group spatial configuration on cyanidin derivatives.

Different anthocyanin/polyphenol ratios between the HRFE and FRFE can explain their spectrophotometric and colorimetric differences. The anthocyanin/phenolic ratio increased from ~20% to 85% from half- to fully-ripened. Therefore, when both extracts were standardized by their anthocyanin concentration, the HRFE contained significant higher polyphenol levels. The main phenolics in the extracts besides anthocyanins were hydroxycinnamic acids and flavonols (data not shown). These compounds can affect the extract color expression via producing yellow colors on their own ($l_{vis\text{-}max}$ ~360 nm), or anthocyanin copigmentation. Though the copigmentation may enhance anthocyanin stability, such interaction is only efficient with non-acylated anthocyanins at a low acidic pH [35], and may alter the color properties simultaneously. With the removal of these interfering compounds, the American elderberry pigments expressed more vivid colors and obtained a sharper spectrum.

Due to the high similarities of their anthocyanin profiles, different cultivars produced similar color with most pairwise $\Delta Es < 5$. All of the extracts were able to express blue hues at pH 8 with h* between 233.1–253.6°, resembling that of FD & C Blue No.2 [33]. About 90% of pigments in American elderberry were cyanidin-3,5-diglycosylated (Figure 1). This glycosidic pattern is characterized by a larger $l_{vis\text{-}max}$ than cyanidin-3-glycosides at all pH, therefore is capable of expressing blue hues at alkaline pH [26]. Common fruit-based anthocyanin sources like European elderberry and chokeberry lack this glycosidic pattern and do not express blue hues at any pH. Currently, natural sources of blue colorants are very limited, thus the glycosidic pattern along with the high acylated anthocyanin content make American elderberry a desirable natural colorant candidate.

4. Materials and Methods

4.1. Reagents

ACS or HPLC grade acetone, chloroform, methanol, trifluoroacetic acid (TFA), ammonium hydroxide (NH_4OH), acetonitrile, potassium hydroxide (KOH), and sodium phosphate dibasic (Na_2HPO_4) were purchased from Fisher Scientific (Fair Lawn, NJ, USA). ACS grade formic acid was purchased from Honeywell (Morris Plains, NJ, USA).

4.2. Collection of Plant Materials

American elderberry fruits of cultivars 'Adams', 'Johns', 'Nova', 'Wyldewood' and 'York' were harvested from plants grown at the South Center, The Ohio State University, near Piketon, Ohio, USA, during two summers in 2015 and 2017. Fruits harvested between mid-August and early-September 2015 were used to investigate the impact of cultivar. 'Nova' and 'Wyldewood' fruits were further harvested in late-August 2017 to determine the impact of ripeness.

Fruit samples were harvested from three plants of each cultivar. After the harvest, samples were placed in polyethylene bags, labeled, and transported to the lab. Fruits were stored at −20°C until further analysis.

4.3. Determination of Maturity Stage

Elderberry fruits on a branch do not mature at the same time (Figure 4); thus, individual fruits were sorted into one of three maturity categories according to their appearance and surface color before analysis: (1) Immature stage: fruits were entirely green or with minimal red color shown on the surface; (2) Half-ripe stage: fruits were overall red with minimal green color shown on the surface; (3) Fully-ripe stage: fruits were entirely dark purple-violet on the surface (Figure 4). Surface color properties (Hunter CIE L*, a*, b*) of samples in categories (2) and (3) were measured by a Minolta handheld colorimeter (Konica Minolta, Osaka, Japan). The L* value indicated the level of lightness with 0 representing the darkest black and 100 representing the brightest white; the a* value indicated the redness (+) and greenness (−) of the object and b* value indicated the yellowish (+) and bluish (−) of the objective, according to the International Commission on Illumination (CIE) [36]. To determine the impact of cultivar, only berries in category (3) were retained to minimize the variance in maturity. To determine the impact of ripeness, berries in category (2) and (3) were stored separately for further analysis.

Figure 4. American elderberry fruits on the same branch at (**1**) Immature, (**2**) Half-ripe, and (**3**) Fully-ripe stages.

4.4. *Fruit Extracts and Their Quality Attributes*

°Brix and pH were measured during sample extraction. Each sample was weighed (~30 g) and blended with liquid nitrogen for 2 min. °Brix was quantified using a handheld refractometer (Atago Co., Ltd., Tokyo, Japan), and pH was measured using a pH meter (Mettler Toledo, Inc., Columbus, OH, US) after the powder was thawed into puree.

Anthocyanins and other phenolics were extracted with acidified acetone and partitioned with chloroform [37]. About 30 mL of acetone acidified with 0.01% HCl was added to the puree and homogenized for 2 min. The blend was then vacuum filtered using a Buchner funnel with Whatman #4 filter paper (Whatman Inc, NJ, US). After filtration, the slurry was re-extracted with 70%(*v*/*v*) aqueous acetone with 0.01% HCl until a pink color was barely visible. The anthocyanin extract was placed in a separatory funnel with 2 volumes of chloroform, and the mixture was gently mixed and left to sit at room temperature until a good separation was achieved. The top layer (anthocyanin and phenolic concentrate) was collected into a flask, while the bottom layer (chloroform and polar solvents) was discarded. Residual acetone was evaporated using a Buchi rotary evaporator at 40 °C. The final volume of sample was documented for quantification purposes.

4.5. *Quantification of Anthocyanin and Phenolic Content*

Monomeric anthocyanin content was estimated using the pH differential method [38]. The absorbance of the anthocyanin extracts at pH 1 and pH 4.5 was measured using a SpectraMax 190 Microplate Reader (Molecular Devices, Sunnyvale, CA, USA) at 520 nm (λ_{max}) and 700 nm with automated 1-cm pathlength correction. The molecular weight (449.2 g mol **) and molar extinction coefficient (29,600 L cm ** mol **) of cyanidin-3-glucoside (C3G) were used for calculation. The total monomeric anthocyanin content was expressed as mg C3GE per 100 g of FW.

Polymeric color was determined by measuring the absorbance of the extracts at 420 nm, 520 nm (λ_{max}), and 700 nm after being treated with sodium bisulfite [38]. The percent polymeric color was expressed as the ratio between polymerized color and overall color density.

Total phenolic content was quantified using the Folin-Ciocalteau method and expressed as gallic acid equivalents [39]. Absorbance was read at 765 nm using the SpectraMax 190 Microplate Reader. Total phenolic content was calculated and expressed as mg gallic acid equivalents (GAE) per 100 g of FW.

4.6. Sample Purification

Anthocyanin extracts were semi-purified using solid phase extraction with a C-18 cartridge. The C-18 cartridge was activated by methanol before loading the elderberry crude extract. The crude extract was then washed with acidified water (0.01% *v/v* HCl) and ethyl acetate to eliminate acids, sugars, and less polar phenolics. The semi-purified anthocyanins were eluted with acidified methanol (0.01% *v/v* HCl) and evaporated until dryness to remove all methanol. Anthocyanins were re-dissolved in acidified water (0.01% *v/v* HCl) for further analysis.

4.7. Anthocyanin Identification

Anthocyanin identification was conducted using a Shimadzu ultra-High-Pressure Liquid Chromatography (uHPLC) system equipped with LC-2040C pumps coupled to a triple-quadrupole Shimadzu LCMS-8040 mass spectrometer using LC-2040 PDA detector (Shimadzu, Columbia, MD, USA). A Restek reverse phase C-18 column (50 × 2.1 mm) with 1.9 μm particle size was used (Restek Corporation, Bellefonte, PA, USA). Samples were filtered through a 0.2 μm RC membrane filter (Phenomenex, Torrance, CA, USA) before injection (10 μL). Samples were run using a flow rate of 0.25 mL/min and solvent A: 4.5% formic acid in HPLC water and solvent B: 100% acetonitrile, at 60 °C. Anthocyanin separation was achieved using a linear gradient from 1% to 3% B in 2 min; 2 to 3 min, 3% to 4.5% B; 3 to 7.5 min, 4.5% to 8.5% B; 7.5 to 13 min, 8.5% to 40% B. Spectra (200–700 nm) was collected. The mass spectrometer was set for positive ion mode, with total ion scan from 100–1000 m/z and precursor ion scan at 271, 287, 301, 303, 317, and 331 m/z. MS data, order of elution, and comparison to literature were used for the anthocyanin identification.

4.8. Buffer and Sample Preparation

Buffer solutions of pH 2–9 were prepared as follows: 0.025 M KCl for pH 2, 0.1 M sodium acetate for pH 3–6, 0.2 M Na_2HPO_4 and 0.2 M NaH_2PO_4 for pH 7–9 [12,40]. The final anthocyanin concentration was adjusted to 100 mM with buffer and kept at 4 °C in dark.

4.9. Spectrophotometric and Colorimetric Analysis

The initial spectral measurement (400–700 nm, 1 nm interval) of each extract at pH 2–9 was taken 1 h after mixing with buffers, when sufficient equilibration was achieved. A 300 mL aliquot was pipetted into a poly-D-lysine coated polystyrene 96-well microplate and read on the SpectraMax 190 Microplate Reader. The spectral data was converted into colorimetric data (L* (lightness), a*, b*, C^*_{ab} (chroma), h^*_{ab} (hue angle)) using the ColorBySpectra software according to CIE 1964 standard observer, D65 illuminant spectral distribution and 10° observer angle [41]. The color difference (ΔE) was calculated using the following equation:

$$\sqrt{(\Delta a^{**} + \Delta b^{**} + \Delta L^{**})}$$

4.10. Statistical Analysis

One-way ANOVA (two-tailed, $a = 0.05$) and post hoc Tukey test were conducted to determine the impact of cultivar. A *t*-test was conducted to evaluate the impact of ripeness. All of the statistical analysis was conducted using Prism software (GraphPad, La Jolla, CA, USA).

5. Conclusions

American elderberry differed on most phytochemical attributes, including pH, anthocyanin content and profile, as well as anthocyanin/phenolic ratio. Those differences led to small but visible color and spectral differences under various pH environments, particularly under alkaline pH. Johns' berries exhibited overall higher anthocyanin content and acylated anthocyanin proportion with relatively low non-anthocyanin phenolic content, possessing more favorable attributes for potential application as natural colorants. All these attributes increased during fruit ripening, except the percentage of polymeric color and acylated anthocyanin. The acylated anthocyanin proportion dropped from 80% at the half-ripe stage to 70% when fully ripened. All the major anthocyanins in American elderberry were cyanidin-derivatives, with both *cis* and *trans*-configured *p*-coumaric acid acylation co-existing. FRFE exhibited a "red-colorless-purple-blue" color expression pattern at pH 2–9 and expressed more vibrant colors and sharper spectra after C-18 semi-purification.

Our results contribute to the selection of proper cultivars and ripeness for specific applications of this North American native plant and expand the potential scientific and industrial applications. The high proportion of acylated anthocyanins, along with blue hues ($\lambda_{vis\text{-}max}$ ~600 nm) expression of its extracts under alkaline pH, make it a promising alternative to synthetic dyes and expands the natural color spectrum. This is particularly attractive for food applications as most fruit sources contain little to no acylated anthocyanins and tend to be less stable. On the other hand, vegetable sources with high levels of acylated anthocyanins typically possess stronger flavor. Moreover, *trans* and *cis*-configured cyanidin-derivatives co-exist in American elderberry, which is rare in nature. Thus, this material can be further utilized to explore the impact of acyl group spatial configuration on anthocyanin color properties.

Author Contributions: Investigation, formal analysis and writing—original draft preparation, Y.Z.; plant selection and growing, sample collection and writing—review and editing, Y.G.G.; writing—review and editing, supervision, project administration and funding acquisition, M.M.G. All authors have read and agreed to the published version of the manuscript.

Funding: This work was supported in part by a Specialty Crop Block Grant from USDA Agricultural Marketing Service through Ohio Department of Agriculture, the USDA National Institute of Food and Agriculture, Hatch Project OHO01423, Accession Number 1014136.

Conflicts of Interest: The authors declare no conflict of interest.

Abbreviations

ACN	Anthocyanin
Cou	Coumaroyl
Cy	Cyanidin
FRFE	Extract from fully-ripe fruits
Glu	Glucoside
HPLC	High-performance liquid chromatography
HRFE	Extract from half-ripe fruits
Mono	Monomeric
MS	Mass spectroscopy
ND	Not detected
Poly	Polymeric
Sam	Sambubioside
TP	Total phenolics

References

1. Lee, J.; Finn, C.E. Anthocyanins and other polyphenolics in American elderberry (*Sambucus canadensis*) and European elderberry (*S. nigra*) cultivars. *J. Sci. Food Agric.* **2007**, *87*, 2665–2675. [CrossRef] [PubMed]
2. Charlebois, D.; Byers, P.L.; Finn, C.E.; Thomas, A.L. Elderberry: Botany, Horticulture, Potential. *Hortic. Rev.* **2010**, *37*, 213–280.
3. Mikulic-Petkovsek, M.; Schmitzer, V.; Slatnar, A.; Todorovic, B.; Veberic, R.; Stampar, F.; Ivancic, A. Investigation of anthocyanin profile of four elderberry species and interspecific hybrids. *J. Agric. Food Chem.* **2014**, *6*, 5573–5580. [CrossRef] [PubMed]
4. Ozgen, M.; Scheerens, J.; Reese, R.; Miller, R. Total phenolic, anthocyanin contents and antioxidant capacity of selected elderberry (*Sambucus canadensis L.*) accessions. *Pharmacogn. Mag.* **2010**, *6*, 198–203. [CrossRef]
5. Byers, P.L.; Thomas, A.L.; Millican, M. "Wyldewood" elderberry. *HortScience* **2010**, *45*, 312–313. [CrossRef]
6. Finn, C.E.; Thomas, A.L.; Byers, P.L.; Serçe, S. Evaluation of American (*Sambucus canadensis*) and European (*S. nigra*) elderberry genotypes grown in diverse environments and implications for cultivar development. *HortScience* **2008**, *43*, 1385–1391. [CrossRef]
7. Iannone, R.; Miranda, S.; Riemma, S.; De Marco, I. Life cycle assessment of red and white wines production in southern Italy. *Chem. Eng. Trans.* **2014**, *39*, 595–600.
8. Wilson, R.J.; Business, R.J.W.; Nickerson, G.; Fried, D.; Tree ER, F.; Nursery, B.; Moonshine, U. *Growing Elderberries: A Production Manual and Enterprise Viability Guide for Vermont and the Northeast*; UVM Center for Sustainable Agriculture, University of Vermont Extension: Burlington, VT, USA, 2016.
9. Sidor, A.; Gramza-Michałowska, A. Advanced research on the antioxidant and health benefit of elderberry (*Sambucus nigra*) in food—A review. *J. Funct. Foods* **2015**, *18*, 941–958. [CrossRef]
10. Torabian, G.; Valtchev, P.; Adil, Q.; Dehghani, F. Anti-influenza activity of elderberry (*Sambucus nigra*). *J. Funct. Foods* **2019**, *54*, 353–360. [CrossRef]
11. Future Market Insight. Natural Antioxidants Market—Towards a Higher Quality Product Consumption. 2020. Available online: https://www.futuremarketinsights.com/reports/natural-antioxidants-market (accessed on 30 November 2020).
12. Tang, P.; Giusti, M.M. Black goji as a potential source of natural color in a wide pH range. *Food Chem.* **2018**, *269*, 419–426. [CrossRef]
13. Giusti, M.M.; Wrolstad, R.E. Acylated anthocyanins from edible sources and their applications in food systems. *Biochem. Eng. J.* **2003**, *14*, 217–225. [CrossRef]
14. Petkovsek, M.M.; Ivancic, A.; Todorovic, B.; Veberic, R.; Stampar, F. Fruit Phenolic Composition of Different Elderberry Species and Hybrids. *J. Food Sci.* **2015**, *80*, 2180–2190. [CrossRef] [PubMed]
15. Dyrby, M.; Westergaard, N.; Stapelfeldt, H. Light and heat sensitivity of red cabbage extract in soft drink model systems. *Food Chem.* **2001**, *72*, 431–437. [CrossRef]
16. Inami, O.; Tamura, I.; Kikuzaki, H.; Nakatani, N. Stability of Anthocyanins of *Sambucus canadensis* and *Sambucus nigra*. *J. Agric. Food Chem.* **1996**, *44*, 3090–3096. [CrossRef]
17. Szalóki-Dorkó, L.; Stéger-Máté, M.; Abrankó, L. Effects of fruit juice concentrate production on individual anthocyanin species in elderberry. *Int. J. Food Sci. Technol.* **2016**, *51*, 641–648. [CrossRef]
18. Szalóki-Dorkó, L.; Stéger-Máté, M.; Abrankó, L. Evaluation of colouring ability of main European elderberry (*Sambucus nigra* L.) varieties as potential resources of natural food colourants. *Int. J. Food Sci. Technol.* **2015**, *50*, 1317–1323. [CrossRef]
19. Perkins-Veazie, P.; Thomas, A.L.; Byers, P.L.; Finn, C.E. Fruit composition of elderberry (*Sambucus spp.*) genotypes grown in Oregon and Missouri, USA. *Acta Hortic.* **2015**, *1061*, 219–224. [CrossRef]
20. Mathieu, F.; Charlebois, D.; Charles, M.T.; Chevrier, N. Biochemical changes in American Elder (*Sambucus canadensis*) fruits during development. *Acta Hortic.* **2015**, *1061*, 61–72. [CrossRef]
21. Guedes, M.N.S.; de Abreu, C.M.P.; Maro, L.A.C.; Pio, R.; de Abreu, J.R.; de Oliveira, J.O. Caracterização química e teores de minerais em frutos de cultivares de amoreira-preta cultivadas em clima tropical de altitude. *Acta Sci. Agron.* **2013**, *35*, 191–196.
22. Matiacevich, S.; Celis Cofré, D.; Silva, P.; Enrione, J.; Osorio, F. Quality parameters of six cultivars of blueberry using computer vision. *Int. J. Food. Sci.* **2013**, *2013*, 419535. [CrossRef]
23. Urgar, D.D.; Duralija, B.; Voća, S.; Vokurka, A.; Ercisli, S. A comparison of fruit chemical characteristics of two wild grown Rubus species from different locations of Croatia. *Molecules* **2012**, *17*, 10390–10398.

24. Charlebois, D. Elderberry as a Medicinal Plant. In *Issues in New Crops and New Uses*; ASHS Press: Alexandria, VA, USA, 2007; pp. 284–292.
25. Elez-Garofulić, I.; Kovačević-Ganić, K.; Galić, I.; Dragovi-Uzelac, V.; Savić, Z. The influence of processing on physico-chemical parameters, phenolics, antioxidant activity and sensory attributes of elderberry (*Sambucus nigra* L.) fruit wine. *Hrvat. Casopis Za prehrambenu Tehnol. Biotehnol. I Nutr.* **2012**, *7*, 9–13.
26. Sigurdson, G.T.; Robbins, R.J.; Collins, T.M.; Giusti, M.M. Impact of location, type, and number of glycosidic substitutions on the color expression of o-dihydroxylated anthocyanidins. *Food Chem.* **2018**, *268*, 416–423. [CrossRef] [PubMed]
27. Macheix, J.J.; Fleuriet, A.; Billot, J. *Fruit Phenolics*; CRC Press: Boca Raton, FL, USA, 2018.
28. Ali, K.; Maltese, F.; Fortes, A.M.; Pais, M.S.; Choi, Y.H.; Verpoorte, R. Monitoring biochemical changes during grape berry development in Portuguese cultivars by NMR spectroscopy. *Food Chem.* **2011**, *124*, 1760–1769. [CrossRef]
29. Wang, S.Y.; Chen, C.T.; Wang, C.Y. The influence of light and maturity on fruit quality and flavonoid content of red raspberries. *Food Chem.* **2009**, *112*, 676–684. [CrossRef]
30. He, J.J.; Liu, Y.X.; Pan, Q.H.; Cui, X.Y.; Duan, C.Q. Different anthocyanin profiles of the skin and the pulp of Yan73 (*muscat hamburg* × *alicante bouschet*) grape berries. *Molecules* **2010**, *15*, 1141–1153. [CrossRef]
31. Zhou, Y.; Giusti, M.M.; Parker, J.; Salamanca, J.; Rodriguez-Saona, C. Frugivory by Brown Marmorated Stink Bug (*Hemiptera: Pentatomidae*) Alters Blueberry Fruit Chemistry and Preference by Conspecifics. *Environ. Entomol.* **2016**, *45*, 1227–1234. [CrossRef]
32. Harakotr, B.; Suriharn, B.; Tangwongchai, R.; Scott, M.P.; Lertrat, K. Anthocyanins and antioxidant activity in coloured waxy corn at different maturation stages. *J. Funct. Foods* **2014**, *9*, 109–118. [CrossRef]
33. Ahmadiani, N.; Robbins, R.J.; Collins, T.M.; Giusti, M.M. Anthocyanins contents, profiles, and color characteristics of red cabbage extracts from different cultivars and maturity stages. *J. Agric. Food Chem.* **2014**, *62*, 7524–7531. [CrossRef]
34. Sigurdson, G.T.; Tang, P.; Giusti, M.M. *Cis-trans* configuration of coumaric acid acylation affects the spectral and colorimetric properties of anthocyanins. *Molecules* **2018**, *23*, 598. [CrossRef]
35. Escribano-Bailón, M.T.; Rivas-Gonzalo, J.C.; García-Estévez, I. Wine Color Evolution and Stability. *Red Wine Technol.* **2018**, 195–205.
36. Schand, J. (Ed.) *Colorimetry: Understanding the CIE System*; John Wiley & Sons, Inc.: New York, NY, USA, 2007.
37. Rodriguez-Saona, L.E.; Wrolstad, R.E. Unit F1.1.1-F1.1.11: Extraction, isolation, and purification of anthocyanins. In *Current Protocols in Food Analytical Chemistry*; John Wiley & Sons, Inc.: New York, NY, USA, 2002.
38. Giusti, M.M.; Wrolstad, R.E. Unit F1.2.1-F1.2.13: Characterization and measurement of anthocyanins by UV-visible spectroscopy. In *Current Protocols in Food Analytical Chemistry*; John Wiley & Sons, Inc.: New York, NY, USA, 2002.
39. Waterhouse, A.L. Unit I1.1: Determination of total phenolics. In *Current Protocols in Food Analytical Chemistry*; John Wiley & Sons, Inc.: New York, NY, USA, 2002.
40. Sigurdson, G.T.; Robbins, R.J.; Collins, T.M.; Giusti, M.M. Molar absorptivities (ε) and spectral and colorimetric characteristics of purple sweet potato anthocyanins. *Food Chem.* **2019**, *271*, 497–504. [CrossRef] [PubMed]
41. Farr, J.E.; Giusti, M.M. ColorBySpectra-Academic License. Available online: https://buckeyevault.com/products/absorbance-to-color-al (accessed on 4 December 2020).

Article

Comparative Study of Anti-Gouty Arthritis Effects of Sam-Myo-Whan according to Extraction Solvents

Yun Mi Lee, Eunjung Son and Dong-Seon Kim *

Herbal Medicine Research Division, Korea Institute of Oriental Medicine, 1672 Yuseong-daero, Yuseong-gu, Daejeon 34054, Korea; candykong@kiom.re.kr (Y.M.L.); ejson@kiom.re.kr (E.S.)
* Correspondence: dskim@kiom.re.kr; Tel.: +82-42-868-9639

Abstract: Sam-Myo-Whan (SMW) has been used in Korean and Chinese traditional medicine to help treat gout, by reducing swelling and inflammation and relieving pain. This study compared the effects of SMW extracted by using different solvents, water (SMWW) and 30% EtOH (SMWE), in the treatment of gouty arthritis. To this end, we analyzed the main components of SMWW and SMWE, using high-performance liquid chromatography (HPLC). Anti-hyperuricemic activity was evaluated by measuring serum uric acid levels in hyperuricemic rats. The effects of SMWW and SMWE on swelling, pain, and inflammation in gouty arthritis were investigated by measuring affected limb swelling and weight-bearing, as well as by enzyme-linked immunosorbent assays, to assess the levels of proinflammatory cytokines and myeloperoxidase (MPO). In potassium oxonate (PO)-induced hyperuricemic rats, SMWW and SMWE both significantly decreased serum uric acid to similar levels. In monosodium urate (MSU)-induced gouty arthritis mice, SMWE more efficiently decreased paw swelling and attenuated joint pain compare to SMWW. Moreover, SMWE and SMWW suppressed the level of inflammation by downregulating proinflammatory cytokines (interleukin-1β, tumor necrosis factor-α, and interleukin-6) and MPO activity. HPLC analysis further revealed that berberine represented one of the major active ingredients demonstrating the greatest change in concentration between SMWW and SMWE. Our data demonstrate that SMWE retains a more effective therapeutic concentration compared to SMWW, in a mouse model of gouty arthritis.

Keywords: Sam-Myo-Whan; traditional medicine; gouty arthritis; inflammation; monosodium urate

Citation: Lee, Y.M.; Son, E.; Kim, D.-S. Comparative Study of Anti-Gouty Arthritis Effects of Sam-Myo-Whan according to Extraction Solvents. *Plants* **2021**, *10*, 278. https://doi.org/10.3390/plants10020278

Academic Editor: Stefania Lamponi

Received: 29 December 2020
Accepted: 28 January 2021
Published: 1 February 2021

1. Introduction

Gout is a metabolic disease caused by increased blood uric acid levels (hyperuricemia) and the deposition of monosodium urate (MSU) crystals in the joints, bone, and subcutaneous tissues. Moreover, gout is closely associated with chronic hyperuricemia, which can markedly reduce patient quality of life due to the severe associated pain [1,2]. Currently, a number of anti-gout agents, including anti-inflammatory drugs (colchicine and indomethacin) as well as urate-lowering drugs (allopurinol and benzbromarone) are often selected as primary therapies for gout. Although these agents are generally effective, they are also associated with various adverse effects, including gastrointestinal, hepatic, and renal toxicity and hypersensitivity [3]. Therefore, it is critical to develop novel agents with fewer associated adverse effects while retaining, or improving, their clinical efficacy. Existing evidence suggests that several natural agents exhibit beneficial efficacy and produce fewer side effects in the treatment of gouty arthritis [4,5]. We have, therefore, focused our research on these candidate natural products.

Sam-Myo-Whan (SMW) has been a common prescription for the treatment of gout and is recorded in traditional Eastern medicine, such as Donguibogam and Chinese Pharmacopoeia. It has good therapeutic efficacy in reducing dampness (edema), decreasing heat and swelling (inflammation), and alleviating pain [6,7]. Moreover, SMW and modified SMW, which is combined with other herbal medicines, are commonly used clinically for the treatment of gouty and rheumatoid arthritis in China [8,9]. SMW is composed of

Phellodendri cortex (*Phellodendron chinense* Schneider), Atractylodes rhizome (the rhizome of *Atractylodes chinensis* Koidzumi), and Achyranthes radix (the root of *Achyranthes japonica* (Miq.) Nakai) in a compatible ratio of 2:3:1. SMW has been shown to inhibit lipopolysaccharide (LPS)-induced inflammatory responses by reducing nitric oxide (NO), tumor necrosis factor-α (TNF-α) production, and inducible nitric oxide synthase (iNOS) expression in RAW264.7 cells and BV2 cells [7]. SMW produced dual hyperuricemic actions by downregulating hepatic XOD to reduce uric acid production and inhibiting renal mURAT1, to decrease urate reabsorption and enhance urate excretion in hyperuricemic mice [10]. In addition, SMW effectively treats osteoarthritis by suppressing chondrocyte apoptosis, cartilage matrix degradation, and the inflammatory response [11]. SMW also modifies the expression of matrix metalloproteinases (MMPs)-3 and aggrecanases (ADAMTSs)-4, which are considered key enzymes in cartilage matrix degradation, and enhances the expression of gouty arthritis-reduced tissue inhibitors of metalloproteinases (TIMPs)-1 and -3, resulting in the effective inhibition of cartilage matrix degradation in gouty arthritis [12]. Several recent studies have also reported that SMW may exhibit therapeutic synergy in gouty arthritis by regulating numerous biological processes and pathways. These include the lipopolysaccharide-mediated signaling pathway, positive regulation of transcription, Toll-like receptor, Janus kinase–signal transducer and activator of transcription (JAK–STAT), nucleotide binding and oligomerization domain (NOD)-like receptor, and mitogen-activated protein kinase (MAPK) signaling pathways [13–15]. In addition, SMW used a modified SMW, adding herbal medicines, to maximize the efficacy of patients with gouty arthritis and to alleviate various symptoms of patients with different phases of gouty pathology. Furthermore, modified SMW has exhibited good results on patients with gout characterized by swelling and edema (dampness-heat type in Chinese medicine) and has been shown to inhibit inflammatory factors in the joint fluid of rats with acute gout arthritis [16–19]. However, according to these previous reports, the SMW was pulverized to a fine powder and suspended in distilled water, or extracted by refluxing with water. While several studies have reported on the efficacy of SMW as a treatment option, there have been no investigations into the differences in composition and efficacy according to the extraction solvent used. Although traditional Chinese and Oriental herbal medicines have used water extracts, ethanol or ethanol/water mixture has recently been introduced as an extraction solvent for pharmaceuticals and dietary supplements. Moreover, the Korea Food and Drug Administration exempts or requires minimum toxicity test data for drug approval of Oriental herbal medicine when using ethanol content up to 30% in mixture with water as an extraction solvent. Thus, this study investigated the differences and changes in the ingredients and efficacy of SMW according to the extraction solvent, namely water (SMWW) and 30% ethanol (SMWE). The quantities of index components and the anti-gouty arthritis activities of two kinds of SMW extract were compared in rat and mouse models.

2. Results

2.1. Chemical Profiling Analysis of SMWW and SMWE

Based on their UV–Vis absorption spectra and retention times, palmatine, armepavine, and berberine, protoberberine groups with quaternary ammonium salt structures, were identified as major components of SMW. SMWW contained 15.2 ± 0.09 mg/g of palmatine, 18.7 ± 0.17 mg/g of armepavine, and 21.1 ± 0.23 mg/g of berberine; while SMWE contained 14.2 ± 0.40 mg/g of palmatine, 21.2 ± 0.26 mg/g of armepavine, and 27.9 ± 0.16 mg/g of berberine. We also identified small amounts of atractylenolides I and III, which are part of the sesquiterpenoid group with three isoprene units, by comparing their retention times and UV–Vis absorption spectra with their reference standards (Figure 1).

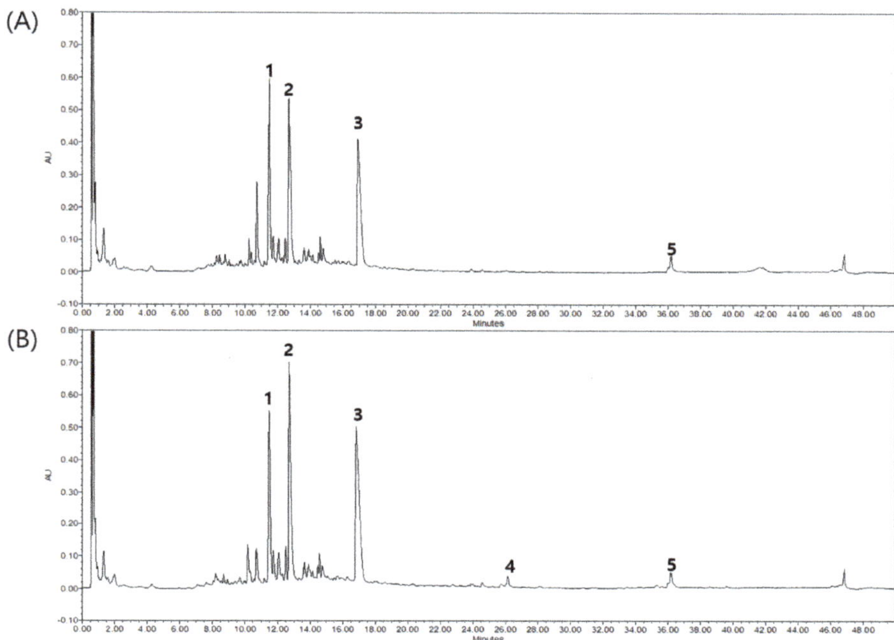

Figure 1. Representative UPLC chromatogram at 200 nm: (**A**) Sam-Myo-Whan (SMW) water extract and (**B**) SMW 30% ethanol extract. (1) Palmatine, (2) armepavine, (3) berberine, (4) atractylenolide III and (5) atractylenolide I.

2.2. Serum Uric Acid Levels of Hyperuricemic Rats Treated with SMHW or SMHE

The effects of SMWW and SMWE on serum uric acid levels in potassium oxonate (PO)-induced hyperuricemic rats are shown in Figure 2. Serum uric acid levels in the PO group rats were significantly increased, compared to those in the Con group ($p < 0.0001$). Treatment with SMWW or SMWE at a 400 mg/kg dose significantly reduced serum uric acid levels by 34.3% and 35.6%, respectively, compared with the PO group (both $p < 0.01$); however, there was no significant difference in efficacy between the two extracts. Rats treated with allopurinol (10 mg/kg) as a positive control showed a 60.4% decrease in their serum uric acid levels ($p < 0.0001$).

2.3. Anti-Inflammatory Effects of SMWW and SMWE on Paw Swelling in MSU-Induced Gouty Arthritis

MSU crystals led to a significant increase in paw thicknesses of injected mice compared with the controls (Figure 3B,C). Meanwhile, treatment with SMWW (100 and 200 mg/kg) or SMWE (50, 100, and 200 mg/kg) significantly suppressed MSU-induced paw swelling compared with the MSU group. At the same dose (200 mg/kg), SMWE caused a greater decrease in paw thickness than SMWW, while the 100 mg/kg SMWE dose showed similar anti-inflammatory effects on paw swelling as the 200 mg/kg SMWW dose.

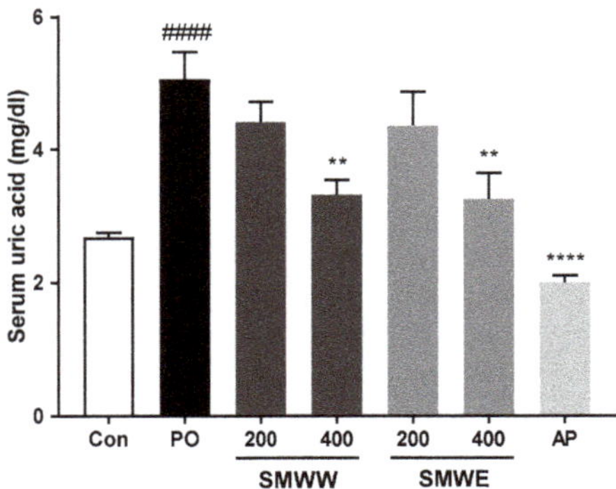

Figure 2. Effects of SMW extracted with water (SMWW) and SMW extracted with 30% EtOH (SMWE) on serum uric acid levels in PO-induced hyperuricemic rats. Con, normal control mice; PO, PO-induced hyperuricemic rat; SMWW, PO rats treated with SMWW; SMWE, PO rats treated with SMWE; AP, PO rats treated with 10 mg/kg of allopurinol. Data are expressed as the mean ± SEM (n = 6). #### $p < 0.0001$ (compared with control group) and ** $p < 0.01$, **** $p < 0.0001$ (compared with PO group).

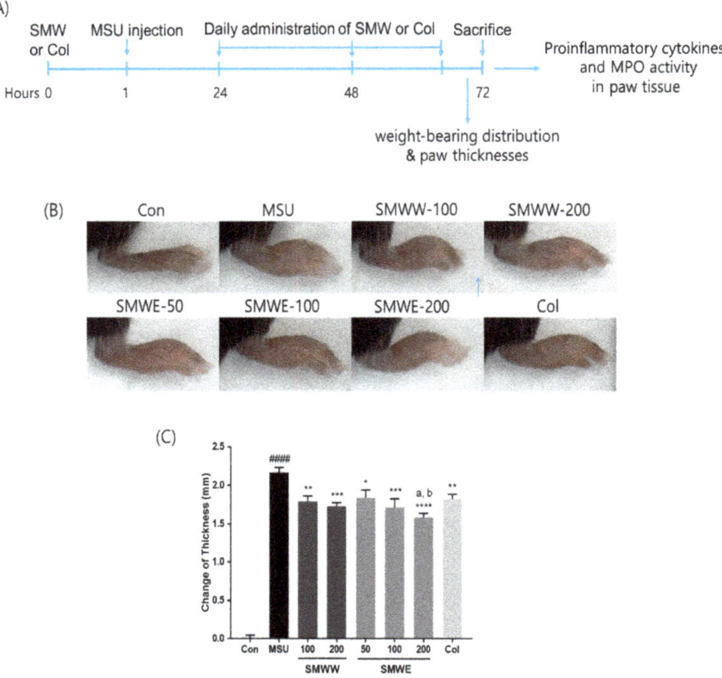

Figure 3. Effect of SMWW and SMWE on paw swelling in mice with monosodium urate (MSU)-crystal-induced gouty arthritis. Con, normal control mice; MSU, MSU-crystal-injected mice; SMWW, MSU mice treated with SMWW; SMWE, MSU mice treated with SMWE; Col, MSU mice treated with 1 mg/kg of colchicine. (**A**) Experimental design. (**B**) Representative images of the right leg from mice in each group. (**C**) Quantification of changes in the thickness of each mouse paw recorded 3 days after the induction of MSU. Data are presented as the mean ± SEM (n = 5). #### $p < 0.0001$ (compared with control group); * $p < 0.05$, ** $p < 0.01$, *** $p < 0.001$, **** $p < 0.0001$ (compared with MSU group); ^{a} $p < 0.05$ (compared with 200 mg/kg of SMWW group); ^{b} $p < 0.01$ (compared with 1 mg/kg of colchicine group).

2.4. Effect of SMWW and SMWE on Hind Paw Weight-Bearing Distribution

In the weight-bearing test, indicating the progressive pain of gouty arthritis, the mice injected with MSU exhibited a clear reduction in weight-bearing on the affected paw, as compared with the control mice (Figure 4). Although hind-paw weight distribution showed no change with a 100 mg/kg SMWW treatment dose, the 200 mg/kg SMWW dose and 50 mg/kg SMWE dose, increased weight distribution to levels similar to the Col treatment group. In particular, the 100 and 200 mg/kg SMWE doses significantly elevated hind-paw weight distribution.

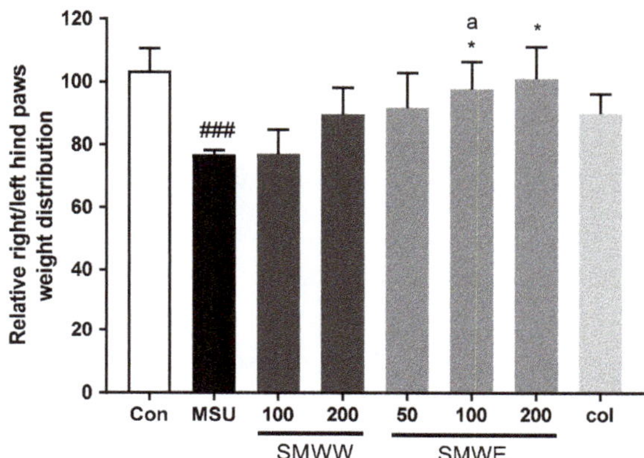

Figure 4. Effect of SMWW and SMWE on hind-paw weight-bearing distribution in mice with MSU-crystal-induced gouty arthritis. The relative right/left hind paw weight-bearing distribution was measured by using a dynamic weight-bearing (DWB) device, compared to that of the MSU-crystal-injected group. Con, normal control mice; MSU, MSU-crystal-injected mice; SMWW, MSU mice treated with SMWW; SMWE, MSU mice treated with SMWE; Col, MSU mice treated with 1 mg/kg of colchicine. Data are presented as the mean ± SEM ($n = 5$). ### $p < 0.001$ (compared with control group); * $p < 0.05$ (compared with MSU group); [a] $p < 0.05$ (compared with 100 mg/kg of SMWW group).

2.5. Effects of SMWW and SMWE on Proinflammatory Cytokines

We investigated the anti-inflammatory effects of MSU-injection by assessing the levels of IL-1β, IL-6, and TNF-α, using ELISA. The results showed that MSU-injected mice had significantly elevated IL-1β, IL-6, and TNF-α levels (Figure 5). However, SMWW and SMWE treatment significantly downregulated IL-1β production by at least 43.9%, at all treatment concentrations, with the 200 mg/kg SMWE dose displaying the greatest efficacy (68.7% reduction), compared with the Col positive control (66.2% reduction). In addition, the 200 mg/kg SMWE dose effectively reduced TNF-α levels by 52%, while the 200 mg/kg SMWW dose and the 100 mg/kg SMWE dose reduced TNF-α to similar levels (29.2% and 30.3%, respectively). Both SMW extracts exhibited a weak dose-dependent decrease in IL-6 production, however, these results were not statistically significant.

2.6. Effects of SMWW and SMWE on MPO Activity

To evaluate the possible cellular infiltration induced by MSU, MPO activity was used as an index of neutrophil accumulation. As shown in Figure 6, MSU injection was found to markedly increase MPO activity in affected paw tissue, compared to the controls ($p < 0.0001$). Meanwhile, SMWW and SMWE both reduced MPO activity, with the highest effect observed following administration of SMWE at a dose of 200 mg/kg ($p < 0.01$). The positive control group, treated with 1 mg/kg Col (which inhibits neutrophil recruitment and activation), also exhibited a significant reduction ($p < 0.05$) in MPO levels, compared to the MSU group.

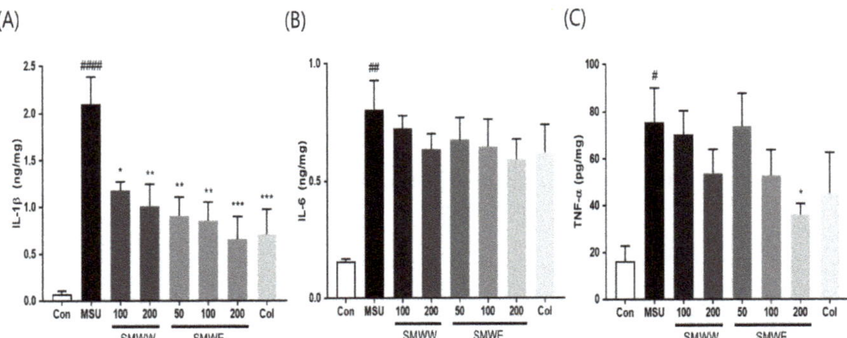

Figure 5. Effects of SMWW and SMWE on proinflammatory cytokines expression in MSU-crystal-injected paw tissue. Con, normal control mice; MSU, MSU-crystal-injected mice; SMWW, MSU mice treated with SMWW; SMWE, MSU mice treated with SMWE; Col, MSU mice treated with 1 mg/kg of colchicine. (**A**) IL-1β, (**B**) IL-6, and (**C**) TNF-α levels measured by ELISA. Data are presented as the mean ± SEM ($n = 5$). # $p < 0.05$, ## $p < 0.01$, #### $p < 0.0001$ (compared with control group); and * $p < 0.05$, ** $p < 0.01$. *** $p < 0.001$ (compared with MSU group).

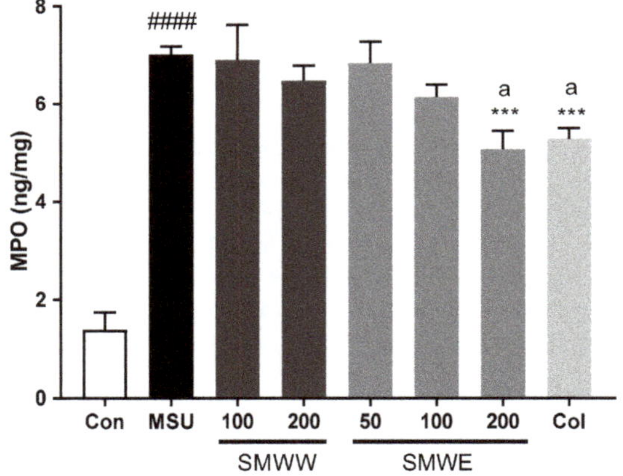

Figure 6. Effects of SMWW and SMWE on myeloperoxidase (MPO) activity in MSU-crystal-injected paw tissue. Con, normal control mice; MSU, MSU-crystal-injected mice; SMWW, MSU mice treated with SMWW; SMWE, MSU mice treated with SMWE; Col, MSU mice treated with 1 mg/kg of colchicine. Data are presented as mean ± SEM ($n = 5$). #### $p < 0.0001$ (compared with control group); *** $p < 0.001$(compared with the MSU group); [a] $p < 0.01$ (compared with 200 mg/kg of SMWW group).

3. Discussion

Gout is a common disease characterized by the deposition of MSU crystals in the joints or subcutaneous tissues, causing acute inflammatory flares or chronic arthritis [20]. Hyperuricemia (high blood uric acid concentration) occurs above the saturation point of MSU, at which point the risk of crystallization increases [21]. MSU crystals result in acute gout attacks characterized by IL-1β-driven acute inflammation, fever, and intense pain caused by neutrophil accumulation and activation in joints. [22]. Therefore, control of hyperuricemia and treatment that reduces inflammation represent the major therapeutic approaches against gouty arthritis [23]. In the present study, we compared the compositional changes as well as treatment efficacy of SMW extracted with water or 30% ethanol. The anti-hyperuricemic effects of SMWW and SMWE in the hyperuricemic animal model, in which serum uric acid levels were increased by intraperitoneal PO injection (to induce hyperuricemia), and the anti-gouty arthritis effects of SMWW and SMWE, were assessed in a gouty arthritis model induced by MSU-crystal injection. In addition, we analyzed the phytochemical contents of SMWW and SMWE, using HPLC.

The ability of SMW to reduce blood uric acid concentration has been demonstrated previously in many animal experiments and clinical studies [10,24,25], and it was confirmed in our study. Moreover, SMWE and SMWW exhibited similar efficacies.

The identification of MSU crystals in joint fluid or synovium is the basis for a clinically definitive diagnosis of gout arthritis, as these crystals have been shown to cause strong inflammatory reactions, leading to acute gout arthritis [26,27]. The most significant symptom of gouty arthritis is swelling and pain, which is observed in the mice injected with MSU [26,28]. In the present study, the MSU-injected mice showed a clear increase in swelling, compared with the controls, and markedly reduced weight-bearing on the affected hind paw, indicating pain. Meanwhile, SMWE treatment markedly prevented the MSU-crystal-induced elevation in paw swelling, compared with that of the SMWW or Col groups. Moreover, the 200 mg/kg SMWE dose elicited excellent pain relief, with hind-paw weight-bearing returning to that similar of the Con group. These results demonstrated that SMWE reduced swelling and pain at dosages of 100–200 mg/kg more effectively than did SMWW at 200 mg/kg.

MSU crystals are one of the most effective proinflammatory stimuli, through their ability to trigger, amplify, and sustain a strong inflammatory reaction in the joint cavity [29]. MSU crystals stimulate the synthesis and release of IL-1β, a key inflammatory cytokine that regulates the differentiation, proliferation, and apoptosis of cells in gout arthritis [30]. In addition, IL-1β induces the expression of a wide range of cytokines, including TNF-α and IL-6, resulting in a large influx of neutrophils into the synovium [31]. In turn, neutrophil interactions with MSU crystals stimulates the synthesis and release of a large variety of pro-inflammatory signals, such as reactive oxygen species, leukotrienes, prostaglandin E2 (PGE2), TNF-α, IL-1, IL-6 and IL-8. This response promotes the vasodilation, erythema and pain associated with acute gout attack [23,32]. Thus, inhibiting MSU-induced recruitment of neutrophils and blocking secretion of inflammatory mediators may prove beneficial for the control and management of acute gouty arthritis [29].

Our results further demonstrated that the levels of IL-1β and TNF-α in the paw tissue were significantly increased in response to MSU, however, became markedly downregulated, in a dose-dependent manner, following SMWW or SMWE treatment. Furthermore, MPO activity was significantly elevated in mice with gouty arthritis, compared to the control group (indicating an influx of neutrophils and acute inflammation), while both SMWW and SMWE effectively decreased MPO activity. Again, SMWE treatment resulted in superior inhibition of MPO activity caopared to SMWW, at a level similar to that of the positive control, colchicine, which is a known regulator of neutrophil activity [33]. These results suggest that SMWE relieves acute gout symptoms caused by MSU crystals by inhibiting the major inflammatory cytokines and suppressing MPO activity, which is a key feature in the initiation and progression of gouty arthritis. Furthermore, our data indicates that SMWE treatment is more effective than SMWW.

Extraction solvents have different abilities to solubilize various biologically active compounds, which can have a significant effect on the content and biological activity of the extract [34,35]. Although SMW has long been used to water extract from herbal medicines consisting of a ratio as 2:3:1 (Phellodendri cortex, Atractylodes rhizome and Achyranthes radix), no studies have reported the specific composition of these compounds. For the single medicinal herb, *Atractylodes japonica*, the extract is reported to contain stigmasterol, hinesol, eudesmol, atractylenolides, atractylon, atractylodin, and sitosterol [36], while methanol extract was reported to contain 0.08% hinesol, 0.09% eudesmol, and 0.02% atractylodin [37]. Moreover, Chikusetsusaponin IVa methyl ester, separated from *Achyranthes japonica* 80% methanol extract, reportedly elicits an anti-inflammatory effect, however, no report has been made on quantity [38]. Additionally, *Phellodendron amurense* is reported to contain alkaloids, such as phellodendrine, magnoflorine, tetrahydropalmatine, columbamine, jatrorrhizine, 8-oxyepiberberine, berberine, palmatine, and bis-[4-(dimethylamino)phenyl] methanone [39]. While most studies of such ingredients are conducted by using non-polar extraction solvents (methanol and ethanol) for a single herb, the only traditional method used includes water extraction in a complex of these three herbs (Phellodendri cortex, Atractylodes rhizome and Achyranthes radix). Alternatively, water and ethanol are commonly used as solvents for the extraction of herbs for preparation of traditional decoction, food ingredients, dietary supplements, etc. Thus, we conducted a study using a 30% ethanol extract, which offers the best efficacy in the range of acceptable ethanol concentrations used in traditional methods.

In this study, SMW was extracted with 30% ethanol or water, and the main ingredients were identified as palmatine, armepavine, and berberine. When SMWE was compared to SMWW, the palmatine content was slightly lower and the armepavine content slightly higher than that of SMWW. However, the berberine content of SMWE was 32.2% higher than that of SMWW. Berberine has been reported to possess a wide range of pharmacological activities, including anti-inflammatory, antimicrobial, antioxidant, hypoglycemic, hypolipidemic, and hepatoprotective properties [40]. Additionally, berberine has been shown to downregulate NLR family pyrin domain-containing protein 3 (NLRP3) and IL-1β expression in MSU-crystal-induced inflammation [41]. Other compounds, such as atractylenolide III (a known anti-inflammatory agent), were only detected in SMWE, albeit in small quantities [42]. It has been shown that extraction using an alcohol/water mixture (versus water alone) increases the content of active components that are insoluble in water-while also extracting water-soluble active ingredients, thus optimizing the extraction of relatively small amounts of active ingredients present in natural products [34]. Therefore, it is suggested that small amounts of compound, atractylenolide III, and 32.2% increased, berberine, are characteristic components of SMWE and are bioactive compounds that may affect the mouse gouty arthritis model. The compounds may contribute to synergistic or additional effects, and our results suggest that SMWE is more effective in reducing swelling, pain, and inflammation in MSU-induced gouty arthritis mouse model than SMWW.

4. Materials and Methods

4.1. Preparation of SMW

The SMW preparation used in this study was purchased from Kwangmyoungdang Pharms (Ulsan, S. Korea). The voucher specimen was deposited at the Korean Herbarium of Standard Herbal Resources of Korea Institute of Oriental Medicine (2-20-0354~2-20-0356, Daejeon, S. Korea). According to Donguibogam, Phellodendri cortex (Phellodendron chinense Schneider) was stir-fried with Makgeolli (1:10, w/v) for 2 h. The Atractylodes rhizome (Atractylodes chinensis Koidzumi) was soaked in rice-washed water for 3 h and then dried. Each sample was ground into a powder. The mixture was prepared with 60 g of Achyranthes radix (Achyranthe japonica Nakai), 180 g of rinsed Atractylodes rhizome, and 120 g of stir-fried Phellodendri cortex, and was extracted with 2 L of water (SMWW) or 30% ethanol (SMWE), for 3 h, by reflux. These extracts were then concentrated under reduced pressure and freeze-dried.

4.2. Components Analysis of SMW

Reference standards, palmatine, armepavine, berberine, atractylenolide III, and atractylenolide I, were purchased from Chemfaces (Hubei, China). After confirming compounds by comparing the retention time and absorption profile of the reference material, each component was quantified through the area comparison.

HPLC analysis was performed on an Acquity UPLC system (Waters, MA, USA) equipped with a quaternary pump, auto-sampler, and photodiode array detector with Acquity UPLC®BEH C18, 100 × 2.1 mm, 1.7 μm. A gradient elution with solvent A (0.1% phosphoric acid) and solvent B (acetonitrile), at a flow rate of 0.5 mL/min, was conducted as follows: 0–2 min, 2–2% B; 2–32 min, 2–50% B; 32–42 min, 50–100% B; 42–45 min, 100–100% B; 45–47 min, 100–2% B; and 47–50 min, 2–2% B. The detection wavelength was set to 200 nm. The column temperature was maintained at 40 °C, and the injection volume was 2 μL.

4.3. Animals

Male Sprague Dawley (SD) rats (7 weeks) and male C57BL6 mice (7 weeks) were purchased from Orient Bio (Seongnam, Korea) and housed at a temperature of 22 ± 2 °C in a 50 ± 10% humidity-controlled room under a 12 h light/dark cycle. The animals were allowed *ad libitum* access to a laboratory diet and water. At the end point of the experiment, the rats were anesthetized using zoletil and sacrificed by cervical dislocation. No systemic adverse effects were observed following treatment with SMWW or SMWE, in any study group. The experimental design was approved by the Committee on Animal Care of the KIOM (approval No. 20-016), and the study was conducted in accordance with the Guide for the Care and Use of Laboratory Animals published by the US National Institutes of Health (Bethesda, MD, United States).

4.4. Hyperuricemia Induction and Sample Treatment.

The uricase inhibitor PO was injected into rats, to induce hyperuricemia [43]. The rats were divided into the following seven groups (n = 6/group): (1) Controls (Con), (2) PO-treated controls, (3) PO+200 mg/kg SMWW, (4) PO+400 mg/kg SMWW, (5) PO+200 mg/kg SMWE, (6) PO+400 mg/kg SMWE, and (7) PO+10 mg/kg allopurinol (AP). Rats in groups (2-7) were injected intraperitoneally with 150 mg/kg PO prepared in 0.5% carboxymethyl cellulose (CMC) with 0.1 M sodium acetate (pH 5.0) to induce hyperuricemia, while the normal control (1) rats were treated with 0.5% CMC with 0.1 M sodium acetate. SMWW, SMWE, and AP were dispersed in 0.5% CMC and administered by oral gavage, 1 h prior to PO injection.

4.5. Analysis of Uric Acid in Serum

Blood samples were collected via cardiac puncture, under anesthesia, 2 h after PO treatment. Serum was obtained by centrifugation at 3000× g for 10 min at 4 °C, after allowing the blood samples to clot for 2 h, at room temperature. The separated serum uric acid levels were determined, using an enzymatic-colorimetric method, using commercial assay kits (Biovision, Milpitas, CA, USA) according to manufacturer's protocols.

4.6. Induction of Gouty Arthritis with MSU Crystals in Mice

MSU was synthesized as previously described [44]. After acclimation, C57BL6 male mice (8 weeks old, 20-22g body weight) were divided into the following eight groups (n = 5/group): (1) normal controls, (2) MSU-crystal-treated, (3) MSU+100 mg/kg SMWW, (4) MSU+200 mg/kg SMWW, (5) MSU+50 mg/kg SMWE, (6) MSU+100 mg/kg SMWE, (7) MSU+200 mg/kg SMWE, and (8) MSU+1 mg/kg colchicine (Col). The right hind paw of each mouse in groups (2–8) was injected intradermally with MSU crystal suspension (4 mg/50 μL) in PBS with 0.5% Tween 80, while the normal control (1) mice were treated with PBS with 0.5% Tween 80. SMWW, SMWE, and Col were dispersed in 0.5% CMC and administered by oral gavage, 1 h before the MSU crystal injection, and then once daily, for 3 days. The experimental design is shown in Figure 3A.

4.7. Assessment of Inflammatory Paw Swelling and Pain

Inflammatory paw swelling was quantified by measuring the thickness of the MSU-injected paw, using a Vernier scale, 3 days after the induction of MSU. The change of thickness (mm) was calculated as follows: Change of thickness (mm) = MSU-treated paw thickness - normal control paw thickness [45]. The pain was measured by right and left hind-limb weight distribution, using a dynamic weight-bearing device (Bioseb, Boulogne, France), which was developed to measure the weight borne by each limb in freely moving animals [44,46]. The mice were placed in a small Plexiglas chamber (11.0 × 19.7 × 11.0 cm) with a floor sensor containing pressure transducer, for 2 minutes, and the analyzer recorded the average weight in grams, for each limb put on the floor. All movements were filmed and validated according to the position of the mouse on the device, and the results were analyzed for the weight of the paw, which touches the floor in grams [47]. The relative right/left hind paws weight-bearing distribution was calculated by using the following equation: (weight on right hind limb / weight on left hind limb) $\times$ 100.

4.8. Measurement of Inflammatory Cytokines and Mediators

The levels of IL-1β, IL-6, TNF-α, and myeloperoxidase (MPO) were measured by using ELISA kits from R&D Systems (Minneapolis, MN, USA) and MyBioSource (San Diego, CA, USA) according to the manufacturers' protocols.

4.9. Statistical Analysis

The results were expressed as the mean $\pm$ standard error of the mean (SEM) and analyzed, using a one-way analysis of variance (ANOVA), followed by Dunnett's tests for multiple comparisons or unpaired Student's t-tests for two-group comparisons. Normality was performed by using Shapiro–Wilk's test. All analyses were performed, using Prism 7.0 (GraphPad Software, San Diego, CA, USA), and p-values < 0.05 were considered significant.

5. Conclusions

In conclusion, this study demonstrated that SMWW and SMWE equally reduced serum uric acid levels in PO-induced hyperuricemic rats. However, in a gouty arthritis animal model, SMWE more efficiently downregulated MSU-crystal-induced swelling and pain, and it exerted anti-inflammatory effects by suppressing proinflammatory cytokines (IL-1β, TNF-α, and IL-6) and MPO activity. Moreover, berberine was found to be one of the most differentially abundant main active ingredients between SMWW and SMWE, while atractylenolide III was identified only in SMWE, both of which are known to elicit anti-inflammatory effects. These observations show that 30% ethanol is an efficient solvent for SMW extraction with anti-gouty arthritis efficacy at the concentrations reduced compared with water extracts. Further studies should be conducted to determine whether SMWE has similar efficacy in clinical trials at lower doses than SMWW.

Author Contributions: Conceptualization, Y.M.L.; methodology, Y.M.L.; software, Y.M.L. and E.S.; validation, Y.M.L. and E.S.; formal analysis, Y.M.L.; investigation, Y.M.L. and E.S.; writing—original draft preparation, Y.M.L.; writing—review and editing, Y.M.L. and E.S.; visualization, Y.M.L. and E.S.; supervision, D.-S.K.; project administration, D.-S.K. All authors have read and agreed to the published version of the manuscript.

Funding: This research was financially supported by grants from the Korea Institute of Oriental Medicine (KSN2012330). The funders have no role in designing the experiment and publication of the manuscript.

Institutional Review Board Statement: The experimental design was approved by the Committee on Animal Care of the KIOM (approval No. 20-016), and the study was conducted in accordance with the Guide for the Care and Use of Laboratory Animals published by the US National Institutes of Health (Bethesda, MD, United States).

Informed Consent Statement: Not applicable.

Data Availability Statement: The data presented in this study is contained within the article.

Acknowledgments: The authors thank all of the colleagues who contributed to this study.

Conflicts of Interest: The authors declare no conflict of interest.

Abbreviations

ADAMTSs	aggrecanases
AP	allopurinol
CMC	carboxymethyl cellulose
Col	colchicine
HPLC	high-performance liquid chromatography
IL-1β	interleukin-1β
IL-6	interleukin-6
iNOS	inducible nitric oxide synthase
JAK–STAT	Janus kinase–signal transducer and activator of transcription
LPS	lipopolysaccharide
MAPK	mitogen-activated protein kinase
MMPs	matrix metalloproteinases
MPO	myeloperoxidase
MSU	monosodium urate
NO	nitric oxide
NOD	nucleotide binding and oligomerization domain
PO	potassium oxonate
PGE2	prostaglandin E2
SMW	Sam-Myo-Whan
SMWE	SMW extracted with ethanol
SMWW	SMW extracted with water
TIMPs	tissue inhibitors of metalloproteinases
TNF-α	tumor necrosis factor-alpha
XOD	xanthine oxidase

References

1. Nielsen, S.M.; Zobbe, K.; Kristensen, L.E.; Christensen, R. Nutritional recommendations for gout: An update from clinical epidemiology. *Autoimmun. Rev.* **2018**, *17*, 1090–1096. [CrossRef] [PubMed]
2. Pascart, T.; Grandjean, A.; Capon, B.; Legrand, J.; Namane, N.; Ducoulombier, V.; Motte, M.; Vandecandelaere, M.; Luraschi, H.; Godart, C.; et al. Monosodium urate burden assessed with dual-energy computed tomography predicts the risk of flares in gout: A 12-month observational study: MSU burden and risk of gout flare. *Arthritis Res. Ther.* **2018**, *20*, 210. [CrossRef] [PubMed]
3. Meng, Z.-Q.; Tang, Z.-H.; Yan, Y.-X.; Guo, C.-R.; Cao, L.; Ding, G.; Huang, W.-Z.; Wang, Z.-Z.; Wang, K.D.; Xiao, W.; et al. Study on the Anti-Gout Activity of Chlorogenic Acid: Improvement on Hyperuricemia and Gouty Inflammation. *Am. J. Chin. Med.* **2014**, *42*, 1471–1483. [CrossRef] [PubMed]
4. Silvestre, S.; Almeida, P.J.S.; El-Shishtawy, R. Natural Products as a Source for New Leads in Gout Treatment. *Evid. Based Complement. Altern. Med.* **2020**, *2020*, 8274975. [CrossRef]
5. Bost, J.; Maroon, A.; Maroon, J.C. Natural anti-inflammatory agents for pain relief. *Surg. Neurol. Int.* **2010**, *1*, 80. [CrossRef]
6. Committee, S.P. *The Pharmacopoeia of People's Republic of China*; China Medical Science Press: Beijing, China, 2015.
7. Lee, J.-H.; Jung, H.-W.; Park, Y.-K. Inhibitory effects of Sam-Myo-San on the LPS-induced production of nitric oxide and TNF−α in RAW 264.7 cells and BV-2 Microglia cells. *Korea Assoc. Herbol.* **2006**, *21*, 59–67.
8. Xu, Y.; Dai, G.J.; Liu, Q.; Liu, Z.L.; Song, Z.Q.; Li, L.; Lin, N. Observation of curative effect on the treatment of acute gouty arthritis with SM in 45 cases. *Yunnan Zhong Yi Zhong Yao Za Zhi* **2003**, *24*, 5–6.
9. Xu, Y.; Dai, G.J.; Liu, Q.; Liu, Z.L.; Song, Z.Q.; Li, L.; Lin, N. The curative effect observation on 68 cases of rheumatoid arthritis treated with Si long San Miao Formula. *Guangming J. Chin. Med.* **2007**, *22*, 86–87.
10. Wang, X.; Wang, C.-P.; Hu, Q.-H.; Lv, Y.-Z.; Zhang, X.; Ouyang, Z.; Kong, L.-D. The dual actions of Sanmiao wan as a hypouricemic agent: Down-regulation of hepatic XOD and renal mURAT1 in hyperuricemic mice. *J. Ethnopharmacol.* **2010**, *128*, 107–115. [CrossRef]
11. Xu, Y.; Dai, G.-J.; Liu, Q.; Liu, Z.-L.; Song, Z.-Q.; Li, L.; Chen, W.-H.; Lin, N. Sanmiao formula inhibits chondrocyte apoptosis and cartilage matrix degradation in a rat model of osteoarthritis. *Exp. Ther. Med.* **2014**, *8*, 1065–1074. [CrossRef]
12. Zhu, F.; Yin, L.; Ji, L.; Yang, F.; Zhang, G.; Shi, L.; Xu, L. Suppressive effect of Sanmiao formula on experimental gouty arthritis by inhibiting cartilage matrix degradation: An in vivo and in vitro study. *Int. Immunopharmacol.* **2016**, *30*, 36–42. [CrossRef] [PubMed]
13. Jiang, T.; Qian, J.; Ding, J.; Wang, G.; Ding, X.; Liu, S.; Chen, W. Metabolomic profiles delineate the effect of Sanmiao wan on hyperuricemia in rats. *Biomed. Chromatogr.* **2017**, *31*, e3792. [CrossRef] [PubMed]
14. Wu, J.; Li, J.; Li, W.; Sun, B.; Xie, J.; Cheng, W.; Zhang, Q. Achyranthis bidentatae radix enhanced articular distribution and anti-inflammatory effect of berberine in Sanmiao Wan using an acute gouty arthritis rat model. *J. Ethnopharmacol.* **2018**, *221*, 100–108. [CrossRef] [PubMed]

15. Qian, H.; Jin, Q.; Liu, Y.; Wang, N.; Chu, Y.; Liu, B.; Liu, Y.; Jiang, W.; Song, Y. Study on the Multitarget Mechanism of Sanmiao Pill on Gouty Arthritis Based on Network Pharmacology. *Evid. Based Complement. Altern. Med.* **2020**, *2020*, 9873739. [CrossRef] [PubMed]
16. Hua, J.; Huang, P.; Zhu, C.-M.; Yuan, X.; Chen-Huan, Y. Anti-hyperuricemic and nephroprotective effects of Modified Simiao Decoction in hyperuricemic mice. *J. Ethnopharmacol.* **2012**, *142*, 248–252. [CrossRef] [PubMed]
17. Liu, Y.; Huang, Y.; Wen, C.-Y.-Z.; Zhang, J.-J.; Xing, G.-L.; Tu, S.-H.; Chen, Z. The Effects of Modified Simiao Decoction in the Treatment of Gouty Arthritis: A Systematic Review and Meta-Analysis. *Evid. Based Complement. Altern. Med.* **2017**, *2017*, 6037037. [CrossRef] [PubMed]
18. Qiu, R.; Shen, R.; Lin, D.; Chen, Y.; Ye, H. Treatment of 60 cases of gouty arthritis with modified Simiao Tang. *J. Tradit. Chin. Med. = Chung i Tsa Chih Ying Wen pan* **2008**, *28*, 94–97.
19. Chi, X.; Zhang, H.; Zhang, S.; Ma, K. Chinese herbal medicine for gout: A review of the clinical evidence and pharmacological mechanisms. *Chin. Med.* **2020**, *15*, 17. [CrossRef]
20. Martillo, M.A.; Nazzal, L.; Crittenden, D.B. The Crystallization of Monosodium Urate. *Curr. Rheumatol. Rep.* **2014**, *16*, 1–8. [CrossRef]
21. Choi, H.K.; Mount, D.B.; Reginato, A.M. Pathogenesis of gout. *Ann. Intern. Med.* **2005**, *143*, 499–516. [CrossRef]
22. Yang, G.; Yeon, S.H.; Lee, H.E.; Kang, H.C.; Cho, Y.-Y.; Lee, J.Y. Suppression of NLRP3 inflammasome by oral treatment with sulforaphane alleviates acute gouty inflammation. *Rheumatology* **2018**, *57*, 727–736. [CrossRef] [PubMed]
23. Liu, X.; Chen, R.; Shang, Y.; Jiao, B.-H.; Huang, C. Lithospermic acid as a novel xanthine oxidase inhibitor has anti-inflammatory and hypouricemic effects in rats. *Chem. Interactions* **2008**, *176*, 137–142. [CrossRef] [PubMed]
24. Kang, D.-H.; Nakagawa, T.; Feng, L.; Watanabe, S.; Han, L.; Mazzali, M.; Truong, L.; Harris, R.; Johnson, R.J. A Role for Uric Acid in the Progression of Renal Disease. *J. Am. Soc. Nephrol.* **2002**, *13*, 2888–2897. [CrossRef] [PubMed]
25. Lam, F.F.; Ko, I.W.; Ng, E.S.; Tam, L.S.; Leung, P.C.; Li, E.K. Analgesic and anti-arthritic effects of Lingzhi and San Miao San supplementation in a rat model of arthritis induced by Freund's complete adjuvant. *J. Ethnopharmacol.* **2008**, *120*, 44–50. [CrossRef]
26. Schlesinger, N. Diagnosing and Treating Gout: A Review to Aid Primary Care Physicians. *Postgrad. Med.* **2010**, *122*, 157–161. [CrossRef]
27. Vaidya, B.; Bhochhibhoya, M.; Nakarmi, S. Synovial fluid uric acid level aids diagnosis of gout. *Biomed. Rep.* **2018**, *9*, 60–64. [CrossRef]
28. Doss, H.M.; Dey, C.; Sudandiradoss, C.; Rasool, M. Targeting inflammatory mediators with ferulic acid, a dietary polyphenol, for the suppression of monosodium urate crystal-induced inflammation in rats. *Life Sci.* **2016**, *148*, 201–210. [CrossRef]
29. Dhanasekar, C.; Kalaiselvan, S.; Rasool, M. Morin, a Bioflavonoid Suppresses Monosodium Urate Crystal-Induced Inflammatory Immune Response in RAW 264.7 Macrophages through the Inhibition of Inflammatory Mediators, Intracellular ROS Levels and NF-κB Activation. *PLoS ONE* **2015**, *10*, e0145093. [CrossRef]
30. Yao, R.; Geng, Z.; Mao, X.; Bao, Y.; Guo, S.; Bao, L.; Sun, J.; Gao, Y.; Xu, Y.; Guo, B.; et al. Tu-Teng-Cao Extract Alleviates Monosodium Urate-Induced Acute Gouty Arthritis in Rats by Inhibiting Uric Acid and Inflammation. *Evid. Based Complement. Altern. Med.* **2020**, *2020*, 3095624. [CrossRef]
31. Pope, R.M.; Tschopp, J. The role of interleukin-1 and the inflammasome in gout: Implications for therapy. *Arthritis Rheum.* **2007**, *56*, 3183–3188. [CrossRef]
32. Prince, S.E.; Nagar, S.; Rasool, M. A Role of Piperine on Monosodium Urate Crystal-Induced Inflammation—An Experimental Model of Gouty Arthritis. *Inflammation* **2011**, *34*, 184–192. [CrossRef]
33. Landis, R.C.; Haskard, D. Pathogenesis of crystal-induced inflammation. *Curr. Rheumatol. Rep.* **2001**, *3*, 36–41. [CrossRef] [PubMed]
34. Altemimi, A.B.; Lakhssassi, N.; Baharlouei, A.; Watson, D.G.; Lightfoot, D.A. Phytochemicals: Extraction, Isolation, and Identification of Bioactive Compounds from Plant Extracts. *Plants* **2017**, *6*, 42. [CrossRef] [PubMed]
35. Silva, R.P.D.; Machado, B.A.S.; Barreto, G.D.A.; Costa, S.S.; Andrade, L.N.; Amaral, R.G.; Carvalho, A.A.; Padilha, F.F.; Barbosa, J.D.V.; Umsza-Guez, M.A. Antioxidant, antimicrobial, antiparasitic, and cytotoxic properties of various Brazilian propolis extracts. *PLoS ONE* **2017**, *12*, e0172585. [CrossRef]
36. Shan, G.-S.; Zhang, L.; Zhao, Q.-M.; Xiao, H.-B.; Zhuo, R.-J.; Xu, G.; Jiang, H.; You, X.-M.; Jia, T.-Z. Metabolomic study of raw and processed Atractylodes macrocephala Koidz by LC–MS. *J. Pharm. Biomed. Anal.* **2014**, *98*, 74–84. [CrossRef]
37. Ishii, T.; Okuyama, T.; Noguchi, N.; Nishidono, Y.; Okumura, T.; Kaibori, M.; Tanaka, K.; Terabayashi, S.; Ikeya, Y.; Nishizawa, M. Antiinflammatory constituents of Atractylodes chinensis rhizome improve glomerular lesions in immunoglobulin A nephropathy model mice. *J. Nat. Med.* **2020**, *74*, 51–64. [CrossRef]
38. Lee, H.-J.; Shin, J.-S.; Lee, W.-S.; Shim, H.-Y.; Park, J.-M.; Jang, D.S.; Lee, K.-T. Chikusetsusaponin IVa Methyl Ester Isolated from the Roots of Achyranthes japonica Suppresses LPS-Induced iNOS, TNF-α, IL-6, and IL-1β Expression by NF-κB and AP-1 Inactivation. *Biol. Pharm. Bull.* **2016**, *39*, 657–664. [CrossRef]
39. Xian, X.; Sun, B.; Ye, X.; Zhang, G.; Hou, P.; Gao, H. Identification and analysis of alkaloids in cortex Phellodendron amurense by high-performance liquid chromatography with electrospray ionization mass spectrometry coupled with photodiode array detection. *J. Sep. Sci.* **2014**, *37*, 1533–1545. [CrossRef]
40. Wang, K.; Feng, X.; Chai, L.; Cao, S.; Qiu, F. The metabolism of berberine and its contribution to the pharmacological effects. *Drug Metab. Rev.* **2017**, *49*, 139–157. [CrossRef]

41. Liu, Y.F.; Wen, C.Y.; Chen, Z.; Wang, Y.; Huang, Y.; Tu, S.H. Effects of Berberine on NLRP3 and IL-1beta Expressions in Monocytic THP-1 Cells with Monosodium Urate Crystals-Induced Inflammation. *Biomed. Res. Int.* **2016**, *2016*, 2503703.
42. Kwak, T.-K.; Jang, H.-S.; Lee, M.-G.; Jung, Y.-S.; Kim, D.-O.; Kim, Y.-B.; Kim, J.-I.; Kang, H. Effect of Orally Administered Atractylodes macrocephala Koidz Water Extract on Macrophage and T Cell Inflammatory Response in Mice. *Evid. Based Complement. Altern. Med.* **2018**, *2018*, 4041873. [CrossRef] [PubMed]
43. Yuk, H.J.; Lee, Y.-S.; Ryu, H.W.; Kim, S.-H.; Kim, D.-S. Effects of Toona sinensis Leaf Extract and Its Chemical Constituents on Xanthine Oxidase Activity and Serum Uric Acid Levels in Potassium Oxonate-Induced Hyperuricemic Rats. *Molecules* **2018**, *23*, 3254. [CrossRef] [PubMed]
44. Lee, Y.-M.; Shon, E.-J.; Kim, O.S.; Kim, D.-S. Effects of Mollugo pentaphylla extract on monosodium urate crystal-induced gouty arthritis in mice. *BMC Complement. Altern. Med.* **2017**, *17*, 1–8. [CrossRef] [PubMed]
45. Aziz, T.A.; Kareem, A.A.; Othman, H.H.; Ahmed, Z.A. The Anti-Inflammatory Effect of Different Doses of Aliskiren in Rat Models of Inflammation. *Drug Des. Dev. Ther.* **2020**, *14*, 2841–2851. [CrossRef]
46. Tétreault, P.; Dansereau, M.-A.; Doré-Savard, L.; Beaudet, N.; Sarret, P. Weight bearing evaluation in inflammatory, neuropathic and cancer chronic pain in freely moving rats. *Physiol. Behav.* **2011**, *104*, 495–502. [CrossRef]
47. Quadros, A.U.; Pinto, L.G.; Fonseca, M.M.; Kusuda, R.; Cunha, F.Q.; Cunha, T.M. Dynamic weight bearing is an efficient and predictable method for evaluation of arthritic nociception and its pathophysiological mechanisms in mice. *Sci. Rep.* **2015**, *5*, 14648. [CrossRef]

Article

Extraction of Anthraquinones from Japanese Knotweed Rhizomes and Their Analyses by High Performance Thin-Layer Chromatography and Mass Spectrometry

Vesna Glavnik and Irena Vovk *

Department of Food Chemistry, National Institute of Chemistry, Hajdrihova 19, SI-1000 Ljubljana, Slovenia; vesna.glavnik@ki.si

* Correspondence: irena.vovk@ki.si; Tel.: +386-1476-0341

Received: 25 November 2020; Accepted: 9 December 2020; Published: 11 December 2020

Abstract: Anthraquinones (yellow dyes) were extracted from Japanese knotweed rhizomes with twelve extraction solvents (water; ethanol$_{(aq)}$ (20%, 40%, 60%, 70% and 80%), ethanol, 70% methanol$_{(aq)}$, methanol, 70% acetone$_{(aq)}$, acetone and dichloromethane). The obtained sample test solutions (STSs) were analyzed using high-performance thin-layer chromatography (HPTLC) coupled to densitometry and mass spectrometry (HPTLC–MS/MS) on HPTLC silica gel plates. Identical qualitative densitometric profiles (with anthraquinone aglycones and glycosylated anthraquinones) were obtained for STSs in all the solvents except for the STS in dichloromethane, which enabled the most selective extractions of anthraquinone aglycones emodin and physcion. The highest extraction efficiency, evaluated by comparison of the total peak areas in the densitograms of all STSs scanned at 442 nm, was achieved for 70% acetone$_{(aq)}$. In STS prepared with 70% acetone$_{(aq)}$, the separation of non-glycosylated and glycosylated anthraquinones was achieved with developing solvents toluene–acetone–formic acid (6:6:1, 3:6:1 and 3:3:1 *v*/*v*) and dichloromethane–acetone–formic acid (1:1:0.1, *v*/*v*). Non-glycosylated anthraquinones were separated only with toluene–acetone–formic acid, among which the best resolution between emodin and physcion gave the ratio 6:6:1 (*v*/*v*). This solvent and dichloromethane–acetone–formic acid (1:1:0.1, *v*/*v*) enabled the best separation of glycosylated anthraquinones. Four HPTLC-MS/MS methods enabled the identification of emodin and tentative identification of its three glycosylated analogs (emodin-8-*O*-hexoside, emodin-*O*-acetyl-hexoside and emodin-*O*-malonyl-hexoside), while only the HPTLC-MS/MS method with toluene-acetone-formic acid (6:6:1, *v*/*v*) enabled the identification of physcion. Changes of the shapes and the absorption maxima (bathochromic shifts) in the absorption spectra after post-chromatographic derivatization provided additional proof for the detection of physcion and rejection of the presence of chrysophanol in STS.

Keywords: Japanese knotweed; *Reynoutria*; *Polygonum*; *Polygonaceae*; anthraquinones; emodin; physcion; HPTLC; HPTLC-MS; densitometry

---

1. Introduction

Japanese knotweed (*Fallopia japonica* Houtt., *Polygonaceae*; synonyms: *Polygonum cuspidatum*, *Polygonum reynoutria* and *Reynoutria japonica*) is on the list of the "100 World's Worst Invasive Alien Species", because it represents huge ecological (biodiversity loss) and economic problems (damage of infrastructure) in Europe and North America. However, in the environment of its origin, which is Eastern Asia, it is used in traditional Chinese and Japanese medicine for healing infections, inflammatory diseases, hyperlipidemia and other diseases [1]. Traditional applications (powder, extracts and herbal infusions) are usually prepared from rhizomes (subterranean stems). Rhizomes contain many secondary metabolites, including stilbenes (especially *trans*-resveratrol and its glycosylated analogs [2–11]); proanthocyanidins (from monomers [2,6,8,12,13], dimers [2,6,8,12,13], oligomers [2,4,8,12,13] and up to polymers [2,8,12,13]); phenolic acids [2,6,8]; phenylpropanoid glycosides [2,4]; flavonoids [2,8]; naphthalenes [2–4,6]; triterpenoids [8] and anthraquinones (especially emodin [2,3,5–7,9–11,14], physcion [2,3,5–7,9–11,14] and their glycosides [2–4,6,9,11,14] and other anthraquinones [2,4,6,7,12]). Leaves of Japanese knotweed rhizomes have been shown to be rich in proanthocyanidins [15] and carotenoids [16].

Anthraquinones are the largest group of natural dyes, with about 700 compounds [17], and can be found in many plant genera, such as *Cassia* [18,19], *Aloe* [17], *Rheum* [20] and *Fallopia* [2], to which Japanese knotweed belongs. Anthraquinones present in Japanese knotweed are traditionally used as laxatives. Besides laxative activities [21], anthraquinones show antibacterial [1,10], antiviral [1,21], antifungal [1], anticancer [22] and estrogenic [1] properties. They also showed a potential to be skin-whitening agents, as they act as tyrosinase inhibitors [9]. Emodin showed antibacterial activities against foodborne bacteria [10] and antifungal activities against phytopathogenic fungi [23]. Physcion has the potential to be an anticancer agent for the treatment of nasopharyngeal carcinoma [22]. Emodin and its glycosides were chosen as marker compounds in Chinese pharmacopoeia [4].

Analyses of anthraquinones, extracted from Japanese knotweed rhizomes with boiling water [6], pure ethanol [9], methanol [3,4,11], aqueous ethanol [7,24–26] and aqueous methanol [5,10], were performed with methods based on high-performance liquid chromatography (HPLC with a fluorescence detector (HPLC-FLD) [9], photo-diode array detector (HPLC-PDA) [24] and (U)HPLC-PDA-MS^n [2–8,10,11]) and thin-layer chromatography (TLC) [25,26]. TLC methods for anthraquinones from Japanese knotweed rhizomes were used for the quantification of emodin [25], for the isolation of fractions [26] and for the testing of Japanese knotweed rhizomes fermentation products [27]. TLC methods for analyses of anthraquinones in other plant materials were applied for the screening [28,29] and quantification of anthraquinones [18,19,30–32]. TLC was also applied for analyzing the mycelium and culture supernatant rich in anthraquinones [29] Separations were mainly performed on a silica gel stationary phase (TLC silica gel plates [19,25,29], TLC silica gel plates F_{254} [28,30,31,33] and TLC silica gel F_{254} G plates [18]) in combination with the following developing solvent mixtures: ethanol–water [28], petroleum ether–ethyl acetate–formic acid [26], toluene–ethyl acetate [19], hexane–ethyl acetate [18], ethyl acetate–methanol–water [30], toluene–ethyl formate–formic acid [33], petroleum ether–ethyl formate–formic acid [29], petroleum–butyl acetate–methanol–acetic acid [25] and toluene–ethyl acetate–formic acid [31]. Additionally, TLC RP-18 F_{254} plates were developed with methanol–water–formic acid [32].

The aims of the study were: (i) selection of extraction solvents for the extraction of anthraquinones from the Japanese knotweed rhizomes, (ii) selection of developing solvents for the separation of non-glycosylated and glycosylated anthraquinones and (iii) development of the first high-performance thin-layer chromatography-mass spectrometry (HPTLC-MS/MS) methods for the separation and identification of anthraquinones from Japanese knotweed rhizomes.

2. Results and Discussion

2.1. Selection of Extraction Solvent

Optimization of extraction of anthraquinones (yellow dyes) and their glycosylated derivatives from the rhizomes of the Japanese knotweed was performed on an analytical scale. Different extraction solvents like water, 20% ethanol$_{(aq)}$, 40% ethanol$_{(aq)}$, 60% ethanol$_{(aq)}$, 70% ethanol$_{(aq)}$, 80% ethanol$_{(aq)}$, ethanol, 70% methanol$_{(aq)}$, methanol, 70% acetone$_{(aq)}$, acetone and dichloromethane were used for the preparation of separate sample test solutions (STSs) from rhizomes. STSs were analyzed on two separate HPTLC silica gel plates. STSs prepared with 70% ethanol$_{(aq)}$, ethanol, 70% methanol$_{(aq)}$, methanol, 70% acetone$_{(aq)}$, acetone and dichloromethane were on the first plate together with standards of physcion, chrysophanol, aloe-emodin, emodin and aloin A (Figure 1).

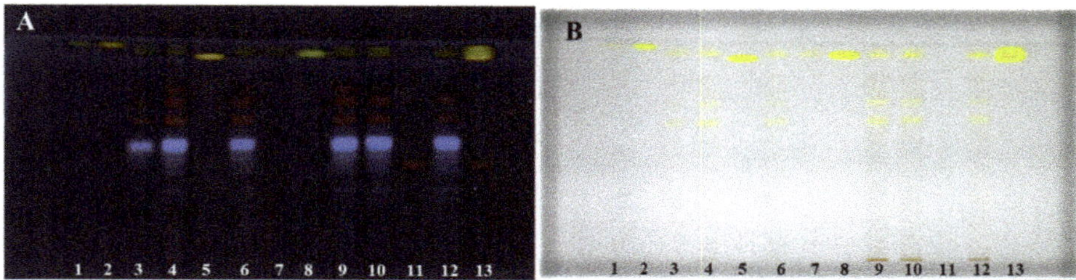

Figure 1. HPTLC chromatograms of physcion (1 μg; track 1), chrysophanol (1 μg; track 2), aloe-emodin (1 μg; track 5), emodin (1 μg; track 8) and aloin A (1 μg; track 11) standards and Japanese knotweed rhizomes STSs (2 μL, 50 mg/mL) prepared with acetone (track 3), methanol (track 4), ethanol (track 6), dichloromethane (track 7), 70% acetone$_{(aq)}$ (track 9), 70% ethanol$_{(aq)}$ (track 10) and 70% methanol$_{(aq)}$ (track 12). HPTLC silica gel plates were developed with toluene–acetone–formic acid (3:6:1, *v*/*v*) and documented after development at 366 nm (**A**) and at white light (**B**).

On the second plate were STSs prepared with 20% ethanol$_{(aq)}$, 40% ethanol$_{(aq)}$, 60% ethanol$_{(aq)}$, 80% ethanol$_{(aq)}$ and ethanol (Figure 2). Both plates were developed with the developing solvent toluene–acetone–formic acid (3:6:1, *v*/*v*). As shown in Figure 1 (tracks 1, 2, 5, 8 and 11), all standards were detected at 366 nm and at white light illumination conditions. At 366 nm standards of physcion, chrysophanol, aloe-emodin and emodin were detected as yellow bands at R_F values in the range of 0.88–0.93, while standard of aloin A (track 11) was detected as orange-brown band at R_F 0.42. At white light, all standards appeared as yellow bands (Figure 1B). At 366 nm, two yellow bands were detected at R_F 0.90 and 0.93 in tracks of STSs prepared with ethanol, 70% ethanol$_{(aq)}$, methanol, 70% methanol$_{(aq)}$, acetone, 70% acetone$_{(aq)}$ and dichloromethane (Figure 1A, tracks 3, 4, 6, 7, 9, 10 and 12). Similarity between the colors of the band of aloin A standard (Figure 1, track 11) and the bands of STSs prepared with ethanol, 70% ethanol$_{(aq)}$, methanol, 70% methanol$_{(aq)}$, acetone and 70% acetone$_{(aq)}$ at R_F values 0.60, 0.69 and 0.74 (Figure 1, tracks 3, 4, 6, 9, 10 and 12) was observed. These bands that appear orange-brown at 366 nm (Figure 1A) are yellow at white light (Figure 1B). Two intensive yellow bands at R_Fs 0.90 and 0.93 were detected at 366 nm and at white light on the second HPTLC plate but only for STSs prepared with 60% ethanol$_{(aq)}$, 80% ethanol$_{(aq)}$ and ethanol (Figure 2, tracks 4–6). The yellow band at R_F 0.90 with comparable intensity was observed also for STSs prepared with 40% ethanol$_{(aq)}$, while a much lower intensity was detected for STSs prepared with water and 20% ethanol$_{(aq)}$ (Figure 2, tracks 1–3). At 366 nm, three bands with the same orange-brown color as was observed for aloin A standard (Figure 1, track 11) were detected for STSs prepared with 60% ethanol$_{(aq)}$, 80% ethanol$_{(aq)}$ and ethanol (Figure 2, tracks 4–6) at R_F values 0.60, 0.69 and 0.74. These bands were yellow at white light.

Evaluation of the chromatograms obtained by different extraction solvents on both plates was performed based on densitometric scanning at 442 nm in absorption/reflectance mode (Figures 3 and 4). The wavelength 442 nm (absorption maximum of emodin, known in the literature to be present in Japanese knotweed rhizomes) was selected based on the absorption spectra of all standards scanned in situ on the developed HPTLC silica gel plate. The absorption maxima for physcion, chrysophanol, aloe–emodin, emodin and aloin A were at 442, 428, 442, 360 and 431 nm, respectively.

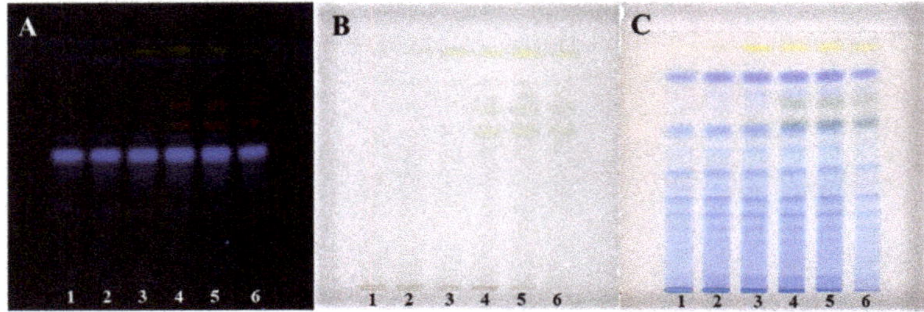

Figure 2. HPTLC chromatograms of Japanese knotweed rhizomes STSs (2 μL, 50 mg/mL) prepared with water (track 1), 20% ethanol$_{(aq)}$ (track 2), 40% ethanol$_{(aq)}$ (track 3), 60% ethanol$_{(aq)}$ (track 4), 80% ethanol$_{(aq)}$ (track 5) and ethanol (track 6). HPTLC silica gel plates were developed with toluene–acetone–formic acid (3:6:1, *v*/*v*). Images of the plates were documented at 366 nm (**A**) and at white light (**B**) after development and at white light (**C**) after derivatization with 4-dimethylaminocinnamaldehyde (DMACA) reagent.

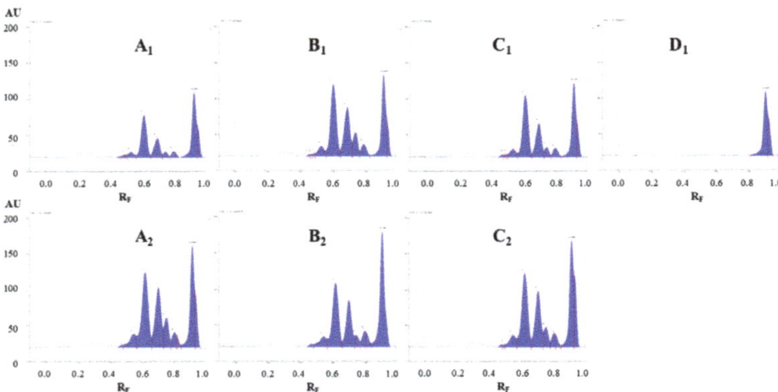

Figure 3. Densitograms of Japanese knotweed rhizomes STSs (2 μL, 50 mg/mL) scanned at 442 nm in absorption/reflectance mode. STSs (2 μL, 50 mg/mL) were prepared with acetone (**A$_1$**), 70% acetone$_{(aq)}$ (**A$_2$**), methanol (**B$_1$**), 70% methanol$_{(aq)}$ (**B$_2$**), ethanol (**C$_1$**), 70% ethanol$_{(aq)}$ (**C$_2$**) and dichloromethane (**D$_1$**). HPTLC silica gel plate was developed with toluene–acetone–formic acid (3:6:1, *v*/*v*).

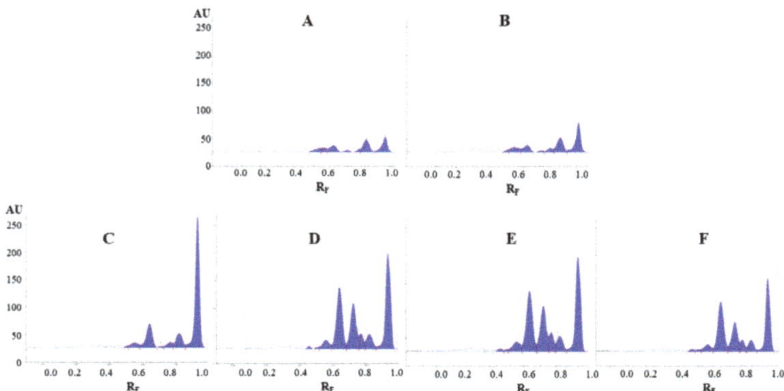

Figure 4. The densitograms of Japanese knotweed rhizomes STSs (2 μL, 50 mg/mL) scanned at 442 nm in absorption/reflectance mode. STSs (2 μL, 50 mg/mL) were prepared with water (**A**), 20% ethanol$_{(aq)}$ (B), 40% ethanol$_{(aq)}$ (**C**), 60% ethanol$_{(aq)}$ (**D**), 80% ethanol$_{(aq)}$ (**E**) and ethanol (**F**). HPTLC silica gel plate was developed with toluene–acetone–formic acid (3:6:1, *v*/*v*).

The densitograms scanned at 442 nm for STSs prepared with different pure organic solvents (acetone, methanol, ethanol and dichloromethane) and mixtures of organic solvents with water (70% $acetone_{(aq)}$, 70% $methanol_{(aq)}$ and 70% $ethanol_{(aq)}$) show similarities and differences in the number and intensities of the peaks (Figure 3). Based on these observations, it can be concluded that dichloromethane is a more selective extraction solvent than all other tested solvents, as no peaks were detected in the densitogram below the R_F value 0.8 (Figure 3D_1). All other extraction solvents show equal qualitative profiles in the densitograms (Figure 3A_1–C_1). The use of extraction solvents that contained water (70% $acetone_{(aq)}$, 70% $methanol_{(aq)}$ and 70% $ethanol_{(aq)}$) resulted in higher peaks compared to pure organic solvents (acetone, methanol and ethanol). Comparison of the densitograms for STSs, prepared with water, 20% $ethanol_{(aq)}$, 40% $ethanol_{(aq)}$, 60% $ethanol_{(aq)}$, 80% $ethanol_{(aq)}$ and ethanol, showed equal qualitative profiles (Figure 4). The increasing of the % of ethanol (from 20% to 80%) in the extraction solvent results in a significant increase in all heights of all the peaks (Figure 4B–E). In the case of STS prepared with pure ethanol, the peak heights were lower than for STS prepared with 80% $ethanol_{(aq)}$ (Figure 4E,F).

The extraction efficiency of all tested extraction solvents was evaluated comparing the total peak areas in the densitograms of STSs prepared with different extraction solvents (Figures 5 and 6). This comparison was performed separately for each HPTLC plate, because peak areas on two different plates cannot directly be compared. On the first HPTLC plate, the highest total peak area was obtained with 70% $acetone_{(aq)}$ and the lowest with dichloromethane (Figure 5). The total peak areas for STSs prepared with other pure organic solvents (acetone, methanol and ethanol) were lower compared to those obtained for STSs prepared with mixtures of organic solvents with water (70% $acetone_{(aq)}$, 70% $methanol_{(aq)}$ and 70% $ethanol_{(aq)}$). On the second HPTLC plate, the highest total peak area was achieved with 80% $ethanol_{(aq)}$ and the lowest with water (Figure 6). Total peak areas increased as the % of ethanol in the extraction solvent was increased from 20% to 80%. However, the total peak area obtained with ethanol was lower compared to that obtained with 60% $ethanol_{(aq)}$. Our results are in agreement with those obtained for the extraction of anthraquinones and other phenolic compounds from Japanese knotweed rhizomes using water and mixtures of water and ethanol (25%, 50%, 75% and 95% $ethanol_{(aq)}$) [24], which showed that 75% $ethanol_{(aq)}$ was the most efficient extraction solvent. Based on our data, 70% $acetone_{(aq)}$ was selected as the extraction solvent, and STS prepared with this solvent was analyzed by HPTLC and HPTLC-MS/MS methods.

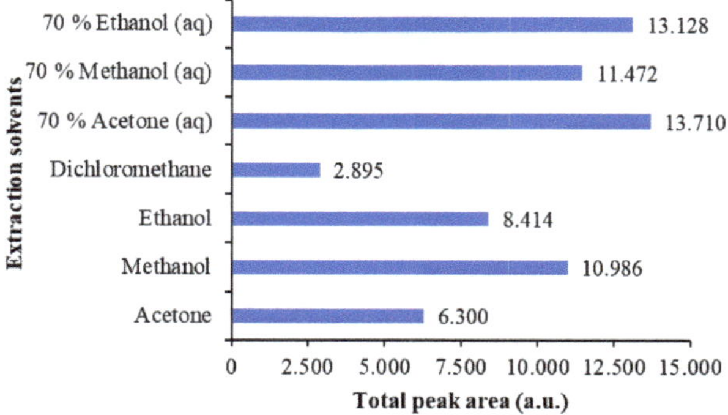

Figure 5. Comparison of the total peak areas obtained from densitograms of Japanese knotweed rhizomes STSs scanned at 442 nm on the HPTLC silica gel plate developed with toluene–acetone-formic acid (3:6:1, *v*/*v*). STSs (2 µL, 50 mg/mL) were prepared with 70% $ethanol_{(aq)}$, ethanol, 70% $methanol_{(aq)}$, methanol, 70% $acetone_{(aq)}$, acetone and dichloromethane.

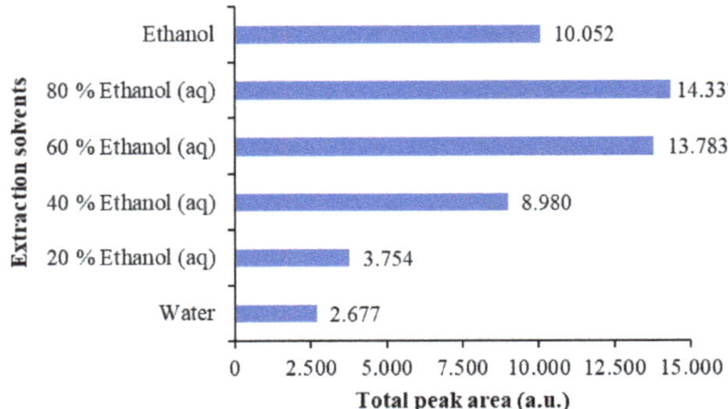

Figure 6. Comparison of the total peak areas obtained from densitograms of Japanese knotweed rhizomes STSs scanned at 442 nm on the HPTLC silica gel plate developed with toluene–acetone-formic acid (3:6:1, *v*/*v*). STSs (2 μL, 50 mg/mL) were prepared with water, 20% ethanol$_{(aq)}$, 40% ethanol$_{(aq)}$, 60% ethanol$_{(aq)}$, 80% ethanol$_{(aq)}$ and ethanol.

2.2. HPTLC Analyses

Developing solvents with different ratios of toluene–acetone–formic acid (6:6:1, *v*/*v*, 3:6:1, *v*/*v* and 3:3:1, *v*/*v*), as well as dichloromethane–acetone–formic acid (1:1:0.1, *v*/*v*) (Figure 7), were tested as possible developing solvents for the separation of anthraquinones from Japanese knotweed rhizomes STSs in 70% acetone$_{(aq)}$. These solvents were used in our previous study for the separation of proanthocyanidins [12]. In the present study, proanthocyanidins were detected by post-chromatographic derivatization with 4-dimethylaminocinnamaldehyde (DMACA) in the STSs prepared with water, 20% ethanol$_{(aq)}$, 40% ethanol$_{(aq)}$, 60% ethanol$_{(aq)}$, 80% ethanol$_{(aq)}$ and ethanol (Figure 2C).

Figure 7. HPTLC chromatograms of Japanese knotweed rhizomes STS (2 μL, 50 mg/mL in 70% acetone$_{(aq)}$) developed up to 9 cm with toluene–acetone–formic acid (6:6:1, *v*/*v*) (**A**), toluene–acetone–formic acid (3:3:1, *v*/*v*) (**B**), toluene–acetone–formic acid (3:6:1, *v*/*v*) (**C**) and dichloromethane–acetone–formic acid (1:1:0.1, *v*/*v*) (**D**) and documented at 366 nm.

As shown in Figure 7, all tested developing solvents enabled separation of non-glycosylated (yellow bands at 366 nm) and glycosylated (orange-brown bands at 366 nm). Only with the developing solvents with different ratios of toluene–acetone–formic acid (6:6:1 (*v*/*v*), Figure 7A, 3:3:1, (*v*/*v*), Figure 7B and 3:6:1, (*v*/*v*), Figure 7C) separation of non-glycosylated anthraquinones was achieved. Among these solvents, the highest resolution between the two yellow bands, later confirmed as emodin and physcion, was achieved with toluene–acetone–formic acid (6:6:1, *v*/*v*) (Figure 7A). Based on the presence of only one yellow band on the plate developed with dichloromethane–acetone–formic acid (1:1:0.1, *v*/*v*), it can be concluded that this developing solvent was not suitable for the separation of non-glycosylated anthraquinones (Figure 7D). The best developing solvents for the separation of glycosylated anthraquinones (orange-brown bands at 366 nm) were dichloromethane–acetone–formic acid (1:1:0.1 (*v*/*v*), Figure 7D) and toluene–acetone–formic acid (6:6:1 (*v*/*v*), Figure 7A). These two solvents enabled separation of six glycosylated anthraquinones, while toluene–acetone–formic acid (3:3:1, *v*/*v*, Figure 7B) and toluene–acetone–formic acid (3:6:1, *v*/*v*, Figure 7C) enabled to separate only four and three glycosylated anthraquinones, respectively. The analysis of STSs (prepared with 70% acetone$_{(aq)}$) together with emodin and aloin A standards on the HPTLC silica gel plates developed with toluene–acetone–formic acid (3:6:1, *v*/*v*) confirmed the presence of emodin by matching the R_F values and the band colors of emodin standard and the corresponding band in the STS track after development and after post-chromatographic derivatization with natural product (NP) and polyethylene glycol (PEG) reagents (Figure 8D,E). Colors of both bands changed from yellow after development and NP reagent to red after PEG reagent. Aloin A was not detected in STS. Additional orange-brown bands with the R_F values lower than emodin and higher than aloin A were detected but were not identified. Some of these bands become red at white light after the application of PEG reagent (Figure 8E).

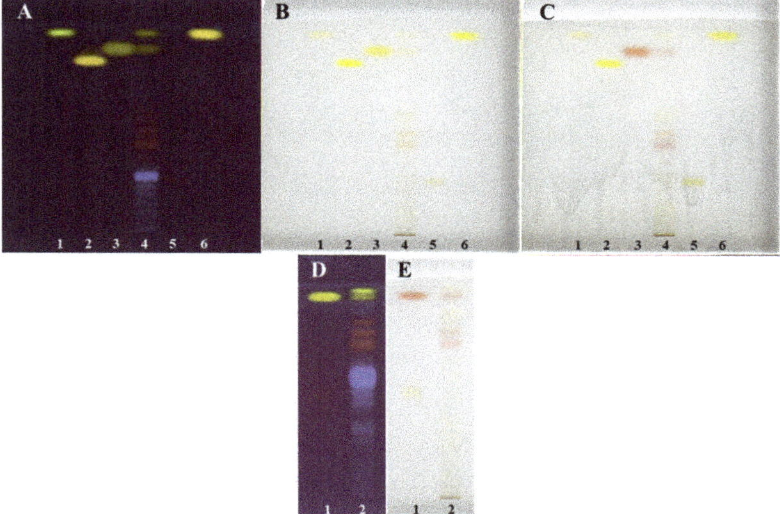

Figure 8. HPTLC chromatograms of standards and Japanese knotweed rhizomes STSs (2 μL, 50 mg/mL) prepared with 70% acetone$_{(aq)}$ (track 4 on the plates (**A**–**C**) and track 2 on the plates (**D**,**E**)). HPTLC silica gel plates were developed with toluene–acetone–formic acid (6:6:1, *v*/*v* (**A**–**C**) or 3:6:1, *v*/*v*; (**D**,**E**)). Documentation was performed at 254 nm after development (**A**,**D**) and at white light after post-chromatographic derivatization with NP reagent (**B**) and 30 min after use of PEGreagent (**C**,**E**). Applications of standards on the plates (**A**–**C**): physcion (1 μg; track 1), aloe-emodin (1 μg; track 2), emodin (1 μg; track 3), aloin A (1 μg; track 5) and chrysophanol (1 μg; track 6). Applications of standards (1 μg; track 1) on the plates: (**D**,**E**) emodin (higher R_F and lower R_F).

Since toluene–acetone–formic acid (6:6:1, *v*/*v*) resulted in the best separation of non-glycosylated and glycosylated anthraquinones among all tested developing solvents, it was chosen for further HPTLC analysis of STS prepared with 70% acetone$_{(aq)}$ on the silica gel plates, together with applied standards of physcion, chrysophanol, aloe-emodin, emodin and aloin A (Figure 8A–C). It is evident that the resolution between the bands of physicon (R_F 0.93, track 1, Figure 8A–C) and chrysophanol (R_F 0.94, track 6, Figure 8A–C) is not good enough. However, there is a difference in shades of yellow colors of the bands and the shapes of their absorption spectra (Figure 9). The presence of emodin and physcion in STS was confirmed based on matching the R_F values and the colors of the bands of the standards (emodin (R_F 0.85) and physicon (R_F 0.93)) with the corresponding bands in the STS track after development and after post-chromatographic derivatization with NP and PEG reagents (Figure 8A–C). Among the standards, only emodin changed color at white light from yellow to red after using the combination of NP and PEG reagents, while physcion, chrysophanol, aloe-emodin and aloin A remained yellow as they were after the development and after the use of NP reagent (Figure 8A–C). The band of aloin A changed its color at 254 nm and 366 nm from orange-brown after development to intensive light green after the application of NP and PEG reagents (data not shown).

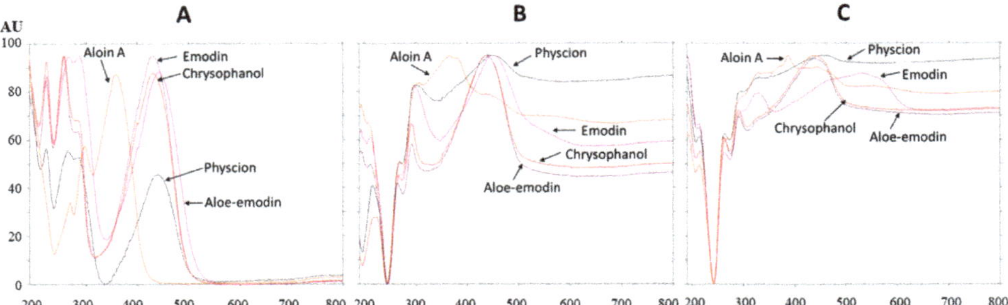

Figure 9. Absorption spectra of standards scanned in situ on the HPTLC silica gel plate developed with toluene–acetone–formic acid (6:6:1, *v*/*v*) after development (**A**), after post-chromatographic derivatization with NP reagent (**B**) and 30 min after use of PEG reagent (**C**). The application of each standard was 1 µg.

The absorption spectra of all standards were scanned after development (Figure 9A), after post-chromatographic derivatization with NP reagent (Figure 9B) and after the application of PEG reagent (Figure 9C). Bathochromic shifts were observed for the absorption maxima of all standards after post-chromatographic derivatization with NP reagent and, also, after use of PEG reagent (Table 1). These shifts were from 4 nm (aloin A) up to 9 nm (aloe–emodin) after the application of NP reagent. Bathochromic shifts were even more pronounced when the application of NP reagent was followed by PEG reagent. The differences between the absorption maxima of the standards on the same plate after development and after post-chromatographic derivatization (with NP reagent, followed by the application of PEG reagent) were compared. The highest difference (86 nm) was obtained for emodin.

Table 1. Absorption maxima of standards determined by densitometric scanning of the HPTLC silica gel plate developed with toluene–acetone–formic acid (3:6:1, *v*/*v*). Spectra were scanned after development, after post-chromatographic derivatization with NP reagent and 30 min after the application of PEGreagent.

	Absorption Maxima of Standards (nm)		
Compounds	**After Development**	**After NP**	**30 min after PEG**
Physcion	442	449	454
Aloe-emodin	428	437	438
Emodin	442	449	528
Aloin A	360	364	387
Chrysophanol	431	437	437

Absorption spectra of emodin (R_F 0.85) and physcion (R_F 0.93) standards and the absorption spectra of compounds in the corresponding bands in the STS track were scanned in situ in the range of 190–800 nm. Comparison of the absorption spectra of emodin (R_F 0.85) and compounds in the corresponding STS band with the same R_F, scanned after development (Figure 10A), after post-chromatographic derivatization with NP reagent (Figure 10B) and after the application of PEG reagent (Figure 10C) confirmed that the spectra matched. This was additional proof of the presence of emodin in the STS. Comparison of the absorption spectra of physcion (R_F 0.93) and compounds in the corresponding STS band with the same R_F scanned after development

(Figure 10D), after post-chromatographic derivatization with NP reagent (Figure 10E) and after the application of PEG reagent (Figure 10F) confirmed that the spectra matched. The possible presence of chrysophanol (R_F 0.94), which was not separated from physcion (R_F 0.93), was rejected. The rejection was based on the comparison of the absorption spectra of chrysophanol, physcion and compounds in the corresponding STS band with the same R_F (Figure 10F). Chrysophanol has a different shape of absorption spectrum and also different absorption maxima after development, after post-chromatographic derivatization and after use of PEG reagent (Table 1).

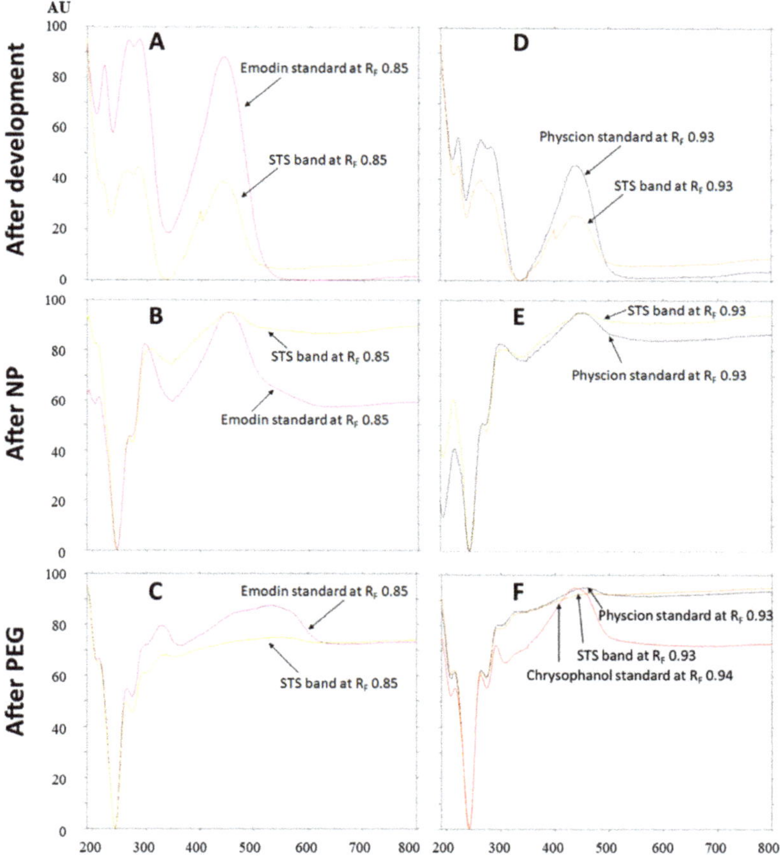

Figure 10. Absorption spectra of emodin (at R_F 0.85; **A–C**) and physcion (at R_F 0.93; **D–F**) standards and compounds in the bands at the same R_Fs (**A–C**: R_F 0.85 and **D–F**: R_F 0.93) in the track of STS (2 μL, 50 mg/mL, 70% acetone$_{(aq)}$). Spectra were scanned in situ on the HPTLC silica gel plate developed with toluene–acetone–formic acid (6:6:1, *v*/*v*) after development, after post-chromatographic derivatization with NP reagent and 30 min after use of PEG reagent. Additional absorption spectrum of chrysophanol scanned after post-chromatographic derivatization with NP reagent and 30 min is presented in (**F**).

Other compounds present in orange-brown bands (at 366 nm; Figure 8A,D) could not be identified by HPTLC analyses due to the lack of appropriate standards. However, they were tentatively identified as glycosylated anthraquinones by HPTLC-MS/MS analyses (Section 2.3) when their fragmentation patterns were compared with those reported in the literature [6,20].

2.3. HPTLC-MS/MS Analyses

Four HPTLC-MS methods (Table 2) were used to analyze anthraquinones from STS of Japanese knotweed rhizomes to obtain MS and MS^n spectra. HPTLC silica gel plates were twice predeveloped (up to the top) before development up to 7 cm with the following developing solvents: toluene–acetone–formic acid (3:6:1 (*v*/*v*), 6:6:1 (*v*/*v*) and 3:3:1 (*v*/*v*)) and dichloromethane–acetone–formic acid (1:1:0.1, *v*/*v*). MS spectra obtained by HPTLC-MS analyses using toluene–acetone–formic acid (6:6:1, *v*/*v*) as the developing solvent are presented in Figure 11.

Table 2. Anthraquinones tentatively identified in Japanese knotweed rhizomes by HPTLC- MS/MS.

Compound	(-)ESI-MS *m*/*z*	(-)ESI-MS^n m/z	Ref.
Emodin [a]	269	MS^2 [269]: 225, 241 MS^3 [269→225]: 181, 210, 197, 207	
Physcion [a]	283	MS^2 [283]: 240, 268 MS^3 [283→240]: 212	
Emodin-8-*O*-hexoside	431	MS^2 [431]: 269, 311 MS^3 [431→269]: 225, 241	[20]
Emodin-*O*-acetyl-hexoside	473	MS^2 [473]: 269, 311 MS^3 [473→269]: 225, 241	[6]
Emodin-*O*-malonyl-hexoside		MS^2 [517]: 473 MS^3 [517→473]: 269, 311 MS4 [517→269]: 225, 241	[20]

[a] Confirmed with a reference standard.

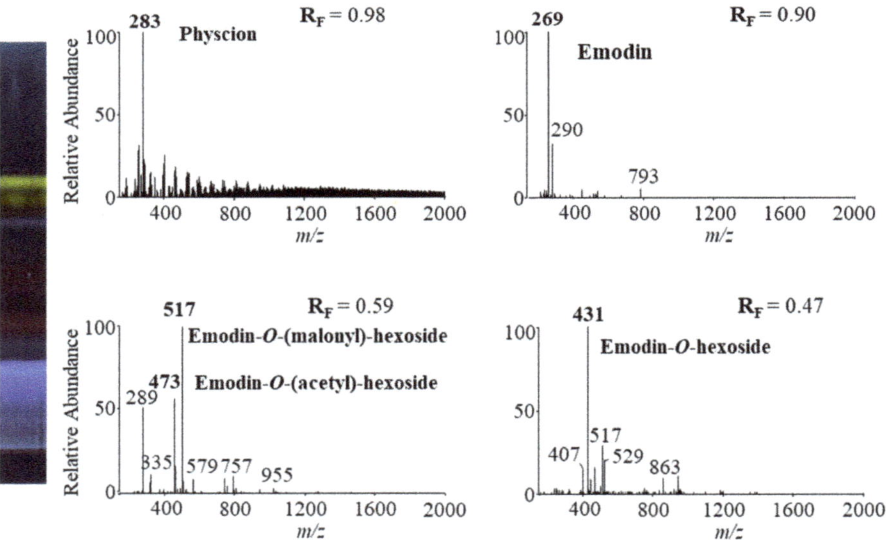

Figure 11. MS spectra of anthraquinones (bold, *m/z*) obtained from Japanese knotweed rhizomes STS from a twice predeveloped HPTLC silica gel plate developed up to 7 cm with toluene–acetone–formic acid (6:6:1, *v*/*v*) as the developing solvent and documented at 366 nm.

The identification of emodin and physcion in Japanese knotweed rhizomes STSs was confirmed by comparing the absorption spectra of yellow bands (Figure 10) and the fragmentation patterns (at R_Fs corresponding to R_Fs for physcion and emodin standards) (Table 2) with absorption spectra and fragmentation patterns of physcion and emodin standards. Other compounds present in orange-brown bands were tentatively identified as glycosylated anthraquinones by comparison of the fragmentation patterns (Table 2) with those reported in the literature. The MS^2 spectrum of the signal at *m*/*z* 431 gave a base ion at *m*/*z* 269 with a neutral loss of 162 Da, and the MS^3 spectrum gave product ions at *m*/*z* 225 and *m*/*z* 241, which were identical to emodin standard. This indicated that signal *m*/*z* 431 could be emodin hexoside. Emodin-1-*O*-glucoside [2,3,6], emodin-6-*O*-glucoside [6] and emodin-8-*O*-glucoside [2,3,6] were previously identified in Japanese knotweed rhizomes. The MS^2 fragmentation pattern of a base peak at *m*/*z* 431 was similar to the fragmentation pattern (product ions at *m*/*z* 269 and *m*/*z* 311) of emodin-8-*O*-glucoside [2,6] and the MS^3 fragmentation pattern was typical for emodin reference standard. Therefore, the signal at *m*/*z* 431 was tentatively assigned to $[M-H]^-$ of emodin-8-*O*-hexoside.

The MS^2 and MS^3 fragmentation patterns of the signal at *m*/*z* 517 were similar to the fragmentation pattern of emodin-8-(6′ malonyl)-glucoside [20], which were reported to be present in Japanese knotweed rhizomes [2,3,6]. Therefore, the signal at *m*/*z* 517 was tentatively identified as emodin–malonyl–hexoside. Another signal at *m*/*z* 473 was obtained at the same R_F value using all HPTLC-MS/MS methods. The MS^2 spectrum of a signal at *m*/*z* 431 gave a base ion at *m*/*z* 269. A loss of 204 Da indicated acetyl and glucosyl residues. MS^3 spectrum of the signal at *m*/*z* 473 gave product ions at *m*/*z* 225 and *m*/*z* 241, which correspond to emodin. Therefore, the signal at *m*/*z* 473 was tentatively assigned as emodin–acetyl–glucoside, which was previously identified in Japanese knotweed rhizomes [6].

HPTLC-MS/MS methods using different ratios of toluene–acetone–formic acid (3:6:1, *v*/*v*, 6:6:1, *v*/*v* and 3:3:1, *v*/*v*) and dichloromethane–acetone–formic acid (1:1:0.1, *v*/*v*) as the developing solvents enabled the detection of emodin and its three glycosylated analogs (Table 3). The method with toluene–acetone–formic acid (6:6:1, *v*/*v*) additionally enabled the detection of physcion.

Table 3. R_F values of anthraquinones identified by HPTLC-MS analyses in Japanese knotweed rhizomes sample test solutions (STSs) (prepared with 70% $acetone_{(aq)}$) on HPTLC silica gel plates developed with different developing solvents.

Compound	*m*/*z*	Developing Solvent			
		DS1	DS2	DS3	DS4
		R_F			
Emodin	269	0.90	0.87	0.97	0.99
Physcion	283	0.98	/	/	/
Emodin-*O*-hexoside	431	0.47	0.62	0.68	0.47
Emodin-*O*-(acetyl)-hexoside	473	0.59	0.74	0.81	0.81
Emodin-*O*-(malonyl)-hexoside	517	0.59	0.74	0.81	0.81

DS1: toluene–acetone–formic acid (6:6:1, *v*/*v*). DS2: toluene–acetone–formic acid (3:3:1, *v*/*v*). DS3: toluene–acetone–formic acid (3:6:1, *v*/*v*). DS4: dichloromethane–acetone–formic acid (1:1:0.1, *v*/*v*).

3. Materials and Methods

3.1. Chemicals

All chemicals used in this study were at least of analytical grade. Toluene, dichloromethane, ethyl acetate hydrochloric acid (37%), formic acid and 4-dimethylaminocinnamaldehyde (DMACA) were purchased from Merck (Darmstadt, Germany). Diphenylboric acid 2-aminoethyl ester (natural product reagent, NP), ethanol (absolute), acetone and HPLC grade methanol and acetonitrile were from Sigma-Aldrich (St. Louis, MO, USA). LC-MS grade methanol and acetonitrile were from Fluka (Buchs, Switzerland). Polyethylene glycol (PEG) 4000 was from Fluka Chemie (Buchs, Switzerland). Bidistilled water was also used.

Standards of physcion, chrysophanol, aloe-emodin, emodin and aloin A were obtained from Extrasynthesè S.A. (Genay, France).

3.2. Preparation of Standard Solutions

Standard solutions (0.2 mg/mL) were separately prepared in methanol and were stored in amber glass storage vials at −80 °C.

3.3. Plant Material and Preparation of Sample Test Solutions (STSs)

Rhizomes (a subterranean stem) of Japanese knotweed (*Fallopia japonica* Houtt.) were collected at Ljubljanica riverside in Ljubljana (Vrhovci, by bridge over Mali graben, N 46°02′33.9″; E 14°27′00.9″ [34]) in August, 2019. The voucher specimen is deposited in the Herbarium LJU (LJU10143477).

The rhizomes were washed with tap water, dried on air, cut into smaller pieces, frozen with liquid nitrogen and lyophilized (Micro Modulyo, IMAEdwards, Bologna, Italy) for 48 h at −50 °C. Freeze-dried rhizomes were again frozen with liquid nitrogen, crushed and pulverized by Mikro-Dismembrator S (Sartorius, Göttingen, Germany) at a frequency of 1700 min^{-1} for 1 min.

Powdered lyophilized plant material (100 mg) was dispersed in 4 mL of extraction solvent. Water, 20% $ethanol_{(aq)}$, 40% $ethanol_{(aq)}$, 60% $ethanol_{(aq)}$, 70% $ethanol_{(aq)}$, 80% $ethanol_{(aq)}$, ethanol, 70% $methanol_{(aq)}$, methanol, 70% $acetone_{(aq)}$, acetone and dichloromethane were used as extraction solvents. Suspensions were vortexed (2 min at 2800 rpm; IKA lab dancer, Sigma-Aldrich) and centrifuged at 6700× *g*. Supernatants were filtered through a 0.45-µm polyvinylidene fluoride (PVDF) membrane filter (Millipore, Billerica, MA, USA). The obtained sample test solutions (STSs) (50 mg/mL) were stored in amber glass storage vials at −80 °C and were used undiluted for the HPTLC and HPTLC-MS analyses.

3.4. HPTLC Analyses

HPTLC analyses were performed on 20 cm × 10 cm glass-backed HPTLC silica gel (Merck, Art. No. 1.05641, Darmstadt, Germany). Standard solutions (1 µg) and STSs, which were prepared in different extraction solvents (2 µL), were applied on the un-predeveloped plates by an automatic TLC Sampler 4 (Camag, Muttenz, Switzerland). The plates were developed up to 9 cm using different ratios of toluene–acetone–formic acid (3:6:1, *v*/*v*, 6:6:1, *v*/*v* and 3:3:1, *v*/*v*) and dichloromethane–acetone–formic acid (1:1:0.1, *v*/*v*) as the developing solvents. The developing solvent (10 mL) was added only in one trough of an unsaturated twin-trough chamber (Camag) for 20 cm × 10 cm plates. Only 5 mL of the developing solvent was used in case of 10 × 10 cm or smaller plates, which were developed in a twin-trough chamber (Camag) for 10 cm × 10 cm plates. The developed plates were dried in a stream of warm air for 3 min.

Post-chromatographic derivatization was performed by heating the plates on a TLC plate heater III (Camag) at 110 °C (3 min), which was immediately followed by dipping the plate for 1 s in the natural product reagent (NP reagent), prepared by dissolving 1 g of NP in 200 mL of ethyl acetate [35]. After drying in a stream of warm air (hair dryer) for 2 min, followed by cooling in the air for 5 min, the plates were dipped in PEG 4000 reagent, prepared by dissolving 10 g of PEG 4000 [35] in 200 mL of dichloromethane. The plates were again dried in a stream of warm air (hair dryer) for 2 min. Additional post-chromatographic derivatization was used to detect proanthocyanidins. In this case, plates were dipped for 1 s in DMACA dipping detection reagent prepared by dissolving 60 mg of DMACA in 13 mL of concentrated hydrochloric acid, which was made up to 200 mL with ethanol [36]. Dipping was followed by drying for 2 min in a stream of warm air. NP reagent, PEG reagent and DMACA reagent were protected from light and stored at 5 °C.

The Camag Digistore 2 Documentation system in conjunction with Reprostar 3 was used to document the images of the chromatographic plates at 254 nm, 366 nm and white light illumination. The plates were documented: (i) immediately after development, (ii) immediately after post-chromatographic derivatization with NP reagent, (iii) immediately after the enhancement and stabilization of fluorescent zones with PEG reagent and (iv) 30 min after the enhancement and stabilization of fluorescent zones with PEG reagent. When DMACA reagent was applied for post-chromatographic derivatization, the images were documented only at white light illumination immediately and 10 min after derivatization. After documentation the developed plates were scanned by the slit-scanning densitometer TLC Scanner 3 (Camag) set in absorption/reflectance mode at 442 nm. The slit length was 6 mm, the slit width 0.30 mm and the scanning speed 20 mm s^{-1}. The absorption spectra (from 190 nm to 800 nm) were scanned in situ before and after post-chromatographic derivatization with NP reagent and also after application of PEG reagent. Both instruments were controlled by winCATS software (Version 1.4.9.2001).

3.5. HPTLC-MS/MS Analyses

The HPTLC silica gel plates were firstly predeveloped with methanol–formic acid (10:1, *v*/*v*) and, secondly, with acetonitrile–methanol (2:1, *v*/*v*)) up to the top and dried for 30 min at 100 °C. Twice predeveloped plates (cut to 5 cm × 10 cm) were used for the application of 20 µL of STS in 70% $acetone_{(aq)}$ as 24 mm band, 10 mm from the bottom edge, 20 mm from the left edge and developed up to 7 cm for the HPTLC-MS/MS analyses of anthraquinones. Toluene–acetone–formic acid (3:6:1 (*v*/*v*), 6:6:1 (*v*/*v*) and 3:3:1 (*v*/*v*)) and dichloromethane–acetone–formic acid (1:1:0.1, *v*/*v*) were used as the developing solvents.

The yellow bands that appeared on the developed plates were used for positioning the oval elution head (4 mm × 2 mm) of the TLC-MS interface (Camag) and were eluted and transferred into a LCQ ion trap system (Thermo Finnigan, San Jose, CA, USA). For evaluation of the collected data Xcalibur 1.3 software was used. HPTLC–MS analyses were performed according to [12,13]. Acetonitrile–methanol (2:1, *v*/*v*) was used as an eluent at 0.2 mL min^{-1} flow rate. A C18 guard column (4 mm × 3 mm ID, Phenomenex, Torrance, CA, USA) was mounted between the TLC–MS interface and MS ion source. Electrospray ionization (ESI) in negative ion mode was used to acquire mass spectra from *m*/*z* 150–2000 scan range in 1 min. The spray voltage was set to 4 kV, capillary temperature to 200 °C, capillary voltage to −38.8 V, tube lens offset to −5, flow rate sheath gas to 95 a.u. (arbitrary units) and flow rate auxiliary gas to 14 a.u. Fragmentation of the parent ion was performed at 45% collision energy.

4. Conclusions

Twelve extraction solvents (water, 20% $ethanol_{(aq)}$, 40% $ethanol_{(aq)}$, 60% $ethanol_{(aq)}$, 70% $ethanol_{(aq)}$, 80% $ethanol_{(aq)}$, ethanol, 70% $methanol_{(aq)}$, methanol, 70% $acetone_{(aq)}$, acetone and dichloromethane) were used for the extraction of anthraquinones (yellow dyes) from Japanese knotweed rhizomes. The obtained sample test solutions (STSs) were analyzed by the HPTLC method on HPTLC silica gel plates developed with toluene–acetone–formic acid (3:6:1, *v*/*v*). Qualitative densitometric profiles scanned at 442 nm (absorption maximum for emodin) for STSs prepared in all the solvents except dichloromethane were identical and included anthraquinone aglycones and glycosylated anthraquinones). The most selective extraction of anthraquinone aglycones emodin and physcion was achieved with dichloromethane. Extraction efficiency, evaluated by comparison of the total peak areas of the densitograms of STSs prepared with all solvents, was the highest with 70% $acetone_{(aq)}$. Therefore, only STS prepared with this solvent was used for further selection of developing solvent for HPTLC and HPTLC-MS analyses.

HPTLC silica gel plates in combination with four new developing solvents were proposed for analyses of anthraquinones by HPTLC and HPTLC-MS. All developing solvents with different ratios of toluene–acetone–formic acid (6:6:1 (*v*/*v*), 3:6:1 (*v*/*v*) and 3:3:1 (*v*/*v*)) and dichloromethane–acetone–formic acid (1:1:0.1, *v*/*v*) enabled the separation of non-glycosylated and glycosylated anthraquinones present in STS prepared with 70% $acetone_{(aq)}$. However, the separation of non-glycosylated was achieved only with the developing solvents with different ratios of toluene–acetone–formic acid (6:6:1 (*v*/*v*), 3:6:1 (*v*/*v*) and 3:3:1 (*v*/*v*)). The best resolution between non-glycosylated anthraquinones emodin and physcion was achieved with toluene–acetone–formic acid 6:6:1 (*v*/*v*). The best separation of glycosylated anthraquinones was achieved with toluene–acetone–formic acid

(6:6:1, *v*/*v*) and dichloromethane–acetone–formic acid (1:1:0.1, *v*/*v*). Changes of the shapes and the absorption maxima (bathochromic shifts) in the absorption spectra of the standards (physcion, chrysophanol, aloe-emodin, emodin and aloin A) scanned in situ on the HPTLC plate after development, after post-chromatographic derivatization with NP reagent and after the application of PEG reagent were observed. These changes were used as additional proof for the presence of emodin and physcion (almost the same R_F but different absorption spectra than chrysophanol) in STS prepared with 70% $acetone_{(aq)}$ analyzed on the same plate as the standards. All four HPTLC-MS/MS methods using different ratios of toluene–acetone–formic acid (3:6:1 (*v*/*v*), 6:6:1 (*v*/*v*) and 3:3:1 (*v*/*v*)) and dichloromethane–acetone–formic acid (1:1:0.1, *v*/*v*) as the developing solvents enabled the identification of emodin and tentative identification of its three glycosylated analogs (emodin-8-*O*-hexoside, emodin-*O*-acetyl-hexoside and emodin-*O*-malonyl-hexoside). Additionally, the HPTLC-MS/MS method with toluene–acetone–formic acid (6:6:1, *v*/*v*) enabled the identification of physcion. The identification of emodin and physcion in Japanese knotweed rhizomes STS was confirmed by comparing the absorption spectra and the fragmentation patterns (at R_Fs corresponding to R_Fs for emodin and physcion standards) with fragmentation patterns of physcion and emodin standards.

Author Contributions: Conceptualization, V.G. and I.V.; data curation, V.G. and I.V.; methodology, V.G. and I.V.; formal analysis, V.G.; investigation, V.G. and I.V.; resources, I.V.; writing—original draft preparation, V.G. and I.V.; writing—review and editing, V.G. and I.V.; visualization, V.G. and I.V.; project administration I.V.; and funding acquisition, I.V. Both authors have read and agreed to the published version of the manuscript.

Funding: The authors acknowledge financial support from the Slovenian Research Agency (research core funding No. P1-0005) and APPLAUSE project (UIA02-228), co-financed by the European Regional Development Fund through the Urban Innovative Action (UIA) initiative.

Acknowledgments: The authors want to express their most sincere gratitude to Andreja Starc for the generous help during the experimental work and Merck KGaA, Germany for donating HPTLC plates.

Conflicts of Interest: The authors declare no conflict of interest.

References

1. Peng, W.; Qin, R.; Li, X.; Zhou, H. Botany, phytochemistry, pharmacology, and potential application of Polygonum cuspidatum Sieb.et Zucc.: A review. *J. Ethnopharmacol.* **2013**, *148*, 729–745. [CrossRef] [PubMed]
2. Nawrot-Hadzik, I.; Ślusarczyk, S.; Granica, S.; Hadzik, J.; Matkowski, A. Phytochemical Diversity in Rhizomes of Three Reynoutria Species and their Antioxidant Activity Correlations Elucidated by LC-ESI-MS/MS Analysis. *Molecules* **2019**, *24*, 1136. [CrossRef] [PubMed]
3. Tang, D.; Zhu, J.-X.; Wu, A.-G.; Xu, Y.-H.; Duan, T.-T.; Zheng, Z.-G.; Wang, R.-S.; Li, D.; Zhu, Q. Pre-column incubation followed by fast liquid chromatography analysis for rapid screening of natural methylglyoxal scavengers directly from herbal medicines: Case study of Polygonum cuspidatum. *J. Chromatogr. A* **2013**, *1286*, 102–110. [CrossRef] [PubMed]
4. Fan, P.; Hay, A.-E.; Marston, A.; Lou, H.; Hostettmann, K. Chemical variability of the invasive neophytes Polygonum cuspidatum Sieb. and Zucc. and Polygonum sachalinensis F. Schmidt ex Maxim. *Biochem. Syst. Ecol.* **2009**, *37*, 24–34. [CrossRef]
5. Chen, H.; Tuck, T.; Ji, X.; Zhou, X.; Kelly, G.; Cuerrier, A.; Zhang, J. Quality Assessment of Japanese Knotweed (Fallopia japonica) Grown on Prince Edward Island as a Source of Resveratrol. *J. Agric. Food Chem.* **2013**, *61*, 6383–6392. [CrossRef]
6. Zhang, Z.; Song, R.; Fu, J.; Wang, M.; Guo, H.; Tian, Y. Profiling of components of rhizoma et radix polygoni cuspidati by high-performance liquid chromatography with ultraviolet diode-array detector and ion trap/time-of-flight mass spectrometric detection. *Pharmacogn. Mag.* **2015**, *11*, 486–501. [CrossRef]
7. Gao, F.; Zhou, T.; Hu, Y.; Lan, L.; Heyden, Y.V.; Crommen, J.; Lu, G.-C.; Fan, G. Cyclodextrin-based ultrasonic-assisted microwave extraction and HPLC-PDA-ESI-$ITMS^n$ separation and identification of hydrophilic and hydrophobic components of Polygonum cuspidatum: A green, rapid and effective process. *Ind. Crop. Prod.* **2016**, *80*, 59–69. [CrossRef]
8. Lachowicz, S.; Oszmiański, J.; Wojdyło, A.; Cebulak, T.; Hirnle, L.; Siewiński, M. UPLC-PDA-Q/TOF-MS identification of bioactive compounds and on-line UPLC-ABTS assay in Fallopia japonica Houtt and

Fallopia sachalinensis (F.Schmidt) leaves and rhizomes grown in Poland. *Eur. Food Res. Technol.* **2018**, *245*, 691–706. [CrossRef]

9. Leu, Y.-L.; Hwang, T.-L.; Hu, J.-W.; Fang, J.-Y. Anthraquinones from Polygonum cuspidatum as tyrosinase inhibitors for dermal use. *Phytother. Res.* **2008**, *22*, 552–556. [CrossRef]
10. Shan, B.; Cai, Y.-Z.; Brooks, J.D.; Corke, H. Antibacterial properties of Polygonum cuspidatum roots and their major bioactive constituents. *Food Chem.* **2008**, *109*, 530–537. [CrossRef]
11. Uddin, Z.; Song, Y.H.; Curtis-Long, M.J.; Kim, J.Y.; Yuk, H.J.; Park, K.H. Potent bacterial neuraminidase inhibitors, anthraquinone glucosides from Polygonum cuspidatum and their inhibitory mechanism. *J. Ethnopharmacol.* **2016**, *193*, 283–292. [CrossRef] [PubMed]
12. Glavnik, V.; Vovk, I.; Albreht, A. High performance thin-layer chromatography–mass spectrometry of Japanese knotweed flavan-3-ols and proanthocyanidins on silica gel plates. *J. Chromatogr. A* **2017**, *1482*, 97–108. [CrossRef] [PubMed]
13. Glavnik, V.; Vovk, I. High performance thin-layer chromatography–mass spectrometry methods on diol stationary phase for the analyses of flavan-3-ols and proanthocyanidins in invasive Japanese knotweed. *J. Chromatogr. A* **2019**, *1598*, 196–208. [CrossRef] [PubMed]
14. Ding, X.-P.; Zhang, C.-L.; Li-Qiong, S.; Sun, L.-Q.; Qin, M.-J.; Yu, B.Y. The Spectrum-Effect integrated fingerprint of Polygonum cuspidatum based on HPLC-diode array detection-flow injection-chemiluminescence. *Chin. J. Nat. Med.* **2013**, *11*, 546–552. [CrossRef] [PubMed]
15. Bensa, M.; Glavnik, V.; Vovk, I. Leaves of Invasive Plants—Japanese, Bohemian and Giant Knotweed-The Promising New Source of Flavan-3-ols and Proanthocyanidins. *Plants* **2020**, *9*, 118. [CrossRef]
16. Metličar, V.; Vovk, I.; Albreht, A. Japanese and Bohemian Knotweeds as Sustainable Sources of Carotenoids. *Plants* **2019**, *8*, 384. [CrossRef]
17. Duval, J.; Pecher, V.; Poujol, M.; Lesellier, E. Research advances for the extraction, analysis and uses of anthraquinones: A review. *Ind. Crop. Prod.* **2016**, *94*, 812–833. [CrossRef]
18. Sihanat, A.; Palanuvej, C.; Ruangrunsi, N.; Rungsihirunrat, K. Estimation of Aloe-emodin Content in Cassia grandis and Cassia garrettiana Leaves Using TLC Densitometric Method and TLC Image Analysis. *Indian J. Pharm. Sci.* **2018**, *80*, 359–365. [CrossRef]
19. Ekambaram, S.P.; Selvan, P.S.; Prakash, P.P. HPTLC Method Development and Validation for Simultaneous Analysis of Emodin and Chrysophanol in Cassia tora Linn Methanolic Extract. *J. Liq. Chromatogr. Relat. Technol.* **2013**, *36*, 2525–2533. [CrossRef]
20. Ye, M.; Han, J.; Chen, H.; Zheng, J.; Guo, D.-A. Analysis of phenolic compounds in rhubarbs using liquid chromatography coupled with electrospray ionization mass spectrometry. *J. Am. Soc. Mass Spectrom.* **2007**, *18*, 82–91. [CrossRef]
21. Patocka, J.; Navrátilová, Z.; Ovando, M. Biologically Active Compounds of Knotweed (Reynoutria spp.). *Mil. Med Sci. Lett.* **2017**, *86*, 17–31. [CrossRef]
22. Pang, M.-J.; Yang, Z.; Zhang, X.-L.; Liu, Z.-F.; Fan, J.; Zhang, H.-Y. Physcion, a naturally occurring anthraquinone derivative, induces apoptosis and autophagy in human nasopharyngeal carcinoma. *Acta Pharmacol. Sin.* **2016**, *37*, 1623–1640. [CrossRef] [PubMed]
23. Kim, Y.-M.; Lee, C.-H.; Kim, A.H.-G.; Lee, H.-S. Anthraquinones Isolated from Cassiatora (Leguminosae) Seed Show an Antifungal Property against Phytopathogenic Fungi. *J. Agric. Food Chem.* **2004**, *52*, 6096–6100. [CrossRef] [PubMed]
24. Gao, F.; Xu, Z.-H.; Wang, W.; Lu, G.; Heyden, Y.V.; Zhou, T.; Fan, G. A comprehensive strategy using chromatographic profiles combined with chemometric methods: Application to quality control of Polygonum cuspidatum Sieb. et Zucc. *J. Chromatogr. A* **2016**, *1466*, 67–75. [CrossRef] [PubMed]
25. Zhao, R.; Liu, S.; Zhou, L.-L. Rapid Quantitative HPTLC Analysis, on One Plate, of Emodin, Resveratrol, and Polydatin in the Chinese Herb Polygonum cuspidatum. *Chromatographia* **2005**, *61*, 311–314. [CrossRef]
26. Zhang, C.; Zhang, X.; Zhang, Y.; Xu, Q.; Xiao, H.; Liang, X. Analysis of estrogenic compounds in Polygonum cuspidatum by bioassay and high performance liquid chromatography. *J. Ethnopharmacol.* **2006**, *105*, 223–228. [CrossRef] [PubMed]
27. Tian, T.; Sun, Q.; Shen, J.; Zhang, T.; Gao, P.; Sun, Q. Microbial transformation of polydatin and emodin-8-β-d-glucoside of Polygonum cuspidatum Sieb. et Zucc into resveratrol and emodin respectively by Rhizopus microsporus. *World J. Microbiol. Biotechnol.* **2008**, *24*, 861–866. [CrossRef]

28. Siva, R.; Mayes, S.; Behera, S.K.; Rajasekaran, C. Anthraquinones dye production using root cultures of *Oldenlandia umbellata* L. *Ind. Crop. Prod.* **2012**, *37*, 415–419. [CrossRef]
29. You, X.; Feng, S.; Luo, S.; Cong, D.; Yu, Z.; Yang, Z.; Zhang, J. Studies on a rhein-producing endophytic fungus isolated from *Rheum palmatum* L. *Fitoterapia* **2013**, *85*, 161–168. [CrossRef]
30. Chewchinda, S.; Ruangwises, N.; Gritsanapan, W. Comparative Analysis of Rhein Content in Cassia fistula Pod Extract by Thin-Layer Chromatographic-Densitometric and TLC Image Methods. *J. Planar Chromatogr. Mod. TLC* **2014**, *27*, 29–32. [CrossRef]
31. Ahmad, S.; Zaidi, S.M.A.; Mujeeb, M.; Ansari, S.H. HPLC and HPTLC Methods by Design for Quantitative Characterization and in vitro Anti-oxidant Activity of Polyherbal Formulation Containing Rheum emodi. *J. Chromatogr. Sci.* **2013**, *52*, 911–918. [CrossRef] [PubMed]
32. Singh, N.P.; Gupta, A.P.; Sinha, A.K.; Ahuja, P.S. High-performance thin layer chromatography method for quantitative determination of four major anthraquinone derivatives in Rheum emodi. *J. Chromatogr. A* **2005**, *1077*, 202–206. [CrossRef] [PubMed]
33. Laub, A.; Sendatzki, A.-K.; Palfner, G.; Wessjohann, L.A.; Schmidt, J.; Arnold, N. HPTLC-DESI-HRMS-Based Profiling of Anthraquinones in Complex Mixtures-A Proof-of-Concept Study Using Crude Extracts of Chilean Mushrooms. *Foods* **2020**, *9*, 156. [CrossRef] [PubMed]
34. Strgulc Krajšek, S.; Dolenc Koce, J. Sexual reproduction of knotweed (Fallopia sect. Reynoutria) in Slovenia. *Preslia* **2015**, *87*, 17–30.
35. Reich, E.; Schibli, A. *High-Performance Thin-Layer Chromatography for the Analysis of Medicinal Plants*; Georg Thieme Verlag KG: Leipzig, Germany, 2007.
36. Glavnik, V.; Simonovska, B.; Vovk, I. Densitometric determination of (+)-catechin and (−)-epicatechin by 4-dimethylaminocinnamaldehyde reagent. *J. Chromatogr. A* **2009**, *1216*, 4485–4491. [CrossRef]

Publisher's Note: MDPI stays neutral with regard to jurisdictional claims in published maps and institutional affiliations.

Article

Biotechnological Potential of *Araucaria angustifolia* Pine Nuts Extract and the Cysteine Protease Inhibitor AaCI-2S

Roberto Carlos Sallai [1,2,†], Bruno Ramos Salu [1,†], Rosemeire Aparecida Silva-Lucca [3], Flávio Lopes Alves [1], Thiago Henrique Napoleão [4], Patrícia Maria Guedes Paiva [4], Rodrigo da Silva Ferreira [1], Misako Uemura Sampaio [1] and Maria Luiza Vilela Oliva [1,*]

1 Departamento de Bioquímica, Universidade Federal de São Paulo, Escola Paulista de Medicina, Rua Três de Maio 100, São Paulo 04044-020, SP, Brazil; roberto.sallai@fsa.br (R.C.S.); brsalu@unifesp.br (B.R.S.); pelopes2@yahoo.com.br (F.L.A.); rodrigobioq@gmail.com (R.d.S.F.); misakosampaio@gmail.com (M.U.S.)

2 Centro Universitário Fundação Santo André, Av. Príncipe de Gales, 821, Santo André 09060-650, SP, Brazil

3 Centro de Engenharias e Ciências Exatas, Universidade Estadual do Oeste do Paraná, Unioeste, Campus de Toledo, R. da Faculdade 645, Toledo 85903-000, PR, Brazil; roselucca@gmail.com

4 Departamento de Bioquímica, Universidade Federal de Pernambuco, Av. Moraes Rego, S/N, Cidade Universitária, Recife 50670-420, PE, Brazil; thiago.napoleao@ufpe.br (T.H.N.); patricia.paiva@ufpe.br (P.M.G.P.)

* Correspondence: mlvoliva@unifesp.br; Tel.: +55-011-5576-4445

† These authors contributed equally to the present work.

Received: 23 October 2020; Accepted: 25 November 2020; Published: 30 November 2020

Abstract: Protease inhibitors are involved in the regulation of endogenous cysteine proteases during seed development and play a defensive role because of their ability to inhibit exogenous proteases such as those present in the digestive tracts of insects. *Araucaria angustifolia* seeds, which can be used in human and animal feed, were investigated for their potential for the development of agricultural biotechnology and in the field of human health. In the pine nuts extract, which blocked the activities of cysteine proteases, it was detected potent insecticidal activity against termites (*Nasutitermes corniger*) belonging to the most abundant termite genus in tropical regions. The cysteine inhibitor (AaCI-2S) was purified by ion-exchange, size exclusion, and reversed-phase chromatography. Its functional and structural stability was confirmed by spectroscopic and circular dichroism studies, and by detection of inhibitory activity at different temperatures and pH values. Besides having activity on cysteine proteases from *C. maculatus* digestive tract, AaCI-2S inhibited papain, bromelain, ficin, and cathepsin L and impaired cell proliferation in gastric and prostate cancer cell lines. These properties qualify *A. angustifolia* seeds as a protein source with value properties of natural insecticide and to contain a protease inhibitor with the potential to be a bioactive molecule on different cancer cells.

Keywords: *Araucaria angustifolia*; bioactive compounds; cysteine protease inhibitor; functional food; insecticide; plant extracts; termites; tumor cells; pine nuts; urban pest

1. Introduction

Araucaria angustifolia is a native gymnosperm of the greatest economic and biological importance in Brazil. It withstood the rigors of the natural selection process for hundreds of millions of years as the planet underwent intense geological and climate change. Because of its wide distribution in Parana state, this species is its state symbol and known for Paraná pine. Uncontrolled logging and the expansion of new agricultural areas harmed the forests harboring these trees to such a critical point that *A. angustifolia* is now on the official list of endangered species of the Brazilian flora and the red list of the International Union for Conservation of Nature [1]. *Araucaria* seeds have a remarkable structure, whose development is controlled directly or indirectly by changes in gene expression patterns and is an interesting biological model for cellular organization studies, protein accumulation, and differential gene expression. In addition to starch (36.28%), proteins (3.57%), lipids (1.26%), carbohydrates (2.43%), and minerals provide high nutritional value [2]. The reddish-brown peel and the thin film are rich in polyphenolic compounds with antioxidant properties that, when transferred to the edible part during the cooking process, make it a very healthy food product. In forests, the pine nuts of Araucária are a key food for various vertebrates such as agouti, squirrels, monkeys, rodents, and various species of birds [3].

The overall seed characteristics and a variety of diverse structures have led researchers to seek new substances with anti-tumor activity and therapeutic effects on various diseases [4,5]. A few studies have been performed with *A. angustifolia*, and lectins having anti-inflammatory, antibacterial, and anti-depressant action on the central nervous system were isolated [6,7].

Plants synthesize numerous proteins that contribute to the protection against attack by microorganisms (fungi and bacteria) and/or invertebrates (insects and nematodes). In most cases, the biological role of these proteins is assigned based on their in vitro activity, as is the case with lectins and enzyme inhibitors. In other cases, their role is confirmed by more direct analysis such as the incorporation of these in artificial diets used in insect feeding, their incorporation into culture media for microorganism culture, or even through the expression of these proteins in transgenic plants [8,9].

Inhibitors of serine proteases have been known when Kunitz and Bowman in 1946, and Birk in 1963 purified and characterized trypsin inhibitors from soybean seeds [10]. Since then, inhibitors have been isolated mainly from reserve organs such as seeds and tubers. Cysteine protease inhibitors or cystatins are reversible protease inhibitors of the papain family and related proteases (e.g., cathepsin B, L, ficin, and bromelain). In plants, orizacystatin from rice seeds was the first inhibitor of cysteine proteases considered a cystatin [11]. Numerous biological functions attributed to phytocystatins have recently been reviewed. It is assumed that they may play a regulatory role in all physiological processes involving cysteine proteases. More systematic studies have led to very promising results, especially the use of such inhibitors as instruments for the study of protease involvement in pathophysiological processes [12,13].

Phytocystatins have been identified in a large variety of monocotyledons such as rice, corn, maize, barley, and sugarcane, and dicotyledons such as beans, potatoes, avocados, kiwis, and nuts [8,14,15]. Fewer reports exist on the purification of cysteine protease inhibitors. In this work, we describe the biotechnological potential of *Araucaria angustifolia* pine nuts on phytopathogenic organisms, extending structural and functional characterization of a cysteine protease inhibitor toxic for human tumor cell lines improving the qualification of the nuts as a functional food.

2. Results

To minimize possible proteolysis, the extract was heated at 60 °C for 15 min, and the inhibitory activity of the cysteine proteases papain, cruzain, and human L-cathepsin was preserved. No inhibitory activity was detected on cathepsin B or serine proteases, such as trypsin, human plasma kallikrein, porcine pancreatic elastase, or human neutrophil elastase.

2.1. *Effect of Extract on Adult Insects*

Protease inhibitors have demonstrated insecticidal activity by interfering with digestion, which leads to poor nutrient absorption and decreased amino acid bioavailability [16]. Thus, we investigate the protective effect of pine nuts on adult termites.

The extract exhibited termiticidal activity on *N. corniger* workers at all tested concentrations (Figure 1a). All the workers died after 10 days in the treatments with extract while 100% mortality in negative control was reached only until the twentieth day. No significant differences ($p > 0.05$)

were detected between the effects of the concentrations tested. Regarding the effect on soldiers, the extract was also able to kill the insects at all tested concentrations (Figure 1b); in negative controls, 100% mortality occurred only on the seventeenth while in the treatment at 1.0 mg/mL, for example, all insets had died on the fifth day.

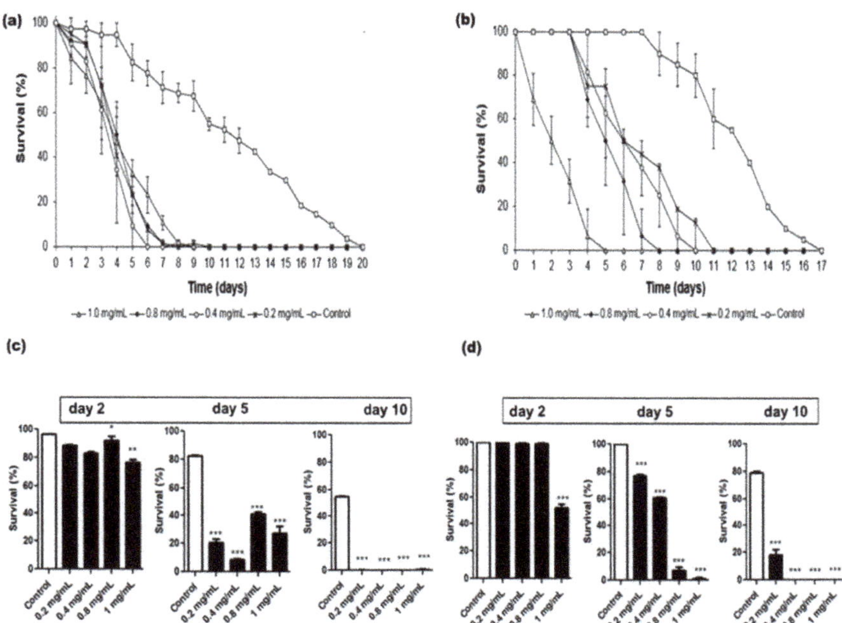

Figure 1. Effect of *A. angustifolia* seed extract on the survival of *Nasutitermes corniger* workers (**a**,**c**) and soldiers (**b**,**d**) during 20 days. Saline solution (0.15 M NaCl) was used in negative control. Each point represents the mean ± standard deviations of three repetitions. (* $p < 0.05$, ** $p < 0.005$, *** $p < 0.0001$; one way-ANOVA, follow Tukey's multiple comparison test).

2.2. Purification of the Cysteine Protease Inhibitor AaCI-2S

The characterization of the cysteine protease inhibitor became the focus of the present study because it is not much studied in gymnosperms, like the inhibitors of serine proteases. The acetone-fractionated proteins from the saline extract (Figure 2a, line 1) were separated by chromatography using a DEAE-Sephadex anion exchange resin followed by the cation exchange chromatography in SP-Sephadex. In the DEAE-Sephadex anion exchange resin, the inhibitor did not bind under the buffer at pH 8. Even with the change in ionic strength and pH parameters, the chromatographic profile was not modified, and most of the inhibitory activity was detected in the non-bonded material eluted with the column equilibration buffer. Using the same buffering conditions, the inhibitor also did not bind to the cationic resin SP-Sephadex. The papain inhibitory activity was detected after the chromatographies were dialyzed, lyophilized, and loaded on a Superdex 30 column in an ÄKTA purifier system (GE Life Sciences, USA). Figure 2b shows the protein profile and the location of the inhibitory activity indicated by the second peak. Fractions with inhibitory activity were pooled and analyzed by SDS-PAGE and reverse-phase chromatography. The estimated molecular mass of the inhibitor was around 18 kDa (Figure 2a, lane 2), and under reducing conditions, it showed a unique band of approximately 9 kDa (Figure 2a, lane 4). Reverse-phase chromatography onto a C-18 column in an HPLC system (Figure 2c) exhibited the presence of a single major peak, indicating the purity of the preparation.

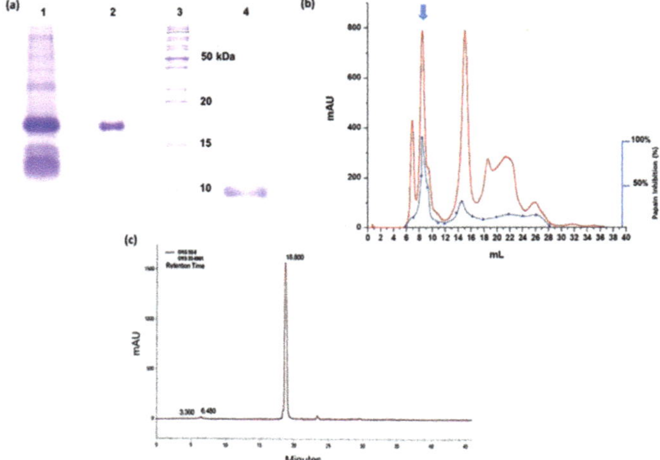

Figure 2. Purification profile of the AaCI-2S inhibitor. (**a**) SDS-polyacrylamide gel electrophoresis (15%). Lane 1, Brazilian pine saline extract (100 μg); Lane 2, non-reduced AaCI-2S (20 μg); Lane 3, molecular mass markers; Lane 4, AaCI-2S (10 μg) under reducing conditions. (**b**) Superdex 30 column equilibrated with 0.05 M Tris-HCl buffer (pH 8.0) containing 0.15 M NaCl at a flow rate of 0.5 mL/min. Absorbance at 280 nm is indicated in red and the inhibitory activity on papain in blue. The arrow indicates the fractions pooled. Sample: protein (2 mg A280) after ion-exchange chromatography. (**c**) Reverse-phase chromatography Vydac C-18 column. The proteins were eluted with an acetonitrile gradient in 0.1% TFA.

2.3. Secondary Structure Estimation and Intrinsic Fluorescence Emission of AaCI-2S

Circular dichroism (CD) spectroscopy was used to characterize the secondary structure of the inhibitory molecule. The spectrum displayed two negative bands, one at 208 and another at 222 nm, a positive band at 192 nm (Figure 3a), and its deconvolution estimated 58% of α-helix, 12% of β-turns, 8% of β-sheets, and 22% of disordered structures. The cluster analysis indicated that AaCI-2S belonged to the α+β class of proteins, which presented a more pronounced band at 208 nm than the one at 222 nm [17], the typical secondary structure of members of the prolamin superfamily [18], similar to napin [19] and 2S albumin isolated from melon *Momordica charantia* [20].

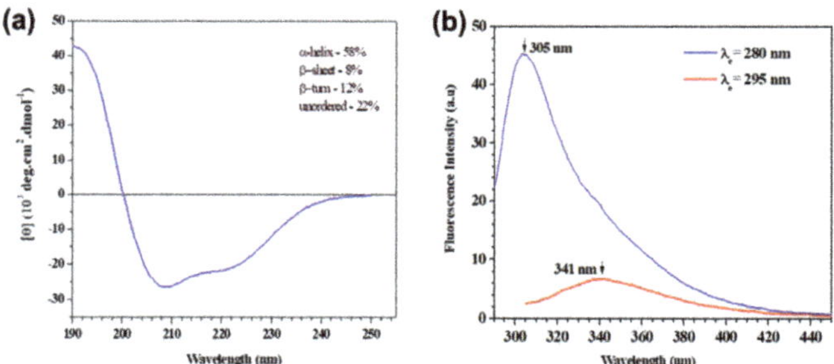

Figure 3. Spectroscopic characteristics of the AaCI-2S. Samples contained a 3 μM concentration of inhibitor, in PBA buffer 10 mM, pH 7.0. (**a**) Far UV-CD spectrum was recorded using a 1 mm cell path length cylindrical cuvette with an average of 8 scans, at 25 °C. The CDPro program was used to estimate the AaCI-2S secondary structure. (**b**) Fluorescence emission spectra of AaCI-2S. The samples were excited at 280 nm and 295 nm, and the fluorescence emission was monitored in the 290–450 and 305–450 nm ranges, respectively, at 25 °C.

The emission fluorescence measurements of the AaCI-2S aromatic amino acids with excitation wavelengths at 280 nm (blue line) and 295 nm (red line) are shown in Figure 3b. The recorded spectra were very different in both the form and location of the emission peak. The intrinsic fluorescence analysis exhibited that with an excitation at 295 nm, the emission peak occurred at 341 nm, which is a characteristic profile of tryptophan class II residues partially exposed to solvent, as in sunflower [21] and buckwheat *Fagopyrum esculentum* 2S albumins [22].

2.4. *Effects of pH and Temperature on the Activity and Structure of the AaCI-2S Inhibitor*

AaCI-2S was stable over a wide pH range (Figure 4a) and temperature (Figure 4b). Its stability was confirmed by CD spectroscopy since no modifications of its secondary structure were observed in the pH range of 2 to 10 (Figure 5a). The results of the intrinsic fluorescence emission of this inhibitor revealed that the microenvironment of Trp residues also did not undergo significant changes in this pH range (Figure 5b), maintaining the emission peak at around 341 nm and subtle variations in intensity. However, as these residues are already partially exposed to the solvent, the ANS extrinsic probe was used to monitor the global conformational changes in the tertiary structure of the inhibitor. This dye has a low fluorescence quantum yield in aqueous environments because it binds preferentially to the hydrophobic sites, promoting a pronounced increase in the fluorescence intensity and a blue shift of the emission peak. Figure 5c showed that significant changes in the fluorescent properties of the probe occur only in acidic environments since the emission peak maximum shifts toward shorter wavelengths (from 510 nm to 478 nm) and the fluorescence intensity increases up to seven-fold, at pH 2, suggesting the exposure of hydrophobic regions at this pH, which was previously inaccessible to the probe. The thermal stability of the secondary structure of the inhibitor can be monitored by the decrease in the CD bands at 208 and 222 nm (Figure 5d). Partial loss of structure was observed after treatment at 100 °C, wherein the inhibitor lost 20% of its activity within 2 h (Figure 4b), but it did not disappear completely even after 3 or 4 h of incubation (Figure 4c).

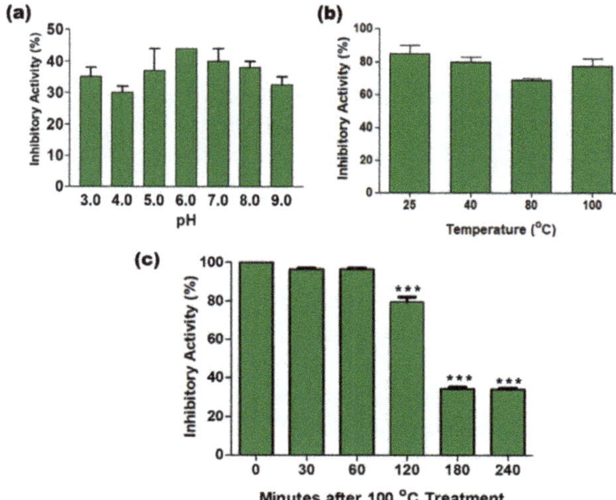

Figure 4. Effects of pH and temperature on the activity of the AaCI-2S inhibitor. (**a**) Functional stability at different pH values. The inhibitor samples were pre-incubated in solutions with different pH values for 30 min, neutralized to an initial pH (8.0), and the inhibitory activity on papain assay was measured. (**b**) Functional stability at different temperatures. The inhibitor was heated at different temperatures for 30 min. (**c**) Functional stability at 100 °C for up to 4 h. After boiling, the inhibitory activity on papain was measured. ((**b**,**c**) after different pretreatment temperatures, the samples were cooled down at room temperature for 30 min before the inhibitory assays). (*** $p < 0.0001$, one-way ANOVA, follow Tukey's multiple comparison test).

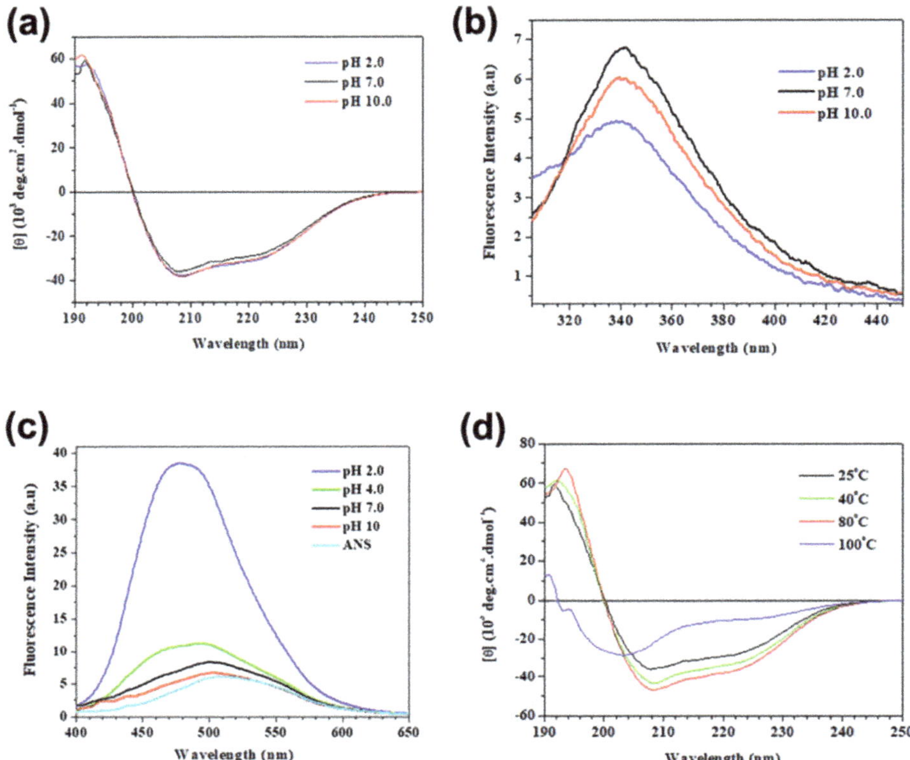

Figure 5. Effects of pH and temperature on AaCI-2S conformation. For pH dependence assays, the inhibitor (4 μM) was incubated in 10 mM PBA buffer for 30 min, at 25 °C: (**a**) Far-UV CD spectra, (**b**) Tryptophan fluorescence spectra of the AaCI-2S at different pH values (blue line, pH 2.0; black line, pH 7.0 and red line, pH 10.0) (**c**) ANS fluorescence spectra in the absence (cyan line) and presence of AaCI-2S as a function of pH. Spectra were taken 30 min after the addition of ANS probe to the protein samples. (**d**) Temperature effects on CD spectra of AaCI-2S (4 μM), in 10 mM PBA buffer, pH 7.0. Before measurements, samples were incubated at their respective temperatures for 30 min and then cooled to 25 °C.

2.5. Inhibitor Sequence

The amino acid sequence of the inhibitor when compared with other protein sequences in the UniProt Knowledgebase database reveals similarities with 2S albumins of conifers and angiosperms. Because of the inhibitory activity on cysteine proteases and the structural similarity with 2S-albumin, the isolated *A. angustifolia* inhibitor was named AaCI-2S. Figure 6 showed the multiple alignments of AaCl-2S with conifer and angiosperm 2S albumin. The highest scores were obtained with conifers *Pinus strobus*, *Picea glauca*, and *Pseudotsuga menziesii* and with the angiosperms *Corylus avellana* (hazel) and *Anacardium occidentale* (cashew tree). A comparison of the AaCl-2S sequence was also performed using the BLAST program with those deposited in the MEROPS database of peptidases and their inhibitors. The search revealed a similar identity with a family of proteins whose inhibitory activity has not yet been demonstrated, denominated in the database by "Family I6 unassigned peptidase inhibitor homolog." The protein sequence data reported in this paper will appear in the UniProt Knowledgebase under the accession number C0HLT8.

```
                        Sinal peptide        N-terminal processing
Angiosperm                                                                          1        =  2
D0PWG2_CORAV    1 --------------MARLATLAALFAALLLVAHAAAFRTTITTVDVDEDIVNQQGRRGESCREQAQRQQNLNQCQRYMRQQSQ  69
B6EU55_BEREX    1 --------------MAKISVAAAALLVLMALGHATAFRATVTTTVVEEE--NQ-----EECREQMQRQQMLSHCRMYMRQQME  62
Q8L694_MOMCH    1 --------------MARLSSMVVLLAVALLLTD--VYAYRTTITTVEVDEDNQG--RHERCHHIRPREQ-LRSCESFLRQSRG  64
B9SA28_RICCO    1 --------------MAKLIPAVALISVLLFIIANASFAYRTTITTVEVDDTN----TQERCFRDLRGKE-FRACQMYLSQSSS  64
Q8H2B8_ANAOC    1 --------------MAKFLLLLSAFAVLLLVANASIYRA---IVEVEEDSGRE-----QSCQRQFEEQQRFRNCQRYVKQEVQ  61
Gimnosperm
Q40997_PINST    1 ERKKSNGCLSSPMSTRLTLKWVTLVAALLFVIHCSTPTVGAHEDTMDGEALQQQ---RRSCDPQR-----LSDCHDYLQRR--  73
O81412_PICGL    1 -----MGVFS-PSTTRLTLKWFSLSVALFLLLHWGIPSVDGHEDNMYGEEIQQQ---RRSCDPQRDPQR-LSSCRDYLERR--  71
O64931_PSEMZ    1 -----MGVFSPPTSTMLRLKWVSLGVALLLLVQWSTPNVDAAGDNMFGEDVVQQQQRRGSCDPQR-----LSSCRDYLERR--  71
AaCI-2S         1 --------------------------------------------------------ERRCDPRR-----LSDCEDFVRGRSK  21
                   internal processing       34  =       5 6  =
D0PWG2_CORAV   70 ---------YGSYDGSNQQ-QQQELEQCCQQLRQMDERCRCEGLRQAV-MQQQGEM--RGEEMR------------------ 120
B6EU55_BEREX   63 ---------ESPYQTMPRRGMEPHMSECCEQLEGMDESCRCEGLRRMMRMMQQKEMQPRGEQMR------------------ 117
Q8L694_MOMCH   65 -------YLEMKGVEENQWERQQGLEECCRQLRNVEEQCRCDALQEIAR---EVQRQERGQEGS------------------ 118
B9SA28_RICCO   65 RRSTDGEVLEMPGEKDQ--QERHQLQECCNELKQVRDECQCEALQVAVEKQIESEQQMQREQYQ------------------ 126
Q8H2B8_ANAOC   62 -------------RGGRYNQRQESLRECCQELQEVDRRCRCQNLEQMVRQLQQQEQ-IKGEEVR------------------ 111
Q40997_PINST   74 ----------------REQP----SERCCEELQRMSPHCRCRAIEQTLDQSLSFDSSTDSDSQDGAPLNQRRRRP-EGRGREE 135
O81412_PICGL   72 ----------------REEP----SERCCEELQRMSPQCRCQAIQQMLDQSLSYDSFMDSDSQEDAPLNQRRRRCREGRGREE 134
O64931_PSEMZ   72 ----------------REQP----SERCCNELERMSPQCRCPAIQQVFDQSSEDLSMVDSHSQN-AAGNQRRREE---RGREE 130
AaCI-2S        22 -------------GG RGEKECRLSERCCTELQKMPRECRCEAVEGMYKEA------ERKERG----------------EG  62
                                                                          <-------------------------
                      =    7 =    8
D0PWG2_CORAV  121 ---EVMETARDLPNQCRLS-PQR-CEIRS--ARF------- 147
B6EU55_BEREX  118 ---RMMRLAENIPSRCNLS-PMR-CPMGGSIAGF------- 146
Q8L694_MOMCH  119 ---QMLQKARMLPAMCGVR-PQR-CDF-------------- 140
B9SA28_RICCO  127 ---EVMQKARSIPSSCGL--PEQ-CQIRTFFF--------- 152
Q8H2B8_ANAOC  112 ---ELYETASELPRICSIS-PSQGCQFQSSY---------- 138
                    --------->
Q40997_PINST  136 EEEEAVERAGELPDRCNVRESPRRCDIRRHSRYSIIG---- 172
O81412_PICGL  135 EE--AMERAAYLPNTCNVREPPRRCDIQRHSRYSMTGSSFK 173
O64931_PSEMZ  131 AEEEMVERAQRLPNTCNVRQPPRHCDIQRHSRYSIIGSSF- 170
AaCI-2S        63 EQRQRLERARALPGLCSIE--PSYCEIRPS----------- 94
```

Figure 6. Comparison between the amino acid sequence obtained from AaCI-2S with precursors of 2S reserve proteins from angiosperms and gymnosperms. The sequences were aligned using ClustaW2. Similar strings include D0PWG2_CORAV: hazelnut (*Corylus avellana*); B6EU55_BEREX: Brazil nut (*Bertollethia excelsa*); Q8L694_MOMCH: São Caetano melon (*Momordica charantia*), B9SA28_RICCO: castor bean (*Ricinus communis*); Q8H2B8_ANAOC: cashew nut (*Anacardium occidentale*); Q40997_PINST: pinus (*Pinus strobus*); O81412_PICGL: spruce (*Picea glauca*); and O64931PSEMZ: Douglas fir (*Pseudotsuga menziesii*). Spaces (-) were introduced to maintain alignment. The eight conserved cysteine residues are indicated numerically in blue. The conserved hydrophobic residues are in red, and the arrows indicate a region rich in arginine and glutamic acid residues.

2.6. Biological Properties of AaCl-2S

The inhibition curves of some cysteine proteases by AaCl-2S was reported in Figure 7. The calculated K_{iapp} of the inhibition of cathepsin L (K_{iapp} = 0.01 nM) was about 20 times lower than that of papain inhibition (K_{iapp} = 0.2 nM), which reflects a greater affinity for cathepsin L. AaCI-2S presented no inhibitory activity on cathepsin B, but it also inhibited other enzymes in the papain family, such as ficin (K_{iapp} = 1.1 nM) and bromelain (K_{iapp} = 8.4 nM). No inhibitory activity was detected on serine proteases (trypsin, human plasma kallikrein, and elastase).

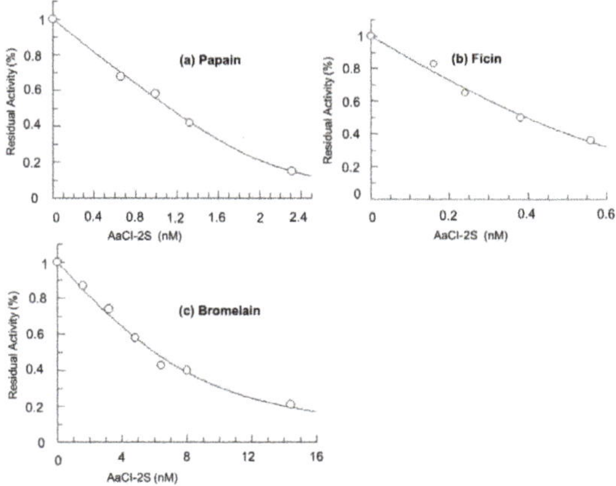

Figure 7. AaCI-2S inhibitory properties. Increasing the concentration of *Araucaria angustifolia* cysteine protease inhibitor was incubated with (**a**) papain (2 nM), (**b**) ficin, and (**c**) bromelain for 20 min, at 40 °C in 0.1 M Na_2PO_4 buffer (pH 6.3) containing 0.4 M NaCl, 0.01 M EDTA, and 8 mM DTT, the enzymatic activities were determined by the hydrolysis of Z-Phe-Arg-pNan (5 mM).

2.7. Effect of AaCI-2S on Predatory Insect Enzymes

Enzymes of the class of cysteine proteases are the major proteolytic enzymes of coleopteran larvae. One of the physiological reasons for the presence of proteins with inhibitory activity in plant seeds is their involvement in the mechanism of seed protection against predatory insects. For this reason, we used larvae from the cowpea bruchid, *Callosobruchus maculatus*, a predator of string bean seeds *Vigna unguiculata*, as a model to investigate whether purified AaCI-2S would decrease the proteolytic activity in the midgut extract of these larvae. Figure 8a indicated the rapid interaction of the inhibitor added to the incubation medium after 40 min and the resulting decrease in proteolytic activity on the colorimetric substrate Z-Phe-Arg-pNan. Figure 8b showed the inhibitory effect of increasing the concentrations of AaCl-2S on the residual activity of cysteine proteases present in the midgut of larvae.

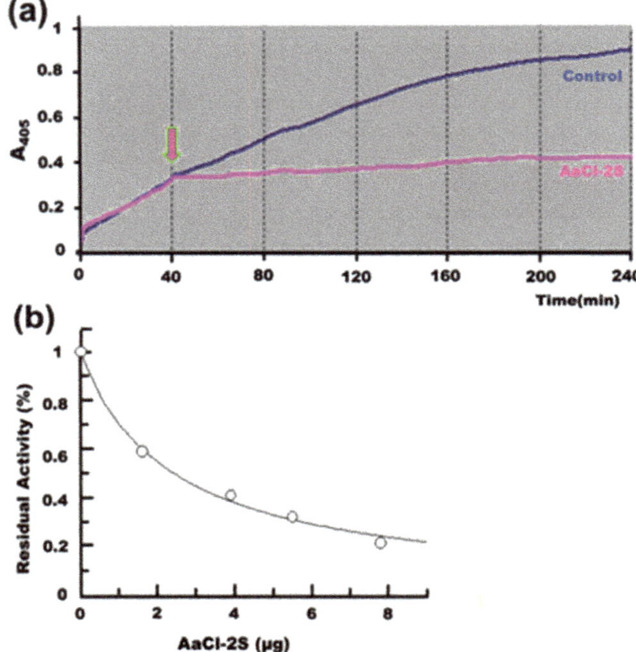

Figure 8. Action of AaCI-2S on the proteolytic activity of *Callosobruchus maculatus* larvae. (**a**) The blue line indicates the increase of proteolytic activity on Z-Phe-Arg-pNan. of the medium intestinal extract containing 43 µg of total proteins. The arrow indicates the time of the addition of 10 µg of the inhibitor to the incubation medium. The pink line indicates a decrease in proteolytic activity. (**b**) Inhibition of the proteolytic activity extracted from the intestine of *Callosobruchus maculatus*. A medium intestinal extract containing 11.5 µg of proteins was preincubated at 37 °C for 10 min with increasing concentrations of AaCl-2S in 0.1 M Na_2PO_4 buffer at pH 6.3, 0.4 M NaCl, 01 M, and 8 mM DTT. Residual activity was determined by the hydrolysis of Z-Phe-Arg-pNan (5 mM).

2.8. Investigation on the Antitumor Activity of AaCI-2S

Pine nuts are used by man as a functional food and, since cysteine proteases are involved in several types of tumors, we were interested in investigating the effect of AaCI-2S on tumor cells, where cathepsin L is recognized to play an important role, as in the models of gastric cancer and prostate cancer.

The effects of the inhibitor on the proliferation of prostate cancer cells (DU-145 and PC3), gastric cancer (Hs746T), and non-tumor human fibroblasts were illustrated in Figure 9. The inhibitor did not affect fibroblast (a) proliferation, while it inhibited the proliferation of both prostate cancer cell lines PC3 (b), DU-145 cells (c), and of the Hs746T cells (d).

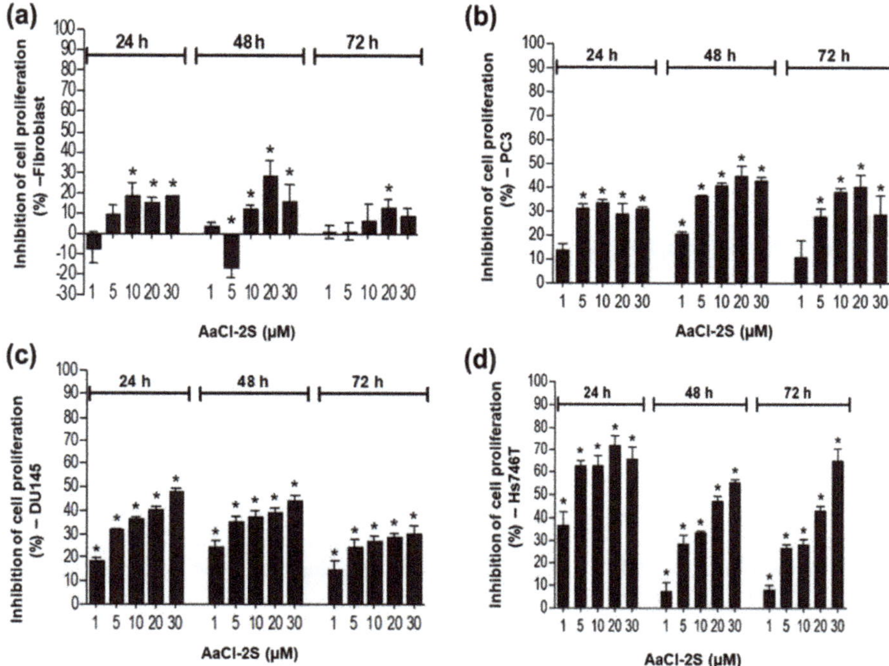

Figure 9. Effects of the inhibitor on the proliferation of prostate cancer cells, gastric cancer, and human fibroblasts. Effect of AaCl-2S on the proliferation of (**a**) fibroblasts, (**b**) PC3, (**c**) DU145, and (**d**) Hs746T cells. Cells were pre-incubated with increasing concentrations of AaCl-2S for 15 min at room temperature and analyzed at different incubation times (* $p < 0.05$, unpaired *t*-test).

3. Discussion

In angiosperms, a considerable fraction of seed proteins includes inhibitors of serine proteases; however, to date, not many protease inhibitors have been purified and characterized in gymnosperms [23]. This was also confirmed by our investigation since, in the present study, the saline extract of *A. angustifolia* seeds did not inhibit trypsin or other serine proteases. A trypsin inhibition has been detected in the embryo tissues only after sample concentration with acetone precipitation [24]. In contrast, the saline extract inhibited two cysteine proteases, papain, and the enzyme cruzain (a recombinant form of the cysteine protease cruzipain from *Trypanosoma cruzi*) [25].

To our knowledge, there are no reports on the effects of cysteine protease inhibitors against termites. However, the deleterious effects found in termites have been attributed to lectins and serine protease inhibitors [16,26–28] and this might not be the case of pine nuts. Although further studies are needed, the description of the termiticidal potential of the seeds is relevant, as it shows that nature has selected an alternative to its century-old forest protection against this pest. This property can be exploited commercially as an alternative for the use of Araucária other than its wood, thus protecting the forest, a world heritage site.

The presence of papain inhibitors has been reported in seeds of some gymnosperms such as *Pinus maritima*, *Picea pungens*, and *A. angustifolia* [29]; however, purification and characterization of the inhibitory activity were not achieved. We did not find any purified gymnosperm phytocystatin in the protein databases, but approximately 200 sequences have been identified through comparative genomic analysis [30]. These analyses have been largely useful for information on the conservation and evolution of proteolytic enzymes and their inhibitors [31].

The concentration of the inhibitor, determined by titration with papain, was 17 mg/kg. The first cysteine protease inhibitor (oryzacystatin) identified in rice during the 1980s occurs at a concentration of 2–3 mg/kg [32] and its structural and functional characterization was investigated only after its recombinant form was reported [33]. In contrast, the concentration of trypsin inhibitors is rather high in legumes [34], for example, the concentration in *Enterolobium contortisiliquum* is approximately 5600 mg/kg [35].

The fact that the inhibitor does not bind to ion exchange resins is an alternative and effective method for use in the processing of large amounts of extract through the batchwise system. Using this strategy, most of the proteins with a molecular weight above 25 kDa were eliminated, thereby favoring size exclusion chromatography in Superdex 30, in which the inhibitory activity was detected only in the second peak. Structurally, the protein database sequences displayed high similarity with conserved proteins of the 2S albumin family and no sequence homology with the typical phytocystatins or with other plant inhibitors of cysteine proteases as the described inhibitor of *B. bauhinioides*, BbCI, which also differs from phytocystatins [36].

From the earliest studies with protease inhibitors in plants, their potential role as a reserve protein has been postulated. The similarity with the 2S albumins of the gymnosperms *Picea glauca*, *Pseudotsuga menziesii* [37], and *Pinus strobus* was 64%, 55%, and 52.6%, respectively. Similarities of 62.5% and 67.5% were also observed with some 2S angiosperm albumins such as *Corylus hazelnut* [38] and *Anacardium occidentale* (cashew nuts) [39], respectively. Thus, based on similarities, sequence alignment, and arrangement of conserved cysteine residues, we can conclude that the *Araucaria angustifolia* inhibitor is a reserve protein of the 2S albumins class which justified the denomination adopted in this work (AaCI-2S). The AaCI-2S also displayed a high content of arginine residues, which is a common feature in conifer reserve proteins. Furthermore, 8 cysteine residues and the hydrophobic residues flanking cysteine residues are conserved positions in relation to the 2S gymnosperm and angiosperm albumins as in the sequence of 2S albumin from *Pseudotsuga menziesii* [37]. Additionally, AaCI-2S displays a molecular mass of 18 kDa and two identical polypeptide chains linked by disulfide bonds similar to the storage protein 2S albumin. Although the reserve function of proteins is usually assigned to 2S albumins, other biological activities have been described such as inhibiting fungal growth [20], hemagglutinating activity [40], the inhibitory activity of trypsin [41], and in the case of AaCI-2S, the inhibitory activity of cysteine proteases.

Many larvae of insects of the order Coleoptera are predators of seeds and the presence of cysteine proteases as digestive enzymes [42] leads to the hypothesis of a possible exogenous protective role of inhibitor of this class of enzymes. Numerous studies evidencing the in vitro and in vivo inhibition of the digestive proteases of these larvae [43,44] and on the growth of fungi [45] support this role. The possible action of the inhibitor to act as a defense protein was confirmed by the rapid and effective inhibition of the cysteine protease activity present in the insect intestine suggesting that this inhibitor may be employed in the functional study of these enzymes. Naturally, these insects do not use the seeds of *Araucaria*, but the result is interesting in the sense of indicating how proteins can be strategic in the composition of compounds involved in the protection of seeds in general.

The presence of protease inhibitors in legume seeds and cereals added to epidemiological studies that identify legumes as potential protective agents in reducing the incidence of some types of cancer in the vegetarian population have stimulated a series of studies involving inhibitory proteases influencing tumor promotion in vivo and in vitro [46–48].

In addition to the inhibition of cruzain and papain, AaCI-2S seems to be selective regarding the two human cathepsins tested, since cathepsin L was inhibited while cathepsin B activity was not altered. Both are medically relevant targets because they are involved in many physiological and pathological processes such as apoptosis, inflammation, and cancer [49]. Among other functions, cathepsins (mainly cathepsins B and L) are involved in extracellular matrix degradation, facilitating the growth, invasion, and metastasis of tumor cells [50,51]. AaCl-2S can interfere with the cellular proliferation of the two cell lines of prostate cancer and shows a more effective inhibitory effect on the proliferation of gastric tumor cells. Increased activity of cathepsins B and L and the reduction of secreted endogenous cystatins have been observed in prostate cancer cell lines PC3 and DU145. The invasive ability of these cell lines was partially inhibited by E-64, a synthetic inhibitor of cysteine proteases [52]. It is worth mentioning that the non-interference in non-tumorigenic cells demonstrate the inhibitory selectivity in cancer cell lines. As the seeds of *Araucaria* are used as food, their antiproliferative effect is of nutritional significance for future studies and provides important information regarding the benefits of including the pine nut in our diet.

The overall yield of the inhibitor by the purification process was low (12%), thus obtaining large amounts of inhibitor is difficult. Notably, this loss is not due to thermal stability, since the inhibitor spectra following treatment at different temperatures exhibited a structure that was quite resistant to thermal denaturation, displaying small conformational changes after heating up to 80 °C. Only the treatment at 100 °C caused a modification of the inhibitor structure with a partial loss of structure. These results are consistent with other observations regarding the thermostability of many members of the prolamin superfamily, such as the presence of intramolecular disulfide bonds, which have been implicated in this thermostability [53]. The maintenance of the inhibitory activity against papain after heating and exposure to extreme pH values indicates that the inhibitor is functionally stable. Many other protease inhibitors purified from plant seeds exhibit high stability at various temperatures [54], but we were surprised that the inhibitor isolated from the *Araucaria* seed maintained its activity even after a 60 min treatment at 100 °C and may implicate in the qualification of pine nuts as a functional food since they are normally consumed roasted or cooked.

4. Material and Methods

4.1. Plant Material

Araucaria angustifolia (Bertol.) O. Kuntze plant material was deposited in the Herbarium of Universidade Estadual da Bahia, UEDB, identified as HUESB 12431. The studies were conducted in accordance with Brazilian legislation (license no. 02/2014, process 02000.003472/2005- 62 Ministério do Meio Ambiente, Coordenação Geral de Autorização de uso da Flora e Floresta, SCEN).

The seeds were purchased from the city of Campos do Jordão—SP from the natural occurrence of *A. angustifolia* located in Campos do Jordão State Park.

4.2. Experimental Reagents

Bovine serum albumin, fibronectin, human neutrophil elastase (EC 3.4.2.37), and porcine pancreas elastase (EC 3.4.21.7) were purchased from Calbiochem®—(EMD Chemicals Inc., Port Wentworth, GA, USA). Bovine trypsin (EC 3.4.21.4), papain (EC 3.4.22.2), bromelain (EC 3.4.22.32), and ficin (3.4.22.3) were obtained from Sigma-Aldrich (Co., St. Louis, MI, USA). Kallikrein (human plasma) (EC 3.4.21.34) was purified according to Oliva [55]. Cruzain, cathepsin B, and L were provided by Prof. Dr. Luís Juliano Neto, Department of Biophysics, UNIFESP. Chromogenic substrates derived from p-nitroanilide (Bz-Arg-pNan, HD-Pro-Phe-Arg-pNan, Suc-Phe-pNan, HD-Val-Leu-Lys-pNan, HD-Phe-L-Pip-L -Arg-pNan, MeO-Suc-Ala-Ala-Pro-Val-pNan, N-Suc-Ala-Ala-Pro-Phe-pNan), and the fluorimetric aminomethyl coumarin substrate Z-Phe-Arg AMC were obtained from Calbiochem® (EMD Chemicals Inc., USA), and starch from Sigma-Aldrich Co. (Saint Louis, MO, USA).

DEAE–Sephadex® A-50; SP-Sephadex® C-50 e Superdex® 30 (GE Healthcare, Chicago, IL, USA)—Biogel® P30 (Bio-Rad Laboratories, Hercules, CA, USA)—C18 column *Protein & Peptide (Vydac® Ultrasphere*—Brea, CA, USA)—Column *Aquapore® RP 300 C* (Varian, Palo Alto, CA, USA). Dinitrosalicylic acid (ADNS), ammonium persulfate, MTT salt, toluidine blue dye, E-64, and 1-anilinonaphthalene-8-sulfonic acid (ANS) probe were obtained from Sigma-Aldrich Co. (USA). Coomassie Brilliant Blue R-250 was obtained from Bio-Rad Laboratories (USA). Fetal bovine serum LB broth, cell culture media, RPMI 1640, DMEM, TEMED, and dithiothreitol from Gibco Invitrogen Co. (Waltham, MA, USA). Acrylamide and N, N, and methylene bisacrylamide were obtained from Serva (Heidelberg, Germany). Molecular weight standards were obtained from Fermentas Inc. (Burlington, ON, Canada) and Bio-Rad Laboratories (USA).

4.3. Cells

The PC3 cell line of prostate adenocarcinoma and the cell line HsT46T of gastric adenocarcinoma were provided by Prof. Dr. Barbara Mayer, from the Klinikum Groβhadern Surgery Department, University of Munich, Germany. The DU145 prostate adenocarcinoma cell line was provided by Prof. Dr. Heloisa Selistre de Araújo, from the Department of Physiological Sciences at the Federal University of São Carlos. The human lineage of fibroblasts, obtained from cells of the amniotic fluid, was provided by Prof. Dr. Leny Toma, from the Biochemistry Department at the Federal University of São Paulo.

4.4. Protein Extraction and Fractionation

The seeds were ground in a blender with a 0.15 M NaCl solution, at a 10% (*w/v*) density, heated at 60 °C for 30 min, cooled in an ice bath for 30 min, stirred at room temperature for 20 min, filtered with cotton and gauze, and centrifuged at 6000× *g* for 15 min at 4 °C. Proteins were estimated spectrophotometrically (A280) as well as by Bradford (1976) [56] assay using bovine serum albumin as the standard.

4.5. Inhibitory Activity

The protein extract of plant seeds as well as the purified inhibitor was tested on proteases. The p-nitroaniline released as a hydrolysis product was measured at 405 nm using a SpectraCount spectrophotometer. In the case of the fluorogenic substrate Z-Phe-Arg-AMC, the Hitachi F-2000 spectrofluorometer was used with excitation and emission wavelengths of 380 and 460 nm, respectively. Different concentrations of the inhibitor solutions were added to the appropriate volumes of activated enzymes in a 100-µL volume of buffer. The volume was topped to 230 µL with a 0.15 M NaCl solution and the mixture was pre-incubated at 37–40 °C for 10 min before the addition of the substrate. The reaction proceeded for 20–30 min at 37–40 °C and was stopped by the addition of 40% acetic acid (*v/v*). The absorbance obtained in the absence of the extracts was considered as 100% of enzymatic activity, and the inhibition was expressed as the reduction of enzyme activity percentage [57].

The concentration of active papain was determined by titration with the synthetic inhibitor E-64 according to Zucker et al. [58]. Once the inhibitory activity of papain was determined, the purified protein was titrated and used to determine the dissociation constants of the enzyme/inhibitor complexes (K_{iapp}) with other cysteine proteases. The determinations were performed following the model suggested by Morrison adapted to an enzymatic kinetics program for computer graphics, and the numerical value was calculated using the GraFit program [59].

4.6. Evaluation of the Extract Effects on Adult Insect Survival

The insecticidal activity of the extract was evaluated by a bioassay based on the method described by Kang [60]. Each assay consisted of a Petri dish (90 × 15 mm) with the bottom covered by a paper filter. Disks (4 cm in diameter) of paper filter were impregnated with 200 µL of extract (0.2, 0.4, 0.8, or 1.0 mg/mL). In the negative control, 200 µL of 0.15 M NaCl was added to the disks. A total of 20 active termites (at a worker-to-soldier ratio of 4:1) were transferred to each dish, which was maintained at 28 °C in the dark. Insect survival was evaluated daily until the death of all insects. The bioassays were carried out in quadruplicate for each tested concentration, and the survival rates (%) were calculated.

4.7. Inhibitor Purification by Ion-Exchange Chromatography (Batchwise)

The proteins from seed extract were precipitated by the slow addition of ice-cold acetone to a final concentration of 80% (*v/v*). After a sedimentation period (30 min), part of the acetone was sucked out with a rubber cannula, and the remaining fraction was centrifuged at 3000× *g* for 15 min at 4 °C. The acetone-precipitated proteins were spread on Petri dishes, dried at 24 °C, and then frozen until use. The acetone-precipitated protein was dissolved in water (1 g/10 mL), centrifuged at 3000× *g* at 4 °C for 15 min, and the conductivity values of the solutions were adjusted by 0.05 M Tris-HCl (pH 8.0) buffer using DEAE–Sephadex®, ion-exchange chromatography. The protein was added to the resins and stirred for 30 min. The mixture was filtered through a funnel with a porous plate, and the resin was washed with an equilibration buffer to remove the non-adsorbed proteins. Elution was performed using 0.15 M or 0.3 M NaCl solution in the equilibration buffer. The non-adsorbed fraction containing papain inhibitory activity was mixed with the SP-Sephadex resin and the procedure was repeated as described above. Papain inhibitory activity was also detected in the non-adsorbed fraction.

4.8. Molecular Exclusion Chromatography

The non-retained ion exchanging fraction was dialyzed in water using a 10 kDa cut off membrane, lyophilized and dissolved in a 0.05 M Tris-HCl buffer (pH 8.0) containing 0.15 M of NaCl, centrifuged at 10,000× *g* for 5 min at room temperature, and loaded onto a Superdex 30 column in a 0.05 M Tris-HCl buffer (pH 8.0) containing a 0.15 M NaCl (equilibrium buffer) under a 0.5 mL/min flow rate in the ÄKTA purifier system (GE Healthcare). The molecular mass of the inhibitor was estimated by standardizing the column with ferritin (440 kDa), SBTI (20 kDa), cytochrome C (12.4 kDa), and aprotinin (6.5 kDa). The protein profile was monitored by absorbance at 280 nm, and fractions (1 mL) with inhibitory activity were pooled, dialyzed, and lyophilized.

4.9. Reverse-Phase Chromatography on an HPLC System

The protein from Superdex 30 chromatography was further purified by C18 reverse phase (Protein & Peptide, 4.6 mm × 14 cm) in an HPLC system (Model SCL-6A—Shimadzu), equilibrated with a 0.1% trifluoroacetic acid (TFA) solution in water (Solvent A). The elution was performed on a gradient of 0.1% TFA in water with 90% acetonitrile (Solvent B) under a constant flow rate of 0.7 mL/min [61].

4.10. Sodium Dodecyl Sulfate-Polyacrylamide Gel Electrophoresis

Denaturing electrophoresis was performed according to the method described by Laemmli [62] on a polyacrylamide gel (15%) in the presence of SDS. The samples were treated with a reducing agent in dithiothreitol-containing sample buffer (200 mg/mL) and heated for 10 min at 100 °C. The proteins were visualized by staining with Coomassie blue R250 solution.

4.11. Estimation of Secondary Structure by Circular Dichroism (CD) Spectroscopy

The far-ultraviolet (UV) CD spectra of the purified inhibitor (an average of 8 scans) were recorded on a J-810 (Jasco Corporation, Tokyo, Japan) spectropolarimeter within the range of 190 to 250 nm in a 1-mm optical path cylindrical quartz cuvette at 25 °C. The inhibitor (3 μM) was dissolved in 10 mM PBA buffer (pH 7.0) and its CD spectra were expressed as molar ellipticity [θ]. The estimated calculation of the secondary structure fractions was performed using the CDPro deconvolution package, with the Selcon3, Continll, and CDSSTR programs [63].

4.12. Intrinsic Fluorescence Measurements

The fluorescence emission measurements were obtained using an F-2500 fluorometer (Hitachi Ltd., Tokyo, Japan) at 25 °C in quartz cuvettes, with an optical path of 1 cm. Analyses were performed with the purified inhibitor (3 μM) in 10 mM PBA buffer (pH 7.0). The sample was excited at 280 or 295 nm, and the fluorescence emission was monitored in the range of 290–450 and 305–450 nm, respectively.

The fluorescence emission spectra of buffers were subtracted from the spectra of the samples to minimize the effect of light scattering and to perform baseline corrections [64].

4.13. Studies on the Influence of pH and Temperature on Structural Stability

Purified inhibitor structural stability was analyzed under different conditions through the measurements of CD and intrinsic and extrinsic fluorescence emission. Samples of the inhibitor (4 μM) in 10 mM PBA buffer (pH 7.0) were incubated at 25, 40, 80, and 100 °C for 30 min and then cooled. Samples with the same concentration were incubated in solutions of different pH (pH 2, pH 4, pH 7, and pH 10) for 30 min. After each treatment, the measurements of CD and intrinsic fluorescence were performed under the same conditions described above. For the extrinsic fluorescence measurements, the treated samples were incubated with the 8-anilino-1-naphthalenesulfonate probe (ANS), 90 μM in 10 mM PBA buffer, pH 7.0 for 15 min, at 25 °C, and the fluorescence spectra were recorded at 400 to 650 nm with a 385 nm excitation, 30 min after adding the probe.

4.14. N-Terminal Sequence Determination

After reverse phase chromatography, the inhibitor (2 nM) was dissolved in 300 μL of buffer 0.25 M Tris-HCl (pH 8.5) containing 6 M guanidine, 1 mM EDTA, and 5 μL of β-mercaptoethanol and then incubated for 2 h at 37 °C, under a nitrogen-saturated atmosphere in the absence of light. Alkylation was performed with the addition of 5 mL of vinylpyridine and re-incubation for 90 min at 37 °C and subjected to HPLC/acetonitrile/isopropanol reverse phase chromatography at a constant flow rate (0.1 mL/min). The N-terminal sequence was obtained automatically by the Edman (1949) [65] degradation method in two independent laboratories. Similarity searches were performed using the FASTA program using the *UniProt Knowledgebase* database Larkin [66] and a *Blossom 80* (EBI) [67] matrix (www.ebi.ac.uk). Multiple alignments of similar sequences were performed using the ClustalW2 program (http://www.ebi.ac.uk/Tools/clustalw2/).

4.15. Effect of pH and Temperature on Inhibitor Activity

To verify the effect of pH, the inhibitor (1 μg) was pre-incubated for 30 min in 50 mM sodium citrate buffer (pH 3 and pH 6) or 50 mM Tris-HCl buffer (pH 7 and pH 9). After pre-incubation, the pH of the samples was adjusted to 8.0 with Tris-HCl, and the ability to inhibit papain was determined. The lability of the inhibitor at different temperatures was investigated by incubating the samples (1 μg) in 50 mM Tris-HCl pH 8.0 buffer, keeping them in a water bath at different temperatures (25, 40, 80, and 100 °C) for 30 min. The resistance of the inhibitor to boiling (100 °C) was also studied by incubation for different durations (30 min, 1, 2, 3, and 4 h). After different heat treatments, the samples were cooled in an ice bath for 5 min and tested for their inhibitory activity.

4.16. Evaluation of the Inhibitor Effect on Insect Enzymes

Twenty larvae of *C. maculatus* with 19–20 days post-hatching (4th instar) were removed from the infested seeds and immersed in 0.15 M NaCl solution for dissection of the intestines with the help of watchmaker tweezers and stereoscopic magnifying glass. Lysis was performed in the intestines by brief sonication wells at 150 μL solution NaCl 0.15 M. The obtained extract was centrifuged at 10,000× *g* at 4 °C for 10 min. The supernatant was collected and frozen at −20 °C. In a preliminary test, 50 μL of crude extract diluted ten times and containing 12 μg of protein was added in duplicates into two wells of a microplate with Na_2PO_4 0.1 M buffer (pH 6.3), 10 mM EDTA, and 0.4 M NaCl. The substrate (20 μL) Z-Phe-Arg-pNan (0.05 M) was added in a final volume of 250 μL. The plate was incubated at 37 °C and hydrolysis was followed photometrically for 40 min. Next, the purified inhibitor (10 μg) was added to one well, and substrate hydrolysis continued to be monitored for up to 4 h. To evaluate the effect of inhibitor concentrations, the extract diluted ten-fold in 0.1 M Na_2 PO_4 buffer (pH 6.3), 10 mM EDTA; 0.4 M NaCl; 8 mM DTT remained at 37 °C for 10 min for enzymatic activation. Next, 50 μL of the activated extract containing 11.5 μg of proteins were preincubated in the absence and presence of different inhibitor concentrations (1 to 8 ug) for 10 min at 37 °C. After the addition of 20 μL of the Z-Phe-Arg-pNan substrate (5 mM) in a final volume of 250 μL, the hydrolysis was monitored for 60 min and then quenched with 50 μL of 40% acetic acid (*v/v*).

4.17. Studies on the Effect of Inhibitors on Tumor Cells and Human Fibroblast Cells

PC3, DU145, and Hs746T tumor cell lines were maintained in RPMI-1640 culture medium and the fibroblast cells were maintained in Dulbecco's modified Eagle's medium (DMEM), pH 7.4. Both culture media were enriched with 10% fetal bovine serum, penicillin (10 UI/mL), and 100 μg/mL streptomycin. Cells were sub-cultured weekly using the following protocol: The medium was removed from the confluent cell flasks (60 × 10 mm) and cells were washed with PBS solution (pH 7.4). For cell detachment, the cells were incubated with 1 mL trypsin solution (0.25%) for 1 min. Next, 1×10^5 cells were resuspended, transferred to a new plate in the appropriate media, cultured at 37 °C under 5% CO_2, and the culture medium was changed every 3 days [46,47].

4.18. Cell Viability Assay

PC3, DU145, and Hs746T cells (5×10^3 cells/100 μL/well) and fibroblast cells (8×10^3 cells/100 μL/well) were incubated at 37 °C and 5% CO_2 for 24 h in RPMI-1640 medium containing 10% fetal bovine serum. A total of 100 μL of inhibitor (2.5–30 μM) diluted in RPMI-1640 medium, previously filtered through a Millipore filter (0.22 μm), was added to the adhered cells and incubated for 24, 48, and 72 h. At the end of each incubation period, 10 μL of MTT (tetrazolium salts) dissolved in PBS (5 mg/mL) was added to each well and the cells were again incubated for 2 h. Subsequently, the medium was removed and 100% DMSO was added to solubilize the formazan crystals and incubated for 20 min at 37 °C. The absorbance was measured at 540 nm using a spectrophotometer (SpectraCount model). Assays were performed in triplicate for each inhibitor concentration and experiments were performed twice as described by Gasperazzo Ferreira et al. [68].

4.19. Statistical Analyses

All assays were performed in triplicate and independently. The statistical analyses were expressed as the mean ± standard deviation (SD) and analyzed using GraphPad Prisma Software. Comparisons among the variables, measured in defined experimental groups, were conducted using one-way ANOVA, followed by Tukey's test. Statistical significance was defined as * $p < 0.05$, ** $p < 0.005$, and *** $p < 0.0001$.

5. Conclusions

These findings provided relevant information about the insecticide and antitumor activity of the pine nuts, the potential biotechnology application in agriculture, and human health that can contribute to the preservation of the Araucária forest. Also, a protein named AaCI-2S, with a molecular mass of 18 kDa composed of two identical polypeptide chains linked by two disulfide bonds was characterized. The studies of the structure–activity relationship at different pH values and temperatures revealed its high functional and structural stability. The inhibitory activity demonstrated on cysteine proteases bromelain, ficin and cathepsin L, and the cysteine proteases of the larval midguts of *C. maculatus* suggests other endogenous roles for 2S albumin. AaCI-2S inhibited cell proliferation of gastric cancer and two lines of prostate cancer and did not affect the proliferation of non-tumorigenic cells. As the seeds of *Araucaria* are used as food, its antiproliferative effect is of nutritional significance for future studies and offers evidence regarding the benefits of including the pine nut as a functional food in our diet.

Author Contributions: All authors participated in this study. R.C.S., B.R.S., R.A.S.-L., F.L.A., T.H.N., and R.d.S.F. performed the assays and wrote the manuscript; M.L.V.O., M.U.S., and P.M.G.P. designed and interpretation of the data and made a critical review of the manuscript. All authors have read and agreed to the published version of the manuscript.

Funding: This study was supported by Fundação de Amparo à Pesquisa do Estado de São Paulo (FAPESP) [2017/07972-9 and 2017/06630-7]; Coordenação de Aperfeiçoamento de Pessoal de Nível Superior-Brasil (CAPES) -Finance Code 001, and Conselho Nacional de Desenvolvimento Científico e Tecnológico (CNPq) [401452/2016-6], and M.L.V.O. received a research fellowship from CNPq, Brazil.

Acknowledgments: Joana G Ferreira, Claudia de Paula, Lucimeire A de Santana and Reinhart Mentele merit our gratitude for the valuable and competent contribution to this work.

Conflicts of Interest: The authors declare no conflict of interest.

References

1. Thomas, P. *Araucaria angustifolia*. The IUCN Red List of Threatened Species. 2013, e.T32975A2829141. Available online: https://www.iucnredlist.org/species/32975/2829141 (accessed on 23 September 2020).
2. Cordenunsi, B.R.; Wenzel, E.M.; Genovese, M.I.; Colli, C.; Souza, A.G.; Lajolo, F.M. Chemical composition and glycemic index of brazilian pine (*Araucaria angustifolia*) seeds. *J. Agric. Food Chem.* **2004**, *52*, 3412–3416. [CrossRef] [PubMed]
3. Iob, G.; Vieira, E.M. Seed predation of *Araucaria angustifolia* (Araucariaceae) in the brazilian Araucária FOREST: Influence of deposition site and comparative role of small and "large" mammals. *Plant Ecol.* **2008**, *198*, 185–196. [CrossRef]
4. Bonturi, C.R.; Motaln, H.; Silva, M.C.C.; Salu, B.R.; De Brito, M.V.; Costa, L.D.A.L.; Torquato, H.F.V.; Nunes, N.N.D.S.; Paredes-Gamero, E.J.; Lah, T.T.; et al. Could a plant-derived protein potentiate the anticancer effects of a stem cell in brain cancer? *Oncotarget* **2018**, *9*, 21296–21312. [CrossRef] [PubMed]
5. Khurshid, Y.; Syed, B.; Simjee, S.U.; Beg, O.; Ahmed, A. Antiproliferative and apoptotic effects of proteins from black seeds (*nigella sativa*) on human breast MCF-7 cancer cell line. *BMC Complement. Med. Ther.* **2020**, *20*, 1–11. [CrossRef]
6. Santi-Gadelha, T.; Gadelha, C.A.A.; Aragao, K.S.; Oliveira, C.C.; Mota, M.R.L.; Gomes, R.C.; Pires, A.F.; Toyama, M.H.; Toyama, D.O.; Alencar, N.M.; et al. Purification and biological effects of *Araucaria angustifolia* (Araucariaceae) seed lectin. *Biochem. Biophys. Res. Commun.* **2006**, *350*, 1050–1055. [CrossRef]
7. Vasconcelos, S.M.M.; Lima, S.R.; Soares, P.M.; Assreuy, A.M.S.; Sousa, F.C.F.; Lobato, R.F.G.; Vasconcelos, G.S.; Santi-Gadelha, T.; Bezerra, E.H.S.; Cavada, B.S.; et al. Central action of *Araucaria angustifolia* seed lectin in mice. *Epilepsy Behav.* **2009**, *15*, 291–293. [CrossRef]
8. Martinez, M.; Santamaria, M.E.; Diaz-Mendoza, M.; Arnaiz, A.; Carrillo, L.; Ortego, F.; Diaz, I. Phytocystatins: Defense proteins against phytophagous insects and acari. *Int. J. Mol. Sci.* **2016**, *17*, 1747. [CrossRef]
9. Shakeel, M.; Ali, H.; Ahmad, S.; Said, F.; Khan, K.A.; Bashir, M.A.; Anjum, S.I.; Islam, W.; Ghramh, H.A.; Ansari, M.J.; et al. Insect pollinators diversity and abundance in *Eruca sativa* Mill. (Arugula) and *Brassica rapa* L. (Field mustard) crops. *Saudi J. Biol. Sci.* **2019**, *26*, 1704–1709. [CrossRef]
10. Ryan, C.A. Plant protease inhibitors in plants: Genes for improving defenses against insects and pathogens. *Annu. Rev. Phytopathol.* **1990**, *28*, 425–449. [CrossRef]
11. Turk, V.; Stoka, V.; Turk, D. Cystatins: Biochemical and structural properties, and medical relevance. *Front. Biosci.* **2008**, *13*, 5406–5420. [CrossRef]
12. Van Wyk, S.G.; Kunert, K.J.; Cullis, C.A.; Pillay, P.; Makgopa, M.E.; Schlüter, U.; Vorster, B.J. The future of cystatin engineering. *Plant Sci.* **2016**, *246*, 119–127. [CrossRef]
13. Siddiqui, M.F.; Ahmed, A.; Bano, B. Insight into the biochemical, kinetic and spectroscopic characterization of garlic (*Allium sativum*) phytocystatin: Implication for cardiovascular disease. *Int. J. Biol. Macromol.* **2017**, *95*, 734–742. [CrossRef] [PubMed]
14. Tremblay, J.; Goulet, M.C.; Michaud, D. Recombinant cystatins in plants. *Biochimie* **2019**, *166*, 184–193. [CrossRef] [PubMed]
15. Mangena, P. Phytocystatins, and their potential application in the development of drought tolerance plants in soybeans (*Glycine max* L.). *Protein Pept. Lett.* **2020**, *27*, 135–144. [CrossRef] [PubMed]

16. Napoleão, T.H.; Albuquerque, L.P.; Santos, N.D.L.; Nova, I.C.V.; Lima, T.A.; Paiva, P.M.G.; Pontual, E.V. Insect midgut structures and molecules as targets of plant-derived protease inhibitors and lectins. *Pest Manag. Sci.* **2019**, *75*, 1212–1222. [CrossRef]
17. Venyaminov, S.Y.; Vassilenko, K.S. Determination of protein tertiary structure class from circular dichroism spectra. *Anal. Biochem.* **1994**, *222*, 176–184. [CrossRef]
18. Breiteneder, H.; Radauer, C. A classification of plant food allergens. *J. Allergy Clin. Immunol.* **2004**, *113*, 821–831. [CrossRef]
19. Schmidt, I.; Renard, D.; Rondeau, D.; Richomme, P.; Popineau, Y.; Axelos, M.A. Detailed Physicochemical Characterization of the 2s Storage Protein from Rape (*Brassica napus* L.). *J. Agric. Food Chem.* **2004**, *52*, 5995–6001. [CrossRef]
20. Vashishta, A.; Sahu, T.; Sharma, A.; Choudhary, S.K.; Dixit, A. In vitro refolded napin-like protein of *Momordica charantia* expressed in *Escherichia coli* displays properties of native napin. *Biochim. Biophys. Acta* **2006**, *1764*, 847–855. [CrossRef]
21. Burnett, G.R.; Rigby, N.; Mills, E.N.; Belton, P.S.; Fido, R.J.; Tatham, A.S.; Shewry, P.R. Characterization of the emulsification properties of 2S albumins from sunflower seed. *J. Colloid Interface Sci.* **2002**, *247*, 177–185. [CrossRef]
22. Tang, C.; Wanga, X. Physicochemical and structural characterization of globulin and albumin from common buckwheat (*Fagopyrum esculentum* Moench) seeds. *Food Chem.* **2010**, *121*, 119–126. [CrossRef]
23. Konarev, A.V.; Lovegrove, A.; Shewry, P.R. Serine Proteinase Inhibitors in Seeds of *Cycas siamensis* and other Gymnosperms. *Phytochemistry* **2008**, *69*, 2482–2489. [CrossRef] [PubMed]
24. Alves, L.P.; Sallai, R.C.; Salu, B.R.; Miranda, A.; Oliva, M.L.V. Identification and characterisation of serine protease inhibitors from *Araucaria angustifolia* seeds. *Nat. Prod. Res.* **2016**, *30*, 2712–2715. [CrossRef] [PubMed]
25. Eakin, A.E.; Mills, A.A.; Harth, G.; McKerrow, J.H.; Craik, C.S. The sequence, organization, and expression of the major cysteine protease (cruzain) from *Trypanosoma cruzi*. *J. Biol. Chem.* **1992**, *267*, 7411–7420.
26. Araújo, R.M.S.; Ferreira, R.S.; Napoleão, T.H.; Carneiro-da-Cunha, M.G.; Coelho, L.C.B.B.; Correia, M.T.S.; Oliva, M.L.V.; Paiva, P.M.G. *Crataeva tapia* bark lectin is an affinity adsorbent and insecticidal agent. *Plant Sci.* **2012**, *183*, 20–26. [CrossRef]
27. Ferreira, R.S.; Brito, M.V.; Napoleão, T.H.; Silva, M.C.C.; Paiva, P.M.G.; Oliva, M.L.V. Effects of two protease inhibitors from *Bauhinia bauhinioides* with different specificity towards gut enzymes of *Nasutitermes corniger* and its survival. *Chemosphere* **2019**, *222*, 364–370. [CrossRef]
28. Da Silva Ferreira, R.; Napoleão, T.H.; Silva-Lucca, R.A.; Silva, M.C.C.; Paiva, P.M.G.; Oliva, M.L.V. The effects of *Enterolobium contortisiliquum* serine protease inhibitor on the survival of the termite *Nasutitermes corniger*, and its use as affinity adsorbent to purify termite proteases. *Pest Manag. Sci.* **2019**, *75*, 632–638. [CrossRef]
29. Galleschi, L.; Capocchi, A.; Ghiringhelli, S.; Saviozzi, F. Some properties of proteolytic enzymes and storage proteins in recalcitrant and orthodox seeds of *Araucaria*. *Biol. Plant.* **2002**, *45*, 539–544. [CrossRef]
30. Benchabane, M.; Schlüter, U.; Vorster, J.; Goulet, M.C.; Michaud, D. Plant cystatins. *Biochimie* **2010**, *92*, 1657–1666. [CrossRef]
31. Rawlings, N.D.; Barrett, A.J.; Thomas, P.D.; Huang, X.; Bateman, A.; Finn, R.D. The MEROPS database of proteolytic enzymes, their substrates, and inhibitors in 2017 and a comparison with peptidases in the PANTHER database. *Nucleic Acids Res.* **2018**, *46*, D624–D632. [CrossRef]
32. Abe, K.; Arai, S. Purification of a cysteine proteinase inhibitor from rice, *Oryza sativa* L. japonica. *Agric. Biol. Chem.* **1985**, *49*, 3349–3350. [CrossRef]
33. Abe, K.; Emori, Y.; Kondo, H.; Suzuki, K.; Arai, S. Molecular cloning of a cysteine proteinase inhibitor of rice (oryzacystatin). Homology with animal cystatins and transient expression in the ripening process of rice seeds. *J. Biol. Chem.* **1987**, *262*, 16793–16797. [PubMed]
34. Janzen, D.H.; Ryan, C.A.; Liener, I.E.; Pearce, G. Potentially defensive proteins in mature seeds of 59 species of tropical leguminosae. *J. Chem. Ecol.* **1986**, *12*, 1469–1480. [CrossRef] [PubMed]
35. Batista, I.F.; Oliva, M.L.; Araujo, M.S.; Sampaio, M.U.; Richardson, M.; Fritz, H.; Sampaio, C.A. Primary structure of a Kunitz-type trypsin inhibitor from *Enterolobium contortisiliquum* seeds. *Phytochemistry* **1996**, *41*, 1017–1022. [CrossRef]
36. De Oliveira, C.; Santana, L.A.; Carmona, A.K.; Cezari, M.H.; Sampaio, M.U.; Sampaio, C.A.; Oliva, M.L.V. Structure of cruzipain/cruzain inhibitors isolated from *Bauhinia bauhinioides* seeds. *Biol. Chem.* **2001**, *382*, 847–852. [CrossRef]

37. Chatthai, M.; Misra, S. Sequence and expression of embryogenesis-specific cDNAs encoding 2S seed storage proteins in *Pseudotsuga menziesii* [Mirb.] Franco. *Planta* **1998**, *206*, 138–145. [CrossRef]
38. Garino, C.; Zuidmeer, L.; Marsh, J.; Lovegrove, A.; Morati, M.; Versteeg, S.; Schilte, P.; Shewry, P.; Arlorio, M.; Van Ree, R. Isolation, cloning, and characterization of the 2S albumin: A new allergen from hazelnut. *Mol. Nutr. Food Res.* **2010**, *54*, 1257–1265. [CrossRef]
39. Robotham, J.M.; Wang, F.; Seamon, V.; Teuber, S.S.; Sathe, S.K.; Sampson, H.A.; Beyer, K.; Seavy, M.; Roux, K.H. Ana o 3, an important cashew nut (*Anacardium occidentale* L.) allergen of the 2S albumin family. *J. Allergy Clin. Immunol.* **2005**, *115*, 1284–1290. [CrossRef]
40. Oguri, S.; Kamoshida, M.; Nagata, Y.; Momonoki, Y.S.; Kamimura, H. Characterization and sequence of tomato 2S seed albumin: A storage protein with sequence similarities to the fruit lectin. *Planta* **2003**, *216*, 976–984. [CrossRef]
41. Mandal, S.; Kundu, P.; Roy, B.; Mandal, R.K. Precursor of the inactive 2S seed storage protein from the Indian mustard *Brassica juncea* is a novel trypsin inhibitor. Characterization, post-translational processing studies, and transgenic expression to develop insect-resistant plants. *J. Biol. Chem.* **2002**, *277*, 37161–37168.
42. Silva, C.P.; Terra, W.R.; Lima, R.M. Differences in midgut serine proteinases from larvae of the bruchid beetles *Callosobruchus maculatus* and *Zabrotes subfasciatus*. *Arch. Insect Biochem. Physiol.* **2001**, *47*, 18–28. [CrossRef] [PubMed]
43. Lalitha, S.; Shade, R.E.; Murdock, L.L.; Bressan, R.A.; Hasegawa, P.M.; Nielsen, S.S. Effectiveness of recombinant soybean cysteine proteinase inhibitors against selected crop pests. *Comp. Biochem. Physiol. Part C Toxicol. Pharmacol.* **2005**, *140*, 227–235. [CrossRef] [PubMed]
44. Sumikawa, J.T.; Brito, M.V.; Macedo, M.L.; Uchoa, A.F.; Miranda, A.; Araujo, A.P.; Silva-Lucca, R.A.; Sampaio, U.M.; Oliva, M.L. The defensive functions of plant inhibitors are not restricted to insect enzyme inhibition. *Phytochemistry* **2010**, *71*, 214–220. [CrossRef] [PubMed]
45. Wang, K.M.; Kumar, S.; Cheng, Y.S.; Venkatagiri, S.; Yang, A.H.; Yeh, K.W. Characterization of inhibitory mechanism and antifungal activity between group-1 and group-2 phytocystatins from taro (*Colocasia esculenta*). *FEBS J.* **2008**, *275*, 4980–4989. [CrossRef] [PubMed]
46. De Paula, C.A.A.; Coulson-Thomas, V.J.; Ferreira, J.G.; Maza, P.K.; Suzuki, E.; Nakahata, A.M.; Nader, H.B.; Sampaio, M.U.; Oliva, M.L.V. *Enterolobium contortisiliquum* trypsin inhibitor (EcTI), a plant proteinase inhibitor, decreases in vitro cell adhesion and invasion by inhibition of Src protein-focal adhesion kinase (FAK) signaling pathways. *J. Biol. Chem.* **2011**, *287*, 170–182. [CrossRef] [PubMed]
47. Bonturi, C.R.; Silva, M.; Motaln, H.; Salu, B.R.; Ferreira, R.; Batista, F.P.; Correia, M.; Paiva, P.; Turnšek, T.L.; Oliva, M.L.V. A bifunctional molecule with lectin and protease inhibitor activities isolated from *Crataeva tapia* bark significantly affects cocultures of mesenchymal stem cells and glioblastoma cells. *Molecules* **2019**, *24*, 2109. [CrossRef]
48. Lobo, Y.A.; Bonazza, C.; Batista, F.P.; Castro, R.A.; Bonturi, C.R.; Salu, B.R.; Sinigaglia, R.C.; Toma, L.; Vicente, C.M.; Pidde, G.; et al. EcTI impairs survival and proliferation pathways in triple-negative breast cancer by modulating cell-glycosaminoglycans and inflammatory cytokines. *Cancer Lett.* **2020**, *491*, 108–120. [CrossRef]
49. Vidak, E.; Javoršek, U.; Vizovišek, M.; Turk, B. Cysteine cathepsins and their extracellular roles: Shaping the microenvironment. *Cells* **2019**, *8*, 264. [CrossRef]
50. Chen, S.; Dong, H.; Yang, S.; Guo, H. Cathepsins in Digestive Cancers. *Oncotarget* **2017**, *8*, 41690–41700. [CrossRef]
51. Vizovišek, M.; Fonović, M.; Turk, B. Cysteine cathepsins in extracellular matrix remodeling: Extracellular matrix degradation and beyond. *Matrix Biol.* **2018**, *75*, 141–159. [CrossRef]
52. Colella, R.; Jackson, T.; Goodwyn, E. Matrigel Invasion by the prostate cancer cell lines, PC3 and DU145, and cathepsin L+B activity. *Biotech. Histochem.* **2004**, *79*, 121–127. [CrossRef]
53. Sharma, G.M.; Mundoma, C.; Seavy, M.; Roux, K.H.; Sathe, S.K. Purification and biochemical characterization of Brazil nut (*Bertholletia excelsa* L.) seed storage proteins. *J. Agric. Food Chem.* **2010**, *58*, 5714–5723. [CrossRef] [PubMed]
54. Lopes, J.L.; Valadares, N.F.; Moraes, D.I.; Rosa, J.C.; Araujo, H.S.; Beltramini, L.M. Physico-chemical and antifungal properties of protease inhibitors from *Acacia plumosa*. *Phytochemistry* **2009**, *70*, 871–879. [CrossRef] [PubMed]

55. Oliva, M.L.V.; Grisolia, D.; Sampaio, M.U.; Sampaio, C.A. Properties of highly purified human plasma kallikrein. *Agents Actions* **1982**, *9*, 52–57.
56. Bradford, M.M. A rapid and sensitive for the quantitation of microgram quantities of protein utilizing the principle of protein-dye binding. *Anal. Biochem.* **1976**, *72*, 248–254. [CrossRef]
57. Nakahata, A.M.; Mayer, B.; Neth, P.; Hansen, D.; Sampaio, M.U.; Oliva, M.L. Blocking the proliferation of human tumor cell lines by peptidase inhibitors from *Bauhinia* seeds. *Planta Med.* **2013**, *79*, 227–235. [CrossRef]
58. Zucker, S.; Buttle, D.J.; Nicklin, M.J.; Barrett, A.J. The proteolytic activities of chymopapain, papain, and papaya proteinase III. *Biochim. Biophys. Acta* **1985**, *828*, 196–204. [CrossRef]
59. Knight, C.G. *The Characterization of Enzymes Inhibition, in Proteinase Inhibitors*; Barrett, A.J., Salvesen, G., Eds.; Elsevier: Amsterdan, The Netherlands, 1986; pp. 23–51.
60. Kang, H.Y.; Matsushima, N.; Sameshima, K.; Takamura, N. Termite resistance tests of hardwoods of Kochi growth. The strong termiticidal activity of kagonoki (*Litsea coreana* Léveillé). *Mokuzai Gakkaishi* **1990**, *36*, 78–84.
61. Nakahata, A.M.; Mayer, B.; Ries, C.; de Paula, C.A.; Karow, M.; Neth, P.; Sampaio, U.M.; Jochum, M.; Oliva, M.L.V. The effects of a plant proteinase inhibitor from *Enterolobium contortisiliquum* on human tumor cell lines. *Biol. Chem.* **2011**, *392*, 327–336. [CrossRef]
62. Laemmli, U.K. Cleavage of structural proteins during the assembly of the head of bacteriophage T4. *Nature* **1970**, *227*, 680–685. [CrossRef]
63. Sreerama, N.; Woody, R.W. Estimation of protein secondary structure from circular dichroism spectra: Comparison of CONTIN, SELCON and CDSSTR methods with expanded reference set. *Anal. Biochem.* **2000**, *287*, 252–260. [CrossRef] [PubMed]
64. Silva-Lucca, R.A.; Tabak, M.; Nascimento, O.R.; Roque-Barreira, M.C.; Beltramini, L.M. Structural and thermodynamic studies of KM+, a D-mannose binding lectin from *Artocarpus integrifolia* seeds. *Biophys. Chem.* **1999**, *79*, 81–93. [CrossRef]
65. Edman, P. A method for the determination of amino acid sequence in peptides. *Arch. Biochem.* **1949**, *22*, 475. [CrossRef] [PubMed]
66. Larkin, M.A.; Blackshields, G.; Brown, N.P.; Chenna, R.; McGettigan, P.A.; McWilliam, H.; Valentin, F.; Wallace, I.M.; Wilm, A.; Lopez, R.; et al. ClustalW and ClustalX version 2. *Bioinformatics* **2007**, *23*, 2947–2948. [CrossRef] [PubMed]
67. UniProt Knowledgebase Database and a Blossom 80 (EBI) Matrix. Available online: http://www.ebi.ac.uk/Tools/clustalw2/ (accessed on 4 April 2020).
68. Ferreira, J.G.; Diniz, P.M.M.; De Paula, C.A.A.; Lobo, Y.A.; Paredes-Gamero, E.J.; Paschoalin, T.; Nogueira-Pedro, A.; Maza, P.K.; Toledo, M.S.; Suzuki, E.; et al. The Impaired Viability of Prostate Cancer Cell Lines by the Recombinant Plant Kallikrein Inhibitor. *J. Biol. Chem.* **2013**, *288*, 13641–13654. [CrossRef] [PubMed]

Article

Three *Scrophularia* Species (*Scrophularia buergeriana*, *S. koraiensis*, and *S. takesimensis*) Inhibit RANKL-Induced Osteoclast Differentiation in Bone Marrow-Derived Macrophages

Hyeon-Hwa Nam, A Yeong Lee, Yun-Soo Seo, Inkyu Park, Sungyu Yang, Jin Mi Chun, Byeong Cheol Moon, Jun-Ho Song * and Joong-Sun Kim *

Herbal Medicine Resources Research Center, Korea Institute of Oriental Medicine, 111, Geonjae-ro, Naju-si 58245, Korea; hhnam@kiom.re.kr (H.-H.N.); lay7709@kiom.re.kr (A.Y.L.); sys0109@kiom.re.kr (Y.-S.S.); pik6885@kiom.re.kr (I.P.); sgyang81@kiom.re.kr (S.Y.); jmchun@kiom.re.kr (J.M.C.); bcmoon@kiom.re.kr (B.C.M.)
* Correspondence: songjh@kiom.re.kr (J.-H.S.); centraline@kiom.re.kr (J.-S.K.)

Received: 20 October 2020; Accepted: 24 November 2020; Published: 26 November 2020

Abstract: Scrophulariae Radix, derived from the dried roots of *Scrophularia ningpoensis* Hemsl. or *S. buergeriana* Miq, is a traditional herbal medicine used in Asia to treat rheumatism, arthritis, and pharyngalgia. However, the effects of *Scrophularia buergeriana*, *S. koraeinsis*, and *S. takesimensis* on osteoclast formation and bone resorption remain unclear. In this study, we investigated the morphological characteristics and harpagoside content of *S. buergeriana*, *S. koraiensis*, and *S. takesimensis*, and compared the effects of ethanol extracts of these species using nuclear factor (NF)-κB ligand (RANKL)-mediated osteoclast differentiation. The harpagoside content of the three *Scrophularia* species was analyzed by high-performance liquid chromatography–mass spectrometry (HPLC/MS). Their therapeutic effects were evaluated by tartrate-resistant acid phosphatase (TRAP)-positive cell formation and bone resorption in bone marrow-derived macrophages (BMMs) harvested from ICR mice. We confirmed the presence of harpagoside in the *Scrophularia* species. The harpagoside content of *S. buergeriana*, *S. koraiensis*, and *S. takesimensis* was 1.94 ± 0.24 mg/g, 6.47 ± 0.02 mg/g, and 5.50 ± 0.02 mg/g, respectively. Treatment of BMMs with extracts of the three *Scrophularia* species inhibited TRAP-positive cell formation in a dose-dependent manner. The area of hydroxyapatite-absorbed osteoclasts was markedly decreased after treatment with the three *Scrophularia* species extracts. Our results indicated that the three species of the genus *Scrophularia* might exert preventive effects on bone disorders by inhibiting osteoclast differentiation and bone resorption, suggesting that these species may have medicinal and functional value.

Keywords: *Scrophularia buergeriana*; *S. koraiensis*; *S. takesimensis*; harpagoside; osteoclast differentiation; RANKL

1. Introduction

An imbalance between osteoclasts and osteoblasts affects bone formation, leading to weakened bone and the development of skeletal diseases such as osteoporosis, rheumatoid arthritis, lytic bone metastases, and chronic obstructive pulmonary disease. The function of osteoclasts, multinucleated giant cells, is bone resorption, but osteoporosis can occur if bone resorption exceeds formation due to an increase in the number of osteoclasts [1]. Most drugs used to treat osteoporosis inhibit osteoclast differentiation to control bone resorption. Receptor activator of nuclear factor (NF)-κB ligand (RANKL), a major osteoclastogenic molecule, is a member of the tumor necrosis factor (TNF) superfamily and is the initial stimulator of osteoclast differentiation, inducing the expression of osteoclast-associated genes, such as tartrate-resistant acid phosphatase (TRAP) [2,3]. Therefore, bisphosphonates and anti-RANKL antibodies that inhibit osteoclast activity are currently used for the treatment of bone resorption diseases.

Scrophulariae Radix, an herbal medicine known as Korean Hyun-Sam, is derived from the dried roots of *Scrophularia ningpoensis* Hemsl. or *S. buergeriana* Miq., plants which are widely distributed throughout the temperate regions of the Northern Hemisphere, including Asia, Europe, and North America [4–6]. Scrophulariae Radix has been traditionally used as a therapeutic agent for blood cooling, yin nourishing, fire pursing, and toxin removal, and is widely used to treat rheumatism and arthritis in Southwest Asia [7–10]. It has also been reported to have neuroprotective, anti-inflammatory, anti-allergy, anti-amnesia, antioxidant, and hepatoprotective effects [11–15]. In Korea, *S. koraiensis* Nakai (Korean: To-Hyun-Sam) has been used as an antipyretic and anti-inflammatory agent in traditional medicine. *S. takesimensis* Nakai (Korean: Seom-Hyun-Sam) is restricted to Ulleung-do Island [16,17]. Although this species is a valuable endemic resource, its medicinal efficacy has not been assessed to date.

The therapeutic potential of *Scrophularia* species is associated with the functions of major secondary metabolites, such as phenylpropanoids and iridoid glycosides, which are present in the plant [18,19]. Harpagoside, an iridoid component present in the *Scrophularia* species, is a bioactive compound of *Harpagophytum procumbens* DC. (Devil's Claw) and has been used in Southern Africa to treat pain, arthritis, and ulcers. Pharmacological effects of harpagoside on RANKL-induced osteoclast differentiation have also been reported [20,21]. However, there are few reports concerning the pharmacological activity of *Scrophularia* species on RANKL-induced osteoclast differentiation, and studies on the biological activity of *S. koraiensis* and *S. takesimensis* have not been reported.

In the current study, we compared the morphological characteristics and harpagoside content of *S. buergeriana*, *S koraiensis*, and *S. takesimensis*, and compared the effects of *Scrophularia* species extracts on RANKL-mediated osteoclast differentiation.

2. Results

2.1. Comparative Morphology of Scrophularia Species

The three species can be distinguished on the basis of leaf shape, apex, margins, pubescence of stems, and calyx shape (Table 1). The leaf blade of *S. buergeriana* is ovate, with an acute apex (Figure 1A), and serrate with a spinose tooth marginal shape (Figure 1D). The stem is glabrous (Figure 1G), and the calyx is ovate with an obtuse apex (Figure 1J). *S. koraiensis* has lanceolate to rarely ovate-shaped leaf blades with an acuminate apex (Figure 1B) and is serrate with a spinose tooth (Figure 1E). The stem of *S. koraiensis* is sparsely pubescent with non-glandular trichomes (Figure 1H), and the calyx is lanceolate with an acute to attenuate apex (Figure 1K). However, *S. takesimensis* has ovate leaf blades with an acute apex (Figure 1C), and has a serrate, almost without spinose tooth, marginal leaf blade (Figure 1F). The stem surface of *S. takesimensis* is glabrous (Figure 1I), and its calyx is semicircular, with a rounded apex (Figure 1L).

Table 1. Major determinants of *Scrophularia* species.

	Leaves			Stems	Calyx	
	Shape	**Apex**	**Margins**	**Surfaces**	**Shape**	**Apex**
S. buergeriana	ovate	acute	serrate with spinose tooth	glabrous	ovate	obtuse
S. koraiensis	lanceolate to rarely ovate	acuminate	serrate with spinose tooth	pubescent	lanceolate	acute to attenuate
S. takesimensis	ovate	acute	serrate almost without spinose tooth	glabrous	semi-circular	rounded

Figure 1. Stereomicroscope micrographs showing the morphology of the three *Scrophularia* species studied. (**A–C**) Apex of leaf blade. (**D–F**) Margin of leaf blade. (**G–I**) Surface of stem. (**J–L**) Calyx of flower and/or fruit. (**A,D,G,J**) *S. buergeriana*. (**B,E,H,K**) *S. koraiensis*. (**C,F,I,L**) *S. takesimensis*. TR, Trichomes. All scale bars = 1 mm.

2.2. Harpagoside Content of Scrophulariae Species

High-performance liquid chromatography (HPLC) chromatograms of *S. buergeriana*, *S. koraiensis*, and *S. takesimensis* are shown in Figure 2A,B. Figure 2A shows the HPLC chromatograms of the harpagoside standard compound and the three species of *Scrophulariae* monitored at 280 nm, because the maximum wavelength of harpagoside is 279.5 nm. Harpagoside was detected at approximately 50.1 min. The harpagoside content of *S. buergeriana*, *S. koraiensis*, and *S. takesimensis* was 1.94 ± 0.24 mg/g, 6.47 ± 0.02 mg/g, 5.50 ± 0.02 mg/g, respectively (Table 2). The total ion chromatography (TIC) of the mass spectrometry (MS) spectrum was confirmed from 190–850 *m*/*z*, and the extracted ion chromatogram (XIC) of harpagoside in the samples was analyzed at 517.11 *m*/*z*, because harpagoside was detected at 517.11 *m*/*z* $[M\text{-}H + Na]^+$ (Figure 2B).

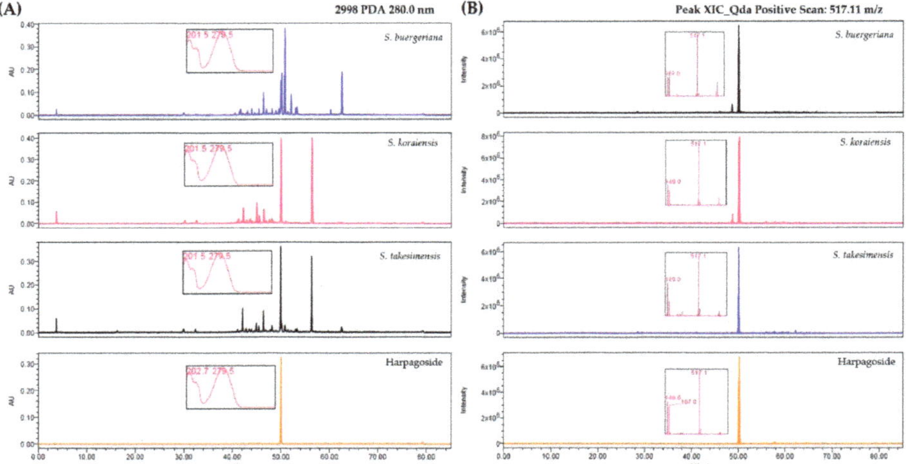

Figure 2. (**A**) Chromatograms of three *Scrophularia* species (*S. buergeriana*, *S. koraiensis*, and *S. takesimensis*) at 280 nm for harpagoside (λ_{max} = 279.5 nm) and (**B**) positive extracted ion chromatogram (XIC) spectrum of harpagoside at 517.11 *m/z*.

Table 2. Harpagoside content of three *Scrophularia* species (*S. buergeriana*, *S. koraiensis*, and *S. takesimensis*) by high-performance liquid chromatography (HPLC) chromatogram analysis.

Scrophulariae Species	Harpagoside (mg/g)
S. buergeriana	1.94 ± 0.24 [(1)]
S. koraiensis	6.47 ± 0.02
S. takesimensis	5.50 ± 0.02

[(1)] Values are expressed as means ± standard deviation (SD) of three samples for each species.

2.3. Cytotoxic Effects on Primary Murine BMM Growth

The XTT assay was conducted to assess cytotoxicity during osteoclast differentiation. *S. buergeriana*, *S. koraiensis*, and *S. takesimensis* did not reduce cell viability at most of the concentrations tested. Treatment with 200 μg/mL *S. koraiensis* reduced cell viability from that of the normal control group (Figure 3A–C), but the difference was not statistically significant.

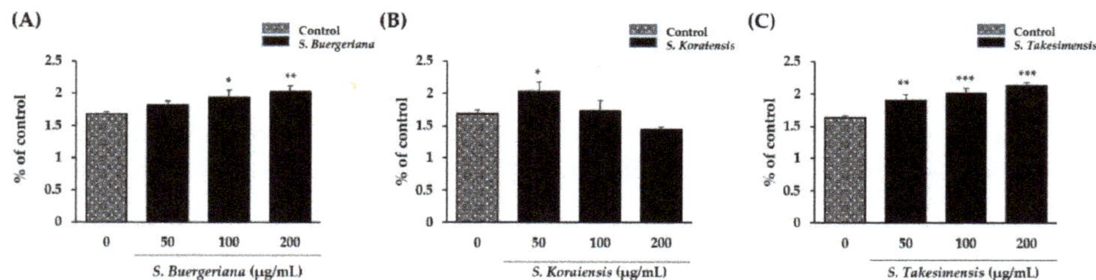

Figure 3. Cell viability affected by ethanol extracts of the three *Scrophularia* species (**A**) *S. buergeriana* (**B**) *S. koraiensis* and (**C**) *S. takesimensis*. * $p < 0.05$, ** $p < 0.01$, and *** $p < 0.001$ vs. control (dimethyl sulfoxide; DMSO).

2.4. Effects on Osteoclast Differentiation

To compare the effect of *S. buergeriana*, *S. koraiensis*, and *S. takesimensis* on osteoclast differentiation, mouse BMMs treated with macrophage colony stimulating factor (M-CSF) and RANKL were cultured in the presence or absence of ethanol extracts of the three *Scrophularia* species. RANKL and M-CSF induced differentiation of BMMs after incubation for 4 days. TRAP-positive osteoclasts were present in higher numbers in the control group, whereas treatment with *Scrophularia* species extracts inhibited the formation of TRAP-positive cells in a dose-dependent manner (Figure 4B). The *S. koraiensis* treatment group showed suppressed formation of RANKL-induced TRAP activity at concentrations of 100 μg/mL and 200 μg/mL.

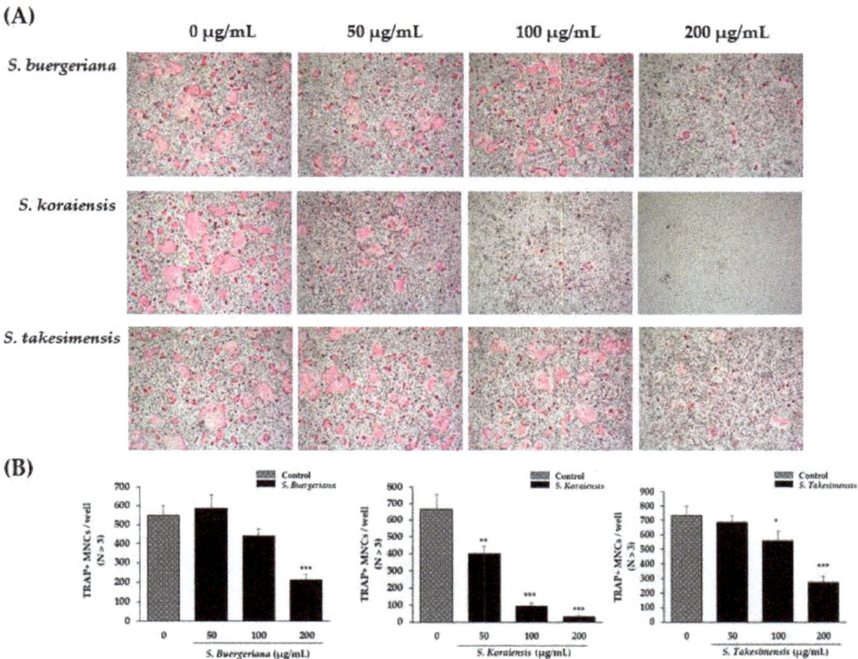

Figure 4. Effects on osteoclast differentiation of ethanol extracts of *Scrophularia buergeriana*, *S. koraiensis*, and *S. takesimensis* at concentrations of 50 μg/mL, 100 μg/mL, and 200 μg/mL. (**A**) Tartrate-resistant acid phosphatase (TRAP)-positive cells photographed (100× magnification) after bone marrow macrophages were cultured with macrophage colony stimulating factor (M-CSF) and nuclear factor (NF)-κB ligand (RANKL) in the presence of ethanol extracts of the three *Scrophularia* species. (**B**) TRAP-positive cells were counted as osteoclasts. * $p < 0.05$, ** $p < 0.01$, and *** $p < 0.001$ vs. control (DMSO).

2.5. Effects on Bone Resorption

To evaluate the effects of *Scrophularia* species on bone resorption, mature osteoclasts and extracts of *S. buergeriana*, *S. koraiensis*, and *S. takesimensis* were applied to plates coated with hydroxyapatite for symbiotic culture of the osteoblasts. Although the area of hydroxyapatite-adsorbed osteoclasts was increased in the control, the resorption area was markedly decreased by treatment with extracts of the three *Scrophularia* species. The resorption inhibition effect was highest for *S. buergeriana*, followed by *S. koraiensis*, and then *S. takesimensis*. (Figure 5A,B).

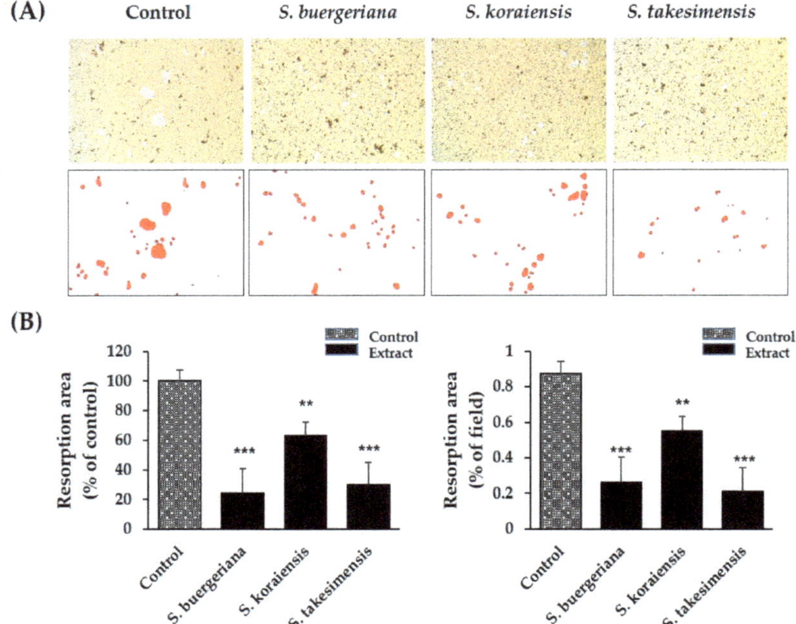

Figure 5. Effects of *S. buergeriana*, *S. koraiensis*, and *S. takesimensis* on bone resorption by mature osteoclasts. (**A**) Hydroxypatite-adherent cells were collected and imaged under a light microscope. (**B**) Resorption areas quantified on hydroxyapatite-coated plates. ** $p < 0.01$ and *** $p < 0.001$ vs. control (DMSO).

3. Discussion

Osteoporosis, one of the major diseases attracting attention worldwide, is associated with lowered bone mass density as a result of an imbalance between the osteoblasts and osteoclasts that influence bone homeostasis [22,23]. Therapeutic agents such as bisphosphonates are currently used to treat bone resorption diseases, but the long-term use of these drugs leads to the suppression of bone formation and osteonecrosis [24]. Therefore, the beneficial pharmacological effects of plant-derived natural compounds have been advocated [10].

Several studies have reported that iridoid glycosides in *Scrophularia* plant extracts demonstrate anti-inflammatory, neuroinflammation, antioxidant, and hepatoprotective effects [11–13,15,25] Harpagoside, an iridoid glycoside, is a main bioactive component of *Harpagophytum procumbens* (family Pedaliaceae) root used to treat chronic rheumatism, osteoarthritis, and arthritis, and is an active constituent of *Scrophularia* species [21,26–28]. We quantified the harpagoside content in *S. buergeriana*, *S. koraiensis*, and *S. takesimensis* using HPLC/MS (Table 2, Figure 2). Harpagoside was detected in three *Scrophularia* species, with the highest content found in *S. koraiensis* extract.

In this study, we investigated the inhibitory effects exerted by three *Scrophularia* species containing harpagoside against osteoclast differentiation and bone resorption without cytotoxic effects in RANKL-induced cells. The cytotoxicity of *S. buergeriana*, *S. koraiensis*, and *S. takesimensis* was evaluated in BMMs and measured by XTT assay during osteoclast differentiation. Treatment with ethanol extracts of the three *Scrophularea* species did not reveal cytotoxic effects on BMMs up to 200 µg/mL, with more than 90% cell viability being observed.

RANKL, a bone formation biomarker, is associated with stimulation of osteoblasts and osteoclast differentiation [22]. The balance of bone formation is influenced by osteoblasts and bone resorption activity by osteoclastic cells [29]. Previous studies have demonstrated that harpagoside improved bone properties by inhibiting the formation of osteoclasts from BMMs and the maturation of osteoblast cells [21,30].

TRAP expression is associated with osteoclast maturation and differentiation and is a standard approach to the detection of osteoclasts [24]. In our study, TRAP staining indicated that the numbers and areas of TRAP-positive cells increased, whereas the *S. buergeriana*, *S. koraiensis*, and *S. takesimensis* treatment groups exhibited considerably fewer TRAP-stained osteoclasts, without any cytotoxicity. The results suggested that the three *Scrophularia* species inhibit osteoclast differentiation and formation in BMMs. The inhibitory effect of harpagoside was increased in a dose-dependent manner by downregulating TRAP expression in BMCs [30]. Therefore, it is presumed that the difference in efficacy of the three Scrophularia species in our results would be affected by the content of harpagoside. To study the inhibitory effects of *Scrophularia* species on bone resorption by osteoclast formation induced by RANKL, we investigated bone resorption pit generation in mature osteoclasts. Although the area of hydroxyapatite-absorbed osteoclasts was increased in DMSO controls, the resorption area was markedly decreased by treatment with the three *Scrophularia* species. The highest resorption inhibition effect was exerted by *S. buergeriana*, followed by *S. takesimensis*, and then *S. koraiensis*.

4. Materials and Methods

4.1. Chemicals

Harpagoside was purchased from Shanghai Sunny Biotech (Shanghai, China). HPLC-grade solvents were obtained from Merck (Darmstadt, Germany). Sodium 3′-[1-(phenyl-aminocarbonyl)-3,4 -tetrazolium]-bis(4-methoxy-6-nitro) benzene sulfonic acid hydrate (XTT reagent) was purchased from Sigma-Aldrich (St. Louis, MO, USA).

4.2. Plant Materials

Samples of the three *Scrophularia* species (*S. buergeriana*, *S. koraiensis*, and *S. takesimensis*) were obtained from the Department of Herbal Crop Research (Eumseong-gun, Chungcheongbuk-do, Korea). The same vouchers were used for both chemical and morphological analyses (*S. buergeriana*; 2-18-0144, *S. koraiensis*; 2-18-0145, *S. takesimensis*; 2-18-0146) from the Korean Herbarium of Standard Resources (KHSR), Korean Institute of Oriental Medicine.

4.3. Comparative Morphology of Scrophularia Species

S. buergeriana and *S. koraiensis* were originally collected from the experimental field of the National Institute of Horticultural and Herbal Science (NIHHS, Korea), and *S. takesimensis* was collected from the natural population on Ulleung-do Island. Vouchers of the studied medicinal materials (*S. buergeriana* (2-18-0144), *S. koraiensis* (2-18-0145), and *S. takesimensis* (2-18-016)) were deposited in the Korean Herbarium of Standard Herbal Resources (Index Herbariorum code: KIOM) at the Korea Institute of Oriental Medicine (Naju, Korea). We observed the voucher specimens of the three species (Figure S1) sing an Olympus SZX16 stereomicroscope (Olympus, Tokyo, Japan) for morphological comparison.

4.4. HPLC/MS Analysis

S. buergeriana (778.14 g), *S. koraiensis* (69.15 g), and *S. takesimensis* (92.5 g) were refluxed in 70% ethanol (*v*/*v*) for 2 h, and these extracts were filtered through filter paper and evaporated in vacuo. The powders of the 70% ethanol extract for *S. buergeriana*, *S. koraiensis*, and *S. takesimensis* were 363.83 g (46.76% of yield, *w*/*w*), 30.40 g (43.96% of yield, *w*/*w*), and 53.91 g (58.28% of yield, *w*/*w*), respectively, and these extracts were stored at 4 °C. Accurately weighed powders of three 70% ethanol extracts (10 mg) were dissolved in 10 mL 70% ethanol and filtered through a 0.2 μm syringe filter prior to HPLC analysis. HPLC was performed using a Waters e2695 Separation Module (Waters Corporation, Milford, MA, USA) and a 2998 photodiode array detector (Waters), combined with an Acquity QDa detector and a micro-splitter (IDEX Health & Science LLC, Oak Harbor, WA, USA) for MS. A Kinetex

Phenyl-hexyl 100A column (4.6 × 250 mm, 5 μm, Phenomenex Inc., Torrance, CA, USA) was utilized. The mobile phase was 0.05% aqueous formic acid (A), methanol (B), and acetonitrile (C) using the following gradient program: 100% A (2 min) → 96% A (3% B and 1% C, 7 min) → isocratic 85% A (11% B and 4% C, from 15 min to 20 min) → 70% A (23% B and 7% C, 35 min) → 50% A (35% B and 15% C, 45 min) → 30% A (50% B and 20% C, 55 min). The following conditions were applied: The flow rate was 0.8 mL/min, injected volume was 10 μL, and UV wavelength was monitored from 195 nm to 400 nm. The QDa mass detector employed the following conditions: Nitrogen carrier gas, 0.8 kV electrospray ionization (ESI) capillary, 600 °C probe temperature, 15 V con voltage, 120 °C source temperature, 20:1 split. TIC was monitored from 150 *m*/*z* to 850 *m*/*z*. Harpagoside was detected at 280 nm and 517.11 *m*/*z*.

4.5. BMM (Bone Marrow Macrophage) Isolation and Culture

Five-week-old ICR mice were sacrificed by cervical dislocation, the thigh and shin bones were aseptically excised, and the soft tissue was removed. After cutting both ends of the iliac crest, bone marrow cells were obtained by flushing both ends of the bone material using a 1 mL syringe. The isolated BMMs were incubated in a culture dish for 1 day in α-MEM medium containing 10% FMS and 1% penicillin/streptomycin, and unsaturated cells were collected. After 3 days, attached macrophages were used in this experiment. Macrophages were treated with M-CSF (30 ng/mL) and RANKL (100 ng/mL), and incubated with *S. takesimensis*, *S. koraiensis*, and *S. buergeriana* extracts at concentrations of 50 μg/mL, 100 μg/mL, and 200 μg/mL. The day following subculture, the cultured cells were stained with a TRAP solution. Cells with three or more nuclei among red-stained cells were considered differentiated osteoclasts, and the degree of differentiation was measured.

4.6. Cell Cytotoxicity

BMMs were seeded in 96-well plates at a density of 1×10^4 cells/well and were treated with M-CSF (30 ng/mL) and 50 μg/mL, 100 μg/mL, and 200 μg/mL ethanol extracts of *Scrophularia* species (*S. takesimensis*, *S. koraiensis*, and *S. buergeriana*) for 3 days. After 3 days, 50 μL XTT reagent (Sigma-Aldrich) was applied for 4 h. The optical density was measured at 450 nm using an ELISA reader (Biotek Instruments, Winooski, VT, USA).

4.7. Bone Resorption Pit Assay

To obtain mature osteoclasts, BMMs obtained from the shin and thigh bones of 5-week-old ICR mice and osteoclasts isolated from the skull of 1-day-old ICR mice were added to a 90 mm culture plate coated with collagen. To produce a symbiotic culture, 1α,25-Dihydroxyvitamin D3 (VitD3) and prostaglandin E2 were added. After incubation for 6 days, cells were removed by treatment with 0.1% collagenase and added to a 96-well plate coated with hydroxyapatite. At the same time, *S. takesimensis*, *S. koraiensis*, and *S. buergeriana* extracts were added to the hydroxyapatite-coated plates at a concentration of 200 ug/mL and cultured for 24 h. The cells were washed with distilled water and observed through an optical microscope (Nikon, Tokyo, Japan). The hydroxyapatite-absorbed portion was imaged using micrography, and the area was measured using ImageJ software version 1.51 (National Institutes of Health, Bethesda, MD, USA).

4.8. Statistical Analysis

Data were expressed as the mean ± standard deviation (SD). Statistical analysis was performed by analysis of variance (ANOVA), followed by a multiple comparison procedure using Dunnett's test. A value of $p < 0.05$ indicated significant difference.

5. Conclusions

Our results suggest that the *Scrophularia* species may prevent bone loss by inhibiting osteoclast differentiation and born resorption without causing cytotoxicity. It was confirmed that similar metabolites were contained in the *Scrophularia* species, and their extracts exerted similar efficacy. *S. buergeriana* and *S. koraiensis* demonstrated inhibitory effects against osteoclast differentiation and bone resorption and had the highest harpagoside content. Therefore, the three *Scrophularia* species, including *S. koraiensis*, may be potentially therapeutic for treating bone disorder diseases, but the mechanism underlying their bioactivity requires further study.

Supplementary Materials: The following are available online at http://www.mdpi.com/2223-7747/9/12/1656/s1.

Author Contributions: Conceptualization, J.-S.K. and J.-H.S.; the morphology study, J.-H.S.; in-vitro analysis, H.-H.N., Y.-S.S. and J.M.C.; resources, I.P. and S.Y.; chemical analysis, A.Y.L.; genetic analysis, B.C.M.; funding acquisition, B.C.M., J.-H.S. and J.-S.K.; writing—review and editing, H.-H.N., J.-H.S. and J.-S.K.; All authors have read and agreed to the published version of the manuscript.

Funding: This research was supported by the Korea Institute of Oriental Medicine, Daejeon, South Korea (grant numbers K18402, KSN2012320) and the National Research Foundation of Korea (NRF) grant funded by the Korea government (MSIT) (NRF-2020R1A2C1004272 and NRF-2020R1A2C1100147).

Acknowledgments: We thank the Korean Herbarium of Standard Herbal Resources (herbarium code KIOM) for the provision of materials. J.-H.S. sincerely thanks Jeong Hoon Lee (Department of Herbal Crop Research, NIHHS) for assistance in material sampling.

Conflicts of Interest: The authors declare no conflict of interest.

Abbreviations

BMM, bone marrow-derived macrophage; RANKL, receptor activator of nuclear factor (NF)-κB ligand; TIC, total ion chromatography; TRAP, tartrate-resistant acid phosphatase; XIC, extracted ion chromatogram; XTT, sodium 3′-[1-(phenyl-aminocarbonyl)-3,4-tetrazolium]-bis(4-methoxy-6-nitro) benzene sulfonic acid hydrate.

References

1. Yu, L.; Jia, D.; Feng, K.; Sun, X.; Xin, H.; Qin, L.; Han, T. A natural compound (LCA) isolated from *Litsea cubeba* inhibits RANKL-induced osteoclast differentiation by suppressing Akt and MAPK pathways in mouse bone marrow macrophages. *J. Ethnopharmacol.* **2020**, *257*, 112873. [CrossRef] [PubMed]
2. Boyle, W.J.; Simonet, W.S.; Lacey, D.L. Osteoclast differentiation and activation. *Nature* **2003**, *423*, 337–342. [CrossRef] [PubMed]
3. Lee, C.H.; Kwak, S.C.; Kim, J.Y.; Oh, H.M.; Rho, M.C.; Yoon, K.H.; Yoo, W.H.; Lee, M.S.; Oh, J. Genipin inhibits RANKL-induced osteoclast differentiation through proteasome-mediated degradation of c-Fos protein and suppression of NF-κB activation. *J. Pharmacol. Sci.* **2014**, *124*, 344–353. [PubMed]
4. Yamazaki, T. Scrophulariaceae. In *Flora of Japan*; Iwatsuki, K., Yamazaki, T., Boufford, D.E., Ohba, H., Eds.; Kodansha: Tokyo, Japan, 1993; Volume IIIa, pp. 326–331.
5. Hong, D.Y.; Yang, H.B.; Jin, C.L.; Fischer, M.A.; Holmgren, N.H.; Mill, R.R. Scrophularia. In *Flora of China*; Wu, Z.Y., Raven, P.H., Eds.; Science Press and Missori Botanical Garden Press: Beijing, China; St. Louis, MO, USA, 1998; Volume 18, pp. 11–20.
6. Choi, H.K. Scrophularia. In *The Genera of Vascular Plants of Korea*; Flora of Korean Editorial Committee, Ed.; Hongreung Publishing Co.: Seoul, Korea, 2018; pp. 1148–1150.
7. Wang, S.; Hua, Y.; Zou, L.; Liu, X.; Yan, Y.; Zhao, H.; Luo, Y.; Liu, J. Comparison of Chemical Constituents in Scrophulariae Radix Processed by Different Methods based on UFLC-MS Combined with Multivariate Statistical Analysis. *J. Chromatogr. Sci.* **2017**, *56*, 122–130. [CrossRef]
8. You-Hua, C.; Jin, Q.; Jing, H.; Bo-Yang, Y. Structural characterization and identification of major constituents in Radix Scrophulariae by HPLC coupled with electrospray ionization quadrupole time-of-flight tandem mass spectrometry. *Chin. J. Nat. Med.* **2014**, *12*, 47–54. [CrossRef]
9. Huang, X.Y.; Chen, C.X.; Zhang, X.M.; Liu, Y.; Wu, X.M.; Li, Y.M. Effects of ethanolic extract from Radix Scrophulariae on ventricular remodeling in rats. *Phytomedicine* **2012**, *19*, 193–205. [CrossRef]

10. Zengin, G.; Stefanucci, A.; Rodrigues, M.J.; Mollica, A.; Custodio, L.; Aumeeruddy, M.Z.; Mahomoodally, M.F. *Scrophularia lucida* L. as a valuable source of bioactive compounds for pharmaceutical applications: In vitro antioxidant, anti-inflammatory, enzyme inhibitory properties, in silico studies, and HPLC profiles. *J. Pharm. Biomed. Anal.* **2019**, *162*, 225–233. [CrossRef]
11. Lee, H.J.; Spandidos, D.A.; Tsatsakis, A.; Margina, D.; Izotov, B.N.; Yang, S.H. Neuroprotective effects of *Scrophularia buergeriana* extract against glutamate-induced toxicity in SH-SY5Y cells. *Int. J. Mol. Med.* **2019**, *43*, 2144–2152. [CrossRef]
12. Shin, N.R.; Lee, A.Y.; Song, J.H.; Yang, S.; Park, I.; Lim, J.O.; Jung, T.Y.; Ko, J.W.; Kim, J.C.; Lim, K.S.; et al. *Scrophularia buergeriana* attenuates allergic inflammation by reducing NF-κB activation. *Phytomedicine* **2020**, *67*, 153159. [CrossRef]
13. Jeong, E.J.; Ma, C.J.; Lee, K.Y.; Kim, S.H.; Sung, S.H.; Kim, Y.C. KD-501, a Standardized Extract of *Scrophularia buergeriana* Has Both Cognitive-Enhancing and Antioxidant Activities in Mice Given Scopolamine. *J. Ethnopharmacol.* **2009**, *121*, 98–105. [CrossRef]
14. Kim, S.R.; Kang, S.Y.; Lee, K.Y.; Kim, S.H.; Markelonis, G.J.; Oh, T.H.; Kim, Y.C. Anti-amnestic Activity of E-p-methoxycinnamic acid from *Scrophularia buergeriana*. *Cogn. Brain Res.* **2003**, *17*, 454–461. [CrossRef]
15. Lee, E.J.; Kim, S.R.; Kim, J.; Kim, Y.C. Hepatoprotective Phenylpropanoids From *Scrophularia Buergeriana* Roots Against CCl(4)-induced Toxicity: Action Mechanism and Structure-Activity Relationship. *Planta Med.* **2002**, *68*, 407–411. [CrossRef] [PubMed]
16. Wang, P.; Wang, D.; Qi, Z.; Li, P.; Fu, C. *Scrophularia koraiensis*, a new synonym to *Scrophularia kakudensis* (Lamiales: Scrophulariaceae). *Phytotaxa* **2015**, *202*, 228–230. [CrossRef]
17. Ma, S.M.; Lim, Y.S.; Na, S.T.; Lee, J.; Shin, H.C. Genetic structure and population differentiation of endangered *Scrophularia takesimensis* (Scrophulariaceae) in Ulleung Island, Korea. *Korean J. Plant Taxon.* **2011**, *41*, 182–193. [CrossRef]
18. Pasdaran, A.; Hamedi, A. The genus *Scrophularia*: A source of iridoids and terpenoids with a diverse biological activity. *Pharm. Biol.* **2017**, *55*, 2211–2233. [CrossRef] [PubMed]
19. García, D.; Fernández, A.; Sáenz, T.; Ahumada, C. Antiinflammatory effects of different extracts and harpagoside isolated from *Scrophularia frutescens* L. *Farmaco* **1996**, *51*, 443–446. [PubMed]
20. Haseeb, A.; Ansari, M.Y.; Haqqi, T.M. Harpagoside suppresses IL-6 expression in primary human osteoarthritis chondrocytes. *J. Orthop. Res.* **2017**, *35*, 311–320. [CrossRef]
21. Kim, J.Y.; Park, S.H.; Baek, J.M.; Erkhembaatar, M.; Kim, M.S.; Yoon, K.H.; Oh, J.; Lee, M.S. Harpagoside inhibits RANKL-induced osteoclastogenesis via Syk-Btk-PLCγ2-Ca(2+) signaling pathway and prevents inflammation-mediated bone loss. *J. Nat. Prod.* **2015**, *78*, 2167–2174. [CrossRef]
22. Zhang, B.; Yang, L.L.; Ding, S.Q.; Liu, J.J.; Dong, Y.H.; Li, Y.T.; Li, N.; Zhao, X.J.; Hu, C.L.; Jiang, Y.; et al. Anti-osteoporotic activity of an edible traditional Chinese medicine Cistanche deserticola on bone metabolism of ovariectomized rats through RANKL/RANK/TRAF6-mediated signaling pathways. *Front. Pharmacol.* **2019**, *10*, 1412. [CrossRef]
23. Kim, K.J.; Lee, Y.; Son, S.R.; Lee, H.; Son, Y.J.; Lee, M.K.; Lee, M. Water Extracts of Hull-less Waxy Barley (*Hordeum vulgare* L.) Cultivar 'Boseokchal' Inhibit RANKL-induced Osteoclastogenesis. *Molecules* **2019**, *24*, 3735. [CrossRef]
24. Kim, J.H.; Kim, M.; Jung, H.S.; Sohn, Y. *Leonurus sibiricus* L. ethanol extract promotes osteoblast differentiation and inhibits osteoclast formation. *Int. J. Mol. Med.* **2019**, *44*, 913–926. [CrossRef] [PubMed]
25. Meng, X.; Xie, W.; Xu, Q.; Liang, T.; Xu, X.; Sun, G.; Sun, X. Neuroprotective Effects of Radix Scrophulariae on Cerebral Ischemia and Reperfusion Injury via MAPK Pathways. *Molecules* **2018**, *23*, 2401. [CrossRef] [PubMed]
26. Che, D.; Cao, J.; Liu, R.; Wang, J.; Hou, Y.; Zhang, T.; Wang, N. Harpagoside-induced anaphylactic reaction in an IgE-independent manner both in vitro and in vivo. *Immunopharmacol. Immunotoxicol.* **2018**, *40*, 173–178. [CrossRef]
27. Su, L.; Deng, Y.; Chen, N.; Zhang, X.; Huang, T. Infrared-assisted extraction followed by high performance liquid chromatography to determine angoroside C, cinnamic acid, and harpagoside content in *Scrophularia ningpoensis*. *BMC Complement. Altern. Med.* **2019**, *19*, 130. [CrossRef] [PubMed]
28. Axmann, S.; Hummel, K.; Nöbauer, K.; Razzazi-Fazeli, E.; Zitterl-Eglseer, K. Pharmacokinetics of harpagoside in horses after intragastric administration of a Devil's claw (*Harpagophytum procumbens*) extract. *J. Vet. Pharmacol. Ther.* **2019**, *42*, 37–44. [CrossRef] [PubMed]

29. Kumagai, M.; Nishikawa, K.; Mishima, T.; Yoshida, I.; Ide, M.; Watanabe, A.; Fujita, K.; Morimoto, Y. Fluorinated Kavalactone Inhibited RANKL-Induced Osteoclast Differentiation of RAW264 Cells. *Biol. Pharm. Bull.* **2020**, *43*, 898–903. [CrossRef]
30. Chung, H.J.; Kim, W.K.; Park, H.J.; Cho, L.; Kim, M.R.; Kim, M.J.; Shin, J.S.; Lee, J.H.; Ha, I.H.; Lee, S.K. Anti-osteoporotic activity of harpagide by regulation of bone formation in osteoblast cell culture and ovariectomy-induced bone loss mouse models. *J. Ethnopharmacol.* **2016**, *179*, 66–75. [CrossRef]

Article

Functional Attributes and Anticancer Potentialities of Chico (*Pachycereus Weberi*) and Jiotilla (*Escontria Chiotilla*) Fruits Extract

Luisaldo Sandate-Flores [1], Eduardo Romero-Esquivel [1], José Rodríguez-Rodríguez [1], Magdalena Rostro-Alanis [1], Elda M. Melchor-Martínez [1], Carlos Castillo-Zacarías [1], Patricia Reyna Ontiveros [2], Marcos Fredy Morales Celaya [3], Wei-Ning Chen [4], Hafiz M. N. Iqbal [1,*] and Roberto Parra-Saldívar [1,*]

1 Tecnologico de Monterrey, School of Engineering and Sciences, Centro de Biotecnología FEMSA, Avenida Eugenio Garza Sada 2501, Monterrey 64849, Mexico; luisaldosf@gmail.com (L.S.-F.); romeroe.ing@gmail.com (E.R.-E.); jrr@tec.mx (J.R.-R.); magda.rostro@tec.mx (M.R.-A.); elda.melchor@tec.mx (E.M.M.-M.); carloscastilloz@tec.mx (C.C.-Z.)

2 Universidad Iberoamericana, Puebla, Avenida Tres Oriente, 615, 6, Centro, Puebla C.P. 72000, Mexico; direccion@rab-gts.com

3 Universidad Tecnológica de los Valles Centrales de Oaxaca, Oaxaca 71270, Mexico; mfmorcel@gmail.com

4 School of Chemical and Biomedical Engineering, College of Engineering, Nanyang Technological University, 62 Nanyang Drive, Singapore 637459, Singapore; wnchen@ntu.edu.sg

* Correspondence: hafiz.iqbal@tec.mx (H.M.N.I.); r.parra@tec.mx (R.P.-S.); Tel.: +52-81-8358-2000 (ext. 5561) (R.P.-S.)

Received: 30 September 2020; Accepted: 30 October 2020; Published: 22 November 2020

Abstract: Mexico has a great diversity of cacti, however, many of their fruits have not been studied in greater depth. Several bioactive compounds available in cacti juices extract have demonstrated nutraceutical properties. Two cactus species are interesting for their biologically active pigments, which are chico (*Pachycereus weberi (J. M. Coult.) Backeb*)) and jiotilla (*Escontria chiotilla* (*Weber) Rose*)). Hence, the goal of this work was to evaluate the bioactive compounds, i.e., betalains, total phenolic, vitamin C, antioxidant, and mineral content in the extract of the above-mentioned *P. weberi* and *E. chiotilla*. Then, clarified extracts were evaluated for their antioxidant activity and cytotoxicity (cancer cell lines) potentialities. Based on the obtained results, Chico fruit extract was found to be a good source of vitamin C (27.19 ± 1.95 mg L-Ascorbic acid/100 g fresh sample). Moreover, chico extract resulted in a high concentration of micronutrients, i.e., potassium (517.75 ± 16.78 mg/100 g) and zinc (2.46 ± 0.65 mg/100 g). On the other hand, Jiotilla has a high content of biologically active pigment, i.e., betaxanthins (4.17 ± 0.35 mg/g dry sample). The antioxidant activities of clarified extracts of chico and jiotilla were 80.01 ± 5.10 and 280.88 ± 7.62 mg/100 g fresh sample (DPPH method), respectively. From the cytotoxicity perspective against cancer cell lines, i.e., CaCo-2, MCF-7, HepG2, and PC-3, the clarified extracts of chico showed cytotoxicity (%cell viability) in CaCo-2 (49.7 ± 0.01%) and MCF-7 (45.56 ± 0.05%). A normal fibroblast cell line (NIH/3T3) was used, as a control, for comparison purposes. While jiotilla extract had cytotoxicity against HepG2 (47.31 ± 0.03%) and PC-3 (53.65 ± 0.04%). These results demonstrated that Chico and jiotilla are excellent resources of biologically active constituents with nutraceuticals potentialities.

Keywords: pachycereus weberi; escontria chiotilla; antioxidant activity; phenolic compounds; betalains; food composition; food analysis; cytotoxicity

1. Introduction

Mexico has a great diversity of cacti [1], but many of the fruits from these plants have not been studied. Cacti fruits have health benefits such as anticlastogenic capacity [2], hepatoprotective effect [3], antimicrobial activity [4], notable antioxidant capacity [5,6], and anticancer activity [7,8]. Currently, the scientific community is interested in two cacti fruits for their pigments, chico (*Pachycereus weberi (J. M. Coult.) Backeb*) and jiotilla (*Escontria chiotilla (Weber) Rose). Pachycereus weberi* is typical in the south-central Mexico region, specifically the arid and semi-arid regions in Puebla, Guerrero, Morelos, and Oaxaca [1]. The fruit of *Pachycereus weberi* is better known as "chico fruit" and "tuna de cardón" [9] (Figure 1). *P Weberi* fruits are ellipsoidal, 6 to 7 cm in diameter [10]. In Santiago Quiotepec, Oaxaca, fruits of this species harvested and commercialized in local markets [11]. The fruit production is 509.3 ton in Tehuacán Valley México from *Pachycereus weberi* [12].

Figure 1. (**A**) *Pachycereus weberi* plant and (**B**) *P. weberi* fruit without prickles (top) with prickles (bottom).

Nowadays, the knowledge about antioxidant activity and phenolic compounds of the fruit from the Mexican plant *Pachycereus weberi* is limited. *Escontria chiotilla,* a columnar cactus is found in Guerrero, Michoacán, Puebla, Oaxaca, and Morelos. The fruit of *Escontria chiotilla* has a red shell with the presence of scales, spherical, and are 5 cm in diameter (Figure 2). The period of growth of the fruit ranged 140 to 175 days [13] and is usually named as "chonostle" and "jiotilla" [14]. The production of this fruit is approximately 99.5 tons in the valley of Tehuacan [12] and is commercialized in local markets, and people from the community of Coxcatlan prepare ice cream and jellies for local trade [15]. Different studies have shown that the pigments present in the jiotilla are vulgaxanthin I, vulgaxanthin II, indicaxanthin, and betanin [16]. These pigments have been extracted using acidified mucilage and microencapsulated for subsequent characterization [17].

Based on the literature, cacti fruits have cytotoxic properties in cancer cell lines [7,8]. For instance, prickle pear juices (*Opuntia spp.*) have shown to produce decreasing viability of cancer cell in vitro, especially colon (HT29/Caco-2) and prostate cancer (PC-3) [7,8]. Indeed, these are promising results as around 70% of cancer deaths occur in low and middle incomes countries [18] and in 2018, 1.8 million colorectal and 1.28 million prostate cases were reported [18]. Furthermore, several studies have shown that fruits growth in semi-arid and arid lands present high mineral content [19,20]. This aspect is relevant as minerals are essential in metabolism and homeostatic of the human body, deficiency of minerals can lead to symptoms of common disorders and diseases as osteoporosis and anemia [21]. One of the essential micronutrients is zinc, and the deficiency of this restricts childhood growth and decreases resistance to infections [22]. In the world, around 20% of children under five years of age are stunted [23]. The principal food with a high concentration of zinc are oysters, red meat, and poultry [24]. The food mentioned before may be more challenging to access for low-income populations [25]. Oaxaca, Guerrero, and Puebla are Mexican states with low-income populations [26]. It is crucial to find other sources of zinc that are accessible to all communities.

Figure 2. (**A**) *Escontria chiotilla* plant, (**B**) *E. chiotilla* flower, and (**C**) *E. chiotilla* fruit.

Studies on potentialities of cacti fruits have gradually increased in recent years, but there is still a lack of information regarding antioxidant capacity, metal content, or phytochemical content of *P. weberi* and *E. chiotilla* fruits pulp and juice extracts. Therefore, this study aimed to evaluate the physiochemical characteristic, antioxidant activities, and anticancer potentialities of *P. weberi* and *E. chiotilla* fruits pulp and/or juice extracts.

2. Materials and Methods

2.1. Chemicals

2,2-Diphenyl-L-picryl-hydrazyl (DPPH) ($C_{18}H_{12}N_5O_6$, Item D9132, Lot 115K1319) was obtained from Sigma-Aldrich (Steinheim, Germany). Methyl alcohol (CH_3OH, Item MS1922, Lot 17020429) was obtained from Tedia High Purity Solvents (Fairfield, OH, USA). Folin&Ciocalteu's Phenol Reagent ($C_7H_6O_5$, Item F9252, Lot SHBD7847V) was obtained from Sigma Aldrich (St. Louis, MO, USA). Sodium Carbonate (Na_2CO_3, Item S7795, Lot BCBP4310V) was obtained from Sigma Aldrich (Steinheim, Germany). Gallic acid monohydrate ($C_7H_6O_5$, Item 398225, Lot MKBD1204) was obtained from Sigma-Aldrich (Shanghai, China). Sodium chloride (NaCl, DEQS290000100, Lot CS180307-03) was obtained from Desarrollo de Especialidades Químicas, S.A. de C.V. (Nuevo Leon, Mexico). Potassium persulfate ($K_2S_2O_8$, Item 216224, Lot MKBR1923V) was obtained from Sigma-Aldrich (St. Louis, MO, USA). Disodium phosphate (Na_2HPO_4, 35902, Lot 536405) obtained from Productos Quimicos Monterrey S.A. de C.V. (Nuevo Leon, Mexico). Potassium chloride (KCl, S32204, Lot 536161) obtained from Productos Químicos Monterrey S.A. de C.V. (Nuevo Leon, Mexico). 2,2′-Azino-bis(3-ethylbenzothiazoline-6-sulfonic acid) ABTS ($C_{18}H_{24}N_6O_6S_4$, Item A9941, Lot SLBT7934) obtained from Sigma-Aldrich (St. Louis, MO, USA). Potassium phosphate monobasic (KH_2PO_4, Item P0662, Lot SLBL9298V) was obtained from Sigma-Aldrich (Tokyo, Japan). Nitric acid (HNO_3, Item UN2031, Lot SCA7227127) was obtained from SPC Science (Baie-d'Urfé, QC, Canada). Ammonium acetate (CH_3COONH_4, Item 0596, Lot L15C82) was obtained from J. T. Baker (Estado de Mexico, Mexico). Milli-Q water purification system is used to obtain the water that was used to perform the procedures (Q-POD, Darmstadt, Germany). Glacial acetic acid (Item AE4001, Lot 1004007) was bought from Tedia (Fairfield, OH, USA). 2,4,6-Tris(2-pyridyl)-s-triazine (TPTZ) ($C_{18}H_{12}N_6$, Item T1253, Lot BCBT8262) was bought from Sigma Aldrich (Buchs, Switzerland). Iron (lll) chloride hexahydrate ($FeCl_3{\cdot}6H_2O$, ItemF2877, Lot MKBR8795V) was bought from Sigma Aldrich (Shanghai, China). Hydrochloric acid (HCl, Item 00636, Lot 71973) was bought from CTR Scientific (Nuevo Leon, Mexico). Gallic acid monohydrate ($C_7H_6O_5{\cdot}H_2O$, item 398225, Lot MKBD1204) was bought from Sigma Aldrich (Shanghai, China). Caffeic acid ($C_9H_8O_4$, item C0625, Lot 089K1114) was obtained from Sigma-Aldrich (Buch, Switzerland). p-Coumaric acid ($C_9H_8O_3$, item C9008, Lot BCBN0412V) was bought from Sigma-Aldrich (United Kingdom). Violuric acid ($C_4H_3O_4{\cdot}H_2O$, item 95120, Lot BCBN0412V) and fumaric acid ($C_4H_4O_4$, item 47910, Lot BCBM2191V) were obtained from Sigma-Aldrich (Vienna, Austria).

2.2. Collection and Preprocessing of Chico and Jiotilla

A single batch of chico fruits (10 kg) was collected from a crop field in Ahuatlán (Puebla, México). The sampling area was situated between 18°34′ latitude (North) and 98°15′ longitude (West). The collected fruit samples were stored at 5 °C for 48 h and processed within 72 h of harvest from the plant. Whereas, a batch of jiotilla fruits (22 kg) was collected from a crop field in Santa María Zoquitlan, Tlacolula, Oaxaca, Mexico 16°33′ latitude (North) and 96°2315′ longitude (West). The collected fruit samples were stored at 5 °C for 12 h and processed within 24 h of harvest from the plant. As collected fruits of both plants, i.e., chico and jiotilla, were preprocessed by washing with tap water and Extran MA05 (Merck, Item 1400001403, Lot Mx1400005004, Estado de Mexico, Mexico). Then, the prickles were removed manually with care using a sterile blade. The electric extractor (Model TU05, Turmix ML, Estado de México) was used to prepare a seed-free pulp sample. This method was used since it is a quick way to eliminate the seeds. Subsequently, around 5 g of seed-free pulp samples were placed in 40 mL vials, 10 mL of deionized water was added, and then the vials were agitated for 15 min. The resultant mix was filtered (Whatman paper grade 4, 150 mm, Item 1009150, GE Healthcare Life Sciences, Little Chalfont, UK) under dark conditions to avoid compounds degradation. The extract obtained (40 mL) was placed in a 100 mL (chico fruit) or 50 mL (jiotilla) volumetric flask and used in the subsequent analysis. Different dilutions were made because a great content of antioxidant capacity was expected in chico fruit. Three extractions for each fruit were carried out following the same conditions.

2.3. Clarified Juice Extract Preparation

Seed-free pulp (30 g) from each fruit was placed in 50-mL polypropylene centrifuge tubes (Corning®, Tewksbury, MA, USA). The tube was subjected to centrifugation (4000× *g*, 4 °C, 10 min, Model SL 40R, Thermo Fisher Scientific, Langenselbold, Germany). The supernatant was filtered in the dark through 150 mm Whatman paper grade 4 (item 1009150, GE Healthcare Life Sciences, Little Chalfont, UK). The pellets were removed, and the clarified supernatant from the pulp of each fruit was used as a clarified juice extract.

2.4. Physiochemical Characterization

Weight of pulp and peel were determined with an analytical balance (OHAUS, model Scout Pro SP4001, China). Equatorial and polar diameters were measured with a caliper (Vernier from Truper, model H87). The pH was determined with Termo Fisher Scientific, ORION 3 STAR pH Benchtop (Singapore, Singapore). Total soluble solids (Brix) were determined with a refractometer HI96811 (HANNA, Smithfield, RI, USA), and moisture content (%) was obtained according to the gravimetric method [27]. A general proximate analysis was applied to the pulps, the method used was AOAC 1999 [28].

2.5. Betacyanins and Betaxanthins Quantification

A spectrophotometric method is used to determine the concentration of betacyanins and betaxanthins in the pulp of the fruit. The spectrophotometer used was Model DR 500 (Hach Lange GmbH, Düsseldorf, Germany), following the methods described by Sandate-Flores et al. [29]. For the dilution of the samples, water Milli-Q was preferred, the dilution was 1:10, and it was carried out in 5 mL volumetric flasks. The extinction coefficients used were the following: for betacyanin (E1% = 60,000 L mol^{-1} cm^{-1}, λ = 540 nm) and for betaxanthin (E1% = 48,000 L mol^{-1} cm^{-1}, λ = 480) [30]. The betalains concentration was showed in milligrams per gram (mg/g, dry weight). The procedure for determining moisture content is described in Section 2.4.

2.6. Vitamin C Quantification

The vitamin C quantification was performed based on the procedure described earlier by Santos et al. [31]. The sample was passed through a nylon filter of 0.20 mL (2 mL of the volumetric extract), placed in vials for HPLC. The HPLC-FLD system (Agilent 1200 HPLC, Santa Clara, CA, USA) was injected with 25 μL of the sample, for which a Zorbax Eclipse XDB C18 column was used (5 μm, 150 9 4.6 mm i.d.) (Agilent, Santa Clara, CA, USA). As eluent solvents, a solution of 10 mM ammonium acetate at a pH of 4.5 was used (phase A), and a methanol solution with 0.1% acetic acid was phase B. Using 80% of phase A, elution was carried out in isocratic mode for 7 min with a flow rate of 1 mL/min. For the measurement of ascorbic acid, the chromatograms were monitored at Exc = 280 and Em = 440 nm. A six-point calibration curve was made using an external standard of ascorbic acid (0.01 to 10 mg/L), the curve used for the quantification of vitamin C.

2.7. Antioxidant Activity

2.7.1. Folin-Ciocalteu Method (Total Antioxidant Capacity)

The Folin-Ciocalteu colorimetric method was used to determine the total antioxidant capacity [32]. This colorimetric method was carried out following the procedure of Singleton et al. [33]. The analysis was performed directly in 96 well plates. Briefly, around 20 μL of the diluted sample was placed in Milli-Q water, then 100 μL of folin reagent at a concentration of 10% was added. After 5 min of the reaction period, around 80 μL of sodium carbonate at a concentration of 7.5% *w*/*v* was added. The above reaction mixture was then incubated at 37 °C for 90 min in the dark. Finally, the sample containing 96-well microplate was placed in a plate reader and scanned at 765 nm. The calibration curve was made using different concentrations (50 to 200 mg/L) of gallic acid. All measurements were made in triplicate, and Milli- Q water was used as blank.

2.7.2. ABTS

For the ABTS method, the reported method of Re et al. [34] was used. ABTS is a single electron transfer (ET) reaction-based assay. The preparation of PBS was carried out to add 0.8 g of NaCl, 0.02 g of KH_2PO_4, 0.115 g of Na_2HPO_4, 0.02 g of KCl, and 0.02 g of NaN_3. Then the solution was filled to 100 mL of Milli Q water. For the ABTS reagent, it was added 38.4 mg of ABTs 1 mM, 6.62 mg of potassium persulfate 2.45 mM, and 10 mL of the solution of PBS. After mixing both solutions, they stirred for 16 h, keep in the dark. To measure the absorbance, the spectrophotometer (Model DR 500, Hach Lange GmbH, Düsseldorf, Germany) was used at 734 nm, a dilution of the initial reagent was needed until it has an absorbance of 0.7 (40 mL of PBS with 3 mL of ABTS solution). The procedure was 20 μL diluted sample, and 2 mL of ABTS solution (absorbance 0.7) were incubated at 30 °C in a water bath for 6 min, then absorbances were read. Trolox was used as standard with concentrations of 5 ppm to 200 ppm. All measurements were made in triplicate. Milli- Q water was used as blank.

2.7.3. Ferric Reducing Antioxidant Power (FRAP)

Based on the procedure of Benzie and Strain [35], the antioxidant capacity of the samples was obtained by the spectrometric method. To prepare the FRAP reagent, three solutions were made. The first one was the acetate buffer 300 mM pH 3.6, for this solution sodium acetate trihydrate, glacial acetic acid, and distilled water were mixed. For the second solution, iron tripyridyltriazine (TPTZ), concentrate HCl and distilled water were used. Then the last solution was $FeCl_3 \cdot 6H_2O$ and Milli-Q water. The three solutions mixed in relation to 10:1:1. FRAP reagent was prepared with 100 mL acetate buffer, 10 mL TPTZ, and 10 mL $FeCl_3 \cdot 6H_2O$. These compounds were mixed at 30 °C for 30 min. After the incubation, 100 μL of the sample was added to 3 mL FRAP reagent. Trolox was used as a standard with concentrations of 10 to 200 ppm. Finally, absorbance was measured at 593 nm against a blank of the reagent. All measurements were made in triplicate.

2.7.4. α-α-. diphenyl-β-picrylhydrazyl (DPPH)

Based on the procedure described by Brand-Williams [36]. Around 0.0148 g of DPPH was weighed and placed in a volumetric flask (25 mL). Then the volumetric flask was filled to the mark with methanol (mother solution). 1 mL of the solution before mention was placed in a volumetric flask (25 mL) and filled to the mark with methanol (diluted solution). The absorbance was measured in a spectrophotometer (Model DR 500, Hach Lange GmbH, Düsseldorf, Germany). The readings were taken in 2.5 mL cuvette, mixing 75 μL of sample and 3 mL of DPPH diluted solution. The reaction takes place 16 min after mixing reactants. Then absorbances were read at 515 nm. The calibration curve was made using Trolox with concentrations of 5 ppm to 200 ppm. All measurements were made in triplicate.

2.8. *Minerals Content Analysis*

Mineral content was evaluated with digestion in nitric acid (HNO_3), ICP from Thermo Scientific, model iCAP 6000 Series (Cambridge, England) was used. The following wavelengths (nm) were used, i.e., Ca (396.8), K (766.4), Fe (259.9), Mg (279.5), Mn (257.6), Na (589.5), P (177.4), Cu (324.7), Se (196.0), and Zn (213.8). The digestion of 1 g of sample was carried out during 4 h at a range 85 to 90 °C. The samples were passed through a paper filter (Whatman paper grade 41,110 mm, Item1441110, GE Healthcare Life Sciences, Little Chalfont, UK). The heating was performed with a hot plate (Model Type 2200, Thermo Fisher Scientific, Dubuque, IA, USA). Three samples for each fruit extract were carried out following the same conditions.

2.9. *Cell Lines and In Vitro Cancer Cell Viability*

Normal fibroblast cell line (NIH/3T3), as a control cell line, and four different mammalian cancer cell lines mammary (MCF-7), prostate (PC3), colon (Caco-2) and hepatic (HepG2) were propagated in DMEM-F12 medium containing 10% FBS (Fetal Bovine Serum) (Gibco, Grand Island, NY, USA) and maintained in 5% of CO_2 atmosphere at 37 °C and 80 % of humidity. The cytotoxicity assay was performed in plates of 96-wells each well was prepared with 100 μL of a cell suspension containing 5×10^5 cells/mL of cancer cells (MCF-7, PC3,Caco-2, and HepG2) and NIH/3T3 after 12 h 100 μL a dilution at 4% of filtrated clarified juice extract in Milli Q water were added to each well in the cell giving a final concentration of 2% of juice in the cell growth media in triplicate (Appendix A). The culture medium without cells was used as a blank. After incubation at 37 °C under 5% CO_2 for 48 h, 20 μL Cell Titer 96®AQueous One Solution Cell Proliferation Assay (Promega, Madison, WI, USA) was used to determine % of cell viability. Absorbance was measured at 490 nm in a microplate reader (Synergy HT, Bio-Tek, Winooski, VT, USA). Cell viability was computed using the average absorbance units obtained from the wells and expressed as a percentage of the untreated cells well.

2.10. *Phenolic Analysis by HPLC*

The determination of phenolic compounds in clarified juices extracts were analyzed by a Perkin Elmer HPLC (Altus 10, Waltham, MA, USA) coupled with a UV-Vis detector. A column Eclipse XDBC18, 5 μm, 150 mm × 4.6 mm (Agilent Technologies, Santa Clara, CA, USA) was used for the analysis. A reversed-phase HPLC was performed, and gradient elution was performed by varying the proportion of solvent A (acidified water with acetic acid pH 2.5) to solvent B (methanol), with a flow rate of 0.8 mL/min. The mobile phase composition started at 100 % solvent A for 3 min, followed by an increase of solvent B to 30% 3 to 8 min, 50% B 8 to 15 min, 30% B 15 to 20 min, and then bring mobile phase composition back to the initial conditions. The reference standards were gallic acid, caffeic acid, coumaric acid, and fumaric acid. The calibration curve range was 10 to 80 mg/L. The aforementioned compounds were selected based on the literature of cacti fruits (*Opuntia ficus Indica and Opuntia littoralis*) [37,38].

3. Results and Discussion

3.1. Physiochemical Characterization

As is shown in Table 1, chico fruit averaged weight was 77.0 g, equatorial and polar diameters were 4.9 and 6.2 cm, respectively (Figure 1b). Regarding its epicarp, this was higher than its mesocarp, and the percentage of edible fruit was 36.6%. With respect to jiotilla, its average weight was 18.7 g, the edible fruit of jiotilla was 51.9%, and the equatorial and polar diameters were 2.8 and 3.4 cm, respectively. Table 2 shows the proximate analysis of chico fruit and jiotilla. Table 3 shows the total soluble solids, pH, and antioxidants activity in the fruits under study. The pH values were 4.5 (chico fruit) and 4.17 (jiotilla). Regarding the total soluble solids, they were 12.6 (chico fruit) and 7.4 (jiotilla) °Brix.

Table 1. Physical characterization of the collected plant samples.

Parameter	Chico Fruit	Jiotilla
Equatorial diameter (cm)	4.87 ± 0.28 [1]	2.81 ± 0.29
Polar diameter (cm)	6.25 ± 0.11	3.48 ± 0.36
Fruit weight (g)	77.0 ± 14.5	18.7 ± 3.12
Mesocarp weight (g)	28.0 ± 9.4	9.21 ± 1.95
Epicarp weight (g)	48.4 ± 10.8	8.53 ± 1.53

[1] Results are shown as average ± standard deviation.

Table 2. General proximate analysis of the collected plant samples.

Parameter	Chico Fruit	Jiotilla
Moisture (%)	71.32 ± 0.46 [1]	85.21 ± 0.54
Carbohydrates (%)	18.84 ± 0.29	12.75 ± 0.19
Protein (%)	4.75 ± 0.10	1.025 ± 0.02
Ethereal extract (%)	1.87 ± 0.06	0.48 ± 0.01
Crude fiber (%)	2.52 ± 0.09	1.31 ± 0.04
Ash (%)	0.7 ± 0.03	0.53 ± 0.02

[1] Results are shown as average ± standard deviation.

Table 3. Chemical characterization of the collected plant samples.

Parameter	Chico Fruit	Jiotilla
Total soluble solids °Brix	12.6	7.4
pH	4.51	4.17
Betacyanins (mg/g DW)	1.55 ± 0.05 [1]	2.32 ± 0.23
Betaxanthins (mg/g DW)	1.46 ± 0.05	4.17 ± 0.35
Vitamin C (mg/100g FS) [2]	27.19 ± 1.95	Not detected
Folin-Ciocalteu method (mg/100g FS) [3]	113.16 ± 5.82	83.40 ± 7.32
ABTS (mmol/100 g FS) [4]	0.864 ± 0.020	0.708 ± 0.124
FRAP (mmol/100g FS) [4]	0.27 ± 0.002	0.315 ± 0.003
DPPH (μmol/g FS) [4]	34.52 ± 3.43	57.17 ± 5.28

[1] Results are shown as average ± standard deviation. [2] Ascorbic acid. [3] Gallic acid. [4] Trolox were used in calibration curves. DW: dry weight; FS: fresh sample.

Comparing the edible percentage, the pitaya (*Stenocereus pruinosus*) fruit has a higher edible percentage (72.5%) [39] than chico fruit and jiotilla. Although the chico fruit (36.6%) and jiotilla (51.9%) have a pulp percentage higher than xoconostle (*Opuntia matudae*) (12% to 18%) [40]. The carbohydrates percentages are higher comparing them with the red pitaya (*Stenocereus pruinosus*) (10.2 ± 0.24%) and orange pitaya (*Stenocereus pruinosus*) (8.5 ± 0.16%) [39], but carbohydrates percentages are lower than presented in other fruits such as Banana (*Musa acuminata*) (22.84%) and Mango (*Mangifera indica*) (17%) [41]. The crude fiber in chico fruit and jiotilla was higher than the percentage in red pitaya (*Stenocereus pruinosus*) (0.67 ± 0.09 %), orange pitaya (0.53 ± 0.02 %) (*Stenocereus pruinosus*) [39].

Furthermore, the protein contents are lower than jackalberry tree fruit (*Diospyros mespiliformis*) (9.28 ± 1.14%) [42], but chico and jiotilla protein percentage are higher than apple (*Malus domestica*) (0.26%) [41].

3.2. Antioxidant Capacity of Pulps

Jiotilla (2.32 ± 0.23 mg/g dw) has a higher betacyanin concentration than chico fruit (1.55 ± 0.05 mg/g DW). The concentrations of betaxanthins in chico fruit and jiotilla were 1.46 ± 0.05 and 4.17 ± 0.35 mg/g DW, respectively. The concentration of vitamin C in the chico fruit was 27.19 ± 1.95 mg/100 g FS, while in jiotilla it was not detected. The total antioxidant capacities (Folin-Ciocalteu method) of chico fruit and jiotilla were 113.16 ± 5.82 and 83.40 ± 7.32 mg/100 g FS, respectively. The antioxidant activities determined as ferric reducing-antioxidant power (FRAP) for chico fruit and jiotilla were 0.27 ± 0.002 and 0.315 ± 0.003 mmol/100 g FS, respectively. In chico fruit and jiotilla, the values obtained in ABTS assay were 0.864 ± 0.023 and 0.708 ± 0.124 mmol/100 g FS, respectively. The antioxidant capacity of the fruits understudy was measured by the DPPH assay, which has been employed to measure the antioxidant capacity as well [43]. The free radical scavenging capacities of chico fruit and jiotilla were 34.52 ± 3.43 and 57.17 ± 5.28 mg/100 g FS, respectively. When compared with other cacti fruits, the betacyanins concentrations were lower than the values reported for red pitaya 2.86 ± 0.38 (*Stenocereus pruinosus*) mg/g DW [39] and prickly pear Rojo cenizo (*Opuntia ficus indica*) 5.95 ± 0.21 mg/g DW [44]. Betaxanthins concentration in jiotilla was higher than red pitaya (*Stenocereus Stellatus*) 1.51 ± 0.06 mg/g DW [45] and orange pitaya (*Stenocereus pruinosus*) 2.67 ± 0.27 mg/g DW [39]. Betalains concentration in chico fruit was lower compared with beetroot (*Beta vulgaris*) cultivar little ball in the flesh 3.6 ± 0.2 mg/g DW of betanin and 1.9 ± 0.1 mg/g DW in betaxanthins [46]. Vitamin C concentration in chico fruit, compared with other fruits is lower than guava (*Psidium guajava*) 131 ± 18.2 mg/100 g FS [47]. The fruits for (*Actinidia spp.*) cultivars and citrus species are an excellent source of vitamin C, comparing the (*Actinidia sp.*) species is reported that kiwi (variety Sanuki Gold) has 156 ± 31.2 mg/100 g FS [48] higher than chico fruit concentration. For citrus fruits like lemon (*Citrus limon*), the amount of vitamin C reported is 34 mg/100 g FS [49] and for orange fruit (*Citrus aurantium*) 36.1 mg/100 g FS [47]. It is important to highlight that chico fruits have a similar concentration of vitamin C like lemon juice and orange fruits. When compared total antioxidant capacity (Folin-Ciocalteu method) with some of the fruits recognized to have the highest values, such as raspberry and blackberry, the concentrations were lower. For instance, raspberry (*Rubus idaeus L.*) is reported to present a value of 1489 ± 4.5 mg GAE/100 g FS [50] and for cultivar Aksu Kırmızısı 1040.9 ± 15.9 milligrams of gallic acid per 100 g of FS [51]. Furthermore, a direct comparison of our results with other fruits such as apple red delicious (*Malus domestica*) (73.96 ± 3.52 mg GAE/100 g fw) [52], papaya (*Carcinia papaya Linn*) (54 ± 2.6 mg GAE/100 g FS) [53] and peach (*Prunus persica*) 27.58 ± 1.57 GAE/100 g FS [52] and it is reported for mango (*Mangifera indica L.*) in dry weight 1.64 ± 0.49 mg GAE/g DW [54] (in consideration of this the concentrations of chico fruit and jiotilla were changed to dry weight 9.19 ± 0.47 and 7.09 ± 0.54 mg GAE/g DW, respectively) Jiotilla and chico, represents higher antioxidant activity than fruits mentioned before.

The antioxidant activity determined as FRAP was lower than reported by red rose grape (*Vitis vinifera*) (0.49 ± 0.04 mmol/100 g FS) [55]. Nevertheless, antioxidant activity in chico fruit and jiotilla was higher than the amount in persimmon (*Spyros kaki*) (0.14 ± 0.03 mmol/100 g FS) and duck pear (*Pyrus bretschneideri*) 0.22 ± 0.03 mmol/100 g FS [55]. Dates (*Phoenix dactylifera*) reported amount between 406.61 ± 14.31 and 818.86 ± 21.91 μmol/100 g FS [56], compared with chico fruit and jiotilla. The antioxidant activities are higher in chico fruit (2201.56 ± 22.01 μmol/100 g DW) and jiotilla (2680.91 ± 26.80 μmol/100 g dry weight) than dates (*Phoenix dactylifera*). Comparing the antioxidant activity concentrations (ABTS method) with purple cactus pear (*Opuntia ficus-indica*) (0.61 ± 0.02 mmol/100 g FS) and orange pulp (0.37 ± 0.02 mmol/100 g FS) [6], the antioxidant activities in chico fruit and jiotilla resulted higher than the presented by *Opuntia ficus-indica* pulps. Antioxidant activity compared with fruits from apricot (*Prunus armeniaca*) for the variety Cöloglu, 0.45 ± 0.09 mmol/g

FS and the variety Zerdali, 0.37 ± 0.03 mmol/g FS [57], the chico fruit and jiotilla have higher activity than apricot (*Prunus armeniaca*). Concentration units of chico fruit and jiotilla (mmol/100 g FS) were changed to μmol/g fresh sample to compared with other fruits. Chico fruit (8.642 ± 0.204 μmol/g FS) and jiotilla (7.083 ± 1.247 μmol/g FS) antioxidant activity is higher than banana (*Musa acuminata*) 3.44 ± 0.29 μmol/g FS, apple red delicious 4.62 ± 0.03 μmol/g FS and pear (*Pyrussp.*) 4.30 ± 0.06 μmol/g FS [52]. Nevertheless, guava (*Psidium guajava*) 15.18 ± 0.81 μmol/g fs and sweetsop 23.60 ± 0.06 μmol/g fs [52] have a higher concentration than chico fruit and jiotilla. Antioxidant activity (DPPH method) reported in raspberry for Reveille *(Rubus idaeus L.)* (695.58 ± 11.56 mg/100 g FS) [58] is higher than chico fruit and jiotilla. Nevertheless, chico fruit (2.81 ± 0.26 mg/g DW) and jiotilla (4.86 ± 0.45 mg/g DW) antioxidant activities are lower than powder of wild plum tree (*Prunus domestica subsp. Insititia L.*) (26.47 ± 0.19 mg/g DW) [59]. Regarding raspberry extracts of (*Rubus Idaeus L.*), the antioxidant activity has been reported to be 29.0 ± 1.1 μmol/g FW [60] which represents a higher antioxidant activity than both fruits in the study, chico fruit (1.38 ± 0.13 μmol/g FW) and jiotilla (2.28 ± 0.21 μmol/g FS).

3.3. Mineral Content Analysis

Table 4 shows mineral content. Magnesium is essential for human metabolism; the enzymes use magnesium as a cofactor. The chico fruit and jiotilla have a magnesium content of 102.26 ± 4.24 mg/100 g and 33.12 ± 2.74, respectively. Regarding the potassium content, the chico fruit has 517.75 ± 16.78 mg/100 g. Comparing magnesium content with the almonds (*Prunus dulcis*) (270 mg/100 g), which represents a high magnesium content [61], chico fruit and jiotilla have lower contents in magnesium than almonds. Likewise, comparing with a cactus, *Opuntia spp.*, it is reported for the pulp fruit content of 76 mg/100 g [62], the recommended intake per day is around 400–420 mg/per day [60]. Regarding the potassium content, the chico fruit has 517.75 ± 16.78 mg/100 g, which is a 159 mg/100 g superior to the value presented by the banana (*Musa acuminate*), a fruit with high potassium content, [61]. A comparison with (*Opuntia ficus indica)* fruits authors reported 559 mg of potassium/100 g [62]. It is important to highlight that the recommended intake per day is around 3400 mg potassium/per day [63]. Calcium is important to maintain strong bones, for the calcium content, the chico fruit and jiotilla have 69.99 ± 4.21 and 40.06 ± 1.43 mg/100 g respectively, these values are higher than reported data for banana (*Musa acuminate*) (26 mg/100 g), the pineapple (*Ananas comosus (L.) Merr*) (21 mg/100 g), papaya (*Carcinia papaya Linn*) (16 mg/100 g) and pitahaya (*Hylocereus undatus*) (31 mg/100 g) [64]. Zinc mineral also is important for children's growth, the chico fruit has 2.46 ± 0.65 mg/100 g, comparing this with reported values of different fruits, that the mango (*Mangifera indica*) has 0.14 mg/100 g, the papaya (*Carica papaya L.)* has 0.09 mg/100 g [64]. Zinc content is higher than fruits before mentioned. The zinc concentration in chico fruit is similar to borojo (*Borojoa sorbilis*) (2.47 mg/100 g) [64]. Nevertheless, the content of chico in zinc is lower than camajon fruit (*Sterculia apetala*) fruit 5.70 mg/100 g (our knowledge the fruit with the highest concentration of zinc) [64].

Table 4. Mineral content analysis profile of chico and jiotilla samples.

Parameter	Chico Fruit	Jiotilla
Ca (mg/100 g)	69.99 ± 4.21 [1]	40.06 ± 1.43
K (mg/100 g)	517.75 ± 16.78	295.43 ± 48.91
Fe (mg/100 g)	2.30 ± 0.80	0.635 ± 0.26
Mg (mg/100 g)	102.26 ± 4.84	33.12 ± 2.74
Mn (mg/100 g)	1.07 ± 0.03	0.69 ± 0.06
Na (mg/100 g)	91.28 ± 22.57	1.90 ± 0.27
P (mg/100 g)	57.78 ± 2.01	16.72 ± 1.49
Cu (mg/100 g)	<2.71 ± 0.02	<0.35
Se (mg/100 g)	<1.41 ± 0.01	<0.1825
Zn (mg/100 g)	2.46 ± 0.65	0.38 ± 0.05

[1] Results are shown as average ± standard deviation.

3.4. Antioxidant Capacity of Clarified Juice Extracts

Clarification is a procedure that removes the solids from a liquid, usually a beverage like wine. The concentration of betacyanins in clarified juice of chico fruit and jiotilla was 1.38 ± 0.01 and 2.93 ± 0.04 mg/g DW. Regarding the betaxanthin concentrations, in chico fruit and jiotilla clarified juice extracts were 1.16 ± 0.01 and 2.52 ± 0.35 mg/g DW. The total antioxidant capacity (Folin–Ciocalteu method) in chico fruit and jiotilla clarified juices were 129.47 ± 1.12 and 195.39 ± 7.82 mg/100 g FS. The antioxidant capacities of clarified juices are shown in Table 5 by methods FRAP, ABTS, and DPPH.

Table 5. Antioxidant activity clarified juices extracts.

Parameter	Chico	Jiotilla
Betacyanins (mg/g DW)	1.38 ± 0.01 [1]	2.93 ± 0.04
Betaxanthins (mg/g DW)	1.16 ± 0.01	2.52 ± 0.35
Folin-Ciocalteu method (mg/100g FS) [2]	129.47 ± 1.12	195.39 ± 7.82
FRAP (mmol/100g FS) [3]	0.665 ± 0.018	3.40 ± 0.539
ABTS (mmol/100 g FS) [3]	2.98 ± 0.07	2.46 ± 0.17
DPPH (mg/100 g FS) [3]	80.01 ± 5.10	280.88 ± 7.62

[1] Results are shown as average ± standard deviation. [2] Gallic acid. [3] Trolox were used in calibration curves. DW: dry weight; FS: fresh sample.

In clarification procedure some antioxidant capacity could be lost [65]. This phenomenon has been reported for green tea (*Camellia sinensis*) [66] and apple juice, where the clarification process reduces 2.5 times its antioxidant capacity [67]. However, in this work, the clarification increased antioxidant capacity. Similar results were observed in freeze-dried cherry laurels (*Prunus laurocerasus*) [68]. The increase in antioxidant capacity after centrifugation in liquids derived from vegetal material is probably due to the fact that antioxidants (polyphenols and betalains) are water-soluble and they are released from the pulp. The concentration of betacyanins in clarified juice of jiotilla (345.52 ± 5.04 µg/g FS) was higher than prickly pear juice (*Opuntia robusta*) betacyanins 300.5 ± 8.8 µg/g FS [7]. Contrary to the concentration found in clarified juice of chico fruit (151.6 ± 1.3 µg/g FS). Furthermore, this clarified juice is lower than the concentration reported for juice of *Opuntia rastrera* 152.6 ± 5.4 µg/g FS [7]. Regarding the betaxanthin concentration, in jiotilla clarified juice (296.8 ± 4.2 µg/g FS) is higher than prickly pear juice (*Opuntia robusta*) 189.9 ± 7.3 µg/g FS [7]. Also, chico fruit clarified juice has a higher betaxanthin concentration (127.5 ± 1.2 µg/g FS) than juice of *Opuntia rastrera* (86.2 ± 22.3 µg/g FS) [7]. Finally, the total antioxidant capacity (Folin-Ciocalteu method) in chico fruit (1294.76 ± 0.01 mg/100 g FS) and jiotilla (1953.39 ± 0.0701 mg/100 g FS) clarified juices are higher than prickly pear juice of Duraznillo rojo (*Opuntia leucotricha*) 226.3 ± 26.4 µg GA/g FS [7]. Comparing with other juices from commonly edible fruits, antioxidant activity in jiotilla juice clarified (3406.6 ± 539.6 µmol/100 mL) is higher than pomegranate (*Punica granatum*) 3281.8 µmol/100 mL and aronia (*Aronia melanocarpa*) and 3277.9 µmol/100 mL juice with FRAP method [69].While in chico fruit juice clarified (665.9 ± 18.1 µmol/100 mL) antioxidant activity is higher than orange juice 203.4 µmol/100 mL [69]. Nevertheless, chico fruit (2983.9 ± 73.4 µmol/100 mL) and jiotilla (2462.6 ± 174.7 µmol/100 mL) juices have lower antioxidant activity than pomegranate 4537.3 µmol/100 mL and aronia 4261.8 µmol/100 mL juices in the ABTS method [69]. Chico fruit 319.7 ± 20.4 µmol/100 mL and Jiotilla 1122.2 ± 30.4 µmol/100 mL have lower antioxidant activity than pomegranate (3138.3 µmol/100 mL) juice in DPPH [69]. Clarified juices extract of chico has a higher antioxidant activity than clarified juice extract of jiotilla, as evident by the ABTS method. This method reacts with any hydroxylated aromatics independently of their real antioxidant activity, including OH-groups, which do not contribute to antioxidant activity [69].

3.5. Cytotoxicity of Clarified Juice Extracts

The cytotoxicity of the juice extracts at 2% is shown in Figure 3. NIH/3T3 cell line was used as a control in order to screen the cytotoxicity of the clarified juices extract of jiotilla and chico on normal cells, as demonstrated in Figure 3, they were not cytotoxic in normal cell lines compared to cancer cell lines. The synergetic effect between betalains and phenolic compounds of cacti juice have been reported with benefits such as anticancer [7,8] and anticlastogenic effects [2]. Chico fruit and jiotilla showed a high concentration of betalains and phenolic compounds respect to different fruits. This contain could be responsible for the antioxidant capacity and cytotoxicity. Furthermore, phenolic acids were analyzed in order to identify the antiproliferative compounds. No cytotoxicity was observed in both juices in cell line NIH/3T3, used as a normal cell in order to evaluate the cytotoxicity effect of the juices relative to cancer cells, these values are congruent compared with other cacti fruits such as red pitahaya (*Hylocereus polyrhizus*) in normal human cell lines (HEK-293/human embryonic kidney and TPH-1/hummonocytes in the concentration of 0.39 to 0.78 mg/mL at 48 h) [4], and *Opuntia spp* Cardón (NIH/3T3 in concentration of 0.5% juice at 48 h) [7]. Chico fruit juice (49.7 ± 0.01%) inhibited the CaCo-2 growth and showed similar cell viability as gavia *Opuntia robusta* 52.50 ± 12.60% [7]. Cacti fruits have been tested in other colorectal cancer cell line HT29 and produced cytotoxicity an effective dose values (ED50 5.8 ± 1.0% *v*/*v* at 96 h) [8]. Cytotoxicity of chico fruit juice was observed (45.56 ± 0.05) in MCF-7, the significant effect was compared to *Opuntia rastrera* 75.40 ± 8.26% [7]. Other foods with a high concentration of betalains such as Beta vulgaris extract was cytotoxic in MCF-7 with IC50 value of 70 μg/mL, and cytotoxicity increase with the combination of Beta vulgaris extract and silver nanoparticles (IC50 47.6 μg/mL) [70]. The combination of chico fruit juice and silver nanoparticles could generate the same effect mentioned above. Jiotilla juice showed a higher value of cytotoxicity (47.31 ± 0.03%) in HepG2 than *Opuntia rastrera* 78.90 ± 9.00% [7]. Betanin from beetroot presented a 49% inhibition of HepG2 cell proliferation [71]. Other benefits that cacti juices have shown are hepatoprotective effect in-vitro and in-vivo [3,72]. Jiotilla juice (53.65 ± 0.04%) diminished the cell viability in PC-3 than moradillo *Opuntia violaceae* 61.20 ± 5.30% [7]. Extract of *Beta vulgaris L.* also has cytotoxicity in PC-3 (IC50 316.0 ± 2.1 μg/mL) [73].

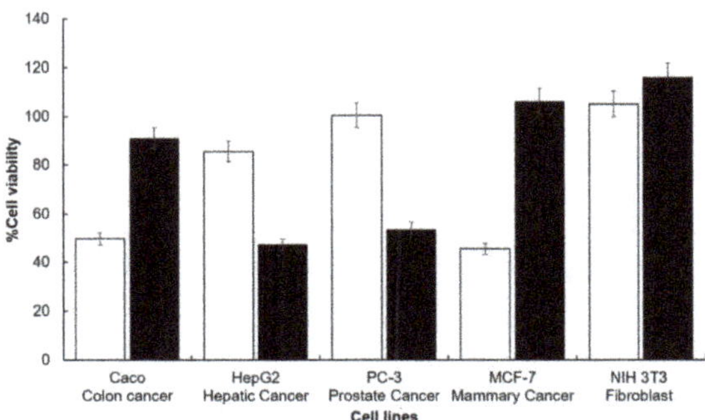

Figure 3. Effect of clarified juices extracts of chico fruit (white) and jiotilla (black) at 2% on the cell viability. The error bars are standard deviations.

3.6. Phenolic Analysis by HPLC

Table 6 shows phenolic acid concentrations of chico fruit and jiotilla that compounds have cytotoxicity in cancer cell lines. *p*-Coumaric acid, a hydroxy derivative of cinnamic acid, is a compound that has a significant antiradical scavenging effect [74]. *p*-Coumaric acid is believed to reduce the risk of stomach cancer by reducing the formation of carcinogenic nitrosamines [75,76]. *p*-Coumaric acid has therapeutic benefits against cancer cell lines (Caco-2) [77]. The *p*-Coumaric acid-induced apoptosis in colon cancer cells (HTC-15) through the ROS-mitochondrial pathway [78]. Furthermore, *p*-Coumaric acid is shown to possess anti-inflammatory, anti-ulcer, anti-cancer, and anti-mutagenic properties [79,80]. Gallic acid, a polyhydroxy phenolic compound that can be found in green tea, grapes, strawberries, and bananas [81]. Gallic acid has demonstrated the potential anticancer activity in vivo and in vitro [82,83]. The anticancer activity of Gallic acid has been reported in various cancer cells, including human ovarian cancer cells (HeLa), leukemia cell lines (C121) [84,85]. It was proven that the anticancer effect of Gallic acid is due to its ability to inhibit cell proliferation and to induce apoptosis [86,87]. Caffeic acid, the primary representative of hydroxycinnamic acids and phenolic acid in general, is widely distributed in plants [88]. Caffeic acid has an action against cervical (HeLa), mammary gland adenocarcinomas (MDA-MB-231), lymphoblastic leukemia (MOLT-3) [88], but it is not cytotoxic to healthy cells [89]. Further, caffeic acid treatment altered the mitochondrial membrane potential on HT-1080 human fibrosarcoma cell line. Other compounds that affect cancer cell lines are the betacyanins, and the main structure is betanin. Kapadia et al. found that betanin has cytotoxic properties in PC-3 cells [73] and Lee et al. observed that the same compound has cytotoxicity in HepG2 [71]. Clarified juice extract of jiotilla shows cytotoxicity in PC-3 and HepG2, this clarified juice has 2.12 times more betacyanins than clarified juice extracts of chico. Vitamin C promotes apoptosis in MCF-7 [90], although the concentration of vitamin C was not analyzed in the clarified juices, it was detected in the chico fruit (27.19 ± 1.95 mg/100 g FS). Probably, it is the main compound in clarified juice extract of chico that produces the cytotoxicity in MCF-7. In this work was identified the main compounds that have been reported in cacti fruits (betalains, p-coumaric acid, caffeic acid, ferulic acid, gallic acid) [37,38,91]. There is a synergy between betalains and phenolic compounds in antioxidant activity and cytotoxicity.

Table 6. Amount of the polyphenolic (mg/100 g fresh sample).

Parameter	Chico Fruit	Jiotilla
p-Coumaric acid	0.32 ± 0.04	N.D.
Gallic acid	0.50 ± 0.01	1.02 ± 0.01
Caffeic acid	0.3 ± 0.00	0.08 ± 0.00
Fumaric acid	N.D.	N.D.

N.D.—Not detected.

4. Conclusions

The presence of a large variety of compounds of high-value in several fruits has been recognized in several studies. In this research, we investigated and demonstrated the nutritional characteristics of two desert fruits, i.e., chico fruit and jiotillain pulp and their clarified juice extracts. Chico fruit was seen to be an excellent resource of vitamin C, potassium, and zinc. Meanwhile, in jiotilla, betacyanins and betaxanthins were the main compounds that deserve to be highlighted, as their concentration was high as compared to earlier reported sources. All these results measured in the pulp of both fruits. On one hand, the determination of the cytotoxicity produced by clarified juice of both fruits in normal cell lines was absent. On the other hand, clarified juice extract of chico fruit showed cytotoxicity in CaCo and MCF-7. Regarding the clarified juice extract of jiotilla, the cytotoxicity activity was showing HepG2 and PC-3 cell lines. For both fruits, in the clarified juice and pulp, the nutritional profiles resulted are high and, in some instances, similar compared with some other and more common edible fruits. Overall, this work increased the knowledge of other different sources of nutrients that can be used to feed the population.

Author Contributions: Conceptualization, L.S.-F., J.R.-R., M.R.-A. and R.P.-S.; data curation, L.S.-F. and J.R.-R.; formal analysis, L.S.-F., E.R.-E., P.R.O. and M.F.M.C.; funding acquisition, R.P.-S.; investigation, L.S.-F., M.R.-A. and E.M.M.-M.; methodology, E.M.M.-M. and C.C.-Z.; project administration, R.P.-S.; resources, C.C.-Z.; supervision, R.P.-S.; validation, L.S.-F. and E.R.-E.; writing—original draft, L.S.-F., E.R.-E., J.R.-R., M.R.-A., E.M.M.-M., C.C.-Z., P.R.O. and M.F.M.C.; writing—review and editing, W.-N.C., H.M.N.I. and R.P.-S. All authors have read and agreed to the published version of the manuscript.

Funding: This research has been supported by Consejo Nacional de Ciencia y Tecnología (CONACYT) Doctoral Fellowship No. 492030 awarded to author L.S.-F. We want to express our profound gratitude to all the support from Opciones de vida para comunidades vulnerables, Water Center, Bioprocess Group, and Synthetic Biology Strategic Focus Group 0821C01004, and Research Chair Funds GIEE 0020209I13 at Tecnologico de Monterrey.

Conflicts of Interest: The authors declare that they have no competing interests.

Appendix A

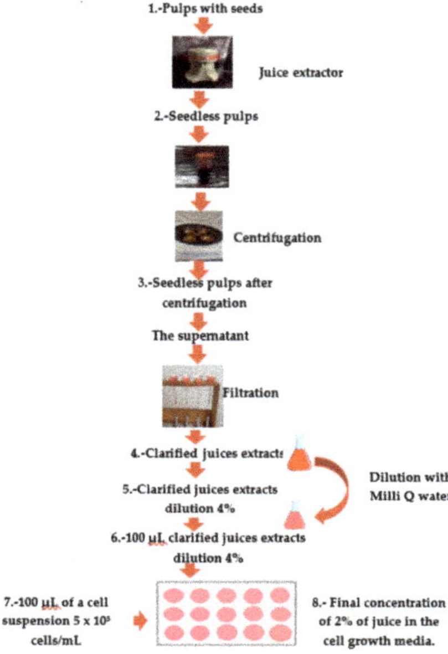

Figure A1. Dilution of clarified juice extract.

References

1. Guzmán, U.; Arias, S.; Dávila, P. *Catálogo de Cactáceas Mexicanas*, 1st ed.; UNAM CONABIO: Mexico City, Mexico, 2003; ISBN 970-9000-20-9.
2. Madrigal-Santillán, E.; García-Melo, F.; Morales-González, J.A.; Vázquez-Alvarado, P.; Muñoz-Juárez, S.; Zuñiga-Pérez, C.; Sumaya-Martínez, M.T.; Madrigal-Bujaidar, E.; Hernández-Ceruelos, A. Antioxidant and anticlastogenic capacity of prickly pear juice. *Nutrients* **2013**, *5*, 4145–4158. [CrossRef] [PubMed]
3. González-Ponce, H.A.; Martínez-Saldaña, M.C.; Rincón-Sánchez, A.R.; Sumaya-Martínez, M.T.; Buist-Homan, M.; Faber, K.N.; Moshage, H.; Jaramillo-Juárez, F. Hepatoprotective effect of Opuntia robusta and Opuntia streptacantha fruits against acetaminophen-induced acute liver damage. *Nutrients* **2016**, *8*, 607. [CrossRef] [PubMed]
4. Yong, Y.Y.; Dykes, G.; Lee, S.M.; Choo, W.S. Effect of refrigerated storage on betacyanin composition, antibacterial activity of red pitahaya (Hylocereus polyrhizus) and cytotoxicity evaluation of betacyanin rich extract on normal human cell lines. *LWT Food Sci. Technol.* **2018**, *91*, 491–497. [CrossRef]
5. Yeddes, N.; Chérif, J.K.; Guyot, S.; Sotin, H.; Ayadi, M.T. Comparative study of antioxidant power, polyphenols, flavonoids and betacyanins of the peel and pulp of three Tunisian Opuntia forms. *Antioxidants* **2013**, *2*, 37–51. [CrossRef] [PubMed]
6. Albano, C.; Negro, C.; Tommasi, N.; Gerardi, C.; Mita, G.; Miceli, A.; De Bellis, L.; Blando, F. Betalains, phenols and antioxidant capacity in cactus pear [Opuntia ficus-indica(L.) Mill.] Fruits from apulia (South Italy) genotypes. *Antioxidants* **2015**, *4*, 269–280. [CrossRef] [PubMed]
7. Chavez-Santoscoy, R.A.; Gutierrez-Uribe, J.A.; Serna-Saldívar, S.O. Phenolic composition, antioxidant capacity and in vitro cancer cell cytotoxicity of nine prickly pear (Opuntia spp.) juices. *Plant Foods Hum. Nutr.* **2009**, *64*, 146–152. [CrossRef] [PubMed]
8. Serra, A.T.; Poejo, J.; Matias, A.A.; Bronze, M.R.; Duarte, C.M.M. Evaluation of Opuntia spp. derived products as antiproliferative agents in human colon cancer cell line (HT29). *Food Res. Int.* **2013**, *54*, 892–901. [CrossRef]

9. Pérez-Negrón, E.; Casas, A. Use, extraction rates and spatial availability of plant resources in the Tehuacán-Cuicatlán Valley, Mexico: The case of Santiago Quiotepec, Oaxaca. *J. Arid Environ.* **2007**, *70*, 356–379. [CrossRef]
10. Valiente-Banuet, A.; Rojas-Martínez, A.; Casas, A.; Arizmendi, M.D.C.; Dávila, P. Pollination biology of two columnar cacti in the tehuacan valley, central Mexico. *J. Trop. Ecol.* **1997**, *84*, 452–455.
11. Llifle Encyclopedias of Living Forms The Encyclopedia of Cacti. Available online: http://www.llifle.com/Encyclopedia/CACTI/Family/Cactaceae/7669/Pachycereus_weberi (accessed on 20 February 2019).
12. Pérez-Negrón, E.; Dávila, P.; Casas, A. Use of columnar cacti in the Tehuacán Valley, Mexico: Perspectives for sustainable management of non-timber forest products. *J. Ethnobiol. Ethnomed.* **2014**, *10*, 1–16. [CrossRef]
13. Ruiz-Huerta, E.A.; Márquez-Guzmán, J.; Pelayo Zaldívar, C.; Barbosa-Martínez, C.; Ponce de León García, L. Escontria chiotilla (Cactaceae): Fruit development, maturation and harvest index. *Fruits* **2015**, *70*, 201–212. [CrossRef]
14. Arellanes, Y.; Casas, A.; Arellanes, A.; Vega, E.; Blancas, J.; Vallejo, M.; Torres, I.; Rangel-Landa, S.; Moreno, A.; Solís, L.; et al. Influence of traditional markets on plant management in the Tehuacán Valley Influence of traditional markets on plant management in the Tehuacán Valley. *J. Ethnobiol. Ethnomed.* **2013**, *9*, 1–15. [CrossRef]
15. Arellano, E.; Casas, A. Morphological variation and domestication of Escontria chiotilla (Cactaceae) under silvicultural management in the Tehuacán Valley, Central Mexico. *Genet. Resour. Crop Evol.* **2003**, *50*, 439–453. [CrossRef]
16. Soriano-Santos, J.; Franco-Zavaleta, M.E.; Pelayo-Zaldivar, C.; Armella-Villalpando, M.A.; Yanez-Lopez, M.L.; Guerrero-Legarreta, I. A partial characterization of the red pigment from the Mexican fruit cactus "jiotilla. " *Rev. Mex. Ing. Química* **2007**, *6*, 19–25.
17. Soto-Castro, D.; Gutiérrez, M.C.; León-Martínez, F.M.; Santiago-García, P.A.; Aragón-Lucero, I.; Antonio-Antonio, F. Spray drying microencapsulation of betalain rich extracts from Escontria chiotilla and Stenocereus queretaroensis fruits using cactus mucilage. *Food Chem.* **2019**, *272*, 715–722.
18. WHO Cancer. Available online: https://www.who.int/news-room/fact-sheets/detail/cancer (accessed on 28 August 2019).
19. Belviranlı, B.; Al-Juhaimi, F.; Özcan, M.M.; Ghafoor, K.; Babiker, E.E.; Alsawmahi, O.N. Effect of location on some physico-chemical properties of prickly pear (Opuntia ficus-indica L.) fruit and seeds. *J. Food Process. Preserv.* **2019**, *43*, 1–9. [CrossRef]
20. Mayer, J.A.; Cushman, J.C. Nutritional and mineral content of prickly pear cactus: A highly water-use efficient forage, fodder and food species. *J. Agron. Crop Sci.* **2019**, *205*, 625–634. [CrossRef]
21. Akhter, S.; Saeed, A.; Irfan, M.; Malik, K.A. In vitro dephytinization and bioavailability of essential minerals in several wheat varieties. *J. Cereal Sci.* **2012**, *56*, 741–746. [CrossRef]
22. Darnton-Hill, I. Zinc Supplementation and Growth in Children. Available online: https://www.who.int/elena/bbc/zinc_stunting/en/ (accessed on 25 May 2019).
23. World Health Organization. Joint Child Malnutrition Estimates. 2019. Available online: http://apps.who.int/gho/tableau-public/tpc-frame.jsp?id=402 (accessed on 19 August 2019).
24. NIH Zinc Fact Sheet for Health Professionals. Available online: https://ods.od.nih.gov/factsheets/Zinc-HealthProfessional/ (accessed on 15 August 2019).
25. WHO Zinc Supplementation to Improve Treatment Outcomes among Children Diagnosed with Respiratory Infections. Available online: https://www.who.int/elena/titles/bbc/zinc_pneumonia_children/en/ (accessed on 18 August 2019).
26. OECD Measuring Well-Being in Mexican States. Available online: http://www.oecd.org/cfe/regional-policy/Mexican-States-Highlights-English.pdf (accessed on 5 August 2019).
27. Association of Official Agricultural Chemists. AOAC Moisture Gravimetric Method. In *AOAC Official Method 934.06*; AOAC International: Rockville, MD, USA, 1990.
28. Association of Official Agricultural Chemists. *AOAC Official Methods of Analysis of AOAC International*; Association of Official Analytical Chemists: Rockville, MD, USA, 1998.
29. Sandate-Flores, L.; Rodríguez-Rodríguez, J.; Calvo-Segura, S.; Mayorga-Martínez, A.; Parra-Saldivar, R.; Chuck-Hernández, C. Evaluation of different methods for betanin quantification in pitaya (Stenocereus spp.). *Agro Food Ind. Hi Tech* **2016**, *27*, 20–25.

30. Stintzing, F.C.; Schieber, A.; Carle, R. Evaluation of colour properties and chemical quality parameters of cactus juices. *Eur. Food Res. Technol.* **2003**, *216*, 303–311. [CrossRef]
31. Santos, J.; Mendiola, J.A.; Oliveira, M.B.P.P.; Ibáñez, E.; Herrero, M. Sequential determination of fat- and water-soluble vitamins in green leafy vegetables during storage. *J. Chromatogr. A* **2012**, *1261*, 179–188. [CrossRef] [PubMed]
32. Everette, J.D.; Bryant, Q.M.; Green, A.M.; Abbey, Y.A.; Wangila, G.W.; Walker, R.B. A thorough study of reactivity of various compound classes towards the Folin-Ciocalteu. *J Agric Food Chem.* **2010**, *58*, 8139–8144. [CrossRef]
33. Singleton, V.L.; Rossi, J.A. Colorimetry of total phenolics with phosphomolybdic-phosphotungstic acid reagents. *Am. J. Enol. Vitic.* **1965**, *16*, 144–158.
34. Re, R.; Pellegrini, N.; Proteggenete, A.; Pannala, A.; Yang, M.; Rice-Evans, C. Antioxidant activity applying an improved ABTS radical cation decolorization assay. *Free Radic. Biol. Med.* **1999**, *26*, 1231–1237. [CrossRef]
35. Benzie, I.F.F.; Strain, J.J. The ferric reducing ability of plasma (FRAP) as a measure of "antioxidant power": The FRAP assay. *Anal. Biochem.* **1996**, *239*, 70–76. [CrossRef] [PubMed]
36. Brand-Williams, W.; Cuvelier, M.E.; Berset, C. Use of a free radical method to evaluate antioxidant activity. *LWT Food Sci. Technol.* **1995**, *28*, 25–30. [CrossRef]
37. Farag, M.A.; Sallam, I.E.; Fekry, M.I.; Zaghloul, S.S.; El-Dine, R.S. Metabolite profiling of three Opuntia ficus-indica fruit cultivars using UPLC-QTOF-MS in relation to their antioxidant potential. *Food Biosci.* **2020**, *36*, 100673. [CrossRef]
38. Abd El-Moaty, H.I.; Sorour, W.A.; Youssef, A.K.; Gouda, H.M. Structural elucidation of phenolic compounds isolated from Opuntia littoralis and their antidiabetic, antimicrobial and cytotoxic activity. *South Afr. J. Bot.* **2020**, *131*, 320–327. [CrossRef]
39. García-Cruz, L.; Valle-Guadarrama, S.; Salinas-Moreno, Y.; Joaquín-Cruz, E. Physical, chemical, and antioxidant activity characterization of pitaya (Stenocereus pruinosus) fruits. *Plant Foods Hum. Nutr.* **2013**, *68*, 403–410. [CrossRef]
40. Guzmán-Maldonado, S.H.; Morales-Montelongo, A.L.; Mondragón-Jacobo, C.; Herrera-Hernández, G.; Guevara-Lara, F.; Reynoso-Camacho, R. Physicochemical, nutritional, and functional characterization of fruits xoconostle (Opuntia matudae) pears from central-México Region. *J. Food Sci.* **2010**, *75*, 485–492. [CrossRef]
41. Mahapatra, A.K.; Mishra, S.; Basak, U.C.; Panda, P.C. Nutrient analysis of some selected wild edible fruits of deciduous forests of India: An explorative study towards non conventional bio-nutrition. *Adv. J. Food Sci. Technol.* **2012**, *4*, 15–21.
42. Hegazy, A.K.; Mohamed, A.A.; Ali, S.I.; Alghamdi, N.M.; Abdel-Rahman, A.M.; Al-Sobeai, S. Chemical ingredients and antioxidant activities of underutilized wild fruits. *Heliyon* **2019**, *5*, 1–8. [CrossRef]
43. Prior, R.L.; Wu, X.; Schaich, K. Standardized methods for the determination of antioxidant capacity and phenolics in foods and dietary supplements. *J. Agric. Food Chem.* **2005**, *53*, 4290–4302. [CrossRef]
44. Jiménez-Aguilar, D.M.; López-Martínez, J.M.; Hernández-Brenes, C.; Gutiérrez-Uribe, J.A.; Welti-Chanes, J. Dietary fiber, phytochemical composition and antioxidant activity of Mexican commercial varieties of cactus pear. *J. Food Compos. Anal.* **2015**, *41*, 66–73. [CrossRef]
45. Pérez-Loredo, M.G.; García-Ochoa, F.; Barragán-Huerta, B.E. Comparative analysis of betalain content in Stenocereus Stellatus fruits and other cactus fruits using principal component analysis. *Int. J. Food Prop.* **2016**, *19*, 326–338. [CrossRef]
46. Kujala, T.S.; Vienola, M.S.; Klika, K.D.; Loponen, J.M.; Pihlaja, K. Betalain and phenolic compositions of four beetroot (Beta vulgaris) cultivars. *Eur. Food Res. Technol.* **2002**, *214*, 505–510. [CrossRef]
47. Leong, L.P.; Shui, G. An investigation of antioxidant capacity of fruits in Singapore markets. *Food Chem.* **2002**, *76*, 69–75. [CrossRef]
48. Vissers, M.C.M.; Carr, A.C.; Pullar, J.M.; Bozonet, S.M. *The Bioavailability of Vitamin C from Kiwifruit*, 1st ed.; Elsevier Inc.: Amsterdam, The Netherlands, 2013; Volume 68, ISBN 9780123942944.
49. Steven, N. Vitamin C contents of citrus fruit and their products: A Review. *J. Agric. Food Chem.* **1980**, *28*, 8–18.
50. Vázquez-Araújo, L.; Chambers, E.; Adhikari, K.; Carbonell-Barrachina, Á.A. Sensory and physicochemical characterization of juices made with pomegranate and blueberries, blackberries, or raspberries. *J. Food Sci.* **2010**, *75*, 398–404. [CrossRef] [PubMed]
51. Sariburun, E.; Saliha, S.; Demir, C.; Türkben, C.; Uylaser, V. Phenolic content and antioxidant activity of raspberry and blackberry cultivars. *J. Food Sci.* **2010**, *75*, 328–335. [CrossRef]

52. Fu, L.; Xu, B.-T.; Xu, X.R.; Gan, R.-Y.; Zhang, Y.; Xia, E.-Q.; Li, H.-B. Antioxidant capacities and total phenolic contents of 62 fruits. *Food Chem.* **2011**, *129*, 345–350. [CrossRef] [PubMed]
53. Patthamakanokporn, O.; Prapasri, P.; Nitithamyong, A.; Sirichakwal, P.P. Changes of antioxidant activity and total phenolic compounds during storage of selected fruits. *J. Food Compos. Anal.* **2008**, *21*, 241–248. [CrossRef]
54. Batista de Medeiros, R.A.; Vieira da Silva Júnior, E.; Fernandes da Silva, J.H.; da Cunha Ferreira Neto, O.; Rupert Brandão, S.C.; Pimenta Barros, Z.M.; Sá da Rocha, O.R.; Azoubel, P.M. Effect of different grape residues polyphenols impregnation techniques in mango. *J. Food Eng.* **2019**, *262*, 1–8. [CrossRef]
55. Guo, C.; Yang, J.; Wei, J.; Li, Y.; Xu, J.; Jiang, Y. Antioxidant activities of peel, pulp and seed fractions of common fruits as determined by FRAP assay. *Nutr. Res.* **2003**, *23*, 1719–1726. [CrossRef]
56. Ramchoun, M.; Alem, C.; Ghafoor, K.; Ennassir, J.; Zegzouti, Y.F. Functional composition and antioxidant activities of eight Moroccan date fruit varieties (Phoenix dactylifera L.). *J. Saudi Soc. Agric. Sci.* **2017**, *16*, 257–264.
57. Güçlü, K.; Altun, M.; Özyürek, M.; Karademir, S.E.; Apak, R. Antioxidant capacity of fresh, sun- and sulphited-dried Malatya apricot (Prunus armeniaca) assayed by CUPRAC, ABTS/TEAC and folin methods. *Int. J. Food Sci. Technol.* **2006**, *41*, 76–85. [CrossRef]
58. Chen, L.; Xin, X.; Zhang, H.; Yuan, Q. Phytochemical properties and antioxidant capacities of commercial raspberry varieties. *J. Funct. Foods* **2013**, *5*, 508–515. [CrossRef]
59. Mousavi, M.; Zaiter, A.; Modarressi, A.; Baudelaire, E.; Dicko, A. The positive impact of a new parting process on antioxidant activity, malic acid and phenolic content of Prunus avium L., Prunus persica L. and Prunus domestica subsp. Insititia L. powders. *Microchem. J.* **2019**, *149*, 1–6. [CrossRef]
60. Mihailovi, N.R.; Mihailovi, V.B.; Ćirić, A.R.; Srećković, N.Z.; Cvijovi, M.R.; Joksovi, L.G. Analysis of wild Raspberries (Rubus idaeus L): Optimization of the ultrasonic-assisted extraction of phenolics and a new insight in phenolics bioaccessibility. *Plant Food Hum. Nutr.* **2019**, *4*, 399–404. [CrossRef]
61. USDA National Nutrient Database for Standard Reference Legacy Release. Available online: https://www.usda.gov/ (accessed on 5 May 2019).
62. Jiménez-Aguilar, D.M.; Mújica-Paz, H.; Welti-Chanes, J. Phytochemical characterization of prickly pear (Opuntia spp) and of its nutritional and functional properties: A Review. *Curr. Nutr. Food Sci.* **2014**, *52*, 57–69. [CrossRef]
63. NIH National Institutes of Health Office of Dietary Supplements. Available online: https://ods.od.nih.gov/factsheets/Magnesium-HealthProfessional/ (accessed on 5 May 2019).
64. Leterme, P.; Buldgen, A.; Estrada, F.; Londoño, A.M. Mineral content of tropical fruits and unconventional foods of the Andes and the rain forest of Colombia. *Food Chem.* **2006**, *95*, 644–652. [CrossRef]
65. Dumitriu, G.D.; López de Lerma, N.; Cotea, V.; Peinado, R.A. Antioxidant activity, phenolic compounds and colour of red wines treated with new fining agents. *Vitis* **2018**, *57*, 61–68.
66. Menezes, M.; Bindes, M.; Luiz, V.; Hespanhol, M.; Reis, M.; Camilla, D. Maximisation of the polyphenols extraction yield from green tea leaves and sequential clarification. *J. Food Eng.* **2019**, *241*, 97–104.
67. Candrawinata, V.; Blades, B.; Golding, J.; Stathopoulos, C.; Roach, P. Effect of clarification on the polyphenolic compound content and antioxidant activity of commercial apple juices. *Int. Food Res. J.* **2012**, *19*, 1055–1061.
68. Cilek Tatar, B.; Sumnu, G.; Oztop, M.; Ayaz, E. Effects of Centrifugation, Encapsulation Method and Different Coating Materials on the Total Antioxidant Activity of the Microcapsules of Powdered Cherry Laurels. *World Acad. Sci. Eng. Technol. Int. J. Nutr. Food Eng.* **2016**, *10*, 901–904.
69. Abountiolas, M.; do Nascimento Nunes, C. Polyphenols, ascorbic acid and antioxidant capacity of commercial nutritional drinks, fruit juices, smoothies and teas. *Int. J. Food Sci. Technol.* **2018**, *53*, 188–198. [CrossRef]
70. Venugopal, K.; Ahmad, H.; Manikandan, E.; Thanigai Arul, K.; Kavitha, K.; Moodley, M.K.; Rajagopal, K.; Balabhaskar, R.; Bhaskar, M. The impact of anticancer activity upon Beta vulgaris extract mediated biosynthesized silver nanoparticles (ag-NPs) against human breast (MCF-7), lung (A549) and pharynx (Hep-2) cancer cell lines. *J. Photochem. Photobiol. B Biol.* **2017**, *173*, 99–107. [CrossRef]
71. Lee, E.J.; An, D.; Nguyen, C.T.T.; Patil, B.S.; Kim, J.; Yoo, K.S. Betalain and betaine composition of greenhouse- or field-produced beetroot (Beta vulgaris L.) and inhibition of HepG2 Cell proliferation. *J. Agric. Food Chem.* **2014**, *62*, 1324–1331. [CrossRef]
72. González-Ponce, H.A.; Rincón-Sánchez, A.R.; Jaramillo-Juárez, F.; Moshage, H. Natural dietary pigments: Potential mediators against hepatic damage induced by over-the-counter non-steroidal anti-inflammatory and analgesic drugs. *Nutrients* **2018**, *10*, 117. [CrossRef]

73. Kapadia, J.G.; Azuine, M.A.; Rao, G.S.; Arai, T.; Iida, A.; Tokuda, H. Cytotoxic effect of the red beetroot (Beta vulgaris L.) extract compared to doxorubicin (Adriamycin) in the Human Prostate (PC-3) and breast (MCF-7) cancer cell lines. *Anticancer. Agents Med. Chem.* **2012**, *11*, 280–284. [CrossRef]
74. Chacko, S.M.; Nevin, K.G.; Dhanyakrishnan, R.; Kumar, B.P. Protective effect of p-coumaric acid against doxorubicin induced toxicity in H9c2 cardiomyoblast cell lines. *Toxicol. Rep.* **2015**, *2*, 1213–1221. [CrossRef]
75. Ferguson, L.R.; Shuo-tun, Z.; Harris, P.J. Antioxidant and antigenotoxic effects of plant cell wall hydroxycinnamic acids in cultured HT-29. *Mol. Nutr. Food Res.* **2005**, *49*, 585–693. [CrossRef]
76. Kikugawa, K.; Hakamada, T.; Hasunuma, M.; Kurechi, T. Reaction of p-hydroxycinnamic acid derivatives with nitrite and its relevance to nitrosamine formation. *J. Agric. Food Chem.* **1983**, *1*, 780–785. [CrossRef]
77. Janicke, B.; Önning, G.; Oredsson, S.M. Differential effects of ferulic acid and p-coumaric acid on S phase distribution and length of S phase in the human colonic cell line Caco-2. *J. Agric. Food Chem.* **2005**, *53*, 6658–6665. [CrossRef] [PubMed]
78. Jaganathan, S.K.; Supriyanto, E.; Mandal, M. Events associated with apoptotic effect of p -Coumaric acid in HCT-15 colon cancer cells. *World J. Gastroenterol.* **2013**, *19*, 7726–7734. [CrossRef]
79. Pragasam, S.J.; Venkatesan, V.; Rasool, M. Immunomodulatory and anti- inflammatory effect of p-coumaric acid: A common dietary polyphenol on experimental inflammation in rats. *Inflammation* **2013**, *36*, 169–176. [CrossRef]
80. Ferguson, L.R.; Lima, I.F.; Pearson, A.E.; Ralph, J.; Harris, P.J. Bacterial antimutagenesis by hydroxycinnamic acids from plant cell walls. *Mutat. Res.* **2003**, *542*, 49–58. [CrossRef]
81. Muhammad, N.; Steele, R.; Isbell, T.S.; Philips, N.; Ray, R.B. Bitter melon extract inhibits breast cancer growth in preclinical model by inducing autophagic cell death. *Oncotarget* **2017**, *8*, 66226–66623. [CrossRef]
82. Bhattacharya, S.; Muhammad, N.; Steele, R.; Peng, G.; Ray, R.B. Immunomodula- tory role of bitter melon extract in inhibition of head and neck squamous cell carcinoma growth. *Oncotarget* **2016**, *7*, 33202–33209. [CrossRef]
83. Sun, J.; Chu, Y.F.; Wu, X.; Liu, R.H. Antioxidant and antiproliferative activities of common Fruits. *J. Agric. Food Chem.* **2002**, *50*, 7449–7454. [CrossRef]
84. De, A.; De, A.; Papasian, C.; Hentges, S.; Banerjee, S.; Haque, I.; Banerjee, S.K. Emblica officinalis extract induces autophagy and inhibits human ovarian cancer cell proliferation, angiogenesis, growth of mouse xenograft tumors. *PLoS ONE* **2013**, *8*, 1–16. [CrossRef]
85. Sourani, Z.; Pourgheysari, B.; Beshkar, P.; Shirzad, H.; Shirzad, M. Gallic acid Inhibits proliferation and induces apoptosis in lymphoblastic leukemia cell line. *Iran. J. Med. Sci.* **2016**, *41*, 525–530.
86. Gonzalez de Mejia, E.; Ramirez-Mares, M.V.; Puangpraphant, S. Bioactive components of tea: Cancer, inflammation and behavior. *Brain. Behav. Immun.* **2009**, *23*, 721–731. [CrossRef]
87. He, Z.; Li, B.; Rankin, G.O.; Rojanasakul, Y.; Chen, Y.C. Selecting bioactive phe- nolic compounds as potential agents to inhibit proliferation and VEGF expression in human ovarian cancer cells. *Oncol. Lett.* **2015**, *9*, 1444–1450. [CrossRef]
88. Touaibia, M.; Doiron, J. Caffeic acid, a versatile pharmacophore: An overview. *Mini-Reviews Med. Chem.* **2011**, *11*, 695–713. [CrossRef]
89. Magnani, C.; Isaac, V.L.B.; Correa, M.A.; Salgado, H.R.N. Caffeic acid: A review of its potential use in medications and cosmetics. *Anal. Methods* **2014**, *6*, 3203–3210. [CrossRef]
90. Sant, D.W.; Mustafi, S.; Gustafson, C.B.; Chen, J.; Slingerland, J.M.; Wang, G. Vitamin C promotes apoptosis in breast cancer cells by increasing TRAIL expressions. *Sci. Rep.* **2018**, *8*, 1–11. [CrossRef]
91. García-Cruz, L.; Santos-Buelgas, C.; Valle-Guadarrama, S.; Salinas-Moreno, Y. Betalains and phenolic compounds profiling and antioxidant capacity of pitaya (Stenocereus spp.) fruit from two species (*S. Pruinosus* and *S. stellatus*). *Food Chem.* **2017**, *234*, 111–118. [CrossRef] [PubMed]

Publisher's Note: MDPI stays neutral with regard to jurisdictional claims in published maps and institutional affiliations.

Article

Euphorbia cuneata Represses LPS-Induced Acute Lung Injury in Mice via Its Antioxidative and Anti-Inflammatory Activities

Hossam M. Abdallah [1,2,*], Dina S. El-Agamy [3,4], Sabrin R. M. Ibrahim [5], Gamal A. Mohamed [1,6], Wael M. Elsaed [7,8], Amjad A. Elghamdi [1], Martin K. Safo [9] and Azizah M. Malebari [10]

1 Department of Natural Products and Alternative Medicine, Faculty of Pharmacy, King Abdulaziz University, Jeddah 21589, Saudi Arabia; gahussein@kau.edu.sa (G.A.M.); aalghamdi2861@stu.kau.edu.sa (A.A.E.)
2 Department of Pharmacognosy, Faculty of Pharmacy, Cairo University, Cairo 11562, Egypt
3 Department of Pharmacology and Toxicology, Faculty of Pharmacy, Mansoura University, Mansoura 35516, Egypt; dagamiabdalla@taibahu.edu.sa
4 Department of Pharmacology and Toxicology, College of Pharmacy, Taibah University, University Street, Al-Madinah Al-Munawarah 30078, Saudi Arabia
5 Department of Pharmacognosy, Faculty of Pharmacy, Assiut University, Assiut 71526, Egypt; sabreen.ibrahim@pharm.au.edu.eg
6 Department of Pharmacognosy, Faculty of Pharmacy, Al-Azhar University, Assiut Branch, Assiut 71524, Egypt
7 Department of Anatomy and Embryology, Faculty of Medicine, Mansoura University, Mansoura 35516, Egypt; wzaarina@taibahu.edu.sa
8 Department of Anatomy and Embryology, College of Medicine, Taibah University, University Street, Al-Madinah Al-Munawarah 30078, Saudi Arabia
9 Department of Medicinal Chemistry, Institute for Structural Biology, Drug Discovery and Development, School of Pharmacy, Virginia Commonwealth University, Richmond, VA 23219, USA; msafo@vcu.edu
10 Department of Pharmaceutical Chemistry, Faculty of Pharmacy, King Abdulaziz University, Jeddah 21589, Saudi Arabia; amelibary@kau.edu.sa
* Correspondence: hmafifi@kau.edu.sa; Tel.: +966 544733110

Received: 14 October 2020; Accepted: 19 November 2020; Published: 21 November 2020

Abstract: *Euphorbia cuneata* (EC; Euphorbiaceae), which widely grows in Saudi Arabia and Yemen, is used traditionally to treat pain and inflammation. This study aimed to evaluate the protective anti-inflammatory effect of a standardized extract of EC against lipopolysaccharide (LPS)-induced acute lung injury (ALI) in mice and the possible underlying mechanism(s) of this pharmacologic activity. ALI was induced in male Balb/c mice using intraperitoneal injection of LPS. A standardized total methanol extract of EC or dexamethasone was administered 5 days prior to LPS challenge. Bronchoalveolar fluid (BALF) and lung samples were collected for analysis. The results demonstrated the protective anti-inflammatory effect of EC against LPS-induced ALI in mice. Standardized EC contained 2*R*-naringenin-7-*O*-β-glucoside (**1**), kaempferol-7-*O*-β-glucoside (**2**), cuneatannin (**3**), quercetin (**4**), and 2*R*-naringenin (**5**) in concentrations of 6.16, 4.80, 51.05, 13.20, and 50.00 mg/g of extract, respectively. EC showed a protective effect against LPS-induced pulmonary damage. EC reduced lung wet/dry weight (W/D) ratio and total protein content in BALF, indicating attenuation of the pulmonary edema. Total and differential cell counts were decreased in EC-treated animals. Histopathological examination confirmed the protective effect of EC, as indicated by an amelioration of LPS-induced lesions in lung tissue. EC also showed a potent anti-oxidative property as it decreased lipid peroxidation and increased the antioxidants in lung tissue. Finally, the anti-inflammatory activity of EC was obvious through its ability to suppress the activation of nuclear factor-κB (NF-κB), and hence its reduction of the levels of downstream inflammatory mediators. In conclusion, these results demonstrate the protective effects of EC against LPS-induced lung injury in mice, which may be due to its antioxidative and anti-inflammatory activities.

Keywords: *Euphorbia cuneata*; Euphorbiaceae; lipopolysaccharide; oxidative stress; NF-κB; lung injury

---

1. Introduction

Acute lung injury (ALI) associated with sepsis is a common clinical problem with a high morbidity rate [1,2]. The pathogenesis of ALI involves disruption of epithelial integrity with massive infiltration of inflammatory cells into the lung tissue, leading to pulmonary edema and severe inflammation [3,4]. Infiltered inflammatory cells, mainly neutrophils and macrophages, release inflammatory mediators such as interleukins (ILs), tumor necrosis factor (TNF)-α, and nitric oxide (NO) [5,6]. Nuclear factor-κB (NF-κB) is a pro-inflammatory transcription factor that regulates and controls inflammatory response during ALI. NF-κB is present in the cytosol in inactive state due to linkage to its inhibitory protein (IκB). Upon stimulation, IκB rapidly degrades to liberate NF-κB, which then migrates into the nucleus where it potentiates gene expression of inflammatory mediators, hence inducing inflammatory responses [7,8].

Lipopolysaccharide (LPS) is bio-active component of the Gram-negative bacterial cell wall [9], and has been used to establish a mouse model of ALI [5]. LPS activates reactive oxygen species (ROS) generation and the release of proteases. Furthermore, LPS induces the activation of NF-κB signaling pathway and cytokine release [10,11]. Inflammatory mediators activate overproduction of ROS, leading to more oxidative damage. Therefore, suppression of oxidative stress and/or inflammation is a potential strategy to improve ALI.

Many people rely on traditional medicine for their primary healthcare, and it is increasingly becoming popular throughout the world. Medicinal plants have also attracted much attention by researchers that are looking for new leads for developing drugs to treat various ailments [12]. Euphorbiaceae is a large plant family comprising about 320 genera with 7950 species, which are distributed mainly in temperate and tropical regions. Some species of this family are of medicinal and economic importance [13]. *Euphorbia* is one of the largest genera of this family, containing approximately 2160 plant species, which are characterized by milky irritant latex [13,14]. Several plants of this genus are used to treat diarrhea, dysentery, gonorrhea, gastric disorders, edema, warts, whooping cough, asthma, and migraine [14–16]. Some of these plants have also been shown to have spasmolytic, diuretic, anti-inflammatory, analgesic, antileukemic, wound healing, hemostatic, and anti-hemorrhoid activities [13,15–17]. This genus is known to possess several phytochemicals, e.g., tannins, phenolic compounds, terpenoids, and flavonoids. *Euphorbia cuneata* Vahl. (EC), one of the plant species in this genus, has various traditional uses in Saudi Arabia and Yemen [15,18,19]. The stem juice when mixed with water or milk is used to treat obesity, food poisoning, and constipation [18]. Moreover, in parts of northern Yemen, the stem juice is used on wound and injuries to reduce external bleeding. It is also used to treat postpartum hemorrhage, and also reported to have analgesic and anti-inflammatory activities [15,18,19]. *E. cuneata* was recently characterized as containing several phytochemicals, including triterpenes and flavonoids [20]. The hemostatic activity of the plant juice is attributed to its flavonoid constituents [15,18]. Unlike several members of the Euphorbiaceae family, *E. cuneata* does not contain the usual toxic diterpenes [21]. On the basis of the traditional anti-inflammatory activity of EC, this study aimed to investigate the possible protective effect of EC against LPS-induced ALI in mice and the possible underlying mechanisms.

2. Materials and Methods

2.1. Chemicals

Acetonitrile, methanol, and formic acid (LC-MS-grade) were obtained from J. T. Baker (Avantor Performance Materials, Radnor, PA, USA). Milli-Q water (Merck Millipore Corporation, Billerica, MA, USA) was used for liquid chromatography analysis. 2*R*-Naringenin, quercetin, cuneatannin, 2*R*-naringenin-7-*O*-*β*-glucoside, and kaempferol-7-*O*-*β*-glucoside were isolated previously from *E. cuneata* [20]. Lipopolysaccharide (LPS; *Escherichia coli* serotype O111:B4) was bought from Sigma-Aldrich (St. Louis, MO, USA) and freshly prepared in normal saline on the challenge day. Ketamine was obtained as ampoules (Tekam, Hikma Pharmaceuticals, Amman, Jordan). Dexamethasone was obtained as ampoules (4 mg/mL, EIPICO, Cairo, Egypt). Other chemicals and reagents were of highest purity. Pierce bicinchoninic acid (BCA) Protein Assay Kit was purchased from Thermofisher Scientific (Waltham, MA, USA). Lactate dehydrogenase (LDH) activity kit was purchased from Human (Wiesbaden, Germany). Malondialdehyde (MDA), catalase, superoxide dismutase (SOD), reduced glutathione (GSH), and total antioxidant capacity (TAC) kits were purchased from Bio-diagnostic Co. (Giza, Egypt). The 4-hydroxynonenal (4-HNE) kit was purchased from My BioSource (San Diego, CA, USA). ELISA kits for NF-κB (ab176648) were purchased form Abcam (Cambridge, MA, USA), while tumor necrosis factor-α (TNF-α, MTA00B), interleukin-1β (IL-1β, MLB00C), and IL-6 (M6000B) were purchased from R&D Systems (Minneapolis, MN, USA).

2.2. Plant Material

E. cuneata aerial parts were collected in March 2016 from Al-Taif City, Saudi Arabia. The plant was kindly identified by a taxonomist at the Department of Natural products and Alternative Medicine, King Abdulaziz University, Saudi Arabia, in addition to its morphological features and the library database [22]. It was confirmed by Dr. Emad Al-Sharif, Associate Professor of Plant Ecology, Department of Biology, Faculty of Science and Arts, Khulais, King Abdulaziz University, Saudi Arabia. A voucher specimen (EC-1036) was archived at the Department of Natural Products and Alternative Medicine herbarium, King Abdulaziz University, Saudi Arabia.

2.3. Extraction Procedures for Pharmacological Study

The air-dried powdered aerial parts of EC (100 g) were extracted with methanol (2 × 500 mL) using an IKA Ultra-Turrax T 25 digital instrument (IKA Labortechnik, Staufen, Germany). The solvent was removed under reduced pressure and the dried total methanolic extract (TEC) (10.9 g) was kept at 4 °C until use in biological tests.

2.4. Extraction Procedures of Plant Material for High Performance Liquid Chromatography Diode-Array Detection

The air-dried powdered aerial parts of EC (1 g) were extracted with methanol (10 mL) as described above. The extract was then vortexed vigorously and centrifuged to remove plant debris. Supernatant was evaporated, and 20 mg of the dry residue (100 mg) was placed on a C18 cartridge preconditioned with methanol and water. The sample was eluted using 3 mL MeOH (100%) and the eluate evaporated. The dry residue was re-suspended in 500 μL methanol, and 3 microliters of the supernatant were used for HPLC analysis.

2.5. HPLC Photodiode Array Determination of Flavonoid Content in TEC

The HPLC system consists of an Agilent 1260 system, solvent delivery module, quaternary pump, autosampler, column compartment, and diode array detector (Agilent Technologies, Germany). The control of the HPLC system and data processing were performed using ChemStation (Rev. B.01.03 SR2 (204)). The separation was performed on Kromasil 100 C18, 5 μm, 250 × 4.6 mm column (Teknokroma, S. Coop. C. Ltd., Barcelona, Spain), maintained at 25 ± 2 °C. The LC system was programmed to deliver the mobile system as follows: Diode array detector (DAD): 284, 330, and 360 nm; **mobile A**: 0.2% formic acid in water; **mobile B**: acetonitrile; gradient elution program: 0–10 min, 5% B; 10–60 min, 5%–38% B; 60–61 min, 38%–100% B; 61–65 min, 100% B; 65–66 min, 5% B; run time, 72 min; flow rate, 0.9 mL/min.

2.6. Calibration Curve of the Isolated Compounds

A weight of 10 mg of each isolated phenolic compound was transferred to a volumetric flask and dissolved in 10 mL methanol. Serial dilutions were prepared in methanol to achieve concentrations of 25, 50, 100, 150, 200, and 250 ng/μL. Each calibration level was analyzed in triplicate. Each compound was injected separately by applying scan mode DAD from 190–500 nm. The UV-VIS scan of each compound was saved and matched with the detected compounds in each sample.

2.7. Biological Study

2.7.1. Animals and Experimental Model

Male Balb/c albino mice (20–25 g) were held in standard conditions in the Animal Facility, College of Pharmacy, King Abdulaziz University, and provided with standard laboratory food and water. All study procedures were approved by the Research Ethical Committee of King Abdulaziz University, Saudi Arabia (reference number PH-116-40), which follows the National Institutes of Health (NIH) guidelines. Mice were divided into 5 groups (n = 8/each group) and were treated as follows: ***control group***: mice were given the vehicle once daily for 5 days; ***LPS group***: to induce acute lung injury, we injected LPS (10 mg/kg) intraperitoneally, as previously described [4]; ***EC + LPS groups***: 2 animal groups that were orally administered EC at 2 different dose levels (25 and 50 mg/kg) for 5 days prior to LPS injection; dexamethasone ***(DEX) + LPS* group**: a positive control group where mice were administered dexamethasone (5 mg/kg) for 5 days prior to LPS injection. The dose of dexamethasone was selected on the basis of previous studies [23,24]

Twenty-four hours after LPS injection, mice were humanely killed under anesthesia using ketamine (50 mg/kg). Right lung was lavaged using 0.9% saline while the left lung was clamped. Bronchoalveolar lavage fluid (BALF) was obtained and centrifuged. Cell pellet was used for the estimation of the cell counts. The supernatants of BALF were stored at −80 °C until further analysis. A small piece of the left lung was weighed, homogenized in phosphate buffer, and centrifuged. The supernatants were stored at −80 °C for further analysis. Another part of the left lung was washed with ice-cold saline and then immersed for 24 h in buffered formalin 10%.

2.7.2. Lung Wet/Dry Weight (W/D) Ratio

W/D ratio is used to estimate the degree of pulmonary edema. It is calculated as the weight of wet piece of the left lung/its weight after drying in an oven (80 °C) for 24 h [4].

2.7.3. Protein Content

Samples of BALF were used for estimation of total protein content according to the manufacturer's kit protocol.

2.7.4. LDH Activity

The LDH activity was determined in BALF samples on the basis of the protocol of the manufacturer kit. In brief, the reaction mixture consisted of nicotinamide adenine dinucleotide phosphate hydrogen (NADPH) (0.8 mmol/L), and sodium pyruvate (1.5 mmol/L) and Tris buffer (50 mmol/L, pH 7.4) was added to the sample. The changes in absorbance were recorded at 340 nm and enzyme activity was calculated and expressed in U/L.

2.7.5. Total and Differential Cell Counts

Cell pellets were resuspended in 0.1 mL sterile saline and then centrifuged onto slides and stained with Wright-Giemsa for 8 min. Total cell counts were determined using a hemocytometer. Differential cell counts were quantified by counting a total of 200 cells per slide at 40 × magnification. Number of each cell type was calculated as the percentage of cell type multiplied by the total number of cells in the BALF.

2.7.6. Lung Histology

Paraffin blocks of lung tissue were obtained from lung samples immersed in formalin, then sectioned (5 μm). Specimens were stained with hematoxylin-eosin (H&E) and examined in random order. Lesions were semi-quantitatively graded as described previously [4].

2.7.7. Immunohistopathology

Immunohistochemistry (IHC) staining was automatically managed using Ventana Benchmark XT system (Ventana Medical Systems, Tucson, AZ, USA). The lung sections were immuno-stained using primary antibodies—rabbit polyclonal antibody to NF-κBp65 following previous procedures [1,25].

2.7.8. Oxidative Stress and Antioxidants

In the supernatants of lung homogenates, lipid peroxidative markers (MDA and 4-HNE) and antioxidants (catalase, SOD, GSH, and TAC) were determined according to instruction of the manufacturer's kits.

Briefly, MDA was quantified by the reaction with thiobarbituric acid in acidic medium at a temperature of 95 °C for 30 min to form thiobarbituric acid-reactive product whose absorption was measured spectrophotometrically at 534 nm. Catalase was determined by its reaction with a known quantity of hydrogen peroxide (H_2O_2). Catalase inhibitor stopped the reaction after 1 min, and then the remaining H_2O_2 reacted with 3,5-dichloro-2-hydroxybenzene sulfonic acid (DHBS) and 4-aminophenazone (AAP) to form a chromophore that was measured at 510 nm. The color intensity was inversely proportional to the amount of catalase in the original sample. Assay of SOD depends on the ability of SOD to inhibit the phenazine methosulphate-mediated reduction of nitroblue tetrazolium dye. The increase in absorbance at 560 nm for 5 min was measured. GSH determination relies on the reaction of GSH with 5,5-dithiobis-2-nitrobenzoic acid. The product was measured spectrophotometrically at 412 nm. The measurement of TAC was performed by the reaction of antioxidants in the sample with a defined amount of H_2O_2. The antioxidants in the supernatant interacted with a specific amount of H_2O_2. The residual H_2O_2 was determined colorimetrically by the conversion of 3,5,dichloro-2-hydroxy benzensulphonate to a colored product that was measured at 510 nm.

2.7.9. NF-κB and Inflammatory Cytokines

Levels of NF-κB, TNF-α, IL-1β, and IL-6 were measured in the supernatants of lung homogenates using ELISA kits.

2.7.10. Statistical Analysis

Presented results are means ± SD ($n = 8$). Statistical analysis was performed using one-way analysis of variance (ANOVA) followed by Tukey's Kramer multiple comparisons test. For non-parametric comparison, Kruskal-Wallis test followed by Dunn's test were used, and a p-value < 0.05 was considered significant.

3. Results

The total methanolic extract (TEC) was standardized for its major phenolic constitutes that were previously isolated [20]. The results showed the presence of 2*R*-naringenin-7-*O*-*β*-glucoside (**1**), kaempferol-7-*O*-β-glucoside (**2**), cuneatannin (**3**), quercetin (**4**), and 2*R*-naringenin (**5**) in concentrations of 6.16, 4.8, 51.05, 13.2, and 50 mg/g of extract, respectively (Figure 1).

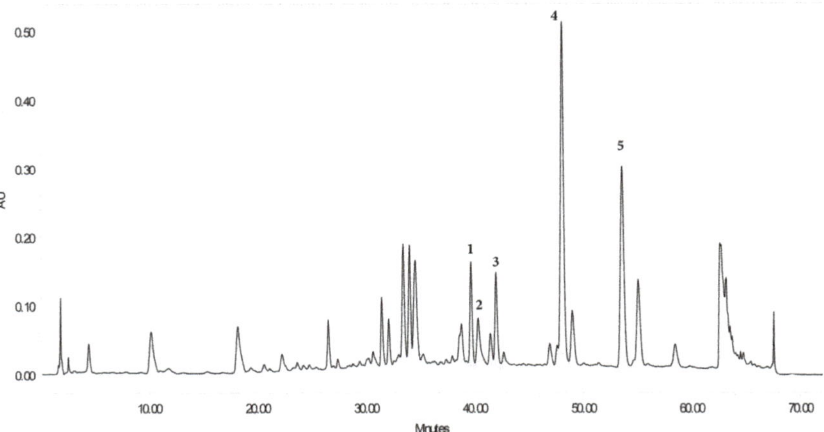

Figure 1. HPLC chromatogram of methanol extract of *Euphorbia cuneata*.

3.1. *Effect of EC on LPS-Induced Lung Edema*

LPS injection to mice resulted in elevation of lung W/D ratio and total protein content of BALF compared to normal mice, indicating development of pulmonary edema (Figure 2). Additionally, LDH activity was remarkably increased in LPS-treated animals compared to the control group. On the contrary, EC pretreatment as well as dexamethasone significantly attenuated W/D ratio, total protein, and LDH activity in comparison with the untreated LPS group. Interestingly, the effect of EC at a high dose was nearly equivalent to the effect of dexamethasone, as there was no significant difference between EC 50 + LPS group compared to the dexamethasone-treated group (Figure 2).

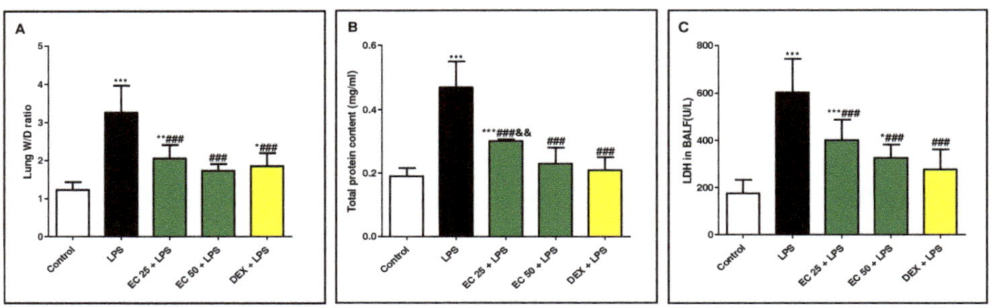

Figure 2. *Euphorbia cuneata* (EC) attenuated lipopolysaccharide (LPS)-induced lung injury. (**A**) Lung wet/dry weight (W/D) ratio, (**B**) protein content, and (**C**) lactate dehydrogenase (LDH) activity in bronchoalveolar lavage fluid (BALF). Mice were administered two different doses of EC (25 and 50 mg/kg) or dexamethasone (5 mg/kg) once daily for 5 days prior to intraperitoneal injection of LPS (10 mg/kg). Samples were collected 24 h after LPS injection. Data are the mean ± SD. (n = 8). * $p < 0.05$, ** $p < 0.01$, *** $p < 0.001$ vs. control group; ### $p < 0.001$ vs. LPS group; && $p < 0.01$ vs. dexamethasone (DEX) + LPS group (one-way ANOVA).

3.2. *Effect of EC on LPS-Induced Increase in the Total and Differential Inflammatory Cell Counts in BALF*

As shown in Figure 3, LPS significantly increased the total and differential cell counts, mainly neutrophils, in the BALF compared to the control group. EC or dexamethasone pretreatment significantly suppressed LPS-induced rise in the total and differential cell counts. The effect of EC at a high dose was not significant compared to that of dexamethasone.

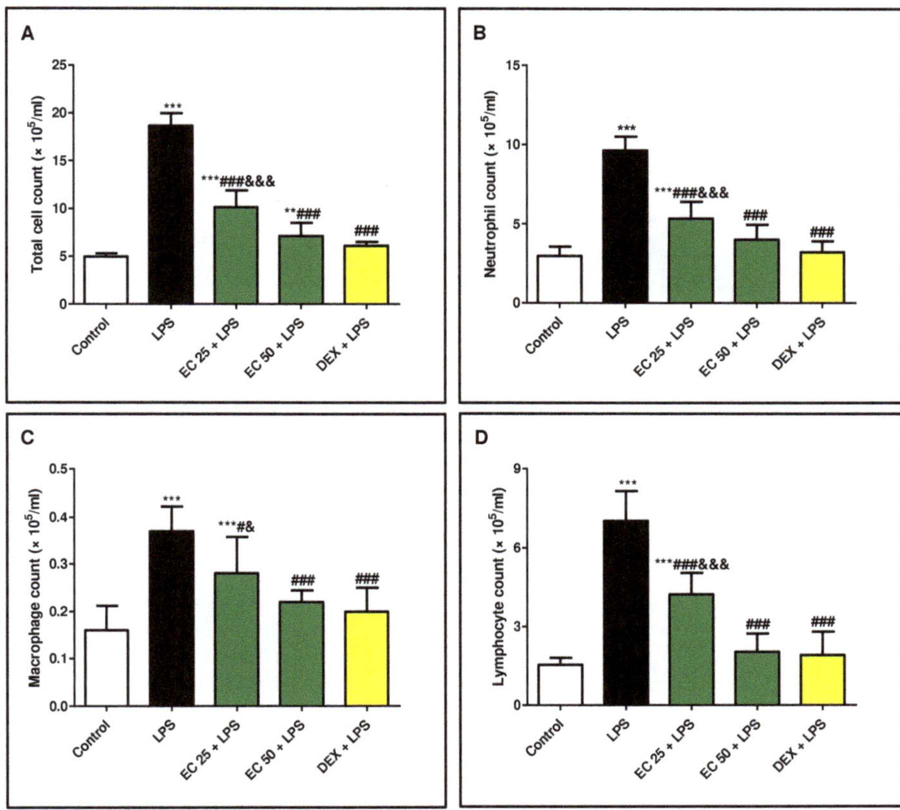

Figure 3. *Euphorbia cuneata* (EC) suppressed lipopolysaccharide (LPS)-induced elevation in total and differential cell counts in bronchoalveolar lavage fluid (BALF). (**A**) Total cell count, (**B**) neutrophil count, (**C**) macrophage count, (**D**) lymphocyte count in lung tissue. Mice were treated with two different doses of EC (25 and 50 mg/kg) or dexamethasone (5 mg/kg) once daily for 5 days prior to intraperitoneal injection of LPS (10 mg/kg). Samples were collected 24 h after LPS injection. Data are the mean ± SD. ($n = 8$). ** $p < 0.01$, *** $p < 0.001$ vs. control group; # $p < 0.05$, ### $p < 0.001$ vs. LPS group; & $p < 0.05$, &&& $p < 0.001$ vs. DEX + LPS group (one-way ANOVA).

3.3. Effect of EC on LPS-Induced Lung Damage

Lung tissue of the control group showed normal histology. There was no sign of lesions in the pulmonary tissue. LPS induced deleterious lung damage in the form of hypertrophied lining epithelium of the pulmonary bronchiole with extravasation of red blood cells (RBCs) and inflammatory cell infiltration in the interalveolar tissue spaces. The thickened inter-alveolar septae were observed with RBC extravasation and excess inflammatory cell infiltration in the interstitial tissue. On the other hand, animals pretreated with EC or dexamethasone exhibited remarkable improvement of the pulmonary lesions compared to the LPS-treated group. Semi-quantitative analysis of LPS-induced lung lesions with regards to the severity and distribution of the lesions indicated significant amelioration of LPS-induced lesions (Figure 4).

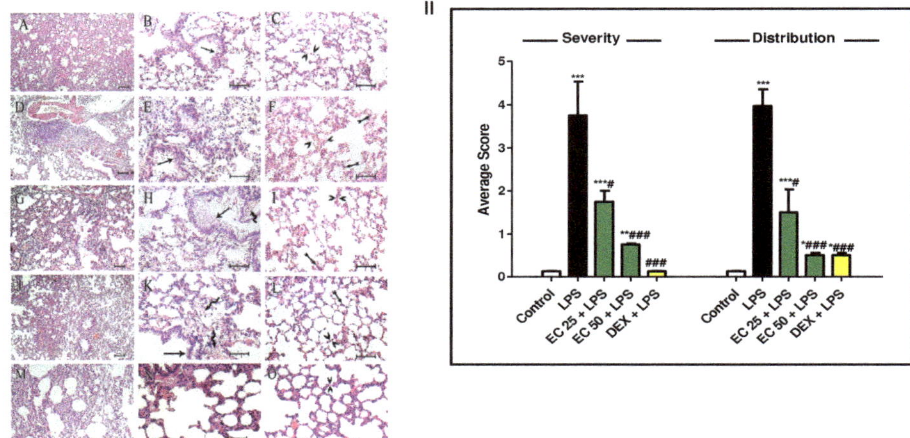

Figure 4. *Euphorbia cuneata* (EC) ameliorated lipopolysaccharide (LPS)-induced histopathological damage of the lung. **I.** Lung specimen of different group stained with hematoxylin-eosin (H&E). (**A–C**) **Control group** where lung specimen displayed normal alveolar bronchioles lined by pseudo-stratified ciliated columnar epithelium (arrow), pulmonary blood vessels, interalveolar septae (between arrow heads), alveolar capillaries, and interstitial tissue. (**D–F**) **LPS** group showing hypertrophied lining epithelium of the pulmonary bronchiole (arrow) with extravasation of red blood cells (RBCs) and inflammatory cell infiltration in the interalveolar tissue spaces, thickened intralveolar septae (arrow heads) with RBCs extravasation, and extensive neutrophil and macrophage infiltration in the interstitial tissue (tailed arrows). (**G–I**) **EC 25 + LPS group,** where the alveolar bronchioles had near normal epithelial lining with interalveolar mucous accumulation (arrow), and lamellae of collagen bundles (curved arrow) is seen close to the bronchiole, with less marked thickened intralveolar septae (arrow heads) with RBC extravasation and scarce neutrophil infiltration in the interstitial tissue (tailed arrow). (**J–L**) **EC 50 + LPS group,** where the alveolar bronchioles had near normal epithelial lining without interalveolar mucous (arrow), and lamellae of collagen bundles (curved arrows) are still seen close to the bronchiole, with no RBCs extravasation nor inflammatory cell infiltration in the interstitial tissue and near normal intralveolar septae (arrow heads) with scarce neutrophil infiltration in the interstitial tissue (tailed arrow). (**M–O**) **DEX + LPS group,** with near normal intralveolar septae (arrow heads) without neutrophil infiltration nor collagen bundle deposition in the interstitial tissue. **II.** Semi-quantitative analysis of LPS-induced lung lesions with regards to the severity and distribution of the lesions. Mice were administered two different doses of EC (25 and 50 mg/kg) or dexamethasone (5 mg/kg) once daily for 5 days prior to intraperitoneal injection of LPS (10 mg/kg). Samples were collected 24 h after LPS injection. Data are the mean ± SD. ($n = 8$). * $p < 0.05$, ** $p < 0.01$, *** $p < 0.001$ vs. control group; # $p < 0.05$, ### $p < 0.001$ vs. LPS group (Kruskal-Wallis).

3.4. Effect of EC on LPS-Induced Lipid Peroxidation and Antioxidants in Lung

LPS injection induced increase in the lipid peroxidative markers, MDA and 4-HNE, in lung in comparison with normal animals (Figure 5A,B). Simultaneously, LPS hindered the antioxidant capacity of the lung due to significant decrease in the endogenous antioxidants such as catalase, SOD, and GSH levels as well as TAC in comparison with the control group (Figure 5C–F). On the other hand, EC or dexamethasone pretreatment significantly augmented the antioxidant activities and diminished the lipid peroxidative parameters in the lung. EC significantly enhanced catalase, SOD, and GSH and significantly decreased MDA and 4-HNE compared to the LPS-treated group. EC at a higher dose exerted a remarkable antioxidant activity compared to dexamethasone.

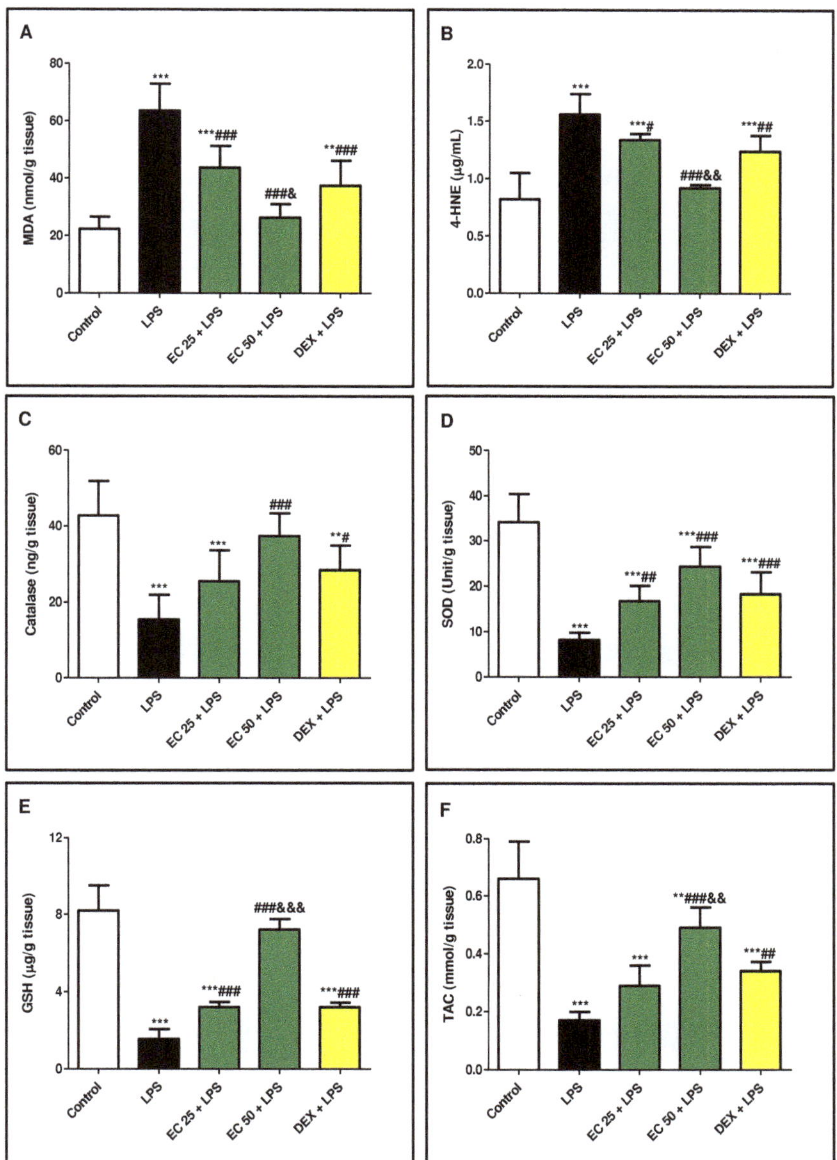

Figure 5. *Euphorbia cuneata* (EC) ameliorated lipopolysaccharide (LPS)-induced lipid peroxidation and increased antioxidant parameters in the lung. (**A**) Malondialdehyde (MDA), (**B**) 4-hydroxynonenal (4-HNE), (**C**) catalase, (**D**) superoxide dismutase (SOD), (**E**) reduced glutathione (GSH), (**F**) total antioxidant capacity (TAC). Mice were treated with two different doses of EC (25 and 50 mg/kg) or dexamethasone (5 mg/kg) once daily for 5 days prior to intraperitoneal injection of LPS (10 mg/kg). Samples were collected 24 h after LPS injection. Parameters were estimated in the supernatants of the lung homogenates. Data are the mean ± SD. (n = 8). ** $p < 0.01$, *** $p < 0.001$ vs. control group; # $p < 0.05$, ## $p < 0.01$, ### $p < 0.001$ vs. LPS group; && $p < 0.01$, &&& $p < 0.001$ vs. DEX + LPS group (one-way ANOVA).

3.5. Effect of EC on LPS-Induced Inflammatory Response in Lung

LPS challenge significantly increased the immuno-expression and the level of NF-κB in the lung in comparison with the control group (Figure 6). In addition, LPS elevated the levels of the inflammatory cytokines TNF-α, IL-1β, and IL-6 (Figure 5II) in the lung compared to the control group. However, EC or dexamethasone reduced immuno-expression and the level of NF-κB simultaneously, with significant reduction in the levels of inflammatory parameters TNF-α, IL-1β, and IL-6 in comparison with the LPS group. Notably, the effect of high dose of EC was not significant from the effect of dexamethasone.

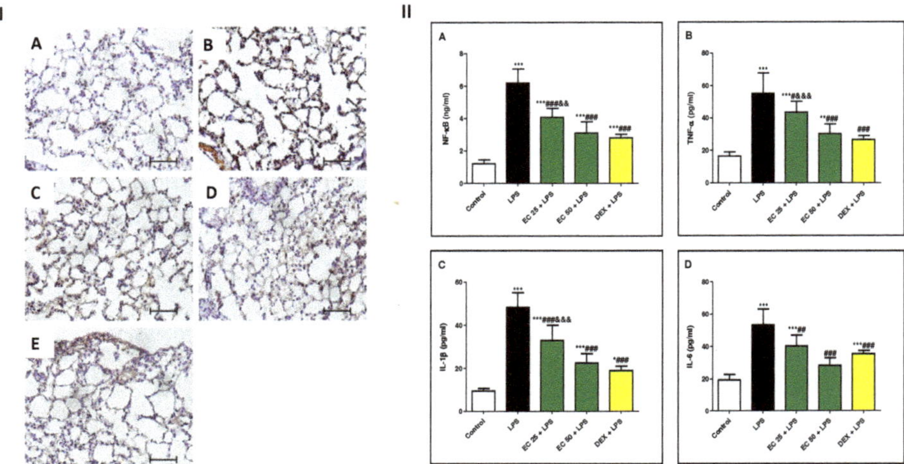

Figure 6. *Euphorbia cuneata* (EC) inhibited lipopolysaccharide (LPS)-induced nuclear factor-κB (NF-κB) activation and cytokine release in lung. **I.** Expression of NF-kB cells in lung tissue determined by immunohistochemistry. (**A**) Control group, the positive NF-kB cells were not observed; (**B**) LPS group, increased expression of NF-kB-positive cells; (**C**) EC 25 + LPS group, there was low staining of NF-kB-positive cells; (**D**) EC 50 + LPS group, very limited expression in the perivascular region, and interstitial lung tissue; (**E**) DEX + LPS group, minor positive NF-kB cells. **II.** Levels of (**A**) NF-κB, (**B**) tumor necrosis factor-α (TNF-α), (**C**) interleukin-1β (IL-1β), (**D**) interleukin-6 (IL-6) in the supernatants of lung homogenates. Mice were treated with two different doses of EC (25 and 50 mg/kg) or dexamethasone (5 mg/kg) once daily for 5 days prior to intraperitoneal injection of LPS (10 mg/kg). Samples were collected 24 h after LPS injection. Data are the mean ± SD. ($n = 8$). * $p < 0.05$, ** $p < 0.01$, *** $p < 0.001$ vs. control group; # $p < 0.05$, ## $p < 0.01$, ### $p < 0.001$ vs. LPS group; && $p < 0.01$, &&& $p < 0.001$ vs. DEX + LPS group (one-way ANOVA).

4. Discussion

ALI is a serious respiratory condition that is characterized by neutrophilia and acute lung inflammation. LPS is an endotoxin derived from Gram-negative bacteria that has been extensively used to establish a model of ALI in rodents. It induces marked pulmonary inflammation after 2–4 h and maximizes at 24–48 h [5]. Thus, in this study, BALF and tissue samples were collected 24 h after LPS exposure. Results of the current study demonstrated the protective antioxidant and anti-inflammatory effects of EC against LPS-induced ALI in mice, which could be related to its ability to modulate the ROS/NF-kB/inflammatory cytokine pathway. The anti-inflammatory efficacy of EC, specifically at the higher dose, was nearly equivalent to that of dexamethasone.

LPS administration results in multiple pathogenic events including massive polymorphonuclear leukocytes (PMN) infiltration in pulmonary tissue, diffuse intravascular coagulation, and profound pulmonary injury [6]. Accumulation of infiltered inflammatory cell in pulmonary tissue exacerbates ALI through the release of multiple toxic mediators including ROS, proteases, and proinflammatory cytokines [4]. Furthermore, recruitment of inflammatory cells contributes to increase of the alveolar-capillary barrier permeability and lung edema. Results of this study were in line with previous research [2,10,26], as LPS induced marked pulmonary edema presented by increased lung W/D ratio. The total protein content in the BALF, as another index of epithelial permeability and pulmonary edema [2], was highly increased in the BALF of mice exposed to LPS. However, EC decreased the lung W/D ratio and the total protein in the BALF, indicating that EC could prohibit the leakage of serous fluid into the lung tissue and attenuate the development of pulmonary edema. Additionally, LDH, as a marker of tissue injury, was increased in BALF upon LPS administration compared with the control group, which was inhibited by EC pretreatment. Inflammatory cell infiltration in the lungs, mainly neutrophil, was noted through increased total and differential cell counts. These biochemical observations were supported by the histopathological examination of the lung, which revealed excessive inflammatory changes such as pulmonary edema, alveolar distortion, and inflammatory cell infiltration lung of LPS-challenged mice. On the other hand, EC attenuated pulmonary edema and inhibited the infiltration of inflammatory cells, as well as restraining the alveolar structural damage. The remarkable attenuation of the biochemical parameters of ALI was parallel to the observed improvement of the histology of the lung in EC-pretreated animals. These findings demonstrated the protective effect of EC on ALI induced by LPS and the ability of EC to prohibit inflammatory cell sequestration and migration into the lung tissue.

Multiple molecular mechanisms mediate the pathogenic events of LPS-induced ALI [27]. Interaction between oxidative stress and inflammation plays a major role in mediating LPS-induced ALI [28]. During the inflammatory response, neutrophils undergo a respiratory burst and produce superoxide. ROS overproduction is extremely toxic to host tissues, and their interactions with various cellular macromolecules result in severe pathophysiological consequences. Excessive LPS-induced ROS release is accompanied by production of lipid peroxides, inactivation of proteins, and DNA mutation [5]. Moreover, LPS-induced oxidative stress is associated with depressed antioxidant activity of the lung, which may aggravate LPS toxicity. Catalase is one of the most important antioxidant enzymes that antagonizes oxidative stress by destroying cellular hydrogen peroxide to produce water and oxygen. GSH acts as a major cellular antioxidant defense system by scavenging free radicals and other ROS. SOD is the only antioxidant enzyme that can scavenge superoxide. Catalase, GSH, and SOD are greatly depressed by LPS [1,2,5]. Our results are consistent with previous studies, as LPS caused marked lipid peroxidation. LPS increased MDA and 4-HNE, which are stable end-products of lipid peroxidation and are frequently used as biomarkers of oxidative stress. Moreover, LPS repressed the activities of the antioxidants (catalase, SOD, GSH, TAC) in lung tissue. Significantly, EC reversed LPS-induced oxidative changes as EC enhanced antioxidant activities and subsequently depressed lipid peroxidation, clearly demonstrating the potent antioxidant activity of EC.

LPS binds and stimulates toll-like receptor 4 (TLR4), which causes activation of NF-κB and subsequent release of inflammatory cytokines (TNFα, IL-1β, and IL-6) [29,30]. Hence, blockade of NF-κB signaling pathways can inhibit the development of ALI induced by LPS. Our study revealed the activation of NF-κB and increased inflammatory cytokines in LPS-challenged mice, consistent with previous studies [4]. Expectedly, pretreatment with EC hindered the activation of this inflammatory pathway, resulting in suppression of inflammatory mediator release. It is worth mentioning that jolkinolides, diterpenoids reported from *Euphorbia* species, have shown a protective effect of LPS-induced ALI via attenuating histological alterations, inflammatory cell infiltration, and lung edema, as well as inhibiting the production of inflammatory mediators, e.g., TNF-α [31,32]. It has also been reported that *Euphorbia* factor L2 improved the survival rate of ALI mice and effectively reduced the pathological changes in the lung by suppression of pro-inflammatory mediators regulated by the NF-κB pathway [33]. As noted above, EC contains several phytochemicals, including quercetin, kaempferol glycoside, and naringenin. Quercetin is known for its antioxidant and anti-inflammatory activity through inhibition of TNF-α, IL-8, IL-4, cyclooxygenase (COX), and lipoxygenase (LOX) [34]. Naringenin has been shown to exhibit anti-inflammatory activity in lung injury in vivo through downregulation of NF-κB, inducing NO synthase, TNF-α, and caspase-3 [35]. Kaempferol glycoside, which is biosynthesized from naringenin, is known for its antioxidant potential, in addition to its anti-inflammatory effect through inhibition of NF-κB, TNF-α, COX, LOX, and expression of IL-1β and IL-8 [36]. Thus, the protective activity of EC against LPS-induced ALI in mice may be attributed to its antioxidant activity due to the presences of the different phenolic constituents.

5. Conclusions

Collectively, this study revealed the potent protective effect of EC against LPS-induced ALI, which may be linked to its antioxidant and anti-inflammatory activities. However, further studies are recommended for better elucidation of the underlying molecular mechanisms of EC.

Author Contributions: Conceptualization, H.M.A.; Data curation, A.M.M.; Funding acquisition, H.M.A.; Methodology, D.S.E.-A., S.R.M.I., G.A.M., W.M.E., A.A.E. and A.M.M.; Writing—original draft, H.M.A., D.S.E.-A., S.R.M.I., G.A.M., W.M.E. and A.M.M.; Writing—review & editing, D.S.E.-A., S.R.M.I., G.A.M., M.K.S. and A.M.M. All authors have read and agreed to the published version of the manuscript.

Funding: This project was funded by the Deanship of Scientific Research (DSR), King Abdulaziz University, Jeddah, under grant no. RG-8-166-38.

Acknowledgments: This project was funded by the Deanship of Scientific Research (DSR), King Abdulaziz University, Jeddah, under grant no. RG-8-166-38. The authors, therefore, acknowledge and thank the DSR's technical and financial support.

Conflicts of Interest: The authors have no conflict of interest to declare.

References

1. El-Agamy, D.S.; Mohamed, G.A.; Ahmed, N.; Elkablawy, M.A.; Elfaky, M.A.; Elsaed, W.M.; Mohamed, S.G.A.; Ibrahim, S.R.M. Protective anti-inflammatory activity of tovophyllin A against acute lung injury and its potential cytotoxicity to epithelial lung and breast carcinomas. *Inflammopharmacology* **2019**, *28*, 153–163. [CrossRef] [PubMed]
2. Ibrahim, S.R.M.; Ahmed, N.; Almalki, S.; Alharbi, N.; El-Agamy, D.S.; Alahmadi, L.A.; Saubr, M.K.; Elkablawy, M.A.; Elshafie, R.M.; Mohamed, G.A.; et al. *Vitex agnus-castus* safeguards the lung against lipopolysaccharide-induced toxicity in mice. *J. Food Biochem.* **2018**, *43*, e12750. [CrossRef] [PubMed]
3. Tseng, T.-L.; Chen, M.-F.; Tsai, M.-J.; Hsu, Y.-H.; Chen, C.-P.; Lee, T.J.F. Oroxylin-A rescues LPS-induced acute lung injury via regulation of NF-κB signaling pathway in rodents. *PLOS ONE* **2012**, *7*, e47403. [CrossRef] [PubMed]
4. Shaaban, A.A.; El-Kashef, D.H.; Hamed, M.F.; El-Agamy, D.S. Protective effect of pristimerin against LPS-induced acute lung injury in mice. *Int. Immunopharmacol.* **2018**, *59*, 31–39. [CrossRef] [PubMed]
5. El-Agamy, D.S. Nilotinib ameliorates lipopolysaccharide-induced acute lung injury in rats. *Toxicol. Appl. Pharmacol.* **2011**, *253*, 153–160. [CrossRef]
6. Guo, S.; Jiang, K.; Wu, H.; Yang, C.; Yang, Y.; Yang, J.; Zhao, G.; Deng, G. Magnoflorine ameliorates lipopolysaccharide-induced acute lung injury via suppressing NF-κB and MAPK activation. *Front. Pharmacol.* **2018**, *9*, 982. [CrossRef]
7. Zhu, T.; Wang, D.-X.; Zhang, W.; Liao, X.-Q.; Guan, X.; Bo, H.; Sun, J.-Y.; Huang, N.-W.; He, J.; Zhang, Y.-K.; et al. Andrographolide protects against LPS-induced acute lung injury by inactivation of NF-κB. *PLOS ONE* **2013**, *8*, e56407. [CrossRef]
8. Lu, R.; Wu, Y.; Guo, H.; Huang, X. Salidroside protects lipopolysaccharide-induced acute lung injury in mice. *Dose-Response* **2016**, *14*, 1559325816678492. [CrossRef]
9. Ahmed, N.; Aljuhani, N.; Salamah, S.; Surrati, H.; El-Agamy, D.S.; Elkablawy, M.A.; Ibrahim, S.R.M.; Mohamed, G.A. Pulicaria petiolaris effectively attenuates lipopolysaccharide (LPS)-induced acute lung injury in mice. *Arch. Biol. Sci.* **2018**, *70*, 699–706.
10. Wu, H.; Yang, Y.; Guo, S.; Yang, J.; Jiang, K.; Zhao, G.; Qiu, C.; Deng, G. Nuciferine ameliorates inflammatory responses by inhibiting the TLR4-mediated pathway in lipopolysaccharide-induced acute lung injury. *Front. Pharmacol.* **2017**, *8*, 939. [CrossRef]

11. Zhang, H.; Chen, S.; Zeng, M.; Lin, D.; Wang, Y.; Wen, X.; Xu, C.; Yang, L.; Fan, X.; Gong, Y.; et al. Apelin-13 administration protects against LPS-induced acute lung injury by inhibiting NF-κB pathway and NLRP3 inflammasome activation. *Cell. Physiol. Biochem.* **2018**, *49*, 1918–1932. [CrossRef] [PubMed]
12. Mohamed, G.A.; Al-Abd, A.M.; El-Halawany, A.M.; Abdallah, H.M.; Ibrahim, S.R.M. New xanthones and cytotoxic constituents from *Garcinia mangostana* fruit hulls against human hepatocellular, breast, and colorectal cancer cell lines. *J. Ethnopharmacol.* **2017**, *198*, 302–312. [CrossRef] [PubMed]
13. Vasas, A.; Hohmann, J. Euphorbia diterpenes: Isolation, structure, biological activity, and synthesis (2008–2012). *Chem. Rev.* **2014**, *114*, 8579–8612. [CrossRef] [PubMed]
14. Rahman, A.; Akter, M. Taxonomy and medicinal uses of Euphorbiaceae (Spurge) family of Rajshahi, Bangladesh. *Res. Plant Sci.* **2013**, *1*, 74–80.
15. Awaad, A.S.; Al-Jaber, N.A.; Moses, J.E.; El-Meligy, R.M.; E Zain, M. Antiulcerogenic activities of the extracts and isolated flavonoids of *Euphorbia cuneata* Vahl. *Phytother. Res.* **2012**, *27*, 126–130. [CrossRef] [PubMed]
16. Awaad, S.A.; Alothman, M.R.; Zain, Y.M.; Alqasoumi, S.I.; Alothman, E.A. Quantitative and qualitative analysis for standardization of *Euphorbia cuneata* Vahl. *Saudi Pharm. J.* **2017**, *25*, 1175–1178. [CrossRef]
17. Das, B.; Alam, S.; Bhattacharjee, R.; Das, B.K. Analgesic and anti-inflammatory activity of *Euphorbia antiquorum* Linn. *Am. J. Pharmacol. Toxicol.* **2015**, *10*, 46–55. [CrossRef]
18. Al-Fatimi, M. Ethnobotanical survey of medicinal plants in central Abyan governorate, Yemen. *J. Ethnopharmacol.* **2019**, *241*, 111973. [CrossRef]
19. Zain, M.E.; Awaad, A.S.; Al-Outhman, M.R.; El-Meligy, R.M. Antimicrobial activities of Saudi Arabian desert plants. *Phytopharmacology* **2012**, *2*, 106–113.
20. Elghamdi, A.A.; Abdallah, H.M.; Shehata, I.A.; Mohamed, G.A.; Shati, A.A.; Alfaifi, M.Y.; Elbehairi, S.E.I.; Koshak, A.E.; Ibrahim, S.R.M. Cyclocuneatol and cuneatannin, new cycloartane triterpenoid and ellagitannin glycoside from *Euphorbia cuneata*. *ChemistrySelect* **2019**, *4*, 12375–12379. [CrossRef]
21. Ghazanfar, S.A. *Handbook of Arabian Medicinal Plants*; CRC Press: Boca Raton, FL, USA; London, UK; New York, NY, USA, 1994.
22. Collenette, S. *Wildflowers of Saudi Arabia; National Commission for Wildlife Conservation and Development*; NCWCD: Riyadh, Saudi Arabia, 1999; p. 167.
23. Al-Harbi, N.O.; Imam, F.; Al-Harbi, M.M.; Ansari, M.A.; Zoheir, K.M.A.; Korashy, H.M.; Sayed-Ahmed, M.M.; Attia, S.M.; Shabanah, O.A.; Ahmad, A.M. Dexamethasone attenuates LPS-induced acute lung injury through inhibition of NF-κB, COX-2, and pro-inflammatory mediators. *Immunol. Investig.* **2016**, *45*, 349–369. [CrossRef] [PubMed]
24. Zhao, M.; Du, J. Anti-inflammatory and protective effects of D-carvone on lipopolysaccharide (LPS)-induced acute lung injury in mice. *J. King Saud Univ. Sci.* **2020**, *32*, 1592–1596. [CrossRef]
25. El-Kholy, A.A.; Elkablawy, M.A.; El-Agamy, D.S. Lutein mitigates cyclophosphamide induced lung and liver injury via NF-κB/MAPK dependent mechanism. *Biomed. Pharmacother.* **2017**, *92*, 519–527. [CrossRef] [PubMed]
26. Wu, K.; Xiu, Y.; Zhou, P.; Qiu, Y.; Li, Y.-H. A New Use for an Old Drug: Carmofur attenuates lipopolysaccharide (LPS)-induced acute lung injury via inhibition of FAAH and NAAA activities. *Front. Pharmacol.* **2019**, *10*, 818. [CrossRef] [PubMed]
27. Liu, H.; Yu, X.; Yu, S.; Kou, J. Molecular mechanisms in lipopolysaccharide-induced pulmonary endothelial barrier dysfunction. *Int. Immunopharmacol.* **2015**, *29*, 937–946. [CrossRef] [PubMed]
28. Liu, Q.; Ci, X.; Wen, Z.; Peng, L. Diosmetin alleviates lipopolysaccharide-induced acute lung injury through activating the Nrf2 pathway and inhibiting the NLRP3 inflammasome. *Biomol. Ther.* **2018**, *26*, 157–166. [CrossRef]
29. Yang, J.; Li, S.; Wang, L.; Du, F.; Zhou, X.; Song, Q.; Zhao, J.; Fang, R. Ginsenoside Rg3 attenuates lipopolysaccharide-induced acute lung injury via MerTK-dependent activation of the PI3K/AKT/mTOR pathway. *Front. Pharmacol.* **2018**, *9*, 850. [CrossRef]
30. Lu, Y.; Xu, D.; Liu, J.; Gu, L. Protective effect of sophocarpine on lipopolysaccharide-induced acute lung injury in mice. *Int. Immunopharmacol.* **2019**, *70*, 180–186. [CrossRef]
31. Yang, H.; Li, Y.; Huo, P.; Li, X.-O.; Kong, D.; Mu, W.; Fang, W.; Li, L.; Liu, N.; Fang, L.; et al. Protective effect of Jolkinolide B on LPS-induced mouse acute lung injury. *Int. Immunopharmacol.* **2015**, *26*, 119–124. [CrossRef]

32. Uto, T.; Qin, G.-W.; Morinaga, O.; Shoyama, Y. 17-Hydroxy-jolkinolide B, a diterpenoid from *Euphorbia fischeriana*, inhibits inflammatory mediators but activates heme oxygenase-1 expression in lipopolysaccharide-stimulated murine macrophages. *Int. Immunopharmacol.* **2012**, *12*, 101–109. [CrossRef]
33. Zhang, Q.; Zhu, S.; Cheng, X.; Lu, C.; Tao, W.; Zhang, Y.; William, B.C.; Cao, X.; Yi, S.; Liu, Y.; et al. Euphorbia factor L2 alleviates lipopolysaccharide-induced acute lung injury and inflammation in mice through the suppression of NF-κB activation. *Biochem. Pharmacol.* **2018**, *155*, 444–454. [CrossRef] [PubMed]
34. Li, Y.; Yao, J.; Han, C.; Yang, J.; Chaudhry, M.T.; Wang, S.; Liu, H.; Yin, Y. Quercetin, inflammation and immunity. *Nutrients* **2016**, *8*, 167. [CrossRef] [PubMed]
35. Nguyen-Ngo, C.; Willcox, J.C.; Lappas, M. Anti-Diabetic, anti-Inflammatory, and anti-Oxidant effects of naringenin in an in vitro human model and an in vivo murine model of gestational diabetes mellitus. *Mol. Nutr. Food Res.* **2019**, *63*, e1900224. [CrossRef] [PubMed]
36. Calderon-Montano, J.M.; Burgos-Moron, E.; Perez-Guerrero, C.; Lopez-Lazaro, M. A Review on the dietary flavonoid kaempferol. *Mini Rev. Med. Chem.* **2011**, *11*, 298–344. [CrossRef] [PubMed]

Publisher's Note: MDPI stays neutral with regard to jurisdictional claims in published maps and institutional affiliations.

Article

Biomolecule from *Trigonella stellata* from Saudi Flora to Suppress Osteoporosis via Osteostromal Regulations

Hairul-Islam Mohamed Ibrahim [1,2,*], Hossam M. Darrag [3,4], Mohammed Refdan Alhajhoj [5] and Hany Ezzat Khalil [6,7]

1 Biological Sciences Department, College of Science, King Faisal University, Al-Ahsa 31982, Saudi Arabia
2 Pondicherry Centre for Biological Sciences and Educational Trust, Kottakuppam 605104, India
3 Research and Training Station, King Faisal University, Al-Ahsa 31982, Saudi Arabia; hdarag@kfu.edu.sa
4 Pesticide Chemistry and Technology Department, Faculty of Agriculture, Alexandria University, Alexandria 21545, Egypt
5 Arid Land Agriculture Department, College of agricultural and Food Sciences, King Faisal University, Al-Ahsa 31982, Saudi Arabia; malhajhoj@kfu.edu.sa
6 Department of Pharmaceutical Sciences, College of Clinical Pharmacy, King Faisal University, Al-Ahsa 31982, Saudi Arabia; heahmed@kfu.edu.sa
7 Department of Pharmacognosy, Faculty of Pharmacy, Minia University, Minia 61519, Egypt
* Correspondence: himohamed@kfu.edu.sa; Tel.: +966-559502963

Received: 7 October 2020; Accepted: 2 November 2020; Published: 20 November 2020

Abstract: *Trigonella stellata* has used in folk medicine as palatable and nutraceutical herb. It also regulates hypocholesterolemia, hypoglycemia, and has showed anti-inflammatory activities as well as antioxidants efficacy. Osteoporosis is a one of bone metabolic disorders and is continuously increasing worldwide. In the present study, caffeic acid was isolated from *Trigonella stellata* and identified using 1 D- and 2 D-NMR spectroscopic data. Caffeic acid was investigated on osteoblast and osteoclast in vitro using mice bone marrow-derived mesenchymal cells. Caffeic acid played reciprocal proliferation between osteoblast and osteoclast cells and accelerated the bone mineralization. It was confirmed by cytotoxicity, alkaline phosphatase (ALP), alizarin red S (ARS), and Tartrate resistant acid phosphatase (TRAP) assay. Caffeic acid regulated the osteogenic marker and upregulated the osteopontin, osteocalcin, and bone morphogenic proteins (BMP). Quantitative real time PCR and Western blot were used to quantify the mRNA and protein markers. It also regulated the matrix metalloprotease-2 (MMP-2) and cathepsin-K proteolytic markers in osteoclast cells. In addition, caffeic acid inhibited bone resorption in osteoclast cells. On the other hand, it upregulate osteoblast differentiation through stimulation of extracellular calcium concentrations osteoblast differentiation, respectively. The results also were confirmed through in silico docking of caffeic acid against cathepsin-B and cathepsin-K markers. These findings revealed that caffeic acid has a potential role in bone-metabolic disorder through its multifaceted effects on osteoblast and osteoclast regulations and controls osteoporosis.

Keywords: *Trigonella stellata*; caffeic acid; osteoporosis; osteoblast; osteoclast; BMP

---

1. Introduction

Osteoporosis is a metabolic bone disease characterized by low bone mass, imbalanced bone cell types that leads to osteoarthritis (OA). Bone is a metabolically active connective tissue, ability to regenerate from the incidence and accident of fractures [1]. Bone reabsorption and osteogenic formation tends the remodeling of bone loss and balancing the bone stereotypes. Osteoclast types of cells efficiently absorb the damaged bone cells and promote osteoconductivity, as in-growth around the bone; induction of osteogenic response promote progenitor cell differentiation and mineral storage in osteoblastic lineages. Balances between osteoblast and osteoclast would be happened by bone marrow derived cells. Bone-marrow derived mesenchymal stem cells (BM-MSCs) are multipotent, differentiated into connective tissue, and develop into mature osteoblasts [2]. Mesenchymal cells directly involved in extracellular matrix composition, mineralization, and coordinate the differentiation of osteostromal cells [3]. Osteoclasts are generated from precursor cells in presence of receptor activator of nuclear factor kappa-B ligand (RANKL) [4]. These inductions activate the inflammatory mediators which given back bone loss. Bone morphogenetic protein (BMP) and IL-10 are examples of osteogenic markers widely used in differentiation of mesenchymal stem cells into osteoblasts. Mitogen activated protein kinase (MAPK) regulates the osteoblast-specific transcription factors for differentiation process [5].

Simple analgesics are now recognized as one of the first line pharmacological treatment of uncomplicated OA. Whereas some non-steroidal anti-inflammatory drugs (NSAIDS) may show some fatal adverse effects particularly if used in long term treatment plans.

Natural products have been recently considered as a source of important therapeutic candidates, that could treat various diseases and considered effective for maintaining good health [6]. The interest in drugs derived from plants is predominantly attributed to the trust that green medicine is safe and dependable in comparison to the synthetic one. Wide ranges of natural candidates have implemented in treatment of chronic and infectious diseases [7].

Currently, a plethora of agents are available for the treatment of inflammatory disorders including OA, but some of the drugs are associated with risk of life-threatening adverse effects leading to its withdrawal from market [8]. Hence, the management of inflammatory disorders using medicines without side effects is still a challenge. In the last decades, hundreds of reports were published regarding the anti-inflammatory activities of plants that were available for alternative therapy [9]. Particularly, phenolic bioactive constituents showed significant attention due to their modulatory activities on inflammasomes [10]. Therefore, it is highly required to find out herbal products or nutraceuticals, which can be used as add-on therapy for long-term management of inflammatory disorders.

Studies have proven that leguminous plants may act as reservoirs of potential secondary metabolites of diverse therapeutic utilities and can produce an anti-inflammatory effect as well as they have significant nutritional value [11,12]. In this aspect, there is a consumption of *Trigonella* species (member leguminous herbs) because of their nutritional value. Particularly, *Trigonella foenum-graecum*, which is a small plant with several benefits, attributed to the diverse array of phytoconstituents such as phenolics and flavonoids [13]. Traditionally *Trigonella foenum-graecum* is reputed to exert anticancer, antidiabetic, antioxidant, antihyperlipidemic, and other various pharmacological effects [14].

Trigonella stellata (*T. stellate*) (Leguminosae) is a member of the genus *Trigonella* that is native and very common to grow wildly in Arabian Peninsula including Saudi Arabia [15]. Traditionally, *T. stellata* is a palatable herb [16] and used to treat abdominal pain, diarrhea, dysentery [17], and as nutraceutical herb [18]. Recent studies reported that *T. stellata* contains isoflavans and saponins and showed antidiabetic and anti-hyperlipidemic activities [19,20].

No previous reports have been performed to illustrate the efficacy of *T. stellata* as anti-erosion agent in vitro and in vivo platforms. Therefore, this study was aimed to investigate the *T. stellata* for the first time for such activities.

2. Results

2.1. *Isolation and Identification of Major Compound*

The methanol extract of shade dried aerial parts (25.0 g) was subjected to several and repeated chromatographic techniques to give the pure phenolic compound; Caffeic acid (CAF) (23.6 mg) [21] (Figure 1). The structure was elucidated by inspection of 1 D- and 2 D-NMR spectroscopic data (Table 1 and Supplementary Material) and compared with literature values [21]. This study represents the first report on the isolation of CAF from *T. stellata*.

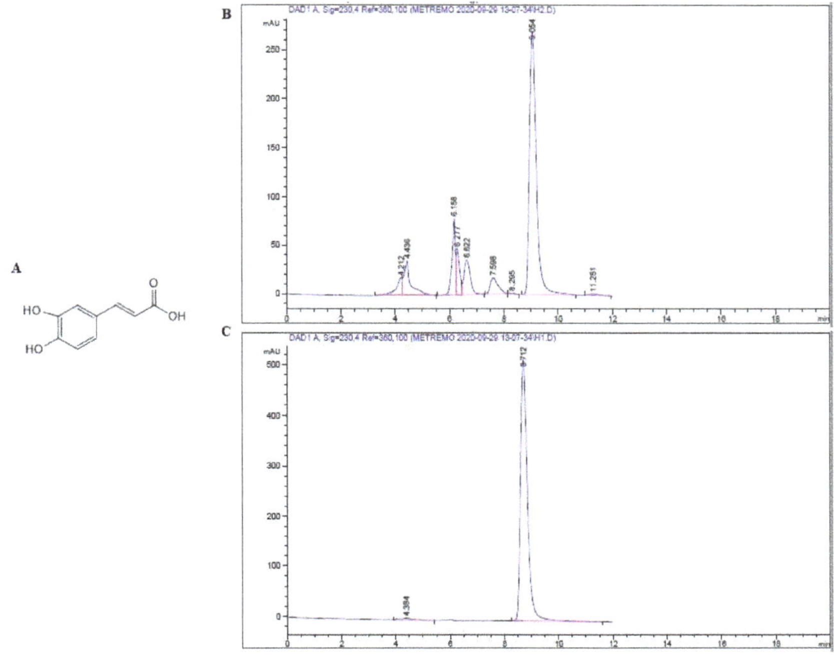

Figure 1. Structure of caffeic acid (CAF) isolated from *T. stellata* (**A**). HPLC chromatogram of SubFr. 2-4-4-1 (**B**). HPLC chromatogram of collected pure CAF (**C**).

Table 1. NMR spectroscopic data of caffeic acid in CD_3OD.

Position	δC	δH, mult. (J in Hz)
1	127.82	-
2	115.10	7.06, d, (2.04)
3	146.82	-
4	149.47	-
5	116.51	6.80, d, (8.16)
6	122.88	6.95, dd, (2.04, 8.16)
7	147.06	7.55, d, (15.88)
8	115.55	6.23, d, (15.88)
9	171.08	-

δC: chemical shift in ppm for ^{13}C-NMR, δH: chemical shift in ppm for ^{1}H-NMR; mult.: multiplicity; J in Hz: coupling constants in Hz; d: doublet, dd: doublet of doublet.

2.2. *Effect of* T. Stellata *Extract Against Mesenchymal Cells*

The biocompatibility of *T. stellata* crude extract (TCE) against murine bone marrow derived mesenchymal cells. Initially, the TCE was evaluated using BM-MSCs viability assay at concentrations (10, 20, 50, 100, 250, and 500 µg/mL) by 3-(4,5-dimethylthiazol-2-yl)-2,5-diphenyltetrazolium bromide (MTT) method (Figure 2A) and neutral red (NR) assay (Figure 2B). This examination revealed that the TCE metabolites play a major role in cell biocompatible. The major moiety isolated from TCE was CAF. The selection of biocompatible concentrations of CAF was carried out using MTT assay. Cell viability was not inhibited up to 50 µg/mL of TCE concentration. The CAF at concentration >10 µM decreases the cell viability with an IC50 with >10 µM concentration. The neutral red assay estimates the uptake of colour from viable cells. NR measurements of the total viable cells was correlated with MTT assay. The NR results revealed that, insignificant cell death was observed at all the tested concentrations. Even at 10 µM CAF treatment showed <25% cell death.

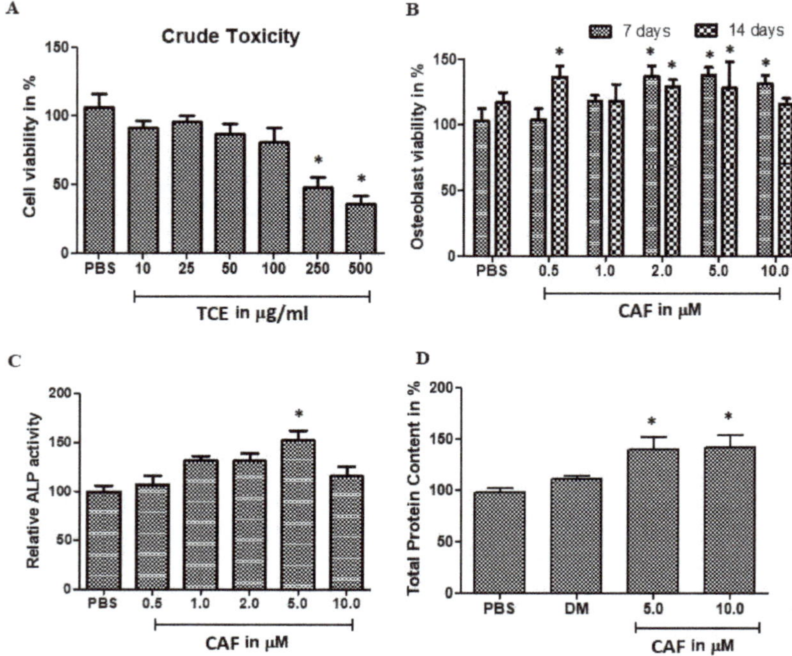

Figure 2. In vitro cytotoxic evaluation of *T. stellata* crude extract (TCE) and its metabolites. (**A**) The osteostromal cells were isolated from murine bone marrow and osteogenic characters were induced. The osteogenic cells were further analyzed for cytotoxic effect against *T. stellata* crude extract. The concentration tested from 10 μM to 500 μM. The crude extract was treated with osteogenic cells for 72 h and evaluate the cell viability using MTT reagent. (**B**) The major metabolites caffeic acid (CAF) was treated with osteogenic cells in concentration of 1.0 μM to 50 μM for 7 days and 14 days cultured cells. (**C**) Alkaline phosphatase-specific activity. (**D**) Total protein content. Bars represent the mean ± SD ($n = 4$). Statistical results are shown as * $p < 0.05$, Values compared between the PBS group with CAF at different concentrations.

Concerning, the effect of bone cell differentiation or proliferation, CAF significantly stimulated alkaline phosphatase (ALP) activity and bone cell differentiation in BM-MSCs cells. Thus, the effect of CAF on osteoblast culture from primary precursors cells was investigated. ALP was estimated at 14 days incubation (Figure 2C). The contents of ALP was increased 50% at 5-μM CAF treatments. ALP results were co-related with total protein content (Figure 2D). Total protein was raised up to 50% in CAF treated osteoblastic cells.

2.3. CAF Regulated In Vitro Mineralization of Osteoblastic Cells

CAF regulated the bone cell differentiation and influenced the loading of calcium on osteoblastic cells. CAF stimulated the proliferation in osteogenic cells. Thus, the effect of CAF was investigated using in vitro osteoblastic cells primary stained with alizarin red S (ARS). As shown in Figure 3A,B, differentiated osteoblasts were stained with ARS for a period of 7, 14, and 21 days, respectively. The contents of ARS staining increased 100–200% at CAF (5 and 10 μM) Figure 3C. Whereas 10 μM showed highly significant amplification of loading mineralization as well as accelerates the ARS staining up to 2-fold compared to PBS group. The inflammatory cytokines from osteoblast differentiation were quantified and found CAF insignificantly regulated the IL-10 and it was not overexpressed by the differentiation cellular modifications Figure 3D. Whereas, TNF-α was reduced significantly at 5 μM of CAF treatment in osteoblast stereotypes.

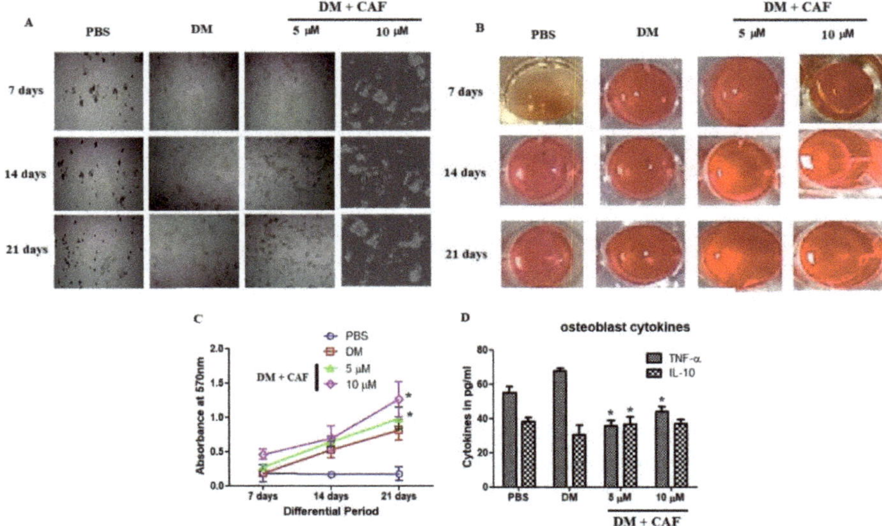

Figure 3. The differentiation and in vitro mineralization of osteoblast was evaluated at CAF treated conditions. (**A**) Representative macroscopic (**B**) Microscopic observation of alizarin red S (ARS) (**C**) ARS quantification for osteoblastic Differentiation at CAF treated conditions. (**D**) Cytokines estimation at 7th day of CAF treated osteoblastic cells. TNF-α and IL-10 were estimation using ELISA kits. Bars represent the mean ± SD (n = 4). Statistical results are shown as * $p < 0.05$. Values compared between the PBS group with CAF at different concentrations.

2.4. Regulation of CAF on Osteoblastic Markers

CAF assessed on mechanistic influences related to the effect on mineralization and differentiation. It was decided to examine the quantification of mRNA and protein markers related to osteoblastic regulations (Figure 4). The protein estimation of cathepsin-B and BMP-2 in CAF treated osteoblastic cells was shown in Figure 4A. The estimation showed that, cathepsin-B protein was significantly increased at 5 µM CAF treated osteoblastic cells. Moreover, BMP-2 was also upregulated parallel to Cath-B protein (Figure 4A). Differential media increased the BMP-2 level 60% compared to control and CAF treatment increased the level up to 300% compared to control groups. These results claim to investigate more markers related to osteoblastic cells. Immunoblot based estimation of cathepsin-B, osteopontin, osteocalcin, and BMP-2 were estimated in CAF treated cells. The protein of Cath-B, osteocalcin, and BMP-2 were increased significantly in CAF treated cells (Figure 4B). On the other hand, osteopontin was not upregulated by CAF treatments. Moreover, 10 µM of CAF treatment upregulate the transcript marker of osteoblast-related morphogenic proteins, osteocalcin (250%), as well as the osteo-inductive protein BMP-2 (230%) (Figure 4C–F). The significance observation was also noted in mRNA estimation of osteoblastic markers. Osteopontin was not significantly increased against differential medium (DM) group, but it was significant against control group.

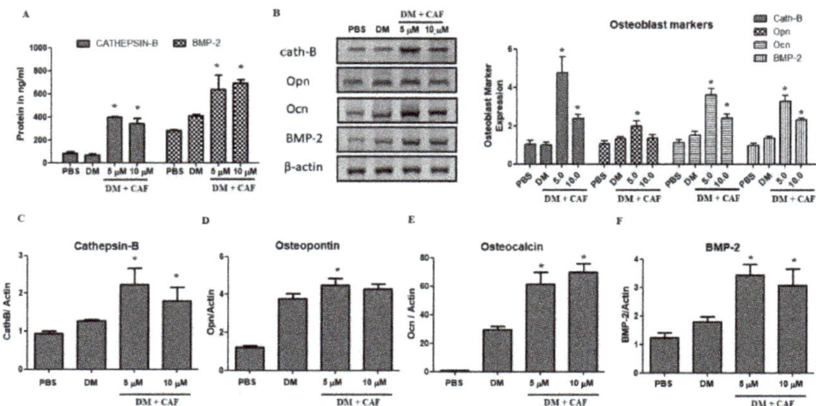

Figure 4. Effect of CAF on bone markers in osteoblast cell organization. (**A**) Cytokines estimation at 7th day of CAF treated osteoblastic cells. TNF-α and IL-10 were estimation using ELISA kits. (**B**) Effect of CAF on the protein expression of bone markers such as cathepsin-B (Cath-B), osteopontin (Opn), steocalcin (Ocn) and Bone morphogenic protein-2 (BMP-2). Protein expression were measured by Western blot analysis. (**C–F**) mRNA expression of bone markers on CAF treated osteoblastic cells and it was quantified by quantitative real time PCR was normalized to β-actin. (**C**) mRNA expression of cathepsin B, (**D**) mRNA expression of osteopontin, (**E**) mRNA expression of osteocalcin and (**F**) mRNA expression of BMP-2 marker, respectively. Bars represent the mean ± SD ($n = 4$). Statistical results are shown as * $p < 0.05$.

2.5. *Effect of CAF on Osteoclastogenic Regulations*

Osteoclasts are major cell type in bone mass and play a critical role in bone rejuvenation and resorption. The negative regulation of osteoclastogenesis has recognized to be a positive treatment for bone loss and bone degenerative diseases. Therefore, it was examined for the anti-osteoclastogenic effect using CAF in RANKL-induced osteoclast cells. As shown in Figure 5A, results showed that CAF treatment concentrations tested from 0.5 to 10 μM regulate the osteoclast cell viability. At 5 and 10 μM concentration plays a 30% and 45% cell death, respectively. CAF treatment strongly inhibited the tartrate resistant acid phosphatase (TRAP) positive osteoclast cells by a dose-dependent manner. The inhibition of osteoclast differentiation was confirmed by TRAP activity (Figure 5B). It was further evaluated using a differentiation platform with RANKL (Figure 5C). As shown in Figure 5D, CAF exhibited significant DNA damage to osteoclast cells. The inhibition was profound nuclear membrane organization and DNA linearization. These results claimed that CAF at 10 μM concentration inhibits the osteoclastogenesis and accelerates RANKL induced DNA damage. Apparently, it was further examined to confirm the osteoclastogenesis mRNA and protein markers.

Osteoclast marker investigation was done using mRNA and Western blot analysis. Western blot based quantification of TRAP, MMP-9, and cathepsin-K were estimated in CAF treated osteoclast cells. The significance observation was noted in mRNA estimation of osteoclast markers. TRAP expression was not appreciated in differential RANKL medium compared to control group, it was reduced at CAF 5 μM and 10 μM treated groups. The protein of TRAP, Cath-K, and MMP-9 were significantly decreased in CAF treated cells (Figure 5E,F). On the other hand, cathepsin-K was elevated at DM group compared to other tested groups. In addition, 10 μM of CAF treatment decreased the protein expression of MMP-9, these results revealed that CAF inhibit the migration and infiltration of bone cells. These results examined, showed that osteoclast markers were negative regulated by CAF treatment. It was significantly regulated by CAF compared to DM group. The inhibition of osteoclast genesis gives a remarkable output of *Trigonella* molecules reciprocally regulated the osteoblast and osteoclast cell types and might control the bone erosion and increases the reabsorption of senescent bone cells.

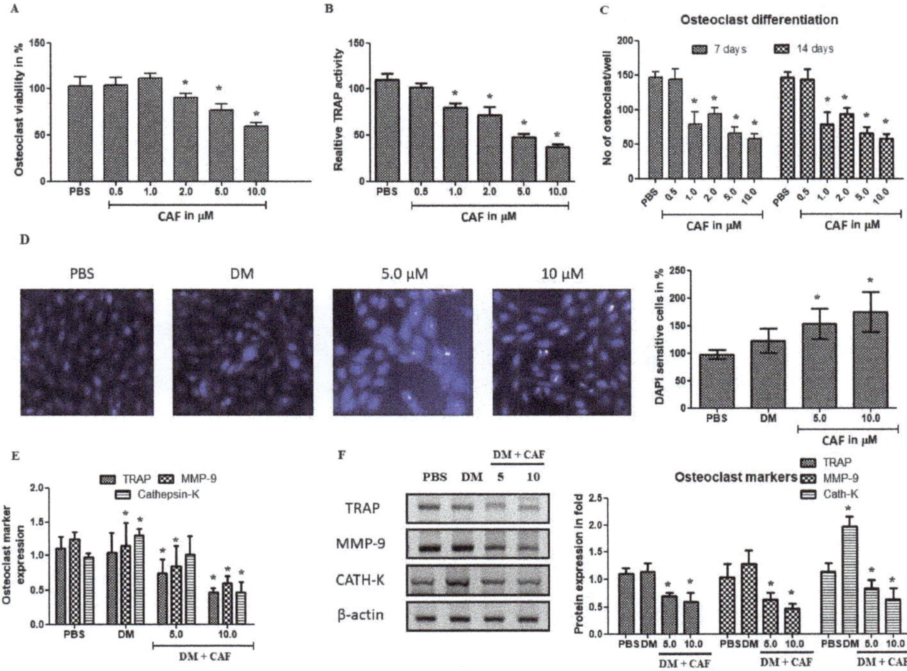

Figure 5. Effects of CAF on receptor activator of nuclear factor kappa-B ligand (RANKL)-induced osteoclast differentiation in mesenchymal cells. Bone-marrow derived mesenchymal stem cells (BM-MSCs) were cultured with vehicle or CAF (5 and 10 μM) in the presence RANKL (100 ng/mL) for 7 days. (**A**) Cultured cells tested against CAF treatment and analysed the cell viability. Cell viability of BM-MSCs was determined using the XTT assay (**B**) The estimation of Tartrate resistant acid phosphatase (TRAP) activity at 450 nm. (**C**) Cell differentiation effect was quantified on two intervals from 7 days to 10 days of incubation with CAF treatment. (**D**) RANKL induced osteoclast cells were treated with CAF and estimate the DAPI positive cells to confirm apoptotic induction as well as nuclear organization by fluorescence. Nuclei were stained by DAPI (blues signal). The area of DNA damage and nuclear organization was measured using ImageJ software. (**E**) Quantification of mRNA expression osteoclastic markers in CAF treated osteoclast cells using quantitative RT-PCR. The markers are TRAP, matrixmettalloprotease-9 (MMP-9), cathepsin-K (Cath-K). (**F**) Quantification of protein expression in osteoclastic cells treated with CAF using Immunoblot methods. The markers are TRAP, matrixmettalloprotease-9 (MMP-9), cathepsin-K (Cath-K). Bars represent the mean ± SD ($n = 4$). Statistical results are shown as * $p < 0.05$.

2.6. In Silico Docking of CAF against Cathepsin-B and Cathepsin-K Markers

Cathepsin is a cysteine residue based protease; ligand selection is based on its hot spot binding in substrate binding site of cathepsins family. These are the key factors for designing effective new ligand based inhibitors. Cathepsin-B structure showed arrangement of 18 amino acids long insertion from (Pro107-Asp124). In current study, orientations of the most potent inhibition of enzymatic activity of cathepsins-B and K were studied. It was found CAF docked similar binding regions with three-dimensional designed crystal structures of cathepsin-B and K (Figure 6A–G). The molecular modeling of CAF identified that a long hydrophobic pocket represents the potential binding site on the surface of cathepsin-K (−6.01) with two hydrogen bonds (Table 2), which is necessary for the peptide binding and excision (Figure 6B–D). The CAF docked the cathepsin-B in four hydrogen bonds, which at the site of hot spot of the receptor and made −4.43 binding energy (Figure 6E–G). Binding energy, ligand efficiency, and intermol energy were noted in Table 2.

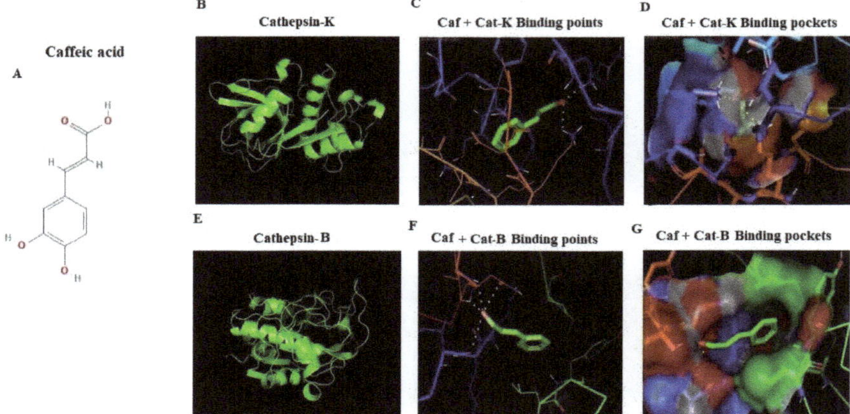

Figure 6. In silico interaction of caffeic acid (CAF) with cathepsin family receptor molecules. Docked orientation of (**A**) CAF (CID_637511), (**B–D**) 3D structure of cathepsin-K (PDB ID: 6QLM), hydrogen bonds between CAF with cathepsin-K, binding pocket of cathepsin-K. (**E–G**) 3D structure of Cathepsin-B (PDB ID:3AI8), hydrogen bonds between CAF with cathepsin-B, binding pocket of cathepsin B with CAF complex. Docking analysis was performed using Autodock tools (ADT) and Autodock v4.2 software.

Table 2. Hydrophobic interaction of caffeic acid and amino acid residues of target proteins.

SL. No	Target Proteins	Binding Energy	Ligand Efficiency	Intermole Energy	Ligand Atoms (Ring)	Docked Amino Acid Residue (Bond Length)
1.	Cat-K	−6.01	−0.6	−6.55	C1-OH	LYS'17/HZ2 (1.7 Å)
					C1-OH	LYS'181/HZ2 (2.3 Å)
					C1-OH	ASN'222/OD1 (2.5 Å)
2	Cat-B	−4.43	−0.44	−4.97	C1-OH	THR'223/HN (2.2 Å)
					C1-OH	THR'223/HG1 (1.7 Å)
					C1-OH	ILE'20/O (3.2 Å)

Cat-K: Cathepsin-K, Cat-B: Cathespsin-B, LYS: Lysine, ASN: Asparagine, THR: Threonine, ILE: isoleucine.

3. Discussion

TCE showed remarkable osteogenic activity and improves bone strength. There is a possibility that the TCE contains active molecules responsible for the potential activity. Further refining and identification of constituents revealed the isolation of CAF. The structure was elucidated by inspection of ^{1}H, ^{13}C, DEPT, HMQC, and HMBC spectroscopic data (Table 1 and Supplementary Material) and compared with the literature values [21]. Whereas the ^{1}H NMR spectrum of CAF (Table 1 and Supplementary Material) revealed the presence of aromatic resonances for an ABX system at σ_H 7.06 (d, J = 2.04 Hz), 6.80 (d, J = 8.16 Hz), and 6.95 (dd, J = 2.04, 8.16 Hz). In addition, a pair of doublet at σ_H 6.23 and 7.55 with coupling constants of 15.88 Hz was observed corresponding to two *trans*-olefin protons. By inspection of the ^{13}C NMR spectrum (Table 1 and Supplementary Material), the obtained resonances were in agreement with the aromatic resonances at σ_C 127.82, 115.10,146.82, 149.47, 116.51, and 122.88, as well as two olefin carbons at σ_C 115.55 and 147.06 and a carboxyl resonance at σ_C 171.08, all of which were characteristic of caffeic acid. This study represents the first report on the isolation of CAF from *T. stellate*.

The obtained results establish the direct stimulatory effect of *T. stellata* metabolites on osteoblast differentiation. The potential stimulation of osteogenic properties of *T. stellata* metabolites has been proposed, but the molecular mechanism of this stimulation remains unclear. Natural products promote the osteoblast activity and suppress the osteoclast activity [22]. There are many studies from plants which showed the osteoblast activation and osteoclast inhibition processes [23–25].

The cytotoxic effects directly explained that the TCE was not toxic up to 100 μg/mL concentration. This biocompatibility on mesenchymal cells triggers us to investigate further on metabolites of *T. stellata*. The cytotoxic activity of CAF also explained its toxic properties and found nontoxic up to 10 μM concentration. The ALP and total protein were also reflecting this viability analysis. The results are comparable with study on medicinal plant's (*Leonurus sibiricus*) ethanol extract and other herbal molecules, which showed osteoblast differentiation and suppress osteoclast differentiation as well as bone resorption in a mouse model [26–29]. One of the studies reported with osteogenic differentiation by the compound leonurine hydrochloride from *Leonurus sibiricus*. Some other studies similar to current study observed with BMP-2 transcription factor, which played key role in bone formation and osteoblast differentiation [23,30]. The presented results demonstrate that CAF from TCE stimulated mineralization in osteoblast by enhancing transcription factors and stimulating cathepsin-B and BMP-2 signaling. Likewise, CAF suppressed RANKL-induced osteoclast differentiation and resorption capacity of cells.

The guava fruit was also effective against osteoporosis according to Chinese medicine due to the presence of polyphenolic compounds and increased bone health [31–35]. The results showed similar properties as *Rumex crispus* extract in osteoblast differentiation through runt related transcription factor 2 and suppress the RANKL induced bone loss through suppressing the RANKL signaling [36,37]. Furthermore, it was observed that the transcription factors were enhanced to activate the osteogenesis [37]. It was found that CAF enhanced the mRNA levels of stimulatory transcription factors to induce osteogenesis. BMP-2 is the osteoblast-regulating factor that induced morphogenic protein involved in differentiation; mineralization and bone strengthen mediators, such as osteopontin and osteocalcin. The mineralization explained the loading of calcium in osteoblast cells improves the bone strength and vitamin-D signaling pathways [38]. The loading of calcium was noted for 21 days and found 50% elevated storage after the 14th day of differentiation. In both tested concentrations got a similar response in differentiation and ARS uptake. These results revealed that, the osteoblast differentiation and mineralization were dose dependent manner.

It is well established that, extracellular signal regulated kinase (ERK) and p38-MAPK are osteogenic mediated signaling pathways that regulate osteoblast markers in unique differentiation. The underlying mechanism of CAF was investigated on the osteogenic activity, the study was examined whether CAF from TCE regulates the BMP-mediated activation of signaling markers and down regulated the TRAP, MMP-9, and cathepsin-K markers. These bone factors considered as major transcription factors required for bone formation, associated with the regulation of osteopontin and osteocalcin mediators [37–39]. In addition, BMP-2-mediated Smad protein stimulation regulate the transcription of osteopontin and osteocalcin markers [40,41]. In this study, CAF synergistically upregulate the BMP-2 and showed reciprocal regulation of cathepsin-B with cathepsin-K, upregulate the osteocalcin as well as bone transcription factors (Figures 5 and 6). These results suggest that the metabolic effect of CAF on BMP-2-dependent osteogenic to enhance the morphogenic protein mediated osteopontin and osteocalcin signaling axis for bone formation. It was showed that CAF treatments at low concentrations expressed an enhancement in differentiation and mineralization of osteoblast in both protein and mRNA level. Understanding the CAF effects on osteogenic markers, which delivers the outlined approach about the therapeutic potential for osteoporotic disorders. Furthermore, the regulations were confirmed by in silico docking analysis. The docking results revealed that, the structures of cathepsins-B and K are very similar, including their cleavage sites containing several "hot-spot" amino acids conserved among all types of cathepsins, along with Lysine 17 for cathepsin K and isoleucine 20 for cathepsin B receptors. The active site contains two histamine residues, which has high affinity towards substrate binding on receptors. These results revealed that CAF docked with cathepsin family receptors and plays a differential expressed ligand role in proteolytic protein osteogenic cell types.

4. Materials and Methods

4.1. General Experimental Procedures and Chemicals

^{1}H-, ^{13}C-, and 2D-NMR spectra were measured on an AVANCE 400 NMR spectrometer (^{1}H-NMR: 400 MHz and ^{13}C-NMR: 100 MHz, Bruker). Silica gel Column Chromatography (SCC) was performed on silica gel 60 (E. Merck, Darmstadt, Germany; 70-230 mesh). Reversed-Phase Silica Gel Column Chromatography (RPCC) was performed on a Cosmosil 75C18-OPN column (Nacalai Tesque, Kyoto, Japan; internal diameter = 50 mm, length = 25 cm, linear gradient: MeOH:H2O). Diaion HP-20 (Mitsubishi Chemical Co., Ltd. Tokyo, Japan). Pre-coated silica gel 60 F254 plates (E. Merck; 0.25 mm and 1000 µm in thickness) were used for Thin Layer Chromatography (TLC) visualization by spraying with p-anisaldehyde reagent and heated to 150 °C on a hotplate. High Performance Liquid Chromatography (HPLC) instrument used (Agilent, 1200 series, Germany) consisted of binary pumps, a PDA detector, and an auto sample injector, with a Chem satiation software module. The column was ZORBAX-SB-C18 (150 mm × 4.6 mm × 5 µm) (Agilent, Folsom, CA, USA). The mobile phase was in an isocratic mode using mixture of MeOH:H2O (45:55 *v*/*v*) with flow rate 1.0 mL/min. The volume of the samples injected into the HPLC system for analysis was set at 10 µL. Chemicals and reagents used were of analytical grade.

4.2. Plant Material

Aerial parts of *T. stellate* were collected from gardens of King Faisal University, Saudi Arabia (September 2016) and was identified by Dr. Mamdouh Shokry, Director of El-Zohria botanical garden, Giza, Egypt. Voucher specimen (9-16-Sept-TS) was deposited at the Herbarium museum of College of Clinical Pharmacy, King Faisal University, Saudi Arabia.

4.3. Extraction and Isolation

Shade-dried aerial parts of *T. stellata* (1.0 kg) were extracted three times with 70.0% methanol (10.0 L) by maceration at room temperature. The well-filtered extracts were combined and concentrated under reduced pressure to give an extract (103.0 g). The extract was partitioned with n-hexane (15.0 L) to give hexane fraction (74.0 g) and remaining mother liquor was concentrated to give (29.0 g) defatted extract. The defatted extract (25.0 g) was subjected to Diaion HP-20 column chromatography (1.0 kg) then it was eluted with water, 50.0% and 100.0% MeOH to obtain the fractions of water (5.0 g), 50.0% MeOH (12.0 g), and 100.0% MeOH (7.0 g). On the basis of the TLC patterns, the 50.0% MeOH-soluble fraction (12.0 g) was subjected to SCC (250.0 g, $CHCl_3$:MeOH:H_2O (15:6:1)(5.0 l)) then washing using 100.0% MeOH (2.0 l) to obtain seven sub-fractions (SubFr. 2-1 to 2-7). Sub-fraction 2-4 (1.3 g) was subjected to RPCC (100.0 g, using MeOH:H_2O gradient elution) to yield six subtractions (SubFr. 2-4-1 (175.3 mg), SubFr. 2-4-2 (26.1 mg), SubFr. 2-4-3 (137.5 mg), SubFr. 2-4-4 (220.9 mg), SubFr. 2-4-5 (56.0 mg), and SubFr. 2-4-6 (330.7 mg)). SubFr. 2-4-4 (220.9 mg) was further fractionated on preparative TLC ($CHCl_3$:MeOH:H_2O (15:6:1)) the spot at R_f value of 0.59 was eluted (SubFr. 2-4-4-1) and was further purified by HPLC to give pure compound; caffeic acid (CAF) (Figure 1B,C).

4.4. Animals and Ethical Aspects

Male C57BL/6j mice 4-week-old were used and weighing between 15 and 20 gm body weight, acquired from the Animal Facility of College of Science of King Faisal University. Animals were housed in cages under standard conditions and room temperature (22 ± 2 °C) under 12 h alternating light/dark conditions. Animal acclimatization were followed by the standard procedure of animal experimentation, and protocols were approved by the Ethics Committee on animal use of the College of Science (No. 180123).

4.5. Culture of BM-MSCs

Primary bone marrow derived mesenchymal cells (BM-MSCs) harvested from bone marrow of male C57BL/6j mice using serum free culture medium [RPMI-1640; UFG, Yanbu, Saudi Arabia). Mice were euthanized by humane control and dissect the femora bones in aseptic conditions following [42]. This isolation procedure was used in all experiments such as osteoblast, osteoclast differentiation studies, respectively.

4.6. In Vitro Cytotoxicity Analyses

4.6.1. MTT Assay

The cytotoxicity of different concentrations of TCE (10, 25, 50, 100, 250, and 500 μg/mL) and CAF (0.5. 1. 2, 5, and 10 μM) were tested in vitro in osteogenic medium. After the 7th day of TCE and the 7th and 14th day of CAF exposures, cell viability was evaluated by the MTT and the NR uptake assays [43] briefly, aspirate the supernatant after CAF treatments, the MTT reagent was added with fresh serum free medium and incubate for 3 h. The formation of formazan crystals were solubilized by adding 0.1 mL dimethyl sulfoxide, and the optical density (OD) was determined at 570 nm.

4.6.2. Neutral Red Assay

The effect of CAF on neutral red uptake using osteogenic cells treated with CAF and cells were aspirate with 0.5 mL of neutral red (3-amino-7-dimethylamino-2-methylphenazine hydrochloride) solution (50 μg/mL). Then, it was incubated for 2 h at 37 °C. Thereafter, the supernatant aspirated with PBS, neutral red dye was extracted using 0.1 % glacial acetic acid, and the OD was determined at 540 nm [44].

4.7. Alkaline Phosphatase Activity

Alkaline phosphatase activity was determined at the 10th day, using the Roy (1970) protocol [45], After CAF treatment, cell layers were scraped off using cell scrapper. Cell pellets were collected, and lysate was prepared by freeze–thaw method. After gently spinning down the cell debris, 20 μL supernatant from each sample were added to an assay mixture of p-nitrophenyl phosphate. Samples were quantified using 405 nm in Biotek elisa plate reader (Biotek Instruments Ltd., England). The specific activity of ALP was quantified by equilibrated total protein concentration.

4.8. Assays of Osteoblast Differentiation

Osteoblast differentiation was induced by addition of differentiation medium (DM), supplemented with 50 μg/mL ascorbic acid, 10 mM β-glycerophosphate, and 10 nM dexamethasone (Sigma-Aldrich, St. Louis, MA, USA). The negative group was kept without additives [42]. Osteoblast differentiation was assessed by mineralization of calcium using the protocols of ARS dye. Osteoblasts were treated with CAF at concentrations of 0, 5, and 10 μM for 21 days. CAF treated osteoblast cells were washed with ice-cold PBS buffer and fixed in ice-cold 10% formalin for 20 min. Then, 1% alcian blue solution was used for fixation. These sections were incubated for 8 min with ARS. Mineralized cell patches observed and counted using an image analyzing system EVOS (Life Technologies, Carlsbad, CA, USA).

4.9. Osteoclast Tartrate-Resistant Acid Phosphatase (TRAP) Activity Estimation

The effect of CAF was examined on osteoclast differentiation isolated according to previously described method [40]. After the RANKL induction, cells were treated with CAF at 5 μM and 10 μM for 7 days, and results were analyzed by TRAP activity assay [26].

Osteoclast Apoptosis

The effect of CAF was evaluated on differentiation of pre-osteoclast cells isolated from bone marrow in mice. To generate osteoclasts, RANKL (150 ng/mL) was used for 4 days. Total TRAP activity was measured at an absorbance of 405 nm after treatment with Substrate (p-nitrophenyl phosphate) as described previously [26]. After treatment with CAF, results were obtained according to previously described protocol [46].

4.10. Gene Expression Analysis by Real-Time Reverse Transcriptase Polymerase Chain Reaction (qRT-PCR)

Total RNA extracted from CAF treated cells by the use of Trizol reagent (Invitrogen, Life Technologies, Grand Island, NY, USA). Extracted RNA were pretreated with DNAse I and quantified using Nanodrop (Thermo scientific, Waltham, MA, USA). Equiliberate the RNA concentration 300 ng/marker and used for complementary DNA preparation using SuperScript™III Reverse Transcriptase (Invitrogen, Waltham, MA, USA). Real time primer details were noted in Table 3. The relative mRNA quantification of cathepsin-B, cathepsin-K, MMP-p, BMP-2, TRAP, osteopontin, osteocalcin were evaluated by quantitative real time PCR (Applied Biosystems, Life Technologies, USA). Relative quantification of target genes were calculated using the $2-\Delta\Delta Ct$ method, as described by [47]. β-actin was used as an internal control.

Table 3. Real time PCR primer details.

Gene Name	Forward Primer	Reverse Primer
MMP-9	GTGCTGGGCTGCTGCTTTGCTG	GTCGCCCTCAAAGGTTTGGAAT
TRAP	GGTCAGCAGCTCCCTAGAAG	GGAGTGGGAGCCATATGATTT
OCN	AGCAAAGGTGCAGCCTTTGT	GCGCCTGGGTCTCTTCACT
OPN	ACATCCAGTACCCTGATGCTACAG	TGGCCTTGTATGCACCATTC
Cath-K	GCCAGGATGAAAGTTGTATG	CAGGCGTTGTTCTTATTCC
Cath-B	GGTTGCAGACCGTACTCCAT	GGAACTGCATCCAAAATGCT
BMP-2	TGCACCAAGATGAACAGC	GTGCCACGATCCAGTCATTC
GAPDH	GTATTGGGCGCCTGGTCACC	CGCTCCTGGAGATGGTGATGG

4.11. Western Blot for Protein Marker Quantification

Protein quantification was done by Western blot analysis as described [48]. Briefly, CAF treated osteostromal cells such as osteoblast and osteoclast cells were washed with cold PBS. Cell pellets were collected by centrifugation and lysed by RIPA cell lysis buffer (Santa cruz biotechnology, USA). Cell lysates were obtained and total protein content in the supernatant was determined by the Bradford assay. It was eluted using SDS-PAGE, and the separated proteins were transferred to nitrocellulose membranes. Primary antibodies against cathepsin-B, cathepsin-K, MMP-p, BMP-2, TRAP, osteopontin, osteocalcin, and β-actin were diluted (1:1000; Cell Signaling, Danvers, MA, USA) Santa Cruz Biotechnology, Santa Cruz, CA, USA) at 4 °C overnight. Membranes were visualized using enhanced luminol-based chemiluminescent (ECL) substrate (Santa Cruz biotechnology, Santa Cruz, CA, USA).

4.12. Cytokine Estimation by ELISA

Production of tumor necrosis factor-α (TNF-α) and interlukin-10 (IL-10) in cell culture supernatant were evaluated. The BMP-2 and cathepsin-B were quantified in whole cell lysate of osteoblast treated with CAF using the manufacture protocol of Abcam validated kits (Abcam, Germany). Assays were performed on 96-well micro titer plates followed by manufacture instructions. The cytokine was quantified by the addition of chromogenic substrate solution (p-Nitrophenol). The reaction was measure at 450 nm. The standard curve was made in regression plot and sample concentrations were calculated in relation to the standard curve.

4.13. Computational Docking Analysis

The potential docking interaction was evaluated CAF against cathepsins-B and K were examined in silico using Autodock tools (ADT) v1.5.4 and Autodock v4.2 program (http://www.scripps.edu/mb/olson/doc/autodock). The respective chemical structure of ligand CAF (CID_637511), was retrieved from Pubchem database (http://www.ncbi.nlm.nih.gov/pccompound). The structures of cathepsin receptors were downloaded from the Protein Data Bank (PDB). The three dimensional structures of cathepsin-B (PDB ID: 6QLM) and cathepsin-K (PDB ID:3AI8) were retrieved from the Protein Data Bank; (http://www.pdb.org). The receptors were prepared by removing polar groups and non-amino acid residues using the protein preparation Wizard of PYMOL and python platform. The active sites of Cath-B and K were identified by Q-site Finder [49]. The competitive inhibition was studied on CAF complexed cathepsin protein family. Ligand docked to the receptor consider as rigid body, and receptor was considered as flexible factor. The ligand efficiency, intermol energy, and binding carbon numbers were recorded.

4.14. Statistical Analysis

The study was designed and evaluated in four independent experiments. The data values were expressed as means ± standard deviation (SD) ($n = 4$). Statistical analysis was conducted by student T test and one-way ANOVA regression plot compared with control group or treatment at different concentrations of TCE and CAF. All analyses were performed using Microsoft Excel office-10 and GraphPad Prism 7.0 (GraphPad Software Inc., San Diego, CA, USA), and statistical results are shown as the corrected p value (* $p < 0.05$).

5. Conclusions

Based on the current findings, it can be concluded that CAF treatment showed promising activation on the bone metabolism via osteoblast differentiation and bone maturation in C57BL/6j bone marrow derived cells. At low concentration of CAF treatments, it enhanced the mineralization and upregulated the osteogenic marker expression in mesenchymal cells. CAF induced the morphogenic protein of bone and cathepsin-B in bone derived precursor cells. On the other hand, CAF attenuated the osteoclast formation and differentiation through the inhibition of osteoclast markers such as TRAP, Cath-K, and MMP-9 in RANKL induced osteoclast cells. Moreover, it regulates apoptotic bone loss in osteoclast cells. These findings recommend, CAF derived for the first time from *T. stellata*, might be used as potential therapeutic alternative for the treatment of bone metabolic diseases.

Supplementary Materials: The following are available online at http://www.mdpi.com/2223-7747/9/11/1610/s1, Figure S1–S8: 1 D- and 2 D-NMR spectroscopic data of CAF.

Author Contributions: H.-I.M.I.: conceptualization, funding acquisition, formal analysis, writing–review and editing and supervision, methodology; H.M.D. and M.R.A.: validation, software, and writing–original draft preparation, H.E.K.; conceptualization, methodology, writing—original draft preparation, formal analysis, writing—review and editing. All authors have read and agreed to the published version of the manuscript.

Funding: The authors extend their appreciation to the Deputyship for Research & Innovation, Ministry of Education in Saudi Arabia for funding this research work through the project number (IFT20094).

Acknowledgments: The authors are thankful to the Deputyship for Research & Innovation, Ministry of Education in Saudi Arabia for funding this research work through the project number (IFT20094).

Conflicts of Interest: The authors declare no conflict of interest.

References

1. Florencio-Silva, R.; Sasso, G.R.d.S.; Sasso-Cerri, E.; Simões, M.J.; Cerri, P.S. Biology of Bone Tissue: Structure, Function, and Factors That Influence Bone Cells. *Biomed Res. Int.* **2015**, *2015*, 421746. [CrossRef] [PubMed]
2. Maniatopoulos, C.; Sodek, J.; Melcher, A. Bone formation in vitro by stromal cells obtained from bone marrow of young adult rats. *Cell Tissue Res.* **1988**, *254*, 317–330. [CrossRef] [PubMed]

3. Komori, M.; Rasmussen, S.W.; Kiel, J.A.; Baerends, R.J.; Cregg, J.M.; van der Klei, I.J.; Veenhuis, M. The Hansenula polymorpha PEX14 gene encodes a novel peroxisomal membrane protein essential for peroxisome biogenesis. *EMBO J.* **1997**, *16*, 44–53. [CrossRef] [PubMed]
4. Suda, T.; Takahashi, N.; Udagawa, N.; Jimi, E.; Gillespie, M.T.; Martin, T.J. Modulation of osteoclast differentiation and function by the new members of the tumor necrosis factor receptor and ligand families. *Endocr. Rev.* **1999**, *20*, 345–357. [CrossRef] [PubMed]
5. Xiao, G.; Jiang, D.; Ge, C.; Zhao, Z.; Lai, Y.; Boules, H.; Phimphilai, M.; Yang, X.; Karsenty, G.; Franceschi, R.T. Cooperative interactions between activating transcription factor 4 and Runx2/Cbfa1 stimulate osteoblast-specific osteocalcin gene expression. *J. Biol. Chem.* **2005**, *280*, 30689–30696. [CrossRef] [PubMed]
6. Razzaq, A.; Ahmad Malik, S.; Saeed, F.; Imran, A.; Rasul, A.; Qasim, M.; Zafar, S.; Kamran, S.K.S.; Maqbool, J.; Imran, M. Moringa oleifera Lam. ameliorates the muscles function recovery following an induced insult to the Sciatic nerve in a mouse model. *Food Sci. Nutr.* **2020**, *8*, 4009–4016. [CrossRef] [PubMed]
7. Vlachogianni, T.; Loridas, S.; Fiotakis, K.; Valavanidis, A. From the traditional medicine to the modern era of synthetic pharmaceuticals natural products and reverse pharmacology approaches in new drug discovery expedition. *Pharmakeftiki* **2014**, *26*, 16–30.
8. Nielsen, O.; Ainsworth, M.; Csillag, C.; Rask-Madsen, J. Systematic review: Coxibs, non-steroidal anti-inflammatory drugs or no cyclooxygenase inhibitors in gastroenterological high-risk patients? *Aliment. Pharmacol. Ther.* **2006**, *23*, 27–33. [CrossRef]
9. Azab, A.; Nassar, A.; Azab, A.N. Anti-inflammatory activity of natural products. *Molecules* **2016**, *21*, 1321. [CrossRef]
10. Owona, B.A.; Abia, W.A.; Moundipa, P.F. Natural compounds flavonoids as modulators of inflammasomes in chronic diseases. *Int. Immunopharmacol.* **2020**, *84*, 106498. [CrossRef]
11. Nunes, C.d.R.; Barreto Arantes, M.; Menezes de Faria Pereira, S.; Leandro da Cruz, L.; de Souza Passos, M.; Pereira de Moraes, L.; Vieira, I.J.C.; Barros de Oliveira, D. Plants as Sources of Anti-Inflammatory Agents. *Molecules* **2020**, *25*, 3726. [CrossRef] [PubMed]
12. Cruz, L.A.; Díaz, M.A.; Gupta, M.P.; López-Pérez, J.L.; Mondolis, E.; Morán-Pinzón, J.; Guerrero, E. Assessment of the antinociceptive and anti-inflammatory activities of the stem methanol extract of *Diplotropis purpurea*. *Pharm. Biol.* **2019**, *57*, 432–436. [CrossRef] [PubMed]
13. Nagulapalli Venkata, K.C.; Swaroop, A.; Bagchi, D.; Bishayee, A. A small plant with big benefits: Fenugreek (*Trigonella foenum-graecum* Linn.) for disease prevention and health promotion. *Mol. Nutr. Food Res.* **2017**, *61*, 1600950. [CrossRef]
14. Sindhu, G.; Ratheesh, M.; Shyni, G.; Nambisan, B.; Helen, A. Anti-inflammatory and antioxidative effects of mucilage of *Trigonella foenum graecum* (Fenugreek) on adjuvant induced arthritic rats. *Int. Immunopharmacol.* **2012**, *12*, 205–211. [CrossRef] [PubMed]
15. Norton, J.; Majid, S.A.; Allan, D.; Al Safran, M.; Böer, B.; Richer, R. *An Illustrated Checklist of The Flora of Qatar*; Browndown Publications: Gosport, UK, 2009.
16. Al-Tabini, R.; Al-Khalidi, K.; Al-Shudiefat, M. Livestock, medicinal plants and rangeland viability in Jordan's Badia: Through the lens of traditional and local knowledge. *Pastor. Res. Policy Pract.* **2012**, *2*, 4. [CrossRef]
17. Alhabri, N. Survey of plant species of medical importance to treat digestive tract diseases in Tabuk Region, Saudi Arabia. *J. King Abdulaziz Univ.* **2017**, *29*, 51–56.
18. Basu, S.K.; Zandi, P.; Cetzal-Ix, W. Fenugreek (*Trigonella foenum-graecum* L.): Distribution, Genetic Diversity, and Potential to Serve as an Industrial Crop for the Global Pharmaceutical, Nutraceutical, and Functional Food Industries. In *The Role of Functional Food Security in Global Health*; Elsevier: Amsterdam, The Netherlands, 2019; pp. 471–497.
19. Shams Eldin, S.M.; Radwan, M.M.; Wanas, A.S.; Habib, A.-A.M.; Kassem, F.F.; Hammoda, H.M.; Khan, S.I.; Klein, M.L.; Elokely, K.M.; ElSohly, M.A. Bioactivity-guided isolation of potential antidiabetic and antihyperlipidemic compounds from *Trigonella stellata*. *J. Nat. Prod.* **2018**, *81*, 1154–1161. [CrossRef] [PubMed]
20. Sheweita, S.A.; ElHady, S.A.; Hammoda, H.M. Trigonella stellata reduced the deleterious effects of diabetes mellitus through alleviation of oxidative stress, antioxidant-and drug-metabolizing enzymes activities. *J. Ethnopharmacol.* **2020**, 112821. [CrossRef]
21. Liu, Q.; Yu, J.; Liao, X.; Zhang, P.; Chen, X. One-step separation of antioxidant compounds from Erythrina variegata by high speed counter-current chromatography. *J. Chromatogr. Sci.* **2015**, *53*, 730–735. [CrossRef]

22. Ogasawara, T.; Kawaguchi, H.; Jinno, S.; Hoshi, K.; Itaka, K.; Takato, T.; Nakamura, K.; Okayama, H. Bone morphogenetic protein 2-induced osteoblast differentiation requires Smad-mediated down-regulation of Cdk6. *Mol Cell Biol* **2004**, *24*, 6560–6568. [CrossRef]
23. Choi, S.-W.; Kim, S.-H.; Lee, K.-S.; Kang, H.J.; Lee, M.J.; Park, K.-I.; Lee, J.H.; Park, K.D.; Seo, W.D. Barley Seedling Extracts Inhibit RANKL-Induced Differentiation, Fusion, and Maturation of Osteoclasts in the Early-to-Late Stages of Osteoclastogenesis. *Evid. Based Complementary Altern. Med.* **2017**, *2017*, 6072573. [CrossRef]
24. Choi, S.-W.; Lee, K.-S.; Lee, J.H.; Kang, H.J.; Lee, M.J.; Kim, H.Y.; Park, K.-I.; Kim, S.-L.; Shin, H.K.; Seo, W.D. Suppression of Akt-HIF-1α signaling axis by diacetyl atractylodiol inhibits hypoxia-induced angiogenesis. *BMB Rep.* **2016**, *49*, 508–513. [CrossRef]
25. Choi, C.-W.; Choi, S.-W.; Kim, H.-J.; Lee, K.-S.; Kim, S.-H.; Kim, S.-L.; Do, S.H.; Seo, W.-D. Germinated soy germ with increased soyasaponin Ab improves BMP-2-induced bone formation and protects against in vivo bone loss in osteoporosis. *Sci. Rep.* **2018**, *8*, 1–12. [CrossRef]
26. Kim, N.; Takami, M.; Rho, J.; Josien, R.; Choi, Y. A novel member of the leukocyte receptor complex regulates osteoclast differentiation. *J. Exp. Med.* **2002**, *195*, 201–209. [CrossRef]
27. Pandey, K.B.; Rizvi, S.I. Plant polyphenols as dietary antioxidants in human health and disease. *Oxidative Med. Cell. Longev.* **2009**, *2*. [CrossRef]
28. Prynne, C.J.; Mishra, G.D.; O'Connell, M.A.; Muniz, G.; Laskey, M.A.; Yan, L.; Prentice, A.; Ginty, F. Fruit and vegetable intakes and bone mineral status: A cross-sectional study in 5 age and sex cohorts. *Am. J. Clin. Nutr.* **2006**, *83*, 1420–1428. [CrossRef]
29. Chin, K.-Y.; Ima-Nirwana, S. Olives and bone: A green osteoporosis prevention option. *Int. J. Environ. Res. Public Health* **2016**, *13*, 755. [CrossRef]
30. Chen, D.; Zhao, M.; Mundy, G.R. Bone morphogenetic proteins. *Growth Factors* **2004**, *22*, 233–241. [CrossRef]
31. Jao, H.-Y.; Hsu, J.-D.; Lee, Y.-R.; Lo, C.-S.; Lee, H.-J. Mulberry water extract regulates the osteoblast/osteoclast balance in an ovariectomic rat model. *Food Funct.* **2016**, *7*, 4753–4763. [CrossRef]
32. Shim, K.-S.; Lee, C.-J.; Yim, N.-H.; Gu, M.J.; Ma, J.Y. Alpinia officinarum stimulates osteoblast mineralization and inhibits osteoclast differentiation. *Am. J. Chin. Med.* **2016**, *44*, 1255–1271. [CrossRef]
33. Mukudai, Y.; Kondo, S.; Koyama, T.; Li, C.; Banka, S.; Kogure, A.; Yazawa, K.; Shintani, S. Potential anti-osteoporotic effects of herbal extracts on osteoclasts, osteoblasts and chondrocytes in vitro. *BMC Complementary Altern. Med.* **2014**, *14*, 1–14. [CrossRef] [PubMed]
34. Devalaraja, S.; Jain, S.; Yadav, H. Exotic fruits as therapeutic complements for diabetes, obesity and metabolic syndrome. *Food Res. Int.* **2011**, *44*, 1856–1865. [CrossRef] [PubMed]
35. Lanham-New, S.A. *Fruit and Vegetables: The Unexpected Natural Answer to the Question of Osteoporosis Prevention?* Oxford University Press: Oxford, UK, 2006.
36. Sharan, K.; Siddiqui, J.A.; Swarnkar, G.; Maurya, R.; Chattopadhyay, N. Role of phytochemicals in the prevention of menopausal bone loss: Evidence from in vitro and in vivo, human interventional and pharmacokinetic studies. *Curr. Med. Chem.* **2009**, *16*, 1138–1157. [CrossRef] [PubMed]
37. Shim, K.S.; Lee, B.; Ma, J.Y. Water extract of Rumex crispus prevents bone loss by inhibiting osteoclastogenesis and inducing osteoblast mineralization. *BMC Complementary Altern. Med.* **2017**, *17*, 1–9. [CrossRef]
38. Ahn, S.H.; Chen, Z.; Lee, J.; Lee, S.-W.; Min, S.H.; Kim, N.D.; Lee, T.H. Inhibitory Effects of 2N1HIA (2-(3-(2-Fluoro-4-Methoxyphenyl)-6-Oxo-1(6H)-Pyridazinyl)-N-1H-Indol-5-Ylacetamide) on Osteoclast Differentiation via Suppressing Cathepsin K Expression. *Molecules* **2018**, *23*, 3139. [CrossRef]
39. Kim, K.-J.; Jeong, M.-H.; Lee, Y.; Hwang, S.-J.; Shin, H.-B.; Hur, J.-S.; Son, Y.-J. Effect of Usnic Acid on Osteoclastogenic Activity. *J. Clin. Med.* **2018**, *7*, 345. [CrossRef]
40. Bikle, D.D. Vitamin D and bone. *Curr. Osteoporos. Rep.* **2012**, *10*, 151–159. [CrossRef]
41. Chen, G.; Deng, C.; Li, Y.-P. TGF-β and BMP signaling in osteoblast differentiation and bone formation. *Int. J. Biol. Sci.* **2012**, *8*, 272. [CrossRef]
42. Maridas, D.E.; Rendina-Ruedy, E.; Le, P.T.; Rosen, C.J. Isolation, culture, and differentiation of bone marrow stromal cells and osteoclast progenitors from mice. *JoVE* **2018**, *131*, e56750. [CrossRef]
43. Freimoser, F.M.; Jakob, C.A.; Aebi, M.; Tuor, U. The MTT [3-(4, 5-dimethylthiazol-2-yl)-2, 5-diphenyltetrazolium bromide] assay is a fast and reliable method for colorimetric determination of fungal cell densities. *Appl. Environ. Microbiol.* **1999**, *65*, 3727–3729. [CrossRef]

44. Borenfreund, E.; Puerner, J.A. A simple quantitative procedure using monolayer cultures for cytotoxicity assays (HTD/NR-90). *J. Tissue Cult. Methods* **1985**, *9*, 7–9. [CrossRef]
45. Roy, A. Rapid method for determining alkaline phosphatase activity in serum with thymolphthalein monophosphate. *Clin. Chem.* **1970**, *16*, 431–436. [CrossRef]
46. Kameda, T.; Mano, H.; Yamada, Y.; Takai, H.; Amizuka, N.; Kobori, M.; Izumi, N.; Kawashima, H.; Ozawa, H.; Ikeda, K. Calcium-sensing receptor in mature osteoclasts, which are bone resorbing cells. *Biochem. Biophys. Res. Commun.* **1998**, *245*, 419–422. [CrossRef] [PubMed]
47. Livak, K.J.; Schmittgen, T.D. Analysis of relative gene expression data using real-time quantitative PCR and the 2− ΔΔCT method. *Methods* **2001**, *25*, 402–408. [CrossRef] [PubMed]
48. Lee, J.; Kim, D.; Choi, J.; Choi, H.; Ryu, J.-H.; Jeong, J.; Park, E.-J.; Kim, S.-H.; Kim, S. Dehydrodiconiferyl alcohol isolated from Cucurbita moschata shows anti-adipogenic and anti-lipogenic effects in 3T3-L1 cells and primary mouse embryonic fibroblasts. *J. Biol. Chem.* **2012**, *287*, 8839–8851. [CrossRef]
49. Laurie, A.T.; Jackson, R.M. Q-SiteFinder: An energy-based method for the prediction of protein–ligand binding sites. *Bioinformatics* **2005**, *21*, 1908–1916. [CrossRef] [PubMed]

Article

Antibacterial Activity of *Arbutus pavarii* Pamp against Methicillin-Resistant *Staphylococcus aureus* (MRSA) and UHPLC-MS/MS Profile of the Bioactive Fraction

Nawal Buzgaia [1,2], Tahani Awin [1], Fakhri Elabbar [1], Khaled Abdusalam [2,3], Soo Yee Lee [2], Yaya Rukayadi [2,4], Faridah Abas [2,4] and Khozirah Shaari [2,5,*]

1 Department of Chemistry, Faculty of Science, University of Benghazi, Benghazi, Libya; nawalbuz@gmail.com (N.B.); awin_org@yahoo.com (T.A.); fakhri.dr@gmail.com (F.E.)
2 Natural Medicines and Products Research Laboratory (NaturMeds), Institute of Bioscience, Universiti Putra Malaysia, UPM Serdang 43400, Selangor, Malaysia; khaledbashirala.79@gmail.com (K.A.); daphne.leesooyee@gmail.com (S.Y.L.); yaya_rukayadi@upm.edu.my (Y.R.); faridah_abas@upm.edu.my (F.A.)
3 Department of Microbiology, Faculty of Science, University of Gharyan, Gharyan, Libya
4 Department of Food Science, Faculty of Food Science and Technology, Universiti Putra Malaysia, UPM Serdang 43400, Selangor, Malaysia
5 Department of Chemistry, Faculty of Science, Universiti Putra Malaysia, 43400 UPM Serdang, Selangor, Malaysia
* Correspondence: khozirah@upm.edu.my

Received: 2 October 2020; Accepted: 3 November 2020; Published: 11 November 2020

Abstract: *Arbutus pavarii* Pamp is a medicinal plant commonly used by local tribes in East Libya for the treatment of many diseases, such as gastritis, renal infections, cancer and kidney diseases. In this study, the antibacterial activity of the leaf and stem bark extracts of the plant against methicillin-resistant *Staphylococcus aureus* (MRSA), as well as the metabolite profiles of the bioactive fractions, was investigated. The antibacterial activity was determined by disc diffusion method, minimum inhibitory concentration (MIC) and minimum bactericidal concentration (MBC), while the microbial reduction by the bioactive fraction was evaluated using time–kill test. The bioactive fraction was further subjected to ultrahigh-performance liquid chromatography–mass spectrometry (UHPLC-ESI-MS/MS) analysis to putatively identify the chemical constituents contained therein. All the extracts and fractions showed different levels of antibacterial activity on the tested MRSA strains. The highest total antibacterial activity, i.e., 4007.6 mL/g, was exhibited by the crude leaf methanolic extract. However, the ethyl acetate fraction of the leaf showed moderate to significant antibacterial activity against MRSA at low MIC (0.08–1.25 mg/mL). Metabolite profiling of this fraction using UHPLC-ESI-MS/MS resulted in the putative identification of 28 compounds, which included phenolic acids, flavan-3-ols and flavonols. The results of this study showed that the ethyl acetate fraction of *Arbutus pavarii* leaf possessed potential antibacterial activity against MRSA and hence can be further explored for pharmaceutical applications as a natural antibacterial agent.

Keywords: *Arbutus pavarii*; antibacterial activity; MRSA; time–kill curves; ultrahigh-performance liquid chromatography; mass spectrometry (UHPLC- ESI-MS/MS)

1. Introduction

Millions of people are affected by contagious bacterial diseases throughout the world. These infectious diseases have persistently caused disability and death throughout mankind's history. According to the World Health Organization (WHO), approximately 50,000 people die from bacterial infectious diseases throughout the world every year [1]. Methicillin-resistant *Staphylococcus aureus* (MRSA) is a group of Gram-positive bacteria that are distinct from other strains of *Staphylococcus aureus* [2]. MRSA is usually found in hospitals, prisons and nursing homes, where the people with open wounds and deteriorated immune systems are at greater risk of hospital-acquired infections. Although MRSA began as a hospital-acquired infection, it can be found in all communities and livestock. The terms HA-MRSA (healthcare-associated or hospital-acquired MRSA), CA-MRSA (community-associated MRSA) and LA-MRSA (livestock-associated) reflect the MRSA infections in a variety of hosts [3]. The MRSA displayed resistance against many antibiotics such as methicillin, a semisynthetic β-lactam antibiotic. Generally, the β-lactam mechanisms of resistance of MRSA strains support cross-resistance to all β-lactam antibiotics [4]. The key mechanism for resistance is the enzyme-catalyzed modification and ultimate destruction of the antibiotic, causing its dynamic efflux from cells and antibiotic target alteration [5]. Therefore, there is a high demand to develop antibiotics from natural sources based on medical plant extracts in a bid to back up the effectiveness and potency of conventional antibiotics [6]. Natural products play an important role in drug discovery, as evidenced by over 50% of all modern clinical drugs being of natural product origin [7].

Medicinal plants are rich sources of secondary metabolites with various biological properties, including antimicrobial properties [6,8]. *Arbutus pavarii* Pamp, an endemic medicinal plant species known locally as Shmar in Libya, is an evergreen shrub belonging to the Ericaceae family [9,10]. In folk medicine, it is used for the treatment of gastritis, renal infections, cancer ailments and kidney diseases [11]. Previous phytochemical studies on *A. pavarii* showed that this plant contains mainly flavonoids, tannins, glycosides, simple phenolics, triterpenes and sterols [11]. In addition, it was also reported that *A. pavarii* demonstrated strong antibacterial activity against several pathogenic bacteria [11]. However, few studies have focused on determining the effect of *A. pavarii* extracts and its fractions against resistant bacterial strains. Thus, the aim of this study was to evaluate the *A. pavarii* leaf and stem bark extracts against methicillin-resistant *Staphylococcus aureus* (MRSA). The anti-MRSA activity of the crude methanolic extract and various solvent fractions were assayed using disc diffusion assay, followed by minimum inhibitory concentration (MIC) and minimum bactericidal concentration (MBC) determinations, as well as time–kill curve analysis. In addition, the active fraction was subjected to ultrahigh-performance liquid chromatography–mass spectrometry (UHPLC-ESI-MS/MS) analysis for the identification of potential bioactive compounds.

2. Results and Discussion

2.1. Antibacterial Activity Of *A. pavarii* Crude Extracts and Solvent Fractions

Disc diffusion test was used to first screen the crude methanolic extracts and solvent fractions for presence of antibacterial activity. The appearance of zones of inhibition produced around the discs was observed, and their diameters were measured and recorded (Table 1). The standard antibiotic, 0.1% CHX, showed inhibition zones ranging from 7.00 to 10.33 mm against the bacterial strains. At the test concentration of 10 mg/mL, the crude methanolic extracts of the leaf and stem bark showed inhibition zones in the ranges of 8.00–9.67 mm and 7.00–10.00 mm, respectively. Among the different solvent fractions of the leaf, the EtOAc fraction showed the greatest activity towards all the bacterial strains, giving inhibition zones of 13.66, 12.00, 13.67 and 13.00 mm against MRSA ATCC 700699, MRSA KCCM 12255, MRSA1 and MRSA2, respectively. The same trend was observed for the stem bark fractions. However, compared to the leaf EtOAc fraction, the stem bark EtOAc fraction showed smaller inhibition zones of 8.00–9.00 mm, indicating that the stem bark either contained different bioactive constituents or lower amounts of the same bioactive constituents [12]. Other solvent fractions of the leaf and stem bark showed no to weak activities against the test bacteria. Previously, Alsabri et al. [13] investigated the antibacterial properties of solvent extracts prepared from the aerial part of *A. pavarii*. They reported that the methanol extract exhibited the highest activity against *S. aureus*, *Escherichia coli* and *Candida albicans*. The chloroform extract was active only against *S. aureus*, while the n-hexane extract showed activity against *C. albicans*. Overall, these results indicated that the polarity of the solvent plays an important role in the extraction of the active ingredients and consequently in their potential antimicrobial activity.

The calculated relative inhibition zone diameter (RIZD) values of the test samples against the MRSA strains varied from 70.99% to 171.43%, as shown in Table 1. The RIZD value provides additional information showing the differential effects of the test extracts and fractions compared to the standard antibiotic used as a positive control. An RIZD value >100% means that the tested extract is more effective than the antibiotic. The leaf EtOAc fraction demonstrated the highest RIZD values, ranging from 125.85% to 171.43% against all MRSA strains. The higher RIZD percentages demonstrated by the leaf EtOAc fraction are a good indication that the leaf of *A. pavarii* contained most, and probably in higher amounts, of the antibacterial compounds of this plant species. It is worthy of note that the local population frequently uses the leaf material for medicinal purposes [11]. Based on the higher biological activity, the leaf and the stem bark EtOAc and n-BuOH fractions were subjected to further evaluation of their MIC, MBC and total activity values.

2.2. Bacteriostatic (MIC) and Bactericidal (MBC) Effects of Bioactive Extracts and Fractions

The antibacterial activity of the bioactive extracts and fractions was further investigated through the determination of the MIC and MBC values, as well as the total activity. The MIC and MBC values are presented in Table 2. The MIC values of the crude leaf methanolic extracts ranged between 0.08 and 1.25 mg/mL, while the MBC values ranged between 0.16 and 2.50 mg/mL. The leaf methanolic extract was more potent against the two standard MRSA strains, i.e., ATCC 700699 (MIC 0.08 mg/mL; MBC 0.16 mg/mL) and KCCM 12255 (MIC 0.63 mg/mL; MBC 1.25 mg/mL), in comparison to the clinical isolates, against which it showed MIC of 1.25 mg/mL and MBC of 2.5 mg/mL for both strains. A similar trend of potency was observed for the leaf fractions, where the standard MRSA strains were more susceptible to the fractions while the clinical isolates were less affected. The MIC and MBC for the EtOAc fractions were in the ranges of 0.08–1.25 mg/mL and 0.16–2.50 mg/mL, respectively; the MIC and MBC for n-BuOH fractions were 0.04–2.50 mg/mL and 0.08–5.00 mg/mL, respectively. In addition, among the activities exhibited by the leaf extract and fractions on the MRSA strains, the n-BuOH fraction showed the highest potency against the ATCC 700699 strain with MIC and MBC values of 0.04 and 0.08 mg/mL, respectively. The antimicrobial activity of an extract is considered very interesting and is of significant scientific value when its MIC values are lower than 100 μg/mL [14]. Hence, the present results revealed that the *A. pavarii* leaf methanolic extract and solvent fractions have moderate to significant activity against the tested MRSA strains.

On the other hand, the MIC values of the crude stem bark methanolic extract were lower than the leaf extract, ranging between 0.63 and 1.25 mg/mL, while the MBC values ranged between 1.25 and 2.50 mg/mL. The stem bark methanolic extract was more potent against the standard MRSA strain, ATCC 700699 (MIC 0.63 mg/mL; MBC 1.25 mg/mL), than against MRSA KCCM 12255 and the two clinical isolates as it showed MIC and MBC values of 1.25 and 2.5 mg/mL, respectively, against these three strains. In comparison, the stem bark EtOAc and n-BuOH fractions were less potent towards all the MRSA strains, except against the clinical isolate MRSA2 (MIC 0.63 mg/mL; MBC 1.25 mg/mL).

Overall, all the extracts and fractions showed different levels of antibacterial activity against the tested MRSA strains. This variation could be due to the different potencies of the bioactive compounds present in the extracts and fractions leading to different bacteriostatic and bactericidal effects on the bacterial strains, as reported by Qaralleh [15] and Oliveira et al. [16]. Several studies investigated the efficacy of plant extracts and their effective compounds as antibacterial agents to control infections by MRSA, suggesting that the bioactive component(s) of the plant extracts interact with enzymes and proteins of the bacterial cell membrane, causing its disruption, to disperse a flux of protons towards the cell exterior, which induces cell death or may inhibit enzymes necessary for the biosynthesis of amino acids [17].

Table 1. Inhibition zones of leaf and stem bark crude methanolic extracts and solvent fractions of *A. pavarii* against methicillin-resistant *Staphylococcus aureus* (MRSA) strains.

MRSA Strains	CHX	CH_3OH		EtOAc		n-BuOH	
		IZD	**% RIZD**	**IZD**	**% RIZD**	**IZD**	**% RIZD**
		Leaf					
ATCC 700699	8.00 ± 0.00	9.33 ± 0.57	120.83 ± 7.22	13.67 ± 0.57	170.83 ± 7.22	8.33 ± 0.57	100.00 ± 5.59
KCCM 12255	7.00 ± 0.00	8.00 ± 0.00	114.29 ± 0.00	12.00 ± 0.00	171.43 ± 0.00	7.00 ± 0.00	100.00 ± 0.00
MRSA1	10.33 ± 0.57	9.67 ± 0.57	93.58 ± 5.59	13.67 ± 1.15	132.30 ± 5.59	n.a	n.d
MRSA2	10.33 ± 0.57	9.33 ± 0.57	83.90 ± 5.59	13.00 ± 0.00	125.85 ± 0.00	n.a	n.d
		Stem bark					
ATCC 700699	8.00 ± 0.00	10.00 ± 0.00	125.00 ± 0.00	9.00 ± 0.00	112.50 ± 0.00	9.00 ± 0.00	112.50 ± 0.00
KCCM 12255	7.00 ± 0.00	7.00 ± 0.00	100.00 ± 0.00	8.00 ± 0.00	114.29 ± 0.00	7.00 ± 0.00	100.00 ± 0.00
MRSA1	10.33 ± 0.57	9.00 ± 0.00	87.12 ± 0.00	9.00 ± 0.00	87.12 ± 0.00	n.a	n.d
MRSA2	10.33 ± 0.57	7.67 ± 0.57	74.22 ± 5.59	8.33 ± 0.57	80.67 ± 5.59	7.33 ± 0.57	70.99 ± 5.59

CH_3OH = methanol extract, EtOAc = ethyl acetate fraction, n-BuOH = butanol fraction. Hexane and chloroform fractions showed no inhibition zones. MRSA1 and MRSA2 are clinical isolates, n.a = no activity (no inhibition zone detected). n.d = not detected. Diameter of inhibition zones in mm (including disc). Positive control: 0.1% CHX; negative control: 10% DMSO. Values are expressed as means ± standard deviation (SD).

Table 2. Minimum inhibitory concentration (MIC) and minimum bactericidal concentration (MBC) values (mg/mL) of leaf and stem bark crude methanolic extracts and solvent fractions of *A. pavarii* against MRSA strains.

MRSA Strains	Parts	CH_3OH		EtOAc		n-BuOH		CHX	
		MIC	MBC	MIC	MBC	MIC	MBC	MIC	MBC
ATCC 700699	Leaf	0.08	0.16	0.08	0.16	0.04	0.08	0.02	0.63
	Stem bark	0.63	1.25	1.25	2.50	2.50	5.00		
KCCM 12255	Leaf	0.63	1.25	0.31	1.25	0.63	1.25	0.02	0.63
	Stem bark	1.25	2.50	1.25	2.50	2.50	5.00		
MRSA1	Leaf	1.25	2.50	1.25	2.50	2.50	5.00	0.03	0.13
	Stem bark	1.25	2.50	1.25	2.50	2.50	5.00		
MRSA2	Leaf	1.25	2.50	1.25	2.50	1.25	2.50	0.03	0.13
	Stem bark	1.25	2.50	0.63	1.25	0.63	1.25		

CH_3OH = methanol extract, EtOAc = ethyl acetate fraction, n-BuOH = butanol fraction, CHX = 0.1% chlorhexidine (standard antibiotic). Hexane and chloroform fractions showed no inhibition zones. MRSA1 and MRSA2 are clinical isolates.

Besides MIC and MBC values, the antibacterial activity against the MRSA strains was also determined based on the total activity of the extracts and fractions. Total activity is defined as the volume to which the biologically active component (extracts, fractions or compounds) present in 1 g of dried plant material can be diluted and still kill the bacteria [18]. Total activity is useful for the selection of sample material for isolating bioactive compounds. Extracts or fractions with large total activity values are considered the best material for isolating potentially bioactive compounds. As shown in Table 3, the total activity values of the extracts and fractions of *A. pavarii* leaf and stem bark demonstrated high variation. The leaf methanolic extract diluted in 4007.60 mL of solvent can still inhibit the growth of MRSA ATCC 700699 (total activity: 4007.60 mL/g). The leaf n-BuOH fraction possessed higher total activity against MRSA ATCC 700699, with a value of 2235.89 mL/g. The leaf EtOAc fraction has higher total activity against MRSA2 and MRSA ATCC 700699, with 2158.97 and 1078.10 mL/g values, respectively. Both the extract and fractions of the stem bark exhibited lower total activity against all the tested MRSA strains as compared to leaf, with values ranging from 54.22 to 452.35.

Table 3. Total activity of leaf and stem bark crude methanolic extracts and solvent fractions of *A. pavarii* against MRSA strains.

MRSA Strains	Total Activity in (mL/g)					
	Leaf			Stem Bark		
	CH_3OH	EtOAc	n-BuOH	CH_3OH	EtOAc	n-BuOH
ATCC 700699	4007.6	1078.1	2235.89	452.35	61.18	54.22
KCCM 12255	500.8	269	139.52	226.16	61.18	54.22
MRSA1	250.4	67.36	34.88	113.08	61.18	54.22
MRSA2	62.6	2158.97	69.76	226.16	122.37	216.90

2.3. Time–Kill Curve for Ethyl Acetate Fraction of the Leaf

A time–kill assay, using the four bacterial strains, was performed for the leaf EtOAc fraction since it exhibited a stronger antibacterial activity in comparison to the other fractions. Although MIC value gives a good indication of the efficacy of an antimicrobial agent, it provides limited information on the kinetics of the antimicrobial action [19]. A better method of assessing the bactericidal or bacteriostatic activity of an antimicrobial agent over time is by using time–kill kinetics assay, where the effect of various concentrations of the antimicrobial agent over time in relation to the growth stages of the bacteria is monitored [20]. The bacterial strains were thus exposed to the EtOAc fraction, at test concentrations of 0, 0.5, 1, 2, 4 and 8 × MIC over a period of 4 h, and the time–kill curve was plotted.

The assay results for MRSA ATCC 700699 (Figure 1A) revealed that the bacteria were completely killed after 4 h when a concentration of 4 × MIC (0.63 mg/mL) was used and after 2 h with the higher concentration of 8 × MIC (1.25 mg/mL). In terms of practical application, the 4 h killing time would be more preferred since the effect was obtained using a lower concentration (0.63 mg/mL) of the disinfecting agent. This condition is similar to that of a drug that exhibits a concentration-dependent bactericidal action, where the bactericidal effect is dependent on the dose of the leaf EtOAc fraction rather than on incubation time [21]. On the other hand, the time–kill curves for MRSA KCCM 12255 (Figure 1B) showed that the time–kill endpoint was achieved after 2 h incubation with a higher concentration of 8 × MIC (2.5 mg/mL). Meanwhile, in the case of the clinical isolates, as illustrated in Figure 1C,D, the time–kill endpoint could only be achieved with a concentration of 4 × MIC (5 mg/mL) after 1 h of incubation.

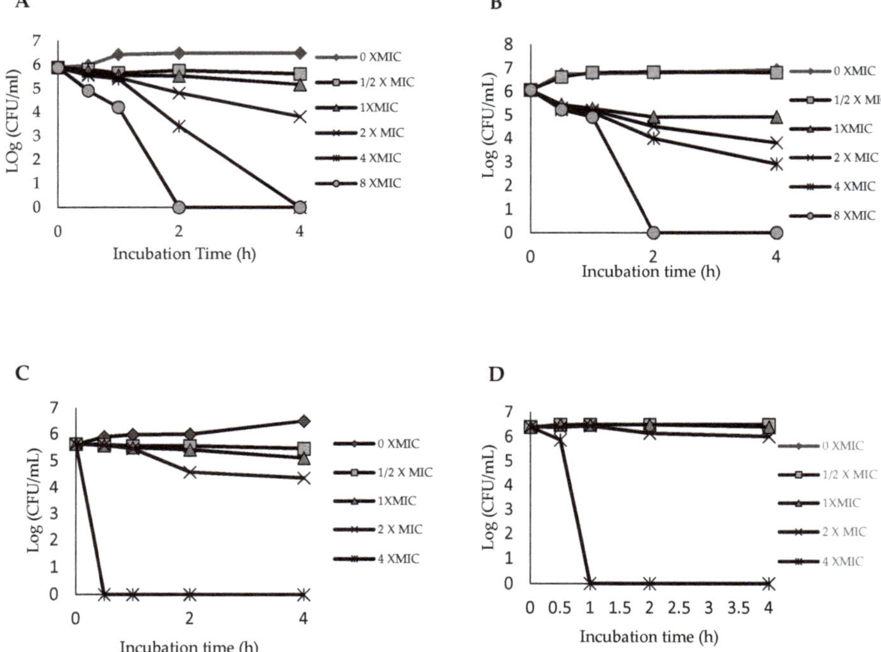

Figure 1. Time–kill curves for leaf EtOAc fraction against (**A**) MRSA ATCC 700699, (**B**) MRSA KCCM 12225, (**C**) MRSA1 and (**D**) MRSA2.

The data demonstrated that the bactericidal ability of the leaf EtOAc fraction is dependent on concentration and the bacterial strain. Generally, the time–kill kinetics results reasserted the expectation that a more concentrated sample will kill the microorganism in a shorter period of time. An increase in concentrations of plant extracts leads to an increase in the diffusion of phytochemicals into the cell membrane of bacteria, thus causing membrane destruction [22]. Furthermore, the bioactive compounds in the fraction may inhibit the synthesis of essential metabolites such as folic acid by preventing the enzymatic reaction. The protein synthesis in the microorganisms also can be inhibited if the bioactive compounds interfere and change the shape of the ribosome, which may lead to misreading of genetic code on mRNA [22]. The results of this time–kill kinetics study, together with the other results presented earlier, including disc diffusion assay, MIC, MBC and total activity determinations, reveal that the *A. pavarii* leaf possesses bacteriostatic and bactericidal effects against the tested MRSA strains, and the bioactive constituents could be largely present in the ethyl acetate fraction. Consequently, the EtOAc fraction was subjected to dereplication using UHPLC-MS/MS in order to gain an insight into the potential bioactive constituents.

2.4. UHPLC-ESI–MS/MS Profile of the EtOAc fraction

Several compounds from the classes of hydroxyquinone (arbutin), phenolic acid (caffeic, ferulic, gallic, rosmarinic, chlorogenic and salicylic acids), flavonoid (catechin, quercetin, dihydroquercetin, isoquercitrin, kaempferol, myricetin, rutin, naringin, neodiosmin, naringenin-7-*O*-glucoside, isovitexin-7-*O*-glucoside and delphinidin-3-*O*-rutinoside) and triterpenoid (oleanolic acid, lupeol and α-amyrin) have been previously reported to be present in *A. pavarii* [11,23,24]. In the present study, 28 compounds were putatively identified from the negative UHPLC-MS/MS spectrum of the leaf EtOAc fraction. The base peak chromatogram is shown in Figure 2, and compounds identified along with their spectral data are shown in Table 4. The results showed that the fraction was rich in phenolic compounds.

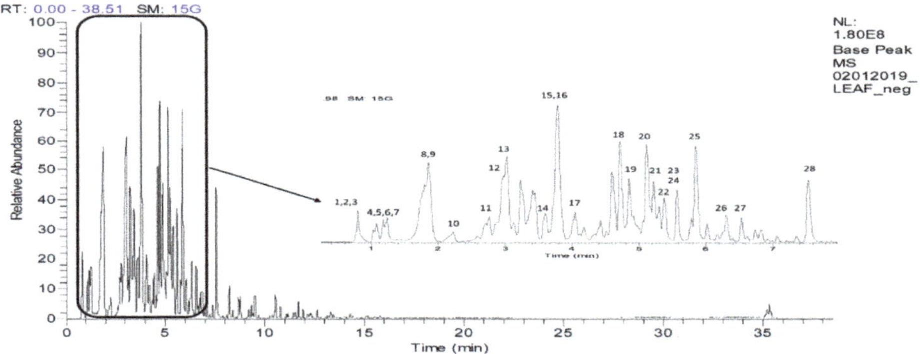

Figure 2. UHPL-ESI-MS/MS base peak chromatogram of the leaf EtOAc fraction of *A. pavarii* in negative ion mode.

Table 4. Compounds identified in the leaf EtOAc fraction of *A. pavarii*.

No	Retention Time (Rt) (min)	[M-H]$^-$ (m/z)	MS/MS Fragment Ions (m/z)	Compound Identity	Molecular Formula
			Phenolic Acids and Derivatives		
2	0.78	331.0668	271.05, 211.02, 169.01	Gallic acid hexoside I	$C_{13}H_{16}O_{10}$
4	1.04	331.0669	271.05, 211.02, 169.01	Gallic acid hexoside II	$C_{13}H_{16}O_{10}$
5	1.18	169.0131	125.02	Gallic acid	$C_7H_6O_5$
6	1.22	331.0668	271.05, 211.02, 169.01	Gallic acid hexoside III	$C_{13}H_{16}O_{10}$
7	1.23	343.0668	191.06, 169.01, 125.02	Galloylquinic acid	$C_{14}H_{16}O_{10}$
8	1.73	315.0720	153.02, 152.01, 109.03, 108.02	Dihydroxybenzoic acid-*O*-hexoside	$C_{13}H_{16}O_9$
11	2.89	483.0774	439.09, 424.54, 331.07, 313.06, 287.08, 271.05, 211.02, 169.01	*Di-O*-galloylhexose	$C_{20}H_{20}O_{14}$
14	3.43	329.0878	167.03, 152.01, 123.04, 108.02	Vanillic acid-*O*-hexoside	$C_{14}H_{18}O_9$
16	3.95	635.0888	465.07, 313.06, 271.05, 211.02, 169.01	*Tri-O*-galloylhexose	$C_{27}H_{24}O_{18}$
			Flavan-3-ol and Derivatives		
9	1.83	305.06638	261.08, 179.03, 138.03, 137.02, 125.02	(Epi)gallocatechin	$C_{15}H_{14}O_7$
10	2.11	451.1254	289.07, 245.08, 151.04, 125.02	(Epi)catechin-3-*O*-hexoside	$C_{21}H_{24}O_{11}$
12	2.90	577.1334	451.10, 425.09, 407.08, 289.07, 287.06, 245.08, 125.02	(Epi)catechin +(epi)catechin I	$C_{30}H_{26}O_{12}$
13	3.11	289.0714	271.06, 245.08, 179.03, 165.02, 150.03, 137.02, 125.02	Catechin	$C_{15}H_{14}O_6$

Table 4. *Cont.*

No	Retention Time (Rt) (min)	[M-H]⁻ (*m/z*)	MS/MS Fragment Ions (*m/z*)	Compound Identity	Molecular Formula
15	3.88	289.0717	271.06, 245.08, 179.03, 165.02, 150.03, 137.02, 125.02	Epicatechin	$C_{15}H_{14}O_6$
17	4.03	729.1458	577.14, 559.13, 451.10, 425.09, 407.08, 289.07, 125.02	(Epi)catechin gallate + (epi)catechin I	$C_{37}H_{30}O_{16}$
20	5.20	441.0823	289.07, 245.08, 203.07, 169.01	(Epi)catecin gallate	$C_{22}H_{18}O_{10}$
21	5.25	729.1453	577.11, 407.08, 425.09, 289.07, 125.02	(Epi)catechin gallate + (epi)catechin II	$C_{37}H_{30}O_{16}$
			Flavonols and Derivatives		
18	4.72	615.0989	463.09, 300.03, 301.03, 271.02, 179.00, 151.00, 169.01	Quercetin-*O*-galloylhexoside	$C_{28}H_{24}O_{16}$
19	4.98	609.1463	301.03, 300.03, 271.02, 255.03	Quercetin-3-*O*-deoxyhexosyl-hexoside	$C_{27}H_{30}O_{16}$
22	5.40	463.0884	317.03, 316.02, 287.02, 271.02, 179.00, 151.00	Myricetin-3-*O*-deoxyhexoside	$C_{21}H_{20}O_{12}$
23	5.55	433.0775	301.03, 300.03, 271.02, 255.03	Quercetin-3-*O*-pentoside	$C_{20}H_{18}O_{11}$
24	5.56	447.0931	285.04, 284.03, 255.03, 227.03	kaempferol-3-*O*-hexoside	$C_{2120}O_{11}$
25	5.99	447.0931	301.03, 300.03, 271.02, 255.03	Quercetin-3-*O*-deoxyhexoside	$C_{21}H_{20}O_{11}$
26	6.30	463.0885	301.03, 300.03, 271.07, 255.03	Quercetin-3-*O*-hexoside	$C_{21}H_{20}O_{12}$
27	6.42	583.1099	463.09, 301.03, 300.03, 271.03, 255.03	Quercetin-*O*-(*p*-hydroxy) benzonylhexoside	$C_{28}H_{24}O_{14}$
28	7.53	301.0354	271.02, 255.03, 179.00, 151.00, 149.02, 121.03, 121.03, 107.01	Quercetin	$C_{15}H_{10}O_7$
			Others		
1	0.76	191.0555	171.03, 127.04, 109.03, 93.03	Quinic acid	$C_7H_{12}O_6$
3	0.80	271.0453	211.02, 108.02	Arbutin	$C_{12}H_{16}O_7$

2.4.1. Identification of Phenolic Acids and Derivatives

Compounds **2**, **4**, **5**, **6**, **7**, **8**, **11**, **14** and **16** were identified as gallic acid and its derivatives based on the presence of the aglycone fragment ion at *m/z* 169 and the characteristic fragment ions at *m/z* 271 and 211 in their MS/MS spectra [25]. Compound **5**, with a pseudomolecular ion at *m/z* 169.0131, was assigned as gallic acid, showing the characteristic base peak at *m/z* 125 for $[M-H-CO_2]^-$. Compounds **2**, **4** and **6**, eluting at three different retention times (0.78, 1.04 and 1.18 min, respectively), were identified as isomers of gallic acid hexoside (I-III). These compounds exhibited pseudomolecular ions at *m/z* 331.0668, 331.0669 and 331.0668, respectively, and all three produced a fragment ion at *m/z* 169 for $[M-H-162]^-$, due to the neutral loss of a hexoxyl moiety. This agrees with previous reports by Mendes et al. [26] and Abu-Reidah et al. [27]. Meanwhile, compound **7** exhibited a pseudomolecular ion at *m/z* 343.0668. The compound was assigned as galloylquinic acid based on the presence of base peak at *m/z* 169 and fragment ion at *m/z* 125 for a further loss of CO_2, all of which were characteristic fragment ions of gallic acid moiety [28]. Compounds **11** and **16** were identified as *di-O*-galloylhexose and *tri-O*-galloylhexose, respectively, based on similar fragmentation pattern showing losses of the corresponding number of galloyl moieties and the presence of a base peak at *m/z* 169 for the gallic acid aglycone.

Compound **8** has a pseudomolecular ion of *m/z* 315.0720, indicative of the molecular formula $C_{13}H_{16}O_9$. It was identified as dihydroxybenzoic acid *O*-hexoside based on fragment ion at *m/z* 153 for $[M-H-162]^-$, due to the loss of a hexoxyl moiety, and fragment ion at *m/z* 109 for $[M-H-162-44]^-$ indicating a further loss of CO_2 moiety, in agreement with Karar and Kuhnert [29]. Meanwhile, compound **14**, which exhibited a pseudomolecular ion of *m/z* 329.0878 and base peak at *m/z* 167 for $[M-H-162]^-$ for a neural loss of a hexoxyl moiety, was assigned as vanillic acid-*O*-hexoside.

The assignment was supported by comparison with the fragmentation pattern previously reported by Morales-Soto et al. [30].

2.4.2. Identification of Flavan-3-ol and Derivatives

Compounds **9**, **10**, **12**, **13**, **15**, **17**, **20** and **21** were identified as (epi)catechin and its derivatives based on the presence of fragment ions at *m/z* 289 and 125, corresponding to the (epi)catechin aglycone [31]. Compound **9**, which displayed a pseudomolecular ion at *m/z* 305.0663, was identified as (epi)gallocatechin based on the fragment ion at *m/z* 179 for $[M\text{-}H\text{-}126]^-$, due to the characteristic loss of the trihydroxybenzene moiety [32]. Compound **10**, with pseudomolecular ion at *m/z* 451.1254, was identified as (epi)catechin-3-*O*-hexoside based on the fragment ion at *m/z* 289, for the loss of a hexoxyl moiety [33]. Compounds **13** (Rt = 5.44 min) and **15** (Rt 7.74 min) showed similar pseudomolecular ions at *m/z* 289.0714 and 289.0717, respectively. By comparison of their elution order with a previous study by Stöggl et al. [34], the compound eluted earlier was identified as catechin while the one eluted later was identified as epicatechin. Both compounds yielded the fragment ions at *m/z* 137 and 151 which were the results of retro-Diels–Alder (RDA) cleavage at ring C of the flavan-3-ol structure.

Three compounds (**12**, **17** and **21**) were identified as the dimeric forms of B-type proanthocyanidins (PAs), which could be differentiated from the A-type Pas with the extra 2 Da in their pseudomolecular ion [35]. Compound **12**, showing pseudomolecular ion at *m/z* 577.1334, was identified as the (epi)catechin + (epi)catechin. The compound also exhibited a fragment ion at *m/z* 425 ($[M\text{–}H\text{-}152]^-$), which was due to the characteristic RDA cleavage at ring C of the dimer top unit [35]. Another fragment ion at *m/z* 407 ($[M\text{-}H\text{-}152\text{-}18]^-$) due to the subsequent loss of a water molecule from the parent molecule was also observed. The presence of two other dimeric derivatives, **17** and **21**, was also indicated by the pseudomolecular ions at *m/z* 729.1458 (Rt = 4.03 min) and 729.1453 (Rt = 5.25 min). These compounds were identified as (epi)catechin gallate + (epi)catechin isomers based on the fragment ion at *m/z* 577 indicative of galloyl moiety losses ($[M\text{-}H\text{-}152]^-$) from the parent ion [36]. Compound **20** at Rt = 5.20 min was identified as (epi)catechin-3-*O*-gallate. It displayed a pseudomolecular ion at *m/z* 441.0823. Its fragmentation pattern showed a fragment ion at *m/z* 289 for $[M\text{-}H\text{-}169]^-$, which corresponded to a loss of gallic acid moiety via cleavage of the ester bond and loss of the (epi)catechin unit [37].

2.4.3. Identification of Flavonols and Derivatives

The ethyl acetate fraction also contained the flavonol quercetin (**28**) and several of its derivatives (**18**, **19**, **23**, **25**, **26**, **27** and **28**). Quercetin (**28**) was identified based on its pseudomolecular ion at *m/z* 301.03 and fragment ions at *m/z* 271, 255, 179 and 151 [36]. Compound **18**, with pseudomolecular ion at *m/z* 615.0997, displayed fragment ions at *m/z* 463 for $[M\text{-}H\text{-}169]^-$, indicating loss of a galloyl moiety, and at *m/z* 301 for $[M\text{-}H\text{-}331]^-$, indicating an additional loss of a hexoxyl moiety. Compound **18** was thus deduced to be quercetin-*O*-galloylhexoside, based on these data and data reported by Mendes et al. [26].

Compounds **19**, **23**, **25** and **26** were assigned as quercetin-3-*O*-deoxyhexosylhexoside, quercetin-3-*O*-pentoside, quercetin-3-*O*-deoxyhexoside and quercetin-3-*O*-hexoside. These compounds exhibited pseudomolecular ions at *m/z* 609.1463, 433.0775, 447.0931 and 463.0885, respectively. The transition of these ions to the aglycone ion (Y_0^-) at *m/z* 301 revealed the losses of the respective sugar moieties [34]. The glycosylation at the C-3 position of these compounds was determined by the higher relative abundance of their radical aglycone ion ($[Y_0\ H]^-$ *m/z* 300) than the Y_0^- ion (*m/z* 301) [38]. Compound **27**, with a pseudomolecular ion at *m/z* 583.1099, was assigned as quercetin-*O*-(*p*-hydroxy)benzonylhexoside. The compound showed a fragment ions at *m/z* 463 for $[M\text{-}H\text{-}120]^-$, indicating a loss of hydroxybenzoyl moiety, and *m/z* 301 ($[M\text{-}H\text{-}282]^-$) for a further loss of hexoxyl moiety, in agreement with data reported by Jaiswal et al. [36].

Compound **22** was identified as myricetin-3-*O*-hexoside based on the presence of fragment ions at *m/z* 317, 316, 179 and 151, corresponding to the aglycone myricetin. The deprotonated aglycone peak observed at *m/z* 316.02 [M-H-162]$^-$ was due to the loss of a hexoxyl moiety [39]. Compound **24** with a pseudomolecular ion at *m/z* 447.093 was identified as kaempferol-3-*O*-hexoside. This compound showed characteristic fragment ion at *m/z* 285 due to the loss of sugar moiety and fragment ions at *m/z* 255 and 227 which are due to the loss of [M-162-CHO]$^-$ and [M-162-H_2O-CO], respectively. Similarly, the fragment ions at *m/z* 179 and 151 were due to RDA cleavage of C-ring [39]. Attachment of the sugar moiety at the C-3 position of these compounds was also determined based on the relative abundance of [Y_0 H]$^-$ and Y_0^- ions [38].

2.4.4. Identification of Other Compounds

Compound **1**, with a pseudomolecular ion at *m/z* 191.0555 ($C_7H_{12}O_6$), was identified as quinic acid. It yielded fragment ions at *m/z* 171 ([M-H-H_2O]$^-$), 127.04 ([M–H-CO_2-H_2O]$^-$) and 109 ([M–H-CO_2-2H_2O]$^-$) [28]. Compound **3**, with pseudomolecular ion at *m/z* 271.0453 ($C_{12}H_{16}O_7$), was identified as arbutin; it yielded a fragment ion at *m/z* 108 for [M-H-162]$^-$ due to loss of the hexose moiety [40].

The UHPLC-MS/MS results showed that the flavonoids and phenolic acid components are major secondary metabolites in the ethyl acetate fraction of *A. pavarii* leaf. Furthermore, among the identified compounds, several of them have been previously reported to possess antibacterial activity against MRSA. Shibata et al. [41] reported that gallic acid has antibacterial activity against MRSA with MIC value of 62.5 µg/mL. Catechins are often linked to antimicrobial effects associated with their interactions with the microbial cell membrane [42]. Cushnie et al. [43] reported that membrane disruption by catechins causes potassium leakage in MRSA strain, which is the first indication of membrane damage in microorganisms [44]. In addition, several studies have shown that the effectiveness of β-lactams can be enhanced by combining them with epigallocatechin gallate [45,46] and epicatechin gallate [47]. Meanwhile, Su et al. [48] reported that quercetin exhibited inhibitory effect against different MRSA strains, with MIC values ranging from 31.25 to 125 µg/mL, while rutin, a quercetin-3-*O*-deoxyhexosylhexoside, was reported to inhibit MRSA with MIC value of 250 µg/mL [49]. Besides, arbutin was reported to exert antibacterial activity against MRSA with MIC value of 10 mg/mL and MBC value of 20 mg/mL [50]. Therefore, the presence of these compounds, especially the flavonoids and phenolic acids, could have contributed significantly to the antibacterial activity of the leaf ethyl acetate fraction of *A. pavarii*.

3. Materials and Methods

3.1. Plant Materials, Extraction and Fractionation

The leaf and stem bark of *A. pavarii* were obtained from Al Jabal Al Akhdar region, Northeast Libya in March 2016 and identified by Dr. Abdulamid Alzerbi, a botanist at Biology Department of Benghazi University, Libya. The leaf and stem bark were dried under shade before being pulverized into a powder using a mechanical grinder (model: MX1100XT11CE, Waring, S/NoB 8643, Atlanta, GA, USA). The powdered plant material was sieved with a steel sieve (80 mesh) to obtain a uniform fine powder. For extraction, 1500 g of the ground leaf and 500 g of the stem bark were separately mixed with methanol at 1:10 solid-to-liquid ratio. The mixtures were sonicated at 35 °C for 60 min with a frequency of 53 kHz using an ultrasonic water bath (Branson, model 8510E-MTH, Danbury, CT, USA). The crude methanolic extracts were filtered (Whatman No. 1 filter paper, USA), and the collected filtrate was concentrated at 45 °C under reduced pressure using a rotary evaporator (Buchi, USA). The crude methanolic extracts were further fractionated using liquid–liquid fractionation to obtain solvent fractions of different polarities, namely hexane, chloroform, ethyl acetate (EtOAc) and n-butanol (n-BuOH) fractions (Merck, Darmstadt, Germany). The yields and physical appearance of the various extracts and fractions are tabulated in Table 5.

Table 5. Yields of extracts and solvent fractions of *Arbutus pavarii*.

Plant Part	Solvent	Weight (g)	Yield %	Physical Appearance
Leaf	CH_3OH	470.00	31.13	Dark greenish brown gum
	Hex	18.60	3.95	Dark green gum
	$CHCl_3$	24.90	5.29	Green gum
	EtOAc	126.48	26.91	Dark orange gum
	n-BuOH	131.00	27.87	Brown gum
Stem Bark	CH_3OH	141.35	28.27	Greenish brown gum
	Hex	11.42	8.08	Dark green gum
	$CHCl_3$	3.65	2.58	Green gum
	EtOAc	38.24	27.05	Dark brown gum
	n-BuOH	67.78	47.95	Dark brown gum

3.2. Bacterial Strains and Preparation of Inoculum

MRSA ATCC 700699 was obtained from the American Type Culture Collection (Rockville, MD, USA) while MRSA KCCM 12255 was obtained from the Korean Culture Center of Microorganisms (Seoul, South Korea). Two clinical isolates (MRSA1 and MRSA2) were collected from the nasal swab of a 4th-year medical student from University Putra Malaysia, Malaysia. The MRSA strains are kept at the Laboratory of Natural Products (Institute of Bioscience, UPM, Malaysia). The MRSA ATCC 700699, MRSA KCCM 12255 MRSA1 and MRSA2 were grown on Mueller Hinton agar (MHA) (Difco, Franklin Lakes, NJ, USA) aerobically for 24 h at 37 °C, whereas inoculum cell suspension was prepared by transferring and incubating a single colony of each bacterial species in 10 mL of Mueller Hinton broth (MHB) at 37 °C overnight with 200 rpm agitation. Then, 1 µL of bacteria suspension was transferred to new MHB in a ratio of 1:10 to yield an inoculum size of 10^6 CFU/mL.

3.3. Disc Diffusion Assay

Antibacterial activity was evaluated using agar diffusion assay, according to Rukayadi et al. [51]. Briefly, an inoculum of the bacterial strain was streaked on the surface of MHA plates using a sterile cotton swab. Sterile 6 mm filter paper discs (Whatman, Germany) were prewetted with 10 µL aliquot of the test extracts or fractions, prepared in DMSO at a concentration of 10 mg/mL. The discs were then placed on the inoculated plates at an appropriate distance from each other. Positive (chlorhexidine, 0.1% CHX, St Louis, MO, USA) and negative (dimethyl sulfoxide, 10% DMSO, Merck, Darmstadt, Germany) control discs were similarly prepared and placed on each test plate. Inoculated plates were subsequently incubated for 24 h at 37 °C and observed for inhibition zones. All experiments were conducted in triplicate, and inhibition zone diameter (IZD) was measured in mm. Antibacterial activity was expressed as the percentage of relative inhibition zone diameter (RIZD) with respect to standard antibiotic (0.1% CHX), according to Alsohaili and Al-fawwaz [52], and calculated using the following formula:

$$\%\ RIZD = [(IZD_{sample} - IZD_{negative\ control})/IZD_{standard\ antibiotic}] \times 100 \quad (1)$$

3.4. Determination of Minimum Inhibitory Concentration (MIC) and Minimum Bactericidal Concentration (MBC) Values

The MIC and MBC values of the test samples against the MRSA strains were established as described by the Clinical and Laboratory Standards Institute (CLSI) [19]. The determination was performed in a 96-well round-bottom microtiter plate (Greiner, Germany) using a 2-fold standard broth microdilution method with an inoculum of about 10^6 CFU/mL. The first well, designated as the negative control, was filled with 100 µL MHB. The second well, designated as the positive control, was filled with 100 µL of the bacterial suspension. A 100 µL aliquot of the test extract or fraction, prepared at a concentration of 10 mg/mL, was then added to the 12th well. Two-fold dilutions were then made from the 12th well down to the 3rd well. Therefore, the 12th well contained the highest concentration (5 mg/mL), while the 3rd well contained the lowest concentration (0.01 mg/mL). The plate was then incubated aerobically for 24 h at 37 °C. After the incubation period, the MIC of the test extract or solvent fraction was determined. The MIC value is defined as the lowest concentration of the test sample that inhibited bacterial growth completely. For determining the MBC value, a 10 μL aliquot of the suspension in each of the 12 wells of the MIC determination was subcultured on an MHA plate. The plate was then incubated for 24 h, at 37 °C. After the incubation period, the plate was observed for bacterial growth and the MBC value was determined. The MBC is defined as the lowest concentration of the test sample that killed the bacterial strain completely. The MIC and MBC values were determined in duplicate. Chlorhexidine (0.1% CHX, St Louis, MO, USA) was used as a positive control. The antimicrobial activity of plant extracts may be expressed in different ways, including total activity values [18]. The total activity of the extract and the fractions was estimated as follows:

$$\text{Total activity} = \text{Quantity of material extracted from 1 g of plant material/MIC} \quad (2)$$

3.5. *Time–Kill Curve*

Time–kill assay against the MRSA strains was performed according to Ramli et al. with slight modifications [22]. Briefly, the inoculum suspension of MRSA was diluted to approximately 10^6 CFU/mL. The ethyl acetate fraction of the leaf was diluted with the MHB medium containing inoculum to obtain final concentrations of 0 × MIC, 0.5 × MIC, 1 × MIC, 2 × MIC, 4 × MIC and 8 × MIC for MRSA ATCC 700699 and MRSA KCCM 12255 and final concentrations of 0 × MIC, 0.5 × MIC, 1 × MIC, 2 × MIC and 4 × MIC for MRSA1 and MRSA2. Cultures (1 mL final volume) were incubated at 30 °C with 200 rpm agitation. At predetermined time points (0, 0.5, 1, 2 and 4 h), 10 μL aliquots were transferred to clean microcentrifuge tubes. The aliquots were serially diluted with 990 μL of 1% phosphate-buffered saline (PBS), and 20 μL was staked onto the MHA plates. The number of colonies formed on the plates after incubation at 30 °C for 24 h was counted and the number of CFU/mL was calculated. Assays were carried out in triplicate. The graph of log CFU/mL versus time was plotted as described by Ramli et al. [22].

3.6. *UHPLC-ESI-MS/MS Analysis*

The bioactive fraction was separated using a Hypersil Gold C18 reversed-phase column (2.1 × 100 mm, 1.9 μm, Thermo, USA) on a Thermos Scientific Ultimate 3000 (Bremen, Germany) with a mobile phase consisting of LCMS grade water (solvent A) and acetonitrile (solvent B), each containing 0.1% formic acid flowing at 0.4 mL/min. The programmed gradient system consisted of 0 min (95% A), 1 min (95% A), 20 min (5% A), 25 min (5% A), 25.1 min (95% A) and 35 min (95% A). The sample of 1 mg/mL (*w*/*v*) was prepared by dissolving 1 mg of a dried sample of the active fraction with 1 mL of methanol. The resultant mixture was then filtered using 0.22 μm Nylon membranes, and then 10 μL of the filtrate was auto-injected. The MS analysis was done on a Q-Exactive Focus Orbitrap LC-MS/MS system. The ESI-MS parameters were set as follows: negative mode, collision energy of 3.5 kV, capillary temperature 350 °C, auxiliary gas heater temperature 0 °C, sheath gas flow rate 40 arbitrary units and auxiliary nitrogen gas (99% pure) flow rate 8 arbitrary units. Then, the mass resolution was set to 70,000 full width at half maximum (FWHM) and a full scan of 150–2000 amu. The identification analysis was carried out by comparing the obtained MS/MS data with the literature.

3.7. *Data Analysis*

Microsoft Excel (Version 2010) was employed to perform the statistical analysis. Disc diffusion results were given as a mean ± standard deviation with three replicates.

4. Conclusions

In this study, the leaf and stem bark of *A. pavarii* were evaluated for anti-MRSA activity. The antibacterial activity was performed using disc diffusion agar test, MIC and MBC assays, in which the methanolic extracts and fractions of leaf and stem bark of *A. pavarii* demonstrated potential antibacterial activity against the tested MRSA strains. Among the extracts and fractions, the EtOAc fraction of *A. pavarii* leaf revealed the highest antibacterial activity against all tested MRSA strains, with activity ranging from moderate to significant (MIC 0.08–1.25 mg/mL). In time–kill analysis, the MRSA strains were found to be completely killed after exposure to this fraction for 30 min to 2 h at 4× MIC and 8× MIC, revealing a remarkable capacity to inhibit or kill the MRSA strains. The UHPLC-ESI-MS/MS profiling of the bioactive fraction revealed that it contains high amounts of polyphenolic compounds. Phenolic acid and flavonoids were the main components and could be responsible for the bioactivity. The present findings add support for the traditional medicinal use of *A. pavarii* and highlight its potential as a source of natural antibacterial agents for future exploitation as natural antibiotics in the fight against MRSA prevalence.

Author Contributions: Conceptualization, K.S., Y.R., F.A. and N.B.; methodology, K.S., Y.R. and N.B.; investigation, N.B.; resources, K.A., F.E. and T.A.; data curation, K.S., Y.R., S.Y.L. and N.B.; Writing—Original draft preparation, N.B.; Writing—Review and editing, K.S. and S.Y.L. All authors have read and agreed to the published version of the manuscript.

Funding: This research received no external funding.

Acknowledgments: The first author gratefully acknowledges support from the Ministry of Higher Education Libya for scholarship.

Conflicts of Interest: The authors declare no conflict of interest.

References

1. Laken, I.I.; Musah Monday, D.M.; Mohammed, S.H.; Paiko, Y.B. Phytochemical and antibacterial activity of *chrysanthellum indicum* (linn) extracts. *Afr. J. Environ. Nat. Sci. Res.* **2019**, *2*, 73–82.
2. Prabhoo, R.; Chaddha, R.; Iyer, R.; Mehra, A.; Ahdal, J.; Jain, R. Overview of methicillin resistant *Staphylococcus aureus* mediated bone and joint infections in India. *Orthop. Rev.* **2019**, *11*, 8070. [CrossRef]
3. Haysom, L.; Cross, M.; Anastasas, R.; Moore, E.; Hampton, S. Prevalence and risk factors for methicillin-resistant *Staphylococcus aureus* (MRSA) infections in custodial populations: A systematic review. *J. Correct. Health Care* **2018**, *24*, 197–213. [CrossRef]
4. Gao, C.; Fan, Y.L.; Zhao, F.; Ren, Q.C.; Wu, X.; Chang, L.; Gao, F. Quinolone derivatives and their activities against methicillin-resistant *Staphylococcus aureus* (MRSA). *Eur. J. Med. Chem.* **2018**, *157*, 1081–1095. [CrossRef]
5. Gonzalez-Bello, C. Antibiotic adjuvants–A strategy to unlock bacterial resistance to antibiotics. *Bioorg. Med. Chem. Lett.* **2017**, *27*, 4221–4228. [CrossRef]
6. Kalan, L.; Wright, G.D. Antibiotic adjuvants: Multicomponent anti-infective strategies. *Expert Rev. Mol. Med.* **2011**, *13*, e5. [CrossRef]
7. Sahgal, G.; Sreeramanan, S.; Sasidharan, S.; Xavier, R.; Ong, M.T. Screening selected medicinal plants for antibacterial activity against Methicillin Resistant *Staphylococcus aureus* (MRSA). *Adv. Nat. Appl. Sci.* **2009**, *3*, 330–338.
8. Oliveira, A.A.; Segovia, J.F.; Sousa, V.Y.; Mata, E.C.; Gonçalves, M.C.; Bezerra, R.M.; Junior, P.O.; Kanzaki, L.I. Antimicrobial activity of Amazonian medicinal plants. *SpringerPlus* **2013**, *2*, 371. [CrossRef]
9. Hegazy, A.; Boulos, L.; Kabiel, H.; Sharashy, O. Vegetation and species altitudinal distribution in Al-Jabal Al-Akhdar landscape, Libya. *Pak. J. Bot.* **2011**, *43*, 1885–1898.
10. Alghazeer, R.; Abourghiba, T.; Ibrahim, A.; Zreba, E. Bioactive properties of some selected Libyan plants. *J. Med. Plant Res.* **2016**, *10*, 67–76.
11. Tenuta, M.C.; Tundis, R.; Xiao, J.; Loizzo, M.R.; Dugay, A.; Deguin, B. Arbutus species (Ericaceae) as source of valuable bioactive products. *Crit. Rev. Food Sci.* **2019**, *59*, 864–881. [CrossRef]
12. Delfin, J.C.; Watanabe, M.; Tohge, T. Understanding the function and regulation of plant secondary metabolism through metabolomics approaches. *Theor. Exp. Plant Phys.* **2019**, *31*, 127–138. [CrossRef]
13. Alsabri, S.G.; El-Basir, H.M.; Rmeli, N.B.; Mohamed, S.B.; Allafi, A.A.; Zetrini, A.A.; Salem, A.A.; Mohamed, S.S.; Gbaj, A.; El-Baseir, M. Phytochemical screening, antioxidant, antimicrobial and anti-proliferative activities study of *Arbutus pavarii* plant. *J. Chem. Pharm. Res.* **2013**, *5*, 2–36.
14. Ríos, J.L.; Recio, M.C. Medicinal plants and antimicrobial activity. *J. Ethnopharmacol.* **2005**, *100*, 80–84. [CrossRef]

15. Qaralleh, H.N. Chemical composition and antibacterial activity of *Origanum ramonense* essential oil on the β-lactamase and extended-spectrum β-lactamase urinary tract isolates. *Bangl. J. Pharmacol.* **2018**, *13*, 280–286. [CrossRef]
16. Oliveira Silva, E.; Batista, R. Ferulic acid and naturally occurring compounds bearing a feruloyl moiety: A review on their structures, occurrence, and potential health benefits. *Compr. Rev. Food Sci. F.* **2017**, *16*, 580–616. [CrossRef]
17. Okwu, M.U.; Olley, M.; Akpoka, A.O.; Izevbuwa, O.E. Methicillin-resistant *Staphylococcus aureus* (MRSA) and anti-MRSA activities of extracts of some medicinal plants: A brief review. *AIMS Microbiol.* **2019**, *5*, 117. [CrossRef] [PubMed]
18. Eloff, J.N. Quantifying the bioactivity of plant extracts during screening and bioassay-guided fractionation. *Phytomedicine* **2004**, *11*, 370–371. [CrossRef]
19. Clinical and Laboratory Standards Institute (CLSI). *Reference Method for Dilution Antimicrobial Susceptibility Tests for Bacteria that Grow Aerobically*, 9th ed.; CLSI document M07-A9; Clinical and Laboratory Standards Institue: Wayne, PA, USA, 2012.
20. Pankey, G.A.; Sabath, L.D. Clinical relevance of bacteriostatic versus bactericidal mechanisms of action in the treatment of Gram-positive bacterial infections. *Clin. Infect. Dis.* **2004**, *38*, 864–870. [CrossRef] [PubMed]
21. Levison, M.E.; Levison, J.H. Pharmacokinetics and pharmacodynamics of antibacterial agents. *Infect. Dis. Clin.* **2009**, *23*, 791–815. [CrossRef]
22. Ramli, S.; Radu, S.; Shaari, K.; Rukayadi, Y. Antibacterial activity of ethanolic extract of *Syzygium polyanthum* L. (Salam) leaf against foodborne pathogens and application as food sanitizer. *Biomed. Res. Int.* **2017**, *2017*. [CrossRef] [PubMed]
23. El-Shibani, F.A. A Pharmacognostical Study of *Arbutus Pavarii* Pampan. Family *Ericaceae* and *Sarcopoterium spinosum* L. Family *Rosaceae* Growing in Libya. Ph.D. Thesis, Cairo University, Giza, Egypt, 2017.
24. Hasan, H.H.; Habib, I.H.; Gonaid, M.H.; Islam, M. Comparative phytochemical and antimicrobial investigation of some plants growing in Al Jabal Al-Akhdar. *J. Nat. Prod. Plant Resour.* **2011**, *1*, 15–23.
25. Kumar, S.; Chandra, P.; Bajpai, V.; Singh, A.; Srivastava, M.; Mishra, D.K.; Kumar, B. Rapid qualitative and quantitative analysis of bioactive compounds from *Phyllanthus amarus* using LC/MS/MS techniques. *Ind. Crops. Prod.* **2015**, *69*, 143–152. [CrossRef]
26. Mendes, L.; de Freitas, V.; Baptista, P.; Carvalho, M. Comparative antihemolytic and radical scavenging activities of strawberry tree (*Arbutus unedo* L.) leaf and fruit. *Food Chem. Toxicol.* **2011**, *49*, 2285–2291. [CrossRef]
27. Abu-Reidah, I.M.; Ali-Shtayeh, M.S.; Jamous, R.M.; Arráez-Román, D.; Segura-Carretero, A. HPLC–DAD–ESI-MS/MS screening of bioactive components from *Rhus coriaria* L. (Sumac) fruits. *Food Chem.* **2015**, *166*, 179–191. [CrossRef]
28. Liu, Y.; Seeram, N.P. Liquid chromatography coupled with time-of-flight tandem mass spectrometry for comprehensive phenolic characterization of pomegranate fruit and flower extracts used as ingredients in botanical dietary supplements. *J. Sep. Sci.* **2018**, *41*, 3022–3033. [CrossRef] [PubMed]
29. Karar, M.; Kuhnert, N. UPLC-ESI-Q-TOF-MS/MS characterization of phenolics from *Crataegus monogyna* and *Crataegus laevigata* (Hawthorn) leaf, fruits and their herbal derived drops (Crataegutt Tropfen). *J. Chem. Bio. Therap.* **2015**, *1*, 102.
30. Morales-Soto, A.; Gómez-Caravaca, A.M.; García-Salas, P.; Segura-Carretero, A.; Fernández-Gutiérrez, A. High-performance liquid chromatography coupled to diode array and electrospray time-of-flight mass spectrometry detectors for a comprehensive characterization of phenolic and other polar compounds in three pepper (*Capsicum annuum* L.) samples. *Food Res. Int.* **2013**, *51*, 977–984. [CrossRef]
31. Munekata, P.E.S.; Franco, D.; Trindade, M.A.; Lorenzo, J.M. Characterization of phenolic composition in chestnut leaf and beer residue by LC-DAD-ESI-MS. *LWT-Food Sci. Technol.* **2016**, *68*, 52–58. [CrossRef]
32. Singh, A.P.; Wang, Y.; Olson, R.M.; Luthria, D.; Banuelos, G.S.; Pasakdee, S.; Vorsa, N.; Wilson, T. LC-MS-MS analysis and the antioxidant activity of flavonoids from eggplant skins grown in organic and conventional environments. *J. Food Sci. Nutr.* **2017**, *8*, 869. [CrossRef]
33. Ismail, B.B.; Pu, Y.; Guo, M.; Ma, X.; Liu, D. LC-MS/QTOF identification of phytochemicals and the effects of solvents on phenolic constituents and antioxidant activity of baobab (*Adansonia digitata*) fruit pulp. *Food Chem.* **2019**, *277*, 279–288. [CrossRef]

34. Stöggl, W.M.; Huck, C.W.; Bonn, G.K. Structural elucidation of catechin and epicatechin in sorrel leaf extracts using liquid-chromatography coupled to diode array-, fluorescence-, and mass spectrometric detection. *J. Sep. Sci.* **2004**, *27*, 524–528. [CrossRef]
35. Gu, L.; Kelm, M.A.; Hammerstone, J.F.; Beecher, G.; Holden, J.; Haytowitz, D.; Prior, R.L. Screening of foods containing proanthocyanidins and their structural characterization using LC-MS/MS and thiolytic degradation. *J. Agric. Food Chem.* **2003**, *51*, 7513–7521. [CrossRef]
36. Jaiswal, R.; Jayasinghe, L.; Kuhnert, N. Identification and characterization of proanthocyanidins of 16 members of the Rhododendron genus (*Ericaceae*) by tandem LC–MS. *J. Mass Spectrom.* **2012**, *47*, 502–515. [CrossRef]
37. Wang, C.; Li, Q.; Han, G.; Zou, L.; Lv, L.; Zhou, Q.; Li, N. LC MS/MS for simultaneous determination of four major active catechins of tea polyphenols in rat plasma and its application to pharmacokinetics. *Chin. Herb. Med.* **2010**, *2*, 289–296.
38. Qing, L.S.; Xue, Y.; Zhang, J.G.; Zhang, Z.F.; Liang, J.; Jiang, Y.; Liu, Y.M.; Liao, X. Identification of flavonoid glycosides in *Rosa chinensis* flowers by liquid chromatography–tandem mass spectrometry in combination with 13C nuclear magnetic resonance. *J. Chromatogr. A* **2012**, *1249*, 130–137. [CrossRef]
39. Zhu, M.Z.; Wu, W.; Jiao, L.L.; Yang, P.F.; Guo, M.Q. Analysis of flavonoids in lotus (*Nelumbo nucifera*) leaf and their antioxidant activity using macroporous resin chromatography coupled with LC-MS/MS and antioxidant biochemical assays. *Molecules* **2015**, *20*, 10553–10565. [CrossRef]
40. De la Luz Cádiz-Gurrea, M.; Fernández-Arroyo, S.; Joven, J.; Segura-Carretero, A. Comprehensive characterization by UHPLC-ESI-Q-TOF-MS from an *Eryngium bourgatii* extract and their antioxidant and anti-inflammatory activities. *Food Res. Int.* **2013**, *50*, 197–204. [CrossRef]
41. Shibata, H.; Kondo, K.; Katsuyama, R.; Kawazoe, K.; Sato, Y.; Murakami, K.; Takaishi, Y.; Arakaki, N.; Higuti, T. Alkyl gallates, intensifiers of β-lactam susceptibility in methicillin-resistant *Staphylococcus aureus*. *Antimicrob. Agents Chemother.* **2005**, *49*, 549–555. [CrossRef] [PubMed]
42. Górniak, I.; Bartoszewski, R.; Króliczewski, J. Comprehensive review of antimicrobial activities of plant flavonoids. *Phytochem. Rev.* **2019**, *18*, 241–272. [CrossRef]
43. Cushnie, T.P.T.; Taylor, P.W.; Nagaoka, Y.; Uesato, S.; Hara, Y.; Lamb, A.J. Investigation of the antibacterial activity of 3-*O*-octanoyl-(−)-epicatechin. *J. Appl. Microbiol.* **2008**, *105*, 1461–1469. [CrossRef]
44. Lambert, P.A.; Hammond, S.M. Potassium fluxes, first indications of membrane damage in micro-organisms. *Biochem. Biophys. Res. Commun.* **1973**, *54*, 796–799.
45. Hu, Z.Q.; Zhao, W.H.; Asano, N.; Yoda, Y.; Hara, Y.; Shimamura, T. Epigallocatechin gallate synergistically enhances the activity of carbapenems against methicillin-resistant *Staphylococcus aureus*. *Antimicrob. Agents Chemother.* **2002**, *46*, 558–560. [CrossRef]
46. Zhao, W.H.; Hu, Z.Q.; Okubo, S.; Hara, Y.; Shimamura, T. Mechanism of synergy between epigallocatechin gallate and β-lactams against methicillin-resistant *Staphylococcus aureus*. *Antimicrob. Agents Chemother.* **2001**, *45*, 1737–1742. [CrossRef]
47. Shiota, S.; Shimizu, M.; Mizushima, T.; Ito, H.; Hatano, T.; Yoshida, T.; Tsuchiya, T. Marked reduction in the minimum inhibitory concentration (MIC) of β-lactams in methicillin-resistant *Staphylococcus aureus* produced by epicatechin gallate, an ingredient of green tea (*Camellia sinensis*). *Biol. Pharm. Bull.* **1999**, *22*, 1388–1390. [CrossRef]
48. Su, Y.; Ma, L.; Wen, Y.; Wang, H.; Zhang, S. Studies of the in vitro antibacterial activities of several polyphenols against clinical isolates of methicillin-resistant *Staphylococcus aureus*. *Molecules* **2014**, *19*, 12630–12639. [CrossRef] [PubMed]
49. Alhadrami, H.A.; Hamed, A.A.; Hassan, H.M.; Belbahri, L.; Rateb, M.E.; Sayed, A.M. Flavonoids as Potential anti-MRSA Agents through Modulation of PBP2a: A Computational and Experimental Study. *Antibiotics* **2020**, *9*, 562. [CrossRef] [PubMed]
50. Ma, C.; He, N.; Zhao, Y.; Xia, D.; Wei, J.; Kang, W. Antimicrobial mechanism of hydroquinone. *Appl. Biochem. Biotechnol.* **2019**, *189*, 1291–1303. [CrossRef]

51. Rukayadi, Y.; Lee, K.; Han, S.; Yong, D.; Hwang, J.K. *In vitro* activities of panduratin A against clinical *Staphylococcus strains*. *AAC* **2009**, *53*, 4529–4532. [CrossRef]
52. Alsohaili, S.A.; Al-fawwaz, A.T. Composition and antimicrobial activity of *Achillea fragrantissima* essential oil using food model media. *Eur. Sci. J.* **2014**, *10*, 156–165.

Publisher's Note: MDPI stays neutral with regard to jurisdictional claims in published maps and institutional affiliations.

Article

Natural Surfactant Saponin from Tissue of *Litsea glutinosa* and Its Alternative Sustainable Production

Jiratchaya Wisetkomolmat [1,2], Ratchuporn Suksathan [3], Ratchadawan Puangpradab [3], Keawalin Kunasakdakul [4], Kittisak Jantanasakulwong [5,6], Pornchai Rachtanapun [5,6] and Sarana Rose Sommano [2,5,7,*]

1 Interdisciplinary Program in Biotechnology, Graduate School, Chiang Mai University, Chiang Mai 50200, Thailand; jiratchaya_wis@cmu.ac.th
2 Plant Bioactive Compound Laboratory, Department of Plant and Soil Sciences, Faculty of Agriculture, Chiang Mai University, Chiang Mai 50200, Thailand
3 Research and Product Development, Department of Research and Conservation, Queen Sirikit Botanic Garden, The Botanical Garden Organisation, Chiang Mai 50180, Thailand; r_spanuchat@yahoo.com (R.S.); ratchadawanp@hotmail.com (R.P.)
4 Department of Entomology and Plant Pathology, Faculty of Agriculture, Chiang Mai University, Chiang Mai 50200, Thailand; kaewalin.k3@gmail.com
5 Cluster of Agro Bio-Circular-Green Industry (Agro BCG), Chiang Mai University, Chiang Mai 50100, Thailand; jantanasakulwong.k@gmail.com (K.J.); pornchai.r@cmu.ac.th (P.R.)
6 Division of Packaging Technology, Faculty of Agro-Industry, Chiang Mai University, Chiang Mai 50100, Thailand
7 Innovative Agriculture Research Centre, Faculty of Agriculture, Chiang Mai University, Chiang Mai 50200, Thailand
* Correspondence: sarana.s@cmu.ac.th; Tel.: +66-53944040

Received: 8 October 2020; Accepted: 6 November 2020; Published: 9 November 2020

Abstract: In this research, we assessed the detergency properties along with chemical characteristic of the surfactant extracts from the most frequently cited detergent plants in Northern Thailand, namely, *Sapindus rarak*, *Acacia concinna*, and *Litsea glutinosa*. Moreover, as to provide the sustainable option for production of such valuable ingredients, plant tissue culture (PTC) as alternative method for industrial metabolite cultivation was also proposed herein. The results illustrated that detergent plant extracts showed moderate in foaming and detergency abilities compared with those of synthetic surfactant. The phytochemical analysis illustrated the positive detection of saponins in *L. glutinosa* plant extracts. The highest callus formation was found in *L. glutinosa* explant cultured with MS medium supplemented with 2.0 mg/L Indole-3-acetic acid (IAA). The callus extract was chemical elucidated using chromatography, which illustrated the presence of saponin similar to those from the crude leaf and Quillaja saponin extracts. Compact mass spectrometry confirmed that the surfactant was of the steroidal diagnostic type.

Keywords: compact mass spectrometry; cleansing properties; detergent plants; phytochemical; tissue culture

1. Introduction

Surfactants are surface-active compounds with significant potentials for food-beverage, medicinal, and pharmaceutical industries. In surfactant manufacturing, they can be either chemically synthesized or found naturally in bio-based forms [1,2]. The first one is non-biodegradable, can be accumulated in the environment, and therefore is known to be toxic to living organisms [2]. Consequently, trends are moving towards finding the biological sources of the surfactants [1]. Biosurfactants can be derived directly from various natural sources including those of microbials, flora, and fauna [3–5]. Plant-based surfactants, such as saponins, have gained increasing attention as they are environmentally safe, less toxic, biodegradable, and renewable [6–10]. Structurally, saponins are amphipathic molecules consisted of triterpenoids, steroids, and glycosides [7,11,12]. Their aglycone parts are known as genin and sapogenin, which covalently bound to one or more polar molecules of sugar moieties [13–16]. Steroidal saponins such as diosgenin had been characterized and found in many plant samples [17–19]. Natural surfactants from plant resources are used mainly as detergents for fabric washing, hair, and body detergents with excellent cleansing properties [20]; for example, *Quillaja Saponaria* bark is used as a detergent for washing hair and clothes [21,22]. *Saponaria officinalis* is known as soap plant [13,23], and *Sapindus saponaria* is manufactured as soap and clothing washing detergent [24–26]. Some natural saponins have been traditionally used in folk medicines [4]. The drawback is that the high value phyto-intermediate raw materials are available naturally with limited abundance. Additionally, destructive harvesting may result in resource exhaustion and even nearly extinction from the nature [27]. More importantly, the degree functionality of active ingredient also varies upon environmental situations [28]. Due to these constraints, the sustainable production of raw material should be considered [27]. Plant tissue culture (PTC) is a potential alternative method to traditional plant propagation for retrieving natural products in the ways that multiplying and maintaining higher consistency and greater true-to-type raw material [29]. Moreover, this in vitro plant regeneration is also simple and time saving [28]. The underlaying advantage is that it can also help reducing the invasion of the natural resources and the destruction of biodiversity [30].

In the context of Northern Thai culture, three plant species namely *Sapindus rarak*, *Acacia concinna*, and *Litsea glutinosa* [31] are frequently mentioned as plants used in cleansing purposes. Among those, *L. glutinosa* was the most potential as the model plant for further conservation studies [32]. Belonging to the family Lauraceae, it also contains different kinds of secondary metabolites, such as alkaloids, tannins, sterols, and flavonoids, with good capability for medicinal purposes [33,34]. Leaf tissue of *L. glutinosa* comprises of numerous secondary metabolites such as megastigmane diglycoside, roseoside, pinoresinol 3-O-b-D-glucopyranoside [35], 2′-oxygenated flavone glycoside [36], and also natural saponin-surfactant with lists of proven active biofunctionalities [37,38]. It was also reported to be an endangered species or very rare in Asia, including in the Philippines, Bangladesh, and India [39]. However, the assessment of its detergency properties along with chemical characterization have not been reported in the literature. Thus, besides the ultimate aim to evaluate the cleaning properties of *L. glutinosa*, we also assessed the PTC as the alternative raw material propagation for extractable ingredients. The outcome of this study is therefore beneficial for ex situ conservation which inhibits the overexploitation of natural resource and the destruction of local forest to obtain natural material.

2. Results and Discussion

2.1. Foaming Ability, Stability, and Detergency Ability

Foaming ability, stability, and detergency ability of crude plant extracts are illustrated in Table 1. Foam produced by mechanical agitation is typically an unsteady thermodynamic system. Although foam generation has little to do with the cleansing ability of the detergent, it is an important criterion to evaluate detergent. Detergent can increase the ability to displace air from a liquid or solid surface [40,41]. By dissolving the crude detergent plant extracts, both of methanol or water extracts, in DI water at the concentration of 5 and 10% (*v/v*), the foaming regeneration of detergent plant extracts showed variable in the results after shaken. The highest foam was found in both types extracts from *S. rarak* (7.1 ± 0.10 cm) and the lowest was found in *L. glutinosa* (1.1 ± 0.06 cm). After 5 min, foam height decreased in all extracts. Chen et al. [40] explained that increase in foam height of detergent was due to the increasing in the concentrations and types of the detergents in aqueous solution. In the same study, foam produced by crude saponin extract of *Camellia oleifera* was less than that from the chemical synthesized products (SLS and Tween80) at 0.5% solution concentration. In the same way with the results from Yang et al. [41] who found that the crude saponin extract from *Sapindus mukorossi* showed lower ability of foaming than 0.5% SLS solution but similar to 0.5% Tween80 solution.

Table 1. Detergent properties of crude detergent plant extracts.

Extracts	Type of Extracts	Concentration (% *v/v*)	Foam Height (cm)		Detergency Ability (%)
			30 s	5 min	
Control	SLS	0.5	7.93 ± 0.06 [a]	7.03 ± 0.06 [a]	88.98 ± 1.92 [a]
Acacia concinna	Methanol	5	2.30 ± 0.10 [g]	1.03 ± 0.21 [f]	71.52 ± 2.70 [d]
	Methanol	10	3.10 ± 0.10 [f]	1.13 ± 0.12 [f]	82.43 ± 1.33 [b]
	Water	5	5.43 ± 0.21 [e]	4.47 ± 0.15 [e]	59.65 ± 2.67 [fg]
	Water	10	6.10 ± 0.20 [d]	5.13 ± 0.15 [d]	78.42 ± 1.10 [c]
Sapindus rarak	Methanol	5	5.93 ± 0.12 [d]	5.30 ± 0.10 [d]	66.60 ± 1.69 [f]
	Methanol	10	7.10 ± 0.10 [b]	6.57 ± 0.25 [b]	74.44 ± 1.90 [d]
	Water	5	6.87 ± 0.06 [c]	5.77 ± 0.15 [c]	57.38 ± 0.66 [fg]
	Water	10	7.13 ± 0.15 [b]	6.74 ± 0.17 [b]	64.29 ± 3.41 [f]
Litsea glutinosa	Methanol	5	1.13 ± 0.06 [i]	0.60 ± 0.10 [g]	54.26 ± 2.96 [g]
	Methanol	10	1.13 ± 0.15 [i]	0.60 ± 0.10 [g]	65.69 ± 2.81 [f]
	Water	5	1.37 ± 0.12 [h]	0.47 ± 0.06 [g]	48.21 ± 1.71 [h]
	Water	10	1.48 ± 0.08 [h]	0.50 ± 0.10 [g]	56.06 ± 1.69 [fg]

Data are expressed as the mean ± SD of three independent experiments; means with the superscription letters (a–i) are significantly different at $p \leq 0.05$.

Different solvents used to extract the active ingredients could obviously play a role in the foaming properties. In our experiment, crude aqueous extract showed the highest content of foam compared to that of the methanol extract. The highest foam was found with 10% *v/v* methanol extract of *S. rarak* (7.1 cm). In the other hand, *L. glutinosa* extracts either by water or methanol gave low content of foam (1.1–1.5 cm). These are similar with Inalegwu and Sodipo [20] found that the crude aqueous extract of *Tephrosia vogelii* leaves illustrated higher foam height than that of the methanolic extract. Structural wise, saponins contain one or more sugar moieties attached to steroid or triterpenoid backbone. The number of sugar molecules could affect foaming characteristics with the lesser the sugar number, the lower foaming ability of the compound produced. It is also possible that some saponins possess no foam forming ability [7]. Böttcher and Drusch [42] also added that saponins from different plant resources illustrated variable in the interfacial properties, which could mask their true functionality. Not only that, crude extract may contain combination of various compounds such as protein that could contribute to the formation of foam [43].

Cleansing as described by the removal of dirt, soil, and grease, is the key important property of the common surfactants and detergency ability. This property is influenced by the content of surfactant ingredients in the extracted samples. From the result in Table 1, it was observed that methanol extracts gave much higher cleansing ability than that of the aqueous extract in all cases. Among the detergent plants tested, the cleansing ability of *A. concinna* was higher at either 5 or 10% than any others detergent plants. Here again, we noticed that such the ability increased with the increasing in the concentrations of all extracts except that of the aqueous extract of *L. glutinosa* that gave the moderate detergent ability (48.21 ± 1.71% detergent ability in 5% *v/v* water extract). Saponins have been proven to be the surfactive molecules that reduce surface tension and enhance cleansing ability [7,44]. Unsurprisingly, plants containing saponins are used as substitutes for soap with the hidden benefits as the treatment of skin lesions caused by fungi [45]. *A. concinna* pod has been traditionally used in India for bathing due to the formation of lather or foam in water. It also gave excellent cleansing actions on dirt similar to the commercial detergents or soaps [46]. Patel and Talathi [44] found that in herbal shampoo comprising of various aqueous plant extracts, the formulation with *A. concinna* extract had higher detergency ability (95.94%) than the extracts of *Aloe vera*, *S. mukorossi*, and *Phyllanthus emblica* because of the higher saponin content. There, nonetheless, is no report on the detergent ability of *L. glutinosa* elsewhere.

2.2. Chemical Properties

2.2.1. Total Saponin Content

Total saponin content showed significant variations within the plant samples and ranged between 0.33–1.55 and 0.52–3.26 mg Diosgenin Equivalents (DE)/g extract, in crude water and crude methanol respectively. By using alcohol as the extracting solvent, *S. rarak* showed the highest amount of total saponin content (3.26 mg DE/g extract), followed by *A. concinna* (1.15 mg DE/g extract) and *L. glutinosa* (0.52 mg DE/g extract). In the same trend with the aqueous extracts, *S. rarak* showed the highest amount of total saponin content (1.88 mg DE/g extract), followed by *A. concinna* (0.52 mg DE/g extract) and *L. glutinosa* (0.33 mg DE/g extract). Notable that the detection of specific compound content in plant extract by using spectrophotometric method may cause an overestimation of the target compound content due to interferences with other components [47]. Moreover, choice of solvent used can also affect the total saponin content, and methanol is usually a much preferable choice for extraction of the compounds of higher polarity [17]. Preliminary phytochemical screening result from Khanpara et al. [48] reported that *A. concinna* fruit powder consisted of saponin (8.04%) and other compounds such as alkaloids, sugars, and flavonoids. Additionally, the study of Chavan et al. [49] revealed that fruits of *A. concinna* gave saponin as high as 10–11.5% and was known as "fruit for hair" that traditionally used as ingredient in traditional shampoo. Saponins exhibit surface properties such as low surface tension and shear visco-elasticity that possess mechanisms associated with foam and emulsion stabilization [50]. Saponin plants are indeed in need for detergent and cosmetic manufacturing. Nonetheless, the limitation of obtaining such the products from the natural resources is that the content is variable upon genotypes, geographical origin [51,52], and maturation state [45].

2.2.2. Fourier Transform Infrared Spectroscopy (FTIR)

Identification of the functional groups can provide useful information on their mechanical and chemical properties [53]. Infrared spectroscopy is used to characterize the specific kind of bounds and the functional groups. The FTIR spectrum of crude plant extracts (prepared in methanol and water) are given in Figure 1A,B. The data of peak values and the possible functional groups of saponins from other literatures are presented in Supplementary Materials. FTIR spectrum of crude methanol or water extracts of the three detergent plants (*A. concinna*, *S. rarak*, and *L. glutinosa*) absorbed light at the wavenumber ranges of 1025–1620 and 2923–3338 cm^{-1}. All detergent plant extracts showed characteristic infrared absorbance of the hydroxyl group (OH) at ~3338.65 cm^{-1} and ~3330.55 cm^{-1} in methanol extracts of *A. concinna* and *S. rarak* and the aqueous extracts gave the peaks at wave numbers at ~3335.52 cm^{-1} and ~3261.94 cm^{-1} of *A. concinna* and *L. glutinosa*, respectively. Carbon–hydrogen (C-H) absorption was found at 2923.23 cm^{-1} only in methanol extract of *L. glutinosa*. The C=C absorbance was observed at 1620.33 cm^{-1} (*A. concinna*) and 1606.13 cm^{-1} (*L. glutinosa*) in the water extracts but was not found in the both extracts of *S. rarak*. Whereas C-O-C absorbance varied from 1035.12 cm^{-1} (*S. rarak*) to 1026.26 cm^{-1} (*A. concinna*) in the methanol extracts and in the crude aqueous of *S. rarak* ~1034.78 cm^{-1} and 1033.22 cm^{-1} in *L. glutinosa*.

This result was similar with previous studies, as follows: Kareru et al. [54] reported characteristic infrared absorbance of phyto-based saponins of the hydroxyl group (OH) ranging from 3429 cm^{-1} to 3316 cm^{-1}. Carbon–hydrogen (C-H) absorption ranged from 2931 cm^{-1} to 2931 cm^{-1} and oligosaccharide linkage absorptions to sapogenins (C-O-C) were at 1074 cm^{-1} and between 1045 cm^{-1} to 1046 cm^{-1}. The study of Almutari and Ali [55] reported that saponins showed characteristic infrared absorbance of the hydroxyl group (OH) in soap nut extract, at 3407 cm^{-1} in the aqueous extract, 3419 cm^{-1} in the 95% ethanol extract, and from 3525 to 3281 cm^{-1} in the standard Quillaja saponin. The results of *Sapindus* extract in the study of Li et al. [56] also found that the stronger absorption of OH, CH_2 and C-O-C in the sapindus extract spectrum, which similar to those of oleanolic acid spectrum (a standard of triterpenoid saponins). As *L. glutinosa* received much attention globally, and several studies were evaluated its chemical compositions and others biological activity [57–59] along with its potential detergency abilities reported herein. *L. glutinosa* was therefore chosen to study the alternative propagation method for bioactive compound induction.

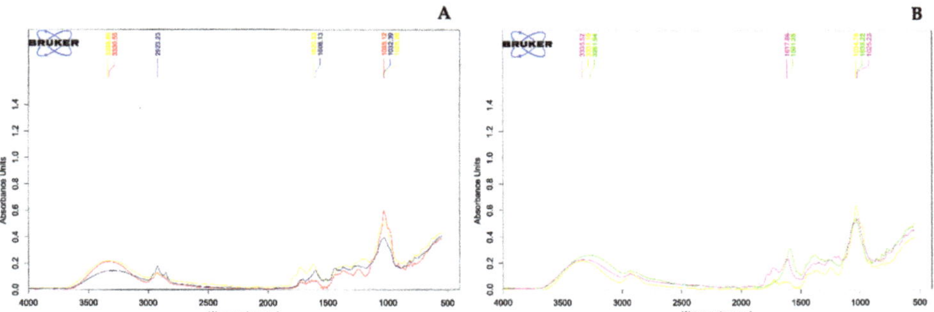

Figure 1. FTIR spectrum of crude detergent plants. (**A**) Crude methanol extracts; *A. concinna* (—) common absorption at 3338.65, 1620.33 and 1026.26 cm^{-1}, *S. rarak* (—) 3330.55 and 1035.12 cm^{-1}, *L. glutinosa* (—) 2923.23, 1606.13 and 1032.39 cm^{-1}; (**B**) Crude water extracts; *A. concinna* (—) common absorption at 3335.52, 1617.86 and 1025.25 cm^{-1}, *S. rarak* (—) 3331.10 and 1034.78 cm^{-1}, *L. glutinosa* (—) 3261.94, 1591.25 and 1033.22 cm^{-1}.

2.3. In Vitro Plant Material Propagation

After six weeks of incubation, the callus induction from the tissue of *L. glutinosa* was observed. The highest callus formation was found in the explants cultured with Murashige and Skoog (MS) medium supplemented with 2.0 mg/L Indole-3-acetic acid (IAA). The greatest fresh weight was observed at 0.84 ± 0.47 g in IAA at concentration of 0.5 mg/L. Meanwhile the dry weight was the highest at 0.13 ± 0.03 g in 2,4-dichlorophenoxyacetic acid (2,4-D) at concentration of 2.0 mg/L (Table 2). Generally, the callus formed were completely friable with greenish and turned to brown colour after four weeks (Figure 2A,B). In addition, Wahyuni et al. [60] stated that the colour of callus becomes brown because they produced phenolic compounds and the variation in callus morphology is due to the explant variant, basic medium, growth regulators, and the biotic and abiotic supplements in the culture. In tissue culture, plant growth regulators play significant role in controlling plant growth and development [61]. Auxins are the major hormones responsible for stem elongation, vascular tissue differentiation, and cell expansion in plant developmental processes [62]. Similarly the cytokinins are required for cell division in a wide variety of plant tissue cultures [63]. During the micropropagation, factors influence the induction of callus are such as plants growth regulator, auxins, and cytokinins, and environmental condition, light, and temperature [64]. Therefore, in studying plant tissue culture for metabolites induction, at least two combination of hormone types are required. The study of callus induction for the production of saponin using different hormones and the combination thereof including 2,4-D and IAA combined with BAP and kinetin (KIN) [65,66]. The total saponin content (TSC) varied in the treatments range between 1.87 and 3.12 mg DE/g extract (Table 2). The callus was used in the evaluation of saponin content and chemical characterization thereafter.

2.4. High Performance Thin Layer Chromatography (HPTLC)

HPTLC result in Figure 3 indicated that crude methanolic extracts of tissues of the detergent plants as well as the callus of *L. glutinosa* (LG callus) gave complex metabolites, while standard diosgenin gave a single band at the rate of flow (Rf) 0.84. For detection of saponins, the blue, violet, yellow, and green should be visible after derivatization at the visible light. They were best resolved at the wavelength UV 366 nm after derivatization [67,68]. Comparing the Rf values with the standard Quillaja saponin, the callus showed the spots at the similar Rfs of 0.2, 0. 39, 0.53 and 0.84, that were comparable with the methanolic extracts of *S. rarak*, *A. concinna*, and *L. glutinosa*, as shown in Figure 3. Other common peaks were also found in the methanolic extracts of detergent plants but were not observed in the standard diosgenin (Table 3).

Table 2. Callus induction in *L. glutinosa* from stem cultured on MS medium treated with various concentrations of auxins (2,4-D and IAA) and 2 cytokinin (kinetin (KIN) and BA) hormones after four weeks of cultured.

Hormone	Concentration (mg/L)	% Callus Induction	Fresh Weight (g)	Dry Weight (g)	TSC
2,4-D	0.5	70 ± 14.14 ^{c}	0.53 ± 0.24 ^{b}	0.03 ± 0.01 ^{e}	1.91 ± 0.03 ^{de}
	1.0	76 ± 16.97 ^{bc}	0.62 ± 0.25 ^{b}	0.09 ± 0.05 ^{b}	2.01 ± 0.04 ^{d}
	2.0	48 ± 5.66 ^{de}	0.61 ± 0.13 ^{b}	0.13 ± 0.03 ^{a}	2.00 ± 0.16 ^{d}
IAA	0.5	90 ± 8.49 ^{ab}	0.84 ± 0.47 ^{a}	0.07 ± 0.03 ^{c}	2.56 ± 0.01 ^{c}
	1.0	92 ± 5.66 ^{ab}	0.56 ± 0.36 ^{b}	0.07 ± 0.05 ^{c}	2.73 ± 0.07 ^{b}
	2.0	100 ± 0.00 ^{a}	0.62 ± 0.29 ^{b}	0.05 ± 0.03 ^{d}	3.12 ± 0.10 ^{a}
KIN	0.5	34 ± 8.49 ^{ef}	-	-	-
	1.0	32 ± 5.66 ^{ef}	-	-	-
	2.0	20 ± 5.66 ^{f}	-	-	-
BA	0.5	64 ± 5.66 ^{cd}	0.06 ± 0.03 ^{c}	0.01 ± 0.00 ^{f}	1.85 ± 0.01 ^{e}
	1.0	66 ± 2.83 ^{cd}	0.06 ± 0.03 ^{c}	0.01 ± 0.00 ^{f}	1.79 ± 0.08 ^{e}
	2.0	38 ± 2.83 ^{ef}	0.05 ± 0.01 ^{c}	0.01 ± 0.00 ^{f}	2.68 ± 0.04 ^{bc}

Data are expressed as the mean ± SD; means with the superscription letters (a–f) are significantly different at $p \leq 0.05$.

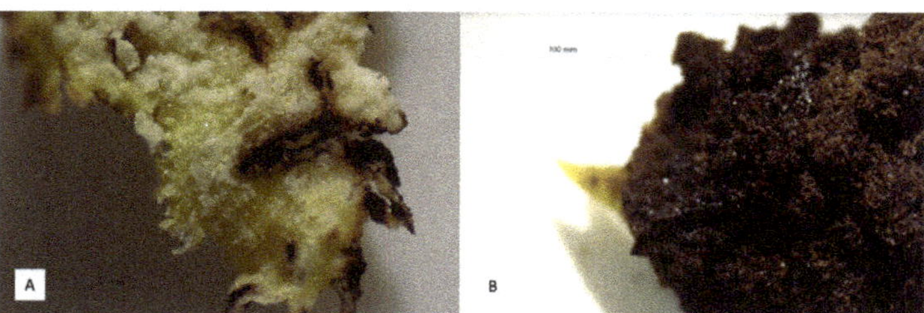

Figure 2. Callus formation; (**A**) Callus from treatment of 2,4-D; (**B**) Callus from treatment of IAA.

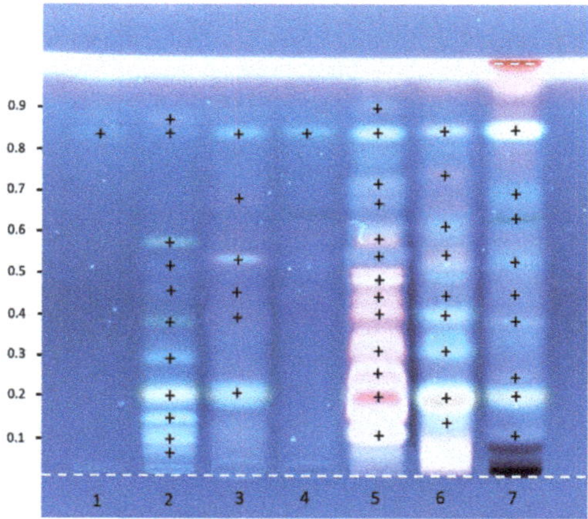

Figure 3. The HPTLC chromatogram of the crude methanolic of detergent plants after derivatized viewed under UV 366 nm. Track 1 = standard Diosgenin, 2 = standard Quillaja saponin, 3 and 4 = Callus of *Litsea* extracts, 5 = *Sapindus* extract, 6 = *Acacia* extract, 7 = *Litsea* extract. The symbol (+) indicate the presence of bands on each track.

Table 3. The rate of flow (Rf) values of the methanolic detergent plant extracts compared with saponin standard from High Performance Thin Layer Chromatography (HPTLC).

Spot No.	Diosgenin	Quillaja	*S. rarak*	*A. concinna*	*L. glutinosa*	LG Callus
1		0.07				
2		0.10	0.12		0.12	
3		0.15		0.14		
4		0.20	0.19	0.19	0.20	0.21
5			0.26			
6		0.29	0.32	0.32		
7		0.38	0.40	0.40	0.38	0.39
8		0.46	0.44	0.44	0.45	0.45
9			0.48			
10		0.52	0.54	0.54	0.53	0.53
11		0.58	0.59	0.62	0.64	
12			0.68			0.69
13			0.72	0.74	0.70	
14	0.84	0.84	0.84	0.84	0.84	0.84
15		0.87	0.90			

In general, saponins from the same plant species can be found in the different forms of molecular structures [42]. Moreover, the variation of the structures could lead to the variety of physicochemical properties and biological activity [50]. The study of Senguttuvan and Subramaniam [68] reported the HPTLC fingerprints of various metabolites in *Hypochaeris radicata* L. and saponins can detected by using the mobile phase consisting of chloroform: glacial acetic acid: methanol: water. The methanolic plant extracts displayed different Rf pattern and the root extract attained the greater number of saponins than the leaf extracts. Further, study of Das et al. [69] showed the variable in the Rf peaks of HPTLC chromatograms of *Diplazium esculentum* Retz. and suggested that the ethanolic extract could yield variety of phenolics, flavonoids, and saponins than those found in the aqueous extract. The type of saponin in the extract was then confirmed by mass spectrometry.

2.5. Compact Mass Spectrometry (CMS)

Mass spectroscopy (MS) is the systematics study of mass spectrum with or without fragmentation which is characterized by a relationship between the mass of a given ion and the number of elementary charges that it carries or mass-to-charge ratios (*m/z*) and relative abundances [70]. Mass spectra of any given substances are different in fragmentation patterns. These fragmentation patterns are useful to determine the molar weight and structural information of the unknown molecule [71]. The mass spectra of the methanolic detergent plant extracts and callus of *L. glutinosa* gave the identical peak with the *m/z* 415.6 in positive ion mode with the standard diosgenin (Table 4). Among the vast array of *m/z* data of the samples, many peaks were also found in common (Figure 4). In the report of Li et al. [72] elucidated that the *m/z* value of 415.3217 was according to the presence protonated aglycone which was the fragmentation of steroidal saponins, 26-*O*-β-D-glucopyranosyl-3β,20α,26-triol-25(R)-Δ-dienofurostan-3-*O*-α-L-rhamnopyranosyl(1→2)-[α-L-rhamnopyranosyl(1→4)]-β-D-glucopyranoside (DFE) (Scheme 1). In another study, fragment of aglycone with the *m/z* value of 415 was detected in the terrestrinin S (or pseudosapogenins of Type IV) which confirmed diagnostic ions of steroidal saponin [73]. Thus, the CMS *m/z* results of crude saponin extracts from all detergent plants along with the callus produced by *L. glutinosa* tissue in our study could be structurally characterized as the steroidal diagnostic type.

Table 4. Mass spectra of detergent plants methanolic extracts.

Label	Maximum Intensity (c/s)	% Peak Area	Base Peak Mass (m/z)
Diosgenin	1.1×10^8	99.6	415.6
Quillaja	3.1×10^5	70.2	415.6
A. concinna	1.4×10^6	81.1	415.6
S. rarak	1.7×10^5	14	415.6
L. glutinosa	1.5×10^6	86.8	415.6
Callus	2.4×10^5	48.5	415.6

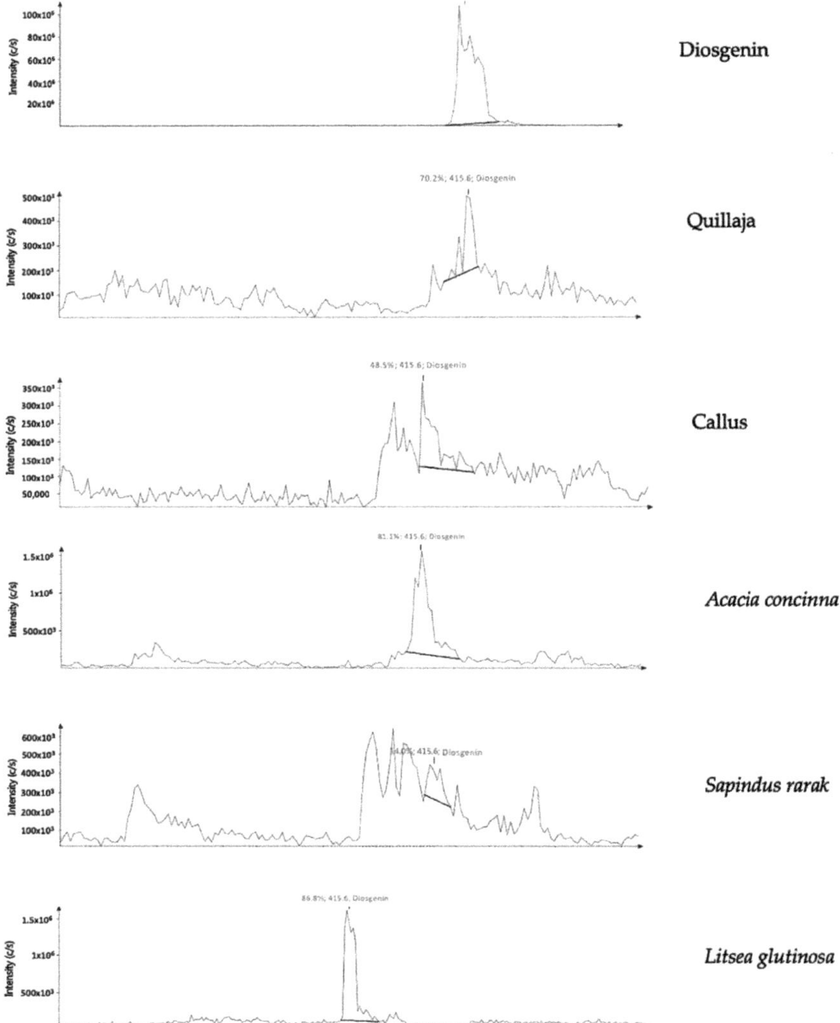

Figure 4. Mass spectra of compound likes diosgenin detected the ion current abundance of m/z 415.6 in methanolic extract of detergent plant, callus compared with standard contained saponin.

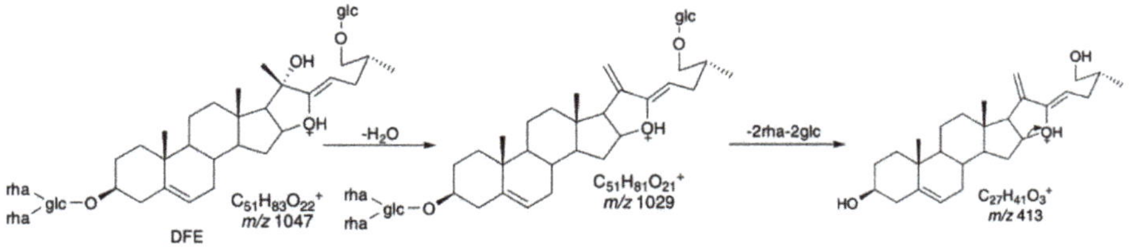

Scheme 1. The proposed of fragmentation patterns of DFE.

3. Materials and Methods

3.1. Chemicals and Standards

Diosgenin (with molecular weight of 414.5 g/mol and purity of ≥93%) was purchased from Sigma-Aldrich (Steinheim, Germany). Saponin from Quillaja Bark pure, 10–14% sapogenin content, was product of PanReac AppliChem (Darmstadt, Germany). The methanol, ethanol, and sulfuric acid (98%) were products of RCI Labscan (Bangkok, Thailand). Vanillin (99%) was product of Sigma-Aldrich (Steinheim, Germany). Sodium Lauryl Sulfate (SLS) was product of KEMAUS (New South Wales, Australia).

3.2. Collection and Authentication of Plant Materials

All plant materials used in this study were collected from the forest reserve restored by the Huai Hong Khrai Royal Development Study Centre (HHKC) during the fields study reported elsewhere [32]. Plant species confirmation was done by comparison the specimen with those deposited at plant collection library of the ethnobotanical laboratory, Department of Biology, Faculty of Science, Chiang Mai University (CMU). The specimens were deposited at Plant Bioactive Compound Laboratory Herbarium (BACH), Faculty of Agriculture with the voucher number JW009-011. The utilized parts, including: pericarp of *Sapindus rarak* (family Sapindaceae), fruit of *Acacia concinna* (family Fabaceae) and Leaves of *Litsea glutinosa* (family Lauraceae) were used in this study. The plant parts were separated, cleaned and dried at 40 °C for 72 h in hot-air oven (WGLL-125BE, Huanghua faithful, Hebei, China). Thereafter, they were grounded into powder using a hand-held grinder (Philips, The Netherlands) at speed level 1 prior to solvent extraction and stored in an air-tight container for further studies. To screen saponin content in plant materials, dry sample powder 0.05 g was suspended in 80% methanol (*v/v*) and extracted using ultrasonication. Unless otherwise stated, the supernatant was used to in colorimetric method for saponin determination later described.

3.3. Determination of Detergent Properties

Unless stated otherwise, 10 g of dried material was extracted with 100 mL of either deionized water (DIH_2O) or 80% methanol at room temperature (25 ± 2 °C) according to the method of Inalegwu and Sodipo [20]. The filtrate was concentrated using automate rotary evaporator under reduced pressure at 40 °C until dried. To evaluate the physiochemical properties of *L. glutinosa*, *A. concinna*, and *S. rarak* were selected to compare their properties based on the ethnobotanical report from our previous study. The detergent properties determined were as followed;

3.3.1. Foaming Ability and Stability

Foaming ability was evaluated according to the methods of Chen et al. [40]; Inalegwu and Sodipo [20] and Pradhan and Bhattacharyya [74] with slight modifications. The extracts were dissolved in water at the concentrations of 5 and 10% (*v/v*). The solution was then vigorously shaken for 1 min by hand in a 10 mL measuring cylinder and the foam height was measured at room temperature. Foam stability and foam height after 30 s and 5 min were also recorded. The measurements were repeated three times from separate set of samples and the values were averaged.

3.3.2. Detergency Ability

The detergent abilities of the crude extract were evaluated following the method of Pradhan and Bhattacharyya [74] with some modifications. The test was conducted using 5 × 5 cm cotton cloth pieces were cleaned with a 5% Sodium Lauryl Sulfate (SLS) solution, then dried. The samples were suspended in artificial sebum (consisting of 1 g coconut oil and 1 g paraffin wax in 100 mL hexane solution) and the mixture was shaken for 15 min. Then samples were removed and left to dryness at room temperature. In the next step, oil-coated materials were divided into two equal portions, the first part was washed with 10 mL of the 5 and 10% crude extracts (both methanol and water fractions) and the second group was cleaned with deionized water (control). After drying, the resided sebum on samples was extracted with 20 mL hexane for 15 min and the materials were removed. After evaporated off to dryness, the sebum content was weighed. Finally, the percentage of detergency ability in both cases was calculated using the following equation;

$$\text{Detergency Ability} = 100 - (T/C \times 100) \quad (1)$$

In which, DA is the percentage of detergency ability. C is the weight of sebum in the controls and T is the weight of sebum in the test sample.

3.4. Chemical Analyses

3.4.1. Determination Saponin Contents

The method of Makkar [75] with slight modification was used to determine the total saponin content in plant materials. Briefly, 50 µL of plant extract was mixed with 2.5 mL sulfuric acid 72% (*v/v*) and 0.1 mL, 8% vanillin solution in ethanol. The mixtures were incubated in a water bath at 60 °C for 10 min and then cooled in cold water. The absorbance of sample was measured at 544 nm using SPECTROstar Nano Microplate Reader (BMG LABTECH, Ortenberg, Germany). Diosgenin was used as the reference standard with saponin content expressed as mg Diosgenin Equivalents (DE) per gram extract (mg DE/g extract). The final quantification of the phytochemicals is the value of mean ± SD of three measurements.

3.4.2. Fourier Transform Infrared Spectroscopy (FTIR)

To identify the presence of saponin functional groups, the FTIR spectra of the extracts were collected by a Bruker model ALPHA II, diamond ATR (Hamburg, Germany) and operating at the basic of 500–4000 wavenumber for averaging 47 scans per spectrum. For the analysis of the different chemical bonds in crude extracts, both methanol and water extracts were used for the FTIR measurement [42].

3.5. *In Vitro Plant Material Propagation*

The lateral bud of plant material was used in this study. For surface sterilization procedures, the explants were soaked in filtered distilled water followed by 95% ethanol for 5 min. Thereafter, the explants were soaked in 25% of sodium hypochlorite solution added with 2–3 drops of tween 20 (Sigma, St. Louis, MO, USA) for 20 min. The explants were then thoroughly rinsed with sterile distilled water to remove any traces of remaining detergents [76].

The explants were inoculated in MS culture medium. The media consisted of 30 g/L sucrose and 7 g/L agar and the pH was adjusted to be within 5.68 after the addition of plant growth regulators. Different concentration, 0.5, 1.0, and 2.0 mg/L of two auxin (2,4-dichlorophenoxyacetic acid (2,4-D) and Indole-3-acetic acid (IAA)) and two cytokinin (6-Benzylaminopurine (BA) and Kinetin (KIN)) were used in the study. The cultures were incubated at 25 ± 2 °C.

The observation of callus was done on a weekly basis. At the end of four weeks, the formed callus was isolated from the explants. The data for callus induction in each treatment was recorded in which the morphology, fresh, and dry weights (g) [76]. The percentage of callus initiation was calculated [30]. Thereafter, the content of total saponin was evaluated and characterization on HPTLC and CMS.

3.6. High Performance Thin Layer Chromatography (HPTLC)

HPTLC finger printing study was carried out according to the methods of Avula et al. [77]; Karthika et al. [67]; and Senguttuvan and Subramaniam [68]. The HPTLC system comprising of Linomat 5 automatic applicator, a twin trough plate development chamber, Camag TLC scanner 3 and winCATS 4 software version 2.5.18262.1 (CAMAG, Muttenz, Switzerland) were used. The methanolic extract of the callus, detergent plants (10 µL) and 3 µL of standards (Diosgenin and *Quillaja* saponins) were spotted separately in form of bands having band width of 5 mm on glass plates (Merck, Darmstadt, Germany) with silica gel 60 F_{254} (20 × 10 cm), using a Hamilton syringe with Linomat 5 applicator attached to CAMAG HPTLC system. The plate was accommodated with 7 tracks and all samples were applied according to the following settings: 8 mm from the bottom of the plate, band width 8 mm; application volume 3–10 µL. All remaining measurement parameters were left at default settings. Chamber saturation was done using 20 × 10 cm Whatman filter paper for 20 min. Development solvent was the lower layer of chloroform: glacial acetic acid: methanol: H_2O (6.4:3.2:1.2:0.8) till 80 mm from the lower edge of the plate. Then chromatogram was developed in the twin trough glass chamber (20 × 10 cm) presaturated with the mobile phase. Developed plates were immersed in Anisaldehyde sulfuric acid reagent, and dried on heat plate for 5 min at 100 °C. The dried plates were kept in a photo documentation chamber and captured the images in visible light, UV 366 nm and UV 254 nm.

3.7. Compact Mass Spectrometry (CMS)

The experiment was performed using the expression Compact Mass Spectrometer and Atmospheric Solids Analysis Probe (ASAP) from Advion, Inc. (Ithaca, NY, USA) to measure the presence of saponin compounds contained in the complex matrix including methanol extracts of detergent plants, callus of selected detergent plant in comparison with the standards following the standard protocol of Perez-Hurtado et al. [78]. The instrument was operated in positive ionization mode. All the spectra were acquired within the range of *m/z* 400–430.

3.8. Statistical Analysis

The extractions of phytochemicals and analyses were of three separate of each sample. The data were subjected to Analysis of Variance (ANOVA) and comparison between the mean values of treatment were confirmed by Duncan's Multiple Range test at the 95% level of significance. A completely randomized design (CRD) with 25 replicates was performed to determine the effect of plant growth regulator on callus induction. The data were presented as mean ± SD. The means comparison and statistical analysis were done using SPSS 17.0 software [76].

4. Conclusions

The recent study assessed the detergency properties along with chemical confirmation of the most cited detergent plant in northern Thailand and its alternative propagation for active ingredient production using plant tissue culture. The results from detergent abilities, physiochemical properties, and phytochemical analysis showed the positive detection of saponins present in *L. glutinosa* plant extracts. Using IAA at 2.0 mg/L concentration in MS media could produce callus that yielded metabolites of different forms. Further study can investigate in-depth on derivatives of saponin, identification, and structure elucidation.

Supplementary Materials: The following is available online at http://www.mdpi.com/2223-7747/9/11/1521/s1, Table S1: FTIR spectra of functional groups of saponins.

Author Contributions: Conceptualization, S.R.S. and K.K.; methodology, S.R.S., K.K., R.S. and R.P.; validation, J.W., R.S. and R.P.; formal analysis, J.W.; investigation, J.W.; resources, K.K., R.S. and R.P.; data curation, J.W.;

writing—original draft preparation, J.W.; writing—review and editing, S.R.S.; visualization, J.W.; supervision, S.R.S.; project administration, K.J. and P.R.; All authors have read and agreed to the published version of the manuscript.

Funding: This research work was partially supported by Chiang Mai University.

Acknowledgments: The authors thank to Taepin Junmahasathien, Faculty of Pharmacy, Chiang Mai University for providing the Quillaja saponin standard. Additional thanks to Angkhana Inta, Faculty of Science, Chiang Mai University for taxonomical identification. We are grateful for the advices from Sila Kittiwachan and Kongkiat Trisuwan in the chemical analyses.

Conflicts of Interest: The authors declare no conflict of interest.

References

1. Akbari, S.; Abdurahman, N.H.; Yunus, R.M.; Fayaz, F.; Alara, O.R. Biosurfactants—A new frontier for social and environmental safety: A mini review. *Biotechnol. Res. Innov.* **2018**, *2*, 81–90. [CrossRef]
2. Fracchia, L.; Ceresa, C.; Franzetti, A.; Cavallo, M.; Gandolfi, I.; Van Hamme, J.; Gkorezis, P.; Marchant, R.; Banat, I.M. Industrial applications of biosurfactants. In *Biosurfactants: Production and Utilization—Processes, Technologies, and Economics*; CRC Press Taylor & Francis Group: Boca Raton, FL, USA, 2014; pp. 245–260.
3. Garai, S. Triterpenoid saponins. *Nat. Prod. Chem. Res.* **2014**, *2*, 1000148.
4. Zhou, W.; Wang, X.; Chen, C.; Zhu, L. Enhanced soil washing of phenanthrene by a plant-derived natural biosurfactant, Sapindus saponin. *Colloids Surf. A Physicochem. Eng. Asp.* **2013**, *425*, 122–128. [CrossRef]
5. Holmberg, K. Natural surfactants. *Curr. Opin. Colloid Interface Sci.* **2001**, *6*, 148–159. [CrossRef]
6. Oleszek, W.; Hamed, A. Saponin-based surfactants. In *Surfactants Renew. Resour*; John Wiley & Sons, Ltd.: Chichester, UK, 2010; Volume 239. [CrossRef]
7. Kregiel, D.; Berlowska, J.; Witonska, I.; Antolak, H.; Proestos, C.; Babic, M.; Babic, L.; Zhang, B. Saponin-based, biological-active surfactants from plants. In *Application and Characterization of Surfactants*; InTech: Rijeka, Croatia, 2017; pp. 183–205.
8. Samal, K.; Das, C.; Mohanty, K. Eco-friendly biosurfactant saponin for the solubilization of cationic and anionic dyes in aqueous system. *Dye. Pigment.* **2017**, *140*, 100–108. [CrossRef]
9. Sahu, S.S.; Gandhi, I.S.R.; Khwairakpam, S. State-of-the-art review on the characteristics of surfactants and foam from foam concrete perspective. *J. Inst. Eng. (India) Ser. A* **2018**, *99*, 391–405. [CrossRef]
10. Basu, A.; Basu, S.; Bandyopadhyay, S.; Chowdhury, R. Optimization of evaporative extraction of natural emulsifier cum surfactant from *Sapindus mukorossi*—Characterization and cost analysis. *Ind. Crops Prod.* **2015**, *77*, 920–931. [CrossRef]
11. Sparg, S.; Light, M.; Van Staden, J. Biological activities and distribution of plant saponins. *J. Ethnopharmacol.* **2004**, *94*, 219–243. [CrossRef]
12. Faizal, A.; Geelen, D. Saponins and their role in biological processes in plants. *Phytochem. Rev.* **2013**, *12*, 877–893. [CrossRef]
13. Osbourn, A. Saponins and plant defence—A soap story. *Trends Plant Sci.* **1996**, *1*, 4–9. [CrossRef]
14. Augustin, J.M.; Kuzina, V.; Andersen, S.B.; Bak, S. Molecular activities, biosynthesis and evolution of triterpenoid saponins. *Phytochemistry* **2011**, *72*, 435–457. [CrossRef] [PubMed]
15. Tippel, J.; Gies, K.; Harbaum-Piayda, B.; Steffen-Heins, A.; Drusch, S. Composition of Quillaja saponin extract affects lipid oxidation in oil-in-water emulsions. *Food Chem.* **2017**, *221*, 386–394. [CrossRef] [PubMed]
16. El Aziz, M.; Ashour, A.; Melad, A. A review on saponins from medicinal plants: Chemistry, isolation, and determination. *J. Nanomed. Res.* **2019**, *8*, 6–12.
17. Le, A.V.; Parks, S.; Nguyen, M.; Roach, P. Improving the vanillin-sulphuric acid method for quantifying total saponins. *Technologies* **2018**, *6*, 84. [CrossRef]
18. Arivalagan, M.; Gangopadhyay, K.; Kumar, G. Determination of steroidal saponins and fixed oil content in fenugreek (*Trigonella foenum*-graecum) genotypes. *Indian J. Pharm. Sci.* **2013**, *75*, 110. [CrossRef]
19. Yang, D.-J.; Lu, T.-J.; Hwang, L.S. Isolation and identification of steroidal saponins in Taiwanese yam cultivar (*Dioscorea pseudojaponica* Yamamoto). *J. Agric. Food Chem.* **2003**, *51*, 6438–6444. [CrossRef]
20. Inalegwu, B.; Sodipo, O. Antimicrobial and foam forming activities of extracts and purified saponins of leaves of *Tephrosia vogelii*. *Eur. J. Exp. Biol.* **2015**, *5*, 49–53.

21. Sarkhel, S. Evaluation of the anti-inflammatory activities of *Quillaja saponaria* Mol. saponin extract in mice. *Toxicol. Rep.* **2016**, *3*, 1–3. [CrossRef]
22. San Martín, R.; Briones, R. Industrial uses and sustainable supply of *Quillaja saponaria* (Rosaceae) saponins. *Econ. Bot.* **1999**, *53*, 302–311. [CrossRef]
23. Koike, K.; Jia, Z.; Nikaido, T. New triterpenoid saponins and sapogenins from *Saponaria officinalis*. *J. Nat. Prod.* **1999**, *62*, 1655–1659. [CrossRef]
24. Albiero, A.L.M.; Sertié, J.A.A.; Bacchi, E.M. Antiulcer activity of *Sapindus saponaria* L. in the rat. *J. Ethnopharmacol.* **2002**, *82*, 41–44. [CrossRef]
25. Tsuzuki, J.K.; Svidzinski, T.I.; Shinobu, C.S.; Silva, L.F.; Rodrigues-Filho, E.; Cortez, D.A.; Ferreira, I.C. Antifungal activity of the extracts and saponins from *Sapindus saponaria* L. *An. Acad. Bras. Cienc.* **2007**, *79*, 577–583. [CrossRef]
26. Damke, E.; Tsuzuki, J.K.; Chassot, F.; Cortez, D.A.; Ferreira, I.C.; Mesquita, C.S.; da-Silva, V.R.; Svidzinski, T.I.; Consolaro, M.E. Spermicidal and anti-Trichomonas vaginalis activity of Brazilian *Sapindus saponaria*. *BMC Complement. Altern. Med.* **2013**, *13*, 196. [CrossRef]
27. Chen, S.-L.; Yu, H.; Luo, H.-M.; Wu, Q.; Li, C.-F.; Steinmetz, A. Conservation and sustainable use of medicinal plants: Problems, progress, and prospects. *Chin. Med.* **2016**, *11*, 37. [CrossRef] [PubMed]
28. Ghaderi, S.; Ebrahimi, S.N.; Ahadi, H.; Moghadam, S.E.; Mirjalili, M.H. In vitro propagation and phytochemical assessment of *Perovskia abrotanoides* Karel. (Lamiaceae)—A medicinally important source of phenolic compounds. *Biocatal. Agric. Biotechnol.* **2019**, *19*, 101113. [CrossRef]
29. Vanisree, M.; Lee, C.-Y.; Lo, S.-F.; Nalawade, S.M.; Lin, C.Y.; Tsay, H.-S. Studies on the production of some important secondary metabolites from medicinal plants by plant tissue cultures. *Bot. Bull. Acad. Sin. Taipei* **2004**, *45*, 1–22.
30. Veraplakorn, V. Micropropagation and callus induction of *Lantana camara* L.—A medicinal plant. *Agric. Nat. Resour.* **2016**, *50*, 338–344. [CrossRef]
31. Wisetkomolmat, J.; Suppakittpaisarn, P.; Sommano, S.R. Detergent plants of Northern Thailand: Potential sources of natural saponins. *Resources* **2019**, *8*, 10. [CrossRef]
32. Wisetkomolmat, J.; Inta, A.; Krongchai, C.; Kittiwachan, S.; Jantanasakulwong, K.; Rachtanapun, P.; Sommano, S.R. Ethnochemometric of plants traditionally utilised as local detergents in the forest dependent culture. *Saudi J. Biol. Sci.* **2020**, Submitted (unpublished).
33. Chowdhury, J.U.; Bhuiyan, M.N.I.; Nandi, N.C. Aromatic plants of Bangladesh: Essential oils of leaves and fruits of *Litsea glutinosa* (Lour.) CB Robinson. *Bangladesh J. Bot.* **2008**, *37*, 81–83. [CrossRef]
34. Sommano, S.; Sirikum, P.; Suksathan, R. Phytochemical screening and ethnobotanical record of some medicinal plants found in Huai Hong Krai royal development study centre, Chiang Mai Thailand. *Med. Plants-Int. J. Phytomedicines Relat. Ind.* **2016**, *8*, 213–218. [CrossRef]
35. Wang, Y.-S.; Liao, Z.; Li, Y.; Huang, R.; Zhang, H.-B.; Yang, J.-H. A new megastigmane diglycoside from *Litsea glutinosa* (Lour.) CB Rob. *J. Braz. Chem. Soc.* **2011**, *22*, 2234–2238. [CrossRef]
36. Wang, Y.-S.; Huang, R.; Lu, H.; Li, F.-Y.; Yang, J.-H. A new 2′-oxygenated flavone glycoside from *Litsea glutinosa* (Lour.) CB Rob. *Biosci. Biotechnol. Biochem.* **2010**, *74*, 652–654. [CrossRef]
37. Pradeepa, K.; Krishna, V.; Santosh, K.; Girish, K.K. Antinociceptive property of leaves extract of *Litsea glutinosa*. *Asian J. Pharm. Clin. Res.* **2013**, *6*, 182–184.
38. Das, D.; Maiti, S.; Maiti, T.K.; Islam, S.S. A new arabinoxylan from green leaves of *Litsea glutinosa* (Lauraeae): Structural and biological studies. *Carbohydr. Polym.* **2013**, *92*, 1243–1248. [CrossRef]
39. Ramana, K.V.; Raju, A.J.S. Pollination ecology of *Litsea glutinosa* (Lour.) CB Robinson (Lauraceae): A commercially and medicinally important semi-evergreen tree species. *Songklanakarin J. Sci. Technol.* **2019**, *41*, 30–36.
40. Chen, Y.-F.; Yang, C.-H.; Chang, M.-S.; Ciou, Y.-P.; Huang, Y.-C. Foam properties and detergent abilities of the saponins from *Camellia oleifera*. *Int. J. Mol. Sci.* **2010**, *11*, 4417–4425. [CrossRef]
41. Yang, C.-H.; Huang, Y.-C.; Chen, Y.-F.; Chang, M.-H. Foam properties, detergent abilities and long-term preservative efficacy of the saponins from *Sapindus mukorossi*. *J. Food Drug Anal.* **2010**, *18*, 155–222.
42. Böttcher, S.; Drusch, S. Interfacial properties of saponin extracts and their impact on foam characteristics. *Food Biophys.* **2016**, *11*, 91–100. [CrossRef]
43. Zayas, J.F. *Functionality of Proteins in Food*; Springer Science & Business Media: Berlin, Germany, 2012.
44. Patel, I.; Talathi, A. Use of traditional Indian herbs for the formulation of shampoo and their comparative analysis. *Int. J. Pharm. Pharm. Sci.* **2016**, *8*, 28–32.

45. Murgu, M.; Rodrigues-Filho, E. Dereplication of glycosides from *Sapindus saponaria* using liquid chromatography-mass spectrometry. *J. Braz. Chem. Soc.* **2006**, *17*, 1281–1290. [CrossRef]
46. Gaikwad, D.; Undale, K.; Kalel, R.; Patil, D. *Acacia concinna* pods: A natural and new bioreductant for palladium nanoparticles and its application to Suzuki–Miyaura coupling. *J. Iran. Chem. Soc.* **2019**, *16*, 2135–2141. [CrossRef]
47. Tenon, M.; Feuillère, N.; Roller, M.; Birtić, S. Rapid, cost-effective and accurate quantification of Yucca schidigera Roezl. steroidal saponins using HPLC-ELSD method. *Food Chem.* **2017**, *221*, 1245–1252. [CrossRef] [PubMed]
48. Khanpara, K.; Renuka, V.; Harisha, C. A detailed investigation on shikakai (*Acacia concinna* Linn.) fruit. *J. Curr. Pharm. Res.* **2012**, *9*, 6–10.
49. Chavan, H.V.; Bandgar, B.P. Aqueous extract of *Acacia concinna* pods: An efficient surfactant type catalyst for synthesis of 3-carboxycoumarins and cinnamic acids via Knoevenagel condensation. *ACS Sustain. Chem. Eng.* **2013**, *1*, 929–936. [CrossRef]
50. Penfold, J.; Thomas, R.; Tucker, I.; Petkov, J.; Stoyanov, S.; Denkov, N.; Golemanov, K.; Tcholakova, S.; Webster, J. Saponin adsorption at the air–water interface—Neutron reflectivity and surface tension study. *Langmuir* **2018**, *34*, 9540–9547. [CrossRef]
51. De Lourdes Contreras-Pacheco, M.; Santacruz-Ruvalcaba, F.; García-Fajardo, J.A.; de Jesús Sánchez, G.J.; Ruíz, L.M.A.; Estarrón-Espinosa, M.; Castro-Castro, A. Diosgenin quantification, characterisation and chemical composition in a tuber collection of *Dioscorea* spp. in the state of J. alisco, M exico. *Int. J. Food Sci. Technol.* **2013**, *48*, 2111–2118.
52. Magedans, Y.V.; Yendo, A.C.; Costa, F.D.; Gosmann, G.; Fett-Neto, A.G. Foamy matters: An update on *Quillaja* saponins and their use as immunoadjuvants. *Future Med. Chem.* **2019**, *11*, 1485–1499. [CrossRef]
53. Ashokkumar, R.; Ramaswamy, M. Phytochemical screening by FTIR spectroscopic analysis of leaf extracts of selected Indian medicinal plants. *Int. J. Curr. Microbiol. Appl. Sci.* **2014**, *3*, 395–406.
54. Kareru, P.; Keriko, J.; Gachanja, A.; Kenji, G. Direct detection of triterpenoid saponins in medicinal plants. *Afr. J. Tradit. Complement. Altern. Med.* **2008**, *5*, 56–60. [CrossRef]
55. Almutairi, M.S.; Ali, M. Direct detection of saponins in crude extracts of soapnuts by FTIR. *Nat. Prod. Res.* **2015**, *29*, 1271–1275. [CrossRef]
56. Li, R.; Wu, Z.L.; Wang, Y.J.; Li, L.L. Separation of total saponins from the pericarp of *Sapindus mukorossi* Gaerten. by foam fractionation. *Ind. Crops Prod.* **2013**, *51*, 163–170. [CrossRef]
57. Haque, T.; Uddin, M.Z.; Saha, M.L.; Mazid, M.A.; Hassan, M.A. Propagation, antibacterial activity and phytochemical profiles of *Litsea glutinosa* (Lour.) CB Robinson. *Dhaka Univ. J. Biol. Sci.* **2014**, *23*, 165–171. [CrossRef]
58. Devi, P.; Meera, R. Study of antioxdant, antiinflammatory and woundhealing activity of extracts of *Litsea glutinosa*. *J. Pharm. Sci. Res.* **2010**, *2*, 155.
59. Agrawal, N.; Pareek, D.; Dobhal, S.; Sharma, M.C.; Joshi, Y.C.; Dobhal, M.P. Butanolides from methanolic extract of *Litsea glutinosa*. *Chem. Biodivers.* **2013**, *10*, 394–400. [CrossRef]
60. Wahyuni, D.K.; Huda, A.; Faizah, S.; Purnobasuki, H.; Wardoyo, B.P.E. Effects of Light, Sucrose concentration and repetitive subculture on callus growth and medically important production in *Justicia gendarussa* Burm. f. *Biotechnol. Rep.* **2020**, *27*, e00473. [CrossRef]
61. Chen, Y.-M.; Huang, J.-Z.; Hou, T.-W.; Pan, I.-C. Effects of light intensity and plant growth regulators on callus proliferation and shoot regeneration in the ornamental succulent Haworthia. *Bot. Stud.* **2019**, *60*, 1–8. [CrossRef]
62. Vanneste, S.; Friml, J. Auxin: A trigger for change in plant development. *Cell* **2009**, *136*, 1005–1016. [CrossRef]
63. D'Agostino, I.B.; Kieber, J.J. Molecular mechanisms of cytokinin action. *Curr. Opin. Plant Biol.* **1999**, *2*, 359–364. [CrossRef]
64. Afshari, R.T.; Angoshtari, R.; Kalantari, S. Effects of light and different plant growth regulators on induction of callus growth in rapeseed (*Brassica napus* L.) genotypes. *Plant Omics J.* **2011**, *4*, 60–67.
65. Jawahar, M.; Ravipaul, S.; Jeyaseelan, M. In vitro regeneration of *Vitex negundo* L.—A multipurpose woody aromatic medicinal shrub. *Plant Tissue Cult. Biotechnol.* **2008**, *18*, 37–42. [CrossRef]
66. Hanafy, M.; Abou-Setta, L.M. Saponins production in shoot and callus cultures of *Gypsophila paniculata*. *J. Appl. Sci. Res.* **2007**, *3*, 1045–1049.
67. Karthika, K.; Jamuna, S.; Paulsamy, S. TLC and HPTLC fingerprint profiles of different bioactive components from the tuber of *Solena amplexicaulis*. *J. Pharmacogn. Phytochem.* **2014**, *3*, 198–206.

68. Senguttuvan, J.; Subramaniam, P. HPTLC Fingerprints of Various Secondary Metabolites in the Traditional Medicinal Herb *Hypochaeris radicata* L. *J. Bot.* **2016**, *2016*, 5429625. [CrossRef]
69. Das, B.; Paul, T.; Apte, K.G.; Chauhan, R.; Saxena, R.C. Evaluation of antioxidant potential & quantification of polyphenols of *Diplazium esculentum* Retz. with emphasis on its HPTLC chromatography. *J. Pharm. Res.* **2013**, *6*, 93–100.
70. Todd, J.F. Recommendations for nomenclature and symbolism for mass spectroscopy (including an appendix of terms used in vacuum technology). (Recommendations 1991). *Pure Appl. Chem.* **1991**, *63*, 1541–1566. [CrossRef]
71. Dass, C. *Fundamentals of Contemporary Mass Spectrometry*; John Wiley & Sons: Hoboken, NJ, USA, 2007; Volume 16.
72. Li, R.; Zhou, Y.; Wu, Z.; Ding, L. ESI-QqTOF-MS/MS and APCI-IT-MS/MS analysis of steroid saponins from the rhizomes of *Dioscorea panthaica*. *J. Mass Spectrom.* **2006**, *41*, 1–22. [CrossRef]
73. Zheng, W.; Wang, F.; Zhao, Y.; Sun, X.; Kang, L.; Fan, Z.; Qiao, L.; Yan, R.; Liu, S.; Ma, B. Rapid characterization of constituents in *Tribulus terrestris* from different habitats by UHPLC/Q-TOF MS. *J. Am. Soc. Mass Spectrom.* **2017**, *28*, 2302–2318. [CrossRef]
74. Pradhan, A.; Bhattacharyya, A. Quest for an eco-friendly alternative surfactant: Surface and foam characteristics of natural surfactants. *J. Clean. Prod.* **2017**, *150*, 127–134. [CrossRef]
75. Makkar, H.P.; Siddhuraju, P.; Becker, K. Saponins. In *Plant Secondary Metabolites*; Springer: Berlin, Germany, 2007; pp. 93–100.
76. Osman, N.I.; Sidik, N.J.; Awal, A. Effects of variations in culture media and hormonal treatments upon callus induction potential in endosperm explant of *Barringtonia racemosa* L. *Asian Pac. J. Trop. Biomed.* **2016**, *6*, 143–147. [CrossRef]
77. Avula, B.; Wang, Y.-H.; Rumalla, C.S.; Ali, Z.; Smillie, T.J.; Khan, I.A. Analytical methods for determination of magnoflorine and saponins from roots of *Caulophyllum thalictroides* (L.) Michx. Using UPLC, HPLC and HPTLC. *J. Pharm. Biomed. Anal.* **2011**, *56*, 895–903. [CrossRef] [PubMed]
78. Perez-Hurtado, P.; Palmer, E.; Owen, T.; Aldcroft, C.; Allen, M.; Jones, J.; Creaser, C.S.; Lindley, M.R.; Turner, M.A.; Reynolds, J.C. Direct analysis of volatile organic compounds in foods by headspace extraction atmospheric pressure chemical ionisation mass spectrometry. *Rapid Commun. Mass Spectrom.* **2017**, *31*, 1947–1956. [CrossRef]

Article

Herbal Combination of *Phyllostachys pubescens* and *Scutellaria baicalensis* Inhibits Adipogenesis and Promotes Browning via AMPK Activation in 3T3-L1 Adipocytes

Yoon-Young Sung [1], Eunjung Son [1], Gayoung Im [2] and Dong-Seon Kim [1,*]

[1] Herbal Medicine Research Division, Korea Institute of Oriental Medicine, 1672 Yuseong-daero, Yuseong-gu, Daejeon 34054, Korea; yysung@kiom.re.kr (Y.-Y.S.); ejson@kiom.re.kr (E.S.)

[2] Nova K Med Co., Ltd., 1646 Yuseong-daero, Yuseong-gu, Daejeon 34054, Korea; imci85@novarex.co.kr

* Correspondence: dskim@kiom.re.kr; Tel.: +82-42-868-9639

Received: 16 September 2020; Accepted: 22 October 2020; Published: 23 October 2020

Abstract: To investigate the anti-obesity effects and underlying mechanism of BS21, a combination of *Phyllostachys pubescens* leaves and *Scutellaria baicalensis* roots was used to investigate the effects of BS21 on adipogenesis, lipogenesis, and browning in 3T3-L1 adipocytes. The expression of adipocyte-specific genes was observed via Western blot, and the BS21 chemical profile was analyzed using ultra-performance liquid chromatography (UPLC). BS21 treatment inhibited adipocyte differentiation and reduced the expression of the adipogenic proteins peroxisome proliferator-activated receptor γ (PPAR-γ), CCAAT/enhancer-binding protein (C/EBP-α), and adipocyte protein 2 (aP2), as well as the lipogenic proteins sterol regulatory element-binding protein 1c (SREBP-1c) and fatty-acid synthase (FAS). BS21 enhanced protein levels of the beta-oxidation genes carnitine palmitoyltransferase (CPT1) and phospho-acetyl-coA carboxylase (p-ACC). BS21 also induced protein expressions of the browning marker genes PR domain containing 16 (PRDM16), peroxisome proliferator-activated receptor gamma co-activator 1-alpha (PGC1α), and uncoupling protein (UCP) 1, and it induced the expression of the thermogenic gene UCP2. Furthermore, BS21 increased adenosine monophosphate-activated protein kinase (AMPK) activation. UPLC analysis showed that BS21 contains active constituents such as chlorogenic acid, orientin, isoorientin, baicalin, wogonoside, baicalein, tricin, wogonin, and chrysin. Our findings demonstrate that BS21 plays a modulatory role in adipocytes by reducing adipogenesis and lipogenesis, increasing fat oxidation, and inducing browning.

Keywords: adipogenesis; AMPK; anti-obesity; triglycerides; UCP1

1. Introduction

Obesity is the main risk factor for several metabolic abnormalities including hyperlipidemia, atherosclerosis, type 2 diabetes, dyslipidemia, and hypertension [1]. Obesity develops following a prolonged period of positive energy imbalance when food intake exceeds total energy expenditure, which must be reversed for treatment to be successful [2,3]. New approaches for promoting energy consumption are emerging to inhibit energy intake (appetite or absorption) [4]. Brown adipocytes in brown adipose tissue (BAT) expend stored energy and produce heat via the expression of brown fat-specific uncoupling protein (UCP) 1 [5]. Beige adipocytes are inducible brown-like fat cells in white adipose tissue (WAT) that can consume stored energy following exposure to adrenergic signaling or chronic cold through UCP1-mediated thermogenesis, and recent treatments for obesity have increased the therapeutic interest in browning of WAT [6].

Adenosine monophosphate-activated protein kinase (AMPK) is an intracellular energy sensor that regulates energy balance and acts as a metabolic switch [7]. AMPK stimulates adenosine triphosphate (ATP)-producing catabolic pathways (e.g., lipolysis or fatty-acid oxidation) and switches off ATP-consuming anabolic pathways (e.g., lipogenesis) that increase energy production and consumption, respectively [8]. Thus, the development of AMPK activators has become a therapeutic target for obesity [9].

Bamboo leaves, usually as tea, have been used for more than 1000 years in Asia for nutrition and as herbal medicine and have various biological functions, including antibacterial, antiviral, and anti-atherosclerosis activities [10]. The antioxidant and anticoagulant effects of *Phyllostachys pubescens* Mazel (Bamboo, Poaceae family) leaves have been reported [11]. It was recently reported that bamboo leaves and its major flavonoid, isoorientin, inhibited adipogenesis in adipocytes [12,13]. *Scutellaria baicalensis* Georgi (Baikal skullcap, family: Lamiaceae family) roots (S) have been used in traditional medicine. These roots have diuretic, anti-diarrhea, and anti-inflammation effects [14] and were recently reported to reduce body weight and improve serum lipid levels in mice [15,16]. Baicalin in *S. baicalensis* exhibited anti-obesity effects through enhanced fatty-acid oxidation [17]. Our recent study suggested that various mixtures of *P. pubescens* leaves and *S. baicalensis* roots (BS) had anti-obesity effects in obese mice fed a high-fat diet, and a 2:1 (*w:w*) mixture of the two plant extracts (BS21) was most effective in decreasing body weight gain and normalizing serum lipid profiles [18]. However, the anti-obesity effects of BS21 and its underlying mechanisms in adipocytes have not been explored. This study examined the effect of BS21—the most effective combination—on adipogenesis and browning in 3T3-L1 adipocytes. Ultra-performance liquid chromatography (UPLC) was performed to determine the constituents of BS21 and assess which constituent inhibited adipogenesis.

2. Results

2.1. Effect of BS21 on Adipocyte Differentiation and Lipolysis in 3T3-L1 Cells

Preadipocytes were stained with fat-specific Oil Red O after inducing adipocyte differentiation in MDI (3-isobutyl-1-methylxanthine, dexamethasone, and insulin) differentiation media (Figure 1A,B). Oil Red O staining quantification demonstrated that treatment with 60, 120, 240, and 480 μg/mL concentrations of BS21 significantly decreased adipocyte differentiation as well as the accumulation and size of lipid droplets, as shown by a ≈50% decrease in fat accumulation with 240 μg/mL treatment. After adipocyte differentiation, the secreted levels of adipokines (leptin and adiponectin) were measured in the supernatant (Figure 1C). BS21 at concentrations of 240 and 480 μg/mL attenuated an increase in leptin levels following differentiation, and 480 μg/mL also decreased adiponectin levels. The glycerol concentration in culture supernatants was used as a marker for adipocyte lipolysis. As shown as Figure 1D, BS21 did not increase lipolysis. Rather, 400 μg/mL of BS21 decreased glycerol levels. None of these extracts affected cell viability after 24 or 48 h treatment at any concentration tested (Figure 1E).

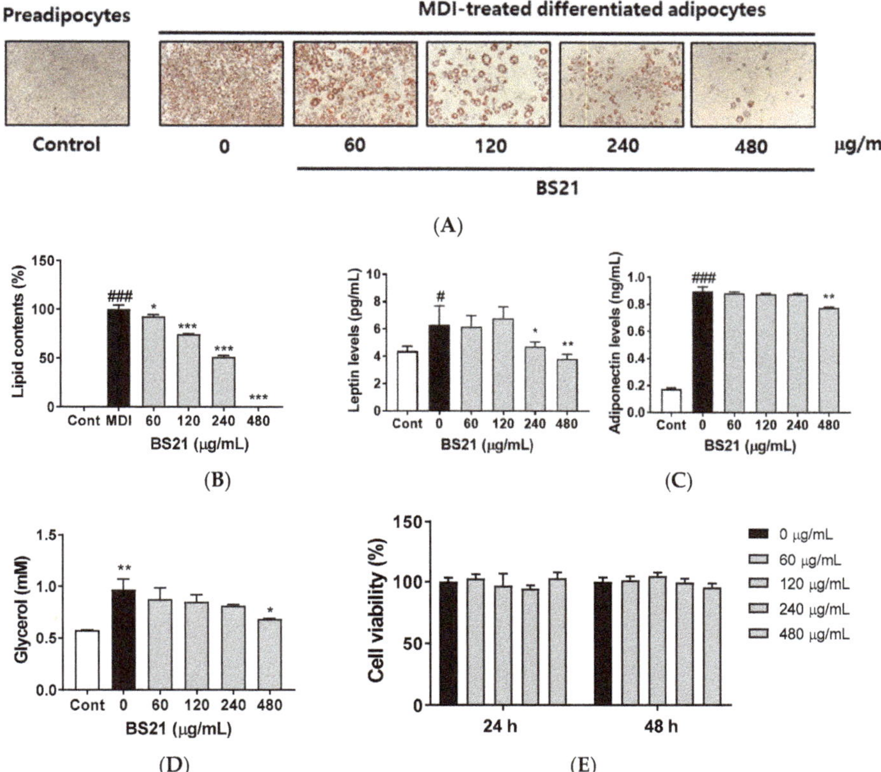

Figure 1. Effect of a 2:1 mixture of *Phyllostachys pubescens* and *Scutellaria baicalensis* (BS21) on adipocyte differentiation in 3T3-L1 cells. (**A**) Oil Red O staining of intracellular triglycerides in 3T3-L1 cells. 3T3-L1 cells were treated with BS21 (60, 120, 240, and 480 μg/mL) during differentiation induction. (**B**) Relative densities of lipid contents. (**C**) Leptin and adiponectin levels in cell supernatants. (**D**) Glycerol levels in differentiated cells. (**E**) Cell viability. Values are expressed as means ± SD ($n = 3$). Significant differences were observed between control (undifferentiated preadipocytes) and differentiated MDI (3-isobutyl-1-methylxanthine, dexamethasone, and insulin) cells: # $p < 0.05$, ## $p < 0.01$, and ### $p < 0.001$. Significant differences were observed between differentiated MDI cells and BS21-treated cells: * $p < 0.05$, ** $p < 0.01$, and *** $p < 0.001$.

2.2. *Effect of BS21 on Adipocyte Marker Protein Expression in 3T3-L1 Cells*

We next confirmed the expression of white adipocyte marker proteins in 3T3-L1 cells. BS21 treatment for 7 days significantly reduced the protein expression of white adipocyte markers (peroxisome proliferator-activated receptor γ (PPARγ), CCAAT/enhancer-binding protein (C/EBPα), and adipocyte protein 2 (ap2)) and lipogenic markers (sterol regulatory element-binding protein (SREBP1c) and fatty-acid synthase (FAS)) in a dose-dependent manner (Figure 2A,B). Expression levels of the β-oxidation proteins (CPT1 and p-ACC) were significantly increased by 400 μg/mL BS21. The phosphorylation of the lipolysis gene HSL decreased with BS2. These results suggest that BS21 can regulate lipid metabolism in 3T3-L1 adipocytes.

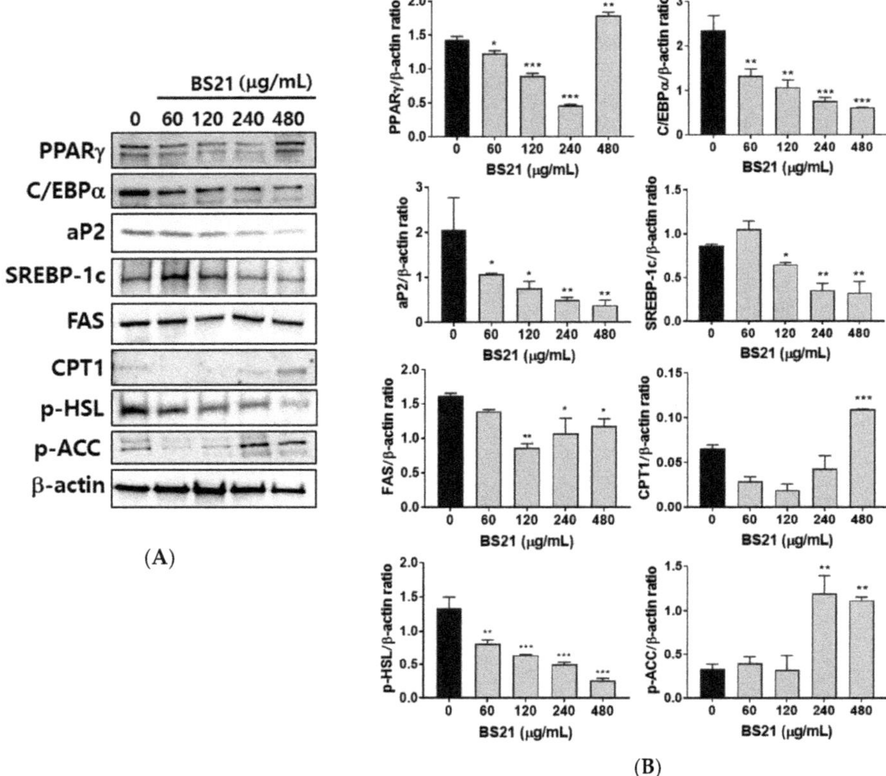

Figure 2. Effect of a 2:1 mixture of *Phyllostachys pubescens* and *Scutellaria baicalensis* (BS21) on lipid metabolism-related protein expression in 3T3-L1 cells. (**A**) Representative bands and (**B**) relative changes in protein expression levels. The relative expression levels of proteins were normalized against a β-actin internal control. Values are expressed as means ± SD ($n = 3$). Significant differences were observed between differentiated MDI (3-isobutyl-1-methylxanthine, dexamethasone, and insulin) cells and BS21-treated cells: * $p < 0.05$, ** $p < 0.01$, and *** $p < 0.001$.

2.3. *Effect of BS21 on Browning and AMPK Signaling in 3T3-L1 White Adipocytes*

To assess whether BS21 regulates browning, we measured the protein levels of browning marker genes. BS21 at concentrations of 240 and 480 μg/mL induced expressions of specific brown fat proteins UCP1, PRDM16, and PGC1α as well as the thermogenesis-related gene UCP2 (Figure 3A,B).

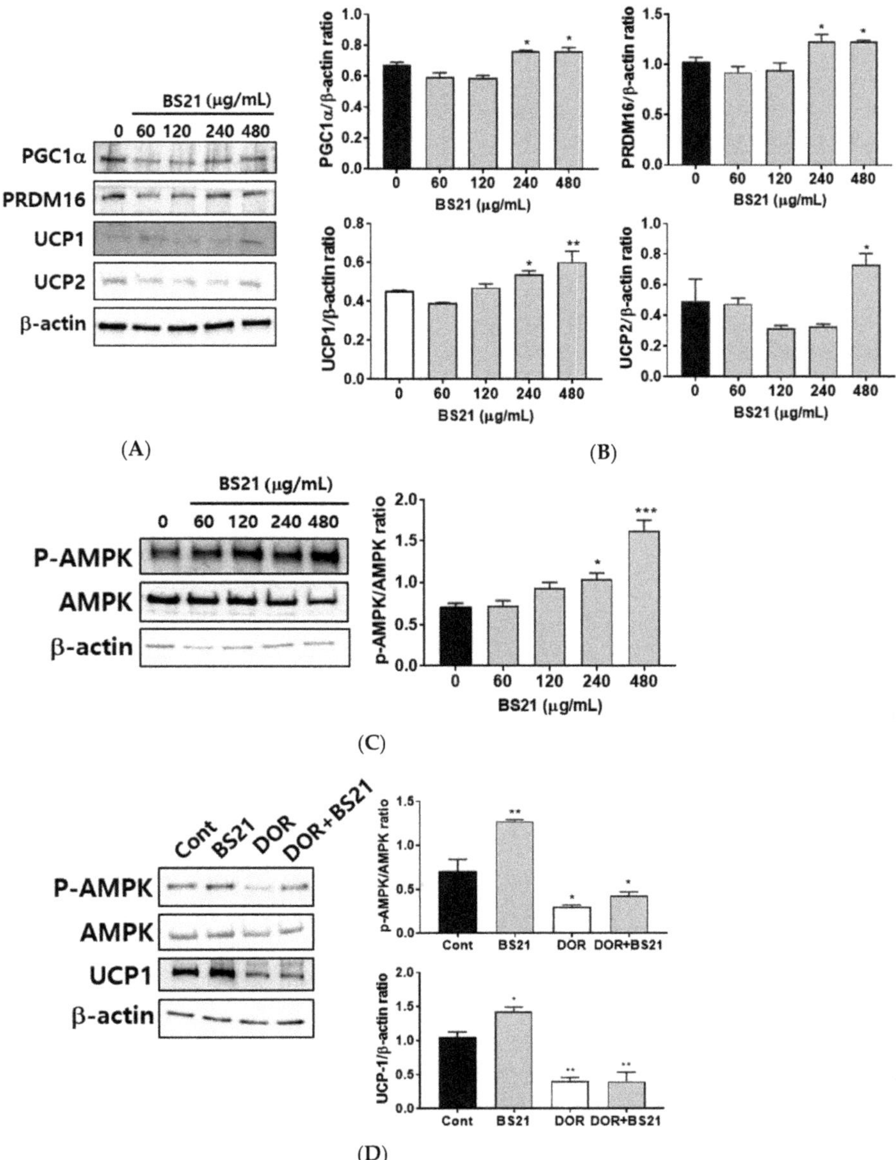

Figure 3. Effect of a 2:1 mixture of *Phyllostachys pubescens* and *Scutellaria baicalensis* (BS21) on brown adipocyte-specific protein expression and adenosine monophosphate-activated protein kinase (AMPK) phosphorylation in 3T3-L1 cells. (**A**) Representative bands and (**B**) relative changes in protein expression levels. The relative expression levels of proteins were normalized against a β-actin internal control. (**C**) Representative bands and relative changes in p-AMPK and AMPK protein levels are shown. Target-protein phosphorylation was normalized to the total protein-expression level. (**D**) 3T3-L1 adipocytes were pretreated with the p-AMPK antagonist dorsomorphin (5 μM) in addition to BS21 at differentiation stages. Values are expressed as mean ± SD ($n = 3$). Significant differences were observed between differentiated MDI (3-isobutyl-1-methylxanthine, dexamethasone, and insulin) cells and BS21-treated cells: * $p < 0.05$, ** $p < 0.01$, and *** $p < 0.001$.

2.4. *Effect of BS21 on AMPK Signaling in 3T3-L1 Cells*

We next investigated whether BS21 affects lipid metabolism and browning via AMPK. BS21 at concentrations of 240 and 480 μg/mL increased AMPK phosphorylation (Figure 3C). This finding suggested that the increased lipid metabolism and browning effect in BS21-treated 3T3-L1 adipocytes could be mediated by AMPK activation. To further examine the AMPK signaling mechanism of the browning effects of BS21, we treated an AMPK antagonist (dorsomorphin). Elevated UCP1 expression was diminished by treatment with dorsomorphin (Figure 3D). These results indicate that BS21 induced the browning of 3T3-L1 adipocytes by enhancing AMPK phosphorylation.

2.5. *Effect of Nine Major BS21 Compounds on 3T3-L1 Adipocyte Differentiation*

UPLC analysis showed that BS21 contained chlorogenic acid, orientin, baicalin, wogonoside, baicalein, tricin, wogonin, and chrysin (Figure 4 and Table 1). Groups of cells were treated with one of the nine compounds at concentrations of 10 and 50 μg/mL, and all compounds significantly reduced adipocyte differentiation and leptin levels except for wogonoside (Figure 5A–C). In particular, baicalein and baicalin at 10 μg/mL inhibited adipocyte differentiation by 50% compared with the differentiated MDI cells. These compounds did not affect cell viability at all concentrations tested (Figure 5D).

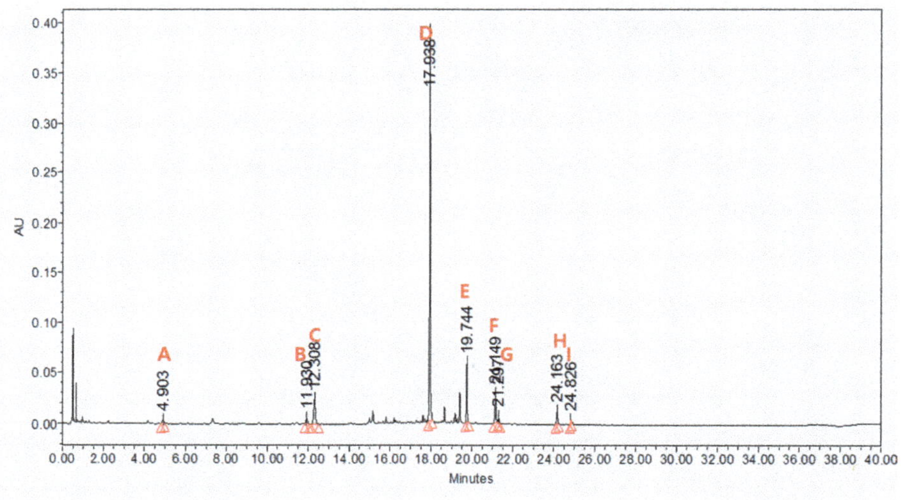

Figure 4. Identification and quantitative analysis of components from BS21: (**A**) cholorogenic acid, (**B**) orientin, (**C**) isoorientin, (**D**) baicalin, (**E**) wogonoside, (**F**) baicalein, (**G**) tricin, (**H**) wogonin, and (**I**) chrysin. Identification was based on comparisons of retention times and UV spectra with those of commercial standards. The components were quantified based on the peak area at 270 nm.

Table 1. BS21 extract constituents.

Chlorogenic Acid	Orientin	Isoorientin	Baicalin	Wogonoside	Baicalein	Tricin	Wogonin	Chrysin
2.9 ± 0.10	2.6 ± 0.06	28.1 ± 0.35	73.2 ± 0.67	8.9 ± 0.27	4.1 ± 0.02	2.9 ± 0.50	2.2 ± 0.07	0.2 ± 0.04

Data are expressed as mean ± standard deviation (SD).

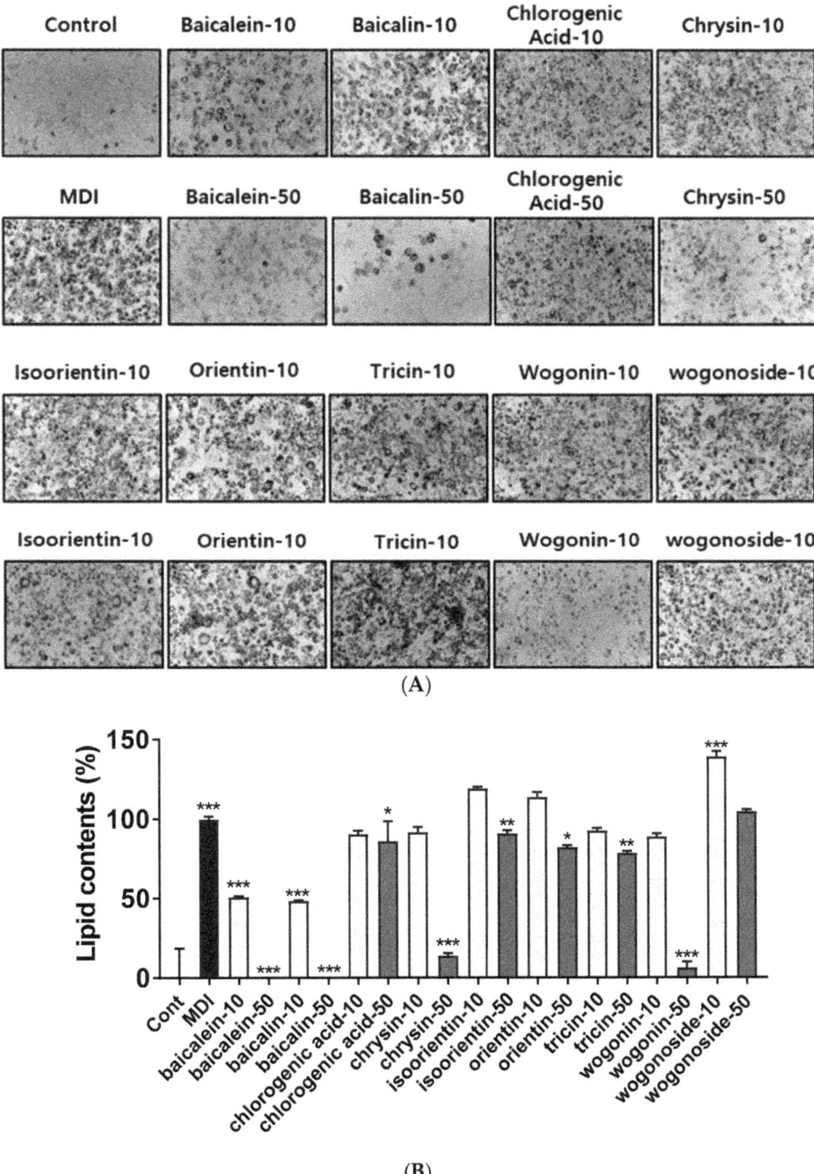

Figure 5. *Cont.*

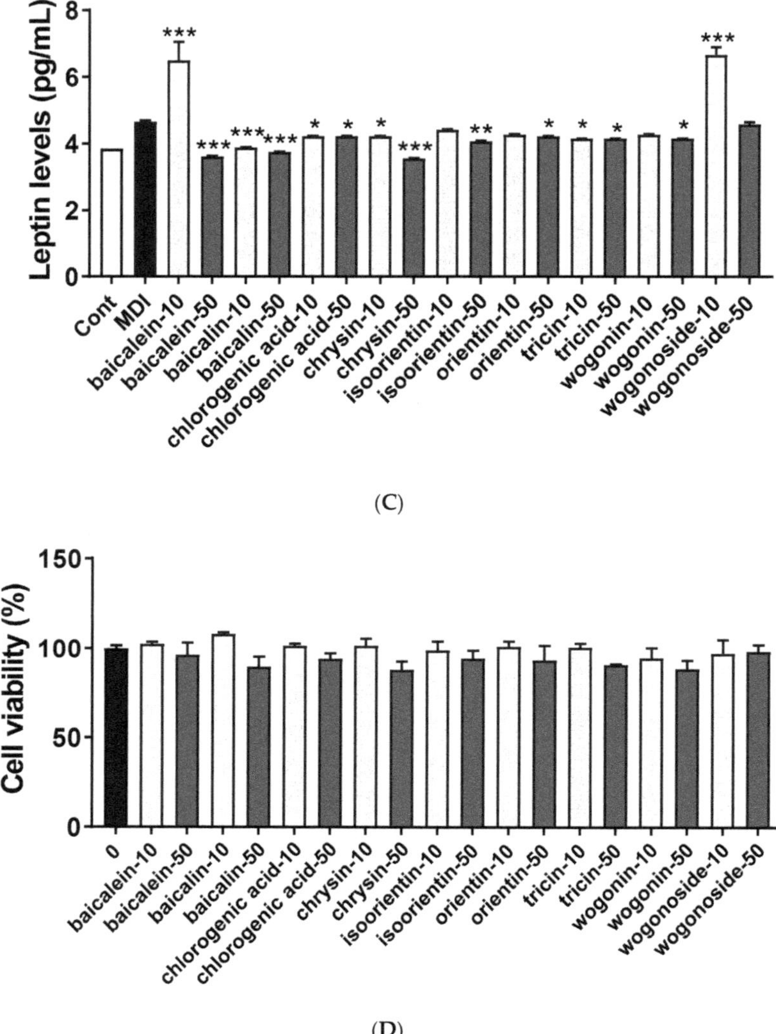

(C)

(D)

Figure 5. Effect of a 2:1 mixture of *Phyllostachys pubescens* and *Scutellaria baicalensis* (BS21) constituents on adipocyte differentiation in 3T3-L1 cells. (**A**) Oil Red O staining of intracellular triglycerides in 3T3-L1 cells. 3T3-L1 cells were treated with compounds (10 and 50 μg/mL) during differentiation induction. (**B**) Relative densities of lipid contents. (**C**) Leptin levels. (**D**) Cell viability. Values are expressed as means ± SD ($n = 3$). Significant differences were observed between control (undifferentiated preadipocytes) and differentiated MDI cells: # $p < 0.05$, ## $p < 0.01$, and ### $p < 0.001$. Significant differences were observed between differentiated MDI (3-isobutyl-1-methylxanthine, dexamethasone, and insulin) cells and BS21-treated cells: * $p < 0.05$, ** $p < 0.01$, and *** $p < 0.001$.

3. Discussion

This study offers the first direct evidence on the anti-obesity effects of BS21 achieved through anti-adipogenesis and the promotion of browning. The results also provide insights into the regulatory mechanisms underlying those effects in 3T3-L1 adipocytes.

PPARγ, C/EBPα, and SREBP-1c are major transcription factors involved in adipocyte differentiation, lipid accumulation, and glucose homeostasis [19,20]. These proteins regulate the expression of adipocyte-specific markers such as FAS, adiponectin, leptin, and aP2 [8,21]. In this study, we found that BS21 decreased the protein levels of transcription factors PPARγ (although it increased at 400 μg/mL concentration), C/EBPα, and SREBP-1c as well as their downstream target proteins FAS, aP2, leptin, and adiponectin in 3T3-L1 adipocytes. BS21 did not enhance lipolysis and the protein levels of phospho-hormone-sensitive lipase (p-HSL), which is an important intracellular neutral lipase that degrades triglycerides by hydrolyzing the ester bond [22]. BS21 also increased protein levels of UCP1 and UCP2, which are mainly associated with thermogenesis and energy expenditure in brown and white adipocytes, respectively [23,24]. These results are supported by decreased fat accumulation in BS21-treated 3T3-L1 adipocytes. Collectively, our findings suggest that BS21 inhibited adipocyte differentiation by down-regulating lipogenesis and up-regulating thermogenesis during adipogenesis.

To further elucidate the mechanism underlying the anti-obesity effects of BS21, we measured marker proteins of BAT (PR domain containing 16 (PRDM16), peroxisome proliferator-activated receptor gamma co-activator 1-alpha (PGC1α), and UCP1) involved in adaptive thermogenesis and the transition of WAT to brown-like adipose tissue [25]. In addition, the activated β-adrenergic signaling cascade (protein kinase A, AMPK, p38, and extracellular signal-regulated kinase) induces white fat browning and brown fat development by elevating PGC1α, PRDM16, PPARγ, and UCP1 [26–29]. In this study, BS21 treatment induced AMPK activation and the expression of brown adipocyte marker proteins in 3T3-L1 adipocytes. In addition, BS21 treatment at 480 ug/mL increased protein levels of PPARγ despite the inhibition of adipogenesis. Previous studies reported that ginsenoside Rb1 and curcumin promote the browning of 3T3-L1 adipocytes through the induction of PPARγ [30,31]. To identify its role in the BS21-induced browning pathway, we treated adipocytes with dorsomorphin, which is a selective inhibitor of AMPK. The arrest of brown fat expression marker protein UCP-1 upon AMPK inhibition reinforced our conclusion that the BS21 mixture induces browning via the AMPK-mediated pathway. Results from previous studies and the present work suggest that BS21 promotes the browning of white adipocytes through the up-regulation of AMPKα-induced white browning proteins PGC1α, PRDM16, and UCP1.

Activated AMPK switches on mitochondrial fatty-acid oxidation and lipolysis [7] and turns off fatty-acid synthesis by inducing the phosphorylation and inactivation of acetyl-CoA carboxylase (ACC), which is an inhibitor of mitochondrial protein CPT1 activity [32]. BS21 treatment led to increased expressions of p-ACC and CPT1. Therefore, the enhanced expression of these proteins is likely to reflect the stimulation of fat oxidation by BS21 treatment. However, even though AMPK activation was increased by BS21, lipolysis and HSL phosphorylation were decreased after BS21 treatment. These data indicate that the anti-obesity effects of BS21 are independent of the HSL enzyme. A recent study showed that AMPK activation inhibited lipolysis [33]. Berberine, a hypoglycemic agent with anti-dyslipidemia and anti-obesity activities, exerts anti-lipolytic effects mainly by reducing the inhibition of phosphodiesterase in 3T3-L1 cells, leading to a decrease in cAMP and HSL phosphorylation independent of the AMPK pathway [34]. These results suggest that the effect of BS21 on AMPK activation may be inhibited or independent of lipolysis.

These results suggest that BS21 may attenuate adipocyte differentiation by suppressing lipid synthesis and promoting fatty-acid oxidation by regulating AMPK activation. UPLC identified nine main compounds (chlorogenic acid, orientin, isoorientin, baicalin, wogonoside, baicalein, tricin, wogonin, and chrysin) in BS21. In our previous study, the two marker compounds isoorientin and baicalin were identified and quantified for the standardization of BS21 [18]. In this study, we identified seven compounds as well as isoorientin and baicalin. Eight constituents (except wogonoside) inhibited adipocyte differentiation and triglyceride accumulation and decreased leptin levels in 3T3-L1 cells. Consistent with our results, previous studies reported that baicalin, baicalein, wogonin, orientin, and isoorientin inhibited adipogenesis in 3T3-L1 adipocytes by decreasing expressions of C/EBPα, SREBP-1c, and PPARγ [13,35–38]. Chlorogenic acid was previously shown to inhibit adipocyte differentiation via AMPK activation [39,40]. Tricin suppressed the expression of transcription factors associated with adipocyte differentiation (PPARγ and C/EBPα) in obese mice fed a high-fat diet [41]. Collectively, the results suggest that these compounds are responsible for the anti-adipogenic activity of BS21. However, further investigation into the effects of these compounds on obesity is required.

The BS2 extract contains two plant extracts and a variety of compounds. A limitation of this study is that it is unclear if one or multiple components of BS21 are responsible for the observed effects on AMPK phosphorylation and thermogenesis. In addition, whether BS2 and its compounds decrease obesity in vivo using these energy expenditure mechanisms needs further investigation.

4. Materials and Methods

4.1. Preparation of BS21

P. pubescens leaves and *S. baicalensis* roots for the BS21 mixture were purchased from Zhenjiang KOC Biotech Co., Ltd. (Jiangsu, China). The dried *P. pubescens* leaves and *S. baicalensis* roots were extracted with 70% ethanol solution for 3 h; then, the extract was prepared by mixing with a 2:1 ratio of *P. pubescens* (B) and *S. baicalensis* root (S) extracts. The lyophilized powder was dissolved in 10% dimethyl sulfoxide and filtered through a 0.22-μm syringe filter to make the stock solution.

4.2. Chemical Profiling of BS21

A Waters ACQUITY UPLC system equipped with a quaternary pump, auto-sampler, and photodiode array detector with an ACQUITY UPLC BEH C18 column (100 × 2.1-mm, 1.7-μm) was used for the chemical profiling of BS21 (Waters Corp., Milford, MA, USA). Elution was performed with 0.1% phosphoric acid solvent A and acetonitrile solvent B. The elution conditions involved holding the starting mobile phase at 94% A and 6% B for 12 min and applying a gradient (89% A: 11% B for 3 min, 80% A: 20% B for 20 min, 40% A: 60% B for 1 min). A wash with 100% B was applied for 2 min, followed by equilibration at 94% A and 6% B for 2 min. The separation temperature was maintained at 40 °C throughout the analysis, with a flow rate of 0.5 mL/min and an injection volume of 3 μL. Each component was analyzed by Quadrupole Dalton (QDa) mass spectrometry, after which the reference standards were purchased from Sigma-Aldrich (St. Louis, MO, USA) and ChemFaces Inc. (Wuhan, Hubei, China), respectively. The retention time and UV spectra were compared and quantified. Components were quantified based on peak areas at 270 nm.

4.3. 3T3-L1 Cell Culture and Differentiation

3T3-L1 cells (ATCC, Rockville, MD, USA) were incubated in Dulbecco's modified Eagle's medium (DMEM) supplemented with calf serum (10%) and penicillin/streptomycin (100 μg/mL) in a humidified atmosphere of 5% CO_2 at 37 °C. To induce the differentiation of 3T3-L1 preadipocytes, cells were incubated with DMEM containing 10% fetal bovine serum (FBS), 3-isobutyl-1-methylxanthine (0.5 mM), dexamethasone (1 μM), and insulin (10 μg/mL) (MDI differentiation media) after reaching confluence in 24-well plates. After 2 days of induction, cells were cultured in DMEM containing FBS (10%) and insulin (10 μg/mL) for 2 days and subsequently maintained in DMEM containing FBS (10%). Various concentrations (60, 120, 240, and 480 μg/mL) of BS21 and its constituents (10 and 50 μg/mL) were added along with the differentiation medium. To test the effects of p-AMPK inhibitor (dorsomorphin, Sigma, St. Louis, MO, USA), cells were supplemented with dorsomorphin (5 μM) in differentiation media until harvest. On day 7, differentiation was assessed based on Oil Red O staining and morphological changes. A cytotoxicity test of 3T3-L1 cells was performed to examine the effect of BS21 on 3T3-L1 cell viability using the 3-(4,5-dimethylthiazol-2-yl)-2,5-diphenyltetrazolium bromide (MTT) assay. The experiment was performed in triplicate, and cell viability was calculated relative to that of the BS21 no-treated control cells.

4.4. Oil Red O Staining

For Oil Red O staining, the differentiated cells were washed gently with phosphate-buffered saline (PBS) and fixed with 10% formalin for 15 min. The fixed cells were stained with Oil Red O solution (in 60 isopropanol and 40% water) for 30 min. After staining, the lipid droplets were visualized using an Olympus CKX41 microscope (Olympus, Tokyo, Japan) and photographed at 100× magnification using the Motic image Plus 2.0 program (Motic, Hong Kong, China).

4.5. Measurement of Lipid Content

To quantify intracellular lipids in Oil Red O-stained cells, the stained lipid droplets were dissolved in isopropanol (500 μL/well) for 30 min, and the extracted dye (100 μL) was transferred to a 96-well plate. Absorbance was measured at 520 nm using a spectrophotometer (BioRad, Hercules, CA, USA). The optical density of the fully differentiated adipocytes was set at 100% relative lipid content.

4.6. Measurement of Adipokines

After differentiation, leptin and adiponectin levels in the medium were measured using a leptin mouse enzyme-linked immunosorbent assay (ELISA) kit (Abcam, Cambridge, UK) and adiponectin mouse ELISA kit (R&D systems, Minneapolis, MN, USA), respectively, following the manufacturer's manuals.

4.7. Measurement of Glycerol Content

Glycerol levels in the medium were measured using a glycerol assay kit (Sigma) with glycerol standards for calibration. Briefly, 100 μL of free glycerol reagent reconstituted in distilled water was mixed with 10 μL of the test sample. Thereafter, the mixtures were incubated at room temperature for 20 min, and the solution absorbance was measured at 570 nm using a microplate reader.

4.8. Western Blot Analysis

Total protein from cells was extracted with PRO-PREP protein extraction solution (Intron, Seoul, Korea) according to the manufacturer's instructions. Total protein samples (15 μg) were separated by sodium dodecyl sulfate polyacrylamide gel electrophoresis and transferred onto nitrocellulose membranes. The membranes were blocked with EzBlock Chemi (ATTO, Tokyo, Japan) and then incubated overnight with primary antibodies at 4 °C. Primary antibodies to peroxisome proliferator-activated receptor (PPAR) γ, CCAAT/enhancer-binding protein (C/EBP) α, sterol regulatory element-binding protein (SREBP)-1c, UCP2, and carnitine palmitoyltransferase (CPT) 1 were obtained from Santa Cruz Biotechnology (Dallas, TX, USA). Antibodies to adipocyte protein 2 (aP2), fatty-acid synthase (FAS), UCP1, AMPKα, phospho-AMPKα, and β-actin were purchased from Cell Signaling (Danvers, MA, USA). Antibodies to phospho-hormone-sensitive lipase (HSL), PR domain containing 16 (PRDM16), and PPARγ co-activator 1-alpha (PGC1α) were obtained from Abcam, and antibodies to phospho-Acetyl-CoA carboxylase (p-ACC) were purchased from Millipore (Billerica, MA, USA). After incubation with primary antibodies, the membranes were incubated with the corresponding horseradish peroxidase-conjugated secondary antibodies (Cell Signaling Technology) for 1 h at room temperature. Then, the membranes were treated with ECL chemiluminescent detection reagent (Amersham Bioscience, Little Chalfont, UK), and protein bands were visualized using LAS-1000 (Fujifilm, Tokyo, Japan). The relative band density was determined by densitometry with Image J software (National Institutes of Health, Bethesda, MD, USA).

4.9. Statistical Analysis

All experiments were repeated at least three times, and each experiment was performed in triplicate. Differences among experimental groups were evaluated using one-way analysis of variance followed by Dunnett's multiple-range tests. All data are presented as mean ± standard deviation (SD). For all tests, $p < 0.05$ was considered statistically significant.

5. Conclusions

The herbal mixture BS21 exerted anti-obesity effects in 3T3-L1 adipocytes by reducing adipogenesis and activating thermogenesis and browning via AMPK activation. Therefore, the data suggest that BS21 might be a potential therapeutic candidate for the prevention and treatment of obesity.

Author Contributions: Conceptualization, Y.-Y.S.; methodology, Y.-Y.S.; software, Y.-Y.S.; validation, Y.-Y.S., E.S. and G.I.; formal analysis, Y.-Y.S.; investigation, Y.-Y.S., E.S. and G.I.; resources, G.I.; data curation, Y.-Y.S.; writing—original draft preparation, Y.-Y.S.; writing—review and editing, Y.-Y.S.; visualization, Y.-Y.S. and E.S.; supervision, D.-S.K.; project administration, D.-S.K.; funding acquisition, D.-S.K. All authors have read and agreed to the published version of the manuscript."

Funding: This research was funded by the INNOPOLIS Foundation of Korea, grant numbers 2019-DD-RD-0039 and 2020-DD-RD-0420-01-201, and the KIOM of Korea, grant number KSN2012330.

Conflicts of Interest: The authors declare no conflict of interest.

References

1. Yumuk, V.; Tsigos, C.; Fried, M.; Schindler, K.; Busetto, L.; Micic, D.; Toplak, H. Obesity Management Task Force of the European Association for the Study of Obesity. *Obes. Facts* **2015**, *8*, 402–424. [CrossRef]
2. Bray, G.A. Drug Treatment of Obesity. *Rev. Endocr. Metab. Disord.* **2001**, *2*, 403–418. [CrossRef] [PubMed]
3. Spiegelman, B.M.; Flier, J.S. Obesity and the Regulation of Energy Balance. *Cell* **2001**, *104*, 531–543. [CrossRef]
4. Kim, G.W.; Lin, J.E.; Blomain, E.S.; Waldman, S.A. Anti-obesity pharmacotherapy: New drugs and emerging targets. *Clin. Pharm.* **2014**, *95*, 53–66.
5. Kajimura, S.; Saito, M. A New Era in Brown Adipose Tissue Biology: Molecular Control of Brown Fat Development and Energy Homeostasis. *Annu. Rev. Physiol.* **2014**, *76*, 225–249. [CrossRef] [PubMed]
6. Kim, K.; Nam, K.H.; Yi, S.A.; Park, J.W.; Han, J.-W.; Lee, J. Ginsenoside Rg3 Induces Browning of 3T3-L1 Adipocytes by Activating AMPK Signaling. *Nutrients* **2020**, *12*, 427. [CrossRef] [PubMed]
7. Hardie, D.G. AMPK: A key regulator of energy balance in the single cell and the whole organism. *Int. J. Obes.* **2008**, *32*, S7–S12. [CrossRef]
8. Son, Y.; Nam, J.-S.; Jang, M.-K.; Jung, I.-A.; Cho, S.-I.; Jung, M.-H. Antiobesity Activity of Vigna nakashimae Extract in High-Fat Diet-Induced Obesity. *Biosci. Biotechnol. Biochem.* **2013**, *77*, 332–338. [CrossRef]
9. Kahn, B.B.; Alquier, T.; Carling, D.; Hardie, D.G. AMP-activated protein kinase: Ancient energy gauge provides clues to modern understanding of metabolism. *Cell Metab.* **2005**, *1*, 15–25. [CrossRef]
10. Kwon, J.H.; Hwang, S.Y.; Han, J.-S. Bamboo (Phyllostachys bambusoides) leaf extracts inhibit adipogenesis by regulating adipogenic transcription factors and enzymes in 3T3-L1 adipocytes. *Food Sci. Biotechnol.* **2017**, *26*, 1037–1044. [CrossRef]
11. Cho, E.-A.; Kim, S.-Y.; Na, I.-H.; Kim, D.-C.; In, M.-J.; Chae, H.-J. Antioxidant and Anticoagulant Activities of Water and Ethanol Extracts of Phyllostachys pubescence Leaf Produced in Geoje. *J. Appl. Biol. Chem.* **2010**, *53*, 170–173. [CrossRef]
12. Yang, J.H.; Choi, M.-H.; Yang, S.H.; Cho, S.S.; Park, S.J.; Shin, H.-J.; Ki, S.H. Potent Anti-Inflammatory and Antiadipogenic Properties of Bamboo (Sasa coreana Nakai) Leaves Extract and Its Major Constituent Flavonoids. *J. Agric. Food Chem.* **2017**, *65*, 6665–6673. [CrossRef]
13. Poudel, B.; Nepali, S.; Xin, M.; Ki, H.-H.; Kim, Y.-H.; Kim, D.-K.; Lee, Y.-M. Flavonoids from Triticum aestivum inhibit adipogenesis in 3T3-L1 cells by upregulating the insig pathway. *Mol. Med. Rep.* **2015**, *12*, 3139–3145. [CrossRef] [PubMed]
14. Yoon, S.-B.; Lee, Y.-J.; Park, S.K.; Kim, H.-C.; Bae, H.; Kim, H.M.; Ko, S.-G.; Choi, H.Y.; Oh, M.S.; Park, W. Anti-inflammatory effects of Scutellaria baicalensis water extract on LPS-activated RAW 264.7 macrophages. *J. Ethnopharmacol.* **2009**, *125*, 286–290. [CrossRef]
15. Yoon, H.-J.; Park, Y.-S. Effects of Scutellaria baicalensis Water Extract on Lipid Metabolism and Antioxidant Defense System in Rats Fed High Fat Diet. *J. Korean Soc. Food Sci. Nutr.* **2010**, *39*, 219–226. [CrossRef]

16. Song, K.H.; Lee, S.H.; Kim, B.-Y.; Park, A.Y.; Kim, J.Y. Extracts ofScutellaria baicalensisReduced Body Weight and Blood Triglyceride in db/db Mice. *Phytother. Res.* **2012**, *27*, 244–250. [CrossRef]
17. Hongxia, L.; Wu, M.; Li, H.; Dong, S.; Luo, E.; Gu, M.; Shen, X.; Jiang, Y.; Liu, Y.; Liu, H. Baicalin Attenuates High Fat Diet-Induced Obesity and Liver Dysfunction: Dose-Response and Potential Role of CaMKKβ/AMPK/ACC Pathway. *Cell Physiol. Biochem.* **2015**, *35*, 2349–2359. [CrossRef]
18. Kim, D.-S.; Kim, S.-H.; Cha, J. Antiobesity Effects of the Combined Plant Extracts Varying the Combination Ratio of Phyllostachys pubescens Leaf Extract and Scutellaria baicalensis Root Extract. *Evid.-Based Complement. Altern. Med.* **2016**, *2016*, 1–11. [CrossRef]
19. Schoonjans, K.; Staels, B.; Auwerx, J. The peroxisome proliferator activated receptors (PPARs) and their effects on lipid metabolism and adipocyte differentiation. *Biochim. Biophys. Acta (BBA) Lipids Lipid Metab.* **1996**, *1302*, 93–109. [CrossRef]
20. Rosen, E.D.; Walkey, C.J.; Puigserver, P.; Spiegelman, B.M. Transcriptional regulation of adipogenesis. *Genes Dev.* **2000**, *14*, 1293–1307.
21. Kliewer, S.A.; Willson, T.M. The nuclear receptor PPARgamma - bigger than fat. *Curr. Opin. Genet. Dev.* **1998**, *8*, 576–581. [CrossRef]
22. Bezaire, V.; Mairal, A.; Ribet, C.; Lefort, C.; Girousse, A.; Jocken, J.; Laurencikiene, J.; Anesia, R.; Rodriguez, A.-M.; Ryden, M.; et al. Contribution of Adipose Triglyceride Lipase and Hormone-sensitive Lipase to Lipolysis in hMADS Adipocytes. *J. Biol. Chem.* **2009**, *284*, 18282–18291. [CrossRef] [PubMed]
23. Dulloo, A.G.; Seydoux, J.; Jacquet, J. Adaptive thermogenesis and uncoupling proteins: A reappraisal of their roles in fat metabolism and energy balance. *Physiol. Behav.* **2004**, *83*, 587–602. [CrossRef] [PubMed]
24. Sung, Y.-Y.; Kim, D.-S.; Kim, H.K. Viola mandshurica ethanolic extract prevents high-fat-diet-induced obesity in mice by activating AMP-activated protein kinase. *Environ. Toxicol. Pharmacol.* **2014**, *38*, 41–50. [CrossRef] [PubMed]
25. Park, S.J.; Park, M.; Sharma, A.; Kim, K.; Lee, H.J. Black ginseng and ginsenosde Rb1 promote browning by inducing UCP1 expression in 3T3-L1 and primary white adipocytes. *Nutrients* **2019**, *11*, 2747. [CrossRef]
26. Yan, M.; Audet-Walsh, É; Manteghi, S.; Dufour, C.R.; Walker, B.; Baba, M.; St-Pierre, J.; Giguère, V.; Pause, A. Chronic AMPK activation via loss of FLCN induces functional beige adipose tissue through PGC-1α/ERRα. *Genes Dev.* **2016**, *30*, 1034–1046. [CrossRef] [PubMed]
27. Yang, Q.; Liang, X.; Sun, X.; Zhang, L.; Fu, X.; Rogers, C.J.; Berim, A.; Zhang, S.; Wang, S.; Wang, B.; et al. AMPK/α-Ketoglutarate Axis Dynamically Mediates DNA Demethylation in the Prdm16 Promoter and Brown Adipogenesis. *Cell Metab.* **2016**, *24*, 542–554. [CrossRef]
28. Lanzi, C.R.; Perdicaro, D.J.; Landa, M.S.; Fontana, A.; Antoniolli, A.; Miatello, R.M.; Oteiza, P.I.; Prieto, M.A.V. Grape pomace extract induced beige cells in white adipose tissue from rats and in 3T3-L1 adipocytes. *J. Nutr. Biochem.* **2018**, *56*, 224–233. [CrossRef]
29. Seo, Y.-J.; Kim, K.-J.; Choi, J.; Koh, E.-J.; Lee, B.-Y. Spirulina maxima Extract Reduces Obesity through Suppression of Adipogenesis and Activation of Browning in 3T3-L1 Cells and High-Fat Diet-Induced Obese Mice. *Nutrients* **2018**, *10*, 712. [CrossRef]
30. Mu, Q.; Fang, X.; Li, X.; Zhao, D.; Mo, F.; Jiang, G.; Yu, N.; Zhang, Y.; Guo, Y.; Fu, M.; et al. Ginsenoside Rb1 promotes browning through regulation of PPARγ in 3T3-L1 adipocytes. *Biochem. Biophys. Res. Commun.* **2015**, *466*, 530–535. [CrossRef]
31. Lone, J.; Choi, J.H.; Kim, S.W.; Yun, J.W. Curcumin induces brown fat-like phenotype in 3T3-L1 and primary white adipocytes. *J. Nutr. Biochem.* **2016**, *27*, 193–202. [CrossRef] [PubMed]
32. Kim, S.-J.; Jung, J.Y.; Kim, H.W.; Park, T. Anti-obesity effects of Juniperus chinensis extract are associated with increased AMP-activated protein kinase expression and phosphorylation in the visceral adipose tissue of rats. *Biol. Pharm. Bull.* **2008**, *31*, 1415–1421. [CrossRef]
33. Daval, M.; Diot-Dupuy, F.; Bazin, R.; Hainault, I.; Viollet, B.; Vaulont, S.; Hajduch, E.; Ferré, P.; Foufelle, F. Anti-lipolytic Action of AMP-activated Protein Kinase in Rodent Adipocytes. *J. Biol. Chem.* **2005**, *280*, 25250–25257. [CrossRef] [PubMed]
34. Zhou, L.; Wang, X.; Yang, Y.; Wu, L.; Li, F.; Zhang, R.; Yuan, G.; Wang, N.; Chen, M.; Ning, G. Berberine attenuates cAMP-induced lipolysis via reducing the inhibition of phosphodiesterase in 3T3-L1 adipocytes. *Biochim. Biophys. Acta (BBA) Mol. Basis Dis.* **2011**, *1812*, 527–535. [CrossRef] [PubMed]
35. Kim, J.; Lee, I.; Seo, J.; Jung, M.; Kim, Y.; Yim, N.; Bae, K. Vitexin, orientin and other flavonoids from Spirodela polyrhiza inhibit adipogenesis in 3T3-L1 cells. *Phytother. Res.* **2010**, *24*, 1543–1548. [CrossRef]

36. Kwak, D.H.; Lee, J.H.; Kim, D.G.; Kim, T.; Lee, K.J.; Ma, J.Y. Inhibitory effects of Hwangryunhaedok-Tang in 3T3-L1 adipogenesis by regulation of Raf/MEK1/ERK1/2 pathway and PDK1/Akt phosphorylation. *Evid.-Based Complement. Alternat. Med.* **2013**, *2013*, 413906. [CrossRef]
37. Kwak, D.H.; Lee, J.H.; Song, K.H.; Ma, J.Y. Inhibitory effects of baicalin in the early stage of 3T3-L1 preadipocytes differentiation by down-regulation of PDK1/Akt phosphorylation. *Mol. Cell Biochem.* **2014**, *385*, 257–264. [CrossRef]
38. Nakao, Y.; Yoshihara, H.; Fujimori, K. Suppression of Very Early Stage of Adipogenesis by Baicalein, a Plant-Derived Flavonoid through Reduced Akt-C/EBPα-GLUT4 Signaling-Mediated Glucose Uptake in 3T3-L1 Adipocytes. *PLoS ONE* **2016**, *11*, e0163640. [CrossRef]
39. Sung, Y.Y.; Kim, D.S.; Kim, H.K. Akebia quinata extract exerts anti-obesity and hypolipidemic effects in high-fat diet-fed mice and 3T3-L1 adipocytes. *J. Ethnopharmacol.* **2015**, *168*, 17–24. [CrossRef]
40. Xu, M.; Yang, L.; Zhu, Y.; Liao, M.; Chu, L.; Li, X.; Lin, L.; Zheng, G. Collaborative effects of chlorogenic acid and caffeine on lipid metabolism via the AMPKα-LXRα/SREBP-1c pathway in high-fat diet-induced obese mice. *Food Funct.* **2019**, *10*, 7489–7497. [CrossRef]
41. Lee, D.; Imm, J.-Y. Antiobesity Effect of Tricin, a Methylated Cereal Flavone, in High-Fat-Diet-Induced Obese Mice. *J. Agric. Food Chem.* **2018**, *66*, 9989–9994. [CrossRef] [PubMed]

Article

Potential Nutraceutical Benefits of In Vivo Grown Saffron (*Crocus sativus* L.) As Analgesic, Anti-inflammatory, Anticoagulant, and Antidepressant in Mice

Asif Khan [1], Nur Airina Muhamad [1,*], Hammad Ismail [2], Abdul Nasir [3], Atif Ali Khan Khalil [4], Yasir Anwar [5], Zahid Khan [6], Amjad Ali [7], Rosna Mat Taha [1], Baker Al-Shara [1], Sara Latif [8], Bushra Mirza [8,9], Yousef Abdal Jalil Fadladdin [5], Isam Mohamed Abu Zeid [5] and Saed Ayidh Al-Thobaiti [10]

1 Institute of Biological Sciences, Faculty of Science, University of Malaya, 50603 Kuala Lumpur, Malaysia; asif.khan.qau@gmail.com (A.K.); rosnataha2020@gmail.com (R.M.T.); Bakearbio@yahoo.com (B.A.-S.)
2 Department of Biochemistry and Biotechnology, University of Gujrat, Gujrat 50700, Pakistan; hammad.ismail@uog.edu.pk
3 Department of Molecular Science and Technology, Ajou University, Suwan 16499, Korea; anasirqau@gmail.com
4 Department of Biological Sciences, National University of Medical Sciences, Rawalpindi 46000, Pakistan; Atif.ali@numspak.edu.pk
5 Department of Biological Sciences, Faculty of Science, King Abdulaziz University, P.O. Box 54229, Jeddah, Saudi Arabia; yanwarulhaq@kau.edu.sa (Y.A.); yfadladdin@kau.edu.sa (Y.A.J.F.); abuzeidmm@yahoo.com (I.M.A.Z.)
6 Department of Pharmacognosy, Faculty of Pharmacy, Federal Urdu University of Arts Science and Technology, Karachi 75300, Pakistan; zahidkhan@fuuast.edu.pk
7 Department of Botany, University of Malakand, Khyber Pakhtunkhwa 18800, Pakistan; amjad1990.aa48@gmail.com
8 Department of Biochemistry, Quaid-i-Azam University, Islamabad 45320, Pakistan; saralatifawan@gmail.com (S.L.); bushramirza@qau.edu.pk (B.M.)
9 Lahore College for Women University, Lahore 54000, Pakistan
10 Department of Biology, University College Turabah, Taif University, Taif 21995, Saudi Arabia; saed@tu.edu.sa
* Correspondence: nurairina@um.edu.my

Received: 24 September 2020; Accepted: 20 October 2020; Published: 22 October 2020

Abstract: *Crocus sativus*, a medicinally important herbaceous plant, has been traditionally used to cure coughs, colds, insomnia, cramps, asthma, and pain. Moreover, the therapeutic applications of saffron include its immunomodulatory and anticancer properties. The current experimental analysis was performed to explore the potential nutraceutical efficacy of corm, leaf, petal, and stigma of saffron ethanolic extracts as analgesic, anti-inflammatory, anticoagulant, and antidepressant using hot plate, carrageenan-induced paw edema, capillary tube and forced swim test, respectively in mice. The results indicated that among all the extracts, stigma ethanolic extract (SEE) represented maximum latency activity (72.85%) and edema inhibition (77.33%) followed by petal ethanolic extract (PEE) with latency activity and edema inhibition of 64.06 and 70.50%, respectively. Corm ethanolic extract (CEE) and leaf ethanolic extract (LEE) displayed mild analgesic activity of 22.40% and 29.07%, respectively. Additionally, LEE (53.29%) and CEE (47.47%) exhibited mild to moderate response against inflammation. The coagulation time of SEE (101.66 s) was almost equivalent to the standard drug, aspirin (101.66 s), suggesting a strong anticoagulant effect followed by PEE (86.5 s). LEE (66.83 s) represented moderate inhibitory effect on coagulation activity while CEE (42.83 s) showed neutral effect. Additionally, PEE and SEE also expressed itself as potential antidepressants with immobility time ≤76.66 s, while CEE (96.50 s) and LEE (106.83 s) indicated moderate to mild antidepressant efficacy. Based on the in vivo activities, saffron extract, particularly SEE and PEE, can be used as a potential nutraceutical and therapeutic agent due to its significant pharmacological activities.

Keywords: *Crocus sativus*; in vivo; carrageenan; analgesic; antidepressant; anticoagulant

1. Introduction

In recent times, traditional medicinal plants are gaining much attention in pharmaceutical industries and served as a part of the primary medical emergencies in the treatment of numerous diseases. *Crocus sativus*, or saffron, is one such stemless herbaceous plant grown as a source of its spice for nearly 3600 years and belongs to the family Iridaceae. Saffron is geographically distributed in Irano-Touranian regions, East Asia, and Mediterranean climates and largely cultivated in countries like Iran, Afghanistan, India, France, Turkey, Italy, Spain, and Morocco [1]. Current estimates for total world annual production of saffron are 418 t per year [2]. Iran, the world's largest producer of saffron with around 90,000 ha harvest area and 336 t annual yield, is said to produce more than 90% of the total saffron produced worldwide [3].

Saffron plant comprises of the corm, foliar structure, and floral organs as the main parts (Figure 1). Commercially, saffron is a dietary spice obtained from three red pungent stigmatic lobes of the *Crocus sativus* flower. According to an estimate, it takes around 60,000 Crocus flowers to produce just 1 kg of this unique spice, which equates to 370 to 470 h of work [4] and needs a cultivation area of approximately 2000 m^2 [5]. At a retail price of almost US$1000–10,000 per kg, saffron is considered the costliest spice of the world [6]. Saffron stigma contains more than 150 phytochemical ingredients, belonging to various divisions of secondary metabolites such as carotenoids, flavonoids, terpenoids, and anthocyanin. Amongst the 50 constituents identified so far, apocarotenoids, including crocin, crocetin, picrocrocin, and safranal, emerge as the major constituents of saffron stigma, primarily responsible for imparting a distinct colorant, flavor, and aroma to the spice [7]. These compounds are derived from lipophilic carotenoid, zeaxanthin by oxidative cleavage. Moreover, a series of flavonoids (all glycosidic derivatives of kaempferol) have also been characterized in the stigma and petals of saffron, which, together with picrocrocin, give the characteristic bitter feature to this spice [8]. Furthermore, some anthraquinones and anthocyanin secondary components are also reported to be extracted from corms and petals of saffron [9].

Conventionally, saffron has a long history of being used as a spice and food additive [10]. Besides its use in culinary [11] and cosmetic preparation [12], saffron has been used as a folk remedy for coughs, colds, insomnia, cramps, asthma and bronchospasms, liver disease, pain, and epilepsy [2]. Saffron tea has been reputed to be a potential complementary treatment for psoriasis in medical nutrition therapy [13]. Many pharmacological studies have shown that this plant and its phytochemicals have emerged as nutraceutical elements, endowed with beneficial effects on health showing antibacterial [14], antioxidant, cytotoxic [15,16], antifungal [17], immunomodulatory [18,19], anti-mutagenic [20], aphrodisiac [21], antitussives [22,23], and antiplatelet effects [24]. Traditional and modern biomedical reports have also proved saffron to cure coronary artery diseases [25], respiratory diseases [26], menstrual disorders [27,28], and neurodegenerative disorders [29]. All these desirable properties of saffron have been attributed to the stigmas, whereas other parts of the plant have been much less studied. Nevertheless, a large quantity of saffron by-products is produced during the stigmatization process with little commercial value but thrown away after harvesting. According to an estimate, approximately 1500 kg of leaves, 350 kg of petals, and several hundred cormlets too small for cultivation or biologically and physically damaged to be regrown are rejected to get spice of only 1 kg [30]. However, this biomass contains a multitude of phytochemical ingredients whose exploitation would significantly enhance the sustainability and profitability of saffron yield. These saffron based by-products need to be tested to assess its pharmacological applications.

Figure 1. Main components of saffron (*Crocus sativus*), (**A**) corms (**B**) flower.

Bioactive compounds from petals have shown antinociceptive, anti-inflammatory [31], cytotoxic, antimicrobial, antioxidant, antidiabetic [32], antiobesity [33], antidepressant [34], anticancer, antityrosinase properties, as well as reduce blood pressure and contractility [35]. Several phenolic compounds from saffron leaves have been identified, showing antibacterial [36,37], anticancer [30], antioxidant, and metal ion chelating activities [37,38]. Likewise, bioactive constituents in saffron corm, such as proteoglycans, revealed to have a cytolytic effect on human tumor and plant cells and triterpenoid saponins on fungicidal and anticancer activities [30]. Polyphenols showing antibacterial, radical scavenging activity [39,40], and a mannan-binding lectin [41] have also been investigated in saffron corms. The existing experimental analysis aimed to explore the anti-inflammatory, analgesic, anticoagulant, and antidepressant potentials of various extracts of *Crocus sativus* L. in mice by carrageenan-induced paw edema test, hot plate assay, capillary method, and forced swim test, respectively. Up to now, ethanolic extracts of stigma, petals, corms, and leaves of in vivo cultivated saffron have not been previously studied for their nutraceutical properties against analgesia, inflammation, coagulation, and depression.

2. Results

2.1. Acute Toxicity Study

During the 7-days study, none of the orally administered saffron ethanolic extracts showed any mortality in the animals treated with 800 mg/kg. Furthermore, tested samples did not produce any notable behavioral changes in mice during the observation period.

2.2. Hot Plate Analgesic Test

The test is based on the use of thermal stimuli to determine the effect of analgesics. For this purpose, an easy and cost-effective method, the hot plate analgesic test, was performed. Results of saffron ethanolic extracts presented in the form of latency time indicated that all the extracts showed significant ($p < 0.05$) analgesic effect in a time-dependent manner (Table 1). Aspirin, used as a reference drug, showed the highest latency (17.51 ± 0.50) at 1 h followed by reduction (16.82 ± 0.45) at 2 h. Among all the extracts of saffron tested, stigma ethanolic extract (SEE) showed the highest latency activity (7.80 ± 0.16, 11.30 ± 0.21, 12.80 ± 0.33, and 13.50 ± 0.28) at 0, 0.5, 1, and 2 h, respectively. On the other hand, leaf ethanolic extract (LEE) exhibited no significant difference at 0 h when compared with negative control (saline) but showed the least activity of 5.92 ± 0.14, 6.22 ± 0.15, and 6.65 ± 0.19 at 0.5, 1, and 2 h, respectively. Based on the percent inhibition (Figure 2), SEE was more prominent in reducing analgesia (44.91%, 63.89%, and 72.85%), followed by petal ethanolic extract (PEE), showing 38.84%, 53.07%, and 64.06% inhibition at 0.5, 1, and 2 h, respectively. Furthermore, corm ethanolic extract (CEE) and LEE responded poorly and showed weak analgesic activity of 22.40% and 29.07% at 2 h, respectively.

Table 1. Latency time of *C. sativus* ethanolic extracts in hot plate assay.

Group	Dose (mg/kg)	Latency Time (s)			
		0 h	0.5 h	1 h	2 h
Saline	1 mL/kg	4.70 ± 0.06	5.00 ± 0.02	5.50 ± 0.01	5.40 ± 0.04
Aspirin	10	9.45 ± 0.28***	16.38 ± 0.42***	17.51 ± 0.50***	16.82 ± 0.45***
CEE	800	6.20 ± 0.17***	6.93 ± 0.16***	7.15 ± 0.13***	7.58 ± 0.14***
LEE	800	5.15 ± 0.15	5.92 ± 0.14**	6.22 ± 0.15*	6.65 ± 0.19**
SEE	800	7.80 ± 0.16***	11.30 ± 0.21***	12.80 ± 0.33***	13.50 ± 0.28***
PEE	800	6.78 ± 0.17***	9.42 ± 0.24***	10.38 ± 0.29***	11.13 ± 0.35***

Each value is represented as mean ± S.D. Where, *$p < 0.05$, ** $p < 0.01$, *** $p < 0.001$ statistically significant relative to control (saline).

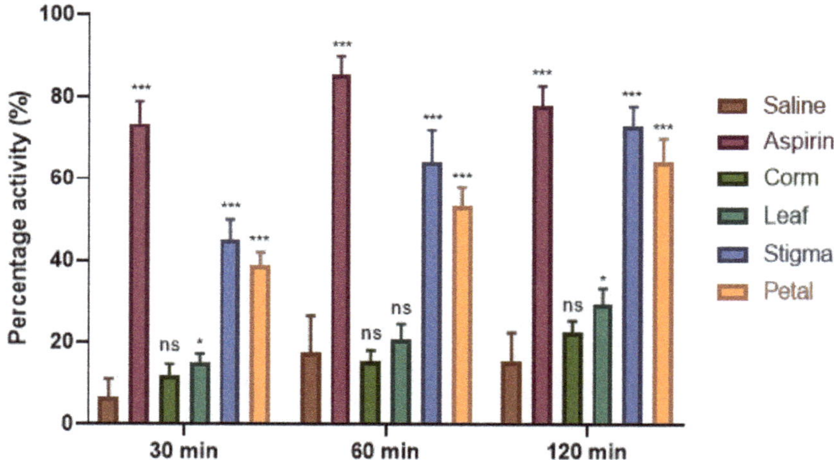

Figure 2. Percentage analgesia of C. *sativus* ethanolic extracts at selected time period in mice. Each value is represented as mean ± S.D. Where, * $p < 0.05$, ** $p < 0.01$, *** $p < 0.001$ statistically significant relative to control (saline).

2.3. *Carrageenan-induced Hind Paw Edema Test*

The anti-inflammatory effects of saffron ethanolic extracts was investigated by the mouse paw-edema test, and the findings are presented in Table 2. Sub planter injection of carrageenan gradually increased edema paw volume of the saline-treated animals. However, as a positive control, diclofenac potassium attenuated paw edema volume by 88.87%, as depicted in Figure 3. Moreover, oral administration of saffron ethanolic extracts showed a significant ($p < 0.05$) decrease in edematous paw volume in a time-dependent manner. SEE (800 mg/kg) produced an anti-inflammatory activity 1 h after administration and continued until the end of the experimentation, with the most prominent inhibition of 77.33% followed by PEE (70.50%) at the fourth h of the study. LEE and CEE exhibited moderate but significant ($p < 0.05$) potential with the percentage edema inhibition of 53.29% and 47.47%, respectively.

Table 2. Anti-inflammatory effect of *C. sativus* ethanolic extracts.

Group	Dose (mg/kg)	Mean Edema Volume (mL)			
		1 h	2 h	3 h	4 h
Saline	1 mL/kg	0.47 ± 0.03	0.56 ± 0.03	0.69 ± 0.04	0.76 ± 0.03
Diclofenac potassium	10	0.25 ± 0.02***	0.17 ± 0.01***	0.08 ± 0.01***	0.04 ± 0.01***
CEE	800	0.43 ± 0.03	0.37 ± 0.03***	0.30 ± 0.01***	0.25 ± 0.03***
LEE	800	0.37 ± 0.03**	0.31 ± 0.02***	0.26 ± 0.01***	0.20 ± 0.03***
SEE	800	0.26 ± 0.01***	0.19 ± 0.02***	0.14 ± 0.01***	0.07 ± 0.01***
PEE	800	0.29 ± 0.02***	0.23 ± 0.02***	0.17 ± 0.01***	0.11 ± 0.01***

Each value is represented as mean ± S.D. Where, * $p < 0.05$, ** $p < 0.01$, *** $p < 0.001$ statistically significant relative to control (saline).

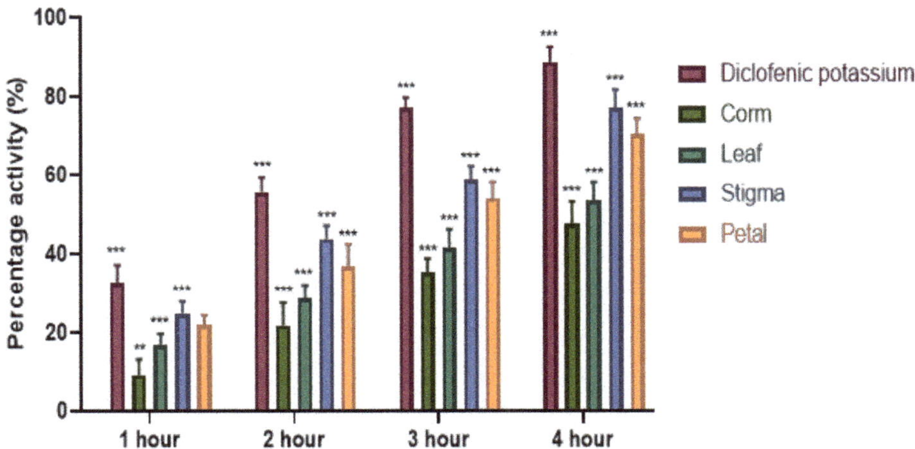

Figure 3. Percentage edema inhibition of *C. sativus* ethanolic extracts at selected time period in mice. Each value is represented as mean ± S.D. Where, * $p < 0.05$, ** $p < 0.01$, *** $p < 0.001$ statistically significant relative to control (saline).

2.4. Anticoagulant Assay

The blood clotting activity of saffron extracts was investigated using the capillary tube method. The results presented in Figure 4 show the effects of oral administration of saffron extracts and aspirin on coagulation time in mice. As a reference drug, the anticoagulant aspirin (10 mg/kg) significantly increased blood clotting time (108.5 ± 8.59 s) compared to the control group, saline (38.33 ± 4.92 s). SEE was the major identified anticoagulant extract showing prominent and significant anticoagulant effect with coagulation time of 101.66 ± 7.20 s followed by PCC (86.50 ± 6.89 s), respectively. LEE, however, had a moderate inhibitory effect on coagulation activity with a clotting time of 66.83 ± 6.17 s. CEE, on the other hand, was unable to prolong the blood clotting time (42.83 ± 6.27 s) and showed an almost equal response to saline, representing the neutral effect of the selected extract.

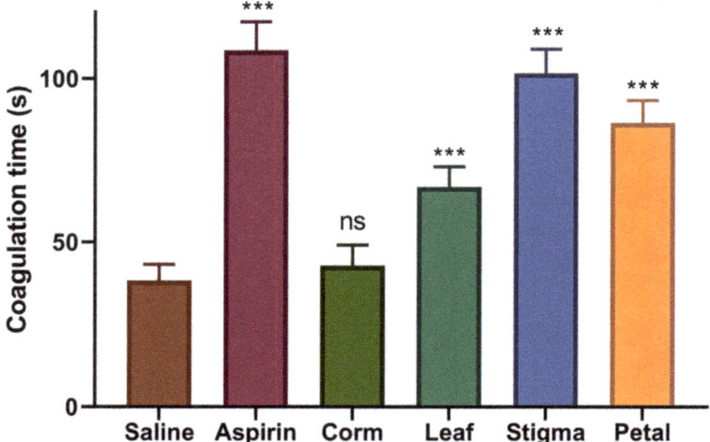

Figure 4. Anticoagulant effect of *C. sativus* ethanolic extracts. Each value is represented as mean ± S.D. Where, * $p < 0.05$, ** $p < 0.01$, *** $p < 0.001$ statistically significant relative to control (saline).

2.5. Antidepressant Activity

The antidepressant activity of orally administered saffron ethanolic extracts was tested by forced swimming test, and findings in the form of immobility time are graphically depicted in Figure 5. Positive control group administered with drug, fluoxetine HCl produced a strong antidepressant effect (41.33 ± 4.71 s) at the concentration of 10 mg/kg against the negative control, saline (141.16 ± 6.40 s). Furthermore, saffron ethanolic extracts significantly attenuated immobility time in mice when compared with the saline-treated control group. At the dose of 800 mg/kg, PEE represented itself as a potential antidepressant candidate showing a significant reduction in immobility time (69.66 ± 7.63 s) with respect to control and equivalent to the standard drug followed by SEE (76.66 ± 6.56 s). CEE significantly declined the mean time spent by mice in the immobile state (96.50 ± 6.28 s), signifying moderate antidepressant effect, whereas LEE exhibited mild activity (106.83 ± 6.24 s) by significantly attenuating immobility time with respect to serine, but not equivalent to fluoxetine HCl.

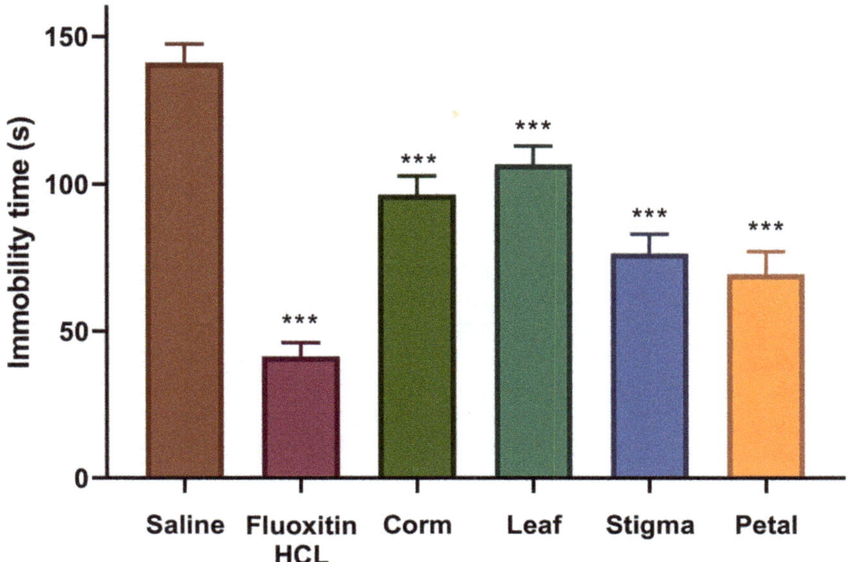

Figure 5. Antidepressant effect of *C. sativus* ethanolic extracts. Each value is represented as mean ± S.D. Where, * $p < 0.05$, ** $p < 0.01$, *** $p < 0.001$ statistically significant relative to control (saline).

3. Discussion

Saffron is an extraordinarily rich source of nutraceutical and pharmaceutical properties that exhibit numerous health benefits and pharmacological effects [42]. In the present study, we have evaluated the analgesic, anti-inflammatory, anticoagulant, and antidepressant activities of saffron ethanolic extracts in mice. The selection of 80% ethanol and/or ethanolic extracts was chosen for the study due to their efficiency against cancer [43,44], blood pressure [45,46], inflammation [47], and nociception [48,49]. None of the orally administered saffron extracts (800 mg/kg) caused any mortality or prominent behavioral abnormalities in mice during the 7-days acute toxicity study. In literature, toxicological reports regarding saffron safety are not uniform. Iranshahi et al. [50] assessed the toxicity level of 800 mg/kg SEE, PEE, stigma aqueous extract (SAE), and petal aqueous extract (PAE) and found no toxicological signs on mice. In another study, the maximum non-toxic dose of SEE and PEE was reported as 2 and 8 g/kg, respectively [48]. Furthermore, LD_{50} values of intraperitoneal administered petal and stigma extracts in mice were 6 and 1.6 g/kg, respectively [51]. Similarly, no mortality of mice was reported within 2 days of study with high oral and intraperitoneal doses (3 g/kg) of the active constituent of saffron, crocin, in mice [52].

The hot plate is one of the oldest and frequently used animal models to quantify "pain-like" behaviors in rodents [53]. Based on the species and strain of rodents used in clinical studies, nearly 12 different behaviors, including grooming, freezing, sniffing, licking, stamping of the legs, leaning, and jumping, have been measured in the hot plate assay [54]. In the present study, hot plate analgesic assay of saffron extracts demonstrated a time-dependent activity. Among all the saffron extracts, SEE exhibited the highest analgesic value in delaying the mean paw licking time (13.50 ± 0.28 s) by suppressing nociception in paws. PEE also delayed the onset time of licking response (11.13 ± 0.35 s). However, CEE and LEE exhibited weak analgesic activity of 22% and 29% at 2 h, respectively. Good analgesic drugs suppress the activity of nociceptors and exhibit the least number of lickings in animals. As per findings, SEE, and PEE thus were found as potent analgesic agents. The maximum analgesic activity of SEE might be due to the presence of carotenoids such as crocetin, crocin, picrocrocin, and safranal, as carotenoids have been reported to suppress the synthesis of prostaglandin synthetase [55].

This speculation is supported by another study where 0.1 and 0.2 g/kg crocin showed significant anti-edematogenic potential in histamine-induced paw edema in rats [56]. However, according to Hosseinzadeh and Younesi [48], aqueous and ethanolic stigma and petal extracts of saffron at any dose exerted no significant analgesic effect in mice. They suggested that the extracts might not act through central mechanisms, although drugs that alter the animal's motor ability may enhance the licking duration on the hot plate method without affecting the central nervous system.

Inflammation serves as a body's defensive biological response to damaged cells and injured tissue. The existence of edema is amongst the major signs of inflammation [57]. Carrageenan-induced paw edema is an established method to investigate the anti-edematous activity of natural products in rodents [58,59]. In the present study, the anti-inflammatory potential of saffron ethanolic extracts was studied after a sub-planter injection of 1% λ-carrageenan into mice. Test samples represented significant anti-edematous potential by regulating biphasic acute inflammatory response induced by carrageenan and showed the highest edema inhibition in mice after 4 h. Carrageenan induces the inflammatory process in two phases. The initial phase, which occurs during the first 2.5 h post-carrageenan injection, attributes to the release of mediators like serotonin, histamine, and bradykinin on vascular permeability. Serotonin and histamine are produced in the first 1.5 h, while bradykinin is produced within 2.5 h post-carrageenan injection [60]. The final phase occurs from 2.5 to 6 h after carrageenan injection, is associated with the overproduction of prostaglandins in tissues [58,61]. In addition to these mediators, Nitric Oxide (NO) is also reported to play a key role in carrageenan-induced acute inflammation [62,63]. Among all the extracts, SEE effectively inhibited the increase in paw volume of carrageenan-induced edema showing a maximum percentage of edema inhibition (77.33%) at the end of 4 h followed by PEE (70.50%) at 800 mg/kg dose. Diclofenac potassium used as a standard drug reduced paw edema volume by 88.87%. Therefore, it can be assumed that the active constituents of the extract might be responsible for the inactivation of inflammation process whose mechanism of action need to be studied. Phytochemicals screening of SAE, SEE, PAE, and PEE on acute inflammation by xylene-induced ear edema indicated that only SAE and SEE at higher doses possessed anti-inflammatory effects in mice. However, SAE, SEE, and PEE showed significant activity against chronic inflammation using formalin-induced edema in rat paw [48]. In another study, saffron aqueous extract suppressed formalin-induced paw edema in the chronic inflammation but failed to show activity against the acute phase of a formalin test [64]. Similarly, intraperitoneal injection of stigma constituent, crocin at concentrations of 0.1 and 0.2 g/kg significantly attenuated paw thickness and infiltration of neutrophils in paw tissues [56]. Kumar et al. [65] examined various petal extracts of *C. sativus* Cashmerianus to assess the anti-inflammatory effect by carrageenan-induced paw edema method. Among all the extracts, methanol extract (400 mg/mL) exhibited 63.16% inhibition of paw volume, followed by aqueous (57.89%) and chloroform (50%) extracts. The phytochemical profile of *C. sativus* petal suggests that the anti-inflammatory properties of PEE might be due to the presence of nutraceutical compounds, particularly flavonoids (kaempferol, 12.6% *w/w*) by modulating the gene and protein expression of inflammatory molecules [66].

The normal hemostatic process is meant to stop a cut or wound on blood vessels through platelet thrombus formation; subsequently, there is an eventual elimination of the plug when healing is complete. It is a delicate multiphase mechanism that requires the interaction of platelets and the coagulation factors with blood vessels. A defect in any of these phases can result in thrombosis or hemorrhage [67]. Extracts of various herbal plants have been used for their hemostatic role in wound healing, anti-infective, and antineoplastic properties [68]. Ankaferd Blood Stopper, an herbal extract from five plants (*Urtica dioica*, *Alpinia officinarum*, *Vitis vinifera*, *Thymus vulgaris*, and *Glycyrrhiza glabra*) is approved for the management of external hemorrhage and post-surgery dental bleedings in Turkey to attenuate blood clotting time effectively [69]. Additionally, crude extracts of *Fagonia cretica* (74.6%), *Hedera nepalensis* (73.8%), and *Phytolacca latbenia* (67.3%) revealed promising anticoagulant effect by delaying blood clotting time to 86.9 s, 84.3 s, and 67.5 s, respectively [70]. In this study, saline and aspirin showed a clear line of difference between the coagulation and anticoagulation by depicting blood clotting time of 38.33 s and 108.5 s, respectively. The most significant anticoagulant effect was reported by SEE with a coagulation time of 101.66 s followed by PEE (86.5 s), respectively. LEE showed a blood clotting time of 66.83 s, representing a moderate inhibitory effect on coagulation activity. CEE was least effective in anticoagulation property and showed an almost equal response to saline, indicating a neutral effect of the selected extract, which can be used in the treatment of blood clotting disorders such as hemophilia. It is said that the corms of *C. sativus* Cartwrightianus have a protein factor involved in human platelet aggregation [71]. However, 5 years later, it was reported that it contains both activator/inducer of platelet aggregation [72]. Crocin delayed blood clotting time and mitigated respiratory distress as a result of pulmonary thrombosis in mice, inhibited thrombosis in rats, and suppressed platelet aggregation in rabbits [73]. Crocetin significantly reduced collagen- and ADP-induced platelet aggregation but failed to reduce arachidonic acid-induced platelet aggregation [74]. Besides that, crocetin significantly reduced dense granule secretion, while neither platelets adhesion to collagen nor cyclic AMP level was affected by crocetin [75]. Saffron stigma tablets (200 and 400 mg/day) assessed for short-term safety and tolerability in a limited number of volunteers showed that only 200 mg of saffron tablets reduced International Normalize Ratio, platelets, and coagulation time [76]. Later, in a double-blind, placebo-controlled clinical study with a large sample size, saffron tablets (200 and 400 mg/day) administration failed to show any major effect on coagulant and anticoagulant system after one month. The authors suggested that the case reports of hemorrhagic complications might be due to the high saffron dose, high period of consumption, or idiosyncrasy activities [77].

Depression is one of the most serious psychiatric disorders affecting approximately 4.7% of the global population and is ranked as the eleventh most frequent cause of disease burden worldwide [78]. Most of the patients have many concerns about commencing synthetic antidepressants in their recommended doses due to the anticipated adverse reactions such as libido, constipation, dry mouth, and dizziness [79]. Hence, extracts of medicinal plants provide the most effective sources of novel drugs showing promising results with minimum side effects in the routine treatment of depression [80,81]. The Forced Swimming Test (FST) or Porsolt swim test, is the most frequently used test to screen antidepressants among all rodents with more reliability, sensitivity, and specificity [82]. Immobility or the flouting response of rodents in FST is traditionally considered an indication of depression and anxiety [83]. During the FST, treatment of antidepressant drugs attenuate immobility, prolong its onset and delay active escape behaviors of the animal [84]. In the present study, a prominent difference in immobility time of saline (141.16 s) and fluoxetine HCl (41.33 s) used as a negative and positive control, respectively, was observed, representing the reliability of the test. Furthermore, the administration of saffron extracts showed a significant effect on reducing the immobility time compared to the saline-treated group. PEE represented itself as a potential antidepressant drug showing a notable diminution in the immobility period (69.66 s) followed by SEE (76.66 s). Moreover, CEE (96.50 s) indicated moderate antidepressant effect, whereas LEE (106.83 s) exhibited mild activity. The potent antidepressant effect exhibited by petal can be strongly linked to the presence of natural flavonoid, kaempferol as it significantly attenuated immobility behaviors in rats and mice and showed an almost similar response to fluoxetine [85]. Similarly, in an 8-week double-blind, randomized clinical study, dried saffron petal (15 mg bid) had similar antidepressant effects as fluoxetine (15 mg bid) in treating patients with mild-to-moderate depression and no significant differences in observed side effects [34]. In an investigation comparing the efficacy of saffron stigma and corms with fluoxetine against depression, petroleum ether, and dichloromethane fractions of saffron stigma and corms significantly reduced the immobility time in the tail suspension and forced swimming test at all doses (150, 300, and 600 mg/kg) without altering the locomotor behavior of mice during the open-field test. The authors highlighted that the antidepressant effect of stigma extracts could be due to crocin analogs, particularly crocin 1. However, HPLC analysis of corm extract revealed the absence of safranal, crocin, crocetin, or kaempferol compounds, assuming the presence of other bioactive compounds in saffron corms showing a potent antidepressant effect which need to be further explored [80]. The promising antidepressant activity of saffron secondary metabolites opens the doors for further studies to understand the mode of action in medicinal plants.

4. Materials and Methods

4.1. Plant Material

Saffron corms obtained from the safranor company, France, were planted in plastic trays and pots containing black soil in August. The soil texture of experimental black soil was a clay-loam structure containing 2.74% of total organic carbon and 0.04% of total nitrogen, in addition to 7 ppm of phosphorus, 331.5 ppm of potassium, 64 ppm of calcium, 75.6 ppm of magnesium, and 25.3 ppm of sodium. The corms were incubated in a temperature-controlled (18 ± 1 °C) room under 16 h photoperiod illuminated with white fluorescent tubes at an irradiance of nearly 40 μmol m^{-2} s^{-1}. Different saffron parts were collected as per life span of the plant, i.e., petal and stigma were handpicked during October, leaves were collected in May, and smaller corms weighing less than 2 g were harvested in June. Individual parts were air-dried under shade for 7 weeks and ground to powder.

4.2. Chemicals

For animal experiments, all the chemicals, including aspirin, diclofenac potassium, fluoxetine HCl, ethanol, and carrageenan were purchased from Sigma. All chemicals used in the experiment were analytical grade.

4.3. Sample Extraction

Powder samples were macerated in ethanol/water (80%, *v/v*) for a week at room temperature. The mixture was subsequently filtered with Whatman filter paper 1 (Sigma-Aldrich, St. Louis, MO, USA) and concentrated by rotary evaporator under reduced pressure at 45 °C. Semi-liquid extracts were further allowed to dry in the fume hood. The crude extracts were freeze-dried and stored in sealed tubes at −20 °C for in vivo studies.

4.4. Standard and Test Drugs Preparation

Ethanolic extracts of saffron were prepared as 0.08 g/mL in 10% DMSO while standard drugs were dissolved in saline water (0.01 g/mL of 0.9% NaCl) and treated orally at 1 mL/100 g of mouse body weight. One optimal dose (800 mg/kg) showing maximum effect with no lethal consequences to mice was selected in this study, as reported previously [50].

4.5. Animals

Adult albino mice (Swiss strain) of either sex between 30–40 g were provided by Veterinary Research Institute Peshawar, Pakistan. All the mice were housed in aluminum cages (grade 304) under controlled natural 12/12 h light/dark cycle, temperature (21 ± 2 °C), and humidity (50–60%), and received tap water ad libitum and standard diet at the Animal House, Quaid-I-Azam University (QAU) Islamabad, Pakistan. Mice were acclimatized in this environment for seven days before the experiments. All the experiments were carried out between 10:00 and 17:00. The study protocol (Bch 0275) for laboratory animals was in accordance with the recommendations of the Institutional Animal Ethics Committee (QAU Islamabad), and all the precautionary measures were followed to reduce animal's fear and suffering.

4.6. Study Design

Mice were weighed, marked with numbers, and split into 6 groups, each containing 6 mice. All the samples were administered by oral gavage. Group 1 received saline (0.9%) and was used as a negative control. Mice in group 2 were given a dose of 10 mg/kg standard drugs (aspirin for analgesic and anticoagulant, diclofenac potassium for anti-inflammatory, and fluoxetine HCl for antidepressant assay) used as a positive control. Group 3, 4, 5, and 6 were administered with a dose of 800 mg/kg CEE, LEE, SEE, and PEE, respectively.

4.6.1. Acute Toxicity Study

The acute toxicity study of each tested sample was undertaken in accordance with the guidelines of the Organization for Economic Cooperation and Development (OECD). Albino mice (n=6) were administered orally with 800 mg/kg of each extract and observed for any changes in behavior and mortality at intervals of 0, 5, 10, 15, 20, and 24 h. Mice were examined for an additional 6 days for any signs of late morbidity and mortality.

4.6.2. Hot Plate Analgesic Assay

The test first reported by Eddy and Leimbach [86], with some modifications, was performed to measure the analgesic activity of saffron extracts. The standard drug, aspirin (10 mg/kg), and saline (1 mL/kg) were served as a positive and negative control, respectively. The parameter recorded was based on the latency time for fore- and hind-paw licking and/or jumping responses by placing the mice on the surface of a hot-plate (IITC Life Science, USA) set at a temperature of 55 ± 1 °C and initial latency time (Ti) was calculated by taking the mean of two readings. Animals with baseline latencies of <5 s or >30 s were ignored from the study. The final latency time (Tf) was noted after administration of the drug for each group at the intervals of 0.5, 1, and 2 h, with a cut off time of 30 s. The tested extracts (0.8 g/kg) were administered orally, and standard (aspirin, 10 mg/kg) was administered subcutaneously. The percentage analgesia (PA) was measured using the equation:

$$\mathrm{PA} = \left(\frac{\mathrm{Tf} - \mathrm{Ti}}{\mathrm{Ti}}\right) \times 100 \quad (1)$$

4.6.3. Carrageenan-induced Paw Edema Test

C. sativus ethanolic extracts were examined for their anti-inflammatory activities using the carrageenan-induced paw edema test as described by Winter et al. [87]. The test is based on the principle to assess acute anti-inflammatory activity in the hind paws of mice by developing inflammatory models with carrageenan, a sulfated polysaccharide agent that triggers inflammation. In this test, diclofenac potassium (10 mg/kg) was used as a positive control, and saline (1 mL/kg) served as a negative control to compare the anti-inflammatory effects. Before experimentation, mice were fasted for 24 h but had access to water. For edema induction, 1% λ-carrageenan (0.1 mL), prepared in 1% saline was injected into the sub planter tissue of mice left hind paw 1 h after administration of the drug. The basal volume of the paw was recorded just before and after injection of λ-carrageenan at 0, 1, 2, 3, and 4 h by digital plethysmometer (UGO Basile, 7140). For every interval, all the data were recorded in triplicate. The degree of swelling was calculated by the delta volume (A–B), where "A" shows the mean volume of the left hind paw after and "B" before the treatment of λ-carrageenan. The percentage edema inhibition (PEI) was measured using the equation:

$$\mathrm{PEI} = \left(\frac{\mathrm{A} - \mathrm{B}}{\mathrm{A}}\right) \times 100 \quad (2)$$

4.6.4. Anticoagulant

Blood clotting activity of saffron extracts was evaluated by the capillary tube method reported by Ismail and Mirza [84]. One hour after oral administration of the dose, the mouse tail was disinfected with spirit and prickled using a lancet. The tail was pressed firmly to get a bigger drop of blood and collected in capillary tubes. The time of appearance of the blood drop on the cut tail was recorded. The tubes were wrapped and maintained at 37 ± 1 °C in a water bath. Small portions of each tube were broken at regular intervals of 10 s, until fibrin thread appeared. The blood coagulation time was measured by considering the appearance of a blood drop as a starting point and thread formation as an endpoint.

4.6.5. Forced Swimming Test

The forced swim test (FST) described by Porsolt et al. [88] was used to screen the extracts of saffron for their antidepressant activity. On day first, mice were forced to swim individually in a glass tank of 40 cm height and 18 cm diameter containing water up to 30 cm under natural light. The water temperature was adjusted to 25 ± 1 °C. The water level was adjusted in such a way that mice could only touch the bottom of the tank with the tip of the tail. After 15 min exposure in the tank, mice were evacuated, dried off with a paper towel, and shifted to their home cage. The next day, saline (1 mL/kg) and fluoxetine HCl (10 mg/kg) served as a negative and positive control, respectively, along with saffron extracts (800 mg/kg) were administered 30 min before the experiment. Mice were allowed to swim freely for 6 min in the tank, and swimming behaviors were recorded with a video camera. Before each test, freshwater was introduced. Animals were placed one by one, and after every 2 min, immobility time was calculated in the last 4 min of swimming practice using a stopwatch. Immobility time is the time when the mouse stopped all additional movements other than those required for survival or balancing of their body.

4.7. Statistical Analysis

Data were expressed as mean along with standard deviation. Significant difference between groups was analyzed with One-way ANOVA in anticoagulant and antidepressant assays and two-way ANOVA in analgesic and anti-inflammatory assays followed by the Post Hoc Dunnett test. Graphs were made in GraphPad Prism 5.0 (La Jolla, CA, USA). The results were considered to be significant at $p < 0.05$.

5. Conclusions

To the best of our knowledge this is the first report of CEE, LEE, SEE, and PEE from in vivo grown *Crocus sativus* L. to determine anti-inflammatory, analgesic, anticoagulant, and antidepressant properties in mice. SEE and PEE were evinced as a safe natural remedy to treat pain, inflammation, depression, and the coagulation system. The above results clearly indicate that saffron is an excellent source of bioactive compounds having great potential as nutraceuticals and health benefits. These properties can be linked to intrinsic active compounds such as carotenoids and flavonoids found in the highest amounts in stigma and petals, respectively. However, further epidemiological investigations, laboratory research, and clinical trials are required to isolate the pharmacologically active molecules that contribute to the observed effects and to explicate the possible mechanism of action and effect of the plant on various critical illnesses and medicinal formulations.

Author Contributions: Conceptualization, A.K., N.A.M. and H.I.; methodology, A.K., H.I. and S.L.; software, A.K. and H.I.; validation, N.A.M. and R.M.T.; formal analysis, H.I., A.A.K.K., Y.A.J.F., I.M.A.Z., S.A.A.T., Z.K., Y.A., A.A., A.N. and B.M.; investigation, H.I., S.L., A.N., A.A.K.K. and B.M.; data curation, A.K. and H.I.; writing—original draft preparation, A.K. and N.A.M.; writing—review and editing, Y.A., I.M.A.Z, S.A.A.T., A.N., A.A.K.K., Z.K., Y.A.J.F., A.A., N.A.M. and R.M.T.; supervision, N.A.M., R.M.T. and B.M. All authors have read and agreed to the published version of the manuscript.

Funding: This research received no external funding.

Conflicts of Interest: The authors declare no conflict of interest.

References

1. Koocheki, A.; Rezvani Moghaddam, P.; Aghhavani-Shajari, M.; Fallahi, H.R. Corm weight or number per unit of land: Which one is more effective when planting corm, based on the age of the field from which corms were selected? *Ind. Crops Prod.* **2019**, *131*, 78–84. [CrossRef]
2. Cardone, L.; Castronuovo, D.; Perniola, M.; Cicco, N.; Candido, V. Saffron (Crocus sativus L.), the king of spices: An overview. *Sci. Hortic.* **2020**, *272*, 109560. [CrossRef]
3. Fallahi, H.R.; Aghhavani-Shajari, M.; Branca, F.; Davarzani, J. Effect of different concentrations of saffron corm and leaf residue on the early growth of arugula, chickpea and fenugreek under greenhouse conditions. *Acta Agric. Slov.* **2018**, *111*, 51–61. [CrossRef]
4. Tsaftaris, A.; Pasentsis, K.; Argiriou, A. Cloning and Characterization of *FLOWERING LOCUS T*-Like Genes from the Perennial Geophyte Saffron Crocus (*Crocus sativus*). *Plant Mol. Biol. Rep.* **2013**, *31*, 1558–1568. [CrossRef]
5. Verma, S.K.; Das, A.K.; Cingoz, G.S.; Uslu, E.; Gurel, E. Influence of nutrient media on callus induction, somatic embryogenesis and plant regeneration in selected Turkish crocus species. *Biotechnol. Rep.* **2016**, *10*, 66–74. [CrossRef]
6. Kafi, M.; Kamili, A.N.; Husaini, A.M.; Ozturk, M.; Altay, V. An Expensive Spice Saffron (*Crocus sativus* L.): A Case Study from Kashmir, Iran, and Turkey. In *Global Perspectives on Underutilized Crops*; Ozturk, M., Hakeem, K.R., Ashraf, M., Ahmad, M.S.A., Eds.; Springer International Publishing: Cham, Switzerland, 2018; pp. 109–149.
7. Chahine, N.; Chahine, R. Protecting Mechanisms of Saffron Extract Against Doxorubicin Toxicity in Ischemic Heart. In *Saffron: The Age-Old Panacea in a New Light*; Sarwat, M., Sumaiya, S., Eds.; Elsevier: Amsterdam, The Netherlands, 2020; pp. 141–154.
8. Hadizadeh, F.; Khalili, N.; Hosseinzadeh, H.; Khair-Aldine, R. Kaempferol from saffron petals. *Iran. J. Pharm. Res.* **2010**, *2*, 251–252.
9. Mykhailenko, O.; Kovalyov, V.; Goryacha, O.; Ivanauskas, L.; Georgiyants, V. Biologically active compounds and pharmacological activities of species of the genus *Crocus*: A review. *Phytochemistry* **2019**, *162*, 56–89. [CrossRef]
10. José Bagur, M.; Alonso Salinas, G.L.; Jiménez-Monreal, A.M.; Chaouqi, S.; Llorens, S.; Martínez-Tomé, M.; Alonso, G.L. Saffron: An Old Medicinal Plant and a Potential Novel Functional Food. *Molecules* **2018**, *23*, 30. [CrossRef]
11. Anuar, N.; Taha, R.M.; Mahmad, N.; Othman, R. Identification of crocin, crocetin and zeaxanthin in *Crocus sativus* grown under controlled environment in Malaysia. *Pigm. Resin Technol.* **2018**, *47*, 502–506. [CrossRef]
12. Mzabri, I.; Addi, M.; Berrichi, A. Traditional and Modern Uses of Saffron (*Crocus Sativus*). *Cosmetics* **2019**, *6*, 63. [CrossRef]
13. Hosseinzadeh, H.; Nassiri-Asl, M. Avicenna's (Ibn Sina) the canon of medicine and saffron (*Crocus sativus*): a review. *Phytother. Res.* **2013**, *27*, 475–483. [CrossRef] [PubMed]
14. Horozić, E. Effects of extraction solvent/technique on the antioxidant and antimicrobial activity of spring saffron (*Crocus vernus* (L.) Hill). *Acta Med. Salin.* **2020**, *49*, 19–22. [CrossRef]
15. Zhang, K.; Wang, L.; Si, S.; Sun, Y.; Pei, W.; Ming, Y.; Sun, L. Crocin improves the proliferation and cytotoxic function of T cells in children with acute lymphoblastic leukemia. *Biomed. Pharmacother.* **2018**, *99*, 96–100. [CrossRef] [PubMed]
16. Sun, C.; Nile, S.H.; Zhang, Y.; Qin, L.; El-Seedi, H.R.; Daglia, M.; Kai, G. Novel Insight into Utilization of Flavonoid Glycosides and Biological Properties of Saffron (*Crocus sativus* L.) Flower Byproducts. *J. Agric. Food Chem.* **2020**, *68*, 10685–10696. [CrossRef]
17. Khoulati, A.; Ouahhoud, S.; Mamri, S.; Alaoui, K.; Lahmass, I.; Choukri, M.; Kharmach, E.Z.; Asehraou, A.; Saalaoui, E. Saffron extract stimulates growth, improves the antioxidant components of *Solanum lycopersicum* L., and has an antifungal effect. *Ann. Agric. Sci.* **2019**, *64*, 138–150. [CrossRef]
18. Boskabady, M.H.; Farkhondeh, T. Antiinflammatory, Antioxidant, and Immunomodulatory Effects of *Crocus sativus* L. and its Main Constituents. *Phytother. Res.* **2016**, *30*, 1072–1094. [CrossRef]

19. Yousefi, F.; Arab, F.L.; Rastin, M.; Tabasi, N.S.; Nikkhah, K.; Mahmoudi, M. Comparative assessment of immunomodulatory, proliferative, and antioxidant activities of crocin and crocetin on mesenchymal stem cells. *J. Cell. Biochem.* **2020**, 1–14. [CrossRef]
20. Bhandari, P.R. *Crocus sativus* L. (saffron) for cancer chemoprevention: A mini review. *J. Tradit. Complement. Med.* **2015**, *5*, 81–87. [CrossRef]
21. Mohammadzadeh-Moghadam, H.; Nazari, S.M.; Shamsa, A.; Kamalinejad, M.; Esmaeeli, H.; Asadpour, A.A.; Khajavi, A. Effects of a topical saffron (*Crocus sativus* L.) gel on erectile dysfunction in diabetics: A randomized, parallel-group, double-blind, placebo-controlled trial. *Evid. Based Complement. Altern. Med.* **2015**, *20*, 283–286. [CrossRef]
22. Hosseinzadeh, H.; Ghenaati, J. Evaluation of the antitussive effect of stigma and petals of saffron (*Crocus sativus*) and its components, safranal and crocin in guinea pigs. *Fitoterapia* **2006**, *77*, 446–448. [CrossRef]
23. Saadat, S.; Shakeri, F.; Boskabady, M.H. Comparative Antitussive Effects of Medicinal Plants and Their Constituents. *Altern. Ther. Health Med.* **2018**, *24*, 36–49. [PubMed]
24. Arimitsu, J.; Hagihara, K.; Ogawa, K. Potential Effect of Saffron as an Antiplatelet Drug in Patients with Autoimmune Diseases. *J. Altern. Complement. Med.* **2014**, *20*, A90. [CrossRef]
25. Moazen-Zadeh, E.; Abbasi, S.H.; Safi-Aghdam, H.; Shahmansouri, N.; Arjmandi-Beglar, A.; Hajhosseinn Talasaz, A.; Salehiomran, A.; Forghani, S.; Akhondzadeh, S. Effects of Saffron on Cognition, Anxiety, and Depression in Patients Undergoing Coronary Artery Bypass Grafting: A Randomized Double-Blind Placebo-Controlled Trial. *J. Altern. Complement. Med.* **2018**, *24*, 361–368. [CrossRef] [PubMed]
26. Hosseini, S.A.; Zilaee, M.; Shoushtari, M.H.; Ghasemi dehcheshmeh, M. An evaluation of the effect of saffron supplementation on the antibody titer to heat-shock protein (HSP) 70, hsCRP and spirometry test in patients with mild and moderate persistent allergic asthma: A triple-blind, randomized placebo-controlled trial. *Respir. Med.* **2018**, *145*, 28–34. [CrossRef] [PubMed]
27. Azimi, P.; Abrishami, R. Comparison of the Effects of Crocus Sativus and Mefenamic Acid on Primary Dysmenorrhea. *J. Pharm. Care* **2018**, *4*, 75–78.
28. Mohammad, N.S.; Atefeh, A.; Mahbobeh, N.S.; Davood, S.; Shakiba, M.; Hamidreza, B.T.; Mehrpour, M. Comparison of Saffron versus Fluoxetine in Treatment of Women with Premenstrual Syndrome: A Randomized Clinical Trial Study. *Indian J. Med. Forensic Med. Toxicol.* **2020**, *14*, 1760–1765.
29. Fernández-Albarral, J.A.; de Hoz, R.; Ramírez, A.I.; López-Cuenca, I.; Salobrar-García, E.; Pinazo-Durán, M.D.; Ramírez, J.M.; Salazar, J.J. Beneficial effects of saffron (*Crocus sativus* L.) in ocular pathologies, particularly neurodegenerative retinal diseases. *Neural Regen. Res.* **2020**, *15*, 1408–1416.
30. Sánchez-Vioque, R.; Santana-Méridas, O.; Polissiou, M.; Vioque, J.; Astraka, K.; Alaiz, M.; Herraiz-Peñalver, D.; Tarantilis, P.A.; Girón-Calle, J. Polyphenol composition and *in vitro* antiproliferative effect of corm, tepal and leaf from *Crocus sativus* L. on human colon adenocarcinoma cells (Caco-2). *J. Funct. Foods* **2016**, *24*, 18–25. [CrossRef]
31. Wali, A.F.; Pillai, J.R.; Al Dhaheri, Y.; Rehman, M.U.; Shoaib, A.; Sarheed, O.; Jabnoun, S.; Razmpoor, M.; Rasool, S.; Paray, B.A.; Ahmad, P. *Crocus sativus* L. Extract Containing Polyphenols Modulates Oxidative Stress and Inflammatory Response against Anti-Tuberculosis Drugs-Induced Liver Injury. *Plants* **2020**, *9*, 167. [CrossRef]
32. Wali, A.F.; Alchamat, H.A.A.; Hariri, H.K.; Hariri, B.K.; Menezes, G.A.; Zehra, U.; Rehman, M.U.; Ahmad, P. Antioxidant, Antimicrobial, Antidiabetic and Cytotoxic Activity of *Crocus sativus* L. Petals. *Appl. Sci.* **2020**, *10*, 1519. [CrossRef]
33. Hoshyar, R.; Hosseinian, M.; Naghandar, M.R.; Hemmati, M.; Zarban, A.; Amini, Z.; Valavi, M.; Beyki, M.Z.; Mehrpour, O. Anti-dyslipidemic properties of Saffron: Reduction in the associated risks of atherosclerosis and insulin resistance. *Iran Red Crescent Med. J.* **2016**, *18*. [CrossRef]
34. Basti, A.A.; Moshiri, E.; Noorbala, A.A.; Jamshidi, A.H.; Abbasi, S.H.; Akhondzadeh, S. Comparison of petal of *Crocus sativus* L. and fluoxetine in the treatment of depressed outpatients: A pilot double-blind randomized trial. *Prog. Neuropsychopharmacol. Biol. Psychiatry* **2007**, *31*, 439–442. [CrossRef]
35. Hosseini, A.; Razavi, B.M.; Hosseinzadeh, H. Saffron (*Crocus sativus*) petal as a new pharmacological target: A review. *Iran. J. Basic Med. Sci.* **2018**, *21*, 1091–1099.
36. Jadouali, S.; Atifi, H.; Bouzoubaa, Z.; Majourhat, K.; Gharby, S.; Achemchem, F.; Elmoslih, A.; Laknifli, A.; Mamouni, R. Chemical characterization, antioxidant and antibacterial activity of Moroccan *Crocus sativus* L. petals and leaves. *J. Mater. Environ. Sci.* **2018**, *9*, 113–118. [CrossRef]

37. Rahaiee, S.; Ranjbar, M.; Azizi, H.; Govahi, M.; Zare, M. Green synthesis, characterization, and biological activities of saffron leaf extract-mediated zinc oxide nanoparticles: A sustainable approach to reuse an agricultural waste. *Appl. Organomet. Chem.* **2020**, *34*, e5705. [CrossRef]
38. Sánchez-Vioque, R.; Rodríguez-Conde, M.; Reina-Urena, J.; Escolano-Tercero, M.; Herraiz-Peñalver, D.; Santana-Méridas, O. *In vitro* antioxidant and metal chelating properties of corm, tepal and leaf from saffron (*Crocus sativus* L.). *Ind. Crops Prod.* **2012**, *39*, 149–153. [CrossRef]
39. Baba, S.A.; Malik, A.H.; Wani, Z.A.; Mohiuddin, T.; Shah, Z.; Abbas, N.; Ashraf, N. Phytochemical analysis and antioxidant activity of different tissue types of *Crocus sativus* and oxidative stress alleviating potential of saffron extract in plants, bacteria, and yeast. *S. Afr. J. Bot.* **2015**, *99*, 80–87. [CrossRef]
40. Esmaeelian, M.; Jahani, M.; Feizy, J.; Einafshar, S. Effects of Ultrasound-Assisted and Direct Solvent Extraction Methods on the Antioxidant and Antibacterial Properties of Saffron (*Crocus sativus* L.) Corm Extract. *Food Anal. Methods* **2020**. [CrossRef]
41. Escribano, J.; Rubio, A.; Alvarez-Ortí, M.; Molina, A.; Fernández, J.A. Purification and Characterization of a Mannan-Binding Lectin Specifically Expressed in Corms of Saffron Plant (*Crocus sativus* L.). *J. Agric. Food Chem.* **2000**, *48*, 457–463. [CrossRef]
42. Mohajeri, S.A.; Hedayati, N.; Bemani-Naeini, M. Available saffron formulations and product patents. In *Saffron*; Elsevier: Amsterdam, The Netherlands, 2020; pp. 493–515.
43. Amin, A.; Hamza, A.; Bajbouj, K.; Ashraf, S.A.; Daoud, S. Saffron: a potential candidate for a novel anti-cancer drug against hepatocellular carcinoma. *Hepatology* **2011**, *54*, 857–867. [CrossRef]
44. Ma, L.; Xu, G.B.; Tang, X.; Zhang, C.; Zhao, W.; Wang, J.; Chen, H. Anti-cancer potential of polysaccharide extracted from hawthorn (*Crataegus.*) on human colon cancer cell line HCT116 via cell cycle arrest and apoptosis. *J. Funct. Foods* **2020**, *64*, 103677. [CrossRef]
45. Fatehi, M.; Rashidabady, T.; Fatehi-Hassanabad, Z. Effects of *Crocus sativus* petals' extract on rat blood pressure and on responses induced by electrical field stimulation in the rat isolated vas deferens and guinea-pig ileum. *J. Ethnopharmacol.* **2003**, *84*, 199–203. [CrossRef]
46. Ahmad Nazri, K.A.; Fauzi, N.M.; Buang, F.; Mohd Saad, Q.H.; Husain, K.; Jantan, I.; Jubri, Z. *Gynura procumbens* Standardised Extract Reduces Cholesterol Levels and Modulates Oxidative Status in Postmenopausal Rats Fed with Cholesterol Diet Enriched with Repeatedly Heated Palm Oil. *Evid. Based Complement. Altern. Med.* **2019**, *2019*, 1–15. [CrossRef] [PubMed]
47. Akbari-Fakhrabadi, M.; Najafi, M.; Mortazavian, S.; Rasouli, M.; Memari, A.H.; Shidfar, F. Effect of saffron (*Crocus sativus* L.) and endurance training on mitochondrial biogenesis, endurance capacity, inflammation, antioxidant, and metabolic biomarkers in Wistar rats. *J. Food Biochem.* **2019**, *43*, 1–10. [CrossRef]
48. Hosseinzadeh, H.; Younesi, H.M. Antinociceptive and anti-inflammatory effects of *Crocus sativus* L. stigma and petal extracts in mice. *BMC Pharmacol.* **2002**, *2*, 7.
49. Li, F.; Huo, J.; Zhuang, Y.; Xiao, H.; Wang, W.; Huang, L. Anti-nociceptive and anti-inflammatory effects of the ethanol extract of *Arenga pinnata* (Wurmb) Merr. fruit. *J. Ethnopharmacol.* **2020**, *248*, 112349. [CrossRef]
50. Iranshahi, M.; Khoshangosht, M.; Mohammadkhani, Z.; Karimi, G. Protective effects of aqueous and ethanolic extracts of saffron stigma and petal on liver toxicity induced by carbon tetrachloride in mice. *Pharmacologyonline* **2011**, *1*, 203–212.
51. Escribano, J.; Alonso, G.L.; Coca-Prados, M.; Fernández, J.A. Crocin, safranal and picrocrocin from saffron (*Crocus sativus* L.) inhibit the growth of human cancer cells in vitro. *Cancer Lett.* **1996**, *100*, 23–30. [CrossRef]
52. Hosseinzadeh, H.; Shariaty, V.M.; Sameni, A. Acute and sub-acute toxicity of crocin, a constituent of *Crocus sativus* L. (Saffron), in mice and rats. *Pharmacologyonline* **2010**, *2*, 943–951.
53. González-Cano, R.; Montilla-García, Á.; Ruiz-Cantero, M.C.; Bravo-Caparrós, I.; Tejada, M.Á.; Nieto, F.R.; Cobos, E.J. The search for translational pain outcomes to refine analgesic development: where did we come from and where are we going? *Neurosci. Biobehav. Rev.* **2020**, *113*, 238–261. [CrossRef]
54. Deuis, J.R.; Dvorakova, L.S.; Vetter, I. Methods used to evaluate pain behaviors in rodents. *Front. Mol. Neurosci.* **2017**, *10*, 284. [CrossRef] [PubMed]
55. Yang, D.J.; Lin, J.T.; Chen, Y.C.; Liu, S.C.; Lu, F.J.; Chang, T.J.; Wang, M.; Lin, H.W.; Chang, Y.Y. Suppressive effect of carotenoid extract of *Dunaliella salina* alga on production of LPS-stimulated pro-inflammatory mediators in RAW264. 7 cells via NF-κB and JNK inactivation. *J. Funct. Foods* **2013**, *5*, 607–615. [CrossRef]
56. Tamaddonfard, E.; Farshid, A.A.; Hosseini, L. Crocin alleviates the local paw edema induced by histamine in rats. *Avicenna J. Phytomed.* **2012**, *2*, 97–104. [PubMed]

57. Gupta, A.K.; Parasar, D.; Sagar, A.; Choudhary, V.; Chopra, B.S.; Garg, R.; Khatri, N. Analgesic and anti-inflammatory properties of gelsolin in acetic acid induced writhing, tail immersion and carrageenan induced paw edema in mice. *PLoS ONE* **2015**, *10*, e0135558. [CrossRef] [PubMed]
58. Masresha, B.; Makonnen, E.; Debella, A. In vivo anti-inflammatory activities of *Ocimum suave* in mice. *J. Ethnopharmacol.* **2012**, *142*, 201–205. [CrossRef] [PubMed]
59. Hossain, K.H.; Rahman, M.A.; Taher, M.; Tangpong, J.; Hajjar, D.; Alelwani, W.; Makki, A.A.; Ali Reza, A.S.M. Hot methanol extract of *Leea macrophylla* (Roxb.) manages chemical-induced inflammation in rodent model. *J. King Saud Univ. Sci.* **2020**, *32*, 2892–2899. [CrossRef]
60. Pandey, K.D.; Singh, Y.; Khatri, N.; Singh, N. Phytochemical Screening and Anti-inflammatory activity of *Murraya koenigii* & *Ficus lacor* roots in Carrageenan, Histamine and Serotonin induced paw edema in albino Wistar rats. *Int. J. Rec. Adv. Sci. Technol.* **2020**, *6*, 1–12.
61. Aoki, T.; Narumiya, S. Prostaglandins in Chronic Inflammation. In *Chronic Inflammation: Mechanisms and Regulation*; Miyasaka, M., Takatsu, K., Eds.; Springer: Tokyo, Japan, 2016; pp. 3–17.
62. Arulmozhi, D.; Veeranjaneyulu, A.; Bodhankar, S.; Arora, S. Pharmacological investigations of *Sapindus trifoliatus* in various in vitro and in vivo models of inflammation. *Indian J. Pharmacol.* **2005**, *37*, 96–102.
63. Yang, Y.; Wei, Z.; Teichmann, A.T.; Wieland, F.H.; Wang, A.; Lei, X.; Zhu, Y.; Yin, J.; Fan, T.; Zhou, L.; Wang, C.; Chen, L. Development of a novel nitric oxide (NO) production inhibitor with potential therapeutic effect on chronic inflammation. *Eur. J. Med. Chem.* **2020**, *193*, 112216. [CrossRef]
64. Arbabian, S.; Izadi, H.; Ghoshouni, H.; Shams, J.; Zardouz, H.; Kamalinezhad, M.; Sahraei, H.; Nourouzzadeh, A. Effect of water extract of saffron (*Crocus sativus*) on chronic phase of formaline test in female mice. *Kowsar Med. J.* **2009**, *14*, 11–18.
65. Kumar, V.; Bhat, Z.A.; Kumar, D.; Khan, N.; Chashoo, I.; Shah, M. Evaluation of anti-inflammatory potential of petal extracts of *Crocus sativus* "cashmerianus". *Int. J. Phytopharm.* **2012**, *3*, 27–31.
66. Kong, L.; Luo, C.; Li, X.; Zhou, Y.; He, H. The anti-inflammatory effect of kaempferol on early atherosclerosis in high cholesterol fed rabbits. *Lipids Health Dis.* **2013**, *12*, 115. [CrossRef] [PubMed]
67. Ogedegbe, H.O. An overview of hemostasis. *Lab. Med.* **2002**, *33*, 948–953. [CrossRef]
68. Scarano, A.; Murmura, G.; di Cerbo, A.; Palmieri, B.; Pinchi, V.; Mavriqi, L.; Varvara, G. Anti-Hemorrhagic Agents in Oral and Dental Practice: An Update. *Int. J. Immunopathol. Pharmacol.* **2013**, *26*, 847–854. [CrossRef] [PubMed]
69. Kurt, M.; Onal, I.; Akdogan, M.; Kekilli, M.; Arhan, M.; Sayilir, A.; Oztas, E.; Haznedaroglu, I. Ankaferd Blood Stopper for controlling gastrointestinal bleeding due to distinct benign lesions refractory to conventional antihemorrhagic measures. *Can. J. Gastroenterol.* **2010**, *24*, 380–384. [CrossRef] [PubMed]
70. Ismail, H.; Rasheed, A.; Haq, I.U.; Jafri, L.; Ullah, N.; Dilshad, E.; Sajid, M.; Mirza, B. Five indigenous plants of Pakistan with Antinociceptive, anti-inflammatory, antidepressant, and anticoagulant properties in Sprague Dawley rats. *Evid. Based Complement. Altern. Med.* **2017**, *2017*, 1–10. [CrossRef]
71. Liakopoulou-Kyriakides, M.; Sinakos, Z.; Kyriakidis, D. A high molecular weight platelet aggregating factor in *Crocus sativus*. *Plant Sci.* **1985**, *40*, 117–120. [CrossRef]
72. Liakopoulou-Kyriakides, M.; Skubas, A. Characterization of the platelet aggregation inducer and inhibitor isolated from *Crocus sativus*. *Biochem. Int.* **1990**, *22*, 103–110.
73. Ma, S.; Liu, B.; Zhou, S.; Xu, X.; Yang, Q.; Zhou, J. Pharmacological studies of glycosides of saffron *Crocus* (*Crocus sativus*). II. Effects on blood coagulation, platelet aggregation and thromobosis. *Chin. Trad. Herbal Drugs* **1999**, *30*, 196–198.
74. Yang, Y.; Qian, Z. Effect of crocetin on platelet aggregation in rats. *Chin. J. Nat. Med.* **2007**, *5*, 374–378.
75. Yang, L.; Qian, Z.; Yang, Y.; Sheng, L.; Ji, H.; Zhou, C.; Kazi, H.A. Involvement of Ca2+ in the Inhibition by Crocetin of Platelet Activity and Thrombosis Formation. *J. Agric. Food Chem.* **2008**, *56*, 9429–9433. [CrossRef] [PubMed]
76. Modaghegh, M.H.; Shahabian, M.; Esmaeili, H.A.; Rajbai, O.; Hosseinzadeh, H. Safety evaluation of saffron (*Crocus sativus*) tablets in healthy volunteers. *Phytomedicine* **2008**, *15*, 1032–1037. [CrossRef] [PubMed]
77. Ayatollahi, H.; Javan, A.O.; Khajedaluee, M.; Shahroodian, M.; Hosseinzadeh, H. Effect of *Crocus sativus* L.(saffron) on coagulation and anticoagulation systems in healthy volunteers. *Phytother. Res.* **2014**, *28*, 539–543. [CrossRef]

78. Haroz, E.E.; Ritchey, M.; Bass, J.K.; Kohrt, B.A.; Augustinavicius, J.; Michalopoulos, L.; Burkey, M.D.; Bolton, P. How is depression experienced around the world? A systematic review of qualitative literature. *Soc. Sci. Med.* **2017**, *183*, 151–162. [CrossRef] [PubMed]
79. Khawam, E.; Laurencic, G.; Malone, D. Side effects of antidepressants: an overview. *Cleve. Clin. J. Med.* **2006**, *73*, 351. [CrossRef]
80. Wang, Y.; Han, T.; Zhu, Y.; Zheng, C.J.; Ming, Q.L.; Rahman, K.; Qin, L.P. Antidepressant properties of bioactive fractions from the extract of *Crocus sativus* L. *J. Nat. Med.* **2009**, *64*, 24. [CrossRef]
81. Hausenblas, H.A.; Saha, D.; Dubyak, P.J.; Anton, S.D. Saffron (*Crocus sativus* L.) and major depressive disorder: a meta-analysis of randomized clinical trials. *J. Integr. Med.* **2013**, *11*, 377–383. [CrossRef]
82. Petit-Demouliere, B.; Chenu, F.; Bourin, M. Forced swimming test in mice: a review of antidepressant activity. *Psychopharmacology* **2005**, *177*, 245–255. [CrossRef]
83. Kayani, W.K.; Dilshad, E.; Ahmed, T.; Ismail, H.; Mirza, B. Evaluation of *Ajuga bracteosa* for antioxidant, anti-inflammatory, analgesic, antidepressant and anticoagulant activities. *BMC Complement. Altern. Med.* **2016**, *16*, 375. [CrossRef]
84. Ismail, H.; Mirza, B. Evaluation of analgesic, anti-inflammatory, anti-depressant and anti-coagulant properties of Lactuca sativa (CV. Grand Rapids) plant tissues and cell suspension in rats. *BMC Complement. Altern. Med.* **2015**, *15*, 199. [CrossRef]
85. Hosseinzadeh, H.; Motamedshariaty, V.; Hadizadeh, F. Antidepressant effect of kaempferol, a constituent of saffron (*Crocus sativus*) petal, in mice and rats. *Pharmacologyonline* **2007**, *2*, 367–370.
86. Eddy, N.B.; Leimbach, D. Synthetic analgesics. II. Dithienylbutenyl- and dithienylbutylamines. *J. Pharmacol. Exp. Ther.* **1953**, *107*, 385–393.
87. Winter, C.A.; Risley, E.A.; Nuss, G.W. Carrageenin-induced edema in hind paw of the rat as an assay for antiinflammatory drugs. *Exp. Biol. Med.* **1962**, *111*, 544–547. [CrossRef] [PubMed]
88. Porsolt, R.D.; Bertin, A.; Blavet, N.; Deniel, M.; Jalfre, M. Immobility induced by forced swimming in rats: Effects of agents which modify central catecholamine and serotonin activity. *Eur. J. Pharmacol.* **1979**, *57*, 201–210. [CrossRef]

Article

Suaeda vermiculata Aqueous-Ethanolic Extract-Based Mitigation of CCl_4-Induced Hepatotoxicity in Rats, and HepG-2 and HepG-2/ADR Cell-Lines-Based Cytotoxicity Evaluations

Salman A. A. Mohammed [1,*], Riaz A. Khan [2,*], Mahmoud Z. El-Readi [3,4], Abdul-Hamid Emwas [5], Salim Sioud [5], Benjamin G. Poulson [6], Mariusz Jaremko [6], Hussein M. Eldeeb [1,7], Mohsen S. Al-Omar [2,8] and Hamdoon A. Mohammed [2,9,*]

1 Department of Pharmacology and Toxicology, College of Pharmacy, Qassim University, Qassim 51452, Saudi Arabia; hu.ali@qu.edu.sa
2 Department of Medicinal Chemistry and Pharmacognosy, College of Pharmacy, Qassim University, Qassim 51452, Saudi Arabia; m.omar@qu.edu.sa
3 Department of Clinical Biochemistry, Faculty of Medicine, Umm Al-Qura University, Makkah 21955, Saudi Arabia; mzreadi@uqu.edu.sa
4 Department of Biochemistry, Faculty of Pharmacy, Al-Azhar University, Assiut 71524, Egypt
5 Core Labs, King Abdullah University of Science and Technology (KAUST), Thuwal 23955-6900, Saudi Arabia; abdelhamid.emwas@kaust.edu.sa (A.-H.E.); salim.sioud@kaust.edu.sa (S.S.)
6 Biological and Environmental Sciences and Engineering Division (BESE), King Abdullah University of Science and Technology (KAUST), Thuwal 23955-6900, Saudi Arabia; benjamingabriel.poulson@kaust.edu.sa (B.G.P); mariusz.jaremko@kaust.edu.sa (M.J.)
7 Department of Biochemistry, Faculty of Medicine, Al-Azhar University, Assiut, 71524, Egypt
8 Medicinal Chemistry and Pharmacognosy Department, Faculty of Pharmacy, JUST, Irbid 22110, Jordan
9 Department of Pharmacognosy, Faculty of Pharmacy, Al-Azhar University, Cairo, 11371, Egypt
* Correspondence: m.azmi@qu.edu.sa (S.A.A.M.); ri.khan@qu.edu.sa (R.A.K.); ham.mohammed@qu.edu.sa (H.A.M.); Tel.: +966-(0)530309899 (S.A.A.M.); +966-(0)508384296 (R.A.K.); +966-(0)566176074 (H.A.M.)

Received: 1 September 2020; Accepted: 22 September 2020; Published: 29 September 2020

Abstract: *Suaeda vermiculata*, an edible halophytic plant, used by desert nomads to treat jaundice, was investigated for its hepatoprotective bioactivity and safety profile on its mother liquor aqueous-ethanolic extract. Upon LC-MS (Liquid Chromatography-Mass Spectrometry) analysis, the presence of several constituents including three major flavonoids, namely quercetin, quercetin-3-*O*-rutinoside, and kaempferol-*O*-(acetyl)-hexoside-pentoside were confirmed. The aqueous-ethanolic extract, rich in antioxidants, quenched the DPPH (1,1-diphenyl-2-picrylhydrazyl) radicals, and also showed noticeable levels of radical scavenging capacity in ABTS (2,2′-azino-bis-3-ethylbenzthiazoline-6-sulphonic acid) assay. For the hepatoprotective activity confirmation, the male rat groups were fed daily, for 7 days ($n = 8$/group, *p.o.*), either carboxyl methylcellulose (CMC) 0.5%, silymarin 200 mg/kg, the aqueous-ethanolic extract of the plant *Suaeda vermiculata* (100, 250, and 500 mg/kg extract), or quercetin (100 mg/kg) alone, and on day 7 of the administrations, all the animal groups, excluding a naïve (250 mg/kg aqueous-ethanolic extract-fed), and an intact animal group were induced hepatotoxicity by intraperitoneally administering carbon tetrachloride (CCl_4). All the animals were sacrificed after 24 h, and aspartate transaminase and alanine transaminase serum levels were observed, which were noted to be significantly decreased for the aqueous-ethanolic extract, silymarin, and quercetin-fed groups in comparison to the CMC-fed group ($p < 0.0001$). No noticeable adverse effects were observed on the liver, kidney, or heart's functions of the naïve (250 mg/kg) group. The aqueous-ethanolic extract was found to be safe in the acute toxicity (5 g/kg) test and showed hepatoprotection and safety at higher doses. Further upon, the cytotoxicity testings in HepG-2 and HepG-2/ADR (Adriamycin resistant) cell-lines were also investigated, and the IC_{50} values were recorded at 56.19 ± 2.55 μg/mL, and 78.40 ± 0.32 μg/mL ($p < 0.001$, Relative Resistance RR 1.39), respectively, while the doxorubicin (Adriamycin) IC_{50} values were found to be 1.3 ± 0.064, and 4.77 ± 1.05 μg/mL ($p < 0.001$, RR 3.67), respectively. The HepG-2/ADR cell-lines when tested in a combination of the aqueous-ethanolic extract with doxorubicin, a significant reversal in the doxorubicin's IC_{50} value by 2.77 folds ($p < 0.001$, CI = 0.56) was noted as compared to the cytotoxicity test where the extract was absent. The mode of action for the reversal was determined to be synergistic in nature indicating the role of the aqueous-ethanolic extract.

Keywords: *Suaeda vermiculata*; halophyte; aqueous-ethanolic extract; antioxidant; liver toxicity; cytotoxicity; hepatoprotective; liver disorders; mass spectrometry; LC-MS; HepG-2; HepG-2/ADR

---

1. Introduction

The liver is a vital organ that regulates both metabolic and detoxification processes. Disorders pertaining to the liver lead to 2 million deaths annually on a global scale [1]. The liver-related mortalities account for the top 15 leading causes of deaths in the United States. In the middle-eastern regions, nearly 10% of the population suffers from the infections of hepatitis-B virus (HBV), and more than 2% of the population is infected with the hepatitis-C virus (HCV), which indicates high prevalence rates of these liver-related infections, which is in addition to other liver-based disorders, including the symptomatic conditions of hyperbilirubinemia-related yellow-coloration, commonly referred to as jaundice. Liver disorders are a serious global healthcare concern [2], and the demand for valuable health-care resources, as well as attention in curtailing them, is huge [3,4]. The common risk factors, such as obesity, diabetes, hypercholesterolemia, and excessive drinking also make the situation further alarming [5]. The available medications for different liver disorders are limited, and some of the currently prescribed chemotherapeutic medications, mainly anti-viral, anti-bacterial, and synthetic steroids, have shown various types of adverse effects in varying proportions to different individuals, and that has the potential to effectively curtail the treatment options, and in turn, negatively impact the desired health outcomes of the subjects [6]. Previous reports on the use of plant-based medications, and investigations of traditionally used herbs, have come back into focus for medicinal purposes, and further drug discovery and development purposes to offer alternative medications to the widely prescribed synthetic-origin drugs for various types of liver disorders. This attention has also provided an impetus to newer strategies in drug development for hepatoprotection purposes [7]. Nonetheless, the use of medicinal plants and their extracts as alternative sources for prophylactic and therapeutic hepatoprotection has always been intriguing, though it has been practiced commonly by traditional healers over time. The majority of the ameliorating claims regarding the hepatoprotective plants, and herbal formulations, remain unproven, vague, and scientifically unsubstantiated, although efforts are still continuously being pushed into the area, and newer plants from varying environments and climatic-condition origins, together with improved testing protocols, are ceaselessly being adopted constantly into the realm of hepatoprotection strategies at global scales [8–11].

Halophytic plants, growing under high-salinity soil conditions, have also been recorded to provide relief from symptomatic liver discomforts [8–10]. The halophytes, under the stress of harsh super-saline habitats in areas with the lowest rainfall and marshy conditions, as well as desert environments, have accustomed to adverse environmental conditions through developing various defense mechanisms for survival. These adoptive techniques include multiple strategies which are morphological, biochemical, physiological, metabolic, and sensory in nature [11,12]. One such strategy of these halophytes is their responsiveness to elevated oxidative stress conditions of the plants, wherein these plants survive high-salinity conditions by producing specific enzymatic, non-enzymatic, and metabolite products, many of which are secondary metabolite antioxidant molecules concentrated abundantly in the plant sap [13]. These products have well-established roles in quenching reactive oxygen species (ROS) [14]. The observation that halophytes are rich in antioxidant constituents lends credence to the hepatoprotective claims regarding this category of the plants. The oxidative stress is implicated in the pathogenesis of various other diseases, including liver disorders [15]. The liver, which is also susceptible to acute and chronic toxicities from chemicals, drugs, plant extracts, and nutraceuticals, as well as to harms from an overdose of food components, along with the inherent adverse metabolic and biochemical reactions within the liver, has been damage-controlled and treated by antioxidants, and antioxidant-rich plants, as well as from other natural product extracts rich in antioxidant materials [16,17]. The pharmacological (or intrinsic) toxicity is prevalent over the idiosyncratic toxicity, and this does not have any dose-response or temporal relationship to the generated toxicity [18–20]. The ROS are unstable and highly reactive entities and cause the majority of the oxidative-stress related damages and injury to the organs, including the liver, which is comparatively at a higher risk and needs antioxidants to avoid damage and maintain homeostasis. The internally produced glutathione and other enzymatic entities counteract the liver injury caused by the ROS, the reactive nitrogen-based species, and the produced oxidative stress. The oral supplementation of antioxidants for liver disorders has been studied, and several non-enzymatic antioxidants, (e.g., ascorbic acid, α-tocopherol, silibinin, naringenin, quercetin, curcumin, resveratrol, products of phenolic, flavonoid, tannin, and carotenoid types) have been exploited [21–23], and the plant-origin secondary metabolite-based antioxidants have revealed promising in vivo therapeutic effects on liver disorders in studied animal models [24], though clinical studies were deemed inconclusive [23]. However, the protective roles of these antioxidants on the liver, in experimentally induced-liver injury and toxication, have been investigated for various pure, single component natural products, and on the naturally-sourced plants' extracts, as well as mixtures of synergistic natural products containing antioxidants of a structurally-varied chemical nature, and were found to be effective [9,25–28].

In this context, the roles of liver enzymes, and other associated biochemical markers of liver injury and amelioration, are extremely important. The liver's diagnostic condition is indicated by measuring these enzymes, and markers in the blood and sera samples. Among the major hallmarks of the diseased conditions of the liver are the damage of hepatic cells, and the changed concentrations of antioxidant-natured serum enzymes (i.e., superoxide dismutase, catalase, glutathione peroxidase, and glutathione transferase). The levels of biomarkers in normal and hepatotoxicity-induced animals, albeit in the absence of external antioxidant compounds, and extracts or concoction, differs significantly. The aspartate aminotransferase, alanine aminotransferase, gamma-glutamyl transpeptidase, and alkaline phosphatase enzyme levels, along with the total protein, total bilirubin, and malondialdehyde, are known as liver damage biomarkers present in serum. The normalization of these serum enzyme levels has been observed to signal amelioration of the diseased conditions of the liver [29–32]. The roles of phenols and polyphenols, flavonoids, tannins, and alkaloid classes of compounds as external antioxidant entities (either as part of the plant-based products, or as a single, pure compound), and the silymarin, a known hepatoprotective agent, are considered crucial in the amelioration process [33–35]. The synergistic actions of the enzymatic antioxidants and their source (S-adenosylmethionine), as well as the plants' extract components and individual compounds, have also been reported participating in the liver protection [36–40].

Suaeda vermiculata, family Amaranthaceae, is a desert halophyte, renowned for its traditionally claimed liver-protecting activity. This ongoing study aimed to investigate the general claims of the treatment of various symptomatic liver disorders, and proposed to set out to evaluate the safety profile of the plant, given its prevalent use by local herbalists and the public in general. The hepatoprotective activity of the aqueous-ethanolic (aq.-ethanolic) extract from leaves and aerial parts, in addition to the safety of plant material at higher doses, were examined. Studies were carried out to find the antioxidant potential, and concentrations of the antioxidant entities in the aq.-ethanolic extract as being reactive to 1,1-diphenyl-2-picrylhydrazyl (DPPH) and 2,2′-azino-bis-3-ethylbenzthiazoline-6-sulphonic acid (ABTS) in their respective assays. The LC-MS analysis was performed to identify the constituent parts of the aq.-ethanolic extract of the plant and identify the major antioxidants of phenolics and flavonoid nature present in the plant extract. Moreover, the effects of the plant extract on the kidneys, liver, blood sugar, and lipid levels in the CCl_4-induced toxicity bearing animal models with lipid peroxidation conditions, and liver damages were investigated in detail. The cytotoxic effects of the antioxidant-rich, high-extractive-value aq.-ethanolic extract were also checked on the hepatic cell-lines, HepG-2, and the resistant HepG-2/ADR.

2. Results

2.1. LC-MS-Based Chemical Fingerprinting of Suaeda vermiculata aq.-Ethanolic Extract

The positive mode electron-spray ionization (ESI) analysis of the aq.-ethanolic extract of *Suaeda vermiculata* led to the identification of nine constituents (Table 1). Of these nine identified compounds, three flavonoids were identified in the plant extract: One flavonol aglycone (quercetin), and two flavonol glycosides (quercetin-3-*O*-rutinoside or rutin, and kaempferol-*O*-(acetyl)-hexoside-pentoside).

Table 1. Identification of constituents from the aq.-ethanolic extract of *Suaeda vermiculata*.

Sr.	RT (min)	Mol. Formula	Calc. Mass	Obsrvd. Mass	Mass Fragments	Error (Δ ppm)	Compound Name	* R%	Refer
1.	16.11	$C_{10}H_{16}O_5$	217.1076	217.10656 [M + H]	239.0883 [M^+ + Na]; 455.1877 [2M + Na]	−2.265	Senecic acid	0.84	[41]
2.	20.24	$C_{28}H_{30}O_{16}$	623.1612	623.1611 $[M + H]^+$	446.9041[M-Pentose-acetyl];582.8751 [M-acetyl + 2H]	−1.007	Kaempferol-*O*- (acetyl) hexoside-pentoside	4.37	[42]
3.	21.66	$C_{15}H_{10}O_7$	303.2443	303.0492 $[M + H]^+$	289.1404 [M-CH_2]	0	Quercetin	12.45	[43]
4.	23.45	$C_{15}H_8O_6$	285.0399	285.04069 $[M + H]^+$	284.99585 [M]; 286.04395 [M + 2H]	0.810	Rhein	1.26	[44]
5.	24.18	$C_{12}O_{14}O_4$	223.0970	223.09634 $[M + H]^+$	177.0543 [M + H-$C_2H_{10}O$]; 245.0782 [M + Na]; 467.16744 [2M + Na]	−0.957	4-Acetyl-6-(2-methylpropionyl)-1,3-resorcinol	2.32	[45]
6.	29.30	$C_{16}H_{22}O_4$	279.1596	279.15900 $[M + H]^+$	301.1407 [M + Na]; 579.2921 [2M + Na]	−0.323	(2E)-Heptyl-3-(3,4-dihydroxyphenyl) acrylate	3.37	[46]
7.	30.15	$C_{16}H_{30}O_4$	287.2222	287.22140 $[M + H]^+$	199.16898 [M-$C_4H_{10}O_2$]; 309.20324 [M + Na]; 595.41675 [2M + Na]	−1.260	Hexadecanedioic acid	1.57	[47]
8.	31.70	$C_{55}H_{74}N_4O_5$	872.5815	872.1408 $[M + H]^+$	256.26268 [Phytyl-C_2H_5 + H]; 280.26241 [phytyl moiety]; 593.15607 [M-phytyl]; 577.12459 [M-phytyl–methyl]; 615.13623 [M-phytyl + Na]	0.240	Pheophytin-a	23.69	[48]
9.	31.75	$C_{27}H_{30}O_{16}$	633.1431	633.14559 $[M + Na]^+$	634.14630 [M + H + Na]	4.719	Quercetin-3-O-rutinoside (Rutin)	7.80	[43]

* Relative % calculated based on the individual compound's peak area and the chromatogram's total peaks area.

2.2. Estimations of the Total Phenolics and Flavonoid Contents in the aq.-Ethanolic Extract, and Fractions of the aq.-Ethanolic Extract Obtained from *Suaeda vermiculata*

The total phenolics and flavonoid contents of the aq.-ethanolic extract of *Suaeda vermiculata* and its fractions were estimated as gallic acid (GAE) and quercetin equivalents (QE) (Table 2). A significant amount of total phenolics contents were present in the aq.-ethanolic extract (200.6 ± 2.14 mg GAE/g of the extract's dry weight). However, the ethyl acetate fraction possessed comparatively very high concentrations of total phenolics (317.5 ± 0.17 mg GAE/g of fraction's dry weight). The majority of the flavonoids' contents were determined to be present in the chloroform fraction (17.47 ± 0.02 mg QE/g of fraction's dry weight), followed by the aq.-ethanolic extract (16.70 ± 1.56 mg QE/g of the extract's dry weight).

Table 2. Total phenolics, total flavonoids of the extracts of *Suaeda vermiculata*.

Extracts	Extractive Values *	Total Phenolic Contents (mg GAE/g)	Total Flavonoid Contents (mg QE/g)
n-Hexane fraction	1.06	15.1 ± 0.60 ^{E}	nd
Ethyl acetate fraction	4.37	317.5 ± 0.17 ^{A}	13.80 ± 0.04 ^{BC}
Chloroform fraction	1.27	140.0 ± 1.02 ^{D}	17.47 ± 0.02 ^{A}
n-Butanolic fraction	6.25	192.6 ± 0.97 ^{C}	11.55 ± 0.02 ^{C}
aq.-Ethanolic extract (mother liquor)	20.00	200.6 ± 2.14 ^{B}	16.70 ±1.56 ^{AB}

* Extractive values calculated as g/100 g of the dried plant powder; nd = not detected; GAE = gallic acid equivalent; QE = quercetin equivalent. The experiment was carried out in triplicates. Values were expressed as mean ± SEM. Means that do not share a letter are significantly different for the relevant column.

2.3. Determination of Antioxidant Potentials Through Measurement of Free-Radical Scavenging Activity

The in vitro radical scavenging activity of the *Suaeda vermiculata* mother liquor aq.-ethanolic extract and its subsequent fractions were estimated by the DPPH and ABTS radical scavenging methods, and IC_{50} values (concentrations causing 50 % inhibitions) were determined based on the requisite percentage inhibitions versus the concentrations (Figure 1). The chloroform and ethyl acetate fractions demonstrated significant concentration-dependent DPPH and ABTS scavenging activities, all being maximum at 5 mg/mL as compared to the standard quercetin at the same concentration (Figure 1). The lowest antioxidant activity was observed for the *n*-hexane fraction in scavenging the DPPH and ABTS free radicals, followed by the antioxidant activity of the *n*-butanolic fraction. The DPPH radical quenching-based antioxidant activity levels were also recorded for the chloroform, and ethyl acetate fractions, observed at about 80 and 65 % of strengths, respectively, while the mother liquor, aq.-ethanolic extract, showed nearly 40 % DPPH radical inhibition strength, all as compared to the referral standard, quercetin (Figure 1A). The ABTS free radical scavenging antioxidant activity was also assayed for the chloroform and ethyl acetate fractions and was found at 72 and 70 % strengths, respectively, followed by the aq.-ethanolic extract antioxidant activity levels at 60 % antioxidant strength in comparison to the standard quercetin (Figure 1B).

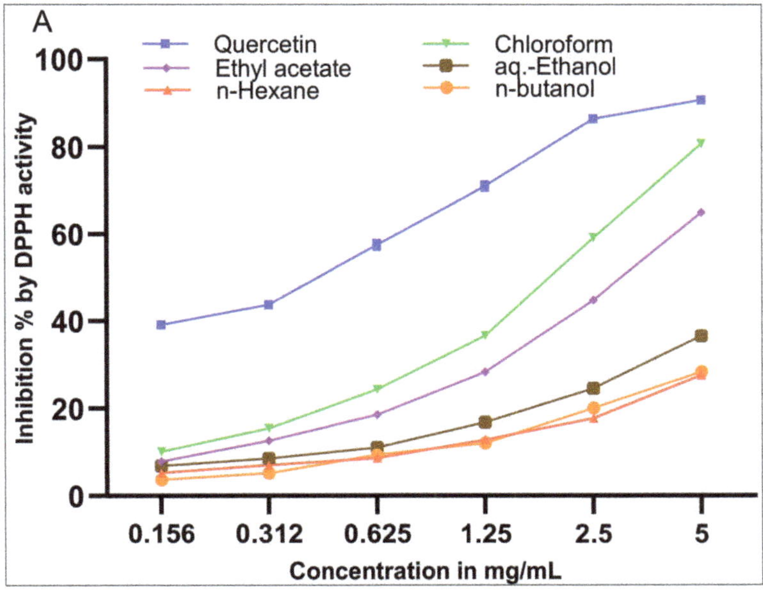

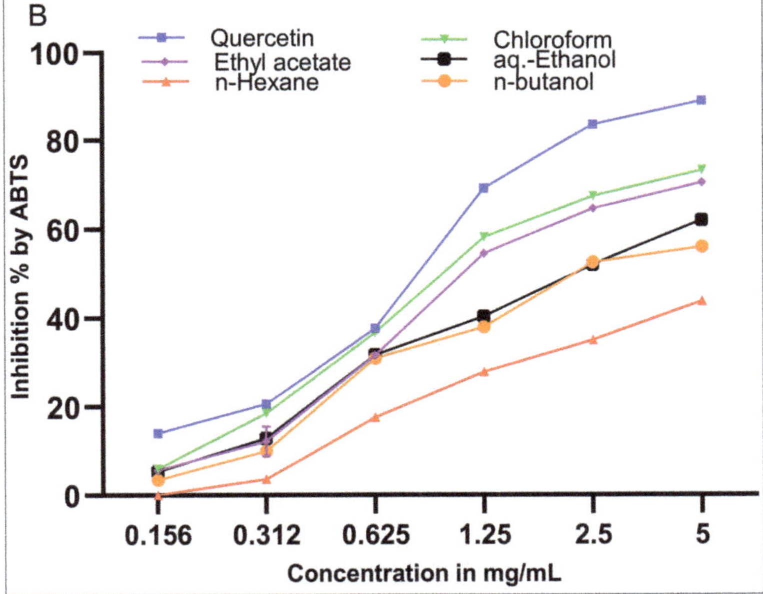

Figure 1. Radical scavenging activity of *Suaeda vermiculata* extracts by (**A**) 1,1-diphenyl-2-picrylhydrazyl (DPPH), and (**B**) 2,2′-azino-bis-3-ethylbenzthiazoline-6-sulphonic acid (ABTS) methods. Values are the mean of three replicates ± SEM. $p < 0.0001$ using two-way ANOVA, and $p < 0.0001$ compared to quercetin for all groups at all the measured concentrations for both methods according to multiple comparisons for Tukey's method.

2.4. Acute Toxicity Study and Dose Selection

Among all the groups (50, 300 mg/kg; 2 and 5 g/kg) checked for acute toxicity, one aq.-ethanolic extract-treated rat (2 g/kg) died on day 2 of the dose administration. The death was attributed to a gavage accident since no behavioral pattern difference was observed among the extract-fed or the control group. According to the Organization for Economic Cooperation and Development (OECD) guidelines, the dose is considered toxic if more than two animals die. All the animal groups were monitored daily for any visible signs of toxicity and mortality for up to two weeks. None of the remaining animals in all the groups displayed any visible signs of toxicity, such as anorexia, hair erection, lacrimation, tremors, convulsions, salivation, or diarrhea, in the first 24 h until 2 weeks of the study period. Based on the data, the given dose was determined to be practically non-toxic according to Hedge and Sterner scale, and 10% (500 mg/kg) of the maximum administered dose (5 g/kg) in the acute toxicity study was selected as the maximum dose for the pharmacological evaluation [49].

2.5. Safety Determination of aq.-Ethanolic Extract on Liver, Heart, and Kidney

The administration of the aq.-ethanolic extract of *Suaeda vermiculata* (250 mg/kg, CCl_4) in normal rats showed no significant effects on the levels of liver enzymes AST (aspartate transaminase) and ALT (alanine transaminase) and was significantly lower ($p < 0.02$) for TP (total protein) contents estimation compared to the intact control group. Moreover, the serum creatinine, triglycerides, cholesterol, and glucose levels showed no significant differences in the 250 mg/kg, CCl_4- group when compared with the intact control group of animals (Tables 3 and 4).

2.6. Hepatoprotective Activity of the aq.-Ethanolic Extract

The AST and ALT values for the CCl_4-induced negative control group treated with CMC were significantly higher ($p < 0.0001$) when compared with all the CCl_4+ induced hepatotoxic groups, i.e., intact, silymarin (200 mg/kg), quercetin (100 mg/kg), and the extract-fed groups (100, 250, and 500 mg/kg) (Table 3). Animals were given *Suaeda vermiculata* aq.-ethanolic extract at daily doses of 100, 250, and 500 mg/kg, and showed a dose-dependent decrease in AST and ALT liver enzyme markers. The 500 mg/kg dose-fed animal group showed maximum hepatoprotective activity (Figure 2), and this group of animals effectively prevented the CCl_4-induced elevation of serum enzyme markers of ALT and AST as compared to the CMC group (CCl_4 induced). The silymarin (200 mg/kg), and quercetin (100 mg/kg)-fed animal groups also showed a significant decrease in AST and ALT liver marker enzymes, while the silymarin group also showed significantly decreased TP contents as compared to the negative control group of animals. No differences were found in the levels of creatinine, or glucose in any CCl_4+ induced hepatotoxic groups pre-treated with various doses of aq.-ethanolic extract as compared to the negative group (Tables 3 and 4). The cholesterol level of the aqueous-ethanol extract-fed group at 250 mg/kg was comparable to the intact control. The extract doses of the 100 and 500 mg/kg-fed animal groups caused a significant decrease in the triglyceride levels as compared to the negative control group of animals (Table 4).

Table 3. Effect of aq.-ethanolic extract of *Suaeda vermiculata* on liver and kidney functions on carbon tetrachloride-induced liver toxicity in experimental rats.

Groups	AST IU/L	ALT IU/L	TP g/dL	Creatinine mg/dL
I. Intact control (no CMC, no drug, no aq.-ethanolic extract, CCl_4-)	62.83 ±2.21 ^{B}	67.42 ± 4.66 ^{B}	7.39 ± 0.17 ^{A}	0.91 ± 0.02 ^{A}
II. aq.-Ethanolic extract, 250 mg/Kg, CCl_4-	54.36 ± 3.64 ^{B}	55.19 ± 3.47 ^{B}	6.14 ± 0.34 ^{BC}	0.80 ± 0.04 ^{A}
III. Negative control (vehicle CMC 0.5%), CCl_4+	169.57 ± 12.68 ^{A}	103.38 ± 4.86 ^{A}	7.23 ± 0.17 ^{AB}	0.82 ± 0.03 ^{A}
IV. Silymarin, 200 mg/kg, CCl_4+	55.40 ± 2.45 ^{B}	55.25 ± 2.91 ^{B}	5.32 ± 0.14 ^{BC}	0.82 ± 0.04 ^{A}
V. aq.-Ethanolic extract, 100 mg/kg, CCl_4+	65.35 ± 3.92 ^{B}	71.77 ± 4.25 ^{B}	7.64 ± 0.23 ^{A}	0.80 ± 0.03 ^{A}
VI. aq.-Ethanolic extract, 250 mg/kg, CCl_4+	56.29 ± 3.30 ^{B}	69.21 ± 3.79 ^{B}	7.23 ± 0.27^{AB}	0.86 ± 0.03 ^{A}
VII. aq.-Ethanolic extract, 500 mg/kg, CCl_4+	54.59 ± 1.67 ^{B}	66.45 ± 7.05 ^{B}	7.56 ± 0.32 ^{A}	0.87 ± 0.02 ^{A}
VIII. Quercetin, 100 mg/kg, CCl_4+	49.54 ± 1.23 ^{B}	52.39 ± 4.56 ^{B}	8.05 ± 0.18 ^{A}	0.82 ± 0.03 ^{A}

Values denoted are mean ± SEM, n = 8 animals/group. AST: Aspartate transaminase, ALT: Alanine transaminase, TP: Total protein, CMC: Carboxyl methylcellulose. CCl_4-: Carbon tetrachloride was not administered. CCl_4+: Carbon tetrachloride was administered. Means that do not share a letter are significantly different for the relevant column.

Table 4. Effect of aq.-ethanolic extract of *Suaeda vermiculata* on blood glucose, triglycerides, and cholesterol on carbon tetrachloride (CCl_4)-induced liver toxicity in experimental rats *.

Groups	Glucose mg/dL	Cholesterol mg/dL	Triglycerides mg/dL
I. Intact control (no CMC, no Drug, no aq.-ethanolic extract, CCl_4-)	122.46 ± 2.66 ^{A}	85.06 ± 1.85 ^{A}	86.26 ± 2.24 ^{BC}
II. aq.-Ethanolic extract, 250 mg/Kg, CCl_4-	101.82 ± 4.47 ^{AB}	78.73 ± 3.87 ^{AB}	74.11 ± 2.95 ^{C}
III.] Negative control (vehicle CMC 0.5%), CCl_4+	111.23 ± 3.23 ^{AB}	61.70 ± 1.42 ^{D}	100.20 ± 6.80 ^{AB}
IV.] Silymarin, 200 mg/kg, CCl_4+	110.43 ± 6.43 ^{AB}	65.94 ± 2.17 ^{CD}	99.54 ± 4.28 ^{AB}
V.] aq.-Ethanolic extract, 100 mg/kg, CCl_4+	99.86 ± 4.15 ^{B}	59.46 ± 1.66 ^{D}	76.56 ± 3.25 ^{C}
VI.] aq.-Ethanolic extract, 250 mg/kg, CCl_4+	99.49 ± 7.00 ^{B}	80.93 ± 2.08 ^{AB}	97.53 ± 3.58 ^{AB}
VII.] aq.-Ethanolic extract, 500 mg/kg, CCl_4+	98.79 ± 6.70 ^{B}	66.66 ± 2.20 ^{CD}	78.16 ± 2.62 ^{C}
VIII.] Quercetin 100 mg/kg, CCl_4+	115.46 ± 3.58 ^{AB}	72.83 ± 1.57 ^{BC}	3.02 ^{A}

* Values are denoted as mean ± SEM. CMC: Carboxyl methylcellulose. CCl_4-: Carbon tetrachloride was not administered. CCl_4+: Carbon tetrachloride was administered. Means that do not share a letter are significantly different for the relevant column.

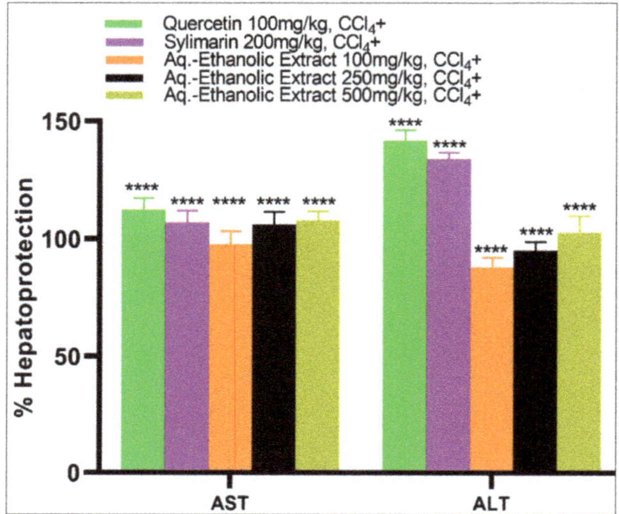

Figure 2. The hepatoprotective effects of *Suaeda vermiculata* aq.-ethanolic extract on CCl_4 induced liver toxicity. Percentage protection of CCl_4 induced elevation of AST and ALT enzymes. The protection percentage formula assumes the enzyme level of the CCl_4 administered negative group (carboxyl methylcellulose) at 0% protection and that of intact group (no exposure to CCl_4) at 100% protection as such they are excluded from the graphical representation. In addition, the aq.-ethanolic extract 250 mg/kg CCl_4- the group which did not receive CCl_4- is also excluded from the graphical representation. **** Data differed significantly at $p < 0.0001$ when compared with the negative control group. Values are expressed as mean ± SEM, n = eight rats per group, CCl_4-: Carbon tetrachloride was not administered. CCl_4+: Carbon tetrachloride-induced liver toxicity.

2.7. Cytotoxicity of Suaeda vermiculata Extract in Wild Type Sensitive and Resistant Hepatic Cell-Lines

To test the cytotoxicity of *Suaeda vermiculata* aq.-ethanolic extract in sensitive HepG-2 and resistant cells, HepG-2/ADR, evaluations were conducted in comparison to the standard cytotoxic drug, doxorubicin (DOX). Figure 3 demonstrates the dose-response curves of the *Suaeda vermiculata* aq.-ethanolic extract group in comparison to the chemotherapeutical agent, DOX. The moderate cytotoxic effect of *Suaeda vermiculata* aq.-ethanolic extract in HepG-2 showed an IC_{50} value of 56.19 µg/mL, and in HepG-2/ADR-resistance cells, the IC_{50} value was 78.40 µg/mL with relative resistance (RR) at 1.39 (Table 5).

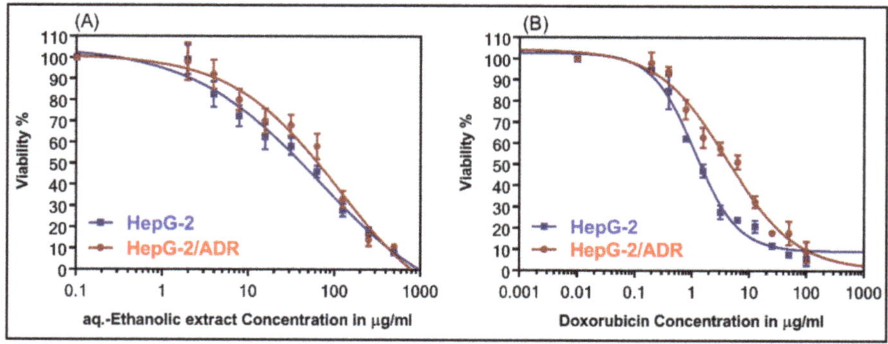

Figure 3. The dose-response curves of *Suaeda vermiculata* aq.-ethanolic extract: (**A**) aq.-ethanolic extract, and (**B**) doxorubicin (DOX) in wild-type HepG-2, and resistant HepG-2/ADR cell-lines.

Table 5. The IC_{50} values of *Suaeda vermiculata* aq.-ethanolic extract compared to DOX in HepG-2 wild types, and HepG-2/ADR resistant cell-lines, and their relative resistance (RR) using MTT [3-(4,5-dimethylthiazol-2-yl)-2,5-diphenyl tetrazolium bromide] assay.

Groups	IC_{50} (μg/mL) *		RR
	HepG-2	HepG-2/ADR	
aq.-Ethanolic extract	56.19 ± 2.55	78.40 ± 0.32 ***	1.39
Doxorubicin	1.3 ± 0.064	4.77 ± 1.05 ***	3.67

* Values expressed as mean ± SEM, *** = $p < 0.001$ using Student's *t*-test.

2.8. Suaeda vermiculata aq.-Ethanolic Extract Reverses DOX Cytotoxicity in Resistant Hepatic Cell-Lines

To examine the reversal effects of the combination of DOX and *Suaeda vermiculata* aq.-ethanolic extract, the DOX-resistant HepG-2/ADR cell lines were selected. The DOX-resistant cells were treated with DOX in the presence and absence of 20 μg/mL *Suaeda vermiculata* aq.-ethanolic extract. As shown in Figure 4, *Suaeda vermiculata* extracts synergistically enhanced the cytotoxicity of DOX, as observed from the dose-response curves of the combination and isobologram. The IC_{50} values of DOX alone or in combination with *Suaeda vermiculata* aq.-ethanolic extract in DOX resistant cells are shown in Table 6. The combination of DOX with *Suaeda vermiculata* extract significantly reduced ($p < 0.001$) the IC_{50} value of DOX, with FR in resistant cell lines to 2.77, and CI at 0.56 (Table 6).

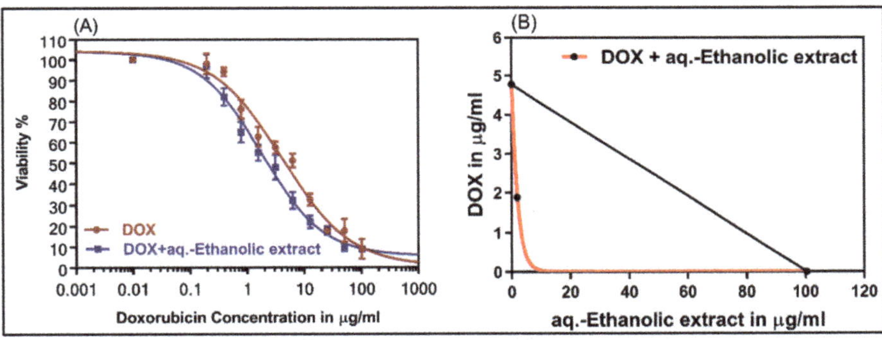

Figure 4. (**A**) The dose-response curves and isobologram analysis of the combination of DOX with 20 μg/mL of *Suaeda vermiculata* extract in resistant HepG-2/ADR. (**B**) The isobologram on the right side of the figure showed the synergistic interaction between DOX and the extract. IC_{50}, DOX, and IC_{50} extract correspond to the IC_{50} for DOX and extract alone. CI < 1 indicates synergism, CI = 1 indicates additive, and CI > 1 indicates antagonism.

Table 6. Synergistic interaction of a combination of DOX with 20 μg/mL of *Suaeda vermiculata* aq.-ethanolic extract in resistant HepG-2/ADR cell-lines *.

Groups	IC_{50} (μg/mL)	Synergistic Parameters			
	HepG-2/ADR	FR	CI	r	IB
Doxorubicin (DOX)	4.77 ± 1.05	–	–	–	–
DOX + aq.-ethanolic extract	1.72 ± 0.07 ***	2.77	0.56	0.99	Synergism

* Values were expressed as mean ± SEM. CI, combination index; FR, fold reversal; IB, isobologram; and medium effect equation (r-value). CI < 1 indicates synergism, CI = 1 indicates additive, and CI > 1 indicates antagonism. *** = $p < 0.001$ using Student's *t*-test.

3. Discussion

The identification of the flavonoid derivatives (quercetin, quercetin-3-*O*-rutinoside or rutin, and kaempferol-*O*-(acetyl)-hexoside-pentoside) in the aq.-ethanolic extract of the *Suaeda vermiculata* is consistent with the halophytic nature of the plant, thereby having excessive antioxidants compounds, especially flavonoids. In addition, the nature of these flavonoids is in conformity with the previously isolated and identified flavonoids from the genus *Suaeda* [43]. For instance, the quercetin (12.45%), and quercetin-3-*O*-rutinoside (7.80%) were previously isolated from *Suaeda japonica, Suaeda salsa,* and *Suaeda physophora* [43], while kaempferol and kaempferol glycoside are also common constituents of the genus *Suaeda* plants, and have been reported both from *Suaeda japonica* and *Suaeda asparagoides* [43]. The LC-MS analysis of the aq.-ethanolic extract showed a base peak at *m/z* 872.1408 $[M + H]^+$ which was identified as the pheophytin-a (23.69%), the reported major chlorophyll-based constituent from *Suaeda vermiculata* species [48]. The fragmentation pattern for pheophytin-a showed mass peaks at *m/z* 256.26268 [phytyl-C_2H_5 + H], 280.26241 [phytyl moiety], 593.15607 [M-phytyl], 577.12459 [M-phytyl–methyl], and *m/z* 615.13623 [M-phytyl + Na], which together confirmed the structure of the compound. The LC-MS analysis of the plant's aq.-ethanolic extract also showed base peaks for the rhein (anthraquinone aglycone) at *m/z* 285.04069 $[M + H]^+$, in addition to the base peaks for senecic acid, 4-acetyl-6-(2-methylpropionyl)-1,3-resorcinol, (2E)-heptyl-3-(3,4-dihydroxyphenyl) acrylate, and hexadecanedioic acid at *m/z* 217.10656 [M + H], 223.09634 $[M + H]^+$, 279.15900 $[M + H]^+$, and *m/z* 287.22140 [M + H], respectively.

The higher contents of phenolics in the ethyl acetate fraction is probably owing to the ability of ethyl acetate as a mid-polarity solvent to dissolve the phenolic contents of the aq.-ethanolic extract much easier than the relatively non-polar, and lipophilic solvents, i.e., *n*-hexane, and chloroform. Besides, the ethyl acetate fraction accumulated more phenolics and flavonoid contents compared to the *n*-butanol fraction, which could be due to the fact that ethyl acetate as a solvent was used in the extraction sequence of aq.-ethanolic mother liquor before the *n*-butanol. Additionally, the majority of the flavonoid contents were present in the chloroform fraction (17.47 ± 0.02 mg QE/g of fraction's dry weight), which was closely followed by the aq.-ethanolic extract (16.70 ±1.56 mg QE/g of extract's dry weight), indicating a lesser degree of hydrophilicity of the products, culminated through the presence of fewer hydroxylation sites and glycosylation of the flavonoids encountered in the chloroform extract, which were able to be extracted in this lipophilic solvent under the presence of aqueous partitioning phase.

The presence of higher concentrations of phenolics and flavonoids is consistent with the previously published data about flavonoid and phenolic contents present in the solvents, such as chloroform and ethyl acetate [50–53]. In the current study, the relatively high mg per gram content values of the phenolics and flavonoids are in the chloroform and ethyl acetate fractions compared to the mother liquor and confirms the higher antioxidant activity of these fractions over the mother liquor. Furthermore, the pheophytin-a, which was reported for the mild antioxidant activity, and isolated previously form *Suaeda vermiculata* chloroform fraction [48] as part of the ongoing study, also played a role in the antioxidant activity of the chloroform fraction. Additionally, the steroidal constituent, β-sitosterol [48], along with other unsaturated oxygenated products [54] identified in the chloroform fraction of the *Suaeda vermiculata* contributed to the higher antioxidant value of the chloroform fraction. Nonetheless, the mother liquor aq.-ethanolic extract was pursued in further studies into the biological evaluations as a hepatoprotective extract, since the traditional practice of the use of plants' aerial parts as a whole for liver disorders provided an equivalence between the prevailing practice and the aq.-ethanolic extract, the mother liquor obtained from the plant's extraction. The mother liquor aq.-ethanolic extract was also selected for the biological evaluations as it represented the largest number of constituents obtained through a comparatively significant high-extractive value (20 g/100 g of the dried plant herbs). Moreover, the statistically insignificant difference (Table 2) in the flavonoid contents of the aq.-ethanolic extract (16.70 ± 1.56 mg/g QE) in comparison to the chloroform fraction's flavonoid contents (17.47 ± 0.02 mg/g QE) supported the choice of the aq.-ethanolic extract for biological activity evaluations.

The current investigation confirmed the *Suaeda vermiculata* plant's richness in antioxidant components compared to other *Suaeda* species [43]. The in vitro radical scavenging activity of the *Suaeda vermiculata* mother liquor aq.-ethanolic extract and its subsequent fractions were estimated by the DPPH and ABTS radical scavenging methods, and the IC_{50} values (concentrations causing 50% inhibitions) were determined based on the requisite percentage of radicals' inhibitions versus the concentrations (Figure 1). Notably, the presence of higher concentrations of antioxidant compounds, measured as radical scavenging capacity in DPPH and ABTS assays is a direct indication of the antioxidant potential of the plant's constituents, but not necessarily the high presence of phenolics and flavonoids, which can only be the reasons for higher antioxidant potential, though these two categories of products, i.e., phenolics and flavonoids, seem to be the reason for the dominating antioxidant potential of the plant. The higher antioxidant presence in the extract and the fractions of the *Suaeda vermiculata* can be attributed to the high-salinity habitat of the plant in comparison to its other counterpart species growing in different climatic locations [43]; the former has been the herb of choice for the local population for treatment of liver disorders, especially jaundice [48,55]. However, the biomechanistics connections between the higher antioxidant concentration levels of compounds in the plant, and its hepatoprotective action have not been conclusively established, though it has been well-observed, and a relationship is proposed at the primary levels (Figure 5), albeit without intricate and basic biomechanistics roles playing, and interconnections, or the feedback responsibilities of the components in the presented cycle. Nonetheless, a plausible mechanistic explanation of the toxicity and its amelioration, based on known information, is proposed herein (Section 3.1).

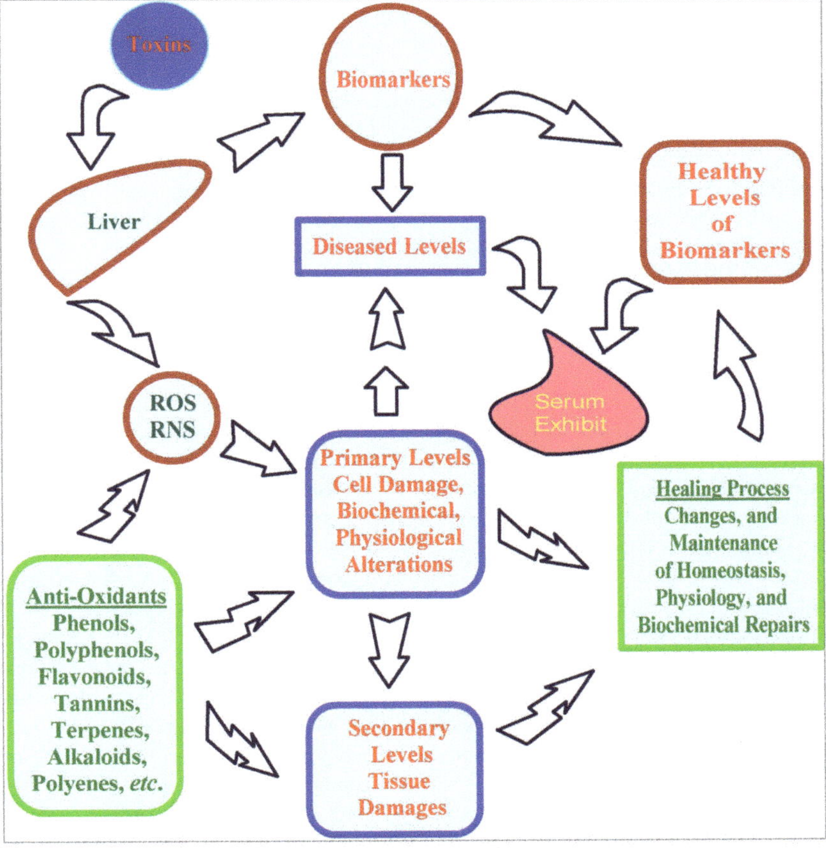

Figure 5. Prospective roles of antioxidants in hepatoprotection.

Moreover, it has to be noted that the present study found no adverse effects of the aq.-ethanolic extract on the liver and kidney enzymes of the normal rats, and the plant extract also did not induce any toxicity at doses as high as 5 g/kg. The study also revealed that the aq.-ethanolic extract of *Suaeda vermiculata* had no significant effects on the blood glucose levels for the treated normal rats as compared to the control group of animals which (blood glucose levels) was measured at 101.82 ± 4.47 and 122.46 ± 2.66 mg/dL, respectively. Additionally, the natural healing in CCl_4 -induced hepatotoxic animals, ingestion of potent hepatotoxins, and evaluations of the hepatoprotective action using several natural products are abundantly available in the literature [56–58]. As for the local *Suaeda* species, where the *Suaeda vermiculata* is concerned, there were no other pharmacological data of any kind, comparative, or otherwise available on the liver-protecting activities of this plant or the plants of the differently-located *Suaeda* species, referring to those located in Iran.

The locally procured *Suaeda vermiculata* aq.-ethanolic extract, as shown in Table 5, exhibited moderate cytotoxicity on HepaG-2 (IC_{50} 56.19 ± 2.55 μg/mL), designated in accordance with the provisions of the previous classification of the cytotoxicity criteria of the natural products extracts on cell lines [59]. As previously reported, the extract was considered active if it had a mean IC_{50} value <100 μg/mL; a moderate level of activity was considered between 10–100 μg/mL; and the strong cytotoxic activity was designated with <10 μg/mL of the IC_{50} values. *Suaeda vermiculata* aq.-ethanolic extract also exhibited moderate cytotoxicity in HepG-2/ADR, the resistance cell-lines (IC_{50} 78.40 ± 0.32 μg/mL with RR 1.39) (Table 5). In resistance cell lines, like HepG-2/ADR, the P-gp is overexpressed as part of the mechanism to escape cell-death induced by anti-cancer chemotherapeutic agents. The cytotoxicity results indicated that *Suaeda vermiculata* aq.-ethanolic extract could be a potential modulator of P-gp, most probably due to the presence of quercetin and other flavonoids (kaempferol, and rutin, Table 1). The flavonoids and phenolics, e.g., kaempferol, rutin, apigenin, daidzein, fisetin, luteolin, silibinin, naringin, proanthocyanidin, (-)epicatechin-3-*O*-gallate, etc., are known P-gp inhibitors [60–62]. The *Suaeda vermiculata* aq.-ethanolic extract's cytotoxicity results were further confirmed from a combination experiment which improved the cytotoxicity of DOX (Table 6) through combination with a low and non-toxic concentration (20 μg/mL) of the aq.-ethanolic extract [63,64]. Again, the presence of the phenolics, together with flavonoids components in the aq.-ethanolic extract (Table 1) could explain these inhibitory effects [63].

3.1. A Plausible Biomechanistic Aspect of Liver Toxicity and Its Amelioration

The induction of liver toxicity by carbon tetrachloride (CCl_4) ensues through free radicals generation [65]. The cytochrome P450 bio-transforms CCl_4 [66] to form trichloromethyl radical ($CCl_3\bullet$), which, in turn, in the presence of oxygen, lipid, and protein, forms trichloromethyl peroxyl radical ($CCl_3OO\bullet$) that eventually leads to lipid peroxidation [67]. These transformations lead to hepatocyte dysfunction through the destruction of the intracellular and the plasma membranes of the organ [68–70].

The hepatotoxicity was induced in animals by CCl_4 injections and led to an increase in serum aminotransferase AST and ALT levels, generally being an AST > ALT relationship in the toxic situations [71], wherein both the AST and ALT were considered as biomarkers of liver damage [72–75]. In the current study, rats with CCl_4-induced hepatotoxicity were confirmed to have significantly increased levels of AST and ALT in the negative control compared to the intact group of animals (Table 3). The administration of the aq.-ethanolic extract induced significant ($p < 0.05$) hepatoprotective action against CCl_4-induced liver-toxicity by improving the liver functions, as indicated by the reductions in the levels of the liver enzymes, ALT, and AST, again as compared to the negative control (Table 3). The aq.-ethanolic extract-treated animals showed liver enzyme levels at comparable levels to the intact animals (as standard controls), and silymarin and quercetin groups of animals. Silymarin is an established antioxidant also used as a standard referral compound in the evaluation of the hepatoprotective activity of nutraceuticals, and other natural products [76–78]. The protective effects of quercetin are associated with a decrease in the oxidative stress in the hepatotoxicity-induced liver tissues [79,80]. The AST values for all the aq.-ethanolic extract dose groups in the current study are comparable with the quercetin. The aq.-ethanolic extract showed protective effects in CCl_4-induced hepatotoxicity conditions, as well as safety from toxicity induction up to 5 g of the aq.-ethanolic extract administration, which showed no adverse effect on the extract-fed animals. The protective effects of the aq.-ethanolic extract indicated the ameliorating effects of the *Suaeda vermiculata* concerning the liver cells in the in vivo conditions (Figure 6).

The comparatively higher concentrations of the flavonoids and phenolics in the halophytes, in comparison to the other normal-conditioned plants, might also be the scientific explanation for their liver-protecting activity [43,48,54,81,82]. Moreover, the hypoglycemic effects reported from other plants of the genus *Suaeda*, for example, *Suaeda fruticosa* produced a significant decrease in the blood glucose levels in normal rats, and a further reduction in the diabetic rat models was observed [83]. The study reported that the aqueous extract of *Suaeda fruticosa* induced a significant decrease in the plasma cholesterol without noticeably affecting the plasma triglyceride levels in the treated rats. The free-radical scavenging potential and the inter-linked antioxidant components in the plant extract reportedly played a critical role in hepatotoxicity which was mediated by the actions of free-radicals on hepatocytes leading to cell necrosis [81]. In this connection, the *Suaeda vermiculata* aq.-ethanolic extract also showed free-radical inhibition properties, and the data reported in this study is following previously published reports demonstrating the hepatoprotective effects in CCl_4-induced extensive liver damage models by the aq.-ethanolic extract containing antioxidant compounds as part of the mother liquor obtained from the plant [56–58,66,84–90].

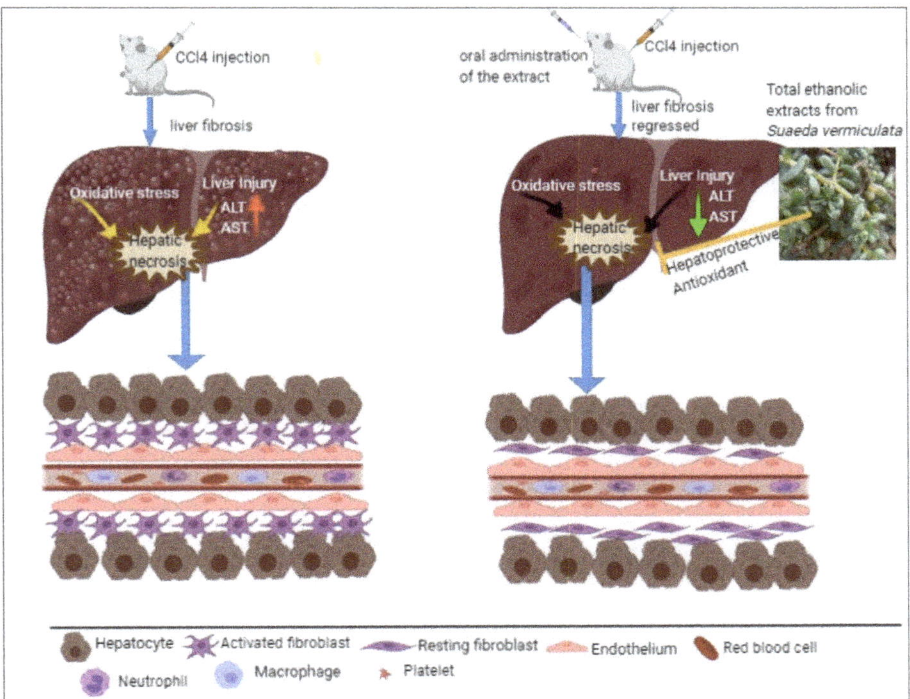

Figure 6. Sketch representing the role of the plant extract in liver protection.

4. Materials and Methods

4.1. Chemicals and Reagents

All chemicals were of analytical grade. Methanol (HPLC grade), carbon tetrachloride, formic acid, (Sigma Aldrich, USA), doxorubicin was also of analytical grade. Silymarin tablets were obtained from Micro Labs Limited, Mumbai, India, and used as obtained.

4.2. Plant Material Collection and Extraction

The *Suaeda vermiculata* plant material (2 kg) was collected (10/2019) from Buraydah City, Qassim, and identified by the institutional botanist at Qassim University (QU). The shade-dried herbs were grinded to a coarse powder (1.4 kg), and extractions performed following the cold maceration method under stirring for 24 h from 70% aqueous-ethanol, repeatedly (3Lx3). The hydroalcoholic extract was filtered and evaporated to dryness under reduced pressure and <40 °C temperature to yield 86.3 g of the dried extract, part of which was stored under −20 °C for further use. The dried aq.-ethanolic extract (50 g) was suspended in 500 mL distilled water and fractionated with *n*-hexane, chloroform, ethyl acetate, and *n*-butanol. The fractions were subjected to dryness under vacuum and <40 °C, and stored under −20 °C for future use.

4.3. LC-MS Experimental Design

Chemical fingerprinting of the aq.-ethanolic extract was performed by C_{18} reverse phase HPLC chromatographic analysis. The compounds were identified by the coupled MS spectrometry through data interpretation and were compared with the literature-reported mass values. The compounds' identification processes were based on the data from the internal library software, and comparison of the available mass-fragments with the literature reports. The relative percentages of the identified compounds were calculated in relation to the total area of the chromatogram. Analytical results are summarized in Table 1.

4.3.1. High-Pressure Chromatography (HPLC): The Chemical Fingerprinting

The separation procedure was carried out using a C_{18} column (C_{18} synchronism 250 × 2.1, 5 μ). The mobile phase solvents were composed of A: 100% water + (0.1 % formic acid), and B: 100% methanol + (0.1 % formic acid). The gradient elution program was followed. The injections [methanol blank, caffeine (QC), and the samples] volume was 20 μL for each, and the flow rate was set at 450 μL/min. UV-Vis detection was performed, and the scanned wavelength was set between 220–600 nm with a scan rate of 20 Hz. Two channels (254 and 268 nm) were also used. The XcaliburTM software (Thermo Scientific) was used for method development and data treatments. All the chromatographic work was carried out at the Faculty of Medicine Laboratories, Umm Al-Qura University, Makkah, Saudi Arabia.

4.3.2. Mass Spectrometer

The analysis was performed using a Thermo LTQ Velos Orbitrap mass spectrometer (Thermo Scientific, Pittsburgh, PA, USA) equipped with a heated ESI ion source. The mass scan range was set to 100–2000 *m/z*, with a resolving power of 100,000. The *m/z* calibration of the LTQ-Orbitrap analyzer was performed in the positive ESI mode using a solution containing caffeine, MRFA (Met-Arg-Phe-Ala) peptide, and Ultramark 1621, in accordance with the manufacturer's guidelines. The ESI was operated with a metal needle operating at 4.5 kV. For all the experiments, the source vaporizer temperature was adjusted to 450 °C, the capillary temperature was set at 275 °C, and the sheath and auxiliary gases were optimized, and set to 30 and 15 arbitrary units, respectively.

4.4. Determination of Total Phenolic Contents

The total phenolic contents of the extract and fractions were determined using the Folin–Ciocalteu method [91]. Briefly, 4 mL of the extract/fraction (40 μg/mL in ethanol) was mixed with 2.5 mL of the Folin–Ciocalteu reagent (diluted with distilled water, 10%, *v/v*) for 5 min, and then 2.5 mL of 20% (*w/v*) sodium carbonate was added. The mixture was allowed to stand for 60 min in the dark, and the absorbance was measured at 760 nm using a V-630, JASCO, Japan, UV-Visible spectrophotometer. The analyses were performed in triplicate, and the total phenolic contents were calculated using the linear regression equation of the calibration curve ($y = 0.002x - 0.014$, $R^2 = 0.998$) of gallic acid performed between the concentrations of 0.05 to 0.5 mg/mL, and expressed as the mg of gallic acid equivalent per gram (mg GAE/g) of the dried extract/fractions weights (Table 2).

4.5. Determination of Total Flavonoid Contents

The total flavonoid contents were estimated using the aluminium chloride complex-forming assay [92]. An aliquot of extract/fraction (1.5 mL of 1 mg/mL in ethanol) was mixed with 500 µL of 10% aluminum chloride solution, and the mixture was allowed to stand for 60 min. Absorbance was measured at 510 nm, and a calibration curve (y = 0.001x + 0.013 R^2 = 0.994) for quercetin was then plotted using 0.05 to 0.5 mg/mL of quercetin in methanol. The equivalents of milligram quercetin per gram (mg QE/g) of the fully concentrated extract/fractions were used to calculate the total flavonoid contents.

4.6. Evaluation of Free-Radical Scavenging Activity

4.6.1. DPPH Assay

The scavenging activity of the extract was estimated using 1,1-diphenyl-2-picrylhydrazyl (DPPH) as a free-radical [93] with slight modification. The DPPH working solution (1.5 mL, 100 µM in methanol) was mixed with 0.5 mL of the extract/fraction/standard (quercetin solution) at different concentrations (0.156 to 5 mg/mL). The mixtures were vortexed thoroughly for 5 min and left in the dark at room temperature (RT) for 20 min, and absorbance was measured at 517 nm to estimate the reduction in DPPH color intensity. The scavenging activity of the extract/fraction and standard control were calculated using the following equation:

$$\text{Scavenging }\% = \left(1 - \frac{S_{ab}}{B_{ab}}\right) \times 100 \quad (1)$$

The $\mathbf{S}_{ab}$ and B_{ab} refer to the sample absorbance and blank absorbance, respectively.

4.6.2. ABTS Assay

The ability of the extract and fractions to scavenge ABTS radical formation was assayed according to the method of Floegel et al. [94] with slight modification. The radical cation of the ABTS (2,2′-azino-bis-3-ethylbenzthiazoline-6-sulphonic acid) was generated by heating the mixture of 2,2-azobis(2-amidinopropane) dihydrochloride (1 mM) and 2.5 mM of the ABTS in phosphate buffer (10 mM, pH 7.4) at 68 °C for 40 min on a water bath. The mixture was allowed to cool to room temperature and filtered. The freshly prepared ABTS solution (980 µL) was mixed with 20 µL of the serially diluted extract/fraction, and quercetin solution (0.156 to 5 mg/mL in ethanol) separately. The mixture was incubated for 10 min at 37 °C before the measurement of ABTS color reduction at 734 nm using a spectrophotometer. The ABTS scavenging activity of the extract and standard quercetin was calculated by defining the percentages of the absorbance reduction of ABTS-extract mixture in comparison to the absorbance of the ABTS blank solution from the following equation:

$$\text{Scavenging }\% = \left(1 - \frac{T_{ab}}{C_{ab}}\right) \times 100\,. \quad (2)$$

The T_{ab} and C_{ab} refer to the sample and blank absorbances, respectively.

4.7. Acute Toxicity Studies and Sample Size

Acute toxicity studies were conducted following the OECD procedure for acute toxicity testing [95,96]. Briefly, ten-week-old female Sprague Dawley rats (n = 20), weighing 200 ± 50 g, overnight fasted, were randomly divided and weighed, and single aq.-ethanolic extract doses were administered (n = 5/group) using oral (p.o.) route with 50 and 300 mg/kg, and 2 and 5 g/kg. The animals were observed for abnormality in behavior and movement for the first three days and mortality for up to two weeks [96].

The required sample size for the experiment was determined using mean ± SEM AST values between the CCl_4-induced injury untreated group and the CCl_4-induced injury aq.-ethanolic extract-treated group, based on a previously published report [58]. The calculated effect size d was 4.27 using a two-tail option on G Power V.3.1.9.4 software, Heinrich Heine University, Düsseldorf, Germany [97]. To achieve a statistical power (1-β err prob) of 80 % and a specific α error probability of 0.05, the minimum required sample size in each group was $n > 3$ mice.

4.8. Effect of aq.-Ethanolic Extract on Liver and Kidney Enzymes in Addition to Lipid and Sugar Levels

The study was performed in accordance with the Animal Research: Reporting In Vivo Experiments (ARRIVE) statement [98].

4.8.1. Experimental Animal Groups

Male ten-week-old naïve Sprague Dawley rats ($n = 64$) weighing 200 ± 50 g, obtained from the animal house facility, College of Pharmacy, Qassim University, Saudi Arabia, were maintained in individual polyacrylic cages housing 2 or 3 rats with a chow diet obtained from the First Milling Company in Qassim, Buraydah, Saudi Arabia, and water ad libitum, 7 days before the start of the experiments. The animals were maintained at RT (~ 25 °C) and relative humidity of ~65% with a controlled light-dark cycle of 12:12 h. The institutional Research Ethics Committee approved the experimental procedure and animal care (Approval ID 2019-CP-8) as per the Guidelines for the Care and Use of Laboratory Animals. Animals were divided randomly into eight groups (n = 8/group). The intact animals (group I) remained untouched during the experimental procedure, while another group received 250 mg/kg (group II) of the aq.-ethanolic extract of *Suaeda vermiculata* orally for seven days to monitor the effect of the extract on the liver and kidney. The remaining groups were administered p.o. once daily with carboxyl methylcellulose 0.5% (CMC, negative control, group III), 200 mg/kg silymarin (positive control, group IV) [99–101], or aq.-ethanolic extract dose of 100, 250, or 500 mg/kg (groups V–VII), and 100 mg/kg quercetin (group VIII) [102,103] for 7 days, followed by induction of hepatotoxicity using single intraperitoneal (i.p.) dose of CCl_4 dissolved in olive oil (1:1, 1.0 mL/kg) [56]. Olive oil (an emulsifying agent) helped to dissolve CCl_4 sufficiently for induction of liver damage and is neither toxic nor possesses any hepatotoxicity related to pharmacological activity [57,58,104]. Twenty-four hours after the administration of CCl_4, blood samples were withdrawn from the orbital vein in ethylene diamine tetraacetic acid (EDTA)-heparinized tubes under mild general anesthesia with ketamine, and samples were centrifuged to separate plasma [105]. The separated plasma was used for the determination of levels of aspartate transaminase (AST) and alanine transaminase (ALT) levels, total protein (TP), creatinine, blood glucose, total cholesterol, and triglycerides with spectrophotometric assays. Percentage hepatotoxic protection was calculated using the following equation [106]:

$$\text{Hepatoprotection } \% = \left[\frac{a-b}{a-c}\right] X\ 100 \tag{3}$$

where a, b, and c refer to the mean value of the marker enzyme level produced by hepatotoxin; toxin plus test sample, and control, respectively.

4.8.2. Determination of Levels AST, ALT, TP, and Creatinine

The levels of ALT and AST in plasma samples were determined using an optimized UV-test, in accordance with the International Federation of Clinical Chemistry (IFCC) procedure. The kit was supplied by the Crescent Diagnostics Company, KSA, (product catalog #CZ902L for ALT, and #CZ904L for AST). The absorbances were measured at 340 nm, and the readings were multiplied by a factor of 952 using the kinetic method [107]. The plasma level of TP was measured using the photometric, colorimetric test. The cupric ions reacted with protein in an alkaline solution to form a purple complex. The absorbance of the complex was proportional to protein concentration in the samples [108]. The plasma creatinine was determined by the kinetic method without deproteinization–Jaffe reaction (Crescent Diagnostics Company, #604). In the Jaffe reaction, the creatinine reacted with alkaline picrate to produce a reddish-orange color, the intensity of which at 490 nm was directly proportional to creatinine concentration [109].

4.8.3. Determination of Plasma Level of Glucose, Cholesterol, and Triglycerides

The glucose level was determined after enzymatic oxidation in the presence of glucose-oxidase (Crescent Diagnostics Company, #605). The formed hydrogen peroxide reacted under the catalysis of peroxidase with phenol and 4-aminoantipyrine to form a red-violet using quinone imine dye as an indicator [110]. The total amount of cholesterol was determined after enzymatic hydrolysis and oxidation (Crescent Diagnostics Company, #603). The colorimetric indicator quinone imine was generated from phenol and 4-aminoantipyrine by hydrogen peroxide under the catalytic action of peroxidase, and the color intensity was then measured at 546 nm [111]. The triglycerides were determined after enzymatic splitting with lipoprotein lipase (Crescent Diagnostics Company, #611) wherein the indicator was quinone imine, generated from 4-chlorophenol and 4-aminoantipyrine by hydrogen peroxide to form a red color quinone imine dye measured at 546 nm [111].

4.9. Cell-Lines

Human cell-lines of hepatocellular carcinoma HepG-2 were cultivated in complete media under standard conditions in 5% CO_2, 37 °C, and were free of mycoplasma [112]. DOX (Adriamycin)-resistant hepatic cell-lines HepG-2/ADR were developed by treating and maintaining the cells in media supplemented with 5 μg/mL DOX in the laboratories of King Abdullah University of Science and Technology, Thuwal at the Core Labs, Biological and Environmental Sciences and Engineering Division (BESE). Using RT-PCR, the development of DOX resistance through P-gp expression as compared to the parent sensitive cell-lines was confirmed. The DOX-free media were applied 7–10 days before conducting any experiments.

4.10. Cytotoxicity and Reversal Assay

Exponentially growing cells (2×10^3 cells/well) were seeded in 96-well plates for MTT cell viability assay [113]. The cells were grown for 24 h, then treated with serial concentrations of *Suaeda vermiculata* aq.-ethanolic extract (up 500 μg/mL) and DOX (100 μg/mL) for 24 h and with MTT solution (0.5 mg/mL) for 4 h. The DMSO dissolved the reaction product (crystals). Absorbance was determined at 570 nm using a SpectraMax M5e Multi-Mode Microplate Reader (Molecular Devices, LLC, USA). The same protocol was used to evaluate the cytotoxicity of the combination of DOX with *Suaeda vermiculata* aq.-ethanolic extract (20 μg/mL, a low, non-toxic dose). All experiments were carried out in triplicate. The IC_{50} values were determined. The Student's *t*-test was applied to determine the degree of significant difference between the sets of data. The relative resistance (RR) for tested samples was calculated using the following equation:

$$\text{Relative resistance} = \frac{\text{IC}_{50}\text{ in resistance cells}}{\text{IC}_{50}\text{ in sensitive cells}} \tag{4}$$

Combination index (CI): The nature of the interaction (synergistic, additive, or antagonistic) between extracts and DOX was determined by the combination index (CI).

$$\text{Combination Index} = \frac{C_{\text{DOX},50}}{\text{IC}_{50,\text{ DOX}}} + \frac{C_{\text{extract},50}}{\text{IC}_{50,\text{extract}}} \tag{5}$$

where $C_{DOX,50}$ is the IC_{50} value for the cytotoxic agent in a two-drug combination, and $C_{extract}$ is the fixed concentration of the extract, herein aq.-ethanolic extract. $IC_{50,DOX}$, and $IC_{50,extract}$ corresponded to the IC_{50} for DOX, and aq.-ethanolic extract alone [114].

4.11. Statistical Analysis

Data were expressed as the mean ± standard error of the mean (SEM). Differences between groups were analyzed using two-way ANOVA followed by a posthoc test using Tukey's multi-group comparison on GraphPad Prism 8.0.2, San Diego, U.S.A. The normality data were checked using GraphPad Prism. The data were considered significant if $p < 0.05$ [115]. The superscripts used to describe significance among the groups in the tables was obtained using Minitab 19.1. A Kolmogorov–Smirnov test was used for the determination of the normality of the data.

5. Conclusions

The current study demonstrated the hepatoprotective potential and ameliorative effects of the *Suaeda vermiculata* aq.-ethanolic extract in CCl_4-induced liver toxicity-bearing rat models. It also demonstrated the extract's effective roles in cell line-based studies in controlling the damages from experimental liver cancer cell-lines. The aq.-ethanolic extract showed no adverse effects on the liver, kidney enzymes, glucose, or on the cholesterol levels in the experimental rats, though it showed a noticeable reducing effect on the triglycerides. The ongoing data on the concentrations of the antioxidant, and comparable liver-protective effects when equated with the silymarin and quercetin, confirmed the beneficial effects of the aq.-ethanolic extract in hepatoprotection, and also the safety of the extract at higher dose administrations. It also suggested that the consumption of the plant material by locals can be regarded as safe. The aq.-ethanolic extract has the potential to offer newer leads, and generate molecular templates upon intensive chemical investigation together with bioassay-guided biological activity pursuance towards finding a potential curative agent as a single molecular entity, or as part of the synergistic action of multiple molecules obtained from the plant's extract for curing liver disorders at par, or higher levels of effectiveness in terms of reduced dose, and speed of recovery in comparison to the available silymarin derivatives, and other hepatoprotective synthetic products utilized for the purpose.

Author Contributions: Conceptualization, S.A.A.M., H.A.M., and R.A.K.; methodology, S.A.A.M., H.A.M., M.S.A., M.Z.E.-R., A.-H.E., S.S., B.G.P., M.J., and H.M.E.; software, S.S., A.-H.E., B.G.P., M.Z.E.-R., M.J., R.A.K., and S.A.A.M.; validation, S.A.A.M., and H.A.M.; formal analysis, S.S., B.G.P., M.Z.E.-R., M.J., S.A.A.M., and H.A.M.; investigation, S.A.A.M., and H.A.M.; resources, H.A.M., R.A.K., S.A.A.M., and H.M.E.; data curation, M.Z.E.-R., A.-H.E., S.S., B.G.P., M.J., S.A.A.M., H.A.M., M.S.A.-O., and H.M.E.; writing—original draft preparation, S.A.A.M., H.A.M., R.A.K., M.S.A., and H.M.E.; writing—review and editing, M.S.A.-O., S.A.A.M., and H.A.M.; visualization, S.A.A.M., H.A.M., and R.A.K.; supervision, S.A.A.M., H.A.M., and R.A.K.; project administration, S.A.A.M., R.A.K., and H.A.M.; funding acquisition, S.A.A.M., H.A.M., and R.A.K. All authors have read and agreed to the published version of the manuscript.

Funding: The authors disclose the receipt of the following financial support for the research and publication of this article. This work was supported by the Qassim University, represented by the Deanship of Scientific Research under the grant (pharmacy-2019-2-2-I-5606) during the academic year 1440 AH/2019 AD.

Acknowledgments: The authors gratefully acknowledge Qassim University, represented by the Deanship of Scientific Research, on the financial support for this research under the number (pharmacy-2019-2-2-I-5606) during the academic year 1440 AH/2019 AD.

Conflicts of Interest: The authors declare no conflict of interest.

References

1. Asrani, S.K.; Devarbhavi, H.; Eaton, J.; Kamath, P.S. Burden of liver diseases in the world. *J. Hepatol.* **2019**, *70*, 151–171. [CrossRef] [PubMed]
2. Everhart, J.E.; Ruhl, C.E. Burden of Digestive Diseases in the United States Part III: Liver, Biliary Tract, and Pancreas. *Gastroenterology* **2009**, *136*, 1134–1144. [CrossRef] [PubMed]

3. Al-Anazi, M.R.; Matou-Nasri, S.; Al-Qahtani, A.A.; Alghamdi, J.; Abdo, A.A.; Sanai, F.M.; Al-Hamoudi, W.K.; Alswat, K.A.; Al-Ashgar, H.I.; Khan, M.Q.; et al. Association between IL-37 gene polymorphisms and risk of HBV-related liver disease in a Saudi Arabian population. *Sci. Rep.* **2019**, *9*. [CrossRef] [PubMed]
4. Al-Johani, J.J.; Aljehani, S.M.; Alzahrani, G.S. Assessment of Knowledge about Liver Cirrhosis among Saudi Population. *Egypt. J. Hosp. Med.* **2018**, *71*, 2443–2446. [CrossRef]
5. Alsabaani, A.; Mahfouz, A.; Awadalla, N.; Musa, M.; Humayed, S. Al Non-Alcoholic Fatty Liver Disease among Type-2 Diabetes Mellitus Patients in Abha City, South Western Saudi Arabia. *Int. J. Environ. Res. Public Health* **2018**, *15*, 2521. [CrossRef]
6. Arhoghro, E.M.; Ekpo, K.E.; Anosike, E.O.; Ibeh, G.O. Effect of Aqueous Extract of Bitter Leaf (Vernonia Amygdalina Del) on Carbon Tetrachloride (CCl4) Induced Liver Damage in Albino Wistar Rats. *Eur. J. Sci. Res.* **2009**, *26*, 122–130.
7. Mohammed, H.A.; Abdel-Aziz, M.M.; Hegazy, M.M. Anti-Oral Pathogens of Tecoma stans (L.) and Cassia javanica (L.) Flower Volatile Oils in Comparison with Chlorhexidine in Accordance with Their Folk Medicinal Uses. *Medicina (B. Aires).* **2019**, *55*, 301. [CrossRef]
8. Jeong, E.; Yang, M.; Kim, N.-H.; Heo, J.-D.; Sung, S. Hepatoprotective effects of Limonium tetragonum, edible medicinal halophyte growing near seashores. *Pharmacogn. Mag.* **2014**, *10*, 563. [CrossRef]
9. Souid, A.; Croce, C.M.D.; Pozzo, L.; Ciardi, M.; Giorgetti, L.; Gervasi, P.G.; Abdelly, C.; Magné, C.; Hamed, K.B.; Longo, V. Antioxidant properties and hepatoprotective effect of the edible halophyte Crithmum maritimum L. against carbon tetrachloride-induced liver injury in rats. *Eur. Food Res. Technol.* **2020**, *246*, 1393–1403. [CrossRef]
10. Gnanadesigan, M.; Ravikumar, S.; Inbaneson, S.J. Hepatoprotective and antioxidant properties of marine halophyte Luminetzera racemosa bark extract in {CCL}4 induced hepatotoxicity. *Asian Pac. J. Trop. Med.* **2011**, *4*, 462–465. [CrossRef]
11. Modarresi, M.; Moradian, F.; Nematzadeh, G. Antioxidant responses of halophyte plant Aeluropus littoralis under long-term salinity stress. *Biologia (Bratisl).* **2014**, *69*. [CrossRef]
12. Boestfleisch, C.; Papenbrock, J. Changes in secondary metabolites in the halophytic putative crop species Crithmum maritimum L., Triglochin maritima L. and Halimione portulacoides (L.) Aellen as reaction to mild salinity. *PLOS ONE* **2017**, *12*, e0176303. [CrossRef] [PubMed]
13. Grigore, M.-N.; Oprica, L. Halophytes as possible source of antioxidant compounds, in a scenario based on threatened agriculture and food crisis. *Iran. J. Public Health* **2015**, *44*, 1153. [PubMed]
14. Soundararajan, P.; Manivannan, A.; Jeong, B.R. Different Antioxidant Defense Systems in Halophytes and Glycophytes to Overcome Salinity Stress. In *Sabkha Ecosystems*; Springer International Publishing: Berlin, Germany, 2019; pp. 335–347.
15. Cichoż-Lach, H. Oxidative stress as a crucial factor in liver diseases. *World J. Gastroenterol.* **2014**, *20*, 8082. [CrossRef]
16. McRae, C.A.; Agarwal, K.; Mutimer, D.; Bassendine, M.F. Hepatitis associated with Chinese herbs. *Eur. J. Gastroenterol. Hepatol.* **2002**, *14*, 559–562. [CrossRef]
17. Pak, E.; Esrason, K.; Wu, V. Hepatotoxicity of herbal remedies: an emerging dilemma. *Prog. Transplant.* **2004**, *14*, 91–96. [CrossRef]
18. Woolbright, B.L.; Williams, C.D.; McGill, M.R.; Jaeschke, H. Liver Toxicity. In *Reference Module in Biomedical Sciences*; Elsevier: Amsterdam, The Netherlands, 2014.
19. Masarone, M.; Rosato, V.; Dallio, M.; Gravina, A.G.; Aglitti, A.; Loguercio, C.; Federico, A.; Persico, M. Role of Oxidative Stress in Pathophysiology of Nonalcoholic Fatty Liver Disease. *Oxid. Med. Cell. Longev.* **2018**, *2018*, 1–14. [CrossRef]
20. Ostapowicz, G. Results of a Prospective Study of Acute Liver Failure at 17 Tertiary Care Centers in the United States. *Ann. Intern. Med.* **2002**, *137*, 947. [CrossRef]
21. Yen, F.-L.; Wu, T.-H.; Lin, L.-T.; Cham, T.-M.; Lin, C.-C. Naringenin-Loaded Nanoparticles Improve the Physicochemical Properties and the Hepatoprotective Effects of Naringenin in Orally-Administered Rats with {CCl}4-Induced Acute Liver Failure. *Pharm. Res.* **2008**, *26*, 893–902. [CrossRef]
22. Chávez, E.; Reyes-Gordillo, K.; Segovia, J.; Shibayama, M.; Tsutsumi, V.; Vergara, P.; Moreno, M.G.; Muriel, P. Resveratrol prevents fibrosis, NF-kappaB activation and TGF-beta increases induced by chronic CCl4 treatment in rats. *J. Appl. Toxicol.* **2007**, *28*, 35–43. [CrossRef]

23. Casas-Grajales, S. Antioxidants in liver health. *World J. Gastrointest. Pharmacol. Ther.* **2015**, *6*, 59. [CrossRef] [PubMed]
24. Gillessen, A.; Schmidt, H.H.-J. Silymarin as supportive treatment in liver diseases: A narrative review. *Adv. Ther.* **2020**, 1–23. [CrossRef] [PubMed]
25. Nwidu, L.L.; Oboma, Y.I.; Elmorsy, E.; Carter, W.G. Hepatoprotective effect of hydromethanolic leaf extract of Musanga cecropioides (Urticaceae) on carbon tetrachloride-induced liver injury and oxidative stress. *J. Taibah Univ. Med. Sci.* **2018**, *13*, 344–354. [CrossRef] [PubMed]
26. Peng, C.; Zhou, Z.; Li, J.; Luo, Y.; Zhou, Y.; Ke, X.; Huang, K. {CCl}4-Induced Liver Injury Was Ameliorated by Qi-Ge Decoction through the Antioxidant Pathway. *Evidence-Based Complement. Altern. Med.* **2019**, *2019*, 1–12. [CrossRef]
27. Chao, W.-W.; Chen, S.-J.; Peng, H.-C.; Liao, J.-W.; Chou, S.-T. Antioxidant Activity of Graptopetalum paraguayense E. Walther Leaf Extract Counteracts Oxidative Stress Induced by Ethanol and Carbon Tetrachloride Co-Induced Hepatotoxicity in Rats. *Antioxidants* **2019**, *8*, 251. [CrossRef]
28. Nwidu, L.L.; Elmorsy, E.; Oboma, Y.I.; Carter, W.G. Hepatoprotective and antioxidant activities of Spondias mombin leaf and stem extracts against carbon tetrachloride-induced hepatotoxicity. *J. Taibah Univ. Med. Sci.* **2018**, *13*, 262–271. [CrossRef]
29. Kuriakose, G.C.; Kurup, M.G. Antioxidant and antihepatotoxic effect of Spirulina laxissima against carbon tetrachloride induced hepatotoxicity in rats. *Food Funct.* **2011**, *2*, 190. [CrossRef]
30. Abdou, R.H.; Basha, W.A.; Khalil, W.F. Subacute Toxicity of Nerium oleander Ethanolic Extract in Mice. *Toxicol. Res.* **2019**, *35*, 233–239. [CrossRef]
31. Alhdad, G.M.; Seal, C.E.; Al-Azzawi, M.J.; Flowers, T.J. The effect of combined salinity and waterlogging on the halophyte Suaeda maritima: the role of antioxidants. *Environ. Exp. Bot.* **2013**, *87*, 120–125. [CrossRef]
32. Pareek, A.; Godavarthi, A.; Issarani, R.; Nagori, B.P. Antioxidant and hepatoprotective activity of Fagonia schweinfurthii (Hadidi) Hadidi extract in carbon tetrachloride induced hepatotoxicity in {HepG}2 cell line and rats. *J. Ethnopharmacol.* **2013**, *150*, 973–981. [CrossRef]
33. Dahiru, D.; Obidoa, O. Evaluation of the antioxidant effects of Ziziphus mauritiana leaf extracts against chronic ethanol-induced hepatotoxicity in rat liver. *African J. Tradit. Complement. Altern. Med.* **2008**, *5*. [CrossRef] [PubMed]
34. Li, A.-N.; Li, S.; Zhang, Y.-J.; Xu, X.-R.; Chen, Y.-M.; Li, H.-B. Resources and Biological Activities of Natural Polyphenols. *Nutrients* **2014**, *6*, 6020–6047. [CrossRef] [PubMed]
35. Chu, X.; Wang, H.; Jiang, Y.; Zhang, Y.; Bao, Y.; Zhang, X.; Zhang, J.; Guo, H.; Yang, F.; Luan, Y.; et al. Ameliorative effects of tannic acid on carbon tetrachloride-induced liver fibrosis in~vivo and in~vitro. *J. Pharmacol. Sci.* **2016**, *130*, 15–23. [CrossRef]
36. Chen, L.; Hu, C.; Hood, M.; Kan, J.; Gan, X.; Zhang, X.; Zhang, Y.; Du, J. An Integrated Approach Exploring the Synergistic Mechanism of Herbal Pairs in a Botanical Dietary Supplement: A Case Study of a Liver Protection Health Food. *Int. J. Genomics* **2020**, *2020*, 1–14. [CrossRef] [PubMed]
37. Rasool, M.; Iqbal, J.; Malik, A.; Ramzan, H.S.; Qureshi, M.S.; Asif, M.; Qazi, M.H.; Kamal, M.A.; Chaudhary, A.G.A.; Al-Qahtani, M.H.; et al. Hepatoprotective Effects of Silybum marianum (Silymarin) and Glycyrrhiza glabra (Glycyrrhizin) in Combination: A Possible Synergy. *Evidence-Based Complement. Altern. Med.* **2014**, *2014*, 1–9. [CrossRef]
38. Ming, Z.; Fan, Y.; Yang, X.; Lautt, W.W. Synergistic protection by S-adenosylmethionine with vitamins C and E on liver injury induced by thioacetamide in rats. *Free Radic. Biol. Med.* **2006**, *40*, 617–624. [CrossRef] [PubMed]
39. Singh, H.; Prakash, A.; Kalia, A.N.; Majeed, A.B.A. Synergistic hepatoprotective potential of ethanolic extract of Solanum xanthocarpum and Juniperus communis against paracetamol and azithromycin induced liver injury in rats. *J. Tradit. Complement. Med.* **2016**, *6*, 370–376. [CrossRef]
40. Yadav, N.P.; Pal, A.; Shanker, K.; Bawankule, D.U.; Gupta, A.K.; Darokar, M.P.; Khanuja, S.P.S. Synergistic effect of silymarin and standardized extract of Phyllanthus amarus against {CCl}4-induced hepatotoxicity in Rattus norvegicus. *Phytomedicine* **2008**, *15*, 1053–1061. [CrossRef]
41. Chen, P.; Wang, Y.; Chen, L.; Jiang, W.; Niu, Y.; Shao, Q.; Gao, L.; Zhao, Q.; Yan, L.; Wang, S. Comparison of the anti-inflammatory active constituents and hepatotoxic pyrrolizidine alkaloids in two Senecio plants and their preparations by LC–UV and LC–MS. *J. Pharm. Biomed. Anal.* **2015**, *115*, 260–271. [CrossRef]

42. Quirós-Guerrero, L.; Albertazzi, F.; Araya-Valverde, E.; Romero, R.M.; Villalobos, H.; Poveda, L.; Chavarría, M.; Tamayo-Castillo, G. Phenolic variation among Chamaecrista nictitans subspecies and varieties revealed through UPLC-ESI (-)-MS/MS chemical fingerprinting. *Metabolomics* **2019**, *15*, 14. [CrossRef]
43. Mohammed, H.A. The Valuable Impacts of Halophytic Genus Suaeda; Nutritional, Chemical, and Biological Values. *Med. Chem.* **2020**. [CrossRef] [PubMed]
44. Hou, M.-L.; Chang, L.-W.; Lin, C.-H.; Lin, L.-C.; Tsai, T.-H. Determination of bioactive components in Chinese herbal formulae and pharmacokinetics of rhein in rats by UPLC-MS/MS. *Molecules* **2014**, *19*, 4058–4075. [CrossRef] [PubMed]
45. Dai, W.; Li, H.; Cong, W.; Zhao, W.; Wang, Q. Two New Eremophilane Sesquiterpenes and One New Resorcinol from Ligularia knorringiana. *Nat. Prod. Commun.* **2016**, *11*, 1934578X1601100201. [CrossRef]
46. Madan, S.; Pannakal, S.T.; Ganapaty, S.; Singh, G.N.; Kumar, Y. Phenolic glucosides from Flacourtia indica. *Nat. Prod. Commun.* **2009**, *4*, 1934578X0900400313. [CrossRef]
47. n-Hexadecanoic Acid. Available online: https://webbook.nist.gov/cgi/cbook.cgi?ID=C57103&Mask=200 (accessed on 14 September 2020).
48. Mohammed, H.A.; Al-Omar, M.S.; El-Readi, M.Z.; Alhowail, A.H.; Aldubayan, M.A.; Abdellatif, A.A.H. Formulation of Ethyl Cellulose Microparticles Incorporated Pheophytin A Isolated from Suaeda vermiculata for Antioxidant and Cytotoxic Activities. *Molecules* **2019**, *24*, 1501. [CrossRef] [PubMed]
49. OECD. *OECD Test No. 425: Acute Oral Toxicity: Up-and-Down Procedure*; OECD: Paris, France, 2008.
50. Jun, H.-I.; Wiesenborn, D.P.; Kim, Y.-S. Antioxidant Activity of Phenolic Compounds from Canola (Brassica napus) Seed. *Food Sci. Biotechnol* **2014**, *23*, 1753–1760. [CrossRef]
51. Kaur, D.; Kaur, A.; Arora, S. Delineation of attenuation of oxidative stress and mutagenic stress by Murraya exotica L. leaves. *Springerplus* **2016**, *5*. [CrossRef] [PubMed]
52. Routray, W.; Orsat, V. Microwave-assisted extraction of flavonoids: a review. *Food Bioprocess Technol.* **2012**, *5*, 409–424. [CrossRef]
53. Stalikas, C.D. Extraction, separation, and detection methods for phenolic acids and flavonoids. *J. Sep. Sci.* **2007**, *30*, 3268–3295. [CrossRef]
54. Mohammed, H.A.; Al-Omar, M.S.; Aly, M.S.A.; Hegazy, M.M. Essential Oil Constituents and Biological Activities of the Halophytic Plants, Suaeda Vermiculata Forssk and Salsola Cyclophylla Bakera Growing in Saudi Arabia. *J. Essent. Oil Bear. Plants* **2019**, 1–12. [CrossRef]
55. Mohammadi Motamed, S.; Bush, S.; Hosseini Rouzbahani, S.; Karimi, S.; Mohammadipour, N. Total phenolic and flavonoid contents and antioxidant activity of four medicinal plants from Hormozgan province, Iran. *Res. J. Pharmacogn.* **2016**, *3*, 17–26.
56. Popoviç, M.; Kaurinoviç, B.; Jakovljeviç, V.; Mimica-Dukic, N.; Bursaç, M.; Popovic, M. PARSLEY ON OXIDATIVE STRESS 717 Effect of Parsley (Petroselinum crispum (Mill.) Nym. ex A.W. Hill, Apiaceae) Extracts on some Biochemical Parameters of Oxidative Stress in Mice treated with CCl 4. *Phytother. Res* **2007**, *21*, 717–723. [CrossRef] [PubMed]
57. Hafez, M.M.; Hamed, S.S.; El-Khadragy, M.F.; Hassan, Z.K.; Al Rejaie, S.S.; Sayed-Ahmed, M.M.; Al-Harbi, N.O.; Al-Hosaini, K.A.; Al-Harbi, M.M.; Alhoshani, A.R.; et al. Effect of ginseng extract on the TGF-β1 signaling pathway in CCl4-induced liver fibrosis in rats. *BMC Complement. Altern. Med.* **2017**, *17*, 45. [CrossRef]
58. Araya, E.M.; Adamu, B.A.; Periasamy, G.; Sintayehu, B.; Gebrelibanos Hiben, M. In vivo hepatoprotective and In vitro radical scavenging activities of Cucumis ficifolius A. rich root extract. *J. Ethnopharmacol.* **2019**, *242*, 112031. [CrossRef] [PubMed]
59. Jabit, M.L.; Wahyuni, F.S.; Khalid, R.; Israf, D.A.; Shaari, K.; Lajis, N.H.; Stanslas, J. Cytotoxic and nitric oxide inhibitory activities of methanol extracts of *Garcinia* species. *Pharm. Biol.* **2009**, *47*, 1019–1026. [CrossRef]
60. Li, S.; Zhao, Q.; Wang, B.; Yuan, S.; Wang, X.; Li, K. Quercetin reversed MDR in breast cancer cells through down-regulating P-gp expression and eliminating cancer stem cells mediated by YB-1 nuclear translocation. *Phyther. Res.* **2018**, *32*, 1530–1536. [CrossRef]
61. Borska, S.; Chmielewska, M.; Wysocka, T.; Drag-Zalesinska, M.; Zabel, M.; Dziegiel, P. In vitro effect of quercetin on human gastric carcinoma: Targeting cancer cells death and MDR. *Food Chem. Toxicol.* **2012**, *50*, 3375–3383. [CrossRef]

62. Kumar, M.; Sharma, G.; Misra, C.; Kumar, R.; Singh, B.; Katare, O.P.; Raza, K. N-desmethyl tamoxifen and quercetin-loaded multiwalled CNTs: A synergistic approach to overcome MDR in cancer cells. *Mater. Sci. Eng. C* **2018**, *89*, 274–282. [CrossRef]
63. Mohana, S.; Ganesan, M.; Rajendra Prasad, N.; Ananthakrishnan, D.; Velmurugan, D. Flavonoids modulate multidrug resistance through wnt signaling in P-glycoprotein overexpressing cell lines. *BMC Cancer* **2018**, *18*, 1168. [CrossRef]
64. Mohana, S.; Ganesan, M.; Agilan, B.; Karthikeyan, R.; Srithar, G.; Beaulah Mary, R.; Ananthakrishnan, D.; Velmurugan, D.; Rajendra Prasad, N.; Ambudkar, S.V. Screening dietary flavonoids for the reversal of P-glycoprotein-mediated multidrug resistance in cancer. *Mol. Biosyst.* **2016**, *12*, 2458–2470. [CrossRef]
65. Qureshi, N.N.; Kuchekar, B.S.; Logade, N.A.; Haleem, M.A. Antioxidant and hepatoprotective activity of Cordia macleodii leaves. *Saudi Pharm. J.* **2009**, *17*, 299–302. [CrossRef] [PubMed]
66. Delgado-Montemayor, C.; Cordero-Pérez, P.; Salazar-Aranda, R.; Waksman-Minsky, N. Models of hepatoprotective activity assessment. *Med. Univ.* **2015**, *17*, 222–228. [CrossRef]
67. Ritesh, K.R.; Suganya, A.; Dileepkumar, H.V.; Rajashekar, Y.; Shivanandappa, T. A single acute hepatotoxic dose of CCl4 causes oxidative stress in the rat brain. *Toxicol. Rep.* **2015**, *2*, 891–895. [CrossRef] [PubMed]
68. Hogade, M.G.; Patil, K.S.; Wadkar, G.H.; Mathapati, S.S.; Dhumal, P.B. Hepatoprotective activity of Morus alba (Linn.) leaves extract against carbon tetrachloride induced hepatotoxicity in rats. *African J. Pharm. Pharmacol.* **2010**, *4*, 731–734.
69. Boll, M.; Weber, L.W.D.; Becker, E.; Stampfl, A. Mechanism of carbon tetrachloride-induced hepatotoxicity. Hepatocellular damage by reactive carbon tetrachloride metabolites. *Zeitschrift Naturforsch. Sect. C J. Biosci.* **2001**, *56*, 649–659. [CrossRef]
70. Recknagel, R.O.; Glende, E.A.; Dolak, J.A.; Waller, R.L. Mechanisms of carbon tetrachloride toxicity. *Pharmacol. Ther.* **1989**, *43*, 139–154. [CrossRef]
71. He, Q.; Guoming, S.; Liu, K.; Zhang, F.; Jiang, Y.; Gao, J.; Liu, L.; Jiang, Z.; Jin, M.; Xie, H. Sex-specific reference intervals of hematologic and biochemical analytes in Sprague-Dawley rats using the nonparametric rank percentile method. *PLoS ONE* **2017**, *12*, e0189837. [CrossRef]
72. Sallie, R.; Michael Tredger, J.; Williams, R. Drugs and the liver part 1: Testing liver function. *Biopharm. Drug Dispos.* **1991**, *12*, 251–259. [CrossRef]
73. Zhou, D.; Ruan, J.; Cai, Y.; Xiong, Z.; Fu, W.; Wei, A. Antioxidant and hepatoprotective activity of ethanol extract of Arachniodes exilis (Hance) Ching. *J. Ethnopharmacol.* **2010**, *129*, 232–237. [CrossRef]
74. Alhassan, A.J.; Sule, M.J.; Aliyu, S.A.; Aliyu, M.D. Ideal hepatotoxicity model in rats using Carbon Tetrachloride (CCl4). *Bayero J. Pure Appl. Sci.* **2011**, *2*. [CrossRef]
75. Singh, A.; Bhat, T.; Sharma, O. Clinical Biochemistry of Hepatotoxicity. *J. Clin. Toxicol.* **2011**, *04*. [CrossRef]
76. Shaker, E.; Mahmoud, H.; Mnaa, S. Silymarin, the antioxidant component and Silybum marianum extracts prevent liver damage. *Food Chem. Toxicol.* **2010**, *48*, 803–806. [CrossRef] [PubMed]
77. Shalan, M.G.; Mostafa, M.S.; Hassouna, M.M.; El-Nabi, S.E.H.; El-Refaie, A. Amelioration of lead toxicity on rat liver with Vitamin C and silymarin supplements. *Toxicology* **2005**, *206*, 1–15. [CrossRef] [PubMed]
78. Flora, K.; Hahn, M.; Rosen, H.; Benner, K. Milk thistle (Silybum marianum) for the therapy of liver disease. *Am. J. Gastroenterol.* **1998**, *93*, 139–143. [CrossRef]
79. Miltonprabu, S.; Tomczyk, M.; Skalicka-Woźniak, K.; Rastrelli, L.; Daglia, M.; Nabavi, S.F.; Alavian, S.M.; Nabavi, S.M. Hepatoprotective effect of quercetin: From chemistry to medicine. *Food Chem. Toxicol.* **2017**, *108*, 365–374. [CrossRef]
80. Elsawaf, A.; Faras, A. El Hepatoprotective activity of quercetin against paracetamol-induced liver toxicity in rats. *Tanta Med. J.* **2017**, *45*, 92. [CrossRef]
81. Oueslati, S.; Trabelsi, N.; Boulaaba, M.; Legault, J.; Abdelly, C.; Ksouri, R. Evaluation of antioxidant activities of the edible and medicinal Suaeda species and related phenolic compounds. *Ind. Crops Prod.* **2012**, *36*, 513–518. [CrossRef]
82. Mohammed, H.A. Behavioral Evaluation for Aqueous and Ethanol Extracts of Suaeda vermiculata Forssk. *Cent. Nerv. Syst. Agents Med. Chem.* **2020**. [CrossRef]
83. Benwahhoud, M.; Jouad, H.; Eddouks, M.; Lyoussi, B. Hypoglycemic effect of Suaeda fruticosa in streptozotocin-induced diabetic rats. *J. Ethnopharmacol.* **2001**, *76*, 35–38. [CrossRef]

84. Issa, M.; Agbon, A.; Balogun, S.; Mahdi, O.; Bobbo, K.; Ayegbusi, F. Hepatoprotective effect of methanol fruit pulp extract of *Musa paradisiaca* on carbon tetrachloride-induced liver toxicity in Wistar rats. *J. Exp. Clin. Anat.* **2018**, *17*, 1. [CrossRef]
85. Khan, R.A.; Khan, M.R.; Sahreen, S. CCl4-induced hepatotoxicity: Protective effect of rutin on p53, CYP2E1 and the antioxidative status in rat. *BMC Complement. Altern. Med.* **2012**, *12*, 178. [CrossRef] [PubMed]
86. Rahman, T.M.; Hodgson, H.J.F. Animal models of acute hepatic failure. *Int. J. Exp. Pathol.* **2000**, *81*, 145–157. [CrossRef] [PubMed]
87. Lima, L.C.D.; Miranda, A.S.; Ferreira, R.N.; Rachid, M.A.; e Silva, A.C.S. Hepatic encephalopathy: Lessons from preclinical studies. *World J. Hepatol.* **2019**, *11*, 173–185. [CrossRef] [PubMed]
88. Tuñón, M.J.; Alvarez, M.; Culebras, J.M.; González-Gallego, J. An overview of animal models for investigating the pathogenesis and therapeutic strategies in acute hepatic failure. *World J. Gastroenterol.* **2009**, *15*, 3086–3098. [CrossRef] [PubMed]
89. Chen, Y.; Chou, P.; Hsu, C.; Hung, J.; Wu, Y.; Lin, J. Fermented Citrus Lemon Reduces Liver Injury Induced by Carbon Tetrachloride in Rats. *Evid. Based. Complement. Alternat. Med.* **2018**, *2018*. [CrossRef]
90. Fouad, D.; Badr, A.; Attia, H.A. Hepatoprotective activity of raspberry ketone is mediated: Via inhibition of the NF-κB/TNF-α/caspase axis and mitochondrial apoptosis in chemically induced acute liver injury. *Toxicol. Res. (Camb).* **2019**, *8*, 663–676. [CrossRef]
91. Kaur, C.; Kapoor, H.C. Anti-oxidant activity and total phenolic content of some Asian vegetables. *Int. J. Food Sci. Technol.* **2002**, *37*, 153–161. [CrossRef]
92. Chantiratikul, P.; Meechai, P.; Nakbanpotec, W. Antioxidant Activities and Phenolic Contents of Extracts from Salvinia molesta and Eichornia crassipes. *Res. J. Biol. Sci.* **2009**, *4*, 1113–1117.
93. Mohammed, H.A.; Al-Omar, M.S.; Mohammed, S.A.A.; Aly, M.S.A.; Alsuqub, A.N.A.; Khan, R.A. Drying Induced Impact on Composition and Oil Quality of Rosemary Herb, Rosmarinus Officinalis Linn. *Molecules* **2020**, *25*, 2830. [CrossRef]
94. Floegel, A.; Kim, D.O.; Chung, S.J.; Koo, S.I.; Chun, O.K. Comparison of ABTS/DPPH assays to measure antioxidant capacity in popular antioxidant-rich US foods. *J. Food Compos. Anal.* **2011**, *24*, 1043–1048. [CrossRef]
95. Organization for Economic Cooperation and Development (OECD) Guidelines for Testing of Chemicals, Guideline 425: Acute Oral Toxicity–Up-And-Down Procedure. *Organ. Econ. Coop. Dev. Paris* **2006**.
96. Ali, E.; Mohamed, H.; Lim, C.P.; Saad-Ebrika, O.; Asmawi, M.Z.; Sadikun, A.; Yam, F.; Mohamed, E.A.H.; Lim, C.P.; Ebrika, O.S.; et al. Toxicity evaluation of a standardised 50% ethanol extract of Orthosiphon stamineus. *J. Ethnopharmacol.* **2011**, *133*, 358–363. [CrossRef]
97. Faul, F.; Erdfelder, E.; Lang, A.G.; Buchner, A. G*Power 3: A flexible statistical power analysis program for the social, behavioral, and biomedical sciences. In *Proceedings of the Behavior Research Methods*; Psychonomic Society Inc.: Madisson, WI, USA, 2007; Volume 39, pp. 175–191.
98. Kilkenny, C.; Browne, W.J.; Cuthill, I.C.; Emerson, M.; Altman, D.G. Improving bioscience research reporting: The arrive guidelines for reporting animal research. *PLoS Biol.* **2010**, *8*. [CrossRef]
99. Wang, L.; Huang, Q.H.; Li, Y.X.; Huang, Y.F.; Xie, J.H.; Xu, L.Q.; Dou, Y.X.; Su, Z.R.; Zeng, H.F.; Chen, J.N. Protective effects of Silymarin on triptolide-induced acute hepatotoxicity in rats. *Mol. Med. Rep.* **2018**, *17*, 789–800. [CrossRef] [PubMed]
100. Freitag, A.F.; Cardia, G.F.E.; Da Rocha, B.A.; Aguiar, R.P.; Silva-Comar, F.M.D.S.; Spironello, R.A.; Grespan, R.; Caparroz-Assef, S.M.; Bersani-Amado, C.A.; Cuman, R.K.N. Hepatoprotective effect of silymarin (silybum marianum) on hepatotoxicity induced by acetaminophen in spontaneously hypertensive rats. *Evidence-based Complement. Altern. Med.* **2015**, *2015*. [CrossRef] [PubMed]
101. Li, C.C.; Hsiang, C.Y.; Wu, S.L.; Ho, T.Y. Identification of novel mechanisms of silymarin on the carbon tetrachloride-induced liver fibrosis in mice by nuclear factor-κB bioluminescent imaging-guided transcriptomic analysis. *Food Chem. Toxicol.* **2012**, *50*, 1568–1575. [CrossRef]
102. Hassan, A. moniem S.; Abo El-Ela, F.I.; Abdel-Aziz, A.M. Investigating the potential protective effects of natural product quercetin against imidacloprid-induced biochemical toxicity and DNA damage in adults rats. *Toxicol. Reports* **2019**, *6*, 727–735. [CrossRef]
103. El-Nekeety, A.A.; Abdel-Azeim, S.H.; Hassan, A.M.; Hassan, N.S.; Aly, S.E.; Abdel-Wahhab, M.A. Quercetin inhibits the cytotoxicity and oxidative stress in liver of rats fed aflatoxin-contaminated diet. *Toxicol. Reports* **2014**, *1*, 319–329. [CrossRef]

104. Szende, B.; Timár, F.; Hargitai, B. Olive oil decreases liver damage in rats caused by carbon tetrachloride (CCl4). *Exp. Toxicol. Pathol.* **1994**, *46*, 355–359. [CrossRef]
105. Bhat, S.H.; Shrivastava, R.; Malla, M.Y.; Iqbal, M.; Anjum, T.; Mir, A.H.; Sharma, M. Hepatoprotective activity of Argemone Mexicana Linn against toxic effects of carbon tetrachloride in rats. *World J. Pharm. Res.* **2014**, *3*, 4037–4048.
106. Meharie, B.G.; Amare, G.G.; Belayneh, Y.M. Evaluation of Hepatoprotective Activity of the Crude Extract and Solvent Fractions of Clutia abyssinica (Euphorbiaceae) Leaf Against CCl 4-Induced Hepatotoxicity in Mice. *J. Exp. Pharmacol.* **2020**, *12*, 137–150. [CrossRef]
107. Thefeld, W.; Hoffmeister, H.; Busch, E.-W.; Koller, P.; Vollmar, J. Referenzwerte für die Bestimmungen der Transaminasen GOT und GPT sowie der alkalischen Phosphatase im Serum mit optimierten Standardmethoden. *DMW Dtsch. Medizinische Wochenschrift* **1974**, *99*, 343–351. [CrossRef] [PubMed]
108. Weichselbaum, C.T.E. An accurate and rapid method for the determination of proteins in small amounts of blood serum and plasma. *Am. J. Clin. Pathol.* **1946**, *10*, 40–49. [CrossRef]
109. Fabiny, D.L.; Ertingshausen, G. Automated Reaction-Rate Method for Determination of Serum Creatinine with the CentrifiChem. *Clin. Chem.* **1971**, *17*. [CrossRef]
110. Trinder, P. Determination of blood glucose using 4-amino phenazone as oxygen acceptor. *J. Clin. Pathol.* **1969**, *22*, 246. [CrossRef] [PubMed]
111. Trinder, P. Enzymatic calorimetric determination of triglycerides by GOP-PAP method. *Ann. ClinBiochem* **1969**, *6*, 24–27.
112. Carmichael, J.; DeGraff, W.G.; Gazdar, A.F.; Minna, J.D.; Mitchell, J.B. Evaluation of a Tetrazolium-based Semiautomated Colorimetric Assay: Assessment of Chemosensitivity Testing. *Cancer Res.* **1987**, *47*.
113. Saito, J.; Okamura, A.; Takeuchi, K.; Hanioka, K.; Okada, A.; Ohata, T. High content analysis assay for prediction of human hepatotoxicity in HepaRG and HepG2 cells. *Toxicol. Vitr.* **2016**, *33*, 63–70. [CrossRef] [PubMed]
114. Danhof, M.; de Jongh, J.; De Lange, E.C.M.; Della Pasqua, O.; Ploeger, B.A.; Voskuyl, R.A. Mechanism-Based Pharmacokinetic-Pharmacodynamic Modeling: Biophase Distribution, Receptor Theory, and Dynamical Systems Analysis. *Annu. Rev. Pharmacol. Toxicol.* **2007**, *47*, 357–400. [CrossRef]
115. Kirkwood, B.R.; Sterne, J.A.C. *Essential Medical Statistics*; Essentials; Wiley: Hoboken, NJ, USA, 2010; ISBN 9781444392845.

Article

Chemical Composition of Cuticular Waxes and Pigments and Morphology of Leaves of *Quercus suber* Trees of Different Provenance

Rita Simões [1], Ana Rodrigues [1], Suzana Ferreira-Dias [2], Isabel Miranda [1,*] and Helena Pereira [1]

[1] Centro de Estudos Florestais (CEF), Instituto Superior de Agronomia, Universidade de Lisboa, Tapada da Ajuda, 1349-017 Lisboa, Portugal; simoes.silva2012@hotmail.com (R.S.); anadr@isa.ulisboa.pt (A.R.); hpereira@isa.ulisboa.pt (H.P.)

[2] Linking Landscape, Environment, Agriculture and Food (LEAF), Instituto Superior de Agronomia, Universidade de Lisboa, Tapada da Ajuda, 1349-017 Lisboa, Portugal; suzanafdias@isa.ulisboa.pt

* Correspondence: imiranda@isa.ulisboa.pt

Received: 30 July 2020; Accepted: 3 September 2020; Published: 9 September 2020

Abstract: The chemical composition of cuticular waxes and pigments and the morphological features of cork oak (*Quercus suber*) leaves were determined for six samples with seeds of different geographical origins covering the natural distribution of the species. The leaves of all samples exhibited a hard texture and oval shape with a dark green colour on the hairless adaxial surface, while the abaxial surface was lighter, with numerous stomata and densely covered with trichomes in the form of stellate multicellular hairs. The results suggest an adaptive role of leaf features among samples of different provenance and the potential role of such variability in dealing with varying temperatures and rainfall regimes through local adaptation and phenotypic plasticity, as was seen in the trial site, since no significant differences in leaf traits among the various specimens were found, for example, specific leaf area 55.6–67.8 cm^2/g, leaf size 4.6–6.8 cm^2 and photosynthetic pigment (total chlorophyll, 31.8–40.4 $\mu g/cm^2$). The leaves showed a substantial cuticular wax layer (154.3–235.1 $\mu g/cm^2$) composed predominantly of triterpenes and aliphatic compounds (61–72% and 17–23% of the identified compounds, respectively) that contributed to forming a nearly impermeable membrane that helps the plant cope with drought conditions. These characteristics are related to the species and did not differ among trees of different seed origin. The major identified compound was lupeol, indicating that cork oak leaves may be considered as a potential source of this bioactive compound.

Keywords: coak oak; chlorophyll; terpenes; lupeol; cuticular permeance

---

1. Introduction

The extracellular surface of plant leaves is covered by a hydrophobic layer known as the cuticle, which is composed primarily of cutin, an insoluble polyester of hydroxyfatty acids, glycerol and complex mixtures of waxes that are deposited within and above the structural cutin matrix [1–3]. Cuticular waxes vary between plant species and are composed of different organic solvent-soluble lipids, consisting of very long chain fatty acids and their derivatives including aldehydes, primary alcohols and alkanes; in some species they also contain significant amounts of pentacyclic triterpenoids [4–6]. They show a high degree of crystallinity, low chemical reactivity and hydrophobicity [7,8]. The composition of cuticular waxes is influenced by plant genotype, leaf side and age. Waxes differ in their functions and responses to biotic and abiotic environments.

Cuticular waxes confer properties upon the leaf surface such as nonstomatal water loss and gas exchange control, as well as protection from the external environment. Wax deposition is often a response to water stress and therefore stress resistant plants adapted to arid conditions often have thicker wax layers than those from more temperate locations or those that are susceptible to stress [9–14]. Content is not the only factor determining the properties and function of the waxy layer; knowledge of its composition is also important. However, the relationship among cuticular wax content and composition, leaf morphology and the response to different environmental conditions remains unclear.

Cork oak (*Quercus suber* L.) is an evergreen sclerophyllous tree species distributed in the western Mediterranean Basin, with a natural range including Algeria, France, Italy, Morocco, Portugal, Spain and Tunisia. Cork oak forests play an important ecological role in terms of carbon sequestration, soil protection, hydrological cycle regulation and ecosystem sustainability, while they are also of critical economic importance due to their production of cork that feeds a dedicated industrial chain [15]. However, climate change scenarios involving enhanced water deficits in the Mediterranean region threatens the ecosystem, even if cork oak shows considerable adaptability to environmental conditions and its genetic variability allows it to cope with climatic variation [16].

An approach to simulating a shift of climatic characteristics is provenance analysis, which enables an assessment of the phenotypic response of various populations and the identification of populations growing well and resistant to adverse environmental factors through the use of indicators or estimators of plant responses to environmental factors, that is, morphological and physiological characteristics [17]. Since the natural distribution of *Q. suber* encompasses significant environmental and geographic gradients, it can be expected that long-term natural selection and genetic drift have resulted in a high level of genetic variation among populations concerning adaptive traits such as leaf morphological area and the accumulation of cuticular wax on leaf surfaces.

However, little information is available for *Q. suber* regarding natural variation in leaf morphology, cuticular features and phytochemical data, despite the fact that it has been demonstrated that the species possesses several mechanisms for drought tolerance including small and thick leaves, sclerophyllicity and deep tap-rooting [18–21]. Only one work was found in the literature regarding the cuticular waxes of *Q. suber* leaves [22]; for a sampling of two trees in spring and summer, the report described a composition including mainly n-alkyl esters (25–45% of the wax extract) and alkanols (18–50%), as well as alkanes, alkanals and alkanoic acids. The paper also noted significant variability in the content of cuticular waxes.

The aim of this study is to develop further knowledge of the features of *Q. suber* leaves and their variability, mainly in terms of cuticular wax quantity and composition, by using samples with different seed geographical origins that represent the area of the species' natural distribution. Sampling was performed in the Cork Oak Provenance and Progeny Trial established in Portugal by the European Network for the Evaluation of Genetic Resources of Cork Oak for Appropriate Use in Breeding and Gene Conservation Strategies [23–25]. Leaves from trees with six different geographical origins (from Portugal, Spain, Italy, France, Morocco and Tunisia) were collected and the morphological parameters, pigments and cuticular wax characteristics were analysed, looking for variations which may be related to genetic or edaphoclimatic issues. This will be the first detailed study of the leaf features and cuticular wax composition of a broad sample of cork oak of different provenance, thereby providing insights into the species' variability.

2. Results

The leaves from all of the *Quercus suber* samples showed a hard texture and oval shape with a dark green colour on the adaxial surface and a lack of trichomes, while the abaxial surface was lighter with numerous stomata and was densely covered with epidermal hairs (trichomas) in the form of stellate multicellular hairs (Figure 1). Figure 2 shows an exemplary *Q. suber* leaf cross-section observed using optical microscopy, showing the thick cuticular membrane covering the adaxial leaf side deposited on the epidermal cell layer with between-cell indentations. The trichomes and stomata are clearly visible on the abaxial side.

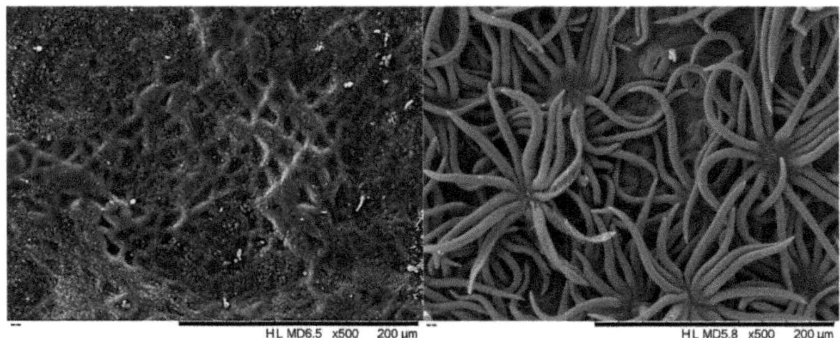

Figure 1. Scanning electron micrographs of the adaxial (**left**) and abaxial (**right**) surface of *Quercus suber* leaves.

The average leaf features of the six cork oak provenances are presented in Table 1. Leaf size (LS) differed significantly among provenances ($p = 0.018$), ranging between 4.6 cm^2 in the French sample (small leaves) and 6.8 cm^2 in the Italian sample (larger leaves), with a mean coefficient of variation of leaf area of 0.39 (between 0.23 and 0.42). The mean specific leaf area (SLA) was 64.1 cm^2/g, ranging between 55.6 cm^2/g (Spanish sample) and 67.8 cm^2/g (Italian sample), with a coefficient of variation in all provenances between 0.15 and 0.9 and lacking statistically significant differences ($p = 0.243$).

Table 1. Morphological and physiological characterization of leaves of cork oak (*Quercus suber*) from six provenances. The provenances are Portugal (PT35), Spain (ES11), Italy (IT3), France (FR3), Morocco (MA27) and Tunisia (TU32). Mean and standard deviation of 20 leaves per provenance.

Leaf Features/Provenance Code	PT35	ES11	IT3	FR3	MA27	TU32
Leaf size (LS; cm^2)	6.2 ± 2.1a	5.1 ± 2.1b	6.8 ± 2.7a	4.6 ± 1.9b	6.4± 2.8a	5.6 ± 2.1b
Specific leaf area (SLA; cm^2/g)	61.5 ± 18.5a	55.6 ± 17.6a	67.8 ± 22.7a	63.1 ± 19.4 a	64.7 ± 27.9a	63.7 ± 28.9a
Sclerophylly index (g/dm^2)	0.9 ± 0.3a	1.0 ± 0.4a	0.8 ± 0.3a	0.9 ± 0.4a	0.9 ± 0.4a	0.9 ± 0.4a
Cuticular wax content (μg/cm^2)	212.5 ± 40.3a	182.2 ± 60.2a	201.0 ± 60.4a	231.5 ± 64.4a	173.0 ± 80.6a	136.4 ± 45.9b
Cuticular wax content (mg/g)	25.9 ± 0.5a	20.7 ± 0.5a	25.7 ± 0.6a	29.7 ± 0.7a	24.2 ± 0.4a	21.6 ± 0.4a
Photosynthetic Pigments						
Chlorophyll a (μg/cm^2)	26.6 ± 1.6a	23.9 ± 3.2b	24.1 ± 2.0b	21.1 ± 1.4b	26.4 ± 2.9a	24.8 ± 0.7b
Chlorophyll b (μg/cm^2)	13.3 ± 2.1a	12.0 ± 1.9a	12.1 ± 1.4a	10.7 ± 1.4b	14.0 ± 1.8a	12.7 ± 0.5a
Chlorophyll a/b ratio	2.0 ± 0.3a	2.0 ± 0.1a	2.0 ± 0.1a	2.0 ± 0.1a	1.9 ± 0.1a	2.0 ± 0.1a
Total chlorophyll (μg/cm^2)	40.0 ± 3.4a	35.9 ± 5.0b	36.3 ± 3.4b	31.8 ± 1.8b	40.4 ± 4.5 a	37.5 ± 1.0a
Total chlorophyll (mg/g)	2.5 ± 0.2a	2.1 ± 0.3a	2.3 ±0.2a	2.1 ± 0.1a	2.5 ±0.3a	2.7 ± 0.1a
Total carotenoids (μg/cm^2)	8.4 ± 0.4 a	7.9 ± 0.7b	7.5 ± 0.8b	7.2 ± 0.1b	8.7 ± 0.1a	8.3 ± 0.2a
Chlorophyll/carotenoids ratio	0.2 ± 0.03a	0.3 ± 0.03a	0.3 ± 0.04a	0.3 ± 0.01a	0.2 ± 0.03a	0.2 ± 0.01a
Cuticular Water Permeability						
$T_{cut} \times 10^{-4}$ (g/m^{2}s)	5.5	2.7	4.9	3.3	4.8	4.0
$P \times 10^{-5}$ (m/s)	2.4	1.2	2.1	1.4	2.1	1.7

(T_{cut}) cuticular transpiration rate at maximum driving force. (P) cuticular permeance. Means in each row followed by the same letter are not significantly different at $p < 0.05$.

The total chlorophyll content expressed on a unit per leaf area basis or per unit dry weight ranged between 40.4 μg/cm^2 (2.53 mg/g) (Moroccan provenance) and 31.8 μg/cm^2 (2.06 mg/g) (French provenance) with similar ratios of chlorophyll *a* to chlorophyll *b* between provenances (1.9 to 2.0). The total carotenoids concentration ranged from 8.7 μg/cm^2 (Moroccan provenance) to 7.2 μg/cm^2 (French provenance). Data on content of total chlorophyll and carotenoids within provenances expressed by mg/g and μg/cm^2 differed significantly ($p < 0.001$). The average values of total chlorophylls and carotenoids in leaves of Portuguese, Moroccan and Tunisian provenance were higher than the average values of other provenances.

The average extracted yield of cuticular wax was 189.4 μg/cm^2, with some variation among provenances, although differences were not statistically significant ($p = 0.449$), ranging from the lowest value of 136.4 μg/cm^2 (Tunisia) to the highest value of 231.5 μg/cm^2 (France), with coefficients of variation between 0.27 and 0.43. On a dry leaf mass basis, the cuticular waxes represented on average 24.6 mg/g, ranging between 20.7 mg/g (Spain) and 29.7 mg/g (France). Based on the driving force, the cuticular transpiration rate of the adaxial leaf surface ranged from 2.7×10^{-4} g/m^{2}s to 5.5×10^{-4} g/m^{2}s and the cuticular water permeance ranged from 1.6×10^{-5} m/s to 2.5×10^{-5} m/s.

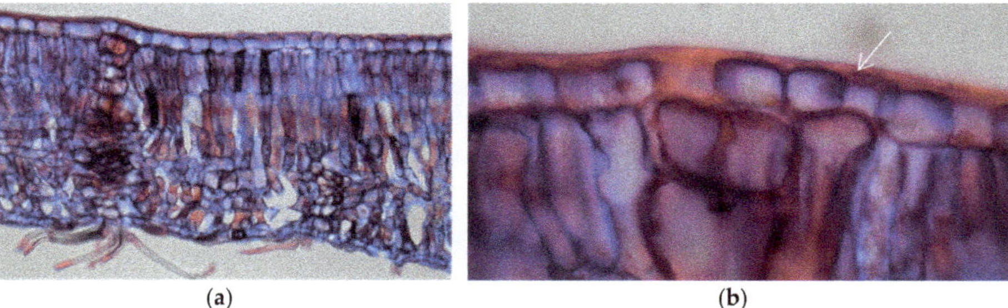

Figure 2. Optical microscopy photographs of a cross-section of a *Quercus suber* leaf (**a**) and an enlargement of the adaxial side showing the cuticular membrane (**b**). Arrows indicate the cuticular structures covering the epidermal cell layer.

The results obtained for the cuticular wax composition of the *Q. suber* leaves are summarized in Table 2 regarding chemical families and detailed for the six provenances of *Q. suber* in Table 3. On average, the cuticular wax extracts were composed of terpenes (60.1 % of peak area), fatty acids (12.7%) and alkanes (6.1%), with a small proportion of alkanols (1.1%) and aromatics compounds (1.9%). The average proportion of identified compounds in the chromatograms was 91.2%.

Table 2. Chemical class composition of the cuticular wax of leaves from six *Quercus suber* provenances, as a percentage of the total peak areas in the GC-MS chromatograms. Mean and standard deviation of 12 samples.

Family	Percent of Total Compounds
n-Alkanols	1.10 ± 0.56
n-Alkanes	6.07 ± 0.72
Fatty acids	12.73 ± 2.41
Aromatic compounds	1.94 ± 0.44
Sterols	5.32 ± 1.23
Terpenes	60.05 ± 3.65
Others	3.76 ± 1.98
Total identified compounds	91.15 ± 3.95

A principal component analysis (PCA) and cluster analysis (CA) were performed on the seven chemical families of epicuticular wax compounds (alkanols, alkanes, fatty acids, aromatic compounds, sterols, terpenes and others) found in 12 *Q. suber* leave samples obtained from six provenances. From the analysis of the eigenvalues of the principal components and of the scree plot, the initial hyperspace defined by seven dimensions (one for each group of compounds) was reduced to a plane defined by the first two principal components. This plane explained 56.5% of the information contained in the original data. Figure 3 shows the projections of both compound groups and samples on this plane. The first dimension (Factor 1) is positively correlated with *n*-alkanols and *n*-alkanes, while terpenes are correlated with the negative side of this axis. This indicates that samples ES-1 and Pt-2 show the highest contents of *n*-alkanols and *n*-alkanes, while samples FR-2 and MA-1 show the highest amounts of terpenes and lowest amounts of *n*-alkanols and *n*-alkanes. The second principal component (Factor 2) is highly correlated with aromatic compounds, which increases along this axis. Therefore, samples ES-2 and TU-1 show the highest content of aromatic compounds. From the analysis of Figure 3, it is difficult to identify groups of samples. Since Figure 3 shows the projections of samples on a plane and not in their original position in the hyperspace, a cluster analysis was performed on the same data to identify groups of samples (Figure 4). Again, no clear group separation was observed by CA. This confirms that there is a similar pattern for the chemical profile of the cuticular waxes of all provenances regarding chemical families.

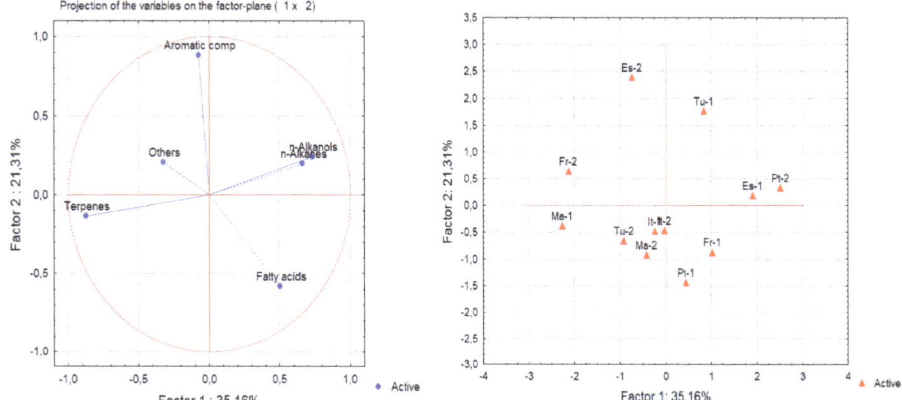

Figure 3. Principal component analysis (PCA) of the chemical classes of epicuticular wax compounds from leaves of 12 samples from six provenances: (left) score plots of the original variables; (right) sample plot (ES—Spain; Fr—France; IT—Italy; MA—Morocco; PT—Portugal; TU—Tunisia).

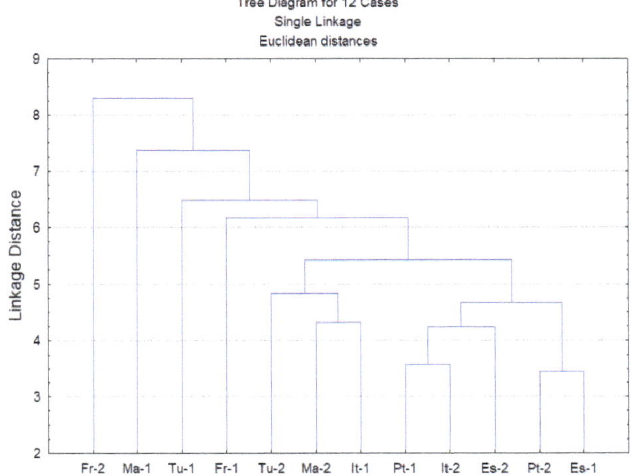

Figure 4. Dendrogram obtained by cluster analysis of the family classes of epicuticular wax compounds from 12 leave samples from six provenances (ES—Spain; FR—France; IT—Italy; MA—Morocco; PT—Portugal; TU—Tunisia).

The chemical compounds identified in the cuticular waxes of leaves from the six *Q. suber* provenances are presented in Table 3 in proportion of the total chromatogram area and are grouped by chemical family.

Table 3. Chemical composition (% of all chromatogram peak areas) of the cuticular waxes of leaves from six *Quercus suber* provenances (mean of two trees per provenance). The provenances are Portugal (PT35), Spain (ES11), Italy (IT3), France (FR3), Morocco (MA27) and Tunisia (TU32).

Compound/Provenance Code	PT35	ES11	IT3	FR3	MA27	TU32	Whole Leaves
n-Alkanols	**2.03**	**1.33**	**0.90**	**0.93**	**0.35**	**1.05**	**1.10 ± 0.56**
hexadecan-1-ol ($C_{16}OH$)	0.21	0.17	0.08	0.00	0.05	0.08	0.10 ± 0.08
octadecan-1-ol ($C_{18}OH$)	0.00	0.17	0.03	0.00	0.00	0.00	0.03 ± 0.07
tretracosan-1-ol ($C_{24}OH$)	1.43	0.75	0.53	0.66	0.15	0.60	0.69 ± 0.42
octacosan-1-ol ($C_{28}OH$)	0.39	0.24	0.27	0.27	0.15	0.37	0.28 ± 0.09
n-Alkanes	**5.78**	**5.92**	**5.57**	**6.09**	**5.55**	**7.46**	**6.06 ± 0.72**
n-pentadecane (C_{15})	0.10	0.06	0.04	0.04	0.08	0.15	0.08 ± 0.04
n-Pentacosane (C_{25})	0.38	0.33	0.36	0.42	0.42	0.44	0.39 ± 0.04
n-heptacosane (C_{27})	0.20	0.28	0.28	0.21	0.10	0.27	0.22 ± 0.07
n-octacosane (C_{28})	3.65	3.73	3.07	3.96	3.56	4.38	3.73 ± 0.44
1-nonacosene ($C_{29(2)}$)	0.06	0.39	0.00	0.10	0.19	0.25	0.17 ± 0.14
n-nonacosane (C_{29})	0.47	0.10	0.33	0.49	0.29	0.40	0.35 ± 0.14
n-hentriacontane (C_{31})	0.93	1.04	1.49	0.87	0.92	1.57	1.14 ± 0.31
Fatty Acids	**15.22**	**11.70**	**14.19**	**14.93**	**11.12**	**9.24**	**12.73 ± 2.41**
Saturated	**14.57**	**11.50**	**13.50**	**13.48**	**9.93**	**8.59**	**11.93 ± 2.33**
octanoic acid ($C_{8:0}$)	0.13	0.17	0.13	0.04	0.08	0.13	0.11 ± 0.05
nonanoic acid ($C_{9:0}$)	0.05	0.21	0.28	0.17	0.03	0.06	0.13 ± 0.10
nonadioc acid ($C_{9(2)}$)	0.04	0.11	0.05	0.40	0.33	1.07	0.33 ± 0.39
decanoic acid ($C_{10:0}$)	0.11	0.09	0.22	0.00	0.05	0.09	0.09 ± 0.07
dodecanoic acid ($C_{12:0}$)	0.13	0.16	0.21	0.12	0.10	0.12	0.14 ± 0.04
tetradecanoic acid ($C_{14:0}$)	0.25	0.15	0.35	0.27	0.28	0.31	0.27 ± 0.07
hexadecanoic acid ($C_{16:0}$)	1.04	1.37	1.01	1.66	1.95	1.62	1.44 ± 0.37
octadecanoic acid ($C_{18:0}$)	0.21	0.23	0.17	0.31	0.27	0.29	0.25 ± 0.05
eicosanoic acid ($C_{20:0}$)	0.35	0.23	0.26	0.43	0.28	0.29	0.31 ± 0.07
docosanoic acid ($C_{22:0}$)	0.67	0.44	0.39	0.69	0.38	0.45	0.50 ± 0.14
tetracoscanoic acid ($C_{24:0}$)	0.55	0.32	0.40	0.82	0.20	0.23	0.42 ± 0.23
hexacosanoic acid ($C_{26:0}$)	0.88	0.53	0.71	0.87	0.64	0.27	0.65 ± 0.23
octacosanoic acid ($C_{28:0}$)	4.03	2.51	3.19	3.39	2.38	1.22	2.79 ± 0.98
triacontanoic acid ($C_{30:0}$)	6.13	4.98	6.13	4.31	2.96	2.44	4.49 ± 1.56
unsaturated	**0.65**	**0.20**	**0.69**	**1.45**	**1.19**	**0.65**	**0.81 ± 0.45**
9,12-octadecadienoic acid ($C_{18:2}$)	0.12	0.06	0.20	0.39	0.37	0.17	0.22 ± 0.13
9,12,15-octadecatrienoic acid ($C_{18:3}$)	0.53	0.14	0.49	1.06	0.82	0.48	0.59 ± 0.32
Aromatic Compounds	**1.39**	**2.57**	**1.77**	**1.77**	**1.41**	**1.74**	**1.78 ± 0.43**
benzoic acid	0.06	0.13	0.11	0.06	0.03	0.08	0.08 ± 0.04
4-hydroxybenzaldehyde	0.11	0.23	0.07	0.17	0.14	0.12	0.14 ± 0.06
vanillin	0.01	0.06	0.06	0.00	0.11	0.03	0.05 ± 0.04
4-(2-hydroxyethy) phenol	0.13	0.20	0.16	0.48	0.35	0.07	0.23 ± 0.15
methyl p-coumarate, trans	0.06	0.16	0.00	0.16	0.12	0.12	0.10 ± 0.06
quercetin/myricetin	0.19	0.05	0.27	0.05	0.21	0.08	0.14 ± 0.09
kaempferol	0.13	0.30	0.55	0.00	0.09	0.49	0.26 ± 0.22
pinoresinol	0.33	0.46	0.32	0.62	0.04	0.50	0.38 ± 0.20
2,3-dihydrobenzofuran	0.26	0.17	0.12	0.17	0.07	0.18	0.16 ± 0.06
hexadecy-(E)-p-coumarate	0.38	0.99	0.23	0.24	0.32	0.25	0.40 ± 0.29
Sterols	**5.11**	**3.16**	**5.14**	**6.36**	**6.62**	**5.54**	**5.32 ± 1.23**
β-tocopherol/γ-tocopherol	0.34	0.42	1.07	0.22	0.29	0.62	0.49 ± 0.31
α-tocopherol	0.58	0.22	0.41	0.66	0.69	0.16	0.45 ± 0.23
β-sitosterol	4.19	2.52	3.66	5.48	5.64	4.76	4.38 ± 1.18
Terpenes	**55.64**	**57.53**	**62.56**	**58.24**	**64.61**	**60.70**	**59.88 ± 3.36**
diterpenes	**1.00**	**0.29**	**1.21**	**1.59**	**1.30**	**0.55**	**0.99 ± 0.49**
phytol	0.46	0.21	1.00	1.18	0.86	0.38	0.68 ± 0.39
α-tocopherolquinone	0.54	0.08	0.21	0.41	0.44	0.17	0.31 ± 0.18
pentacyclic triterpenes	**54.64**	**57.24**	**61.35**	**56.65**	**64.31**	**60.15**	**59.06 ± 3.54**
α-amyrin	1.01	0.92	0.97	1.00	1.09	0.73	0.95 ± 0.12
β-amyrin	5.14	4.82	4.63	5.68	8.26	5.30	5.64 ± 1.34
lupeol	37.76	34.37	38.59	32.91	37.22	40.49	36.89 ± 2.79
friedooleanan-3-ol	3.95	7.59	8.25	7.29	7.85	5.82	6.79 ± 1.62
friedelin	2.94	5.39	3.79	4.60	4.86	3.12	4.12 ± 0.99
betulin	1.42	1.40	1.44	2.02	2.25	1.75	1.71 ± 0.36
oleanolic acid	0.43	0.63	0.61	0.92	0.42	0.76	0.63 ± 0.19
betulinic acid	1.27	1.31	1.78	1.73	1.94	1.83	1.64 ± 0.28
ursolic acid	0.72	0.81	1.29	0.50	0.42	0.35	0.68 ± 0.35
Others	**3.48**	**3.59**	**2.66**	**8.21**	**3.29**	**3.48**	**4.12 ± 2.03**
myo-inositol/ Scyllo-inositol	1.55	1.30	1.60	3.46	1.06	1.01	1.66 ± 0.91
D (-) fructofuranose	0.39	0.33	0.00	0.79	0.39	0.38	0.38 ± 0.25
D (-) fructopyranose	0.43	0.36	0.00	0.79	0.28	0.51	0.40 ± 0.26
glycerol	0.46	0.75	0.45	0.56	0.59	0.29	0.52 ± 0.16
6,10,14 trimethtylpentadecan-2-one	0.09	0.52	0.23	1.91	0.40	0.82	0.66 ± 0.66
1-Linoleylglycerol	0.08	0.01	0.14	0.25	0.27	0.11	0.14 ± 0.10
Total identified compounds	**88.65**	**85.80**	**92.77**	**96.52**	**93.94**	**89.21**	**91.15 ± 3.95**

Bold is highlighted the family of compounds.

Triperpenes represented between 54.7% (Portuguese provenance) and 64.3% (Moroccan provenance) of the total peak area. Lupeol was the major triterpene found in all the provenances (33–40% of the compounds). Other abundant triterpenes were friedooleanan-3-ol (6.8% on average), friedelin (4.1%) and β-amyrin (5.7%), followed by smaller amounts of betulin and betulinic acid. Oleanolic and ursolic acids were also identified but in small amounts.

Sterols were present in amounts varying between 3.2% and 6.6% in the Spanish and Maroccan provenances, respectively. β-Sitosterol was the major compound from this family in all cases, while tocopherols (α, β and γ tocopherol) were present in smaller amounts.

The aliphatic compounds constituted an abundant group including fatty acids (9.2–15.2% of the compounds), n-alkanes (5.6–7.5%) and primary alcohols (0.4–2.0%), which together accounted for 17–23% of the total compounds. Among the fatty acids, C30, C28 and C16 acids were predominant, representing 36.8%, 22.7% and 12.9% of the total fatty acids, respectively. The homologous series of saturated acids with an even number of C was present from C8 to C30. Unsaturated acids only represented on average 6.9% of the total fatty acids. The chain length of the n-alkanes varied between C15 and C31, with n-octacosane (C28) and hentriacontane (C31) as the major compounds. Alkanols comprised only 0.4–1.3% of all compounds, with tretracosanol ($C_{24}OH$) and octacosanol ($C_{28}OH$) as the major compounds in the wax extracts. Glycerol was present in minor amounts, as well as one monoacylglycerol. Aromatic compounds comprised 1.4–2.6% of all compounds, with kaempferol, pinoresinol and hexadecy-(E)-p-coumarate as the major compounds.

3. Discussion

The study was conducted in a *Q. suber* provenance trial where 35 cork oak provenances covering the natural distribution range are represented, which is of utmost importance to assess the effect of seed geographic origin on adaptive traits [26,27]. In the present study, six provenances were selected that cover the broad geographic area of cork oak distribution (Portugal, Spain, France and Italy in southern Europe and Morocco and Tunisia in northern Africa).

The cork oak leaves in the trees of all the provenances showed a clear schlerophytic character (Figures 1 and 2) and their morphological features (Table 1) were within the range of values reported for the species. Leaf sizes (4.6–6.8 cm^2) were similar to those reported by Mediavilla et al. [28] for *Q. suber* leaves taken from different orientations in the canopy (5.5–7.4 cm^2) and the 7.1 cm^2 reported by Prats et al. [20]. The specific leaf area values (55.6–67.8 cm^2/g) were of the same order of magnitude as those obtained in adult leaves of *Q. suber* growing under contrasting environments and located in different positions and orientations of the canopy (50.0 to 126.0 cm^2/g) [28–32].

When comparing the six cork oak provenances, two sets could be defined in relation to leaf area (Table 1), with ES, FR and TU provenances characterized by smaller leaves and PT, IT and MA provenances characterized by larger leaves. Specific leaf area (SLA) for the six provenances showed a low coefficient of variation (CV 0.14), suggesting that they developed a similar degree of leaf sclerophylly. This finding is in line with most of the few available studies on *Q. suber*. Rzigui et al. [31] observed that there was no difference in SLA values between two provenances in Tunisia (Gafour and Feija) with contrasting environments (125 and 126 cm^2/g) and Daoudi et al. [33] reported that SLA did not differ significantly between three humid, semi-arid and sub-humid provenances in Algeria, although SLA was lower under conditions of water deficiency. Lobo-do-Vale et al. [32] reported an adaptive response to drought with an SLA reduction (85 cm^2/g in a mild year and 65 cm^2/g in a dry year), while Aranda et al. [34] observed that, while irrigation had no significant effect, SLA decreased with increasing irradiance, thereby improving the potential for carbon uptake relative to transpiration water loss. SLA plays an important role in linking plant carbon and water cycles and is sensitive to environmental change [35,36], with leaf traits being influenced by a combination of species, climate and soil factors [37–39].

In the present study, the observed amounts of the total chlorophyll, expressed per unit of leaf area or per unit of dry weight (Table 1), were consistent with those reported in the available studies for *Q. suber*, for example, the results of Ramírez-Valiente et al. [30] for leaves of open-pollinated trees from populations in Morocco, Portugal and Spain (2.68 mg/g, 2.64 mg/g and 2.74 mg/g respectively) and 55.7 μg/cm^2 and 56.0 μg/cm^2 total chlorophyll in sun and shade leaves of 40-year-old *Q. suber* trees [40]. Mediavilla et al. [28] also observed that the chlorophyll content differed among canopy orientations, with lower values on the west side (80 vs. 110 μg/cm^2), probably due to a direct effect of excess radiation at the time of the day with highest leaf temperatures and lowest water potential on the west facing leaves.

A substantial average wax layer of 189.4 μg/cm^2 (or 24.6 mg/g) covered the leaves of the six *Q. suber* provenances (Table 1) without any provenance effect. The values are slightly above the 125 μg/cm^2 reported by Martins et al. [22] for the cuticular wax of young leaves of *Q. suber*. A higher amount of cuticular wax is observed in *Q. suber* leaves than in other *Quercus* species. For instance, young leaves of *Q. ilex* (holm oak, a persistent leaf oak) have 71 μg/cm^2 of cuticular wax [22], *Q. robur* (pedunculate oak, deciduous tree growing in cooler climates with less hydric stress) has 59 μg/cm^2 [41] and *Q. petraea* (sessile oak) has 101.5–134.5 μg/ cm^2 [42]. Similar values were also reported for *Q. polymorpha* (Mexican white oak) leaves, with a wax layer of 199.4 μg/cm^2 [43].

The high wax content of the cork oak leaves suggests that cuticular characteristics may be associated with adaptation to the local environmental conditions of high temperatures and water deficit. Accumulation of cuticular wax is considered an important strategy against drought in many plant species, providing an essential barrier to protect plants from drought stress [44,45]. In this study, the cork oak did not differ in their responses to environmental conditions by provenance, indicating that the accumulation of cuticular waxes on the leaves, albeit depending on the species, mainly responds to the specific environmental conditions.

Despite the role of the cuticle as a water barrier, there is still a movement of water through the cuticle between the outer cell wall of the epidermis and the surrounding atmosphere, giving the cuticle some permeability [46]. This movement is based on a simple diffusion process along a gradient of the water's chemical potential with the water molecules being absorbed at one interface, following a random path through the cuticle in a mainly lipophilic chemical environment and desorbed at the other interface [47]. The degree to which cuticles transferred water was measured in the adaxial leaf surface of the *Q. suber* samples, with an average cuticular permeance of 1.8×10^{-5} m/s (Table 2). No information is available on the cuticular permeance of *Q. suber* leaves but comparison with other *Quercus* species shows that this value is quite low, in agreement with its adaptation to hot and dry environments. The available values for cuticular permeances range from 3.6×10^{-5} m/s for *Q. ilex*, 7.3×10^{-5} m/s for *Q. rubra*, 10×10^{-5} m/s for *Q. coccifera*, to 27×10^{-5} m/s for *Q. sessiliflora* [47]. The permeances determined with isolated cuticular membranes were from 3.8×10^{-5} m/s to 7.4×10^{-5} m/s for *Q. petraea* [48].

Our results on the chemical composition of the dichloromethane extract of *Q. suber* leaves (Tables 2 and 3) show that the majority of compounds are pentacyclic triterpenoids (mainly lupeol) and that long-chain aliphatic components are mainly fatty acids (mainly in C30, C28 and C16). The extraction method may have a determining role in the extent of compound solubilization, namely, on the amount of terpenes in relation to aliphatic compounds. Since the objective in the present study was to extract the total wax layer, including epicuticular and intracuticular waxes from both sides of the leaf, an intensive 6 h extraction with dichloromethane was conducted.

The only work to our knowledge regarding the composition of cuticular waxes in young *Q. suber* leaves [22] applied a quick surface extraction method by dipping and shaking the leaves for a few seconds with chloroform and this explains the compositional differences to our results. In fact, this extract contained only small amounts of triterpenoids (triterpenone, friedelin), although the aliphatic composition was similar to that found in the present study: 4–27% *n*-alkanes; 18–50% even chain amphiphilic compounds (*n*-alkan-1-ols); up to 25% *n*-alkanals; <5% *n*-alkanoic acids; and 25–45% *n*-alkyl esters [22]. It was reported that triterpenoids are located almost exclusively in the intracuticular wax compartment and therefore require a more intensive extraction procedure [49,50].

The cuticular wax composition appears to be related to the species, since no statistical significant differences were found between cork oak provenances and there was no clustering of provenances by chemical families, based on PCA and cluster analysis (Figures 3 and 4).

Q. suber leaf wax contains the same lipid classes as *Q. ilex* and *Q. robur* but with different distribution patterns. In a *Q. ilex* leaf, the most abundant wax components are C22 and C24 *n*-alkanoic acids (38%) and *n*-alkan-1-ols (43–54%), with small amounts of the triterpenols α- and β-amyrin [30]. In a *Q. robur* leaf wax, the dominating classes are alcohols (about 70% of the wax with chain lengths ranging from C16 to C34, with tetracosanol as a main component), fatty acids (20% of the wax, especially with chain length of C14 and C22), aldehydes (28%, of the wax with chain lengths from C20 to C32 with C26 and C28 as the main components) and several triperpenoids (8% of the wax, taraxerol, β-amyrin, α-amyrin and lupeol were identified) [41,51].

A comparison with *Fagus* and *Castanea* species, also members of the Fagaceae family to which *Q. suber* belongs, shows that the epicuticular leaf wax of *Castanea sativa* consists of a homologous series of wax lipids (wax esters, aldehydes, primary alcohols and fatty acids) and large amounts of triterpenoids (α- and β-amyrin and lupeol) [52], while that of *Fagus sylvatica* contains only wax lipids, without any triterpenoids [53].

The cuticular wax of *Q. suber* leaves is rich in triterpenoids that can be obtained as an extract after solubilization. Over the last three decades, extensive research has revealed important pharmacological applications of triterpenoids with potential uses in new functional foods, drugs, cosmetics and healthcare products. Lupeol was studied for the treatment of various diseases, including skin wounds and various medicinal properties of lupeol have been reported, including anti-inflammatory, antioxidant, anti-diabetic and anti-mutagenic effects [54–56].

Quercus suber leaves are a promising and highly available source of triterpenes, since large quantities of leaves are generated each year from silvicultural practices (e.g., pruning), which can be used to produce chemicals using environmentally friendly extraction processes [57]. Our results estimate a potential extraction yield per kg of dry leaves of *Q. suber* of 15 g triterpenes, of which 9 g is lupeol.

4. Material and Methods

4.1. Sampling

The study was carried out on a provenance trial of *Quercus suber* L. at Herdade do Monte Fava, Santiago do Cacém, near Setúbal, in central Portugal (38° 00′ N, 08°07′ W, altitude 79 m). The site has a Mediterranean climate, with hot and dry summers (total year rainfall, 556.6 mm; summer accumulated rainfall, 19.4 mm; annual average temperature, 15.8 °C; average minimum temperature of the coldest month, 4.3 °C; average maximum temperature of the hottest month, 31.3 °C) and the soil has a sandy texture [26]. The trial was established in March 1998, as part of an international network on cork oak genetic resources funded by an EU Concerted Action Fair 202 [23]. The plant material included in this field trial resulted from a seed collection conducted during the autumn of 1996 from 35 cork oak populations that covered the species' natural range; seedlings were raised from these seeds with a common protocol in one nursery, planted in the trial field and the trees were allowed to grow until leaves were sampled [23,26].

The sampling carried out for the present study included 21-year-old trees from six provenances: Portugal, Spain, Italy, France, Morocco and Tunisia. Detailed information on the location and climate data from the original seed collection sites of the studied provenances is given in Table 4 as well as that of the trees' growing site. Leaves were randomly sampled from two trees of each provenance in March 2019. The leaves were collected from different branches on the southern exposed side of the crown, in the lower part of the canopy up to a height of approximately 2 m, making up a total sample per tree of about 100 leaves.

Table 4. Characterization of the trial site and identification of the *Quercus suber* provenances with collection site location, T_m annual average air temperature (°C) and the long-term annual average precipitation (PPT, mm) (adapted from Varela [23]) and characterization of the provenance trial of *Quercus suber* (Herdade do Monte Fava).

Provenance Code	Country of Seed Collection	Latitude	Longitude	Altitude (m)	T_m (°C)	PPT (mm)
PT35	Portugal, Ermidas do Sado	38° 00′ N	8° 70′ W	79	15.8	557
ES11	Spain, Alpujarras	36° 50′ N	3° 18′ W	1300	13.0	742
IT13	Italy, Puglia	40° 34′ N	17° 40′ E	45	16.6	588
FR3	France, Landes	43° 45′ N	1° 20′ W	20	12.3	870
MA27	Morocco, Rif Occidental I.2	35° 07′ N	5° 16′ W	300	n.a.	1280
TU32	Tunisia, Mekna	36° 57′ N	8° 51′ W	12	17.9	948
Monte da Fava	Portugal, Santiago do Cacém	38° 00′ N	8° 07′ W	79	15.8	557

n.a.: not available.

4.2. Morphological Variables

The morphological variables were measured in 20 leaves that were randomly selected from leaves sampled for each of the two trees of each provenance. Leaves were digitalized and analysed using WinSEEDLE^{TM} 2011. The leaves were oven-dried at 70 °C to a constant mass and the total dry mass per leaf was determined. Specific leaf area (SLA, cm^2/g) was calculated as the ratio between the measured leaf area and dry weight; this was used as an indirect index of sclerophylly. The sclerophylly index was calculated as (IE) = leaf dry mass (g)/2 × leaf area (dm^2) and defines sclerophylly as IE > 0.6 and mesophylly as IE < 0.6, according to Rizzini [58].

4.3. Determination of Leaf Pigment Contents

Chlorophyll *a* and *b* were determined in six disks of known area taken from leaves of each of the two trees per provenance. The tissue samples were homogenized in the dark during 24 to 36 h, with 3 mL of dimethylformamide solution. Chlorophyll *a* and *b*, carotenoids and total chlorophyll were quantified in the supernatant phase by using a spectrophotometer and readings were taken at 663.8, 646.8 and 480.0 nm, respectively, according to the methodology described by Lichthenthaler [59] and using the equations and specific absorption in the wavelength reported by Wellburn [60]. The leaf pigments chlorophyll a and b, total chlorophyll and carotenoids were expressed as $\mu g/cm^2$ of leaf area.

4.4. Cuticle Permeability Assay

The cuticular permeance was determined by the measurement of water loss through the adaxial, astomatous leaf surface. Leaf samples were rehydrated by submerging the petioles of the leaves in water at room temperature for 12 h. The leaf petioles and the abaxial leaf surface were sealed with paraffin wax to ensure that water transpiration only occurred via the stomata-free adaxial leaf surface. Leaf weight was measured immediately afterwards. Leaf samples were then placed in boxes over silica gel and placed in an incubator with control the surrounding temperature (25 °C) and weighed repeatedly over a 4 h period. The transpiration rate (flux of water vapor; J, in g m^{-2} s^{-1}) was obtained from the change in fresh weight of the samples (ΔW, in g) over time (Δt, in s) and surface area (A, in m^2):

$$J = \frac{\Delta W}{\Delta t \times A}.$$

The leaf surface area (*A* leaf) of the adaxial surface was obtained by scanning the leaf surface.

The cuticular permeance P (in m s^{-1}) and minimum leaf conductance g_{min} can be obtained from the transpiration rate J and the driving force Δc of water across the cuticle according to

$$P = g_{min} = \frac{J}{c_{WV}\left(a_{leaf} - a_{air}\right)} = \frac{J}{c_{WV}\,\Delta c}.$$

The driving force (Δc) is the difference between the concentration of water vapour in the leaf interior and the surrounding atmosphere (g m^{-3}), resulting in permeability parameters in m s^{-1}. Since during cuticular transpiration the diffusion of water occurs in the solid phase of the cuticle, atmospheric pressure has no influence and, consequently, using the concentration-based driving force is appropriate. The resulting permeability parameters were described in m s^{-1}. The water activity of leaf samples (a_{leaf}) was assumed to be unity [49]. The humidity of the air was controlled by silica gel, resulting in a water activity (a_{air}) close to zero. Therefore, the driving force for water loss through transpiration was identical to the density of water vapour at saturation in the air (c_{WV} at 25 °C was 23.07 g m^{-3}; [61]) [14,62].

4.5. Extraction of Cuticular Waxes

Cuticular wax was extracted from whole fresh leaves with dichloromethane in a Soxhlet apparatus over 6 h using a sample of 20 leaves per tree. After extraction, the solvent was evaporated. This extraction method yields a total wax mixture containing both epicuticular and intracuticular waxes that cover both sides of the leaf.

The amount of soluble cuticular lipids was determined from the mass difference of the extracted leaves after drying at 105 °C and was expressed on a leaf surface area and dry weight basis (the ratio between wax in μg and the two-sided leaf surface area in cm^2, obtained by digitalization).

4.6. Cuticular Wax Composition

The cuticular wax (obtained as dichloromethane extracts from 20 leaves per tree) was analyzed using gas-chromatography mass spectrometry (GC-MS). Two mg of each leaf extract was taken and derivatized in 120 μL of pyridine; the compounds with hydroxyl and carboxyl groups were trimethylsilylated into trimethylsilyl (TMS) ethers and esters, respectively, by adding 80 μL of bis(trimethylsily)-trifluoroacetamide (BSTFA). The reaction mixture was heated at 60 °C for 30 min in an oven. The derivatized extracts (1 μL) were immediately analyzed by GC-MS (EMIS, Agilent 5973 MSD, Palo Alto, CA, USA), with an ionization energy of 70 eV and the MS source was kept at 220 °C under the following GC conditions: Zebron 7HGG015-02 column (30 m, 0.25 mm; ID, 0.1 μm film thickness), with injector at 280 °C. The column temperature was initially held at 50 °C for 1 min, raised to 150 °C at a rate of 10 °C min^{-1}, then to 300 °C at 4 °C min^{-1}, to 370 °C at 5 °C min^{-1} and at 8 °C min^{-1} until it reached 380 °C; this was followed by an isothermal period of 5 min. Compounds were identified as TMS derivatives by matching their mass spectra with a GC-MS spectral library (Wiley, NIST) and by comparing their fragmentation profiles with published data [63,64]. Two replicates were made per extract.

4.7. Structural and Anatomical Observations

For scanning electron microscopy (SEM), small pieces of fresh- and air-dried leaves were fixed to sample holders and observed with a scanning electron microscope (TM 3030 Plus Hitachi).

The cuticular membrane was observed in a thin leaf cross section by optical microscopy. The leaves were impregnated with DP 1500 polyethylene glycol and cross sections of approximately 10 μm thickness were cut with a rotary microtome (Medite M530). The sections were stained with safranin and astral blue and were mounted in Euparal. Observations were made using a light microscope (Leica DM LA) and the photomicrographs were taken with a Nikon Microphot-FXA.

4.8. Statistical Analyses

To compare leaf functional traits among provenances, one-way analysis of variance (ANOVA) was performed. Duncan's post-hoc tests were used to analyse pairwise differences between provenances. Statistical significance was set at $p < 0.05$. All statistical analyses were performed using the Sigmaplot® (Version 11.0, Systat Software, Inc., Chicago, IL, USA).

Principal component analysis (PCA) and cluster analysis (CA) were performed to analyse the chemical composition of cuticular wax from leaves of *Quercus suber* trees of different provenance. Using PCA, the hyperspace was defined by the original variables and dimension reduction was carried out using the significant principal components (new axes). These new axes were correlated with the original variables. Thus, the samples were plotted onto a reduced space where similar samples could be grouped. Agglomerative hierarchical CA, using the Euclidean distance and single-linkage method, was carried out to assess the existence of groups of samples suggested by PCA [65]. PCA and CA were performed with the Statistica™ software, version 6, from Statsoft (Tulsa, OK, USA).

5. Conclusions

The results obtained for *Quercus suber* leaves from six different provenances suggest the adaptive value of leaf features under a Mediterranean climate, namely, specific leaf area, leaf size and photosynthetic pigment, highlighting their potential role in dealing with varying temperatures and rainfall regimes through local adaptation and phenotypic plasticity. Cork oak leaves possess a substantial cuticular wax layer that forms a nearly impermeable membrane which is a tool to cope with adverse drought related with environmental conditions. The composition of the cuticular wax also favours the hydrophobicity of the layer by inclusion of very long-chain alkanes and alkanoic acids (e.g., C28 and C30). These characteristics are species specific and did not differ across provenances.

Triterpenes are the major component of the cuticular wax complex and contain a high proportion of lupeol. Thus, the leaves of cork oak represent a potential source for this interesting bioactive compound. The chemical composition of the cuticular wax did not differ among provenances, which may be ascribed to similar water stress conditions in the Mediterranean region, which probably has a stronger effect than genetic variability in *Q. suber*.

Author Contributions: H.P. and I.M. conceived the study and carried out the experimental design; S.F.-D. performed data curation; R.S. and A.R. performed the experiments and data interpretation; I.M. wrote the first manuscript draft; H.P., I.M. and S.F.-D. were responsible for data validation and manuscript revisions. All authors have read and agreed to the published version of the manuscript.

Funding: This work was funded by the Portuguese Foundation for Science and Technology (FCT) through the funding of Centro de Estudos Florestais (UID/AGR/00239/2019) and LEAF (UID/AGR/04129/2020). Rita Simões acknowledges a doctoral scholarship from FCT with the SUSFOR Doctoral Programme (PD/BD/128259/2016). The cork oak provenance field trials were funded by the European Commission (FAIR1-CT-95-0202) and national programs (PBIC/AGR/2282/95, PAMAF 4027, PRAXIS/3/3.2/Flor/2110/95).

Acknowledgments: We are particularly indebted to Helena Almeida in providing plant materials.

Conflicts of Interest: The authors declare no conflict of interest.

References

1. Pollard, M.; Beisson, F.; Li, Y.; Ohlrogge, J.B. Building lipid barriers: Biosynthesis of cutin and suberin. *Trends Plant Sci.* **2008**, *13*, 236–246. [CrossRef] [PubMed]
2. Domínguez, E.; Heredia-Guerrero, J.A.; Heredia, A. The biophysical design of plant cuticles: An overview. *New Phytol.* **2011**, *189*, 938–949. [CrossRef] [PubMed]
3. Ingram, G.; Nawrath, C. The Roles of the Cuticle in Plant Development: Organ Adhesions and Beyond. *J. Exp. Bot.* **2017**, *68*, 5307–5321. [CrossRef] [PubMed]
4. Samuels, L.; Kunst, L.; Jetter, R. Sealing Plant Surfaces: Cuticular Wax Formation by Epidermal Cells. *Annu. Rev. Plant Biol.* **2008**, *59*, 683–707. [CrossRef] [PubMed]
5. Kunst, L.; Samuels, L. Plant cuticles shine: Advances in wax biosynthesis and export. *Curr. Opin. Plant Biol.* **2009**, *12*, 721–727. [CrossRef]
6. Jetter, R.; Kunst, L.; Samuels, A.L. Composition of plant cuticular waxes. In *Biology of the Plant Cuticle, Annual Plant Reviews*; Riederer, M., Müller, C., Eds.; Blackwell: Oxford, UK, 2006; Volume 23, pp. 145–181.
7. Reynhardt, E.C.; Riederer, M. Structures and molecular dynamics of plant waxes. II. Cuticular waxes from leaves of *Fagus sylvatica* L. and *Hordeum vulgare* L. *Eur. Biophys. J.* **1994**, *23*, 59–70.

8. Zeisler-Diehl, V.; Müller, Y.; Schreiber, L. Epicuticular wax on leaf cuticles does not establish the transpiration barrier, which is essentially formed by intracuticular wax. *J. Plant Physiol.* **2018**, *227*, 66–74. [CrossRef]
9. Sharma, P.; Kothari, S.L.; Rathore, M.S.; Gour, V.S. Properties, variations, roles and potential applications of epicuticular wax: A review. *Turk. J. Bot.* **2018**, *42*, 135–149. [CrossRef]
10. Shepherd, T.; Griffiths, D.W. The effects of stress on plant cuticular waxes. *New Phytol.* **2006**, *171*, 469–499. [CrossRef]
11. Kosma, D.K.; Jenks, M. Eco-physiological and molecular-genetic determinants of plant cuticle function in drought and salt stress tolerance. In *Advances in Molecular Breeding toward Drought and Salt Tolerant Crops*; Jenks, M.A., Hasegawa, P.M., Jain, S.M., Eds.; Springer: Dordrecht, The Netherlands, 2007; pp. 91–120.
12. Serrano, M.; Coluccia, F.; Torres, M.; L'Haridon, F.; Métraux, J.-P. The cuticle and plant defense to pathogens. *Front. Plant Sci.* **2014**, *5*, 274. [CrossRef]
13. Domínguez, E.; Heredia-Gerrero, J.A.; Heredia, A. The plant cuticle: Old challenges, new perspectives. *J. Exp. Bot.* **2017**, *68*, 5251–5255. [CrossRef] [PubMed]
14. Schuster, A.C.; Burghardt, M.; Riederer, M. The ecophysiology of leaf cuticular transpiration: Are cuticular water permeabilities adapted to ecological conditions? *J. Exp. Bot.* **2017**, *68*, 5271–5279. [CrossRef] [PubMed]
15. Pereira, H. *Cork: Biology, Production and Uses*; Elsevier: Amsterdam, The Netherlands, 2007.
16. Leite, C.; Oliveira, V.; Lauw, A.; Pereira, H. Cork rings suggest how to manage *Quercus suber* to mitigate the effects of climate changes. *Agric. For. Meteorol.* **2019**, *266*, 12–19. [CrossRef]
17. Pérez-Harguindeguy, N.; Díaz, S.; Garnier, E.; Lavorel, S.; Poorter, H.; Jaureguiberry, P.; Bret-Harte, M.S.; Cornwell, W.K.; Craine, J.M.; Gurvich, D.E.; et al. New handbook for standardised measurement of plant functional traits worldwide. *Aust. J. Bot.* **2013**, *61*, 167–234. [CrossRef]
18. Oliveira, G.; Correia, O.; Martins Loução, M.; Catarino, F.M. Phenological and growth-patterns of the Mediterranean oak *Quercus suber* L. *Trees Struct. Funct.* **1994**, *9*, 41–46. [CrossRef]
19. Vaz, M.; Pereira, J.S.; Gazarini, L.C.; David, T.S.; David, J.S.; Rodrigues, A.; Moroco, J.; Chaves, M.M. Drought-induced photosynthetic inhibition and autumn recovery in two Mediterranean oak species (*Quercus ilex* and *Quercus suber*). *Tree Physiol.* **2010**, *30*, 946–956. [CrossRef] [PubMed]
20. Prats, K.A.; Brodersen, C.R.; Ashton, M.S. Influence of dry season on *Quercus suber* L. leaf traits in the Iberian Peninsula. *Am. J. Bot.* **2019**, *106*, 656–666. [CrossRef]
21. Costa-e-Silva, F.; Correia, A.C.; Piayda, A.; Dubbert, M.; Rebmenn, C.; Cuntz, M.; Werner, C.; David, J.S.; Pereira, J.S. Effects of an extremely dry winter on net ecosystem carbon exchange and tree phenology at a cork oak woodland. *Agric. For. Meteorol.* **2015**, *204*, 48–57. [CrossRef]
22. Martins, C.M.C.; Mesquita, S.M.M.; Vaz, W.L.C. Cuticular Waxes of the Holm (*Quercus ilex* L. subsp. ballota (Desf.) Samp.) and Cork (*Q. suber* L.) Oaks. *Phytochem. Anal.* **1999**, *10*, 1–5. [CrossRef]
23. Varela, M.C. *European Network for the Evaluation of Genetic Resources of Cork Oak for Appropriate Use in Breeding and Gene Conservation Strategies*; Handbook; INIA: Lisbon, Portugal, 2000.
24. Eriksson, G. *Quercus suber—Recent Genetic Research*; European Forest Genetic Resources Programme (EUFORGEN), Bioversity International: Rome, Italy, 2017; 30p, Available online: http://www.euforgen.org/fileadmin/templates/euforgen.org/upload/Publications/Thematic_publications/Quercus_suber_OpenSourceCR_web.pdf (accessed on 25 June 2020).
25. Sampaio, T.; Branco, M.; Guichoux, E.; Petit, R.J.; Pereira, J.S.; Varela, M.C.; Almeida, M.H. Does the geography of cork oak origin influence budburst and leaf pest damage? *For. Ecol. Manag.* **2016**, *373*, 33–43. [CrossRef]
26. Sampaio, T.; Gonçalves, E.; Patrício, M.S.; Cota, T.M.; Almeida, M.H. Seed origin drives differences in survival and growth traits of cork oak (*Quercus suber* L.) populations. *For. Ecol. Manag.* **2019**, *448*, 267–277. [CrossRef]
27. Varela, M.C.; Tessier, C.; Ladier, J.; Dettori, S.; Filigheddu, M.; Bellarosa, R.; Vessella, F.; Almeida, M.H.; Sampaio, T.; Patrício, M.S. Characterization of the International Network Fair 202 of provenance and progeny trials of cork oak on multiple sites for further use on forest sustainable management and conservation of genetic resources. In Proceedings of the Second International Congress of Silviculture, Designing the Future of the Forestry Sector, Florence, Italy, 26–29 November 2014.
28. Mediavilla, S.; Martín, I.; Babiano, J.; Escudero, A. Foliar plasticity related to gradients of heat and drought stress across crown orientations in three Mediterranean Quercus species. *PLoS ONE* **2019**, *14*, e0224462. [CrossRef] [PubMed]

29. Vaz, M.; Moroco, J.; Ribeiro, N.; Gazarini, L.C.; Pereira, J.S.; Chaves, M.M. Leaf-level responses to light in two co-occurring Quercus (Quercus ilex and Quercus suber): Leaf structure, chemical composition and photosynthesis. *Agrofor. Syst.* **2011**, *82*, 173–181. [CrossRef]
30. Ramírez-Valiente, J.A.; Valladares, F.; Delgado, A.; Granados, S.; Aranda, I. Factors affecting cork oak growth under dry conditions: Local adaptation and contrasting additive genetic variance within populations. *Tree Genet. Genomes* **2011**, *7*, 285–295. [CrossRef]
31. Rzigui, T.; Jazzar, L.; Ben Baaziz, K.; Fkiri, S.; Nasr, Z. Drought tolerance in cork oak is associated with low leaf stomatal and hydraulic conductances. *iForest* **2018**, *11*, 728–733. [CrossRef]
32. Lobo-do-Val, R.; Besson, C.K.; Caldeira, M.C.; Chaves, M.M.; Pereira, J.S. Drought reduces tree growing season length but increases nitrogen resorption efficiency in a Mediterranean ecosystem. *Biogeosciences* **2019**, *16*, 1265–1279. [CrossRef]
33. Daoudi, H.; Derridj, A.; Hannachi, L.; Mévy, J.P. Comparative drought responses of Quercus suber seedlings of three algerian provenances under greenhouse conditions. *Revue d'Ecologie (Terre et Vie)* **2018**, *73*, 57–70.
34. Aranda, I.; Pardos, M.; Puértolas, J.; Jiménez, M.D.; Pardos, J.A. Water use efficiency in cork oak (*Quercus suber* L.) is modified by the interaction of water and light availabilities. *Tree Physiol.* **2007**, *27*, 671–677. [CrossRef]
35. Gunn, S.; Farrar, J.F.; Collis, B.E.; Nason, M. Specific leaf area in barley: Individual leaves versus whole plants. *New Phytol.* **1999**, *143*, 45–51. [CrossRef]
36. Pierce, L.L.; Running, S.W.; Walker, J. Regional-scale relationships of leaf-area index to specific leaf-area and leaf nitrogen-content. *Ecol. Appl.* **1999**, *4*, 313–321. [CrossRef]
37. Reich, P.B.; Wright, I.J.; Lusk, C. Predicting leaf physiology from simple plant and climate attributes: A global glopnet analysis. *Ecol. Appl.* **2007**, *17*, 1982–1988. [CrossRef] [PubMed]
38. Han, Q.; Kabeya, D.; Günter, H. Leaf traits, shoot growth and seed production in mature Fagus sylvatica trees after 8 years of CO_2 enrichment. *Ann. Bot.* **2011**, *107*, 1405–1411. [CrossRef] [PubMed]
39. Liu, C.; Li, Y.; Li, X.; Chen, Z.; He, N. Variation in leaf morphological, stomatal, and anatomical traits and their relationships in temperate and subtropical forests. *Sci. Rep.* **2019**, *9*, 5803. [CrossRef] [PubMed]
40. Faria, T.; García-Plazaola, J.I.; Abadía, A.; Cerasoli, S.; Pereira, J.S.; Chaves, M.M. Diurnal changes in photoprotective mechanisms in leaves of cork oak (*Quercus suber* L.) during summer. *Tree Physiol.* **1996**, *16*, 115–123. [CrossRef]
41. Prasad, R.B.N.; Gülz, P.G. Surface structure and chemical composition of leaf waxes from *Quercus robur* L., *Acer pseudoplatanus* L. and *Juglans regia* L. *Z. Naturforsch. C* **1990**, *45*, 813–817. [CrossRef]
42. Bahamonde, H.A.; Gil, L.; Fernández, V. Surface properties and permeability to calcium chloride of *Fagus sylvatica* and *Quercus petraea* leaves of different canopy heights. *Front. Plant Sci.* **2018**, *9*, 494. [CrossRef]
43. Maiti, R.; Rodriguez, H.G.; Sarkar, N.C.; Kumari, A. Biodiversity in leaf chemistry (Pigments, epicuticular wax and leaf nutrients) in woody plant species in north-eastern Mexico, a synthesis. *Forest Res.* **2016**, *5*, 170.
44. Sánchez, F.J.; Manzanares, M.; de Andrés, E.F.; Tenorio, J.L.; Ayerbe, L. Residual transpiration rate, epicuticular wax load and leaf colour of pea plants in drought conditions. Influence on harvest index and canopy temperature. *Eur. J. Agron.* **2001**, *15*, 57–70. [CrossRef]
45. Xue, D.; Zhang, X.; Lu, X.; Chen, G.; Chen, Z.-H. Molecular and Evolutionary Mechanisms of Cuticular Wax for Plant Drought Tolerance. *Front. Plant Sci.* **2017**, *8*, 621. [CrossRef]
46. Riederer, M.; Schreiber, L. Protecting against water loss: Analysis of the barrier properties of plant cuticles. *J. Exp. Bot.* **2001**, *52*, 2023–2032. [CrossRef]
47. Kerstiens, G. Cuticular water permeability and its physiological significance. *J. Exp. Bot.* **1996**, *47*, 1813–1832. [CrossRef]
48. Burghardt, M.; Riederer, M. Ecophysiological relevance of cuticular transpiration of deciduous and evergreen plants in relation to stomatal closure and leaf water potential. *J. Exp. Bot.* **2003**, *54*, 1941–1949. [CrossRef] [PubMed]
49. Jetter, R.; Schäffer, S. Chemical composition of the *Prunus laurocerasus* leaf surface. Dynamic changes of the epicuticular wax film during leaf development. *Plant. Physiol.* **2001**, *126*, 1725–1737. [CrossRef] [PubMed]
50. Buschhaus, C.; Jetter, R. Composition differences between epicuticular and intracuticular wax substructures: How do plants seal their epidermal surfaces? *J. Exp. Bot.* **2011**, *62*, 841–853. [CrossRef]
51. Gülz, P.G.; Müller, E. Seasonal variation in the composition of epicuticular waxes of *Quercus robur* leaves. *Z. Naturforsch. C* **1992**, *47*, 800–806. [CrossRef]

52. Gülz, P.G.; Müller, E.; Herrmann, T. Chemical composition and surface structures of epicuticular leaf waxes from *Castanea sativa* and *Aesculus hippocastanum*. *Z. Naturforsch. C* **1992**, *47*, 661–666. [CrossRef]
53. Gülz, P.G.; Prasad, R.B.N.; Müller, E. Surface structures and chemical composition of epicuticular waxes during leaf development of *Fagus sylvatica* L. *Z. Naturforsch. C* **1992**, *47*, 190–196. [CrossRef]
54. Fernández, M.A.; de las Heras, B.; García, M.D.; Sáenz, M.T.; Villar, A. New insights into the mechanism of action of the anti-inflammatory triterpene lupeol. *J. Pharm. Pharmacol.* **2001**, *53*, 1533–1539. [CrossRef]
55. Zhang, L.; Zhang, Y.; Zhang, L.; Yang, X.; Lv, Z. Lupeol, a dietary triterpene, inhibited growth and induced apoptosis through down-regulation of DR3 in SMMC7721 cells. *Cancer Investig.* **2009**, *27*, 163–170. [CrossRef]
56. Beserra, F.P.; Xue, M.; Maia, G.L.A.; Leite Rozza, A.L.; Pellizzon, C.H.; Jackson, C.J. Lupeol, a Pentacyclic Triterpene, Promotes Migration, Wound Closure and Contractile Effect In Vitro: Possible Involvement of PI3K/Akt and p38/ERK/MAPK Pathways. *Molecules* **2018**, *23*, 2819. [CrossRef]
57. Paulo, J.A.; Crous-Duran, J.; Firmino, P.N.; Faias, S.P.; Palma, J.H.N. *System Report: Cork Oak Silvopastoral Systems in Portugal*; The AGFORWARD Research Project WP 2; AGFORWARD: Bedfordshire, UK, 2006.
58. Rizzini, C.T. *Tratado de Fitogeografia do Brasil*; HUCITEC, USP: São Paulo, Brazil, 1976.
59. Lichtenthaler, H.K. Chlorophylls and carotenoids: Pigments of photosynthetic biomembranes. *Meth. Enzymol.* **1987**, *148*, 350–382.
60. Wellburn, A.R. The spectral determination of chlorophylls a and b, as well as total carotenoids, using various solvents with spectrophotometers of different resolution. *J. Plant Physiol.* **1994**, *144*, 307–313. [CrossRef]
61. Nobel, P.S. *Physicochemical and Environmental Plant Physiology*; Academic Press: Oxford, UK, 2009.
62. Bueno, A.; Alfarhan, A.; Arand, K.; Burghardt, M.; Deininger, A.-C.; Hedrich, R.; Leide, J.; Seufert, P.; Staiger, S.; Riederer, M. Temperature effects on the cuticular transpiration barrier of two desert plants with water-spender and water-saver life strategies. *J. Exp. Bot.* **2019**, *70*, 1613–1625. [CrossRef] [PubMed]
63. Kolattukudy, P.E.; Agrawal, V.P. Structure and composition of aliphatic constituents of potato tuber skin (suberin). *Lipids* **1974**, *9*, 682–691. [CrossRef]
64. Ferreira, J.P.; Miranda, I.; Sen, A.; Pereira, H. Chemical and cellular features of virgin and reproduction cork from *Quercus variabilis*. *Ind. Crops Prod.* **2016**, *94*, 638–648. [CrossRef]
65. Alvin, C.R. *Methods of Multivariate Analysis*; Wiley: New York, NY, USA, 2002.

Article

Phytochemical, Cytotoxicity, Antioxidant and Anti-Inflammatory Effects of *Psilocybe Natalensis* Magic Mushroom

Sanah M. Nkadimeng [1,*], Alice Nabatanzi [1,2,3], Christiaan M.L. Steinmann [4] and Jacobus N. Eloff [1]

1 Phytomedicine Programme, Paraclinical Sciences Department, University of Pretoria, P/Bag X04, Onderstepoort, Pretoria, Gauteng 0110, South Africa; alice2nabatanzi@gmail.com (A.N.); kobus.eloff@up.ac.za (J.N.E.)

2 Future Africa, University of Pretoria, Hatfield, Pretoria, Gauteng 0186, South Africa

3 Department of Plant Sciences, Microbiology and Biotechnology, College of Natural Sciences, Makerere University, Kampala 00256, Uganda

4 Physiology Department, Sefako Makgatho Health Sciences University, Ga-Rankuwa, Gauteng 0208, South Africa; chris.steinmann@smu.ac.za

* Correspondence: sanah.nkadimeng@up.ac.za

Received: 4 August 2020; Accepted: 25 August 2020; Published: 31 August 2020

Abstract: Psilocybin-containing mushrooms, commonly known as magic mushrooms, have been used since ancient and recent times for depression and to improve quality of life. However, their anti-inflammatory properties are not known. The study aims at investing cytotoxicity; antioxidant; and, for the first time, anti-inflammatory effects of *Psilocybe natalensis*, a psilocybin-containing mushroom that grows in South Africa, on lipopolysaccharide-induced RAW 264.7 macrophages. Macrophage cells were stimulated with lipopolysaccharide and treated with different concentrations of *Psilocybe natalensis* mushroom extracted with boiling hot water, cold water and ethanol over 24 h. Quercetin and N-nitro-L-arginine methyl ester were used as positive controls. Effects of extracts on the lipopolysaccharide-induced nitric oxide, prostaglandin E_2, and cytokine activities were investigated. Phytochemical analysis, and the antioxidant and cytotoxicity of extracts, were determined. Results showed that the three extracts inhibited the lipopolysaccharide-induced nitric oxide, prostaglandin E_2, and interleukin 1β cytokine production significantly in a dose-dependent manner close to that of the positive controls. A study proposed that ethanol and water extracts of *Psilocybe natalensis* mushroom were safe at concentrations used, and have antioxidant and anti-inflammatory effects. Phytochemical analysis confirmed the presence of natural antioxidant and anti-inflammatory compounds in the mushroom extracts.

Keywords: depression; antioxidant; cytokines; anti-inflammatory; cytotoxicity; medicinal mushroom; *Psilocybe natalensis*

1. Introduction

Studies have shown an association between pathological inflammation and various chronic diseases, such as cancer and cardiovascular diseases [1]. Inflammation is known as a normal protective response to injury and/or infection. It involves complex processes that limit tissue injury in normal circumstances; however, in chronic inflammation, immune cells are dysregulated and tend to lose their self-limiting nature as a result thereof [1].

A number of inflammatory cytokines are involved in the pathogenesis of chronic inflammation [2]. Inflammatory cytokines are cell-signalling protein molecules that are released during inflammation and introduce signalling cascades that are able to activate the immune system [3]. Type 1 cytokines include tumour necrosis factor (TNF-α), and interferon-y and interleukin (IL) 1β, which play a primary role in enhancing the cellular immune response. Type 2, on the other hand, includes IL6, IL10 and IL13, which are more linked to antibody responses [4]. These cytokine also set in motion the activation of acute-phase proteins like C-reactive protein that further activate the immune system to release more cytokines and sustain the pathological inflammation state [4]. As a result, there is a higher uncontrolled release of pro-inflammatory cytokine such as TNF-α and IL 1β observed in many chronic diseases associated with pathological inflammation [2]. Oxidative stress also plays a significant role in the pathophysiology of inflammation via the actions of free radicals, non-radical molecules, and reactive oxygen and nitrogen species [5]. Lowered antioxidant concentrations and increased oxidative stress are also associated with mitochondrial dysfunctions and cell death [2]. In macrophages, lipopolysaccharide (LPS), a well-known endotoxin, induces productions of inflammatory mediators like inducible nitric oxide (iNO) and prostaglandinE_2, which are synthesised by inducible nitric oxide synthase (iNOS) and cyclooxygenase-2, also known as prostaglandin endoperoxide sythase-2 (PGE_2), respectively, in addition to pro-inflammatory cytokine activation [2,6]. As a result, LPS is commonly used as a potent inflammatory agent in different experimental models of inflammation studies [7].

Psilocybin-containing mushrooms, commonly known as magic mushrooms, are reported to have been used for centuries for their sacred, healing and mind-manifestation hallucinogenic powers within various indigenous societies [8,9]. Fatal intoxications due to exposure to magic mushrooms are rare and often reported to be mainly in combination with other drugs [10]. Furthermore, the lethal dose of magic mushrooms in rats is 280 mg/kg, and for humans is 17 kg/70 kg, which is very low, and magic mushrooms are therefore not normally considered toxic [10].

The crude water and sometimes ethanol extracts of psilocybin-containing mushrooms are the main source and method of treatment taken by many people. Many species of mushrooms are known to contain various molecules and scavenger-free radicals such as polysaccharides and phenol compounds [11]. Consequently, many naturally occurring substances in plants and mushrooms are perceived to possess antioxidant activities [11]. Studies have been performed on magic mushrooms with regards to their antidepressant properties; however, very little is known of their antioxidant potential and, to the best of our knowledge, there are no scientific reports on their anti-inflammatory potentials or properties.

This study aimed at investigating the cytotoxicity; antioxidant; and, for the first time, anti-inflammatory effects of *Psilocybe natalensis* mushroom, commonly known as "Natal Super strength", which is one of the well-known psilocybin mushrooms in the family Strophariaceae and genera of *Psilocybe*, that grow in South Africa on LPS-induced RAW264.7 macrophages [9]. The *Psilocybe natalensis* mushroom spores were identified and authenticated with an SKU number NSS-1 by the Spore Spot Company, Durban, South Africa. We hypothesize that the water and ethanol extracts of *Psilocybe natalensis* mushroom have anti-inflammatory properties. These results will reveal for the first time the potential anti-inflammatory mechanism offered by *Psilocybe natalensis* magic mushroom.

2. Results

2.1. ABTS Free Radical Scavenging Activity of P. natalensis Extracts

The extracts and positive controls, trolox and ascorbic acid, had concentration-dependent effect on the antioxidant activity. The positive controls showed inhibitory concentration (IC_{50}) very low under 1 μg/mL, as shown in Table 1. The ethanol displayed an IC_{50} less than 50 μg/mL, while the hot water and cold water showed IC_{50} greater than 50 μg/mL but less than 100 μg/mL.

2.2. Cytotoxicity of P. natalensis Extracts

The lethal concentrations (LC_{50}) of the extracts were higher than the positive control—doxorubicin—on measurements of toxicity on normal Vero cells, as shown on Table 1.

The ethanol extracts were mostly safe on the viability of cells with $LC_{50} > 100$ µg/mL, followed by hot water and cold water, which had the lowest LC_{50} of 25 µg/mL, Table 1.

Table 1. 2,2′-azinobis-(3-ethylbenzothiazoline-6-sulfonic acid) (ABTS) radical scavenging activity of the extracts and their cytotoxicity effect on Vero cells.

Sample	ABTS IC_{50} (µg/mL)	Vero LC_{50} (µg/mL)
Hot-water	86.233 ± 2.370	49.080 ± 3.340
Cold-water	90.154 ± 8.748	25.046 ± 0.460
70% Ethanol	26.586 ± 5.378	182.190 ± 11.860
Ascorbic acid	0.026 ± 0.003	not applicable
Trolox	0.928 ± 0.006	not applicable
Doxorubicin	not applicable	10.000 ± 1.327

2.3. *Anti-Inflammatory Effects of the Extracts*

2.3.1. Inhibitory Effects of *P. natalensis* Extracts on Inducible NO Production and % Cell Viability

The control LPS-induced cells significantly increase ($p < 0.001$) the nitrite content compared to the normal cells, as shown in Figure 1. The positive controls—quercetin and LNAME—both inhibited the inducible NO significantly ($p < 0.001$ and $p < 0.001$, respectively) compared to the control cells. Figure 1 also shows the hot-water extract significantly decreased the inducible NO with all the concentrations 50 µg/mL ($p < 0.001$), 25 µg/mL ($p < 0.001$) and 10 µg/mL ($p < 0.001$) in comparison to the control cells. The ethanol extract significantly decreased the iNO with the 50 µg/mL ($p < 0.001$) very close to LNAME values and 25 µg/mL ($p < 0.001$) but non-significantly with 10 µg/mL ($p = 0.299$) compared to the control cells. Meanwhile the cold-water significantly inhibited the iNO only with the lowest concentration 10 µg/mL ($p < 0.001$). The results also showed that the extracts were not toxic to the macrophage cells in comparison to doxorubicin, the toxic positive control, Figure 1. Furthermore, the % cell viability of treated cells where iNO production was significantly inhibited were safe at ≥80% cell viability.

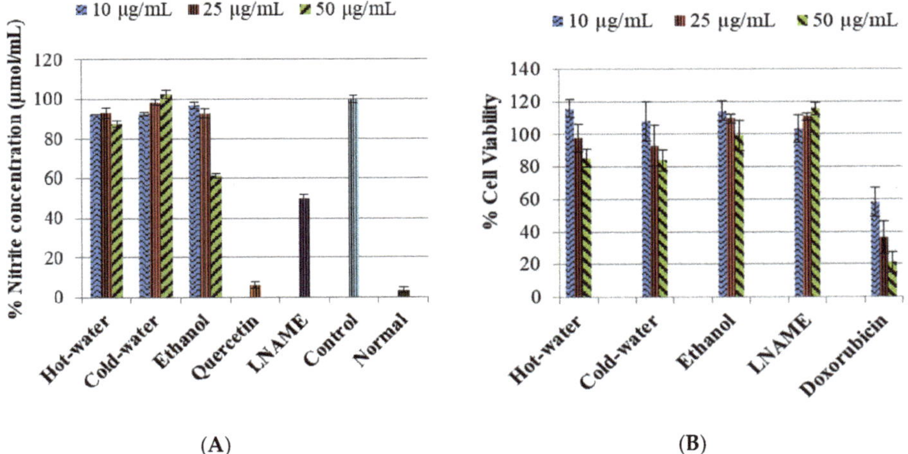

Figure 1. The inhibitory effect of *P. natalensis* extracts on LPS-induced NO production (**A**) and % cell viability (**B**) on RAW 264.7 macrophages treated with different concentrations (10, 25 and 50 µg/mL) and positive controls; for Figure 1A: quercetin (50 µg/mL), N-Nitro-L-Arginine Methyl Ester (LNAME) (100 µM) and for Figure 1B: LNAME (25, 50 and 100 µM) and doxorubicin (4, 10 and 20 µM) over 24 h.

2.3.2. Effects of *P. natalensis* Extracts on PGE_2 Production

The control LPS-induced cells significantly ($p < 0.001$) increased the concentration of PGE_2 in comparison to the normal cells, as shown in Figure 2. The positive controls—quercetin and LNAME—significantly reversed and inhibited the PGE_2 concentrations $p < 0.001$ and $p < 0.001$ respectively compared to the control cells. The three extracts also significantly inhibited the PGE_2 concentrations in comparison to the control cells, see Figure 2. The ethanol had an accelerating dose-dependent inhibition that increased as concentration increased, and potent inhibition was with 50 µg/mL ($p < 0.001$) concentration, which was very close to quercetin and lower than LNAME. The hot-water and cold-water, on the other hand, had a deceleration dose-dependent response where PGE_2 inhibition was weakened as concentrations increases. The potent inhibition for both hot-water and cold-water was with the lowest concentration 10 µg/mL ($p < 0.001$ and $p < 0.001$, respectively), which was lower than both LNAME and quercetin and very close to the normal cells.

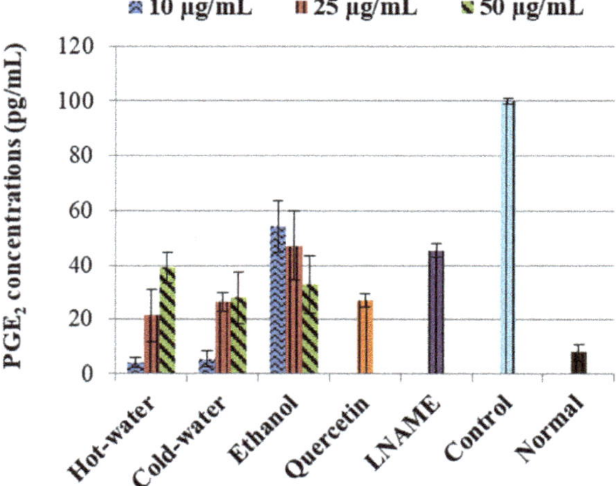

Figure 2. Inhibitory effects *of P. natalensis* extracts on LPS-induced PGE_2 production on RAW 264.7 macrophages treated with different concentrations (10, 25 and 50 µg/mL) and positive controls; quercetin (50 µg/mL) and LNAME (100 µM) over 24 h.

2.3.3. Effects of *P. natalensis* Extracts on Cytokine Production

The LPS-induced control cells increased the TNF-α non-significantly while IL1β ($p < 0.001$) and IL10 ($p < 0.001$) were significantly increased in comparison to the normal cells (Figure 3). Moreover, the three extracts significantly decreased the IL1β production very closed to LNAME. The ethanol extract significantly increased the IL10 production with 50 µg/mL ($p < 0.001$) same as LNAME ($p < 0.001$) in comparison to the control, while the hot-water and cold-water extracts decreased it in a dose-dependent response but were still higher than normal non-induced cells, Figure 4.

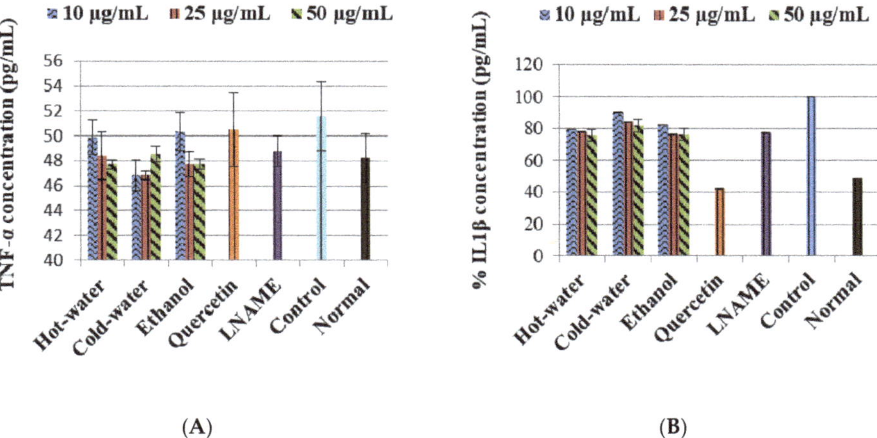

Figure 3. Effects of *P. natalensis* extracts on LPS-induced TNF-α (**A**) and IL1β (**B**) production in RAW 264.7 macrophages treated with different concentrations (10, 25 and 50 µg/mL), and positive controls quercetin (50 µg/mL) and LNAME (100 µM) over 24 h.

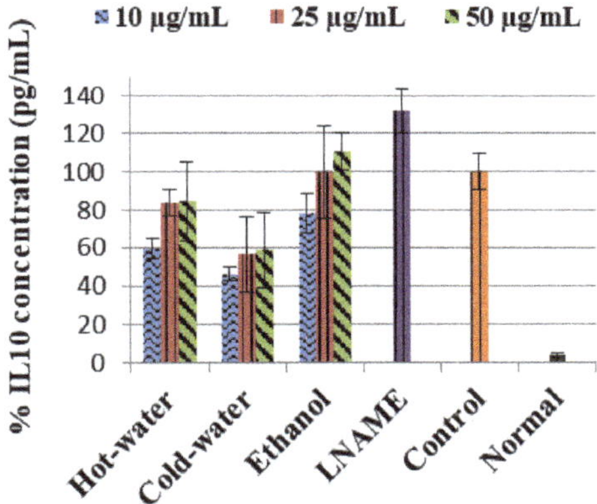

Figure 4. Effects of *P natalensis* extracts on LPS-induced IL10 production in RAW 264.7 macrophages treated with different concentrations (10, 25 and 50 µg/mL) and positive control LNAME (100 µM) over 24 h.

2.3.4. Phytochemical Determination

Figure 5 shows the gas chromatography-mass spectrometry (GCMS-MS) chromatograms of the hot-water, cold-water and ethanol extracts of *P. natalensis* mushrooms, respectively. The compounds with known antioxidant and anti-inflammatory from the chromatograms of the three extracts are tabulated in Table 2 with their molecular weight, formulas and area % per extracts. There were six known compounds with natural antioxidant and anti-inflammatory activities extracted from the three *P. natalensis* mushroom extracts. Compound n-hexadecoid acid with known antioxidant and anti-inflammatory activity was found to be present in all the three extracts with different degrees of area percentage. Ethanol extracts had the highest area % of compounds like nonadecane and tetradecane, which are known to possess anti-inflammatory and antioxidant properties along with other medicinal benefits that were not derived from the water extracts.

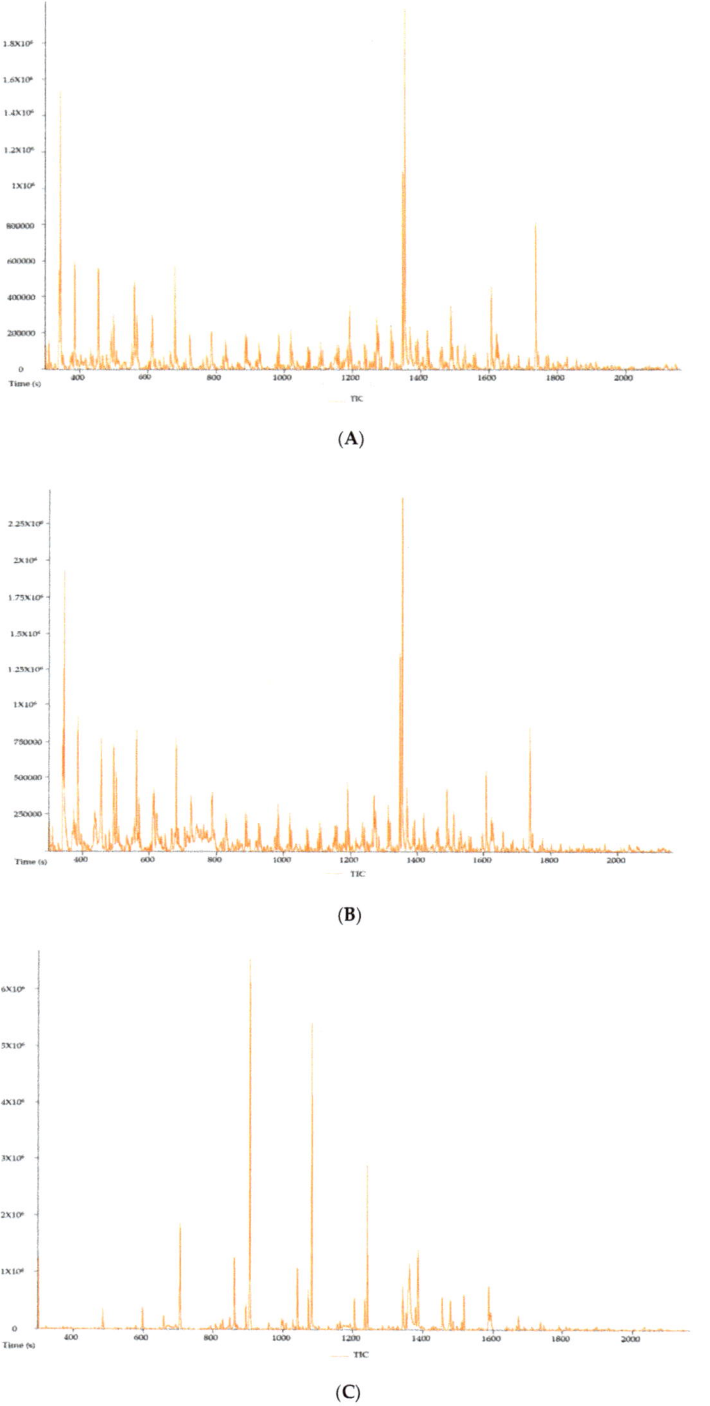

Figure 5. Gas chromatography-mass spectrometry (GCMS-MS) chromatogram of hot-water (**A**), cold-water (**B**) and 70% ethanol (**C**) extracts of *P. natalensis* mushroom.

Table 2. Compounds identified in the cold-water, hot-water and 70% ethanol extracts of *P. natalensis* mushroom with antioxidant and anti-inflammatory activities.

Compound Name	MW	Formula	Area %			Activity	Reference
			Cold	Hot	Ethanol		
n-Hexadecanoic acid	256	$C_{16}H_{32}O_2$	1.7129	2.0765	2.313	Anti-inflammatory Antioxidant	[12] [13]
4H-Pyran-4-one, 2,3-dihydro-3,5-dihydroxy-6-methyl-	144	$C_6H_8O_4$	2.0452			Antioxidant Anti-inflammatory	[13]
3-Octanone	128	$C_8H_{16}O$	3.6742	3.2977		Antioxidant Anti-inflammatory	[14]
Dibutyl phthalate	278	$C_{16}H_{22}O_4$		12.383		Anti-inflammatory	[15]
Nonadecane	268	$C_{19}H_{40}$			19.7244	Antioxidant Anti HIV Antibacterial Antimalarial	[16,17]
Tetradecane	198	$C_{14}H_{30}$			17.1872	Anti-inflammatory Antimicrobial Anti-diarrhoeal	[18]

MW: Molecular weight.

3. Discussion

The antioxidant potential of the extracts was determined using ABTS assay, which is one of the most commonly used radical scavenging assays methods for evaluating different natural product and functional food materials because of its quick, reproducible and inexpensive properties [19]. Phongpaichit et al. [20] state that the radical scavenging activity of extracts with $IC_{50} < 100$ μg/mL has good antioxidant potential while the ones with $IC_{50} < 50$ μg/mL are considered potent antioxidant agents. Accordingly, the hot-water and cold-water extracts of *P. natalensis* mushroom had good antioxidant potential while the ethanol had potent activity. Since increased oxidative stress and lowered antioxidant concentrations play a great role in the genesis and progression of inflammation, the scavenging properties of *P. natalensis* mushroom extracts will be of beneficial in chronic inflammation treatment [2]. Cytotoxicity results showed that the extracts were safe on normal Vero cells compared to positive control doxorubicin. Moreover, the results also indicated that the ethanol and water extracts of *P. natalensis* mushroom will be regarded as safe in relation to the American National Cancer Institution guidelines, which state that an extract with $LC_{50} \leq 20$ μg/mL is toxic [21].

Macrophage cells are reported as key immune cells in the initial stage of inflammation [22]. According to [23], RAW 264.7 macrophage cells provide an excellent model in drug screening for their potential use in inflammatory diseases and to evaluate potential inhibitors of pathways leading to inducible NO production. As a result, when the RAW cells are induced with LPS, they lead to a series of responses that include synthesis and production of prostanoids and pro-inflammatory cytokines present in pathological inflammation illnesses. Our study showed that the control LPS-induced cells significantly increase in iNO, PGE_2, IL1β and IL10 productions and TNF-α non-significantly in comparison to the normal cells. This was in agreement with other studies confirming the activation of signalling pathways, which leads to the release of pro-inflammatory cytokines, including IL1 β, TNF-α and mediators like NO and PGE_2 and anti-inflammatory cytokines like IL10 induced by LPS in RAW macrophage cells in the study [24–26]. The positive controls LNAME and quercetin in our study significantly reversed these effects.

The three extracts of *P. natalensis* magic mushroom dose-dependently inhibited iNO production, and the effects were more pronounced with the highest concentration of the ethanol extract. Furthermore, these inhibition effects were not due to cytotoxicity, as shown in Figure 1, where the percentage cell viability of the extracts was ≥80% in safe margins in all the concentrations where significant iNO inhibition was observed. Since iNO accumulation is a major macrophage-derived inflammatory mediator involved in the pathogenesis of various inflammation diseases, inhibiting or controlling its production is significant in anti-inflammatory investigation [25,27]. By suppressing iNO production, the extracts demonstrated an important healing potential effect in pathological inflammation.

The three extracts inhibited the LPS-induced PGE_2 production, which is another mediator of inflammation significantly close to the LNAME and quercetin. During inflammation, induced increase in PGE_2 production is also associated with the occurrence of pathological pain; as a result, the inhibitory properties of *P. natalensis* mushrooms extracts on PGE_2 demonstrated their potential ability to alleviate inflammation and physical chronic pains associated with pathological inflammation [28,29].

With regards to the cytokine activities, the hot-water, cold-water and ethanol extracts of *P. natalensis* mushroom decreased significantly the LPS-induced pro-inflammatory cytokines IL1β and reduced TNF-α non-significantly. Only the ethanol extract at the highest concentration used significantly increased the anti-inflammatory cytokine IL10 above the control in line with LNAME, while the other extracts down-regulated it in a dose-dependent manner. It is known that high levels of pro-inflammation cytokines like IL1β and TNF-α are present in the pathogenesis of many chronic diseases associated with chronic inflammation, and that they contribute to cell injury and damage and also initiate or sustain the inflammation state [4,30]. By suppressing the LPS-induced production of IL1β and TNF-α cytokines, the *P. natalensis* mushroom extracts demonstrate another positive potential benefit in chronic inflammation treatment.

These *P. natalensis* mushroom extracts effects were confirmed by the GCMS-MS analysis, which showed that the three mushroom extracts contained compounds known to induce natural antioxidant and anti-inflammatory effects such as n-hexadecanoic acid; 4h-Pyran-4-one, 2,3-dihydro-3,5-dihydroxy-6-methyl-; 3-octanone; and dibutyl phthalate. The GCMS-MS results also showed that the ethanol extract had very high percentage area of compounds known to have antioxidant and anti-inflammatory effects such as nonadecane and tetradecane, as shown in Table 2. As a result, the phytochemical analysis supported the greater positive effects that were observed with the ethanol extracts in comparison to the water extracts in the study.

In summary, these results indicated that the water and ethanol extracts of *Psilocybe natalensis* mushroom have potential antioxidant activity, an important factor in oxidative stress dysfunctions associated with inflammation. The results also indicated that the extracts will be considered safe in concentrations used. The study showed that the three extracts inhibited induced NO and PGE_2 productions, the two inflammatory mediators involved in pathogenesis of inflammatory diseases. Furthermore, the results proposed that the hot-water, cold-water and ethanol extract lowered LPS-induced TNF-α and inhibited induced pro-inflammatory cytokine IL1β, the well-known pro-inflammatory cytokines reported to be high in chronic inflammations, and their reduction was associated with improvement in the diseases [31]. Moreover, the ethanol extract increased the concentrations of anti-inflammatory cytokine IL10. The GCMS-MS analysis supported these findings by showing presence of compounds known to induce natural antioxidant and anti-inflammatory compounds in the three extracts.

4. Materials and Methods

4.1. Ethical Clearances

The protocol for this study was approved by University of Pretoria Research Ethics Committee, and the protocol number REC045-18 was assigned. Since psilocybin mushrooms are schedule 7 substances in South Africa, approval by the South African Department of Health Medical Control Council (MCC) was applied for, and a permit license POS 223/2019/2020 was given for the project.

4.2. Mushroom Growth and Making Extracts

The spore prints syringe of *Psilocybe natalensis* (*P. natalensis*) mushroom also known as "Natal super strength" were identified and authenticated with an SKU number NSS-1by the Spore Spot Company, Durban, South Africa, together with a growing sterile substrate kit (SSK-2), and were both purchased from the Spore Spot Company. Upon arrival, the spores were inoculated in a sterile substrate in sterile conditions in a triple locked laboratory under strict supervision as required by the MCC department. As soon as mycelium started to colonise about 60% of the substrate, it was transferred into a sterile monotub designed and donated by Mr L. Morland for the project. The temperature was controlled with an air conditioner and humidity supplied using a Clicks humidifier. As soon as the mushrooms fruit and mature, they were harvested and dried in an open oven at 35–36 °C over one to two days. The dried mushrooms were grounded into fine powder using a grinder. Extracts were made by measuring 5 g of dried powder into small sterile beaker and dissolved with 50 mL of boiling distilled water, cold water and 70% ethanol. The mixture was stirred using a stirrer for 5 min and allowed to stand for 24 h. After 24 h, the mixtures were filtered into small vials no 6 (Lasec, Johannesburg, South Africa) that were previously weighed and dried over night at 30 °C in an open oven. The extracts' yield was calculated and stored in dark in a fridge until use.

4.3. *Free Radical Scavenging Activity on ABTS*

The 2,2′-azinobis-(3-ethylbenzothiazoline-6-sulfonic acid) (ABTS) (Sigma-Aldrich, Johannesburg, South Africa) scavenging activity of the three extracts was assessed using methods of [32]. Ascorbic acid and trolox were used as positive controls. Briefly, freshly prepared ABTS was added into the two wells of each extract sample, while methanol was added into the other two wells to be used as blank, and the ability of the mushroom extracts to scavenge ABTS radicals was determined using a microplate reader (Biotek, Synergy HT, Analytical & Diagnostic Products CC, Johannesburg, South Africa) at a wavelength of 734 nm. The experiments were repeated three times, and inhibition concentration(IC) measured as IC_{50} values were calculated according to the formula: % ABTS inhibition = ((Absorbance control − Absorbance sample)/Absorbance control) × 100. The antioxidant ability was expressed as IC_{50} value, which is the concentration of the sample necessary to inhibit ABTS by 50%.

4.4. *Cytotoxicity of Extracts on Vero Normal Cells*

Viability of cells was determined using the tetrazolium-based colorimetric (MTT) assay described by [33] with modifications on normal African green monkey kidney (Vero) cells purchased from American Type Culture Collection (ATCC, Manassas, VA, USA) [34]. When cells had reached confluent culture, they were harvested and centrifuged at 200× *g* for 5 min and then resuspended in growth medium to 1 × 10^4 cells/mL into each well of columns 2 to 12 of a sterile 96-well microtitre plate, and column 1 was used as blank (no cells). The growth medium used was Minimal Essential Medium (MEM), (PAN Biotech, Biocom Africa, Johannesburg South Africa) supplemented with 0.1% gentamicin (Virbac, Johannesburg, South Africa) and 5% foetal calf serum (separation scientific, Johannesburg, South Africa). The plates were incubated for 24 h at 37 °C in a 5% CO_2 incubator. Then, MEM was aspirated from the cells and replaced with 200 µL of a serial prepared concentration range of extract. The serial dilutions of the test extracts were prepared in sera-free MEM. The microtitre plates were incubated at 37 °C in a 5% CO_2 incubator for 48 h with extracts. Untreated cells (negative control) and positive control (doxorubicin chloride, Pfizer Laboratories, Johannesburg, South Africa) were included.

After incubation, the media (with and without treatment) were aspirated, and cells were washed with 200 µL phosphate buffered saline (PBS) (Whitehead Scientific, Johannesburg, South Africa). Then, 100 µL of media was added to all the wells and 30 µL MTT (Inqaba biotec, Pretoria, South Africa, stock solution of 5 mg/mL in PBS) was added to each well and the plates incubated for a further 4 h at 37 °C. After incubation with MTT, the medium in each well was carefully removed without disturbing the MTT crystals in the wells. The MTT formazan crystals were dissolved by adding 50 µL DMSO to each well. The plates were shaken gently until the MTT solution was dissolved. The amount of MTT reduction was measured immediately by detecting absorbance in a microplate reader (Biotek, Synergy HT, Analytical & Diagnostic Products CC, Johannesburg, South Africa) at a wavelength of 540 nm and a reference wavelength of 630 nm. The wells in column 1, containing medium and MTT but no cells, were used to blank the plate reader. Viability of cells in percentages was calculated using the formula:

% Viability = ((Sample Absorbance/control Absorbance) × 100). The experiments were performed in triplicate and repeated twice in different times. The lethal concentrations (LC_{50}) values were calculated as the concentration of test extract resulting in a 50% reduction of absorbance compared to untreated cells. The 50% lethal concentration of the samples and positive control doxorubicin were obtained by linear regression analysis of concentration-response curve plotting between percentage of viability and sample concentration of two independent assays.

4.5. Anti-Inflammatory Effects of the Extracts

4.5.1. RAW 264.7 Macrophage Cell Culture

The RAW 264.7 macrophage cells purchased from American Type Culture Collection (ATCC, Manassas, VA, USA USA) were used in this study and maintained in Dulbecco's modified Eagle's medium (DMEM) (Pan Biotech, Separations scientific) supplemented with 10% Foetal bovine serum (Gibco, Sigma-Aldrich, Johannesburg, South Africa) and 1% of penicillin (100 units/mL) and streptomycin (100 μg/mL) (Celtic Molecular Diagnostics) at 37 °C in a 5% CO_2 atmosphere (HERAcell 150, Thermo Electron Corp., Separation scientific, Johannesburg, South Africa). The cells were allowed to grow, and reached 80% confluence before being used in the experiments.

4.5.2. Cytotoxicity of Extracts on LPS-Induced RAW 264.7 Macrophages

The RAW 264.7 cells were seeded at a density of 4×10^4 cells/well into each well of column 2 to 11 of sterile tissue culture treated 96 well plates (NEST, Whitehead scientific, Johannesburg, South Africa) and incubated for 24 h at 37 °C in a 5% CO_2. Then, media was aspirated from all the wells and replaced with fresh medium. The cells were treated with lipopolysaccharide (LPS) (Sigma-Aldrich, Johannesburg, South Africa) at a concentration of 1 μg/mL in the presence of different concentrations of the three extracts (10, 25 and 50 μg/mL) and incubated for 24 h. Cytotoxicity was measured using MTT assay same as with the Vero cells above. Viability of cells in percentages was calculated using the formula: % Viability = ((Sample Absorbance/control Absorbance) × 100). The experiments were repeated two times on different occasions.

4.5.3. Treatment with the Extracts

The RAW 254.7 macrophage cells were plated 1×10^6 cells per 25 cm^2 tissue culture flasks (NEST, Whitehead scientific, Johannesburg, South Africa) over 24 h. Then, the medium was removed, fresh media added and the cells were stimulated with LPS 1 μg/mL and treated with the extracts at 10, 25 and 50 μg/mL concentrations. Quercetin (Sigma-Aldrich, Johannesburg, South Africa), a well-known antioxidant and a flavonol found in many fruits and plants, and N-Nitro-L-Arginine Methyl Ester (LNAME) (Sigma-Aldrich, Johannesburg, South Africa), a nitric oxide synthase (NOS) inhibitor, were used as positive controls. Control cells were stimulated with LPS but not treated. Normal cells were cells that were neither stimulated with LPS nor treated with extracts. The cells were exposed to LPS and treated over 24 h. After 24 h, medium was removed and stored in −80 freezer until day of analysis.

4.5.4. Nitrite Content Measurements

To measure nitric oxide production in cell culture media after 24 h of treatments in LPS-induced macrophage cells, nitrite content was measured in cell culture supernatant as an indication of NO production based on the Griess reaction (Sigma-Aldrich, Johannesburg, South Africa). Briefly, 100 μL of cell culture supernatant was added to 100 μL of Griess reagent incubated for 15 min and absorbance measured at 540 nm. The concentration of nitrite was determined from the serial diluted standard curve.

4.5.5. PGE_2 Activity Measurements

The effects of the extracts on PGE_2 were determined using the mouse $PTGS_2$/COX-2 Prostaglandin endoperoxide synthase 2 (PGE_2) ELISA kit (Elabscience, Biocom Africa, Johannesburg South Africa) according to the manufacture manual protocol. Concentrations of mouse PGE_2 in the cell culture media samples were calculated from the standard curve. The absorbance was directly proportional to the concentrations of PGE_2 in the sample medium.

4.5.6. Cytokine Activity Measurements

The effects of the extracts on levels of TNF-α, IL1 β and IL10 were determined and quantified using the mouse ELISA kits with catalogue numbers E-EL-M0049, E-EL-M0037 and E-EL-M0046, (Elabscience, Biocom Africa, Johannesburg South Africa)), respectively, following the same protocol as above using the instructor manual on the cell culture medium.

4.6. Phytochemical Determination of the Extracts

Phytochemical determination of extracts was performed using the gas chromatography-mass spectrometry (GCMS-MS) by the LC-MS (Synapt Waters, Johannesburg, South Africa) facility at the Chemistry Department, University of Pretoria. The water and ethanol extracts of *Psilocybe natalensis* mushrooms were dissolved in methanol (1 mg/mL). Chromatograms and presence of compounds in the three extracts were produced.

4.7. Statistical Analysis

Results are expressed as mean ± standard deviations, and statistically significant values were compared using one-way ANOVA analysis of variance using an interactive statistical program (Sigmastat, SPSS version 26, San Jose, CA, USA) and pairwise multiple comparison procedures using Holm—Sidak method. Normality testing was done using Shapiro—Wilk and equal variance test using Brown—Forsythe. The p-value of ≤ 0.050 was considered statistically significant.

5. Conclusions

The results proposed for the first time that the water and ethanol extracts of *Psilocybe natalensis* are safe in the concentrations used, and have antioxidant and anti-inflammatory properties. The study recommends further investigations of the mushroom effects in vivo in this field.

Author Contributions: Conceptualization, S.M.N.; Methodology, S.M.N.; writing—original draft preparation, S.M.N.; writing—review and editing S.M.N., A.N., C.M.L.S., J.N.E. All authors have read and agreed to the published version of the manuscript.

Funding: This study (project REC045-18) was funded by the Health and Welfare Sector Education and Training Authority (HWSETA) granted to SM Nkadimeng and Future Africa (University of Pretoria) grant to A Nabatanzi.

Acknowledgments: We appreciate the support of Llewelyn Morland, who assisted with the growing of mushrooms, and Lebogang E Moagi, for her assistance with statistics. We kindly acknowledge the LC-MS Synapt Facility (Department of Chemistry, University of Pretoria) for chromatography and mass spectrometry services provided by Madelien Wooding.

Conflicts of Interest: The authors declare no conflict of interest. The funders had no role in the design of the study; in the collection, analyses, or interpretation of data; in the writing of the manuscript; or in the decision to publish the results.

References

1. Zhong, J.; Shi, G. Editorial: Regulation of Inflammation in Chronic Disease. *Front. Immunol.* **2019**, *10*. [CrossRef]
2. Kanwar, J.R.; Kanwar, R.K.; Burrow, H.; Baratchi, S. Recent advances on the roles of NO in cancer and chronic inflammatory disorders. *Curr. Med. Chem.* **2009**, *16*, 2373–2394. [CrossRef] [PubMed]

3. Amodeo, G.; Trusso, M.A.; Fagiolini, A. Depression and Inflammation: Disentangling a Clear Yet Complex and Multifaceted Link. *Neuropsychiatry* **2017**, *7*, 448–457. [CrossRef]
4. Lucey, D.R.; Clerici, M.; Shearer, G.M. Type 1 and type 2 cytokine dysregulation in human infectious neoplastic, and inflammatory diseases. *Clin. Microbiol. Rev.* **1996**, *9*, 532–562. [CrossRef] [PubMed]
5. Leonard, B.; Maes, M. Mechanistic explanations how cell-mediated immune activation, inflammation and oxidative and nitrosative stress pathways and their sequels and concomitants play a role in the pathophysiology of unipolar depression. *Neurosci. Biobehav. Rev.* **2012**, *36*, 764–785. [CrossRef] [PubMed]
6. Vuolteenaho, K.; Moilanen, T.; Knowles, R.G.; Moilanen, E. The role of nitric oxide in osteoarthritis. *Scand. J. Rheumatol.* **2007**, *36*, 247–258. [CrossRef]
7. Gasparrini, M.; Forbes-Hernández, T.Y.; Giampieri, F.; Afrin, S.; Alvarez-Suarez, J.M.; Mazzoni, L.; Mezzetti, B.; Quiles, J.L.; Battino, M. Anti-inflammatory effect of strawberry extract against LPS-induced stress in RAW 264.7 macrophages. *Food Chem. Toxicol.* **2017**, *102*, 1–10. [CrossRef]
8. Johnson, M.W.; Griffiths, R.R. Potential Therapeutic Effects of Psilocybin. *Neurotherapeutics* **2017**, *14*, 734–740. [CrossRef]
9. Guzman, G.; Allen, J.W.; Gartz, J. A worldwide geographical distribution of the neurotropica fungi, an analysis and discussion. *Ann. Mus. Civ. Rovereto* **1998**, *14*, 189–280.
10. Van Amsterdam, J.G.; Opperhuizen, A.; Brink, W.V.D. Harm potential of magic mushroom use: A review. *Regul. Toxicol. Pharmacol.* **2011**, *59*, 423–429. [CrossRef]
11. Dore, C.M.P.G.; Alves, M.G.C.F.; Santos, M.G.L.; de Sauza, L.A.R.; Baseia, I.G.; Leite, E.L. Antioxiant and anti-inflammatory properties of an extract rich in polysaccharides of the mushroom *Polyporus dermoporus*. *Antioxidants* **2014**, *3*, 730–744. [CrossRef]
12. Aparna, V.; Dileep, K.V.; Mandal, P.K.; Karthe, P.; Sadasivan, C.; Haridas, M. Anti-Inflammatory Property of n-Hexadecanoic Acid: Structural Evidence and Kinetic Assessment. *Chem. Boil. Drug Des.* **2012**, *80*, 434–439. [CrossRef] [PubMed]
13. Kumar, P.P.; Kumaravel, S.; Latlitha, C. Screening of antioxidant activity, total phenolics and GC-MS study of Vitex negundo. *Afr. J. Biochem. Res.* **2010**, *4*, 191–195. [CrossRef]
14. Palariya, D.; Singh, A.; Dhami, A.; Pant, A.K.; Kumar, R.; Prakash, O. Phytochemical analysis and screening of antioxidant, antibacterial and anti-inflammatory activity of essential oil of *Premna mucronata* Roxb. leaves. *Trends Phytochem. Res.* **2019**, *3*, 275–286.
15. Maestre-Batlle, D.; Pena, O.M.; Huff, R.D.; Rhandhawa, A.; Carlisten, C.; Bolling, A.K. Dibutyl phthalate modulate phenotype of gradunulocytes in human blood in response to inflammation. *Toxicol. Lett.* **2018**, *296*, 23–30. [CrossRef] [PubMed]
16. Mahmoodreza, M.; Younes, G.; Forough, K.; Hossein, T. Composition of the essential oil of Rosa damascene mill from South of Iran. *Iran. J. Psychiatry Bahav. Sci.* **2010**, *6*, 59–62.
17. Ombito, O.J.; Matasyoh, C.J.; Vulule, M.J. Chemical composition and larvicidal activity of *Zantoxylem gilletii* essential oil against *Anopheles gambie*. *Afr. J. Biotech.* **2014**, *13*, 2175–2180. [CrossRef]
18. Gurudeeban, S.; Sathyavani, K.; Ramanathan, T. Bitter apple (Citrulus colocynthis): An overview of chemical composition and biomedical potentials. *Asian J. Plant Sci.* **2010**, *9*, 394–401.
19. Joo, T.; Sowndhararajan, K.; Hong, S.; Lee, J.; Park, S.-Y.; Kim, S.; Jhoo, J.-W. Inhibition of nitric oxide production in LPS-stimulated RAW 264.7 cells by stem bark of Ulmus pumila L. *Saudi J. Boil. Sci.* **2014**, *21*, 427–435. [CrossRef]
20. Phongpaichit, S.; Nikom, J.; Rungjindamai, N.; Sakayaroj, J.; Hutadilok-Towatana, N.; Rukachaisirikul, V.; Kirtikara, K. Biological activities of extracts from endophytic fungi isolated fromGarciniaplants. *FEMS Immunol. Med. Microbiol.* **2007**, *51*, 517–525. [CrossRef]
21. Mahavorasirikul, W.; Viyanant, V.; Chaijaroenkul, W.; Itharat, A.; Na-Bangchang, K. Cytotoxic activity of Thai medicinal plants against human cholangiocarcinoma, laryngeal and hepatocarcinoma cells in vitro. *BMC Complement. Altern. Med.* **2010**, *10*, 55. [CrossRef] [PubMed]
22. Jeong, D.; Yang, W.S.; Yang, Y.; Nam, G.; Kim, J.H.; Yoon, D.H.; Noh, H.J.; Lee, S.; Kim, T.W.; Sung, G.H.; et al. In vitro and in vivo anti-inflammatoty effect of *Rhodomyrtus tomentosa* methanol extract. *J. Ethnopharmacol.* **2014**, *146*, 205–213. [CrossRef] [PubMed]
23. An, H.-J.; Kim, I.-T.; Park, H.-J.; Kim, H.-M.; Choi, J.-H.; Lee, K.-T. Tormentic acid, a triterpenoid saponin, isolated from *Rosa rugosa*, inhibited LPS-induced iNOS, COX-2 and TNF-α expression through inactivation of the nuclear factor-κb pathway in RAW 264.7 macrophages. *Int. Immunopharmacol.* **2011**, *11*, 504–510. [CrossRef] [PubMed]

24. Arulselvan, P.; Tan, W.S.; Gothal, S.; Muniandy, K.; Fakurazi, S.; Esa, N.M.; Alarfaj, A.A.; Kumar, S.S. Anti-inflammatory potential of ethyl acetate fraction of *Moringa oleifera* in downregulating the NF-κb signalling pathways in lipopolysaccharide-stimulated macrophages. *Molecules* **2016**, *21*, 1452. [CrossRef] [PubMed]
25. Muniadndy, K.; Gothai, S.; Badran, K.M.H.; Kumar, S.S.; Esa, N.M.; Arulselvan, P. Suppression of proinflammatory cytokine and mediators in LPS-induced RAW 264.7 macrophages by stem extract of Alternathera sessilis via the inhibition of the NF-κb pathway. *J. Immunol. Res.* **2018**, *2018*, 1–12. [CrossRef]
26. Sanjeewa, K.K.A.; Nagahawatta, D.P.; Yang, H.-W.; Oh, J.Y.; Jayawardena, T.U.; Jeon, Y.-J.; De Zoysa, M.; Whang, I.; Ryu, B. Octominin inhibits LPS-induced chemokine and pro-inflammatory cytokines secretion from RAW 264.7 macrophages via blocking TLRs/NF-κβ signal transduction. *Biomolecules* **2020**, *10*, 511. [CrossRef]
27. Xie, Q.; Cho, H.; Calaycay, J.; Mumford, R.; Świderek, K.; Lee, T.; Ding, A.; Troso, T.; Nathan, C. Cloning and characterization of inducible nitric oxide synthase from mouse macrophages. *Science* **1992**, *256*, 225–228. [CrossRef]
28. Wang, Q.; Liang, J.; Brennan, C.; Ma, L.; Li, Y.; Lin, X.; Liu, H.; Wu, J. Anti-inflammatory effect of alkaloids extracted from Dendrobium aphyllum on macrophage RAW 264.7 cells through NO production and reduced IL-1, IL-6, TNF-α and PGE2 expression. *Int. J. Food Sci. Technol.* **2019**, *55*, 1255–1264. [CrossRef]
29. Park, T.; Park, J.-S.; Sim, J.H.; Kim, S.-Y. 7-Acetoxycoumarin inhibits LPS-induced inflammatory cytokine synthesis by IκBα degradation and MAPK activation in macrophage cells. *Molecules* **2020**, *25*, 3124. [CrossRef]
30. Kobelt, D.; Zhang, C.; Clayton-Lucey, I.A.; Glauben, R.; Voss, C.; Siegmund, B.; Stein, U. Pro-inflammatory TNF-α and IFN-γ Promote Tumor Growth and Metastasis via Induction of MACC1. *Front. Immunol.* **2020**, *11*, 1–15. [CrossRef]
31. Kuriakose, T.; Kanneganti, T.-D. Regulation and functions of NLRP3 inflammasome during influenza virus infection. *Mol. Immunol.* **2017**, *86*, 56–64. [CrossRef] [PubMed]
32. Re, R.; Pellegrini, N.; Proteggente, A.; Pannala, A.; Yang, M.; Rice-Evans, C. Antioxidant activity applying an improved ABTS radical cation decolourization assay. *Free Rad. Biol. Med.* **1999**, *26*, 1231–1237. [CrossRef]
33. Mosmann, T. Rapid colorimetric assay for cellular growth and survival: Application to proliferation and cytotoxicity assays. *J. Immunol. Methods.* **1983**, *65*, 55–63. [CrossRef]
34. Mwinga, J.L.; Asong, J.A.; Van Staden, J.; Nkadimeng, S.M.; McGaw, L.J.; Aremu, A.O.; Mbeng, W.O. In vitro antimicrobial effects of Hypoxis hemerocallidea against six pathogens with dermatological relevance and its phytochemical characterization and cytotoxicity evaluation. *J. Ethnopharmacol.* **2019**, *242*, 112048. [CrossRef] [PubMed]

Article

Anti-Melanoma Activities and Phytochemical Compositions of *Sorbus commixta* Fruit Extracts

Sora Jin [1], Kyeoung Cheol Kim [2], Ju-Sung Kim [2], Keum-Il Jang [3,*] and Tae Kyung Hyun [1,*]

1 Department of Industrial Plant Science and Technology, College of Agricultural, Life and Environmental Sciences, Chungbuk National University, Cheongju 28644, Korea; thfkrhehd123@naver.com
2 College of Agriculture & Life Sciences, SARI, Jeju National University, Jeju 63243, Korea; cheolst0516@gmail.com (K.C.K.); aha2011@jejunu.ac.kr (J.-S.K.)
3 Department of Food Science and Biotechnology, College of Agricultural, Life and Environmental Sciences, Chungbuk National University, Cheongju 28644, Korea
* Correspondence: jangki@chungbuk.ac.kr (K.-I.J.); taekyung7708@chungbuk.ac.kr (T.K.H.); Tel.: +82-(43)-2612569 (K.-I.J.); +82-(43)-2612520 (T.K.H.)

Received: 13 July 2020; Accepted: 19 August 2020; Published: 21 August 2020

Abstract: *Sorbus commixta* Hedl. (Rosaceae family) has a long history as a medicinal plant in East Asian countries. In this study, we evaluated the effect of *S. commixta* fruit extracts prepared with different ethanol concentrations on anti-melanoma activity, and the extraction yield of phenolic compounds and flavonoids. Using the partitioned fractions from the EtOH extract, we found that the butanol fraction (BF) possessed strong cytotoxic activity against SK-MEL-2 cells (human melanoma cells) but not against HDFa cells (human dermal fibroblast adult cells). Additionally, BF-induced cell death was mediated by the inhibition of the mitogen-activated protein kinase/extracellular regulated kinase (MEK/ERK) signaling pathway, coupled with the upregulation of caspase-3 activity in SK-MEL-2 cells. Furthermore, HPLC analysis of polyphenolic compounds suggested that *S. commixta* fruits contained several active compounds including chlorogenic acid, rutin, protocatechuic acid, and hydroxybenzoic acid, all of which are known to possess anti-cancer activities. Although this study has been carried out by cell-based approach, these results suggest that *S. commixta* fruits contain promising anti-melanoma compounds.

Keywords: *Sorbus commixta* fruit; anti-melanoma activity; caspase-3; polyphenolic compounds

---

1. Introduction

Malignant melanoma, an aggressive and fatal form of skin cancer, is a malignant melanocyte tumor that accounts for approximately 75% of skin cancer-related deaths worldwide [1–3]. In recent years, melanoma incidence has continued to increase worldwide, and the onset of malignant melanomas is reportedly affected by various environmental and genetic factors, such as ultraviolet exposure and carcinogenic BRAF mutations [4–6]. Despite the wide variety of therapies available to treat melanoma, including surgery, radiation therapy, immunotherapy, and chemotherapy [7–9], phytochemicals have been recognized as better anti-cancer therapies to prevent or inhibit carcinogenesis [10]. Camptothecin, vincristine, vinblastine, taxol, topotecan, podophyllotoxin, and irinotecan are good examples of plant-based anticancer molecules [11]. Growing evidence supports that plant-derived natural products are important sources of novel candidates, with pharmacological applicability.

Sorbus commixta Hedl. (Rosaceae family) is a well-known medicinal plant used traditionally in East Asian countries including Japan, China, and Korea for the treatment of asthma and other bronchial disorders [12,13]. Additionally, *S. commixta* extracts exhibit a range of biological effects, such as anti-inflammatory [14], antioxidative [15], anti-atherogenic [13,16], vasorelaxant [17], and anti-lipid peroxidation activities [18]. Furthermore, the *S. commixta* fruit also has been used for the treatment of bronchitis and gastrointestinal disorders, as well as for its anti-inflammatory, anti-diabetic, diuretic, and vasorelaxant properties [12,19]. Phytochemical analysis of *S. commixta* fruits has revealed the presence of active ingredients often used in cosmetics, including rutin, isoquercitrin, caffeoylquinic acid, dicaffeoylquinic acid, neosakuranin, chlorogenic acid, neochlorogenic acid, carotenoids, and ascorbic acid [20–22], suggesting the potential applicability of *S. commixta* fruits for the production of herbal cosmetics with anti-melanoma activity. Nonetheless, the anti-melanoma properties of *S. commixta* fruits are yet to be comprehensively characterized.

Therefore, our study analyzed the effect of *S. commixta* fruit extracts prepared with different ethanol (EtOH) concentrations on anti-melanoma activity, as well as the role of the mitogen-activated protein kinase/extracellular regulated kinase (MEK/ERK) pathways in the activation of cytotoxicity. Additionally, seven polyphenolic compounds were identified in *S. commixta* fruit extracts via HPLC. Therefore, we expect our study to motivate further interest in the use of *S. commixta* fruits as a natural anti-melanoma source in the cosmetic and pharmaceutical industries.

2. Results and Discussion

2.1. *Effects of* S. commixta *Fruit Extracts and Solvent Fractions on Human Melanoma SK-MEL-2 Cell*

Extraction efficiencies are affected by the solvent type and concentration [23]. Particularly, EtOH is considered an excellent solvent and is therefore often used to recover polyphenols from plant matrices, as it is also safe for human consumption [24]. In addition, EtOH is completely miscible with water. Therefore, our study conducted further experiments to define the effect of *S. commixta* fruit extracts prepared with different EtOH concentrations on anti-melanoma activity. As shown in Figure 1A, incubation with 100 µg/mL of EtOH extracts significantly inhibited the proliferation of SK-MEL-2 cells (28.85 ± 3.38%), although ethanol concentration did not have a substantial impact on SK-MEL-2 cell cytotoxicity. Among the bioactive compounds of *S. commixta* fruit [22], phenolic and flavonoid compounds have been proposed. Binary-solvent systems have been found to be more effective for the extraction of these compounds from plants compared to mono-solvent systems [25,26]. However, in the case of *S. commixta* fruit, EtOH extract contained higher TPC (total phenolic content) levels than water extracts, but solvent concentrations did not significantly affect the extraction of phenolic compound yields. TFC (total flavonoid content) in the extracts increased with increasing ethanol concentrations, and the mono-solvent system examined herein (EtOH) was more effective in extracting phenolic and flavonoid compounds than EtOH/water solvents (Table 1). Additionally, higher concentrations of chlorogenic acid, rutin, protocatechuic acid, and hydroxybenzoic acid in EtOH extracts were observed compared to other extracts, suggesting that the differences in the cytotoxic activity of *S. commixta* fruit extracts were due to their concentration and composition of polyphenolic compounds.

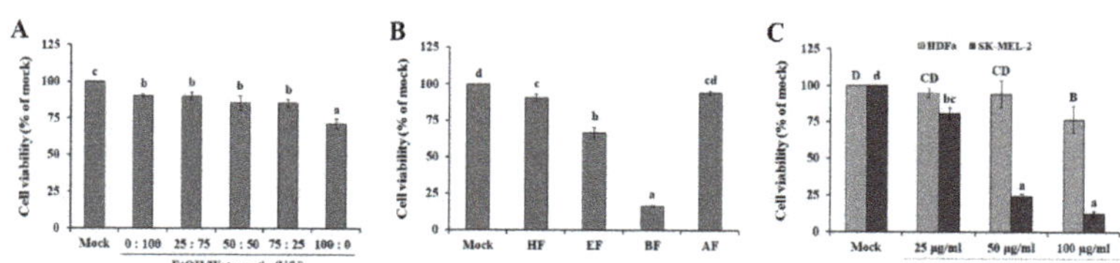

Figure 1. Anti-melanoma activities of *Sorbus commixta* fruit extracts. (**A**) Cytotoxic effects of ethanol concentration on SK-MEL-2 cells. (**B**) Effect of solvent fractions of *Sorbus commixta* fruit ethanol extract on anti-melanoma activity. SK-MEL-2 cells were treated with 100 µg/mL of each extract or solvent fraction. After incubation for 48 h, cell viability was determined via the MTT assay. (**C**) Dose-dependent effect of butanol fraction (BF) on the viability of SK-MEL-2 cells and HDFa human dermal fibroblast adult cells. Dimethyl sulfoxide (DMSO) treated samples served as mock. Values with different letters were found to be significantly different ($p < 0.05$). Values are reported as the mean ± SE. Hexane fraction, HF; ethyl acetate fraction, EF; butanol fraction, BF; aqueous fraction, AF.

Table 1. Effect of ethanol concentration on polyphenolic compound extraction from *Sorbus commixta* fruits.

Polyphenolic Compounds	Water Extract	25% EtOH Extract	50% EtOH Extract	75% EtOH Extract	EtOH Extract
Total phenol content [1]	50.27 ± 4.2 [a]	55.71 ± 3.44 [a]	60.02 ± 9.35 [a]	64.36 ± 6.35 [a]	67.98 ± 4.28 [a]
Total flavonoid content [2]	2.49 ± 0.25 [a]	3.53 ± 0.23 [a]	4.91 ± 0.25 [b]	6.09 ± 0.21 [c]	8.65 ± 0.5 [d]
Chlorogenic acid [3]	111.81 ± 2.54 [a]	127.14 ± 1.56 [b]	150.70 ± 3.40 [c]	176.72 ± 4.58 [d]	344.70 ± 7.18 [e]
Ferulic acid [3]	0.14 ± 0.02 [b]	0.09 ± 0.01 [a]	0.06 ± 0.002 [a]	0.08 ± 0.01 [a]	0.13 ± 0.01 [b]
Hydroxybenzoic acid [3]	0.13 ± 0.01 [a]	0.13 ± 0.02 [a]	0.10 ± 0.01 [a]	0.10 ± 0.01 [a]	0.20 ± 0.01 [b]
Protocatechuic acid [3]	0.79 ± 0.01 [a]	0.78 ± 0.09 [a]	0.92 ± 0.05 [ab]	1.05 ± 0.03 [bc]	1.22 ± 0.06 [c]
Rutin [3]	ND	ND	0.28 ± 0.06 [a]	0.48 ± 0.08 [b]	0.60 ± 0.04 [b]

[1] Total phenolic content analyzed as gallic acid equivalent (GAE) μg/mg of extract; values are the average of triplicate experiments. [2] Total flavonoid content analyzed as quercetin equivalent (QE) μg/mg of extract; values are the average of triplicate experiments. [3] μg/g of extract values are the average of triplicate experiments. Values with different superscripted letters are significantly different ($p < 0.05$). ND = Not detectable.

Organic solvents such as ethyl acetate (relative polarity 0.228), butanol (relative polarity 0.586), and hexane (relative polarity 0.009) were used to partition the crude extract via the liquid–liquid extraction technique, which is a method to separate compounds, based on their relative solubilities in two different immiscible liquids with different polarities. Therefore, this method is commonly used for the separation of a substance from mixture [27]. It is well known that ethanol (relative polarity 0.654) can dissolve polar compounds, such as sugar, amino acid, glycoside compounds, phenolic compounds with medium polarity, aglycon flavonoids, anthocyanins, terpenoids, flavones, and polyphenols [28], indicating that EtOH extract is a complex mixture of organic compounds. Based on the liquid–liquid extraction technique, highly polar substances, such as organic acids, polysaccharides, free sugars, and proteins stay in the aqueous phase, whereas other relatively less polar compounds, such as terpenoids and polyphenols, are more soluble in organic solvents [29,30]. To enrich the active compounds in *S. commixta* fruit EtOH extract, the extracts were partitioned with different solvents, after which the cytotoxicity of each solvent fraction against SK-MEL-2 cells was evaluated. As shown in Figure 1B, the anti-melanoma activities of solvent fractionated EtOH extracts exhibited the following order: BF > EF > HF > AF (83.47%, 33.57%, 9.43%, and 5.56%, respectively). Although several drugs have been used for the treatment of cancer, their cytotoxicity towards normal cells remains a major drawback, which results in secondary malignancy risk [31]. To investigate the effect of BF on the viability of normal cells, HDFa and SK-MEL-2 cells were treated with different concentrations (25 μg/mL, 50 μg/mL, and 100 μg/mL) of BF (Figure 1C). BF exhibited very little cytotoxicity against HDFa cells (5.98%) up to 50 μg/mL, whereas BF showed cytotoxic effects on SK-MEL-2 cells at concentrations of 50 μg/mL, indicating that BF can be a potent source of anticancer agents due to its selective cytotoxicity against melanoma cells. Further experiments were performed at exposure concentrations below (≤50 μg/mL).

Plant-derived polyphenolic compounds (i.e., including flavonoids) have been reported to possess a wide range of pharmacological properties and many polyphenolic compounds have been shown to induce cell cycle arrest and apoptosis in various types of cancer cells [32,33]. Moreover, the anticancer activity of plant extracts has been widely linked to their TPC and TFC. Therefore, our study analyzed TPC and TFC in solvent fractions of *S. commixta* fruit ethanol extracts. As shown in Table 2, BF contained the highest amount of phenolic compounds (87.15 ± 4.66 μg GAE/mg of extract), whereas the lowest level of TPC was observed in HF (19.89 ± 2.65 μg GAE/mg of extract). Additionally, total flavonoid levels were measured in the following order: BF (22.25 ± 0.77 μg QE/mg of extract) > EF (13.54 ± 0.43 μg QE/mg of extract) > HF (6.25 ± 1.25 μg QE/mg of extract) > AF (5.96 ± 1.11 μg QE/mg of extract). These results suggested that the inhibitory effect of *S. commixta* fruit solvent fractions on SK-MEL-2 cell proliferation was significantly related to their TFC and TPC. To further identify the active substance that caused SK-MEL-2 cell cytotoxicity, the polyphenolic compounds of the solvent fractions of *S. commixta* fruit were analyzed and quantified via HPLC. We identified and quantified seven polyphenolic compounds, including chlorogenic acid, ferulic acid, gallic acid, hydroxybenzoic acid, protocatechuic acid, rutin, and sinapinic acid (Table 2). BF contained the highest amount of chlorogenic acid (912.72 ± 68.04 μg/g of extract), which is a known antioxidant, anti-inflammatory, and anti-cancer polyphenol compound [34,35].

Table 2. Polyphenolic compounds in solvent fractions of *Sorbus commixta* fruit ethanol extract.

Polyphenolic Compounds	Hexane Fraction	Ethyl Acetate Fraction	Butanol Fraction	Aqueous Fraction
Total phenol content [1]	19.89 ± 2.65 ^{a}	72.59 ± 5.34 ^{c}	87.15 ± 4.66 ^{d}	44.07 ± 2.77 ^{b}
Total flavonoid content [2]	6.25 ± 1.25 ^{a}	13.54 ± 0.43 ^{b}	22.25 ± 0.77 ^{c}	5.96 ± 1.11 ^{a}
Chlorogenic acid [3]	2.20 ± 0.03 ^{a}	93.12 ± 14.54 ^{b}	912.72 ± 68.04 ^{c}	74.20 ± 2.88 ^{ab}
Ferulic acid [3]	0.02 ± 0.004 ^{a}	1.24 ± 0.05 ^{b}	0.11 ± 0.01 ^{a}	0.02 ± 0.001 ^{a}
Gallic acid [3]	ND	0.17 ± 0.02 ^{b}	0.04 ± 0.02 ^{a}	ND
Hydroxybenzoic acid [3]	0.36 ± 0.02 ^{b}	1.66 ± 0.11 ^{c}	0.13 ± 0.01 ^{a}	0.04 ± 0.003 ^{a}
Protocatechuic acid [3]	0.05 ± 0.003 ^{a}	2.20 ± 0.22 ^{b}	2.86 ± 0.01 ^{c}	0.26 ± 0.05 ^{a}
Rutin [3]	ND	0.12 ± 0.03 ^{a}	2.59 ± 0.17 ^{b}	0.12 ± 0.02 ^{a}
Sinapinic acid [3]	ND	0.13 ± 0.01 ^{b}	0.07 ± 0.01 ^{a}	ND

[1] Total phenolic content analyzed as gallic acid equivalent (GAE) μg/mg of extract; values are the average of triplicate experiments. [2] Total flavonoid content analyzed as quercetin equivalent (QE) μg/mg of extract; values are the average of triplicate experiments. [3] μg/g of extract values are the average of triplicate experiments. Values with different superscripted letters are significantly different ($p < 0.05$). ND = Not detectable.

2.2. BF Blocks the MEK/ERK Signaling Pathway

The mitogen-activated protein kinase (MAPK) cascade is an important signaling pathway involved in cellular processes, such as proliferation, and apoptosis. Importantly, dysregulation of MAPK cascades has been linked to several cancers and other diseases [36]. For instance, the MEK/ERK signaling pathway has been shown to play an important role in tumorigenesis and cancer progression [37], suggesting that MEK and ERK are key protein kinases to target for the discovery of anticancer drugs. Therefore, to investigate the involvement of the MEK/ERK signaling pathway in cell death induced by BF, we analyzed the activation level of the MEK/ERK pathway after treatment with BF. As shown in Figure 2A, activation of MEK1/2 and ERK1/2 was inhibited by BF treatment in a dose-dependent manner, indicating that BF reduces the survival of SK-MEL-2 cells by inhibiting the activation of MEK and ERK.

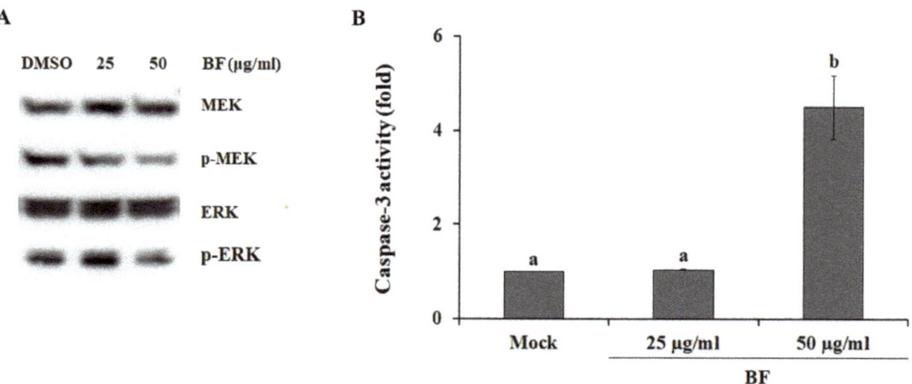

Figure 2. Molecular mechanisms of BuOH fraction (BF)-induced cytotoxicity against SK-MEL-2 cells. (**A**) Effect of the mitogen-activated protein kinase/extracellular regulated kinase (MEK/ERK) signaling pathway on BF-induced cell death. (**B**) Effect of BF on caspase-3 activity in SK-MEL-2 cells. All values are reported as the mean ± SE. DMSO treated samples served as mock. Mean separation within columns was determined via Duncan's multiple range test at a 0.05% level.

Caspases (aspartate-specific cysteine proteases) are a family of protease enzymes with fate-determining roles involved in many cellular processes, including programmed cell death, differentiation, neuronal remodeling, and inflammation [38]. During apoptosis, caspase-3 (i.e., a major executioner caspase) is cleaved at an aspartate residue to yield a p12 and a p17 subunit to form the active caspase-3 enzyme [39], resulting in the cleavage of key structural proteins, cell cycle proteins, and DNase proteins, such as poly (ADP-ribose) polymerase, gelsolin, ICAD/DFF, and DNA-dependent kinase [40]. When SK-MEL-2 cells were treated with 50 µg/mL BF, caspase-3 activity was strongly increased (Figure 2B), indicating that BF induced cell death via caspase-3 activation. ERK has been found to directly phosphorylate pro-caspase-9 to inhibit caspase-9 processing and caspase-3 activation, resulting in the inactivation of the caspase cascade during apoptosis [41], suggesting that BF increased caspase-3 activity by inhibiting ERK activation in SK-MEL-2 cells. Additionally, chlorogenic acid has been shown to inactivate ERK in hepatocellular carcinoma cells [42]. Furthermore, rutin, which exhibits several promising pharmacological properties including antitumor activity [43], was approximately 22 times more abundant in BF (2.59 ± 0.17 µg/g of extract) than in EF (0.12 ± 0.03 µg/g of extract) and AF (0.12 ± 0.02 µg/g of extract) (Table 2). Taken together, these relationships suggest that rutin and chlorogenic acid might be potential anti-melanoma active compounds in *S. commixta* fruit.

In cancer biology, reactive oxygen species (ROS) are known as a double-edged sword, because the imbalance or accumulation of ROS lead to both the survival and death of cancer cells, respectively [44,45]. Particularly, it has been shown that ROS enhances the activation of the Raf/MEK/ERK signaling pathways to promote cancer cell survival, cell proliferation, cell migration, and differentiation [46], whereas ROS accumulation via pro-oxidants under severe oxidative stress conditions leads to apoptosis and cell death [47]. BF-treated SK-MEL-2 cells exhibited similar levels of ROS as compared with mock control (Figure 3), suggesting that inactivation of MEK by BF-treatment (Figure 2A) was not mediated by ROS levels.

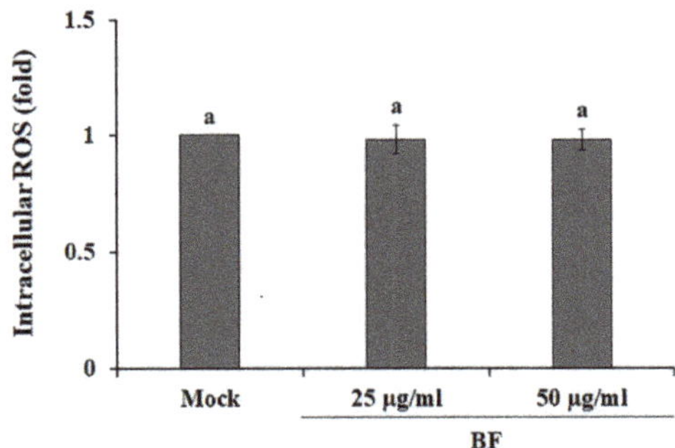

Figure 3. Effects of the BuOH fraction (BF) on reactive oxygen species (ROS) production in SK-MEL-2 cells. ROS level was analyzed using the fluorescent probe DCFH-DA. DMSO treated samples served as mock. Values are reported as the mean ± SE. Values with different letters were found to be significantly different ($p < 0.05$).

3. Materials and Methods

3.1. Plant Materials and Sample Preparation

Fresh *S. commixta* fruits were harvested from the research forest at Chungbuk National University, after which the air-dried fruits were ground into a fine powder using a blender. The ground materials were soaked in water, EtOH/water mixtures (25:75 *v/v* (EtOH 25%), 50:50 *v/v* (EtOH 50%), and 75:25 *v/v* (EtOH 75%)), or absolute EtOH for 24 h at room temperature, then sonicated in an ultrasonic bath (1 h × 3 times). The suspension was then filtered and evaporated under reduced pressure, and lyophilized to produce dried powder extract. Additionally, 150 g of the crude EtOH extract was suspended in 1 L of water and sequentially partitioned with hexane (HF, 1 L × 2 times), ethyl acetate (EF, 1 L × 2 times), and n-butanol (BF, 1 L × 2 times). The remaining aqueous extract was used as an aqueous fraction (AF). After filtration, each fraction was evaporated using a vacuum evaporator.

3.2. Cell Viability Assay

SK-MEL-2 human melanoma cells (ATCC® HTB-68™) and HDFa human dermal fibroblast adult cells (ATCC® PCS-201-012™) were purchased from the American Type Culture Collection (ATCC). Cells were cultured in RPMI 1640 medium or DMEM medium supplemented with 10% fetal bovine serum (FBS), 100 U/mL penicillin, and 100 μg/mL streptomycin at 37 °C in a humidified chamber containing 5% CO_2.

The cytotoxicity of each sample was determined via the MTT assay. Briefly, cells were seeded into 96-well plates at a 5×10^3 cells/well seeding density and incubated at 37 °C in a humidified chamber containing 5% CO_2 for 24 h. The cells were then treated with the above-described extracts or fractions at a 100 μg/mL concentration. After incubation for 48 h, the medium was replaced with 20 μL of 3-(4,5-dimethylthiazol-2-yl)-2,5-diphenyltetrazolium bromide (MTT) solution (1 mg/mL in PBS) for 4 h. Formazan crystals were dissolved in dimethyl sulfoxide (DMSO), and the absorbance was measured at 520 nm using an iMARK microplate reader (Bio-Rad Laboratories GmbH, Munich, Germany).

3.3. Western Blot

After treatment with various BF concentrations for 48 h, SK-MEL-2 cells were analyzed via immunoblotting. The cells were washed with ice-cold PBS and lysed in RIPA buffer (50 mM Tris-HCl pH 7.5, 150 mM NaCl, 1% Triton X-100, 0.1% SDS, 0.5% sodium deoxycholate, 1 mM EDTA, and 10 mM NaF). Protein concentrations were quantified with the Pierce™ BCA Protein Assay Kit (Thermo Fisher Scientific, Waltham, MA, USA). Afterward, 20 μg protein samples were separated via 10% SDS-PAGE and transferred to a PVDF membrane (Millipore, Burlington, MA, USA). After blocking with 5% non-fat dried milk, the membranes were hybridized with MEK1/2, phospho-MEK1/2 (Ser217/221), ERK1/2, and phospho-ERK1/2 (Thr202/Tyr204) (Cell Signalling Technology, Beverly, MA, USA). The signal was detected and visualized via the ECL reagent (SuperSignal™ West Pico PLUS Chemiluminescent Substrate; Thermo Fisher Scientific, Waltham, MA, USA) with an Azure c280 imaging system (Azure Biosystems, Inc., Dublin, CA, USA).

3.4. Caspase-3 Activity and Intracellular ROS Measurement

Protein extraction and the analysis of caspase-3 activity were performed using a Caspase-3/CPP32 Fluorometric Assay Kit (BioVision, Milpitas, CA, USA), according to the manufacturer's instructions. In addition, Intracellular ROS was detected using the DCFH-DA fluorogenic probe as described by Yoo et al. [48]. Caspase-3 activity (400 nm excitation and 505 nm emission) and DCF fluorescence (485 nm excitation and 525 nm emission) were analyzed using a SpectraMax Gemini EM microplate reader (Molecular Devices, San Jose, CA, USA).

3.5. Determination of TPC and TFC

TPC and TFC were measured according to the Folin–Ciocalteu and colorimetric methods, respectively, as described by Jin et al. [49] TPC was expressed in milligrams of gallic acid equivalents (mg GAE/g extract) using an equation obtained from a standard gallic acid graph. TFC was determined as milligrams of quercetin equivalents (QE) per gram of extract (mg QE/g extract).

3.6. HPLC Analysis

HPLC analysis was performed using a Shimadzu liquid chromatography system (LC-10ADvp) coupled with an ultraviolet-visible detector (SPD-10A; Shimadzu, Kyoto, Japan). The samples were separated with PerkinElmer Brownlee SPP columns (2.7 μm C18 2.1 × 150 mm) at 40 °C at a 0.3 mL/min of flow rate. The mobile phases consisted of water containing 0.1% formic acid (mobile phase A) and acetonitrile containing 0.1% formic acid (mobile phase B). The following gradient elution was performed: 10% B in 0–1.6 min, 20% B in 1.6–11.3 min, 25% B in 11.3–17.8 min, 30% B in 17.8–22.7 min, 60% B in 22.7–25.9, and holding at 80% B in 25.9–28.2 min, the samples were finally reconditioned to the initial conditions. Concentrations were calculated by comparing the sample peak areas with the standard calibration curve.

3.7. Statistical Analyses

All experiments were conducted in independent triplicates and significant differences between groups were determined via Duncan's multiple range test. Values of $p < 0.05$ were considered significant.

4. Conclusions

Our study analyzed the effect of different ethanol concentrations on the polyphenolic compound composition and anti-melanoma activity of *S. commixta* fruit extracts. The overall results of the present study suggest that EtOH extract contained a high level of polyphenolic compounds, such as chlorogenic acid, rutin, protocatechuic acid, and hydroxybenzoic acid, all of which are well-known pharmaceutically active compounds, indicating that EtOH extract of *S. commixta* fruits could be a useful source of natural anti-melanoma agents. Additionally, we found that BF induced SK-MEL-2 cell death by increasing caspase-3 activity and inhibiting the MEK/ERK pathway. Taken together, our results provide valuable information for the development of novel anticancer drugs based on *S. commixta* fruit, as well as the optimization of extraction conditions in future studies.

Author Contributions: Conceptualization, S.J., K.-I.J. and T.K.H.; Investigation, S.J., K.C.K. and J.-S.K.; Writing—Original Draft Preparation, S.J., K.-I.J. and T.K.H.; Writing—Review and Editing, S.J., J.-S.K., K.-I.J. and T.K.H. All authors have read and agreed to the published version of the manuscript.

Funding: This work was carried out with the support of "Cooperative Research Program for Agriculture Science and Technology Development (Project No. PJ015285012020)" Rural Development Administration, Republic of Korea.

Conflicts of Interest: The authors confirm that there is no conflict of interest to declare.

References

1. Miller, K.D.; Siegel, R.L.; Lin, C.C.; Mariotto, A.B.; Kramer, J.L.; Rowland, J.H.; Stein, K.D.; Alteri, R.; Jemal, A. Cancer treatment and survivorship statistics, 2016. *CA Cancer J. Clin.* **2016**, *66*, 271–289. [CrossRef]
2. Cummins, D.L.; Cummins, J.M.; Pantle, H.; Silverman, M.A.; Leonard, A.L.; Chanmugam, A. Cutaneous malignant melanoma. *Mayo Clin. Proc.* **2006**, *81*, 500–507. [CrossRef]
3. Hussein, M.R.; Haemel, A.K.; Wood, G.S. Apoptosis and melanoma: Molecular mechanisms. *J. Pathol.* **2003**, *199*, 275–288. [CrossRef] [PubMed]
4. Markovic, S.N.; Erickson, L.A.; Rao, R.D.; Weenig, R.H.; Pockaj, B.A.; Bardia, A.; Vachon, C.M.; Schild, S.E.; McWilliams, R.R.; Hand, J.L.; et al. Malignant melanoma in the 21st century, part 1: Epidemiology, risk factors, screening, prevention, and diagnosis. *Mayo Clin. Proc.* **2007**, *82*, 364–380. [CrossRef]
5. Kato, M.; Liu, W.; Akhand, A.A.; Hossain, K.; Takeda, K.; Takahashi, M.; Nakashima, I. Ultraviolet radiation induces both full activation of Ret kinase and malignant melanocytic tumor promotion in RFP-RET-transgenic mice. *J. Investig. Dermatol.* **2000**, *115*, 1157–1158. [CrossRef] [PubMed]
6. Libra, M.; Malaponte, G.; Navolanic, P.M.; Gangemi, P.; Bevelacqua, V.; Proietti, L.; Bruni, B.; Stivala, F.; Mazzarino, M.C.; Travali, S.; et al. Analysis of BRAF mutation in primary and metastatic melanoma. *Cell Cycle* **2005**, *4*, 1382–1384. [CrossRef] [PubMed]
7. Guevara-Canales, J.O.; Gutiérrez-Morales, M.M.; Sacsaquispe-Contreras, S.J.; Sánchez-Lihón, J.; Morales-Vadillo, R. Malignant melanoma of the oral cavity. Review of the literature and experience in a Peruvian Population. *Med. Oral Patol. Oral Cir. Bucal* **2012**, *17*, e206–e211. [CrossRef] [PubMed]
8. Vikey, A.K.; Vikey, D. Primary malignant melanoma, of head and neck: A comprehensive review of literature. *Oral Oncol.* **2012**, *2012*, 399–403. [CrossRef] [PubMed]
9. Xiao, P.; Zheng, B.; Sun, J.; Yang, J. Biochanin a induces anticancer effects in SK-Mel-28 human malignant melanoma cells via induction of apoptosis, inhibition of cell invasion and modulation of Nf-κB and MAPK signaling pathways. *Oncol. Lett.* **2017**, *14*, 5989–5993. [CrossRef]

10. Srivastava, V.; Negi, A.S.; Kumar, J.K.; Gupta, M.M.; Khanuja, S.P.S. Plant-based anticancer molecules: A chemical and biological profile of some important leads. *Bioorganic Med. Chem.* **2005**, *13*, 5892–5908. [CrossRef]
11. Dar, K.B.; Bhat, A.H.; Amin, S.; Anees, S.; Zargar, M.A.; Masood, A. Herbal compounds as potential anticancer therapeutics: Current. *Ann. Pharmacol. Pharm.* **2017**, *2*, 1106.
12. Lee, T.K.; Roh, H.S.; Yu, J.S.; Kwon, D.J.; Kim, S.Y.; Baek, K.H.; Kim, K.H. A novel cytotoxic activity of the fruit of Sorbus commixta against human lung cancer cells and isolation of the major constituents. *J. Funct. Foods* **2017**, *30*, 1–7. [CrossRef]
13. Sohn, E.J.; Kang, D.G.; Mun, Y.J.; Woo, W.H.; Lee, H.S. Anti-atherogenic effects of the methanol extract of Sorbus cortex in atherogenic-diet rats. *Biol. Pharm. Bull.* **2005**, *28*, 1444–1449. [CrossRef] [PubMed]
14. Yu, T.; Lee, Y.J.; Jang, H.J.; Kim, A.R.; Hong, S.; Kim, T.W.; Kim, M.Y.; Lee, J.; Lee, Y.G.; Cho, J.Y. Anti-inflammatory activity of Sorbus commixta water extract and its molecular inhibitory mechanism. *J. Ethnopharmacol.* **2011**, *134*, 493–500. [CrossRef]
15. Bae, J.T.; Sim, G.S.; Kim, J.H.; Pyo, H.B.; Yun, J.W.; Lee, B.C. Antioxidative activity of the hydrolytic enzyme treated Sorbus commixta Hedl. and its inhibitory effect on matrix metalloproteinase-1 in UV irradiated human dermal fibroblasts. *Arch. Pharm. Res.* **2007**, *30*, 1116–1123. [CrossRef]
16. Sohn, E.J.; Kang, D.G.; Choi, D.H.; Lee, A.S.; Mun, Y.J.; Woo, W.H.; Kim, J.S.; Lee, H.S. Effect of methanol extract of Sorbus cortex in a rat model of L-NAME-induced atherosclerosis. *Biol. Pharm. Bull.* **2005**, *28*, 1239–1243. [CrossRef]
17. Yin, M.H.; Kang, D.G.; Choi, D.H.; Kwon, T.O.; Lee, H.S. Screening of vasorelaxant activity of some medicinal plants used in Oriental medicines. *J. Ethnopharmacol.* **2005**, *99*, 113–117. [CrossRef]
18. Lee, S.-O.; Lee, H.W.; Lee, I.-S.; Im, H.G. The pharmacological potential of Sorbus commixta cortex on blood alcohol concentration and hepatic lipid peroxidation in acute alcohol-treated rats. *J. Pharm. Pharmacol.* **2006**, *58*, 685–893. [CrossRef]
19. Shikov, A.N.; Pozharitskaya, O.N.; Makarov, V.G.; Wagner, H.; Verpoorte, R.; Heinrich, M. Medicinal Plants of the Russian Pharmacopoeia; Their history and applications. *J. Ethnopharmacol.* **2014**, *154*, 481–536. [CrossRef]
20. Raudonis, R.; Raudone, L.; Gaivelyte, K.; Viškelis, P.; Janulis, V. Phenolic and antioxidant profiles of rowan (*Sorbus* L.) fruits. *Nat. Prod. Res.* **2014**, *28*, 1231–1240. [CrossRef]
21. Zymone, K.; Raudone, L.; Raudonis, R.; Marksa, M.; Ivanauskas, L.; Janulis, V. Phytochemical profiling of fruit powders of twenty Sorbus L. Cultivars. *Molecules* **2018**, *23*, 2593. [CrossRef] [PubMed]
22. Sołtys, A.; Galanty, A.; Podolak, I. Ethnopharmacologically important but underestimated genus Sorbus: A comprehensive review. *Phytochem. Rev.* **2020**, *19*, 491–526. [CrossRef]
23. Che Sulaiman, I.S.; Basri, M.; Fard Masoumi, H.R.; Chee, W.J.; Ashari, S.E.; Ismail, M. Effects of temperature, time, and solvent ratio on the extraction of phenolic compounds and the anti-radical activity of Clinacanthus nutans Lindau leaves by response surface methodology. *Chem. Cent. J.* **2017**, *11*, 54. [CrossRef] [PubMed]
24. Do, Q.D.; Angkawijaya, A.E.; Tran-Nguyen, P.L.; Huynh, L.H.; Soetaredjo, F.E.; Ismadji, S.; Ju, Y.H. Effect of extraction solvent on total phenol content, total flavonoid content, and antioxidant activity of Limnophila aromatica. *J. Food Drug Anal.* **2014**, *22*, 296–302. [CrossRef] [PubMed]
25. Nawaz, H.; Shi, J.; Mittal, G.S.; Kakuda, Y. Extraction of polyphenols from grape seeds and concentration by ultrafiltration. *Sep. Purif. Technol.* **2006**, *48*, 176–181. [CrossRef]
26. Chew, K.K.; Khoo, M.Z.; Ng, S.Y.; Thoo, Y.Y.; Aida, W.M.W.; Ho, C.W. Effect of ethanol concentration, extraction time and extraction temperature on the recovery of phenolic compounds and antioxidant capacity of Orthosiphon stamineus extracts. *Int. Food Res. J.* **2011**, *18*, 1427–1435.
27. Khoo, K.S.; Leong, H.Y.; Chew, K.W.; Lim, J.-W.; Ling, T.C.; Show, P.L.; Yen, H.-W. Liquid biphasic system: A recent bioseparation technology. *Processes* **2020**, *8*, 149. [CrossRef]
28. Widyawati, P.S.; Budianta, T.D.W.; Kusuma, F.A.; Wijaya, E.L. Difference of solvent polarity to phytochemical content and antioxidant activity of Pluchea indicia less leaves extracts. *Int. J. Pharmacogn. Phytochem. Res.* **2014**, *6*, 850–855.
29. Haminiuk, C.W.I.; Plata-Oviedo, M.S.V.; de Mattos, G.; Carpes, S.T.; Branco, I.G. Extraction and quantification of phenolic acids and flavonols from Eugenia pyriformis using different solvents. *J. Food Sci. Technol.* **2014**, *51*, 2862–2866. [CrossRef]
30. Chua, L.S.; Lau, C.H.; Chew, C.Y.; Dawood, D.A.S. Solvent fractionation and acetone precipitation for crude saponins from eurycoma longifolia extract. *Molecules* **2019**, *24*, 1416. [CrossRef]

31. Khazir, J.; Mir, B.A.; Pandita, M.; Pilcher, L.; Riley, D.; Chashoo, G. Design and synthesis of sulphonyl acetamide analogues of quinazoline as anticancer agents. *Med. Chem. Res.* **2020**, *29*, 916–925. [CrossRef]
32. Gào, X.; Schöttker, B. Reduction-oxidation pathways involved in cancer development: A systematic review of literature reviews. *Oncotarget* **2017**, *8*, 51888–51906. [CrossRef]
33. Afanas'ev, I. Reactive oxygen species signaling in cancer: Comparison with aging. *Aging Dis.* **2011**, *2*, 219–230. [PubMed]
34. Hadi, S.M.; Asad, S.F.; Singh, S.; Ahmad, A. Putative mechanism for anticancer and apoptosis-inducing properties of plant-derived polyphenolic compounds. *IUBMB Life* **2000**, *50*, 167–171. [CrossRef] [PubMed]
35. Khan, H.Y.; Hadi, S.M.; Mohammad, R.M.; Azmi, A.S. Prooxidant anticancer activity of plant-derived polyphenolic compounds: An underappreciated phenomenon. *Funct. Foods Cancer Prev. Ther.* **2020**, 221–236. [CrossRef]
36. Roberts, P.J.; Der, C.J. Targeting the Raf-MEK-ERK mitogen-activated protein kinase cascade for the treatment of cancer. *Oncogene* **2007**, *26*, 3291–3310. [CrossRef]
37. Guo, Y.; Pan, W.; Liu, S.; Shen, Z.; Xu, Y.; Hu, L. ERK/MAPK signalling pathway and tumorigenesis (Review). *Exp. Ther. Med.* **2020**, *19*, 1997–2007. [CrossRef]
38. Julien, O.; Wells, J.A. Caspases and their substrates. *Cell Death Differ.* **2017**, *9*, 738–742. [CrossRef]
39. O'Donovan, N.; Crown, J.; Stunell, H.; Hill, A.D.K.; McDermott, E.; O'Higgins, N.; Duffy, M.J. Caspase 3 in breast cancer. *Clin. Cancer Res.* **2003**, *9*, 738–742.
40. Ponder, K.G.; Boise, L.H. The prodomain of caspase-3 regulates its own removal and caspase activation. *Cell Death Discov.* **2019**, *28*, 56. [CrossRef]
41. Cheung, E.C.C.; Slack, R.S. Emerging role for ERK as a key regulator of neuronal apoptosis. *Sci. STKE* **2004**, *2004*, 45. [CrossRef] [PubMed]
42. Yan, Y.; Li, J.; Han, J.; Hou, N.; Song, Y.; Dong, L. Chlorogenic acid enhances the effects of 5-fluorouracil in human hepatocellular carcinoma cells through the inhibition of extracellular signal-regulated kinases. *Anti-Cancer Drugs* **2015**, *26*, 540–546. [CrossRef] [PubMed]
43. Caparica, R.; Júlio, A.; Araújo, M.E.M.; Baby, A.R.; Fonte, P.; Costa, J.G.; de Almeida, T.S. Anticancer activity of rutin and its combination with ionic liquids on renal cells. *Biomolecules* **2020**, *10*, 233. [CrossRef] [PubMed]
44. Fiaschi, T.; Chiarugi, P. Oxidative stress, tumor microenvironment, and metabolic reprogramming: A diabolic liaison. *Int. J. Cell Biol.* **2012**, *2012*, 762825. [CrossRef]
45. Liu, D.; Qiu, X.; Xiong, X.; Chen, X.; Pan, F. Current updates on the role of reactive oxygen species in bladder cancer pathogenesis and therapeutics. *Clin. Transl. Oncol.* **2020**, *22*, 1687–1697. [CrossRef]
46. Naveed, M.; Hejazi, V.; Abbas, M.; Kamboh, A.A.; Khan, G.J.; Shumzaid, M.; Ahmad, F.; Babazadeh, D.; FangFang, X.; Modarresi-Ghazani, F.; et al. Chlorogenic acid (CGA): A pharmacological review and call for further research. *Biomed. Pharmacother.* **2018**, *97*, 67–74. [CrossRef]
47. Santana-Gálvez, J.; Villela Castrejón, J.; Serna-Saldívar, S.O.; Jacobo-Velázquez, D.A. Anticancer potential of dihydrocaffeic acid: A chlorogenic acid metabolite. *CYTA J. Food* **2020**, *18*, 245–248. [CrossRef]
48. Yoo, T.K.; Kim, J.S.; Hyun, T.K. Polyphenolic composition and anti-melanoma activity of white forsythia (Abeliophyllum distichum nakai) organ extracts. *Plants* **2020**, *9*, 757. [CrossRef]
49. Jin, S.; Eom, S.H.; Kim, J.S.; Jo, I.H.; Hyun, T.K. Influence of ripening stages on phytochemical composition and bioavailability of ginseng berry (Panax ginseng C.A. Meyer). *J. Appl. Bot. Food Qual.* **2019**, *92*, 130–137. [CrossRef]

Article

Identification of Bioactive Phytochemicals in Leaf Protein Concentrate of Jerusalem Artichoke (*Helianthus tuberosus* L.)

László Kaszás [1], Tarek Alshaal [1,2,*], Hassan El-Ramady [1,2], Zoltán Kovács [1], Judit Koroknai [1], Nevien Elhawat [1,3], Éva Nagy [1], Zoltán Cziáky [4], Miklós Fári [1] and Éva Domokos-Szabolcsy [1]

1. Department of Agricultural Botany, Plant Physiology and Biotechnology (MEK), Debrecen University, Böszörményi Street 138, 4032 Debrecen, Hungary; kaszas.laszlo@agr.unideb.hu (L.K.); ramady2000@gmail.com (H.E.-R.); kovacs.zoltan@agr.unideb.hu (Z.K.); koroknaij@agr.unideb.hu (J.K.); nevienelhawat@gmail.com (N.E.); nagyeva0116@gmail.com (É.N.); fari@agr.unideb.hu (M.F.); szabolcsy@agr.unideb.hu (É.D.-S.)
2. Soil and Water Department, Faculty of Agriculture, Kafrelsheikh University, Kafr El-Sheikh 33516, Egypt
3. Department of Biological and Environmental Sciences, Faculty of Home Economic, Al-Azhar University, Tanta 31732, Egypt
4. Agricultural and Molecular Research and Service Institute, University of Nyíregyháza, 4407 Nyíregyháza, Hungary; cziaky.zoltan@nye.hu

* Correspondence: alshaaltarek@gmail.com; Tel.: +36-2035-404-38

Received: 24 June 2020; Accepted: 13 July 2020; Published: 14 July 2020

Abstract: Jerusalem artichoke (JA) is widely known to have inulin-rich tubers. However, its fresh aerial biomass produces significant levels of leaf protein and economic bioactive phytochemicals. We have characterized leaf protein concentrate (JAPC) isolated from green biomass of three Jerusalem artichoke clones, Alba, Fuseau, and Kalevala, and its nutritional value for the human diet or animal feeding. The JAPC yield varied from 28.6 to 31.2 g DM kg^{-1} green biomass with an average total protein content of 33.3% on a dry mass basis. The qualitative analysis of the phytochemical composition of JAPC was performed by ultra-high performance liquid chromatography-electrospray ionization-Orbitrap/mass spectrometry analysis (UHPLC-ESI-ORBITRAP-MS/MS). Fifty-three phytochemicals were successfully identified in JAPC. In addition to the phenolic acids (especially mono- and di-hydroxycinnamic acid esters of quinic acids) several medically important hydroxylated methoxyflavones, i.e., dimethoxy-tetrahydroxyflavone, dihydroxy-methoxyflavone, hymenoxin, and nevadensin, were detected in the JAPC for the first time. Liquiritigenin, an estrogenic-like flavanone, was measured in the JAPC as well as butein and kukulkanin B, as chalcones. The results also showed high contents of the essential amino acids and polyunsaturated fatty acids (PUFAs; 66-68%) in JAPC. Linolenic acid represented 39–43% of the total lipid content; moreover, the ratio between ω-6 and ω-3 fatty acids in the JAPC was ~0.6:1. Comparing the JA clones, no major differences in phytochemicals, fatty acid, or amino acid compositions were observed. This paper confirms the economic and nutritional value of JAPC as it is not only an alternative plant protein source but also as a good source of biological valuable phytochemicals.

Keywords: circular economy; green biorefinery; polyunsaturated fatty acids; phytochemicals; amino acids; food and feed; UHPLC-ESI-ORBITRAP-MS/MS

---

1. Introduction

The global protein demand continuously grows as the world population exponentially increases. In Europe, the increasing protein dependency particularly obtained from soybean has triggered an urgent need for alternative production systems. Locally grown green biomass crops represent an alternative protein source. Due to high green biomass yield and regrown capacity, clover, alfalfa, and grasses are the most common and prospective plant species for leaf protein isolate. However, digestion of green biomass by monogastric animals is difficult because of its high fiber content [1]. Green biorefinery is a complex processing system with a dedicated goal of making a commercially viable production system of added-value protein based on green biomass [2]. Separating fresh green biomass into two fractions is a key step in the green biorefinery. The fibrous pulp contains insoluble and fiber-bound protein, while the other fraction (green juice) is soluble protein-rich [1]. Soluble proteins in green juice can be precipitated by different techniques. Recovered protein concentrate is separated from brown juice fraction by filtration. Moreover, the quality of leaf protein concentrate as a main product is very important. Based on extended qualitative and quantitative analysis, alfalfa leaf protein concentrate can be directed towards feed and/or food [3,4]. However, besides the well-known herbaceous species, a range of agro-industrial crops is constantly expanding, which can be utilized in the green biorefinery [5].

Jerusalem artichoke (JA), a perennial plant, belongs to the Asteraceae family. Cultivation of JA has many advantages as it is tolerant to biotic stress, i.e., pests and diseases [6]. It can grow normally in a wide range of soils including salt-affected soil, sandy soil, and marginal lands with nearly zero levels of fertilization [7–9]. Moreover, it showed potential resistance to drought, frost, and high temperatures [10]. It yields a huge green biomass almost 120 tons ha^{-1} fresh mass [11]. These aspects are important when avoiding competition with food production on arable lands. The recognized nutritional value of JA is mainly due to the high inulin and fructose contents in its tubers, which additionally contain protein, nutrients, and vitamins [12]. Additionally, JA is well-known as multipurpose use crop where its aerial part has attracted the interest of many researchers, firstly, concerning bioenergy production due to its high lignocellulosic content, high biomass yield, and low inputs [6]. Among the phytochemicals, sesquiterpene lactones, phenolic acids, flavone glucosides (kaempferol 3-*O*-glucoside and quercetin 7-*O*-glucoside), chlorophylls, and carotenoids have been described by several authors in the whole plant or different organs such as tubers, leaves, or flowers [13–19]. These isolated phytochemicals are known as potential anticancer, antidiabetic, antioxidant, antifungal, and antimicrobial in addition to their other medical uses [13,17].

Despite green leafy shoot of JA can be utilized directly as fresh forage, silage, or food pellets for animal feeding [9,12], most of the animal species do not prefer it because of trichome-rich leaves and stems [8]. Considering its high green biomass, regeneration capacity, and chemical composition, leafy shoots of JA can be alternatively used in the green biorefinery practice; however, there is a shortage of knowledge in this area [20].

The objectives of the present work were to produce and characterize the biological value of JAPC originating from the green biomass of JA. We aimed to provide detailed insights into the extraction efficiency and biochemical composition of JAPC. Therefore, three clones of JA representing different climatic zones were grown under low input conditions in Hungary. In addition to total protein, amino acid composition, and fatty acids profile the biochemical composition and qualitative determination of phytochemicals in the JAPCs from these clones were measured using ultra-high performance liquid chromatography-electrospray ionization-Orbitrap/mass spectrometry analysis (UHPLC-ESI-ORBITRAP-MS/MS).

2. Materials and Methods

2.1. Experimental Installation

A field experiment was conducted in 2016 at the Horticultural Demonstration garden of the University of Debrecen, Hungary (47°33′ N; 21°36′ E). Three different clones of JA (i.e., Alba, Fuseau, and Kalevala) were compared for their fresh aerial biomass, phytochemical content, and biochemical traits of the JAPC, under low input conditions. Tubers of JA clones representing three climatic zones were obtained from different sources as follows: Alba was obtained from a Hungarian market; Fuseau was obtained from Ismailia, Egypt; and Kalevala was obtained from Helsinki, Finland. The experiment was set up in a randomized complete block design with six replicates. The area of the experimental plot was 0.8 × 0.6 m2; the row was 3.5 m in length and 0.8 m in width, and within-row spacing was 0.6 m. The cultivation of the JA clones started on 5 April 2016, using identically sized tubers (60–80 g/tuber). Neither irrigation nor fertilization was applied. The chemical characteristics of the experimental soil were as follows: total N (555 ± 2 mg kg^{-1}); total P (6793 ± 17 mg kg^{-1}); total K (1298 ± 7 mg kg^{-1}); and humus (1.9% ± 0.02%).

2.2. *Harvest of Above-Ground Biomass*

Due to the ability of JA plants to regrow, the green biomass of the three clones was harvested twice during the growing season, when young shoots reached 1.3–1.5 m in height from the soil surface. The first harvest was conducted on 27 June 2016, and the second on 8 August 2016. The fresh yield of the aerial parts was measured.

2.3. *Fractionation of Harvested Green Biomass*

The harvest of JA plants was conducted early in the morning and they were immediately transferred to the laboratory in an icebox to prevent the chemical compounds from degrading. The plants were harvested 15–20 cm above the soil surface. A 1 kg harvest of green biomass was mechanically pressed and pulped using a twin-screw juicer (Green Star GS 3000, Toronto, ON, Canada) in three replicates. Thereafter, the green juice was thermally coagulated at 80 °C in one step to obtain the JAPC. The JAPC was separated from the brown-colored liquid fraction using cloth filtration. Both the fresh and dry masses of the JAPC were measured before it was lyophilized using an Alpha 1–4 LSC Christ lyophilizer.

2.4. *Biochemical Composition of JAPC*

2.4.1. Crude Protein Content

The total protein content of the JAPC was measured as total N content using the Kjeldahl method [21]. Briefly, 1 g lyophilized sample was weighed in a 250 mL Kjeldahl digestion tube, then 15 mL concentrated sulfuric acid (99%, VWR Ltd., Debrecen, Hungary) and two catalyst tablets were added. The Kjeldahl digestion tubes were placed in a Tecator Digestor (VELT, VWR Ltd, Debrecen, Hungary) at 420 °C for 1.5 h. The total N content in the digested samples was measured by titration and calculated based on the weight of the titrated solution and the sample weight. The total protein content of the sample was calculated using the following equation: Total protein % = total N content × 6.25.

2.4.2. Quantification of Amino Acid Composition in JAPC Using an Amino Acid Analyzer

Lyophilized and ground samples of JAPC were digested with 6 M HCl at 110 °C for 23 h. Since the digested sample was designed to contain at least 25 mg N, the measured weights of the samples were variable. Alternating application of inert gas and a vacuum using a three-way valve was conducted to remove air. Following hydrolysis, the sample was filtered into an evaporator flask and the filtrate was evaporated under 60 °C to achieve a syrup-like consistency. Thereafter, distilled water was added to the sample and evaporation was conducted twice more under the same conditions. The evaporated sample was washed with citrate buffer pH 2.2. For the analysis of amino acid composition an INGOS AAA500 (Ingos Ltd., Prague, Czech Republic) amino acid analyzer was used. The separation was based on ionic exchange chromatography with post-column derivatization of ninhydrin. A UV/VIS detector was used at 440/570 nm.

2.4.3. Determination of Fatty Acid Composition in JAPC Using Gas Chromatography

The esterification of fatty acids in the JAPC fraction into methyl esters was conducted using a sodium methylate catalyst. Lyophilized homogeneous sample (70 mg) was weighed into a 20 mL tube;

3 mL of n-hexane, 2 mL of dimethyl carbonate and 1 mL of sodium methylate in methanol were added. The contents of the test tube were shaken for 5 min (Janke and Kunkel WX2) and then 2 mL of distilled water was added before the tube was shaken again. The samples were centrifuged at 3000 rpm for 2 min (Heraeus Sepatech, UK). A 2.0 mL sample of supernatant (hexane phase) was transferred into a container through filter paper, which contained anhydrous sodium sulfate. The prepared solution contained approximately 50–70 mg cm−3 fatty acid methyl ester (FAME) and was suitable for analysis by gas chromatography. Gas chromatography was performed using an Agilent 6890 N coupled to an Agilent flame ionization detector. A Supelco Omegawax capillary column (30 m, 0.32 mm i.d., 0.25 μm film thickness) was used to separate FAMEs. The oven temperature was 180 °C and the total analysis time was 36 min. An Agilent 7683 automatic split/splitless injector was used with an injector temperature of 280°C and a 100:1 split ratio. The injection volume was 1 μL. The carrier gas was hydrogen with a flow rate of 0.6 mL min−1 and the makeup gas was N with a flow rate of 25.0 mL min^{-1}. The components were identified from retention data and standard addition.

2.5. *Screening of Phytochemicals in JAPC by UHPLC-ESI-ORBITRAP-MS/MS*

2.5.1. Sample Preparation

To prepare the hydro-alcoholic extracts, 0.5 g ground JAPC powder was extracted with 25 mL methanol:water solution. The mixture was stirred at 150 rpm for 2 h at room temperature. The hydro-alcoholic extracts were filtered using a 0.22 μm PTFE syringe filter.

2.5.2. UHPLC-ESI-ORBITRAP-MS/MS Analysis

Phytochemical analyses were performed using UHPLC-ESI-ORBITRAP-MS/MSwith a Dionex Ultimate 3000RS UHPLC system (Thermo Fisher, Waltham, MA, USA) coupled to a Thermo Q Exactive Orbitrap hybrid mass spectrometer equipped with a Thermo Accucore C18 analytical column (2.1 mm × 100 mm, 2.6 μm particle size). The flow rate was maintained at 0.2 mL/min and the column oven temperature was set to 25 °C ± 1 °C. The mobile phase consisted of methanol (A) and water (B) (both acidified with 0.1% formic acid). The gradient program was as follows: 0–3 min, 95% B; 3–43 min, 0% B; 43–61 min, 0% B; 61–62 min, 95% B; and 62–70 min, 95% B. The injection volume was 2 μL.

2.5.3. Mass Spectrometry Conditions

A Thermo Q Exactive Orbitrap hybrid mass spectrometer (Thermo Fisher, Waltham, MA, USA) was equipped with an ESI source. The samples were measured in both positive and negative ionization modes separately. The capillary temperature was 320 °C and spray voltages were 4.0 kV in positive ionization mode and 3.8 kV in negative ionization mode, respectively. The resolution was 35,000 for MS1 scans and 17,500 for MS2 scans. The scanned mass interval was 100–1500 *m*/*z*. For the tandem MS (MS/MS) scans, the collision energy was set to 30 nominal collision energy units. The difference between measured and calculated molecular ion masses was less than 5 ppm in each case. The data were acquired and processed using Thermo Trace Finder 2.1 software based on own and internet databases (Metlin, Mass Bank of North America, *m*/*z* Cloud). After processing, the results were manually checked using Thermo Xcalibur 4.0 software (ThermoFisher, Waltham, MA, USA).

2.6. *Quality Assurance of Results*

The glass- and plastic-ware used for analyses were usually new and were cleaned by soaking in 10% (v/v) HNO_3 for a minimum of 24 h, followed by thorough rinsing with distilled water. All chemicals were analytical reagent grade or equivalent analytical purity. All equipment was calibrated, and uncertainties were calculated. Internal and external quality assurance systems were applied at the Central Laboratory of the University of Debrecen, according to MSZ EN ISO 5983-1: 2005 (for Total N), and the Bunge Private Limited Company Martfű Laboratory, according to MSZ 190 5508: 1992 (for fatty acid composition).

2.7. Statistical Analysis

Before the ANOVA test, Levene's Test for Equality of Variances was performed. The Levene's test for different variables at all treatments was negative, $p < 0.05$, showing homogeneity of the variances. The experimental design was established as a randomized complete block design with six replicates. The data obtained from the experiments were subjected to one-way ANOVA by 'R-Studio' software and the means were compared by Duncan's Multiple Range Test at $p < 0.05$ [22].

3. Results

3.1. Green Biomass of Jerusalem Artichoke Clones

The yield of the aerial fresh biomass of different JA clones is presented in Table 1. Clones displayed almost the same fresh biomass yield. Hence, no significant differences among the clones (i.e., Alba, Fuseau, and Kalevala) were noticed, especially during the first harvest. The harvest time largely influenced the yield. The average fresh biomass yield was approximately 5.3 kg m^{-2} for the first harvest, while for the second harvest the yield was significantly reduced to 2.4 kg m^{-2} (Table 1). The total aerial fresh biomass yield—as an average—was estimated to be 7.7 kg m^{-2}.

Table 1. Aerial fresh biomass, dry mass, and total protein content of Jerusalem artichoke leaf protein concentrate (JAPC) isolated from green biomass of different clones.

Clones	Fresh Biomass Yield (kg m^{-2})		JAPC (g kg^{-1} Fresh Biomass)		Total Protein %	
	1st Harvest	2nd Harvest	1st Harvest	2nd Harvest	1st Harvest	2nd Harvest
Alba	5.0 ± 0.43 a	1.8 ± 0.22 b	31.9 ± 0.63 a	30.4 ± 0.59 a	35.3 ± 0.8 a	31.6 ± 0.8 b
Fuseau	5.2 ± 0.28 a	2.6 ± 0.19 ab	28.3 ± 0.04 a	28.8 ± 0.25 a	33.3 ± 0.9 a	35.2 ± 0.8 a
Kalevala	5.6 ± 0.65 a	2.8 ± 0.57 a	32.3 ± 0.53 a	28.0 ± 0.13 a	33.8 ± 0.7 a	33.4 ± 0.7 ab

Means followed by different letters in the same column show significant differences according to Duncan's test at $p < 0.05$.

3.2. JAPC Yield

The yield of JAPC, extracted using thermal coagulation, from 1 kg fresh green biomass of the JA clones is displayed in Table 1. No significant differences were seen between the JA clones in either the first or the second harvests. The JAPC yield ranged from 28.3 (Fuseau) to 32.3 (Kalevala) g kg^{-1} fresh biomass for the first harvest, while for the second harvest it varied from 28 (Kalevala) to 30.4 (Alba) g kg^{-1} fresh biomass (Table 1). However, the results showed that the average JAPC dry yield from the first and second harvests was 30.8 and 29.1 g kg^{-1} fresh biomass, respectively. Therefore, 1 kg of green biomass of JA was estimated to yield approximately 30 g JAPC dry mass as an annual average.

3.3. Total Protein Content of JAPC

The total protein content (m/m%) of JAPC generated from fresh green biomass of JA clones ranged between 33.3 m/m% (Fuseau) and 35.3 m/m% (Alba) in the first harvest, while in the second, it varied from 31.6 m/m% (Alba) to 35.2 m/m% (Fuseau). Statistically, no significant differences were calculated either between the clones or harvests (Table 1). The average total protein content in the first harvest was 34.1 m/m% and 33.4 m/m% in the second based on the dry weight. The annual average total protein content of the JAPC extracted from the JA fresh biomass was estimated to be 33.8 m/m% (Table 1).

3.4. Amino Acid Composition of JAPC

The amino acid composition of the JAPC obtained from the green biomass of the JA clones is presented in Table 2. Essential amino acids (i.e., lysine, histidine, isoleucine, leucine, phenylalanine, methionine, threonine, and valine) play a major nutritional role in feed; therefore, they are of special interest. Among the investigated JA clones, Kalevala displayed the highest content of five essential amino acids (i.e., phenylalanine, histidine, isoleucine, threonine, and valine). Additionally, the content of aspartic acid, glycine, glutamic acid, proline, and serine was the highest in Kalevala, with values of 4.23, 2.13, 4.82, 2.20, and 1.90 m/m%, respectively (Table 2). Lysine is particularly important in animal feed and its content in Alba, Fuseau, and Kalevala ranged between 2.19 and 2.32 m/m% in the first harvest. Lysine content in the clones was similar regardless of the harvest time with higher value in the second harvest (2.35–2.54 m/m%) than the first harvest. Methionine is another limiting essential amino acid. The methionine content in Alba and Fuseau clones ranged between 0.82 and 0.95 m/m% in both harvests (Table 2). A reduction in methionine content was found in the second harvest for all clones except Fuseau.

Table 2. Amino acid profile (m/m%) of Jerusalem artichoke leaf protein concentrate (JAPC) extracted from green biomass of different clones.

Amino Acid	1st Harvest			2nd Harvest		
	Alba	Fuseau	Kalevala	Alba	Fuseau	Kalevala
Lysine	2.32 ± 0.02 ‡ a	2.19 ± 0.02 c	2.25 ± 0.02 b	2.35 ± 0.03 c	2.54 ± 0.01 a	2.46 ± 0.02 b
Histidine	0.80 ± 0.20 a	0.71 ± 0.01 b	0.83 ± 0.03 a	0.72 ± 0.02 c	0.76 ± 0.02 b	0.82 ± 0.02 a
Isoleucine	1.72 ± 0.03 a	1.64 ± 0.02 b	1.77 ± 0.02 a	1.72 ± 0.02 bc	1.86 ± 0.02 a	1.78 ± 0.02 ab
Leucine	3.25 ± 0.05 b	3.08 ± 0.02 c	3.31 ± 0.01 a	3.19 ± 0.02 b	2.46 ± 0.02 c	3.30 ± 0.10 a
Phenylalanine	2.12 ± 0.02 b	1.96 ± 0.02 c	2.19 ± 0.01 a	2.03 ± 0.03 b	2.20 ± 0.10 a	2.18 ± 0.02 a
Methionine	0.87 ± 0.03 a	0.84 ± 0.02 a	0.79 ± 0.03 b	0.82 ± 0.02 b	0.95 ± 0.01 a	0.77 ± 0.02 c
Threonine	1.96 ± 0.01 b	1.87 ± 0.02 c	2.33 ± 0.03 a	1.95 ± 0.02 c	2.12 ± 0.02 b	2.33±0.03 a
Valine	2.05 ± 0.05 a	2.02 ± 0.02 a	2.06 ± 0.02 a	2.10 ± 0.02 b	2.34 ± 0.01 a	2.09 ± 0.01 b
Alanine	2.36 ± 0.05 a	2.20 ± 0.10 b	2.35 ± 0.02 a	2.32 ± 0.02 b	2.47 ± 0.02 a	2.34 ± 0.02 b
Arginine	2.08 ± 0.04 a	1.88 ± 0.02 b	1.86 ± 0.01 b	1.87 ± 0.02 c	1.97 ± 0.02 b	2.21 ± 0.01 a
Aspartic acid	3.81 ± 0.01 b	3.63 ± 0.03 c	4.23 ± 0.03 a	3.89 ± 0.02 b	4.23 ± 0.03 a	4.24 ± 0.04 a
Cysteine	0.24 ± 0.02 a	0.22 ± 0.02 a	0.22 ± 0.02 a	0.24 ± 0.02 ab	0.26 ± 0.02 a	0.23 ± 0.03 bc
Glycine	2.04±0.04 b	1.93 ± 0.03 c	2.13 ± 0.01 a	1.99 ± 0.01 b	2.14 ± 0.01 a	2.14 ± 0.02 a
Glutamic acid	4.29 ± 0.01 bc	4.14 ± 0.02 c	4.82 ± 0.02 a	4.38 ± 0.02 c	4.74 ± 0.02 b	4.79 ± 0.02 a
Proline	1.92 ± 0.03 b	1.82 ± 0.02 c	2.20 ± 0.10 a	2.04 ± 0.02 b	2.18 ± 0.01 a	2.19 ± 0.01 a
Serine	1.74 ± 0.04 b	1.67 ± 0.02 b	1.90 ± 0.10 a	1.77 ± 0.02 c	1.89 ± 0.01 b	1.93 ± 0.01 a
Tyrosine	1.48 ± 0.02 a	1.38 ± 0.02 c	1.46 ± 0.02 ab	1.42 ± 0.02 c	1.61 ± 0.01 a	1.55 ± 0.01 b
Ammonia	0.49 ± 0.01 ab	0.47 ± 0.02 b	0.52 ± 0.02 a	0.52 ± 0.02 a	0.48 ± 0.02 b	0.54 ± 0.02 a

‡ Standard deviation. Means followed by different letters in the same row and same harvest show significant differences according to Duncan's test at $p < 0.05$.

3.5. *Qualitative Analysis of JAPC Fatty Acid Composition*

Both saturated (SFA) and unsaturated fatty acids (UFA) were detected in the JAPC. Polyunsaturated fatty acids (PUFA) including linoleic acid (C18:2ω –6) and linolenic acid (C18: 3ω –3) predominated (66%–68%) in all of the JA clones (Figures 1 and 2). Among these fatty acids, linolenic acid (38.6%–42.7%) exhibited a narrow range of content that was present in the highest amount regardless of harvest time or clone. Linoleic acid was present in the second-highest concentration, at a minimum of 23.4% in the first harvest JAPC of Kalevala and a maximum of 26.9% in the second harvest JAPC of Alba. All of the analyzed JAPC samples exhibited a low concentration of unknown fatty acid, which comprised 0.3–0.6% of the total fatty acid content (Figure 1). Among the monounsaturated fatty acids (MUFAs), oleic acid (C18:1ω–9) was detected at a high value (6.6–11.6%), whereas the content of palmitoleic acid (C16:1ω–7) was significantly lower and ranged from 0.7% to 1.1% (Figure 1). The saturated fatty acids (SFA), myristic acid (C14:0), palmitic acid (C16:0), and stearic acid (C18:0) were also identified. Palmitic acid was the most abundant saturated component with no significant differences (16.4–17.9%) either between clones or time of harvest. The percent composition of myristic acid (2.5–6.9%) and stearic acid (1.5–1.8%) in the JAPC fractions were markedly lower than that of palmitic acid. Opposing tendencies were found for the oleic and myristic acid contents between the first and second harvests. The myristic acid content in JAPC was higher in the first harvest in all the three JA clones, while the oleic acid content was higher in the second harvest JAPC of Alba and Kalevala (Figure 1).

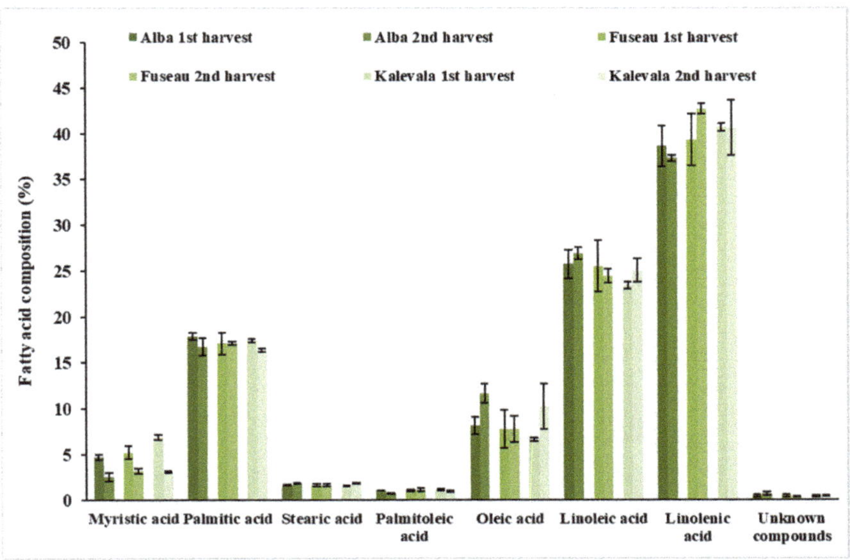

Figure 1. Fatty acid composition (%) of Jerusalem artichoke leaf protein concentrate (JAPC) extracted from the green biomass of three clones (Alba, Fuseau, and Kalevala).

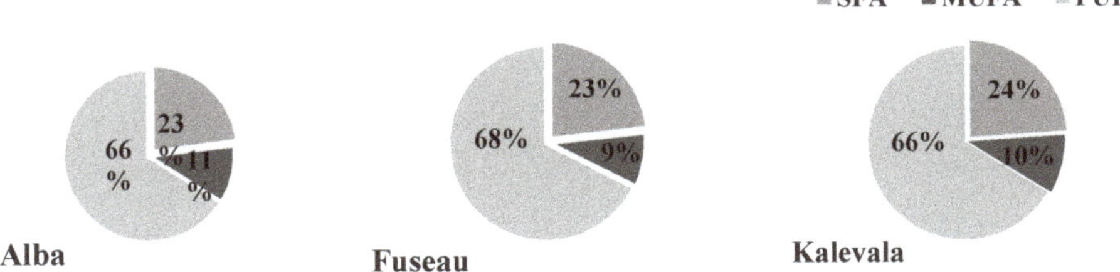

Figure 2. Distribution of saturated fatty acids (SFA), monounsaturated fatty acids (MUFA), and polyunsaturated fatty acids (PUFA) in Jerusalem artichoke leaf protein concentrate (JAPC) extracted from the green biomass of three clones (Alba, Fuseau, and Kalevala).

3.6. Screening JAPC Phytochemicals Using UHPLC-ESI-ORBITRAP-MS/MS

The profiles of the phytochemicals in the JAPCs isolated from the JA clones Alba, Fuseau, and Kalevala, exhibited negligible differences between them. Up to 61 phytochemicals were defined, based on specific retention time, accurate mass, isotopic distribution, and fragmentation pattern, and by screening the following MS databases: Metlin, mzCloud, MoNA-MassBank of North America, and our own database. Table 3 indicates that phenolic compounds comprised a significant component of the compounds identified. Regardless of JA clones, three caffeoylquinic acid isomers: chlorogenic acid (3-*O*-caffeoylquinic acid), neochlorogenic acid (5-*O*-caffeoylquinic acid), and cryptochlorogenic acid (4-*O*-caffeoylquinic acid), respectively, were identified in the JAPCs with a characteristic $[M-H]^-$ ion at *m*/*z* 353.0873. Considering the area of the peak of extracted ion chromatogram of isomers the 3-*O*-caffeoylquinic acid is the dominant one, while a lower ratio of neochlorogenic acid (5-*O*-caffeoylquinic acid) and cryptochlorogenic acid (4-*O*-caffeoylquinic acid) were detected (Figure 3). Additionally, three di-*O*-caffeoylquinic acid isomers ($[M-H]^-$ ion at *m*/*z* 515.1190), four coumaroylquinic acid isomers ($[M-H]^-$ ion at *m*/*z* 337.0924), and a 5-*O*-feruloylquinic acid ($[M-H]^-$ ion at *m*/*z* 367.1029) were identified in the hydro-alcoholic extracted JAPC. The investigation also revealed a compound with a $[M-H]^-$ ion at *m*/*z* 299.0767 in all of the JAPC extracts. The ion scan experiment of this ion showed corresponding fragment ions at m/z values of 137.0233; 113.0229; 93.0331; 85.0281; and 71.0122. After comparison with the databases, this compound was identified as salicylic acid 2-*O*-β-D-glucoside.

Table 3. Chemical composition of Jerusalem artichoke leaf protein concentrate (JAPC) extracted from green biomass.

No.	Compound	Formula	Retention Time	Measured Mass (m/z)		Fragments 1	Fragments 2	Fragments 3	Fragments 4	Fragments 5
				$[M+H]^+$	$[M-H]^-$					
1	γ-Aminobutyric acid	$C_4H_9NO_2$	1.25	104.07116		87.0446	86.0607	69.0342	58.0658	
2	Quinic acid	$C_7H_{12}O_6$	1.27		191.05557	173.0447	171.0289	127.0388	93.0331	85.0280
3	Betaine (Trimethylglycine)	$C_5H_{11}NO_2$	1.28	118.08681		59.0737	58.0659			
4	Malic acid	$C_4H_6O_5$	1.33		133.01370	115.0024	89.0230	87.0075	72.9916	71.0123
5	Nicotinic acid (Niacin)	$C_6H_5NO_2$	1.51	124.03986		96.0450	80.0501	78.0347		
6	Citric acid	$C_6H_8O_7$	1.73		191.01918	173.0082	129.0182	111.0075	87.0073	85.0280
7	Neochlorogenic acid (5-*O*-Caffeoylquinic acid)	$C_{16}H_{18}O_9$	10.14		353.08726	191.0557	179.0344	173.0448	135.0441	
8	Salicylic acid-2-*O*-glucoside	$C_{13}H_{16}O_8$	13.56		299.07670	137.0234	113.0229	93.0331	85.0280	71.0123
9	Chlorogenic acid (3-*O*-Caffeoylquinic acid)	$C_{16}H_{18}O_9$	14.83		353.08726	191.0556	179.0344	173.0443	161.0234	135.0441
10	Cryptochlorogenic acid (4-*O*-Caffeoylquinic acid)	$C_{16}H_{18}O_9$	16.11		353.08726	191.0555	179.0344	173.0447	161.0232	135.0441
11	4-*O*-(4-Coumaroyl) quinic acid	$C_{16}H_{18}O_8$	16.14		337.09235	191.0555	173.0447	163.0390	119.0489	93.0331
12	Vanillin (4-Hydroxy-3-methox ybenzaldehyde)	$C_8H_8O_3$	16.22	153.05517		125.0600	111.0445	110.0366	93.0341	65.0393
13	5-*O*-(4-Coumaroyl)quinic acid	$C_{16}H_{18}O_8$	17.38		337.09235	191.0556	173.0447	163.0391	119.0490	93.0332
14	Indole-3-acetic acid	$C_{10}H_9NO_2$	17.98		174.05551	146.0601	144.0440	130.0651	128.0492	
15	4-*O*-(4-Coumaroyl)quinic acid cis isomer	$C_{16}H_{18}O_8$	18.04		337.09235	191.0556	173.0447	163.0391	119.0489	93.0331
16	Isoscopoletin (6-Hydroxy-7-meth oxycoumarin)	$C_{10}H_8O_4$	18.33	193.05009		178.0264	165.0550	149.0598	137.0600	133.0287
17	5-*O*-Feruloylquinic acid	$C_{17}H_{20}O_9$	18.42		367.10291	193.0503	191.0556	173.0447	134.0362	93.0331
18	Riboflavin	$C_{17}H_{20}N_4O_6$	19.03	377.14611		359.1352	243.0879	200.0824	172.0872	69.0342
19	Scopoletin (7-Hydroxy-6-meth oxycoumarin)	$C_{10}H_8O_4$	19.08	193.05009		178.0263	165.0546	149.0597	137.0601	133.0287
20	Azelaamic acid (9-Amino-9-oxononanoic acid)	$C_9H_{17}NO_3$	19.21		186.11302	125.0959	97.0647			
21	6-Methylcoumarin	$C_{10}H_8O_2$	19.44	161.06026		133.0651	115.0547	105.0704	91.0547	79.0549
22	5-*O*-(4-Coumaroyl)quinic acid cis isomer	$C_{16}H_{18}O_8$	19.63		337.09235	191.0555	173.0446	163.0390	119.0491	93.0330
23	Indole-4-carbaldehyde	C_9H_7NO	19.67	146.06059		118.0655	117.0574	91.0548		
24	Fraxidin or Isofraxidin	$C_{11}H_{10}O_5$	19.72		221.04500	206.0219	190.9983	163.0030		
25	Loliolide	$C_{11}H_{16}O_3$	20.05	197.11777		179.1069	161.0962	135.1171	133.1015	107.0860
26	4-Hydroxy-3-meth oxycinnamaldehyde (Coniferyl aldehyde)	$C_{10}H_{10}O_3$	20.59	179.07082		161.0599	147.0442	133.0652	119.0495	55.0186
27	7-Deoxyloganic acid isomer	$C_{16}H_{24}O_9$	22.36		359.13421	197.0815	153.0909	135.0805	109.0643	89.0230
28	Di-*O*-caffeoylquinic acid isomer 1	$C_{25}H_{24}O_{12}$	22.61		515.11896	353.0884	191.0556	179.0342	173.0447	135.0441
29	Di-*O*-caffeoylquinic acid isomer 2	$C_{25}H_{24}O_{12}$	22.77		515.11896	353.0884	191.0556	179.0342	173.0446	135.0440
30	Salvianolic acid derivative isomer 1	$C_{27}H_{22}O_{12}$	22.80		537.10331	375.0705	201.0165	179.0343	161.0234	135.0440
31	Butein (2′,3,4,4′-Tetrah ydroxychalcone)	$C_{15}H_{12}O_5$	23.00	273.07630		255.0656	227.0699	209.0602	163.0391	137.0235
32	Quercetin-3-*O*-glucuronide	$C_{21}H_{18}O_{13}$	23.26		477.06692	301.0359	178.9980	163.0028	151.0026	121.0281
33	Isoquercitrin (Hirsutrin, Quercetin-3-*O*-glucoside)	$C_{21}H_{20}O_{12}$	23.47		463.08765	301.0358	300.0283	271.0253	255.0300	
34	Chrysoeriol-*O*-glucoside	$C_{22}H_{22}O_{11}$	23.87		461.10839	299.0560	298.0484	270.0537	255.0292	227.0346
35	Salvianolic acid derivative isomer 2	$C_{27}H_{22}O_{12}$	24.60		537.10331	375.0705	201.0166	179.0343	161.0236	135.0440
36	Di-*O*-caffeoylquinic acid isomer 3	$C_{25}H_{24}O_{12}$	24.62		515.11896	353.0884	191.0557	179.0342	173.0447	135.0440

Table 3. *Cont.*

No.	Compound	Formula	Retention Time	Measured Mass (*m*/*z*)		Fragments 1	Fragments 2	Fragments 3	Fragments 4	Fragments 5
				$[M + H]^+$	$[M - H]^-$					
37	Azelaic acid	$C_9H_{16}O_4$	25.05		187.09704	169.0863	143.1070	125.0959	123.0803	
38	Kaempferol-3-*O*-glucuronide	$C_{21}H_{18}O_{12}$	25.18		461.07200	285.0410	229.0505	113.0231		
39	Apigenin-*O*-malonylglucoside	$C_{24}H_{22}O_{13}$	25.21		517.09822	473.1116	269.0461	268.0376		
40	Astragalin (Kaempferol-3-*O*-glucoside)	$C_{21}H_{20}O_{11}$	25.26		447.09274	285.0410	284.0331	255.0302	227.0350	
41	Isorhamnetin-3-*O*-glucoside	$C_{22}H_{22}O_{12}$	25.48		477.10330	315.0524	314.0437	285.0406	271.0248	243.0292
42	Kukulkanin B (2′,4′,4-Trihydroxy-3′-methoxyxchalcone)	$C_{16}H_{14}O_5$	25.50	287.09195		269.0810	241.0864	177.0548	145.0286	137.0235
43	Isorhamnetin-3-*O*-glucuronide	$C_{22}H_{20}O_{13}$	25.70		491.08257	315.0517	300.0275	271.0249		
44	Dihydroactinidiolide	$C_{11}H_{16}O_2$	27.16	181.12286		163.1119	145.1014	135.1171	121.1015	107.0860
45	Dimethoxy-tetrahydroxyflavone	$C_{17}H_{14}O_8$	28.38		345.06105	330.0386	315.0153	287.0204	215.0347	178.9978
46	Dihydroxy-methoxyflavone	$C_{16}H_{12}O_5$	29.89		283.06065	268.0381	267.0305	240.0427	239.0350	211.0396
47	Dimethoxy-trihydr oxyflavone isomer 1	$C_{17}H_{14}O_7$	30.09		329.06613	314.0439	299.0197	283.0869	271.0247	255.0913
48	Trihydroxy-trime thoxyflavone	$C_{18}H_{16}O_8$	30.36		359.07670	344.0541	329.0307	314.0075	301.0358	286.0129
49	Dimethoxy-trihyd roxyflavone isomer 2	$C_{17}H_{14}O_7$	30.38		329.06613	314.0439	299.0201	283.0871	271.0252	253.0763
50	Liquiritigenin (4′,7-Dihydroxyflavanone)	$C_{15}H_{12}O_4$	30.56		255.06574	153.0183	135.0077	119.0489	91.0175	
51	Hymenoxin (5,7,Dihydroxy-3′,4′, 6,8-tetramethoxyflavone)	$C_{19}H_{18}O_8$	32.11	375.10800		360.0840	345.0606	342.0736	330.0367	317.0659
52	Epiafzelechin trimethyl ether	$C_{18}H_{20}O_5$	33.32	317.13890		167.0704	163.0755	155.0705	137.0598	121.0651
53	Nevadensin (5,7-Dihydroxy-4′, 6,8-trimethoxyflavone)	$C_{18}H_{16}O_7$	33.91	345.09743		330.0736	315.0501	312.0631	287.0554	

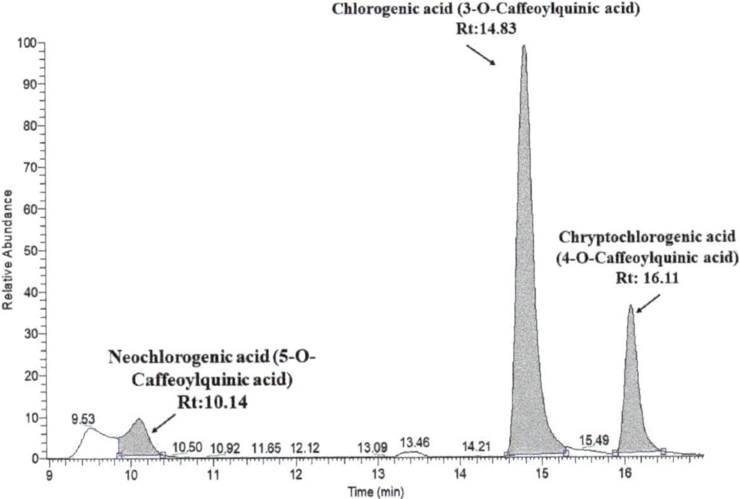

Figure 3. Extracted ion chromatogram of chlorogenic acid structural isomers.

Among flavonoids, isorhamnetin-3-*O*-glucoside with *m*/*z* 477.1033, kaempferol 3-glucuronide (kaempferol 3-*O*-β-D-glucopyranosiduronic acid) with *m*/*z* 461.0720, and astragaline (kaempferol 3-*O*-β-D-glucopyranoside) with *m*/*z* 447.0927 was found in the JAPC. However, to our knowledge, this is the first time glucuronide derivatives of isorhamnetin (isorhamnetin-3-*O*-glucuronide) and isoquercetin (quercetin 3-*O*-β-D-glucopyranoside) with *m*/*z* 463.0877 (Table 3 and Figure 4) have been identified. In addition to flavonols, most of the identified flavonoids belonged to the flavones. As far as we are aware, none of these has been identified previously in JAPC. For instance, we identified two dimethoxy-trihydroxyflavone isomers ([M − H]$^-$ ion at *m*/*z* 329.0661), dimethoxy-tetrahydroxyflavone ([M − H]$^-$ ion at *m*/*z* 345.0611), dihydroxy-methoxyflavone ([M − H]$^-$ ion at *m*/*z* 283.0607), and trihydroxy-trimethoxyflavone ([M − H]$^-$ at *m*/*z* 359.0767). Hymenoxin (5,7-dihydroxy-3′,4′,6,8-tetramethoxyflavone) at *m*/*z* 375.1080 and nevadensin (5,7-hydroxy-4′,6,8-trimethoxyflavone) at *m*/*z* 317.1389 were identified in positive ESI mode (Table 3). Within flavonoids, Butein (2′,3,4,4′-tetrahydroxychalcone) and kukulkanin B (3′-methoxy-2′,4,4′-methoxychalcone) which related to chalcones subgroup were identified. Finally, liquiritigenin (4′,7-dihydroxyflavanone; [M − H]$^-$ at *m*/*z* 255.0657) was the only flavanone found in this study (Figure 4).

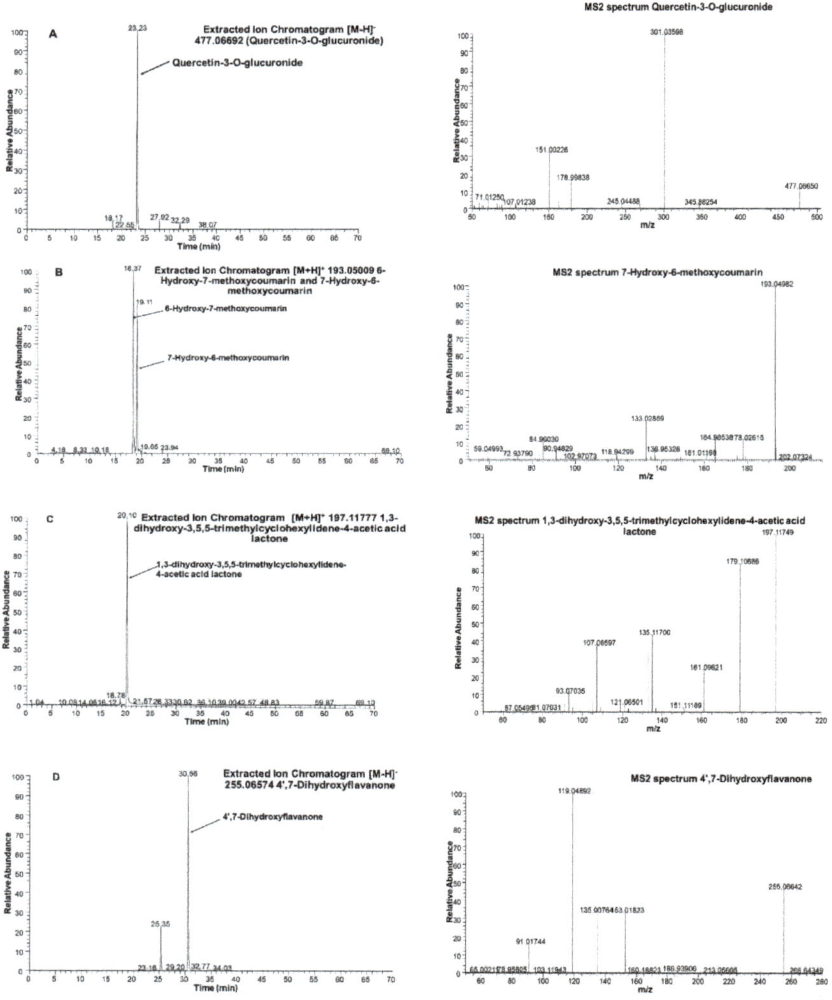

Figure 4. Extracted Ion Chromatograms (XIC) and MS spectra of selected phytoconstituents from Jerusalem artichoke leaf protein concentrate: (**A**): quercetin-3-*O* -glucuronide; (**B**): 7-Hydroxy-6-methoxycoumarin (Scopoletin); (**C**): 1,3-dihydroxy-3,5,5-trimethylcyclohexylidene -4-acetic acid lactone (Loliolide); and (**D**): 4′,7-Dihydroxyflavanone (Liquiritigenin).

In addition to polyphenols, three different terpenes consistently appeared in the JAPC of the JA clones. Loliolide (1,3-dihydroxy-3,5,5-trimethylcyclohexylidene-4-acetic acid lactone) is a C_{11} monoterpenoid lactone, which was observed with a $[M + H]^+$ ion at *m*/*z* 197.1178 (Figure 4). Dihydroactinidiolide as a volatile monoterpene with a $[M + H]^+$ ion at *m*/*z* 181.1229, and 7-deoxyloganic acid isomer, an iridoid monoterpene with a $[M - H]^-$ ion at *m*/*z* 359.1342, were recognized. Several proteinogenic amino acids were also identified (Table 3). In terms of vitamins, vitamin B molecules such as nicotinic acid (niacin; $[M + H]^+$ ion at *m*/*z* 124.0399) and riboflavin ($[M + H]^+$ ion at *m*/*z* 377.1461) were seen, while organic acids, i.e., malic acid and citric acid, and plant hormones such as indole acetic acid, were also identified in the JAPC.

4. Discussion

One important aspect of the biorefinery to become a competitive process is to produce at least one product of high value. The quantitative analysis of crude protein content of JAPC is a priority. The protein content of JAPC is influenced by plant type and also by the processing method. The average total protein content of the JAPC produced from Alba, Fuseau, and Kalevala was 33.4 m/m%; however, most of the isolated protein was found in the leaves, as these organs contain 3-fold higher total protein than the stem [23]. The JAPC comprised parenchyma tissues (80–87%) containing easily released cytoplasmic and chloroplast proteins such as Rubisco, which is of high nutritional value [24]. The time of harvest is critical to the quantity and quality of the JAPC produced from the aerial parts of the JA. Rashchenko [25] reported that the N content of older leaves is ~50% less than that in young leaves and Seiler [26] reported that the total protein content fell by 32.6% between the vegetative and flowering stages of JA growth. Knowing this, the shoots were harvested at the point of the maximum green leaf; ahead of senescence and before the bottom leaves turn dry. Ultimately, there was no significant difference in protein content between the two harvests.

In terms of an ideal protein source, the amino acid profile cannot be ignored, because among the 20 proteinogenic amino acids, nine cannot be synthesized by most animal species [20]. The content of these essential amino acids is, therefore, of particular interest. Among the green biomass fractions, the JAPC, as a dedicated protein enriched product for feed, was examined thoroughly. Several indispensable amino acids, i.e., lysine, isoleucine, leucine, methionine, and threonine, were present in high concentrations in the JAPC. However, even higher amino acid contents were found in JAPC by Rawate and Hill [27]; this may be attributed to different extraction methods and varieties. Additionally, the amino acid profiles exhibited minor differences between the two harvests, which may be due to differences in weather and plant age, as has previously been documented [11,25,26].

Considering the scientific literature about phytoconstituents of different JA organs, it was assumed that the green biomass-originated JAPC can be more than an alternative protein source. Qualitative analysis of phytochemicals in JAPC was performed by UHPLC-ESI-MS in both negative and positive ESI modes. The negative mode was used to identify flavonoid and phenolic acid (hydroxycinnamic acid and benzoic acid) derivatives, as it provided better sensitivity. The easy protonation of N in the positive mode made it suitable for identifying terpenes, amino acids, coumarins, and coumaroylquinic acids.

Phenolic compounds are one of the largest groups of plant secondary metabolites. Among them, phenolic acids are an important subgroup and their presence is characteristic of the Asteraceae family. The most revealed phenolic acids are the mono- and di, and even tri-hydroxycinnamic acid (p-coumaric, caffeic, and ferulic acids) esters of quinic acids in the tuber and shoot organs of JA [15,17,19]. Our measurements confirmed 13 different "phenolic acids" from green biomass originated hydro-alcoholic extracted JAPC. The three structural isomers of caffeoylquinic acid were identified with a similar degree of ionization, and the same molecular weight and fragmentation pattern. Hence, the area of the peak of extracted ion chromatogram of isomers is comparable and the 3-*O*-caffeoylquinic acid seemed to be the dominant one (Figure 3). However, neochlorogenic acid (5-*O*-caffeoylquinic acid) displayed the lowest ratio. Chlorogenic acid (3-*O*-caffeoylquinic acid) is known as the most abundant isomer in plants, whereas cryptochlorogenic acid (4-*O*-caffeoylquinic acid) and neochlorogenic acid (5-*O*-caffeoylquinic acid) are present in much lower concentration [27]. Yuan et al. [15] cited 3-*O*-caffeoylquinic acid and 1,5-dicaffeoylquinic acid in high concentrations in JA leaves. However, Liang and Kitts [28] mentioned that 5-*O*-caffeoylquinic acid is the predominant isomer in fruits and vegetables. The presence of these phenolic acids is interesting from the aspect of both humans and animals, as several biological roles are attributed to caffeoylquinic acid isomers including antioxidant and antibacterial activities, hepato- and cardio-protection, anti-inflammatory and antipyretic activities, neuroprotection, anti-obesity, antiviral, and anti-hypertension activities, and central nervous system stimulation. Additionally, these compounds modulate lipid metabolism and glucose levels in both genetic metabolism-related disorders and healthy people [15,29]. Based on their health-promoting effects, caffeoylquinic acid isomers are increasingly recommended as natural and safe food additives, in place of synthetic antibiotics and immunity boosters.

The four different coumarins have also been revealed in the JAPC. Coumarins are widely distributed non-flavonoid polyphenols in the plant kingdom (Figure 5B). However, "simple coumarins" as coumarin subgroup is mainly present in the Asteraceae family. Therefore, each coumarin subclass-related compounds are used for the chemotaxonomic approach, too. Scopoletin and ayapin were already described in tubers of JA and assumed the presence of them in aerial part as in the case of *Helianthus annuus* [30]. Our measurement confirmed the presence of scopoletin along with isoscopoletin, 6-methyl coumarin, and fraxidin from green biomass originated product of JA. Some of simple coumarins are known as phytoalexins. At the same time, fraxidin and scopoletin have also shown potent antiadipogenic activity against the preadipocyte cell line in vitro assay systems [31]. Within non-flavonoid phenolics, two salvianolic acid derivatives and a salicylic acid-2-*O*-glucoside were also in detectable amounts.

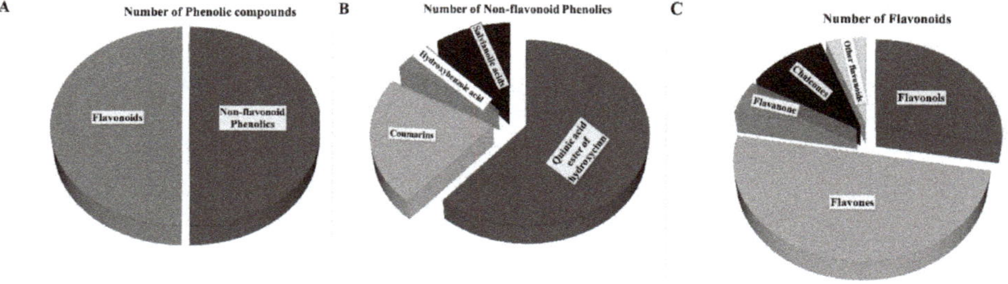

Figure 5. Identified phenolic compounds from Jerusalem artichoke leaf protein concentrate: (**A**) ratio of flavonoid and non-flavonoid phenolic compounds; (**B**) number of identified compounds within non-flavonoid phenolics subgroup; and (**C**) number of identified compounds within flavonoid phenolics subgroup.

Flavonoids are widespread secondary metabolites that occur as part of the phenolic constituents of plants. However, only a few of these have been described in the aerial part of JA including isorhamnetin glucoside, kaempferol glucuronide, and kaempferol-3-*O*-glucoside [14]. Based on the present qualitative analysis, 18 flavonoid compounds were revealed in the JAPC as green biomass originated product (Figure 5). Generally, cell vacuoles are the main storage places for soluble flavonoids. The JPAC is mostly made up of content released from the cytoplasm and vacuoles cell fractions, which may be the reason for the relatively high proportion of identified flavonoids. Within flavonoids, five flavonols were detected in the JAPC, in which all of them occurred as glycosides. Primarily, the solubility of flavonoids is due to their sugar substitutions. Among the sugars, glucose and glucuronic acid at a single position are probably the most common substituents [32]. The importance of flavonoid glucuronides is related to their health-promoting activities such as the anti-inflammatory and neuroprotective activities of quercetin-3-*O*-glucuronide [33]. Most of the identified flavonoid compounds belong to the flavones (Figure 5C). All of the flavone compounds were hydroxylated methoxyflavones, which contain one or more methoxy groups instead of a hydroxyl group on a flavone framework. The substitution of a methoxy group for a hydroxyl group in flavones has significant importance. One side the hydroxyl groups of flavones have free radical scavenging activity, but extensive conjugation of free hydroxyl groups to flavones results in low oral bioavailability; hence, they undergo rapid sulfation and glucuronidation in the small intestine and liver by phase II enzymes. Consequently, conjugated metabolites, but not the original compounds, can be found in plasma [34]. However, if one or more hydroxyl groups are capped by methylation, the substitution of a methoxy group by the hydroxyl group induces an increase in metabolic stability and improves transport and absorption. Considering the biological properties and chemical characteristics of hydroxyl and methoxy groups together, the hydroxylated methoxyflavones combine many advantages from both functional groups, improving their potential for application in human health [34]. Therefore, the presence of several hydroxylated methoxyflavones such as dimethoxy-trihydroxyflavone isomers, dimethoxy-tetrahydroxyflavone, dihydroxy-methoxyflavone, trihydroxy-trimethoxyflavone, hymenoxin, and nevadensin, increase the value of JAPC.

From minor flavonoids, two chalcones were detected in JAPC. Butein (2′,3,4,4′ -Tetrahydroxychalcone) is one of them which is widely biosynthesized in plants; however, no reference has been found citing it in JA. Based on preclinical studies, butein exhibits significant therapeutic potential against various diseases. In vitro and in vivo studies support that butein can suppress proliferation and trigger apoptosis in various human cancer cells with no or only minimal toxicity inducing in normal cells [35].

Liquiritigenin (4′,7-dihydroxyflavanone) as the only flavanone was measured in JA flowers by Johansson et al. [13]. Our results confirmed the presence of liquiritigenin in JAPC, too. Liquiritigenin is known to be a promising active estrogenic compound and is a highly selective estrogen receptor β agonist, which may be helpful to women who suffer from menopausal symptoms [36].

Three terpenes consistently appeared in the tested JAPC from all the JA clones. Loliolide, a C_{11} monoterpenoid lactone, is considered to be a photo-oxidative or thermally degraded product of carotenoids [37]. Similarly, we identified dihydroactinidiolide, a volatile monoterpenoid, which is a flavor component of several plants such as tobacco and tea. According to Yun et al. [38], thermal treatment induces the formation of dihydroactinidiolide from β-carotene. Kaszás et al. [8] confirmed that the green juice of the JAPC contains a marked number of carotenoids, which may be able partially to convert to loliolide or dihydroactinidiolide, causing the number of detectable terpenes to increase. Studies have confirmed that loliolide inhibits growth and germination, while also being phytotoxic, repelling leaf-cutter ants and having antitumor and antimicrobial activities in animals and microorganisms [37,39]. Dihydroactinidiolide has a carbonyl group that can react with nucleophilic structures in macromolecules, providing high potential reactivity to the molecules. It also shows cytotoxic effects against cancer cell lines [38]. The 7-Deoxyloganic acid isomer is the third terpene, which is known to be an intermediate in the secoiridoid pathway in plants.

The fatty acid and lipid contents of JA tubers have been reported by several authors [11,40]; however, little information is available about the fatty acid composition of its leaves and JAPC [41]. Recently, rapidly growing interest is for PUFAs, because humans and other mammals are incapable of synthesizing omega-6 and omega-3 PUFAs, due to the lack of Δ12 and Δ15 desaturase enzymes, which insert a cis double bond at the n-6 and n-3 positions [42]. Hence, linolenic and linoleic acids are essential nutrients converted from oleic acid in the endoplasmic reticulum of plant cells. Linolenic acid is the precursor of longer-chain PUFAs such as eicosapentaenoic acid (EPA: C20:5ω −3) and docosahexaenoic acid (DHA: C22:6ω −3), which can be synthesized in humans. Similarly, linoleic acid is an essential precursor to dihomo-γ-linoleic acid (DGLA: C20:3ω −6) and arachidonic acid (C20:4ω −6). As they are essential to life, linolenic and linoleic acids must be supplied to animals and humans through diet. In the JAPC of all JA clones, the highest contribution to the fatty acid profile was made by linolenic acid (38.6–42.7%) and linoleic acid (23.4–26.9%) as shown in Figures 1 and 2. The correct proportions of linoleic and linolenic acids are emphasized by anthropological and epidemiological studies. The required ratio of omega-6 to omega-3 essential fatty acids is ~1:1, according to the evolutionary history of the human diet. In contrast, in the current Western diet, this ratio has shifted to 10–20:1, which is not beneficial to health and promotes the pathogenesis of many diseases [43]. We found a ratio of ~0.6:1 for omega-6 to omega-3 essential fatty acids in JAPC, which is very favorable and close to Paleolithic nutrition levels.

Concerning the harvest time, Alba and Kalevalas' JAPC exhibited higher oleic acid contents in the second harvest (when the nights were cooler). According to Barrero-Sicilia et al. [44], plants often respond to low temperature by increasing the levels of unsaturated fatty acids in the membrane and increasing membrane fluidity and stabilization. We found the opposing tendency in the saturated myristic acid (C14:0) contents of the JAPC, in which levels were higher in the first harvest (when the nights were warmer).

In summary, this study delivers deeper insights into JAPC originating from the fractionated green biomass of different JA clones focusing on its content of different phytochemicals that are potentially bioactive compounds and have several important uses. Future studies should investigate the anti-nutritional ingredients of JAPC, analyze the chemical composition of other fractions such as the brown juice and fiber, and calculate the economic viability of JA crops.

5. Conclusions

This paper discusses two important points related to JA. Firstly, we examined the potential production of leaf protein from its aerial parts. Secondly, we aimed to determine the quality of produced JAPC as a promising protein source that could be directed to human consumption and/or animal feeding. Biochemical analyses revealed that the JAPC is not only a good source of protein with a favorable amino acid composition but also a repository of essential fatty acids, flavonoid and non-flavonoid phytonutrients. The saturated palmitic acid (C16:0), stearic acid (C18:0), and the monosaturated form of stearic acid, oleic acid (C18:1ω–9), are often referred to as common fatty acids. They are biosynthesized in the plastids and partially incorporated into the cell and subcellular membranes [45]. The JAPC originates mainly from crushed cells of vegetative tissues containing membrane debris, which explains the relatively higher proportion of palmitic (16.4–17.9%) and oleic (6.6–11.6%) acids. Moreover, several important viable compounds were detected in JAPC. These compounds are known for their antibacterial, anti-inflammatory, and antipyretic activities, neuroprotection, anti-obesity, antiviral, and anti-hypertension activities, hepato- and cardio- protection, and central nervous system stimulation. However, the quantity and quality of the phytochemicals are specific to the species and can vary with the bioanalytical technology used. Hence, a quantitative analysis of identified phytocompounds needs to confirm the nutritional value of JAPC. Overall, the present results confirm that the green aerial parts of this underestimated plant can be a source of marketable products involving into green biorefinery concept.

Author Contributions: Conceptualization, L.K., M.F., and É.D.-S.; data curation, L.K. and É.D.-S.; formal analysis, É.D.-S.; funding acquisition, M.F.; methodology, Z.K.; project administration, É.D.-S.; software, É.N. and Z.C.; supervision, É.D.-S.; validation, J.K.; visualization, T.A.; writing—original draft, H.E.-R. and N.E.; and writing—review and editing, T.A. All authors have read and agreed to the published version of the manuscript.

Funding: This research received no external funding.

Acknowledgments: The research was financed by the Higher Education Institutional Excellence Programme (NKFIH-1150-6/2019) of the Ministry of Innovation and Technology in Hungary, within the framework of the Biotechnology thematic programme of the University of Debrecen. This research was financed also by the "Complex Rural Economic and Sustainable Development, Elaboration of its Service Networks in the Carpathian Basin (Project ID: EFOP-3.6.2-16-2017-00001, Hungary)" research project. This paper was also supported by the János Bolyai Research Scholarship of the Hungarian Academy of Sciences. The authors thank Mohamed E. Ragab (Horticulture Department, Faculty of Agriculture, Ain Shams University, Egypt) for providing them with tubers of Fuseau.

Conflicts of Interest: The authors declare no conflict of interest.

References

1. La Cour, R.; Schjoerring, J.K.; Jørgensen, H. Enhancing Protein Recovery in Green Biorefineries by Lignosulfonate-Assisted Precipitation. *Front. Sustain. Food Syst.* **2019**, *3*, 112. [CrossRef]
2. Kamm, B.; Schönicke, P.; Hille, C.H. Green biorefinery–industrial implementation. *Food Chem.* **2016**, *197*, 1341–1345. [CrossRef] [PubMed]
3. Grela, E.R.; Pietrzak, K. Production technology, chemical composition and use of alfalfa protein xanthophyll concentrate as dietary supplement. *J. Food Process. Technol.* **2014**, *5*, 10.
4. European Food Safety Authority, EFSA. Scientific Opinion of the Panel on Dietetic Products Nutrition and Allergies on a request from the European Commission on the safety of 'alfalfa protein concentrate' as food. *EFSA J.* **2009**, *997*, 1–19.
5. Tenorio, A.T.; Kyriakopoulou, K.E.; Suarez-Garcia, E.; van den Berg, C.; van der Goot, A.J. Understanding differences in protein fractionation from conventional crops, and herbaceous and aquatic biomass-consequences for industrial use. *Trends Food Sci. Technol.* **2018**, *71*, 235–245. [CrossRef]
6. Gunnarsson, I.B.; Svensson, S.-E.; Johansson, E.; Karakashev, D.; Angelidaki, I. Potential of Jerusalem artichoke (*Helianthus tuberosus* L.) as a biorefinery crop. *Ind. Crops Prod.* **2014**, *56*, 231–240. [CrossRef]
7. Fang, Y.R.; Liu, J.A.; Steinberger, Y.; Xie, G.H. Energy use efficiency and economic feasibility of Jerusalem artichoke production on arid and coastal saline lands. *Ind. Crops Prod.* **2018**, *117*, 131–139. [CrossRef]
8. Kaszás, L.; Kovács, Z.; Nagy, E.; Elhawat, N.; Abdalla, N.; Domokos-Szabolcsy, E. Jerusalem artichoke (Helianthus tuberosus L.) as a potential chlorophyll source for humans and animals nutrition. *Environ. Biodivers. Soil Secur.* **2018**, *2*, 1–20. [CrossRef]
9. Razmkhah, M.; Rezaei, J.; Fazaeli, H. Use of Jerusalem artichoke tops silage to replace corn silage in sheep diet. *Anim. Feed Sci. Technol.* **2017**, *228*, 168–177. [CrossRef]
10. Niu, L.; Manxia, C.; Xiumei, G.; Xiaohua, L.; Hongbo, S.; Zhaopu, L.; Zed, R. Carbon sequestration and Jerusalem artichoke biomass under nitrogen applications in coastal saline zone in the northern region of Jiangsu, China. *Sci. Total Environ.* **2016**, *568*, 885–890. [CrossRef]
11. Kays, S.J.; Nottingham, S.F. *Biology and Chemistry of Jerusalem Artichoke: Helianthus tuberosus L.*; CRC Press, Taylor & Francis Group: New York, NY, USA, 2008.
12. Long, X.H.; Shao, H.B.; Liu, L.; Liu, L.P.; Liu, Z.P. Jerusalem artichoke: A sustainable biomass feedstock for biorefinery. *Renew. Sustain. Energy Rev.* **2016**, *54*, 1382–1388. [CrossRef]
13. Johansson, E.; Prade, T.; Angelidaki, I.; Svensson, S.E.; Newson, W.R.; Gunnarsson, I.B.; Hovmalm, H.P. Economically viable components from Jerusalem artichoke (*Helianthus tuberosus* L.) in a biorefinery concept. *Int. J. Mol. Sci.* **2015**, *16*, 8997–9016. [CrossRef]
14. Chen, F.; Long, X.; Liu, Z.; Shao, H.; Liu, L. Analysis of phenolic acids of Jerusalem artichoke (*Helianthus tuberosus* L.) responding to salt-stress by Liquid chromatography/tandem mass spectrometry. *Sci. World J.* **2014**, *2014*, 568043. [CrossRef]
15. Yuan, X.; Gao, M.; Xiao, H.; Tan, C.; Du, Y. Free radical scavenging activities and bioactive substances of Jerusalem artichoke (*Helianthus tuberosus* L.) leaves. *Food Chem.* **2012**, *133*, 10–14. [CrossRef]
16. Chae, S.W.; Lee, S.H.; Kang, S.S.; Lee, H.J. Flavone glucosides from the leaves of Helianthus tuberosus. *Nat. Prod. Sci.* **2002**, *8*, 141–143.
17. Oleszek, M.; Kowalska, I.; Oleszek, W. Phytochemicals in bioenergy crops. *Phytochem. Rev.* **2019**, *18*, 893–927. [CrossRef]
18. Pan, L.; Sinden, M.R.; Kennedy, A.H.; Chai, H.; Watson, L.E.; Graham, T.L.; Kinghorn, A.D. Bioactive constituents of Helianthus tuberosus (Jerusalem artichoke). *Phytochem. Lett.* **2009**, *2*, 15–18. [CrossRef]
19. Showkat, M.M.; Falck-Ytter, A.B.; Strætkvern, K.O. Phenolic Acids in Jerusalem Artichoke (*Helianthus tuberosus* L.): Plant Organ Dependent Antioxidant Activity and Optimized Extraction from Leaves. *Molecules* **2019**, *24*, 3296. [CrossRef]
20. Boisen, S.; Hvelplund, T.; Weisbjerg, M.R. Ideal amino acid profiles as a basis for feed protein evaluation. *Livest. Prod. Sci.* **2000**, *64*, 239–251. [CrossRef]
21. Sparks, D.L.; Page, A.L.; Helmke, P.A.; Loppert, R.H. *Methods of Soil Analysis: Chemical Methods, Part 3*; ASA and SSSA: Madison, WI, USA, 1996.
22. Duncan, D.B. Multiple range and multiple F-tests. *Biometrics* **1955**, *11*, 1–42. [CrossRef]

23. Malmberg, A.; Theander, O. Differences in chemical composition of leaves and stem in Jerusalem artichoke and changes in low-molecular sugar and fructan content with time of harvest. *Swed. J. Agric. Res.* **1986**, *16*, 7–12.
24. Lamsal, B.P.; Koegel, R.G.; Gunasekaran, S. Some physicochemical and functional properties of alfalfa soluble leaf proteins. *LWT-Food Sci. Technol.* **2007**, *40*, 1520–1526. [CrossRef]
25. Rashchenko, I.N. Biochemical investigations of the aerial parts of Jerusalem artichoke. *Tr. Kazakh Sel'skokhoz Inst* **1959**, *6*, 40–52.
26. Naveed, M.; Hejazi, V.; Abbas, M.; Kamboh, A.A.; Khan, G.J.; Shumzaid, M.; Ahmad, F.; Babazadeh, D.; Xia, F.; Modarresi-Ghazani, F.; et al. Chlorogenic acid (CGA): A pharmacological review and call for further research. *Biomed. Pharmacother.* **2018**, *97*, 67–74. [CrossRef]
27. Rawate, P.D.; Hill, R.M. Extraction of a high-protein isolate from Jerusalem artichoke (Helianthus tuberosus) tops and evaluation of its nutrition potential. *J. Agric. Food Chem.* **1985**, *33*, 29–31. [CrossRef]
28. Liang, N.; Kitts, D.D. Role of chlorogenic acids in controlling oxidative and inflammatory stress conditions. *Nutrients* **2016**, *8*, 16. [CrossRef]
29. Zhang, L.T.; Chang, C.Q.; Liu, Y.; Chen, Z.M. Effect of chlorogenic acid on disordered glucose and lipid metabolism in db/db mice and its mechanism. *Zhongguo Yi Xue Ke Xue Yuan Xue Bao Acta Acad. Med. Sin.* **2011**, *33*, 281–286.
30. Cabello-Hurtado, F.; Durst, F.; Jorrin, J.V.; Werck-Reichhard, D. Coumarins in Helianthus tuberosus: Characterization, induced accumulation and biosynthesis. *Phytochemistry* **1998**, *49*, 1029–1036. [CrossRef]
31. Venugopala, K.N.; Rashmi, V.; Odhav, B. Review on natural coumarin lead compounds for their pharmacological activity. *BioMed Res. Int.* **2013**, 963248. [CrossRef]
32. Docampo, M.; Olubu, A.; Wang, X.; Pasinetti, G.; Dixon, R.A. Glucuronidated flavonoids in neurological protection: Structural analysis and approaches for chemical and biological synthesis. *J. Agric. Food Chem.* **2017**, *65*, 7607–7623. [CrossRef]
33. Ho, L.; Ferruzzi, M.G.; Janle, E.M.; Wang, J.; Gong, B.; Chen, T.Y.; Lobo, J.; Cooper, B.; Wu, Q.L.; Talcott, S.T.; et al. Identification of brain-targeted bioactive dietary quercetin-3-*O*-glucuronide as a novel intervention for Alzheimer's disease. *FASEB J.* **2013**, *27*, 769–781. [CrossRef] [PubMed]
34. Lai, C.S.; Wu, J.C.; Ho, C.T.; Pan, M.H. Disease chemopreventive effects and molecular mechanisms of hydroxylated polymethoxyflavones. *BioFactors* **2015**, *41*, 301–313. [CrossRef] [PubMed]
35. Yang, P.Y.; Hu, D.N.; Lin, I.C.; Liu, F.S. Butein Shows Cytotoxic Effects and Induces Apoptosis in Human Ovarian Cancer Cells. *Am. J. Chin. Med.* **2015**, *43*, 769–782. [CrossRef]
36. Mersereau, J.E.; Levy, N.; Staub, R.E.; Baggett, S.; Zogovic, T.; Chow, S.; Ricke, W.A.; Tagliaferri, M.; Cohen, I.; Bjeldanes, L.F.; et al. Liquiritigenin is a plant-derived highly selective estrogen receptor beta agonist. *Mol. Cell. Endocrinol.* **2008**, *283*, 49–57. [CrossRef]
37. Seiler, G.J. Nitrogen and mineral content of selected wild and cultivated genotypes of Jerusalem artichoke. *Agron. J.* **1988**, *80*, 681–687. [CrossRef]
38. Murata, M.; Nakai, Y.; Kawazu, K.; Ishizaka, M.; Kajiwara, H.; Abe, H.; Takeuchi, K.; Ichiro, M.; Mochizuki, A.; Seo, S. Loliolide, a carotenoid metabolite, is a potential endogenous inducer of herbivore resistance. *Plant Physiol.* **2019**, *179*, 1822–1833. [CrossRef] [PubMed]
39. Yun, N.; Kang, J.W.; Lee, S.M. Protective effects of chlorogenic acid against ischemia/reperfusion injury in rat liver: Molecular evidence of its antioxidant and anti-inflammatory properties. *J. Nutr. Biochem.* **2012**, *23*, 1249–1255. [CrossRef]
40. Islam, M.S.; Iwasaki, A.; Suenaga, K.; Kato-Noguchi, H. Isolation and identification of two potential phytotoxic substances from the aquatic fern Marsilea crenata. *J. Plant Biol.* **2017**, *60*, 75–81. [CrossRef]
41. Chernenko, T.V.; Glushenkova, A.I.; Rakhimov, D.A. Lipids of Helianthus tuberosus tubers. *Chem. Nat. Compd.* **2008**, *44*, 1–2. [CrossRef]
42. Kaszás, L.; Alshaal, T.; Kovács, Z.; Koroknai, J.; Elhawat, N.; Nagy, É.; El-Ramady, H.; Fári1, M.; Domokos-Szabolcsy, É. Refining high-quality leaf protein and valuable co-products from green biomass of Jerusalem artichoke (Helianthus tuberosus L.) for sustainable protein supply. *Biomass Convers. Biorefinery* **2020**. [CrossRef]
43. Simopoulos, A.P. Evolutionary aspects of diet, the omega-6/omega-3 ratioand genetic variation: Nutritional implications for chronic diseases. *Biomed. Pharmacother.* **2006**, *60*, 502–507. [CrossRef] [PubMed]

44. Barrero-Sicilia, C.; Silvestre, S.; Haslam, R.P.; Michaelson, L.V. Lipid remodelling: Unraveling the response to cold stress in Arabidopsis and its extremophile relative Eutrema salsugineum. *Plant Sci.* **2017**, *263*, 194–200. [CrossRef] [PubMed]
45. Sadali, N.M.; Sowden, R.G.; Ling, Q.; Jarvis, R.P. Differentiation of chromoplasts and other plastids in plants. *Plant Cell Rep.* **2019**, *38*, 803–818. [CrossRef] [PubMed]

Article

Evaluation of Alpha-Amylase Inhibitory, Antioxidant, and Antimicrobial Potential and Phytochemical Contents of *Polygonum hydropiper* L.

Abdul Nasir [1,2], Mushtaq Khan [1], Zainab Rehman [3], Atif Ali Khan Khalil [4], Saira Farman [1], Naeema Begum [1], Muhammad Irfan [5], Wasim Sajjad [4,*] and Zahida Parveen [1,*]

1 Department of Biochemistry, Abdul Wali Khan University Mardan, Mardan 23200, Pakistan; anasirqau@gmail.com (A.N.); mushtaqkhanbiochemist@gmail.com (M.K.); sairafarman@awkum.edu.pk (S.F.); yzai_taurus@yahoo.com (N.B.)
2 Department of Molecular Science and Technology, Ajou University, Suwan 16499, Korea
3 Laboratory of Animal and Human Physiology, Department of Animal Sciences, Quaid-i-Azam University, Islamabad 45320, Pakistan; abbassids@outlook.com
4 Department of Biological Sciences, National University of Medical Sciences, Rawalpindi 46000, Pakistan; atif.ali@numspak.edu.pk
5 College of Dentistry, Department of Oral Biology, University of Florida, Gainesville, FL 32610, USA; irfanmuhammad@ufl.edu
* Correspondence: sajjadw@numspak.edu.pk (W.S.); zahida@awkum.edu.pk (Z.P.)

Received: 10 June 2020; Accepted: 3 July 2020; Published: 6 July 2020 

Abstract: *Polygonum hydropiper* L. is a traditionally used medicinal plant. The present study was designed to explore the α-amylase inhibitory, antioxidant, and antimicrobial activities of *Polygonum hydropiper* L. Polarity-based solvent extracts (*n*-hexane, acetone, chloroform, methanol, ethanol, and water) of *Polygonum hydropiper* leaves and stem were used. Antioxidant activity was assessed by free radical scavenging assay (FRAP) and 2,2-diphenylpicrylhydrazyl (DPPH) free radical scavenging activity methods. Quantitative phytochemical analyses suggested that the stem of *Polygonum hydropiper* L. contains higher levels of bioactive compounds than its leaves ($p < 0.05$). The results suggested that stem-derived extracts of *Polygonum hydropiper* L. are more active against bacterial species, including two Gram-positive and three Gram-negative strains. Moreover, our results showed that the bioactive compounds of *Polygonum hydropiper* L. significantly inhibit α-amylase activity. Finally, we reported the polarity-based solvent extracts of *Polygonum hydropiper* L. and revealed that the stem, rather than leaves, has a high antioxidant potential as measured by FRAP and DPPH assay with IC_{50} values of 1.38 and 1.59 mg/mL, respectively. It may also be deducted from the data that the *Polygonum hydropiper* L. could be a significant candidate, which should be subjected to further isolation and characterization, to be used as an antidiabetic, antimicrobial and antioxidant resource in many industries, like food, pharmaceuticals and cosmetics.

Keywords: water-pepper; FRAP; DPPH; polarity-based solvent extraction; α-amylase inhibition

1. Introduction

The Polygonaceae family is comprised of 48 genera and 1200 species [1,2]. Among the 60 species of the *Polygonum* distributed throughout the world, approximately 20 species are found in Pakistan [3]. They grow in moist places and shallow water. *Polygonum hydropiper* L. is an important medicinal plant of Polygonaceae family. It is commonly grown annually in the wet areas and is recognized as a common weed which is native to Southeast Asia [4]. Morphologically, the stem is round; its length usually varies from 40 to 70 cm [5].

This species has anti-inflammatory potential [6] and insecticidal properties [7], along with anticholinesterase, phytotoxic, anthelmintic, antiangiogenic, anticancer, antimicrobial, and antioxidant potential [8–11]. This plant is used in treating broad range of disorders, including gastrointestinal disturbances, neurological disorders, inflammation, and diarrhea [12]. Moreover, it is used to treat other diseases, like dyspepsia, itching skin, excessive menstrual bleeding, hemorrhoids, and cancer (particularly colon, breast, and prostate cancer) [13]. Furthermore, sap of the leaves is used to treat headache, pain, toothache, liver enlargement, gastric ulcer, dysentery, and loss of appetite. The juice of this plant is considered useful for treating wounds, skin diseases, and painful carbuncles [14]; it can also be used as an anthelminthic herb, in the treatment of snake-bites, and as a diuretic [15].

The phytochemicals of *P. hydropiper* L. include catechins, procyanidins, condensed tannins, and antitumor agents that are flavanols in nature [16]. Flavonoids with antioxidant activity and aldose reductase and tyrosinase inhibitory activities have also been found in *P. hydropiper* L. [4,17]. Due to the presence of drimane-type sesquiterpenes, *P. hydropiper* L. has some insecticidal and antifungal activities [18,19]. The antifungal activities of *P. hydropiper* L. are due to the presence of confertifolin, a natural antimicrobial compound [20]. Other bioactive compounds of *P. hydropiper* L. include gallic acid; ellagic acid; 3,3′-di-O-methyl ether; anthraquinone; aromatic 6-lactone; and flavonoids such as viscosumic acid, oxymethyl-anthraquinones, rutin, hyperin, isoquercitrin, epicatechin, quercetin, kaempferol, and isorhamnetin [21]. It has been reported that this species contains pyrocatechol, 4-methyloxazole, caryophyllene, succinimide, vanillic acid, myristic acid, farnesol, arachidic acid, methyl ester, and capsaicin [10].

The excessive and long-term use of synthetic drugs causes several side effects [22]. In the human body, numerous cellular processes produce free radicals (FR) and reactive oxygen species (ROS). Such toxics may also be produced exogenously by pollutants, radiation, smoke, drugs, and xenobiotics [23]. Overproduction of different chemically reactive oxygen plagiaristic molecules such as hydrogen peroxide (H_2O_2), superoxide (O_2^-), and hydroxyl radicals (OH^-) is highly toxic and leads to biomolecular damage, which in turn causes diabetes mellitus, cancer, atherosclerosis, and heart and neurodegenerative diseases [24]. In contrast, the overproduction of FR and ROS species can be averted by antioxidant substances. It is reported that secondary metabolites of plants like phenols, flavonoids, and alkaloids have antioxidant, antidiabetic, anthelminthic, lipid-lowering, anticoagulant, and antimicrobial properties [25].

Diabetes mellitus is characterized by hyperglycemia, lipedema, and oxidative stress and predisposes affected individuals to long-term complications afflicting the eyes, skin, kidneys, nerves, and blood vessels. Recently, it has been estimated that the prevalence of diabetes by 2025 will increase from 143 million to 300 million patients [26]. Various studies have indicated that dietary supplementation with combined antiglycation and antioxidant nutrients might be a safe and simple complement to traditional therapies targeting diabetic complications [27]. Hyperglycemia, through both enzymatic and non-enzymatic mechanisms, produces oxidative stress by producing free radicals. α-Amylase hydrolyzes (1,4)-α-D-glucosidic linkages in polysaccharides containing three or more (1,4)-α-linked D-glucose units, which ultimately increases the blood sugar level; α-amylase-inhibiting drugs such as acarbose, miglitol, and voglibose are frequently used in diabetes management [28,29]. However, α-amylase-inhibiting drugs cause severe side effects like bloating and abdominal uneasiness [30]. Therefore, the administration of natural resources to treat diabetes seems to be a promising strategy. To this end, targeting α-amylase inhibition by natural remedies may be an ideal platform in diabetes prevention [31,32]. In this context, the current study was designed to determine the bioactive compounds of *P. hydropiper*. Furthermore, the extracts of *P. hydropiper* were evaluated for their potential to inhibit α-amylase activity and reduce oxidative stress. Finally, the antimicrobial potential of *P. hydropiper* was demonstrated.

2. Results

2.1. Determination of Bioactive Compounds

In first series of experiments, class of secondary metabolites including alkaloids, tannins, flavonoids, β-carotene, and lycopene from the leaves (PHL) and stem (PHS) of *P. hydropiper* were extracted with their respective solvents and quantified. The comparative analyses found that concentrations of alkaloids and tannins were significantly higher in *P. hydropiper* stem than in leaves ($t = -3.22$, $p = 0.032$; $t = -3.61$, $p = 0.023$). However, flavonoids were found in comparable amounts in both stem and leaves ($t = -0.427$, $p = 0.691$) (Table 1). We also noticed that the leaves of *P. hydropiper* contained significantly higher concentrations of β-carotene and lycopene than those found in the stem of *P. hydropiper* ($t = 2.90$, $p = 0.044$; $t = 4.31$, $p = 0.013$).

Table 1. Phytochemical constituents of *P. hydropiper* leaves and stem.

Phytochemical	PHL Concentration (mg/mL) [a]	PHS Concentration (mg/mL) [a]
Alkaloids	4.32 ± 0.354	8.17 ± 1.13 *
Tannins	1.76 ± 0.287	3.54 ± 0.402 *
Flavonoids	6.11 ± 0.344	6.33 ± 0.392
β-Carotene	0.461 ± 0.075 *	0.194 ± 0.053
Lycopene	0.762 ± 0.138 *	0.136 ± 0.043

[a] Values are means of triplicate determination (n = 3) ± standard deviations; * indicates significant difference ($p < 0.05$) between leaves and stem; PHL, *P. hydropiper* leaves; PHS, *P. hydropiper* stem.

2.2. In Vitro Evaluation of α-Amylase Inhibition

The different solvent extracts of *P. hydropiper* stem and leaves were investigated for their potential to inhibit α-amylase activity at six different concentrations (0.46, 0.94, 1.88, 3.75, 7.50, and 15 mg/mL). The dose–response calibration curves for n-hexane, acetone, chloroform, ethanol, methanol, and water extracts of *P. hydropiper* leaves and stem were constructed individually. Percent α-amylase inhibition and IC_{50} values were determined from the dose–response calibration curves for each type of extract (Figures 1 and 2).

The α-amylase inhibitory activities of the leaf extracts were ranked in the following order: n-hexane (IC_{50} 1.03 mg/mL; R^2 0.566) > chloroform (IC_{50} 1.53 mg/mL; R^2 0.6492) > methanol (IC_{50} 2.32 mg/mL; R^2 0.7255) > acetone (IC_{50} 4.70 mg/mL; R^2 0.9919) > water (IC_{50} 4.85 mg/mL; R^2 0.7629) > ethanol (IC_{50} 13.89 mg/mL; R^2 0.3249). Differently from those of the leaf extracts, the α-amylase inhibitory activities of the stem extracts were ranked in the following order: chloroform (IC_{50} 2.599 mg/mL; R^2 0.8232) > methanol (IC_{50} 3.517 mg/mL; R^2 0.8375) > ethanol (IC_{50} 5.672 mg/mL; R^2 0.4736) > n-hexane (IC_{50} 6.910 mg/mL; R^2 0.4399) > acetone (IC_{50} 11.86 mg/mL; R^2 0.5608) > water (IC_{50} 13.12 mg/mL; R^2 0.6824).

2.3. Antioxidant Capacity

The antioxidant potential of the different extract types was analyzed at six concentrations (0.46, 0.94, 1.88, 3.75, 7.50, and 15 mg/mL) by using the free radical scavenging assay (FRAP) and the 2,2-diphenylpicrylhydrazyl (DPPH) assay. Table 2 shows the antioxidant potential measured by two assays.

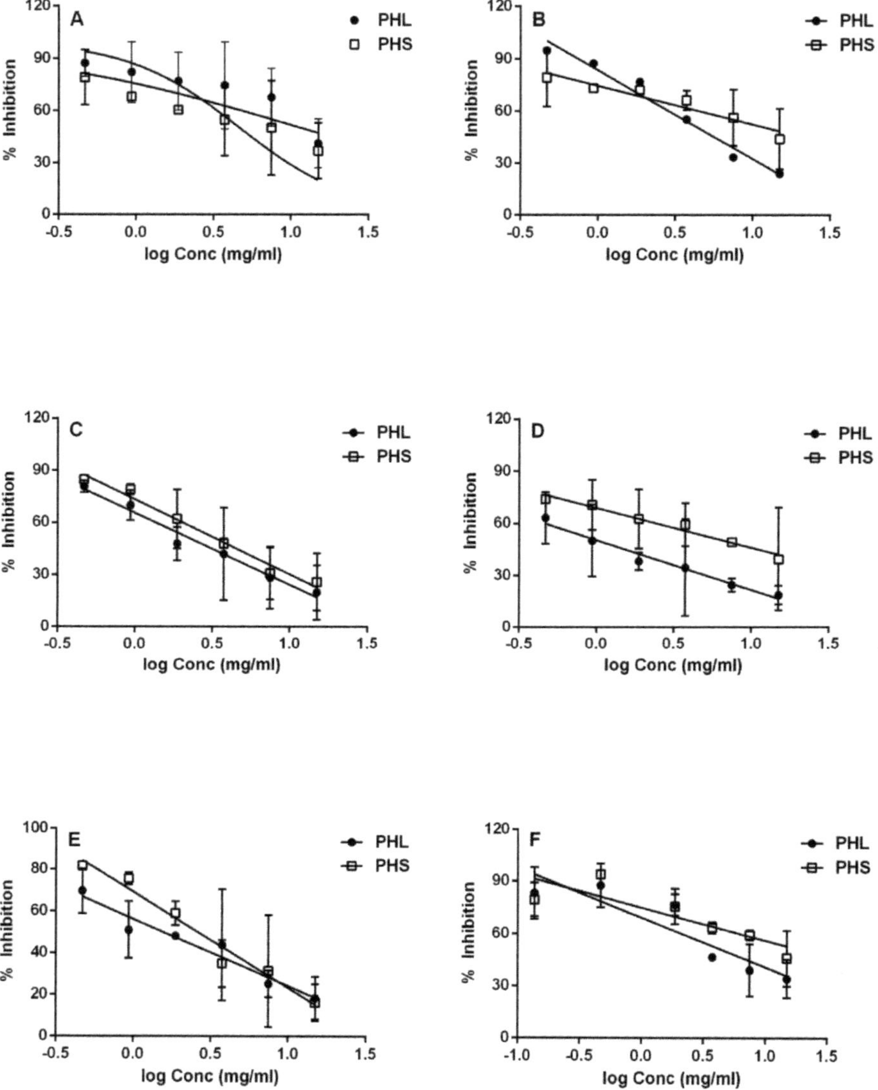

Figure 1. Inhibition of α-amylase activity by different extracts (ethanol, (**A**); acetone, (**B**); methanol, (**C**); *n*-hexane, (**D**); chloroform, (**E**) and water, (**F**)) of *P. hydropiper* leaves (PHL) and stem (PHS). Solid lines represent hyperbolic dose–response curves, which were generated in GraphPad Prism. Values represent the means of triplicate measurements (n = 3).

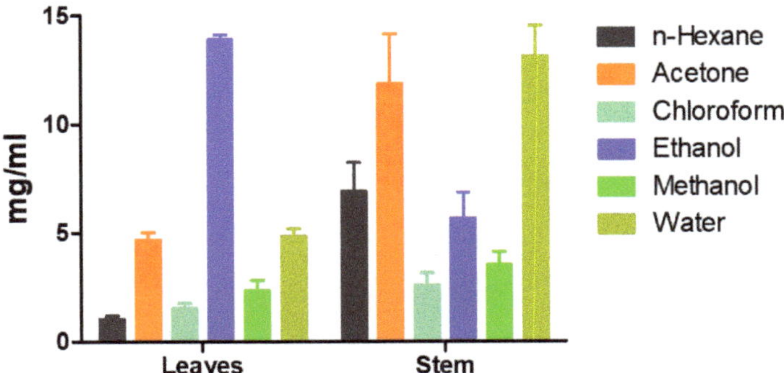

Figure 2. α-Amylase inhibitory activity of the different extracts tested: comparison of IC_{50} values. The IC_{50} values were calculated from dose-dependent percent inhibition. Values represent the means of triplicate measurements (n = 3). Bars represent the standard deviation.

Table 2. Antioxidant activity of *P. hydropiper* extracts in different solvents at different concentrations.

FRAP Activity	**% Activity**						
	Concentration (mg/mL) [a]						
	0.46	**0.94**	**1.88**	**3.75**	**7.50**	**15**	**IC_{50}**
PHL n-hexane	83.33 ± 7.2 **	79.28 ± 6.1 **	74.07 ± 3.5 **	63.86 ± 6.4 *	44.38 ± 11	32.87 ± 7.8 **	5.52
PHL acetone	74.72 ± 1.8 **	71.95 ± 4.1 **	51.53 ± 4.7 **	19.27 ± 3.1 **	8.43 ± 1.9	3.95 ± 0.8	2.29
PHL ethanol	68.37 ± 7.3 **	62.46 ± 3.2 **	56.68 ± 8.0 **	42.18 ± 4.3 **	29.35 ± 3.3	26.48 ± 4.8 *	2.99
PHL methanol	65.53 ± 3.2 **	59.78 ± 2.5 **	49.43 ± 3.8 **	25.97 ± 3.9 **	25.17 ± 3.3	12.87 ± 4.6	2.30
PHS n-hexane	72.15 ± 3.2 **	67.87 ± 5.3 **	45.94 ± 5.3 **	42.88 ± 1.6 *	31.45 ± 3.3	29.10 ± 6.1 *	1.50
PHS acetone	67.23 ± 12 **	52.70 ± 3.6 **	40.04 ± 6.5 **	16.03 ± 4.0 **	7.39 ± 1.4	3.57 ± 0.8 *	1.81
PHS ethanol	74.52 ± 3.32 **	69.66 ± 5.5 **	46.91 ± 5.1 **	43.67 ± 6.0 *	34.41 ± 4.8	28.66 ± 8.2	1.38
PHS methanol	69.31 ± 3.19 **	65.96 ± 5.3 **	44.33 ± 5.3 **	37.62 ± 4.0 **	27.61 ± 4.5	22.55 ± 6.0	1.73
DPPH Activity							
PHL acetone	72.91 ± 5.7	68.06 ± 3.6 **	65.16 ± 3.6 **	32.93 ± 6.0 *	22.03 ± 5.1 *	16.88 ± 2.2	2.94
PHL ethanol	82.68 ± 5.6	80.52 ± 4.4 **	72.02 ± 4.7 **	59.93 ± 5.6 **	42.19 ± 2.5	30.06 ± 1.3 *	5.14
PHS acetone	77.93 ± 3.2	71.95 ± 5.5 **	50 ± 5.1 **	46.21 ± 3.8 *	32.60 ± 3.3 *	27.57 ± 9.1	1.59
PHS ethanol	82.93 ± 9.4	78.13 ± 4.6 **	72.55 ± 5.8 **	64.78 ± 3.4 **	47.88 ± 0.2 *	36.72 ± 5.3 *	6.88

[a] Values are means of triplicate determination (n = 3) ± standard deviations; *p* values less than 0.05 were considered to be statistically significant; * $p < 0.05$, ** $p < 0.01$; PHL, *P. hydropiper* leaves; PHS, *P. hydropiper* stem.

The FRAP activities for different extract types of both stem (PHS) and leaves (PHL) were observed to be ranked in the following order: PHS ethanol (IC_{50} 1.38 mg/mL; R^2 0.7847) > PHS n-hexane (IC_{50} 1.50 mg/mL; R^2 0.8407) > PHS methanol (IC_{50} 1.73 mg/mL; R^2 0.8483) > PHS acetone (IC_{50} 1.81 mg/mL; R^2 0.8814) > PHL acetone (IC_{50} 2.29 mg/mL; R^2 0.977) > PHL methanol (IC_{50} 2.30 mg/mL; R^2 0.9121) > PHL ethanol (IC_{50} 2.99 mg/mL; R^2 0.8090) > PHL n-hexane (IC_{50} 5.52 mg/mL; R^2 0.7588). All types of extracts of the stem showed higher FRAP activity than those of the leaves of *P. hydropiper*.

The DPPH activities of both stem and leaves were also analyzed for different extract types at different concentrations. The acetonic extracts of both stem and leaves were more effective than ethanolic extracts. The activities was observed to be ranked in the following order: PHS acetone (IC_{50} 1.59 mg/mL; R^2 0.8330) > PHL acetone (IC_{50} 2.94 mg/mL; R^2 0.9244) > PHL ethanol (IC_{50} 5.14 mg/mL; R^2 0.9102) > PHS ethanol (IC_{50} 6.88 mg/mL; R^2 0.8126).

2.4. Antimicrobial Activity

Our results showed strong antibacterial activities for both acetonic and ethanolic extracts at all six concentrations (0.46, 0.94, 1.88, 3.75, 7.50, and 15 mg/mL) of *P. hydropiper* stem and leaves. Zones of inhibition of the tested extracts are depicted in Table 3. The acetonic stem extract showed antibacterial activity against five different microbial species (*Escherichia coli*, *Staphylococcus aurous*, *Klebsiella pneumoniae*, *Morganella morganii*, and *Haemophilus influenzae*); however, the leaves did not show antibacterial activity against *E. coli and S. aureus*. The ethanolic extracts of stem and leaves showed strong antiproliferative activities against all the microbial species tested (Table 3).

Table 3. Antimicrobial activity of different solvent extracts of *P. hydropiper* leaves and stem at different concentrations.

	Clear Zone of Inhibition (mm) of HP Extracts [a]												Ant Ag (mm) [a]	
Microorganisms	**Leaf Concentration (mg/mL)**						**Stem Concentration (mg/mL)**						**DMSO**	**A**
	0.46	**0.94**	**1.88**	**3.75**	**7.50**	**15**	**0.46**	**0.94**	**1.88**	**3.75**	**7.50**	**15**		
Acetonic extract														
E. coli	ND	ND	ND	ND	ND	ND	9	10	12	14	16	17	ND	20
S. aureus	ND	ND	ND	ND	ND	ND	9	11	13	15	17	18	ND	22
K. pneumonia	9	10	12	13	14	16	9	10	11	13	14	15	ND	17
H. influenzae	9	10	12	13	14	16	9	11	12	14	16	18	ND	20
M. morganii	10	12	13	14	15	17	10	12	14	15	17	19	ND	17
Ethanolic extract														
E. coli	9	10	11	13	14	15	9	11	12	13	14	16	ND	20
S. aureus	9	10	12	13	15	17	9	10	12	14	15	17	ND	22
K. pneumonia	7	8	9	13	14	15	9	10	12	13	14	16	ND	17
H. influenzae	7	11	12	14	15	17	10	11	12	13	15	17	ND	20
M. morganii	9	10	11	13	15	17	9	10	12	13	15	16	ND	17

[a] Values are means of triplicate determination (n = 3); ND, not detected at this concentration; A, ampicillin (10 µg); DMSO, dimethyl sulfoxide (100%); Ant Ag, antimicrobial agent.

3. Discussion

The excessive use of synthetic drugs causes several side effects and may lead to conflict [22]. In contrast, traditional medicines are harmless, effective, and inexpensive drug candidates. In order to evaluate the biological activities of plant extracts with respect to traditional uses, *P. hydropiper* was previously screened for its antioxidant potency and antimicrobial and antipathogenic activities at low doses and with few solvent extractions [8,10]. We evaluated the antioxidant and antibacterial potential at high doses and with several different solvent extractions; additionally, we screened the α–amylase inhibitory potential of *P. hydropiper*.

Our results show that *P. hydropiper* leaves are rich in tannins and flavonoids, which is in strong agreement with the results of Nakao et al. (1999), who they reported condensed tannins, procyanidins, and catechins which are flavones in nature [16]. Yang et al. (2011) also reported the presence of flavonoid in *P. hydropiper* leaves [21]. Previously, it was reported that *P. hydropiper* species contained 4-methyloxazole (flavonoid in nature) and succinimide [8]. The therapeutic application of β-carotene showed a strong association with a lower risk of lung cancer [33]. β-Carotene along with phenytoin has great antiepileptic activity and can be used as a therapeutic agent in epilepsy management [34]. Tannins have been reported as anticancer and anti-inflammatory agents and can be used for ulcerated tissues [35–37]. Polyphenols are known not only to ease the oxidative stress status, but also to act on cellular signaling pathways, including vascular endothelial growth factor (VEGF)-mediated angiogenesis, endoplasmic reticulum (ER) stress, nitric oxide (NO$^{\cdot}$) signaling, and nuclear factor E2-related factor 2 antioxidant pathways, thus preventing vascular complications in diabetes. Resveratrol, one of the most studied polyphenols, has been reported to restore the insulin receptor substrate 1 (IRS-1) and endothelial nitric oxide synthase (eNOSx) signaling pathway in endothelial cells under palmitate-induced insulin resistance [38,39]. As a source of tannins, *P. hydropiper* may be used to treat cancer, inflammation, and ulcerated tissues. Furthermore, our findings revealed that the leaves of *P. hydropiper* contain a significantly high level of lycopene. The intake of lycopene reduces the risk of prostate cancer [40] and plays a protective role against nephrotoxicity [41]. Lycopene consumption can also regulate endothelial function, thereby reducing oxidative stress in healthy humans [42].

Enzymes that are primarily involved in increasing blood glucose, such as α-amylase and α-glucosidase, have been targeted as a therapeutic approach in postprandial hyperglycemia [43] because their inhibition can cause reduction in postprandial hyperglycemia [44]. Our results indicate that α-amylase activity was inhibited by the extracts of *P. hydropiper* leaves only. Specifically, the *n*-hexane extract of the leaves showed the most potent antiamylase activity. Therefore, we suggest that leaf extracts of *P. hydropiper* could be used as a future therapeutic candidate in treating hyperglycemia.

The antioxidant activity of leaf and stem extracts was determined by two methods: the free radical scavenging assay (using ferric ion reducing agent) using and the DPPH scavenging assay. The results in our study indicate that this plant species was potently active, which suggests that all different extracts of *P. hydropiper* leaves and stem contained compounds that are capable of hydrogen donation to the free radical for the purpose of eliminating the odd electron, thus reducing the radicals' activity. This also implies that this plant species may be beneficial, especially at high concentrations, for treating the pathological damage caused by radicals' activities. All of the extracts of both stem and leaves at different concentrations were also active against the Gram-negative and Gram-positive bacteria. However, such activity was not shown by the acetonic extract of leaves against *E. coli* (which might be due to outer membrane in Gram-negative bacteria acting as a permeability barrier) and *S. aureus* [45]. Hence, we suggest that ethanolic extracts of *P. hydropiper* stem and leaves could be used as antioxidant and antibacterial agents and should be studied further.

4. Materials and Methods

4.1. Chemical and Reagents

Soluble starch, potassium ferrocyanide, trichloroacetic acid, ferric chloride, and solvents used for polarity-based extraction (n-hexane, acetone, chloroform, ethanol, and methanol) were purchased from Sigma-Aldrich (Lahore, Pakistan). Dimethylsulfoxide, quercetin, Folin's phenol, and tannic acid were obtained from Merck (Karachi, Pakistan). Porcine pancreatic α-amylase, ascorbic acid, and DPPH solution were purchased from Sigma-Aldrich (Lahore, Pakistan) for measurement of α-amylase, FRAP, and DPPH assay, respectively. All reagents were biochemical reagent grade.

4.2. Sample Collection

Polygonum hydropiper was collected from Mardan District, Khyber Pakhtunkhwa, Pakistan, in April 2016. The plant was identified as *P. hydropiper* and deposited in the Department of Botany, Abdul Wali Khan University, Mardan, Pakistan.

4.3. Sample Preparation

Fresh leaves and stem of *P. hydropiper* were separated, washed with tap water to remove dust, and then air-dried at 25 °C for 30 days in an air flux drying oven. The plant parts were crushed to fine powder (80 mashes) with the help of an electric blender. Powdered samples were placed in sterile sealed bags each containing a damp paper towel and kept at 4 °C for further analyses.

4.4. Solvent–Solvent Extraction

The extracts of powdered samples were isolated using the solvent–solvent extraction method. Several solvents (n-hexane, acetone, chloroform, ethanol, methanol, and water) were used to ensure the polarity-based extraction. Initially, each sample was extracted by shaking with n-hexane at a ratio of 1:10 (*w/v*) for 24 h at room temperature. The corresponding sample was then filtered by Whatman filter paper. The pellet was separated, and the supernatant was transferred to a preweighed Falcon tube. The residual pellet was re-extracted with next solvent, which was slightly polar than n-hexane, using the same ratio, temperature, and time. The same procedure was repeated with all solvents, and all extracts were allowed to dry in oven at 37 °C. The dried sample was weighed and redissolved in dimethylsulfoxide (DMSO) at a final concentration of 15 mg/mL.

4.5. Phytochemical Determination

4.5.1. Determination of Flavonoids

Flavonoid estimation was carried out by spectrophotometric assay as previously described [46]. Five grams of air-dried powdered sample was dissolved in 50 mL of 80% aqueous ethanol and incubated for 24 h in a shaker incubator. The extract was centrifuged at 10,000 rpm and 25 °C for 15 min. The pellet was discarded and the supernatant containing flavonoids was stored in a 50 mL Falcon tube at 4 °C. The flavonoid extract (250 µL) was mixed with 1.25 mL of sterile distilled water and 75 µL of 5% $NaNO_2$ solution. After 5 min, 150 µL of 10% $AlCl_3.H_2O$ was added, and the mixture was incubated for 6 min. Thereafter, 500 µL of 1 M NaOH and 275 µL of sterile distilled water were added to the mixture. The solution was mixed, and absorbance was measured at λ of 415 nm. Different concentrations of quercetin (15–500 µg) were used as standards to calculate the standard curve, while 80% aqueous ethanol was used as blank.

4.5.2. Determination of β-Carotene and Lycopene

β-Carotene and lycopene were extracted and quantified according to the method described by Lillian et al. [47]. Briefly, methanol extract was prepared by dissolving 10 g of air-dried powdered sample in 100 mL of methanol and incubating in a shaker incubator for 24 h. The extract was filtered by Whatman filter paper, the pellet was discarded, and supernatant was isolated. Thereafter, the methanol contents were evaporated by heating in a water bath. The dried sample was dissolved in acetone and n-hexane mixture (4:6). Finally, the reaction mixture containing β-carotene and lycopene was stored at 4 °C. The spectrophotometric analysis was carried out by measuring absorbance at λ's of 453, 505, 645, and 663 nm. β-Carotene and lycopene contents were calculated [48] by using the following equations:

$$\text{Lycopene } (\text{mg}/50\text{ mL}) = 0.0458\text{A}_{663} + 0.372\text{A}_{505} - 0.0806\text{A}_{453} \tag{1}$$

$$\beta - \text{carotene } (\text{mg}/50\text{ mL}) = 0.216\text{A}_{663} + 0.304\text{A}_{505} - 0.452\text{A}_{453} \tag{2}$$

4.5.3. Determination of Tannins

Tannin contents of *P. hydropiper* were extracted using the method of Makkar et al. [49]. Briefly, 0.5 g of each air-dried powdered sample was dissolved in 100 mL of 70% acetone and incubated while shaking for 6 h. The sample was filtered by Whatman filter paper, the pellet was discarded, and the supernatant was stored in a 50 mL Falcon tube at 4 °C. Different concentrations of tannic acid (3–50 mg) were prepared by serial dilution from stock solution (50 mg/100 mL of 70% acetone). The tannin extracts (50 µL) was mixed with 950 µL of sterile distilled water. Thereafter, 0.5 mL of Folin's phenol reagent (mixture of phosphomolybdate and phosphotungstate), used for phenolic and polyphenolic antioxidants detection, and 2.5 mL of 20% $NaCO_3$ solution were added and vortexed. The solution was incubated at room temperature for 40 min. Finally, the absorbance was measured at λ of 725 nm. During the experiment, 70% acetone was used as blank and treated as positive control.

4.5.4. Determination of Alkaloids

Alkaloids were extracted by using the acid–base shifting method [46]. Briefly, dried powdered sample was dissolved in ethanol at a ratio of 1:10 (w/v) and left to shake for 24 h at room temperature. The extract was concentrated while drying in an oven. The dried sample was redissolved in ethanol with 1% HCl. The mixture was then precipitated in a refrigerator for 3 days. The solution was filtered, and pH was maintained at 8–10 by the addition of ammonium hydroxide. This basic solution was extracted with chloroform by using a separating funnel. The chloroform layer containing alkaloids was recovered, and ethanol layer was discarded. The chloroform was evaporated by heating in a water bath. Finally, the sample was dried in an oven and alkaloid contents were measured.

4.6. α-Amylase Inhibition Assay

Screening of extracts for α-amylase inhibition was carried as previously described for the starch iodine assay [5]. In brief, the assay mixtures composed of 120 μL of 0.02 M sodium phosphate buffer (pH 6.9 containing 6 mM sodium chloride), 1.5 mL of porcine pancreatic α-amylase (PPA) solution (0.05 mg/2 mL H_2O) and *P. hydropiper* extracts (sample from solvent–solvent extraction) at concentrations ranging from 0.46 to 15 mg/mL (w/v) were incubated at 37 °C for 10 min. Afterward, soluble starch (1%, 0.1 g/10 mL (w/v)) was added to each reaction test tube and incubated at 37 °C for 15 min. Then, 1 M HCl (60 μL) was added to stop the enzymatic reaction, followed by the addition of 300 μL of iodine reagent (5 mM I_2 and 5 mM KI). The color change was noted, and the absorbance was recorded at λ of 620 nm on a spectrophotometer (721 2C50811136 Shimadzu, Japan). The control reaction representing 100% enzyme activity did not contain any *P. hydropiper* extract. To eliminate the absorbance produced by extracts, appropriate extract controls without α-amylase were also included. A dark blue color indicates the presence of starch, a yellow color indicates the absence of starch, and a brownish color indicates partially degraded starch in the reaction mixture. In the presence of extracts, the starch will not degrade upon its addition to the enzyme assay mixture, hence giving a dark blue color complex. In contrast, the absence of color complex indicates the lack of inhibitor and means that starch is completely hydrolyzed by α-amylase. The percentage inhibition of α-amylase was calculated by the following formulas:

$$I_{\alpha-\text{amylase}}\% = \frac{\Delta A_{\text{control}} - \Delta A_{\text{sample}}}{\Delta A_{\text{control}}} \times 100 \quad (3)$$

$$\Delta A_{\text{control}} = A_{\text{test}} - A_{\text{blank}} \quad (4)$$

$$\Delta A_{\text{sample}} = A_{\text{test}} - A_{\text{blank}} \quad (5)$$

4.7. Antioxidant Activity Determination

4.7.1. Free Radical Scavenging Assay (FRAP)

Herein, the ability of extracts to reduce ferric ions was determined by using a previously described method [50], with modifications. The extract (750 μL) of each sample was mixed with an equal amount of phosphate buffer (0.2M, pH 6.6) and 1% potassium ferricyanide (a source of ferric ions). The mixture was incubated at 50 °C for 20 min. After incubation, an equal amount of trichloroacetic acid (10%) was added to stop the reaction. The sample was then centrifuged at 3000 rpm for 10 min. The upper layer (1.5 mL) was separated and mixed with an equal amount of distilled water and 0.1 mL of $FeCl_3$ solution (0.1%). A blank was also prepared by using same procedure, and the absorbance was measured at λ of 700 nm as the reducing power. In parallel, ascorbic acid (Vitamin C) was used as standard positive control. The assay was repeated in triplicate, and percentage inhibition was calculated using the following formula:

$$\%\text{Scavenging effect} = \frac{\text{Control absorbance} - \text{Sample absorbance}}{\text{Control absorbance}} \times 100 \quad (6)$$

4.7.2. 2,2-Dipheny-1-Picrylhydrazyl (DPPH) Scavenging Assay

The DPPH (2,2-dipheny-1-picrylhydrazyl) free radical scavenging capacity of the extract was determined by using a previously described method [51], with modifications. The solution was prepared by dissolving the 0.006 g of DPPH in 100 mL dimethylsulfoxide (DMSO). The extract (1 mL) of each sample was mixed with an equal amount of DMSO and used to prepare desired concentrations of sample (i.e., six concentrations from 0.468 to 15 mg/mL). The solutions of the required concentrations were then transferred into a test tube by taking 1 mL of sample and 2 mL of DPPH solution. The blank was prepared by mixing 1 mL of DMSO with 2 mL of DPPH. All solutions were then incubated for 30 min at 37 °C. The absorbance of each concentration was taken at λ of 517 nm. The percent scavenging activity was calculated as:

$$\%\text{Scavenging activity} = \frac{A_0 - A_1}{A_0} \times 100 \tag{7}$$

where A_0 represents absorbance of the control and A_1 is the absorbance of the sample. Each experiment was performed in triplicate.

4.8. Antimicrobial Activity Determination

4.8.1. Selection of Microorganisms

Antimicrobial activity was evaluated against *Escherichia coli*, *Staphylococcus aurous*, *Klebsiella pneumoniae*, *Morganella morganii*, and *Haemophilus influenzae*. Microorganisms were obtained from the National Institute of Health (NIH) in Islamabad, Pakistan. The stock inoculums were subcultured using the streaking method. Inoculums of all microbes were prepared in sterilized LB-broth medium (Miller's LB Broth, Sigma-Aldrich, St. Louis, MO, USA). Twenty grams of LB-broth powder was added to 1 L of distilled water and autoclaved for 15 min at 121 °C. The autoclaved liquid medium (5 mL) was then poured into separate test tubes and settled to cool at 50 °C. Bacterial inoculums were transferred to the medium-filled test tube and incubated while shaking at 37 °C for 24 h. Later, optical density of each culture was taken at λ of 660 nm and an absorbance level of 0.5–1.0 was considered for the optimal determination of antimicrobial activity.

4.8.2. Preparation of the Culture Medium

The culture medium was prepared by dissolving 20 g of LB-broth medium in 1000 mL of distilled water. A turbid solution was obtained, which was heated until it became a clear transparent solution, using continuous shaking to dissolve the agar completely. The medium was sterilized at 121 °C for 15 min at 15 pounds of pressure. Sterilized medium (35 mL) was poured into petri dishes under the laminar flow hood and left to solidify at room temperature.

4.8.3. Antimicrobial assay

Antimicrobial activity was evaluated by agar well diffusion method. Seventy-five microliters of each microbial culture was spread individually on separate plates. A sterile cork-borer was used to bore wells (9 mm) in each inoculum-spread plate. Acetone and ethanol extracts (100 μL of each) were pipetted into individual wells. In each plate, a negative control (DMSO 100%) and a positive control (ampicillin 10 μg) were treated as standard. The plates were incubated at 37 °C for 24 h. For each plate, zones of inhibition were measured in millimeters.

4.9. Statistics

All the statistical analyses were performed using IBM SPSS Statistics 20.0 software (Armonk, NY, USA). The graphs were created in GraphPad Prism 5.0 software (La Jolla, CA, USA). One-way ANOVA test was performed for the comparison of groups, and an independent samples *t*-test was performed for differences between the two groups. All the data are presented as mean and standard error of triplicate measurements. The IC_{50} values were calculated from dose-dependent percent inhibition using GraphPad Prism. Statistically significant difference was considered for $p < 0.05$.

5. Conclusions

The present study highlights that *P. hydropiper* possesses a strong α-amylase inhibitory potential and reveals its potency to be used as a strong source of future therapeutic agents in diabetes. Our study also indicates that this plant species may be beneficial for treating the pathological damage caused by radicals' activities and bacterial infections. Future studies are required to unveil the novel bioactive compounds of *P. hydropiper*, which might be helpful in studying the precise mechanisms of α-amylase inhibition, antioxidant potential, and antimicrobial activity.

Author Contributions: Conceptualization, A.N. and M.K.; methodology, A.N., M.K., and Z.R.; software, A.N., M.K., and Z.R.; validation, A.N. and M.I.; formal analysis, A.N., A.A.K.K., S.F., N.B., M.I., and W.S.; investigation, Z.P., W.S., and A.A.K.K.; resources, Z.P.; writing—original draft preparation, A.N. and Z.P.; writing—review and editing, A.A.K.K., S.F., N.B., M.I., W.S., and Z.P.; visualization, A.N., A.A.K.K., S.F., N.B., M.I., W.S., and Z.P.; supervision, Z.P. All authors have read and agreed to the published version of the manuscript.

Funding: This study was supported by Higher Education Commission Pakistan (Grant No: 20-3589) and Directorate of Science and Technology.

Conflicts of Interest: The authors declare no conflict of interest.

References

1. Craig, C.; Freeman, J.L.R. *Polygonaceae*; Oxford University Press: Oxford, UK, 2005; Volume 5, pp. 216–221.
2. Sanchez, A.; Kron, K.A. Phylogenetics of polygonaceae with an emphasis on the evolution of eriogonoideae. *Syst. Bot.* **2008**, *33*, 87–96. [CrossRef]
3. Qaiser, M. *Flora of Pakistan*; Missouri Botanical Garden Press: St. Louis, MO, USA, 2001; Volume 205.
4. Peng, Z. Antioxidant flavonoids from leaves of *Polygonum hydropiper* L. *Phytochemestry* **2003**, *62*, 219–228. [CrossRef]
5. Xiao, H.; Ravu, R.R.; Jacob, M.R.; Khan, S.; Tekwani, B.; Khan, I.A.; Wang, W.; Li, X.C. Structure identification and biological evaluation of compounds isolated from *Polygonum hydropiper*. *Planta Med.* **2016**, *82*, PC65. [CrossRef]
6. Yang, Y.; Yu, T.; Jang, H.-J.; Byeon, S.E.; Song, S.-Y.; Lee, B.-H.; Rhee, M.H.; Kim, T.-W.; Lee, J.; Hong, S.; et al. In vitro and in vivo anti-inflammatory activities of *Polygonum hydropiper* methanol extract. *J. Ethnopharmacol.* **2012**, *139*, 616–625. [CrossRef] [PubMed]
7. Mollah, J.U.; Islam, W. Effect of *Polygonum hydropiper* L. extracts on the oviposition and egg viability of Callosobruchus cinensis F.(Coleoptera: Bruchidae). *J. Bio Sci.* **2005**, *12*, 101–109.
8. Ayaz, M.; Junaid, M.; Ahmad, J.; Ullah, F.; Sadiq, A.; Ahmad, S.; Imran, M. Phenolic contents, antioxidant and anticholinesterase potentials of crude extract, subsequent fractions and crude saponins from *Polygonum hydropiper* L. *BMC Complement. Altern. Med.* **2014**, *14*, 145. [CrossRef] [PubMed]
9. Ayaz, M.; Junaid, M.; Subhan, F.; Ullah, F.; Sadiq, A.; Ahmad, S.; Imran, M.; Kamal, Z.; Hussain, S.; Shah, S.M. Heavy metals analysis, phytochemical, phytotoxic and anthelmintic investigations of crude methanolic extract, subsequent fractions and crude saponins from *Polygonum hydropiper* L. *BMC Complement. Altern. Med.* **2014**, *14*, 465. [CrossRef]
10. Ayaz, M.; Junaid, M.; Ullah, F.; Sadiq, A.; Ovais, M.; Ahmad, W.; Ahmad, S.; Zeb, A. Chemical profiling, antimicrobial and insecticidal evaluations of *Polygonum hydropiper* L. *BMC Complement. Altern. Med.* **2016**, *16*, 502. [CrossRef]
11. Ayaz, M.; Junaid, M.; Ullah, F.; Sadiq, A.; Subhan, F.; Khan, M.A.; Ahmad, W.; Ali, G.; Imran, M.; Ahmad, S. Molecularly characterized solvent extracts and saponins from *Polygonum hydropiper* L. show high anti-angiogenic, anti-tumor, brine shrimp, and fibroblast NIH/3T3 cell line cytotoxicity. *Front. Pharmacol.* **2016**, *7*, 18. [CrossRef]
12. Sharma, R. *Medicinal Plants of India: An Encyclopaedia*; Daya Pub: New Delhi, India, 2003.
13. Nawab, A.; Yunus, M.; Mahdi, A.A.; Gupta, S. Evaluation of anticancer properties of medicinal plants from the Indian sub-continent. *Mol. Cell. Pharmacol.* **2011**, *3*, 21–29.

14. Ghani, A. *Medicinal Plants of Bangladesh: Chemical Constituents and Uses*; Asiatic Society of Bangladesh: Dhaka, Bangladesh, 1998.
15. Loi, D.T. *The Glossary of Vietnamese Medicinal Plants and Items*; Hanoi Medicine Publishing House: Hanoi, Vietnam, 2000.
16. Nakao, M.; Ono, K.; Takio, S. The effect of calcium on flavanol production in cell suspension cultures of *Polygonum hydropiper*. *Plant Cell Rep.* **1999**, *18*, 759–763. [CrossRef]
17. Miyazawa, M.; Tamura, N. Components of the essential oil from sprouts of *Polygonum hydropiper* L. ('Benitade'). *Flavour Fragr. J.* **2007**, *22*, 188–190. [CrossRef]
18. Hasan, M.F.; Das, R.; Khan, A.; Hasan, M.S.; Rahman, M. The determination of antibacterial and antifungal activities of *Polygonum hydropiper* (L.) Root Extract. *Adv. Biol. Res.* **2009**, *3*, 53–56.
19. Kundu, B.; Ara, R.; Begum, M.; Sarker, Z. Effect of Bishkatali, *Polygonum hydropiper* L. plant extracts against the red flour beetle, Tribolium castaneum Herbst. *Univ. J. Zool. Rajshahi Univ.* **2008**, *26*, 93–97. [CrossRef]
20. Duraipandiyan, V.; Indwar, F.; Ignacimuthu, S. Antimicrobial activity of confertifolin from *Polygonum hydropiper*. *Pharm. Biol.* **2009**, *48*, 187–190. [CrossRef]
21. Yang, X.; Wang, B.-C.; Zhang, X.; Yang, S.-P.; Li, W.; Tang, Q.; Singh, G.K. Simultaneous determination of nine flavonoids in *Polygonum hydropiper* L. samples using nanomagnetic powder three-phase hollow fibre-based liquid-phase microextraction combined with ultrahigh performance liquid chromatography–mass spectrometry. *J. Pharm. Biomed. Anal.* **2011**, *54*, 311–316. [CrossRef]
22. Abas, F.; Lajis, N.H.; Kalsom, Y.U. Antioxidative and radical scavenging properties of the constituents isolated from Cosmos caudatus Kunth. *Nat. Prod. Sci.* **2003**, *9*, 245–248.
23. Armstrong, D. Introduction to free radicals, inflammation, and recycling. In *Oxidative Stress and Antioxidant Protection: The Science of Free Radical Biology and Disease*; John Wiley & Sons, Inc.: Hoboken, NJ, USA, 2016; pp. 1–10.
24. Van Kiem, P.; Nhiem, N.X.; Cuong, N.X.; Hoa, T.Q.; Huong, H.T.; Huong, L.M.; Van Minh, C.; Kim, Y.H. New phenylpropanoid esters of sucrose from *Polygonum hydropiper* and their antioxidant activity. *Arch. Pharmacal. Res.* **2008**, *31*, 1477–1482. [CrossRef]
25. Teoh, E.S. *Medicinal Orchids of Asia*; Springer: Berlin/Heidelberg, Germany, 2016.
26. Mitra, A.; Bhattacharya, D.; Roy, S. Dietary influence on TYPE 2 Diabetes (NIDDM). *J. Hum. Ecol.* **2007**, *21*, 139–147. [CrossRef]
27. Elosta, A.; Ghous, T.; Ahmed, N. Natural products as anti-glycation agents: Possible therapeutic potential for diabetic complications. *Curr. Diabetes Rev.* **2012**, *8*, 92–108. [CrossRef]
28. Agarwal, P.; Gupta, R. Alpha-amylase inhibition can treat diabetes mellitus. *Res. Rev. J. Med. Health Sci.* **2016**, *5*, 1–8.
29. Mahmood, N. A review of α-amylase inhibitors on weight loss and glycemic control in pathological state such as obesity and diabetes. *Comp. Clin. Path.* **2016**, *25*, 1253–1264. [CrossRef]
30. Tripathi, B.K.; Srivastava, A.K. Diabetes mellitus: Complications and therapeutics. *Med. Sci. Monit.* **2006**, *12*, RA130–RA147.
31. Pokhum, C.; Chawengkijwanich, C.; Kobayashi, F. Enhancement of non-thermal treatment on inactivation of glucoamylase and acid protease using CO_2 microbubbles. *J. Food Process. Technol.* **2015**, *6*, 1–5.
32. Alborzi, Z.; Zibaee, A.; Sendi, J.J.; Ramzi, S. Effects of the agglutinins extracted from Rhizoctonia solani (Cantharellales: Ceratobasidiaceae) on Pieris brassicae (Lepidoptera: Pieridae). *J. Econ. Èntomol.* **2016**, *109*, 1132–1140. [CrossRef] [PubMed]
33. Donaldson, M. Nutrition and cancer: A review of the evidence for an anti-cancer diet. *Nutr. J.* **2004**, *3*, 19. [CrossRef] [PubMed]
34. Jagannath, C.V. Potentiation of antiepileptic activity of phenytoin using beta carotene against maximal electroshock induced convulsions in mice. *World J. Pharm. Pharm. Sci.* **2017**, 1574–1585. [CrossRef]
35. Ruch, R.J.; Cheng, S.-J.; Klaunig, J.E. Prevention of cytotoxicity and inhibition of intercellular communication by antioxidant catechins isolated from Chinese green tea. *Carcinogenesis* **1989**, *10*, 1003–1008. [CrossRef]
36. Ferhi, S.; Santaniello, S.; Zerizer, S.; Cruciani, S.; Fadda, A.; Sanna, D.; Dore, A.; Maioli, M.; D'Hallewin, G. Total phenols from grape leaves counteract cell proliferation and modulate apoptosis-related gene expression in MCF-7 and HepG2 human cancer cell lines. *Molecules* **2019**, *24*, 612. [CrossRef]
37. Ogawa, S.; Yazaki, Y. Tannins from acacia mearnsii De Wild. Bark: Tannin determination and biological activities. *Molecules* **2018**, *23*, 837. [CrossRef]

38. Cruciani, S.; Santaniello, S.; Garroni, G.; Fadda, A.; Balzano, F.; Bellu, E.; Sarais, G.; Fais, G.; Mulas, M.; Maioli, M.; et al. Myrtus polyphenols, from antioxidants to anti-inflammatory molecules: Exploring a network involving cytochromes P450 and vitamin D. *Molecules* **2019**, *24*, 1515. [CrossRef] [PubMed]
39. Suganya, N.; Bhakkiyalakshmi, E.; Sarada, D.V.L.; Ramkumar, K.M. Reversibility of endothelial dysfunction in diabetes: Role of polyphenols. *Br. J. Nutr.* **2016**, *116*, 223–246. [CrossRef]
40. Giovannucci, E.; Rimm, E.B.; Liu, Y.; Stampfer, M.J.; Willett, W.C. A prospective study of tomato products, lycopene, and prostate cancer risk. *J. Natl. Cancer Inst.* **2002**, *94*, 391–398. [CrossRef] [PubMed]
41. Atessahin, A.; Yilmaz, S.; Karahan, I.; Ceribasi, A.O.; Karaoglu, A. Effects of lycopene against cisplatin-induced nephrotoxicity and oxidative stress in rats. *Toxicology* **2005**, *212*, 116–123. [CrossRef] [PubMed]
42. Kim, J.Y.; Paik, J.K.; Kim, O.Y.; Park, H.W.; Lee, J.H.; Jang, Y.; Lee, J.H. Effects of lycopene supplementation on oxidative stress and markers of endothelial function in healthy men. *Atherosclerosis* **2011**, *215*, 189–195. [CrossRef] [PubMed]
43. Shim, Y.-J.; Doo, H.-K.; Ahn, S.-Y.; Kim, Y.-S.; Seong, J.-K.; Park, I.-S.; Min, B.-H. Inhibitory effect of aqueous extract from the gall of Rhus chinensis on alpha-glucosidase activity and postprandial blood glucose. *J. Ethnopharmacol.* **2003**, *85*, 283–287. [CrossRef]
44. Kwon, Y.-I.; Apostolidis, E.; Kim, Y.-C.; Shetty, K. Health benefits of traditional corn, beans, and pumpkin: In vitro studies for hyperglycemia and hypertension management. *J. Med. Food* **2007**, *10*, 266–275. [CrossRef]
45. Chew, Y.L.; Chan, E.W.L.; Tan, P.L.; Lim, Y.; Stanslas, J.; Goh, J.K. Assessment of phytochemical content, polyphenolic composition, antioxidant and antibacterial activities of Leguminosae medicinal plants in Peninsular Malaysia. *BMC Complement. Altern. Med.* **2011**, *11*, 12. [CrossRef]
46. Barros, L.; Calhelha, R.C.; Vaz, J.; Ferreira, I.C.F.R.; Baptista, P.; Estevinho, L.M.; Estevinho, L.M. Antimicrobial activity and bioactive compounds of Portuguese wild edible mushrooms methanolic extracts. *Eur. Food Res. Technol.* **2006**, *225*, 151–156. [CrossRef]
47. Barros, L.; Ferreira, M.-J.; Queirós, B.; Ferreira, I.C.F.R.; Baptista, P. Total phenols, ascorbic acid, β-carotene and lycopene in Portuguese wild edible mushrooms and their antioxidant activities. *Food Chem.* **2007**, *103*, 413–419. [CrossRef]
48. Nagata, M.; Yamashita, I. Simple method for simultaneous determination of chlorophyll and carotenoids in tomato fruit. *J. Jpn. Soc. Food. Sci.* **1992**, *39*, 925–928.
49. Makkar, H.P.S.; Blümmel, M.; Borowy, N.K.; Becker, K. Gravimetric determination of tannins and their correlations with chemical and protein precipitation methods. *J. Sci. Food Agric.* **1993**, *61*, 161–165. [CrossRef]
50. Xiao, Z.; Storms, R.; Tsang, A. A quantitative starch? Iodine method for measuring alpha-amylase and glucoamylase activities. *Anal. Biochem.* **2006**, *351*, 146–148. [CrossRef] [PubMed]
51. Brand-Williams, W.; Cuvelier, M.-E.; Berset, C. Use of a free radical method to evaluate antioxidant activity. *LWT-Food Sci. Technol.* **1995**, *28*, 25–30. [CrossRef]

Review

Phytochemicals and Biological Activity of Desert Date (*Balanites aegyptiaca* (L.) Delile)

Hosakatte Niranjana Murthy [1], Guggalada Govardhana Yadav [1], Yaser Hassan Dewir [2,3,*] and Abdullah Ibrahim [2]

1 Department of Botany, Karnatak University, Dharwad 580003, India; hnmurthy60@gmail.com (H.N.M.); govardhanyadavgs@gmail.com (G.G.Y.)
2 Plant Production Department, P.O. Box 2460, College of Food and Agriculture Sciences, King Saud University, Riyadh 11451, Saudi Arabia; adrahim@ksu.edu.sa
3 Faculty of Agriculture, Kafrelsheikh University, Kafr El-Sheikh 33516, Egypt
* Correspondence: ydewir@ksu.edu.sa

Abstract: Many underutilized tree species are good sources of food, fodder and possible therapeutic agents. *Balanites aegyptiaca* (L.) Delile belongs to the Zygophyllaceae family and is popularly known as "desert date", reflecting its edible fruits. This tree grows naturally in Africa, the Middle East and the Indian subcontinent. Local inhabitants use fruits, leaves, roots, stem and root bark of the species for the treatment of various ailments. Several research studies demonstrate that extracts and phytochemicals isolated from desert date display antioxidant, anticancer, antidiabetic, anti-inflammatory, antimicrobial, hepatoprotective and molluscicidal activities. Mesocarp of fruits, seeds, leaves, stem and root bark are rich sources of saponins. These tissues are also rich in phenolic acids, flavonoids, coumarins, alkaloids and polysterols. Some constituents show antioxidant, anticancer and antidiabetic properties. The objective of this review is to summarize studies on diverse bioactive compounds and the beneficial properties of *B. aegyptiaca*.

Keywords: bioactive compounds; polysterols; polyphenols; saponins; therapeutic properties

Citation: Murthy, H.N.; Yadav, G.G.; Dewir, Y.H.; Ibrahim, A. Phytochemicals and Biological Activity of Desert Date (*Balanites aegyptiaca* (L.) Delile). *Plants* **2021**, *10*, 32. https://dx.doi.org/10.3390/plants10010032

Received: 28 November 2020
Accepted: 22 December 2020
Published: 25 December 2020

1. Introduction

Balanites aegyptiaca (L.) Delile (Family: Zygophyllaceae) is an underutilized fruit-yielding tree (Figure 1A) native to Africa and distributed in tropical and subtropical regions of Africa, from Senegal in the west (16 °W) to Somali in the East (49 °E) and Jordan in the north (35 °N) to Zimbabwe in the south (19 °S). *B. aegyptiaca* is also distributed in India, Myanmar, Iran, Jordan, Oman, Palestine, Saudi Arabia, Syria and Yemen [1]. Young leaves (Figure 1B) and tender shoots are used as vegetables. Leaves and fruits are used as fodder for livestock [1]. Fiber obtained from tender bark and older dried bark is used for the preparation of medicines (Figure 1C). Unripe and ripe fruits (Figure 1D,E) are edible and popularly known as "desert date". The fruits are processed into beverages and liquor. Timber is suitable for the construction of furniture, domestic items and musical instruments. The wood produces high-quality charcoal fuel and industrial activated charcoal. Gum or resin produced from stems are used as glue. Seeds contain about 49% edible oil (Figure 1F,G),which is also used in the production of biodiesel fuel [1].

B. aegyptiaca is used in African and Indian traditional medicine. Roots and bark are purgative and anthelmintic. A decoction of roots is used to treat malaria. The bark is used to deworm cattle, and the roots are boiled into a soup and used to treat edema and stomach pains. Roots are also used as an emetic [1]. The fruit is usedto treat jaundice in Sudan [2]. Seed oil is used as a laxative and for the treatment of hemorrhoids, stomach aches, jaundice, yellow fever, syphilis and epilepsy [3]. Bark extracts are used to kill freshwater snails and copepods. A decoction of bark is also used as an abortifacient and antidote in West African traditional medicine [4].

Figure 1. Morphology of *Balanitesaegyptiaca* (L.) Delile: (**A**) habit, (**B**) leaves, (**C**) stem bark, (**D**) ripened fruits, (**E**) rind (left) and pulp (right), (**F**) seed kernels and (**G**) seed oil.

2. Nutritional Composition of Fruits, Seeds and Leaves

Ripe fruits display a thin brownishepicarp (Figure 1E), dark brown and fleshy mesocarp (Figure 1E) and thick endocarp nut (Figure 1F). The edible partsof the pulp and kernel yield oil. The pulp is rich in carbohydrates (62.63%) and protein (9.19%; Table 1) [5]. Fruit pulp shows lesser amounts of fat (2.58%) and dietary fiber (2.93%). The overall energy value is 346.82 kcal/100 g. Fruits are also rich in minerals, including calcium, magnesium, phosphorus, potassium and sodium (Table 1) [6]. Iron, copper, manganese, lead, chromium, cobalt, cadmium and selenium are reported in lower concentrations (Table 1). Major fatty acids in fruit pulp are oleic (37.17%), linoleic (27.73%) and palmitic (22.02%; Table 2) [7]. The fruit pulp also exhibits amino acids (Table 3) [8] and vitamins (Table 3). Antinutritional factors are comparatively less (Table 4) [5].

Seeds are rich in fixed oil content (49.00%) with a significant content of proteins (32.40%) and carbohydrates (8.70%; Table 1) [9,10]. Seed oil is used for edible purposes; major fatty acids arelinoleic (47.84%), oleic (22.80%), palmitic (16.68%) and stearic (11.67%) (Table 2) [11]. It has been demonstrated that biodiesel from seed oil meets all international biodiesel standards [11]. Seeds contain minerals, such as potassium, phosphorus and calcium in higher concentrations (Table 1) and amino acids (Table 3) [9]; seed cake isused for animal feed. However, seeds also contain oxalate (8.51 mg/g DW), antinutrient and possibly toxic constituents (Table 4).

Young leaves and shoots are used as vegetables in African countries. Leaves and shoots are also popular livestock fodder [1]. Leaves are a good source of carbohydrates (28.12%) and proteins (15.86%) and contain ash (9.26%) and dietary fiber (30.75%; Table 1) [12]. Leaves also provide minerals (Table 1), fatty acids (Table 2), amino acids (Table 3) and vitamins (Table 3) [12,13]. Leaves contain antinutrients in meager concentrations (Table 4).

Table 1. Nutritional and mineral composition of desert date pulp, seeds and leaves.

Proximate (%)			
Components	**Pulp [5,6]**	**Seeds [9,10]**	**Leaves [12,13]**
Moisture	18.27	5.20	13.11
Protein	9.19	32.40	15.86
Fat	2.58	49.00	2.90
Ash	4.40	3.30	9.26
Carbohydrate	62.63	8.70	28.12
Dietary fiber	2.93	1.40	30.75
Energy (kcal/100g)	346.82	605.40	202.02
Mineral composition (mg/g DW)			
Calcium (Ca)	1.41	1.51	0–2.65
Iron (Fe)	0.0494	0.0484	NR *
Magnesium (Mg)	0.73	0.887	0.23–0.77
Phosphorus (P)	0.48	3.60	1.51–5.32
Potassium (K)	22.20	6.36	1.76–4.81
Sodium (Na)	0.48	0.053	NR
Zinc (Zn)	0.0065	0.0286	NR
Copper (Cu)	0.0039	0.0118	0.05–0.65
Manganese (Mn)	0.0033	0.0192	0.02–0.08
Lead (Pb)	0.0030	0.0050	NR
Chromium (Cr)	0.0040	0.0060	NR
Cobalt (Co)	0.0107	0.0120	NR
Cadmium (Cd)	0.0347	0.0163	NR
Selenium (Se)	0.0005	NR	NR

* NR = not reported.

Table 2. Fatty acid composition of desert date pulp, seeds and leaves.

Fatty Acid	Pulp (% in oil) [7]	Seeds (% in oil) [11]	Leaves (μg/g DW) [13]
Saturated fatty acids (SFA)			
Lauric (C12:0)	ND *	ND	0.17–0.28
Myristic (C14:0)	0.1	0.05	0.043–0.074
Pentadecylic (C15:0)	ND	0.046	ND
Palmitic (C16:0)	22.02	16.683	0.29–1.98
Margaric (17:0)	ND	0.106	ND
Stearic (C18:0)	ND	11.67	0.008–0.049
Nonadecylic(19:0)	ND	0.032	ND
Arachidic (C20:0)	ND	0.34	0.274–0.439
Behenic (C22:0)	ND	0.059	0.026–0.038
Tricosylic (C23:0)	ND	0.012	0–0.003
Lignoceric (C24:0)	ND	0.042	0.028–0.298
Hyenic (C25:0)	ND	ND	0.078–0.121
Monounsaturated fatty acids (MUSFA)			
Pentadecenoic (C15:1)	ND	0.003	0.08–0.31
Palmitoleic (C16:1)	ND	0.027	ND
Oleic (18:1)	37.17	22.807	0.03–0.061
Nonadecenoic (19:1)	ND	0.175	ND
Gadoleic (20:1)	ND	0.061	ND
Tetracosenoic (C24:1)	ND	ND	0.077–0.266

Table 2. *Cont.*

Fatty Acid	Pulp (% in oil) [7]	Seeds (% in oil) [11]	Leaves (µg/g DW) [13]
	Polyunsaturated fatty acids (PUSFA)		
Hexadecadienoic (C16:2)	ND	ND	0.406–1.835
Linoleic (C18:2)	27.73	47.847	0.025–0.642
Eicosadienoic (C20:2)	ND	ND	0.116–0.296
Hexadecatrienoic (C16:3)	ND	ND	0.761–2.142
Octadecatrienoic (C18:3)	ND	ND	0.20–0.525
Total SFA (%)	22.16	29.04	1.84–3.38
Total MUSFA (%)	37.17	23.073	0.179–0.637
Total PUSFA (%)	27.73	47.847	1.727–5.174

* ND = not detected.

Table 3. Amino acid and vitamin composition of desert date pulp, seeds and leaves.

Amino Acid	Pulp (mg/g DW) [5,8,14]	Seeds (g/100 g of Protein) [9]	Leaves (g/100 g of Protein) [12,13]
Alanine	2.90	3.50	1.80
Aspartic acid	4.43	8.29	7.86
Arginine	3.00	2.70	4.20
Cystine	1.65	2.52	0.79
Glutamic acid	7.10	8.91	10.80
Glycine	2.52	4.10	9.65
Histidine	0.80	1.99	2.83
Isoleucine	1.87	3.47	3.50
Leucine	3.04	6.47	6.23
Lysine	1.64	5.00	4.51
Methionine	0.60	0.75	0.73
Phenylalanine	1.90	4.61	4.80
Proline	30.80	2.78	1.85
Serine	1.80	4.29	2.01
Threonine	2.17	4.25	2.88
Tyrosine	1.84	2.75	3.16
Valine	2.23	3.29	4.07
Tryptophan	0.70	NR *	NR
Vitamin (mg/g DW)			
	Vitamin A		
α-carotene	NR	NR	0.33–0.54
β-carotene	0.6484	NR	0.25–0.81
β-cryptoxanthin	NR	NR	0.02–1.14

Table 3. *Cont.*

Amino Acid	Pulp (mg/g DW) [5,8,14]	Seeds (g/100 g of Protein) [9]	Leaves (g/100 g of Protein) [12,13]
	Vitamin B		
Thiamine	0.0027	NR	0.24–0.51
Riboflavin	0.0007	NR	NR
Niacin(B3)	0.0174	NR	NR
Vitamin B6	0.0021	NR	NR
	Vitamin C		
Ascorbic acid	1.05	NR	0.57–2.05
	Vitamin E		
α-tocopherol	NR	NR	0.08–0.57
β-tocopherol	NR	NR	0.01–0.04
γ-tocopherol	NR	NR	0.01–0.063
δ-tocopherol	NR	NR	0.13–0.96
	Vitamin K		
Phylloquinone	NR	NR	0.21–1.37

* NR = not reported.

Table 4. Antinutritional components of desert date pulp, seeds and leaves.

Antinutritional Factors (mg/g DW)			
Components	**Pulp [5]**	**Seeds [9]**	**Leaves [12]**
Oxalate	0.38	8.51	0.75
Phytate	1.82	0.2133	0.0297
Tannins	0.40	0.275	0.041
Saponin	0.62	4.279	ND
Nitrate	ND *	0.39	ND
Cyanogenic glycosides	ND	0.0096	ND

* ND = not detected.

3. Phytochemicals Isolated from Desert Date

B. aegyptiaca produces a variety of secondary metabolites, such as polyphenols (phenolic acids, flavonoids and coumarins), alkaloids, steroids, saponins (spirostanolsaponins, furostanolsaponins and open-chain steroidal saponins) and pregnane glycosides, isolated from plant tissues, such as fruit, seeds, leaves, stem bark, roots and galls (Table 5).

3.1. Polyphenols

Polyphenols exhibit phenolic structural features with one or more aromatic rings, each with one or more hydroxyl groups [15]. Polyphenols are grouped into phenolic acids, flavonoids, stilbenes, lignans and tannins. These compounds are important as natural therapeutic agents involved in the prevention of degenerative diseases, particularly cancers, cardiovascular diseases and neurodegenerative diseases [16].

Phenolic acids are nonflavonoid polyphenolic compounds of benzoic acid and cinnamic acid. Major phenolic acids, which are isolated from tissues of *B. aegyptiaca*, include caffeic acid (**1**), ferulic acid (**2**), gentisic acid (**3**), p-coumaric acid (**4**), sinapic acid (**5**), syringic acid (**6**), vanillicacid (**7**), 2-methoxy-4-vinylphenol (**8**), 2,6-dimethoxyphenol (**9**), 2-methoxy-3(-2-propenyl)-phenol (**10**), 2-methoxy-4-(1-propenyl)-phenol (**11**), 2,4-di-tert-butyl-phenol (**12**), 2,6-di-tert-butyl-phenol (**13**) and 3-hydroxy-1-(4-hydroxy-3-methoxyphenyl)-1-propanone (**14**) (Table 5; Figure 2) [2,17–19].

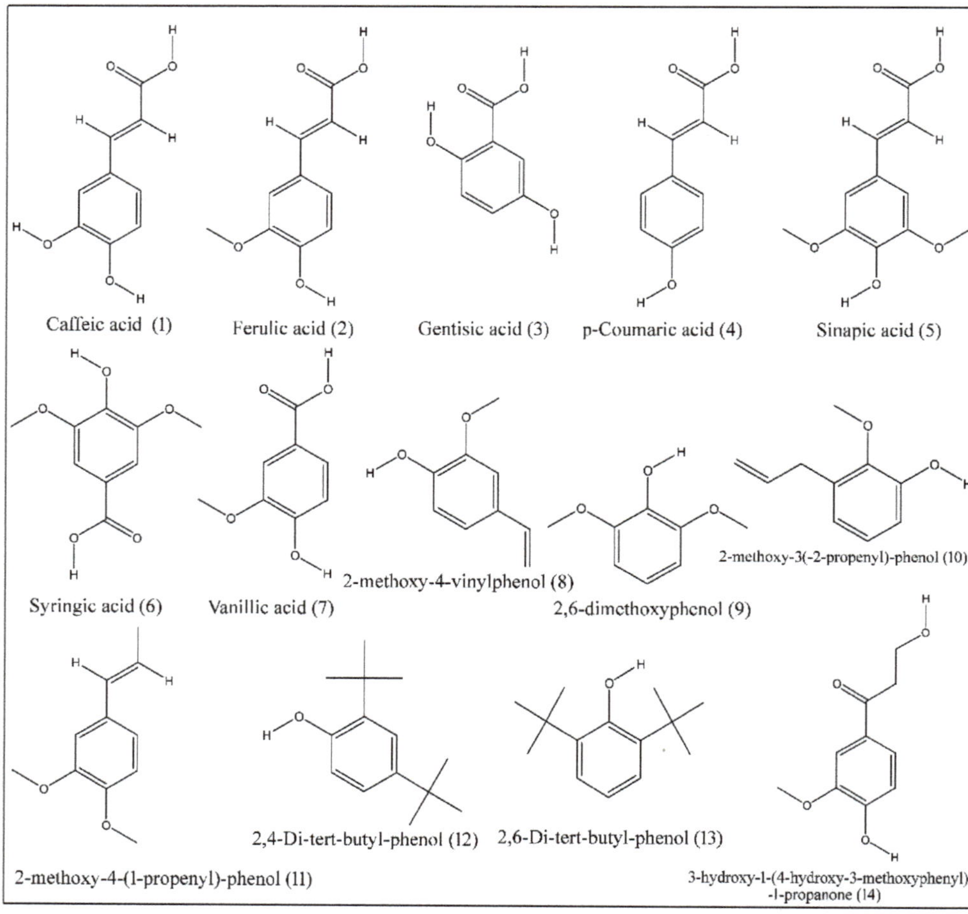

Figure 2. Structures of phenolic compounds isolated from desert date.

3.2. Flavonoids

Flavonoids exhibit a diphenyl propane–flavone skeleton with a three-carbon bridge between phenyl groups and commonly cyclized with oxygen. Epicatechin O-glucoside (**28**), hyperoside (**19**), isorhamnetin (**18**), isorhamnetin-3-O-glucoside (**23**), isorhamnetin 3,7-diglucoside (**25**), isorhamnetin 3-O-galactoside (**27**), isorhamnetin 3-O-robinobioside (**26**), isorhamnetin 3-rutinoside (**24**), kaempferol (**15**), myricetin (**16**), quercetin (**17**), quercetin 3-glucoside (**21**), quercetin 3-rutinoside (**22**) and quercitrin (**20**) are isolated from different tissues of *B. aegyptiaca* (Table 5; Figure 3) [19–22].

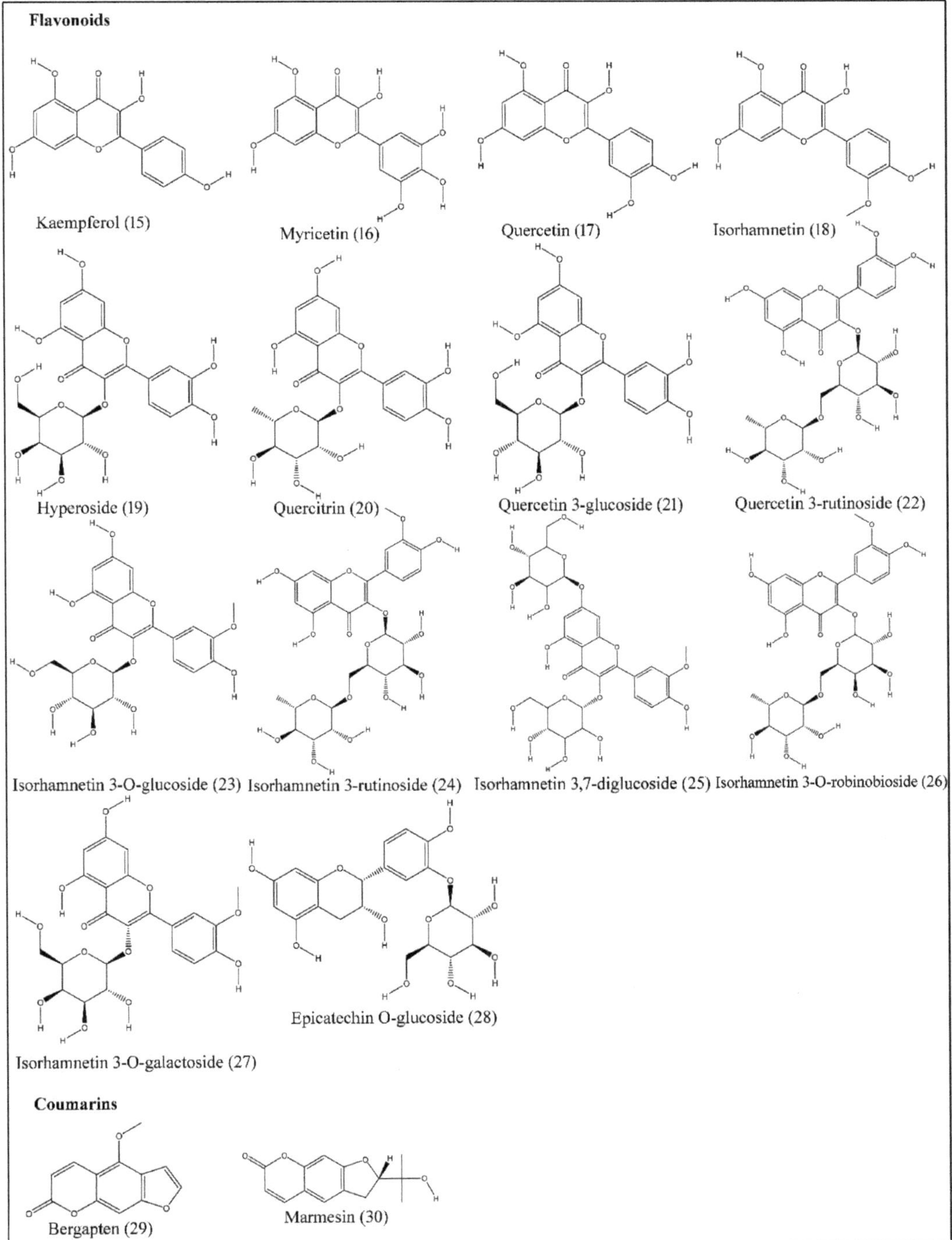

Figure 3. Structures of flavonoids and coumarins isolated from desert date.

Table 5. Phytochemicals isolated from various parts of desert date.

No.	Compound Name	Plant Parts	Reference
	PHENOLICS		
1	Caffeic acid	Gall, leaf	[19]
2	Ferulic acid	Gall, leaf	[19]
3	Gentisic acid	Gall, leaf	[19]
4	p-Coumaric acid	Gall, leaf	[19]
5	Sinapic acid	Gall, leaf	[19]
6	Syringic acid	Stem bark	[2]
7	Vanillic acid	Stem bark	[2]
8	2-methoxy-4-vinylphenol	Fruit, leaf	[18]
9	2,6-dimethoxyphenol	Leaf	[18]
10	2-methoxy-3(-2-propenyl)-phenol	Leaf	[18]
11	2-methoxy-4-(1-propenyl)-phenol (Isoeugenol)	Leaf	[18]
12	2,4-di-tert-butyl-phenol	Seed	[17]
13	2,6-di-tert-butyl-phenol	Seed	[17]
14	3-hydroxy-1-(4-hydroxy-3-methoxyphenyl)-1-propanone	Stem bark	[2]
	FLAVONOIDS		
15	Kaempferol	Leaf	[19]
16	Myricetin	Leaf	[19]
17	Quercetin	Fruit, leaf, seed	[19,20,22]
18	Isorhamnetin	Fruit, seed	[20,22]
19	Hyperoside	Gall, leaf	[19]
20	Quercitrin	Leaf	[19]
21	Quercetin 3-glucoside (isoquercetin)	Leaf, seed	[19,21,22]
22	Quercetin 3-rutinoside (Rutin)	Fruit, gall, leaf, seed	[19–22]
23	Isorhamnetin-3-O-glucoside	Fruit, leaf, seed	[20–22]
24	Isorhamnetin 3-rutinoside	Fruit, leaf, seed	[20–22]
25	Isorhamnetin 3,7-diglucoside	Leaf, seed	[21,22]
26	Isorhamnetin 3-O-robinobioside	Seed	[22]
27	Isorhamnetin 3-O-galactoside	Seed	[22]
28	Epicatechin O-glucoside	Fruit	[20]
	COUMARINS		
29	Bergapten	Stem bark	[23]
30	Marmesin	Stem bark	[23]
	ALKALOIDS		
31	N-trans-Feruloyltyramine	Stem bark	[2]
32	N-cis-feruloyltyramine	Stem bark	[2]
33	Trigonelline	Fruit	[20]
	STEROIDS		
34	Diosgenin	Fruit	[24]
35	Yamogenin	Fruit, root, stem bark	[25]
36	6-Methyldiosgenin	Fruit	[24]
37	Rotenone	Root	[26]
38	β-Sitosterol	Seed (oil)	[27]
39	Cholesterol	Seed (oil)	[27]
40	Campesterol	Seed (oil)	[27]
41	Stigmasterol	Seed (oil)	[27]

Table 5. *Cont.*

No.	Compound Name	Plant Parts	Reference
	PREGNANE GLYCOSIDES		
42	Pregn-5-ene-3β,16β,20(R)-triol 3-O-(2,6-di-O-α-L-rhamnopyranosyl)-β-D-glucopyranoside	Fruit	[28]
43	Pregn-5-ene-3β,16β,20(R)-triol 3-O-β-D-glucopyranoside	Fruit	[28]
	SAPONINS		
	SPIROSTANOL SAPONINS		
44	Balanitin 1	Root, stem bark	[29]
45	Balanitin 2	Root, stem bark	[29]
46	Balanitin 3	Root, stem bark	[29]
47	Balanitin 4	Seed	[30]
48	Balanitin 5	Seed	[30]
49	Balanitin 6	Seed	[30,31]
50	Balanitin 7	Fruit, root, seed	[30–32]
51	Deltonin	Seed	[33]
52	(3β,20S,22R,25R)-spirost-5-en-3-yl β-D-xylopyranosyl-(1→3)-β-D-glucopyranosyl-(1→4)[α-L-rhamnopyranosyl-(1→2)]-β-D-glucopyranoside	Root	[34]
53	(3β,20S,22R,25S)-spirost-5-en-3-yl β-D-xylopyranosyl-(1→3)-β-D-glucopyranosyl-(1→4)[α-L-rhamnopyranosyl-(1→2)]-β-D-glucopyranoside	Root	[34]
	FUROSTANOL SAPONINS		
54	Balanitesin	Fruit	[35]
55	Balanitoside	Fruit	[24,36]
56	22R and 22S epimers of 26-(O-β-D-glucopyranosyl)-3-β-[4-O-(β-D-glucopyranosyl)-2-O-(α-L-rhamnopyranosyl)-β-D-glucopyranosyloxy]-22,26-dihydroxyfurost-5-ene	Fruit	[37]
57	Xylopyranosyl derivative of 26-(O-β-D-glucopyranosyl)-3-β-[4-O-(β-D-glucopyranosyl)-2-O-(α-L-rhamnopyranosyl)-β-D-glucopyranosyloxy]-22,26-dihydroxyfurost-5-ene	Fruit	[37]
58	(3β,20S,22R,25R)-26-(β-D-glucopyranosyloxy)-22-methoxyfurost-5-en-3-yl β-D-xylopyranosyl-(1→3)-β-D-glucopyranosyl-(1→4) [α-L-rhamnopyranosyl-(1→2)]-β-D-glucopyranoside	Root	[34]

Table 5. *Cont.*

No.	Compound Name	Plant Parts	Reference
59	(3β,20S,22R,25S)-26-(β-D-glucopyranosyloxy)-22-methoxyfurost-5-en-3-yl β-D-xylopyranosyl-(1→3)-β-D-glucopyranosyl-(1→4)[α-L-rhamnopyranosyl-(1→2)]-β-D-glucopyranoside	Root	[34]
60	26-O-β-D-glucopyranosyl-(25R)-furost-5-ene-3β,22,26-triol 3-O-(2,4-di-O-α-L-rhamnopyranosyl)-β-D-glucopyranoside	Fruit	[38]
61	22-methyl ether of 26-O-β-D-glucopyranosyl-(25R)-furost-5-ene-3β,22,26-triol 3-O-(2,4-di-O-α-L-rhamnopyranosyl)-β-D-glucopyranoside	Fruit	[38]
62	26-O-β-D-glucopyranosyl-(25R)-furost-5-ene-3β,22,26-triol 3-O-[α-L-rhamnopyranosyl-(1→2)]-[β-D-xylopyranosyl (1→3)]-[α-L-rhamnopyranosyl-(1→4)]-β-D-glucopyranoside	Fruit	[38]
63	22-methyl ether of 26-O-β-D-glucopyranosyl-(25R)-furost-5-ene-3β,22,26-triol 3-O-[α-L-rhamnopyranosyl-(1→2)]-[β-D-xylopyranosyl (1→3)]-[α-L-rhamnopyranosyl-(1→4)]-β-D-glucopyranoside	Fruit	[38]
64	Balanin B2	Stem bark	[39]
65	26-(O-β-D-glucopyranosyl)-22-O-methylfurost-5-ene-3β,26-diol-3-O-β-D-glucopyranosyl-(1→4)-[α-L-rhamnopyranosyl-(1→2)]-β-D-glucopyranoside	Fruit	[40]
66	25R and 25S epimers of 26-O-β-D-glucopyranosyl-furost-5-ene-3,22,26- triol 3-O-[α-L-rhamnopyranosyl-(1→3)-β-D-glucopyranosyl-(1→2)]-α-L-rhamnopyranosyl-(1→4)-β-D-glucopyranoside	Fruit	[41]

Table 5. *Cont.*

No.	Compound Name	Plant Parts	Reference
67	26-O-β-D-glucopyranosyl-(25R)-furost-5-ene-3,22,26-triol 3-O-[β-D-glucopyranosyl-(1→2)]- α-L-rhamnopyranosyl-(1→4)-β-D-glucopyranoside	Fruit	[41]
68	26-O-β-D-glucopyranosyl-(25R)-furost-5,20-diene-3,26-diol 3-O-[α-L-rhamnopyranosyl-(1→3)- β-D-glucopyranosyl-(1→2)]- α-L-rhamnopyranosyl-(1→4)- β-D-glucopyranoside	Fruit	[41]
69	25R and 25S epimers of 26-O-β-D-glucopyranosyl-furost-5,20-diene-3,26-diol 3-O-[β-D-glucopyranosyl-(1→2)]- α-L-rhamnopyranosyl-(1→4)- β-D-glucopyranoside	Fruit	[41]
	OPEN-CHAIN STEROIDAL SAPONINS		
70	(3β,12α,14β,16β)-12-hydroxycholest-5-ene-3,16-diyl bis (β-D-glucopyranoside)	Root	[34]
71	Balanin B1	Stem bark	[39]
72	β-Sitosterol glucoside	Stem bark	[42]
73	Stigmasterol-3-O-β-D-glucopyranoside	Stem bark	[40]

3.3. Coumarins

Coumarins are phenolic compounds displaying fused benzene and α-pyrone rings and are known for anti-inflammatory, anticoagulant, antimicrobial, anticancer, antioxidant and neuroprotective properties [43]. Bergapten (**29**) and marmesin (**30**) are coumarins extracted from stem bark (Table 5; Figure 3) [23].

3.4. Alkaloids

Alkaloids are compounds that contain basic nitrogen atoms [44] and show varied biological activities. They are especially useful for cancer treatment. N-cis-feruloyltyramine (**32**), N-trans-feruloyltyramine (**31**) and trigonelline (**33**) are some of the alkaloids isolated from stem bark and fruit (Table 5; Figure 4) [2,20].

3.5. Phytosterols

Phytosterols are bioactive compounds found naturally in food with chemical structures similar to cholesterol. Various clinical studies consistently show that intake of phytosterols, such as beta-sitosterol, campesterol and stigmasterol, is associated with a significant reduction in levels oflow-density lipoprotein in humans. *B. aegyptiaca* produces several steroids, such as campesterol (**40**), cholesterol (**39**), diosgenin (**34**), 6-methyldiosgenin (**36**), rotenone (**37**), β-sitosterol (**38**), stigmasterol (**41**) and yamogenin (**35**) (Table 5; Figure 4) [24–27].

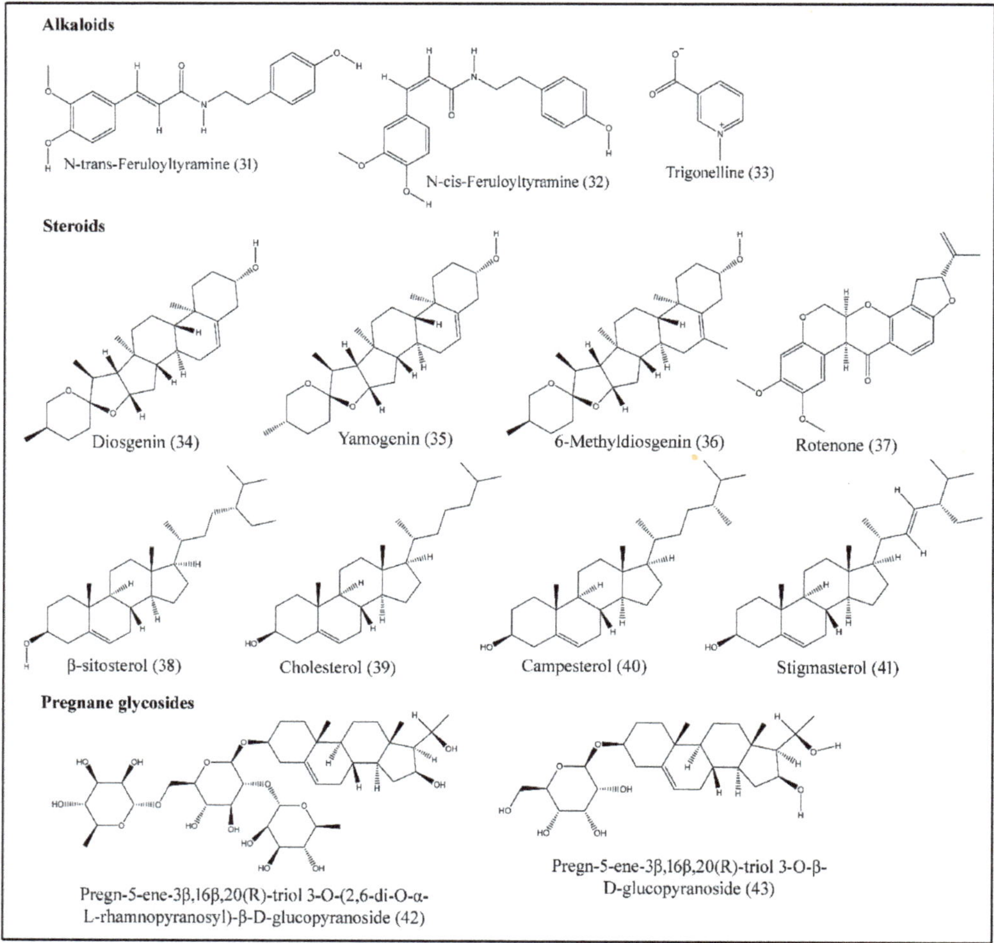

Figure 4. Structures of alkaloids, steroids and pregnane glycosides isolated from desert date.

Pregnane glycosides are naturally occurring sugar conjugates of C_{21} steroidal compounds, isolated from various plants and many show anticarcinogenic properties [45]. Pregn-5-ene-3β,16β,20(R)-triol 3-O-(2,6-di-O-α-L-rhamnopyranosyl)-β-D-glucopyranoside (**42**) andpregn-5-ene-3β,16β,20(R)-triol 3-O-β-D-glucopyranoside (**43**) were extracted from the fruits of desert date (Table 5; Figure 4) [28].

3.6. *Saponins*

Saponins are bioorganic compounds that exhibit triterpenoid or steroidal skeletons that are glycosylated by varying numbers of sugar moieties attached at different positions. Steroidal saponins are further classified intospirostanol, furostanol and open-chain steroidal saponins [46]. Saponins exhibit a widerange of biological properties, including hemolytic factorsand anti-inflammatory, antimicrobial, insecticidal, anticancer and molluscicidal activities [47]. Various spirostanol, furostanol and open-chain steroidal saponins, which are isolated from fruits, seeds, roots and stem bark are presented in Table 5 and Figures 5 and 6 [24,29–42].

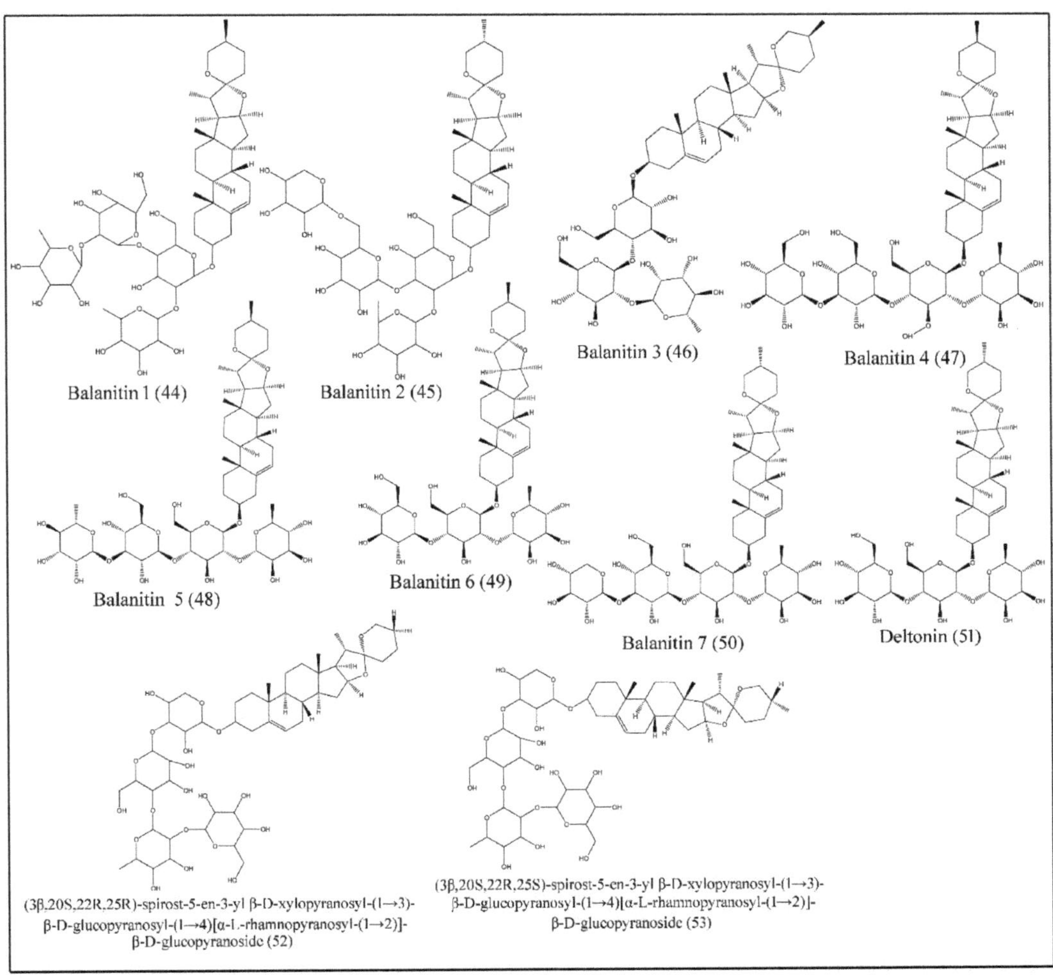

Figure 5. Structures of spirostanolsaponins isolated from desert date.

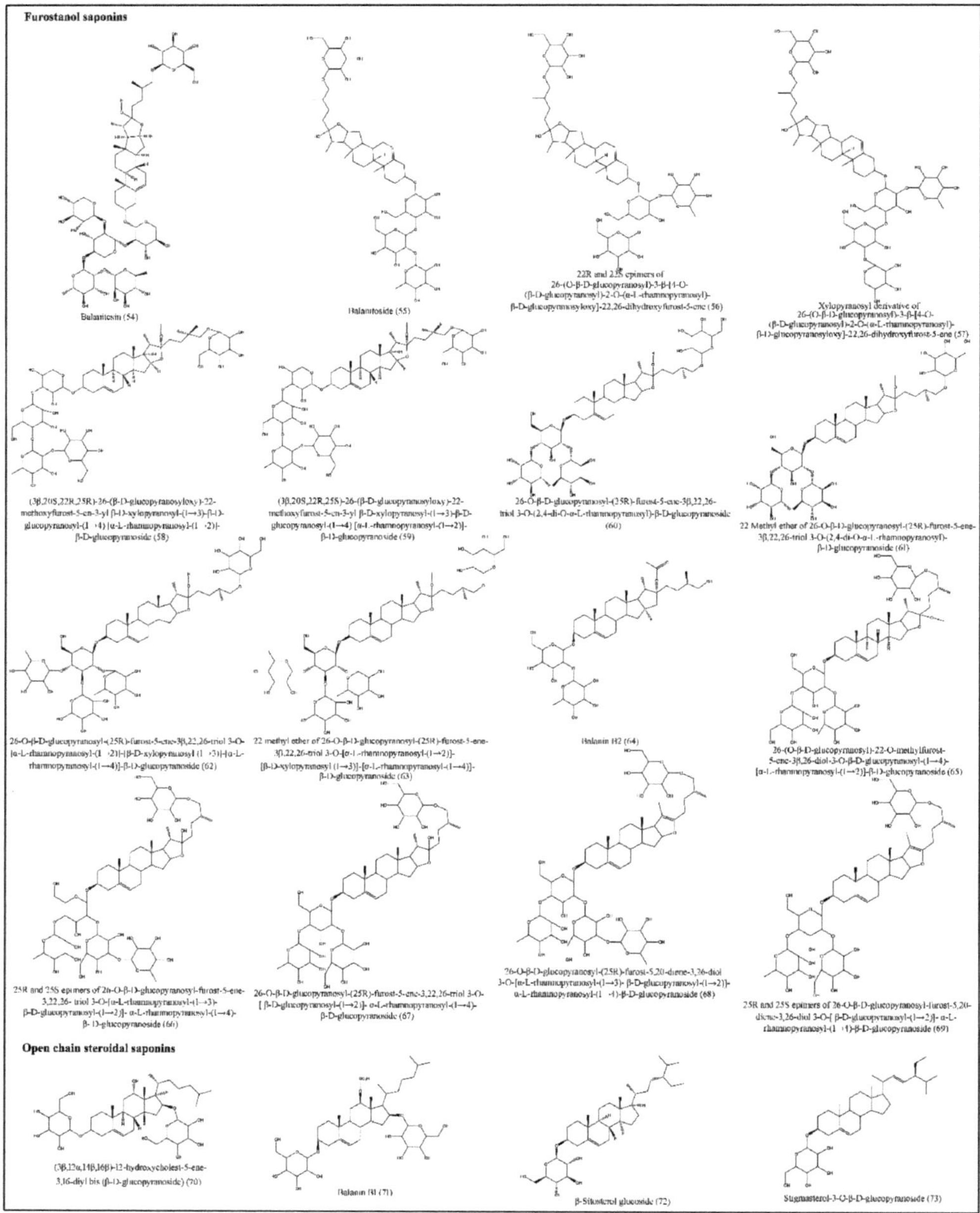

Figure 6. Structures of furostanoland open-chain steroidal saponins of desert date.

4. Biological Activity

Extracts and compounds from extractionsof *B. aegyptiaca* todate exhibita wide range of biological activity (Table 6).

4.1. Antioxidant Properties

Various kinds of physical and physiological stresses lead to the overproduction of oxidants in the human body, which can cause oxidative damage of DNA, proteins and lipids. Furthermore, this damage is responsible for several disorders in the human body such as cardiovascular diseases, cancer and aging. It was reported that minor fruits and nuts possess abundant antioxidant phytochemicals, and the consumption of minor fruits and nuts is beneficial to the human body [48]. The antioxidant effects ofmethanol extracts ofstembark on 1,1-diphenyl-2-picrylhydrazyl (DPPH) and 2,2'-azino-bis(3-ethylbenzothiazoline-6-sulfonic acid (ABTS) scavenging is demonstrated and accounted for in the total soluble phenolic and flavonoid contents [49]. Furthermore, Hassan et al.'s [49] results show that methanol extracts display the highest phenolic content (35.17 mg gallic acid equivalents/g) and considerable flavonoid content (112.83 mg quercetin equivalents/g). Methanol extract showed the highest free radical scavenging activity atIC_{50} = 40 μg/mL and IC_{50} = 125.85 μg/mL in DPPH and ABTS assays, respectively. The antioxidant properties of aqueous fruit extracts were assessed in streptozotocin-induced diabetic rats [50]. Oral administration produced a significant ($p < 0.01$) increase in mean plasma total antioxidant levels and a significant ($p < 0.01$) decrease in malondialdehyde levels. The antioxidant properties of leaf and root extracts were also demonstrated [51]. Balanitin 1 and balanitin 2 (saponins) were isolated from bark extracts and demonstrated antioxidant propertiesin vitro, using a method based on the Briggs–Rauscher oscillating reaction [39]. Polyphenols such as quercetin and kaempferol are the major components responsible for antioxidant activities [52]. In addition, phytosterols including ß-sitoterol, stigmasterol and campesterolhave been reported to exert antioxidant activity [53]. Polyphenols, phytosterols and saponins together might be responsible for the antioxidant activity of desert date.

4.2. Antimicrobial Properties

Plants synthesize several antimicrobial compounds, including phenolics such as simple phenols, phenolic acids, quinones, flavonoids, flavones, flavonols, tannins, coumarins, terpenoids, essential oils and alkaloids [54]. The mechanism of action of these compounds ranges from membrane disruption, substrate deprivation, intercalation into the cell wall/or DNA and enzyme inhibition. Desert date is rich in all these phytochemicals and demonstrates potent antimicrobial activity. The bark of *B. aegyptiaca* is widely used in African folk medicine for the treatment of wounds and skin diseases. The effects of aqueous ethanolic extracts of bark on bacteria isolated from wounds have been reported [55]. These extracts inhibited the growth of *Pseudomonas aeruginosa* and *Staphylococcus aureus* in vitro. The in vitro antifungal activity of saponin-rich extracts of fruit mesocarpwas explored against phytopathogenic fungi [56]. These extracts were moderately active (34.7%) against *Alternariasolani* and highly active (89.01%) against *Pythium ultimum*, and activity was significantly higher compared to the fungicide, metalaxyl (15 μg/mL). The antifungal activity of ethanolic and methanolic extracts of root bark and fruit have been demonstratedagainst *Aspergillus niger*, *Candida albicans*, *Penicilliumcrustosum* and *Saccharomyces cerevisiae* [57].

4.3. Hepatoprotective Properties

A methanolic extract of leaves was evaluated for hepatoprotective activity against carbon tetrachloride (CCl_4)-induced hepatic damage in rats [58]. Administration of the extract (200 and 400 mg/kg per os) markedly reduced the CCl_4-induced elevation of serum marker enzymes, such as glutamate pyruvate transaminase, glutamate oxaloacetate transaminase, alkaline phosphatase and bilirubin. Similarly, fruit mesocarp and stem bark aqueous extracts amelioratedCCl_4-induced hepatotoxicity in rats, as measured by liver enzyme activity, blood parameters and histopathology [59]. Ethanolic extracts of bark protected hepatocytes against paracetamol-and CCl_4-induced hepatotoxicity in rats, analogous to silymarin [60]. Bioactive compounds, primarily obtained from dietary sources, contain a wide range of free radical scavenging constituents, including polyphenols, alkaloids and phytosterols, which are responsible for hepatoprotective effects [61]. Desert date is rich in polyphenols, phytosterols and saponins. It has depicted very good antioxidant potential and thus increased the cellular antioxidant defense system, which may be responsible for the hepatoprotective effects of desert date.

4.4. Anticancer Properties

Cancer is a major health problem. Radiotherapy, chemotherapy and surgical removal are the current treatment methods. However, these methods have varied disadvantages such as drug resistance and toxic effects on nontargeted tissues. Therefore, researchers are searching for naturally available plant-based bioactive compounds for cancer therapy [62]. Among the plant-based bioactive compounds, saponins and phytosterols have significant importance in reducing the risk of cancer [63,64]. Variaoussteroidal saponins isolated from various tissues of *B. aegyptiaca* are reported to display anticancer activities. For example, a mixture of balanitin-6 and balanitin-7 (28:72) isolated from kernels showgrowth inhibition in human cancer cell lines in vitro [31]. Balanitin-6/balanitin-7 exhibited higher antiproliferative activity than well-known natural cancer therapeutic agents, such as etoposide and oxaliplatin. Balanitin-6/balanitin-7 displayed its highest activity against A549 nonsmall cell lung cancer (IC_{50}, 0.3 μM) and U373 glioblastoma (IC_{50}, 0.5 μM) cell lines. Balanitoside extracted from the fruit also showed anticancer activity against Ehrlich ascites carcinoma (EAC)-bearing Swiss albino mice [36]. Mice injected intraperitoneally with balanitoside (10 mg/kg body weight) displayed decreases in liver and serum enzyme levels. Issa et al. [65] studied an aqueous extract of pulp on the development and growth of EAC and metastasis to the liver and spleen. Treatment with the extract (400 mg/kg) inhibited tumor growth and proliferation in ascetic fluid, inducing a significant decrease in tumor volume, total cell volume and viable cell count and prolongedmouse survival. The authors also recorded significant decreases in levels of lipid peroxidation and increased superoxide dismutase and catalase activity and P53 (a tumor suppressor protein) expression. The saponin, balanitin-7 isolated from seed kernels, showed antiproliferative activity [32]. These agents showed potent antiproliferative activity against MCF-7 human breast cancer cells and HT-29 human colon cancer cells, with IC_{50} values of 2.4 and 3.3 μM, respectively.

4.5. Anti-Inflammatory Properties

Inflammation is a pattern of response to injury, which involves the accumulation of cells, exudates in irritated tissue, which allows protection from further damage. A variety of in vitro and in vivo experiments has shown that certain flavonoids and saponins possess anti-inflammatory activity [66]. The mechanism by which flavonoids and saponins exert their anti-inflammatory effects involves the inhibition of cyclooxygenase and lipoxygenase activities [67]. Desert date exhibited potent anti-inflammatory activity; for example, Speroni et al. [39] studied the in vivoanti-inflammatory activity of methanol and butanol extracts and two saponins, viz. balanin-B1 and balanin-B2, isolated from *B. aegyptiaca* bark in rats with edema induced bycarrageenin. Both extracts exhibited a significant reduction of rat paw edema. The inhibition produced by methanolextract, butanol extract, balanin-B1 and balanin-B2 were 32%, 68%, 62% and 59%, respectively. Likewise, the influence of seed oil on liver and kidney fractions in rat serum was evaluated [68]. Seed oil (100 mg/kg) in the rat dietdecreased nitrogen oxide and lipid peroxidation. Further, mRNA and protein expression of tumor necrosis factor-α and interleukin-6 were downregulated, leading to a reduction of cyclooxygenase-2, reflecting anti-inflammatory activity.

4.6. Antidiabetic Activity

Diabetes is a chronic disease that occurs either when the pancreas does not produce enough insulin or when the body cannot effectively use the insulin it produces. Several medicinal plants have demonstrated hypoglycemic and hyperglycemic activities; these activities seem to be mediated through increased insulin secretion via stimulation of pancreatic cells, interfering with dietary glucose absorption or through insulin-sensitizing action [69]. Kamel et al. [38] demonstrated the antidiabetic effect of an aqueous extract of fruit in streptozotocin (STZ)-induced diabetic mice after oral administration. They also identified steroidal saponins, 26-O-ß-D-glucopyranosyl-(25R)-furost-5-ene-3ß,22,26-triol-3-O-[α-L-rhamnophyranosyl-(1→1)]-[ß-D-xylopyranosyl-(1→3)]-[α-L-rhamnopyronosyl-(1→4)]-ß-D-glucopyranoside and its 22-methyl ether in the extract and recognized two additional saponins, 26-O-ß-D-glycopyranosyl-(25R)-furost-5-ene-3ß,22,26,-triol-3-O-[2,4-di-O-α-L-rhamnopyranosyl)-ß-D-glucopyranoside and its methyl ether. A combination of saponinsexhibited greater antidiabetic activity than individual saponins. Gad et al. [70] administered fruit extracts (1.5 g/kg body weight) to STZ-induced diabetic rats and studied the glycogencontent of liver and kidney and on some key enzymes of liver involved in carbohydrate metabolism. STZ (50 mg/kg body weight) caused a five-fold increase in blood glucose level, an 80% reduction in serum insulin level, a 58% decrease in liver glycogen and a seven-fold increase in kidney glycogen content. A marked increment in the activity of glucose-6-phosphatase activity and decreasedactivity of glucose-6-phosphate dehydrogenase and phosphofructokinase were recorded. Treatment of rats with fruit extract reduced blood glucose levels by 24% and significantly decreased liver glucose-6-phosphatase activity. The authors also demonstrated that the extract inhibited α-amylase activity in vitro. The major component in the extract was diosgenin, based on high-performance thin-layer chromatography. Additionally, Al-Malki et al. [71] showed that ethyl acetate extract containing β-sitosterol modulatedoxidative stress induced by streptozotocin.

Table 6. Biological activities of compounds isolated from various parts of desert date.

Compound	Part	Activity	Model/Method	Reference
Balanitin 1	Root and stem bark	Molluscicide	*Biomphalaria glabrata*	[29]
Balanitin 2	Root and stem bark	Molluscicide	*Biomphalaria glabrata*	[29]
Balanitin 3	Root and stem bark	Molluscicide	*Biomphalaria glabrata*	[29]
Balanitin 4	Seed	Anticancer	P-388 Lymphocytic leukemia cell line	[30]
Balanitin 5	Seed	Anticancer	P-388 Lymphocytic leukemia cell line	[30]
Balanitin 6	Seed	Anticancer	Different cancer cell lines including the P-388 Lymphocytic leukemia cell line and female mice injected with L1210 syngeneic murine leukemia cells	[30,31]
Balanitin 7	Fruit, root, seed	Anticancer	1.Different cancer cell lines and female mice injected with L1210 syngeneic murine leukemia cells 2. P-388 Lymphocytic leukemia cell line 3.Human breast cancer cells (MCF-7) and human colon cancer cells (HT-29)	[30–32]
Deltonin	Seed	Nematocidal	*Caenorhabditiselegans*	[72]
		Molluscicidal	*Biomphalaria glabrata*	[33]
Balanitoside	Fruit	Anticancer	Ehrlich ascites carcinoma bearing Swiss albino mice	[36]
		Antidiabetic	Streptozotocin-induced diabetes in Wistar rats	[73]
Balanin B2	Stem bark	Anti-inflammatory	Carrageenin-induced paw edema in male Sprague Dawley rats	[39]
26-(O-β-D-glucopyranosyl)-22-O-methylfurost-5-ene-3β,26-diol-3-O-β-D-glucopyranosyl-(1→4)-[α-L-rhamnopyranosyl-(1→2)]-β-D-glucopyranoside	Fruit	Antidiabetic	α-Glucosidase and aldose reductase inhibitory activities (in vitro) and streptozotocin-induced diabetes in male albino Wistar rats (in vivo)	[40]

Table 6. *Cont.*

Compound	Part	Activity	Model/Method	Reference
25R and 25S epimers of 26-O-β-D-glucopyranosyl-furost-5-ene-3,22,26- triol 3-O-[α-L-rhamnopyranosyl-(1→3)- β-D-glucopyranosyl-(1→2)]-α-L-rhamnopyranosyl-(1→4)-β-D-glucopyranoside	Fruit	Aldose reductase inhibitor	Aldose reductase inhibition activity on rat liver homogenate	[41]
26-O-β-D-glucopyranosyl-(25R)-furost-5-ene-3,22,26-triol 3-O-[β-D-glucopyranosyl-(1→2)]- α-L-rhamnopyranosyl-(1→4)-β-D-glucopyranoside	Fruit	Aldose reductase inhibitor	Aldose reductase inhibition activity on rat liver homogenate	[41]
26-O-β-D-glucopyranosyl-(25R)-furost-5,20-diene-3,26-diol 3-O-[α-L-rhamnopyranosyl-(1→3)-β-D-glucopyranosyl-(1→2)]-α-L-rhamnopyranosyl-(1→4)-β-D-glucopyranoside	Fruit	Aldose reductase inhibitor	Aldose reductase inhibition activity on rat liver homogenate	[41]
25R and 25S epimers of 26-O-β-D-glucopyranosyl-furost-5,20-diene-3,26-diol 3-O-[β-D-glucopyranosyl-(1→2)]-α-L-rhamnopyranosyl-(1→4)-β-D-glucopyranoside	Fruit	Aldose reductase inhibitor	Aldose reductase inhibition activity on rat liver homogenate	[41]
Balanin B1	Stem bark	Anti-inflammatory	Carrageenin-induced paw edema in male Sprague Dawley rats	[39]
		Antioxidant	ROS scavenging activity by Briggs-Rauscher oscillating reaction	[39]

Hassanin et al. [74] tested a crudeethanolic fruit extract and itsbutanolic and dichloromethane fractions on stress-activated protein kinase/c-Jun N-terminal kinase (SAPK-JNK) signaling in experimental diabetic rats. Six groups of male Wistar rats were used: normal control, diabetic, diabetic rats treated with crude, butanol or dichloromethane factions (50 mg/kg body weight), and diabetic rats were treated with gliclazide as a reference drug. Treatments continued for one month. Extract treatments produced a reduction in plasma glucose, hemoglobin A1c, lactic acid, lipid profile and malondialdehyde levels, which induced an increase in insulin and reduced glutathione (GSH) levels and catalase and superoxide dismutase activities. Moreover, the authors observed the down-regulation of apoptosis signal-regulating kinase 1, c-Jun N-terminal kinase 1 and protein 53 and the upregulation of insulin receptor substrate 1 in rat pancreas. Glucose transporter 4 was upregulated in rat muscle. Liquid chromatography and high-resolution mass spectrometry (LC-HRMS) analysis identified balanitin-2, hexadecenoic acid, methyl protodioscin and 26-(O-β-D-glucopyranosyl)-3-β-[4-O-(β-D-glucopyranosyl)-2-O-(α-L-rhamnopyranosyl)-β-D-glucopyranosyloxy]-22,26-dihydroxyfurost-5-ene in crude extract and balanitin-1 and trigonelloside C in butanoland dichloromethane fractions of crude extract. Ezzat et al. [40] isolated several compounds from pericarp, including stigmasterol-3-O-β-D-glucopyranoside (**a**), a pregnane glucoside: pregn-5-ene-3β,16β,20(R)-triol-3-O-β-D-glucopyranoside (**b**); a furostanolsaponin: 26-(O-β-D-glucopyranosyl)-22-O-methylfurost-5-ene-3β,26-diol-3-O-β-D-glucopyranosyl-(1→4)-[α-L-rhamnopyranosyl-(1→2)]-β-D-glucopyranoside (**c**). The latter component possessed significant α-glucodidase (AG) and aldose reductase inhibitory activities in streptozotocin-induced diabetic Wistar rats. Compound (**c**) also caused a significant increment in insulin and C-peptide levels.

4.7. Molluscicidal Activity

Regarding the effects of fruit extracts on juvenile and adult *Bulinusglobosus* and *B.truncatus*, two Planorbid (ramshorn) freshwater snails have been reported [75]. LC_{95} values were 16.9 and 19.7 μg/mLand 14.2 and 12.0μg/mLforjuvenile and adults of *B. globosus* and *B. truncatus*, respectively. Seed, endocarp, mesocarp and whole fruit extracts were assessed against adult *Biomphalariapfeifferi*, another Planorbid snail, and *Lymnaeanatalensis*, a Lymnaeid pond snail [76]. LC_{90} values were 77.70, 120.04, 89.50 and 97.55 mg/L against *Biomphalariapfeifferi* for seed, endocarp, mesocarp and whole fruit extracts, respectively, and 102.30, 138.21, 115.42 and 127.69 mg/L against *Lymnaeanatalensis*. Furthermore, the molluscicidal activity of seed oil on *Monachacartusiana*, aHygromiid land snail, has been demonstrated [77]. Bioactive compounds were identified assaponins, such as diosgenin, yamogenin and 3,5-spirostadiene.

4.8. Other Activities

Several studies demonstrate the antinematode andantiplasmodial activities of *B. aegyptiaca* extracts. Shalaby et al. [78] showed the effects of methanolicfruit extracts on enteral and parenteral stages of *Trichinellaspiralis* (pork worm). The authors also evaluated the effectiveness of methanolic extract against preadult migrating larvae and encysted larvae of *Trichinellaspiralis* in rats and compared them with the commonly used anthelmintic chemical, albendazole. Methanolic extract (1000 mg/kg body weight) for five successive days throughout the parasite lifecycle led to a marked reduction in migrating and encysted larvae by 81.7% and 61.7%, respectively. In another study, the efficacy of a methanolicextract on *Toxocaravitulorum* (roundworm), a major parasite in cattle and buffalo, was assessed [79]. They incubated parasites in a ringer solution containing 10, 30, 60, 120 and 240 μg/mL of ethanolic extract for 24 h. The most prominent activity at 240 μg/mLcausedthe disorganization of body cuticle musculature. Kusch et al. [17] evaluated a crude extract of seeds for antiplasmodial activity. An IC_{50} value for chloroquine-susceptible *Plasmodium falciparum* NF54 was 68.26 μg/μL. The compound responsible for this activity was6-phenyl-2(H)-1,2,4-triazin-5-one oxime. The authorsalso showed that two phenolic compounds, 2,6-di-*tert*-butyl-phenol and 2,4-di-*tert*-butylphenol, displayedantiplasmodial activity at IC_{50} values of 50.29 and 47.82 μM, respectively.

5. Conclusions

B. aegyptiaca or desert date is an underutilized tree species. The nutritional status of the fruits, leaves and seeds indicate that this species could be exploited as a food source. Seed oil might also be a good source of biodiesel. Leaves and young shoots are nutritionally rich and could be exploited as cattle feed. Furthermore, fruits, leaves, roots and the bark of stem and roots are substantial sources of bioactive phytochemicals that display a host of possibly useful biological properties. *B. aegyptiaca* might prove to be a valuable source of bioactive agents for use in human and veterinary medicine.

Author Contributions: Conceptualization and methodology, H.N.M. and G.G.Y.; validation, H.N.M. and G.G.Y.; investigation, H.N.M., G.G.Y., Y.H.D. and A.I.; formal analysis H.N.M. and G.G.Y.; resources, H.N.M. and Y.H.D.; data curation, H.N.M., G.G.Y. and Y.H.D.; writing—original draft preparation, H.N.M. and G.G.Y.; writing—review and editing, H.N.M., Y.H.D. and A.I.; visualization, H.N.M., G.G.Y., Y.H.D. and A.I. All authors have read and agreed to the published version of the manuscript.

Funding: The authors extend their appreciation to the Deputyship for Research and Innovation, "Ministry of Education" Saudi Arabia for funding this research work through the project number IFKSURP-59. H.N.M. is thankful to the University Grants Commission-Basic Scientific Research (UGC-BSR) mid-career award (Grant No. F.19-223/2018/ (BSR)).

Institutional Review Board Statement: Not applicable.

Informed Consent Statement: Not applicable

Data Availability Statement: Please refer to suggested Data Availability Statements in section "MDPI Research Data Policies" at https://www.mdpi.com/ethics.

Acknowledgments: The authors thankful to the Researchers Support and Services Unit (RSSU) for their technical support.

Conflicts of Interest: The authors declare no conflict of interest.

References

1. Orwa, C.; Mutua, A.; Kindt, R.; Simons, A.; Jamnadass, R.H. Agroforestree Database: A Tree Reference and Selection Guide Version 4.0. World Agroforestry Centre, Kenya. 2009. Available online: http://apps.worldagroforestry.org/treedb2/speciesprofile.php?Spid=279 (accessed on 10 November 2020).
2. Sarker, S.D.; Barholomew, B.; Nash, R.J. Alkaloids from *Balanitesaegyptiaca*. *Fitoterapia* **2000**, *71*, 328–330. [CrossRef]
3. Ojo, O.O.; Nadro, M.S.; Tell, I.O. Protection of rats by extracts of some common Nigerian trees against acetaminophen-induced hepatotoxicity. *Afr. J. Biotechnol.* **2006**, *5*, 755–760.
4. IwuMaurice, M. *Handbook of African Medicinal Plants*; CRC Press: Boca Raton, FL, USA, 1993; p. 129.
5. Achaglinkame, M.A.; Aderibigbe, R.O.; Hensel, O.; Sturm, B.; Korese, K. Nutritional characteristics of four underutilized wild fruits of dietary interest in Ghana. *Foods* **2019**, *8*, 104. [CrossRef] [PubMed]
6. Sagna, M.B.; Diallo, A.; Sarr, P.S.; Ndiaye, O.; Goffner, D.; Guisse, A. Biochemical composition and nutritional value of *Balanitesaegyptiaca*(L.) Del fruit pulp from Northern Ferlo in Senegal. *Afr. J. Biotechnol.* **2014**, *13*, 336–342.
7. Amadou, I. Date fruits: Nutritional composition of dates (*Balanitesaegyptiaca* Delile and *Phoenix dactylifera* L.). In *Nutritional Composition of Fruit Cultivars*; Simmonds, M.S.J., Preedy, V.R., Eds.; Elsevier Inc.: Amsterdam, The Netherlands, 2016; pp. 215–233.
8. Cook, J.A.; VanderJagt, D.J.; Pastuszyn, A.; Mounkaila, G.; Glew, R.S.; Glew, R.H. Nutrient content of two indigenous plant foods of western Sahel: *Balanitesaegyptiaca* and *Maeruacrassifolia*. *J. Food Comp. Anal.* **1998**, *11*, 221–230. [CrossRef]
9. Samuel, A.L.; Temple, V.J.; Ladeji, O. Chemical and nutrition evaluation of the seed kernel of *Balanites aegyptiaca*. *Niger. J. Biotechnol.* **1997**, *8*, 57–63.
10. Mohamed, A.M.; Wolf, W.; Spieb, W.E.L. Physical, morphological and chemical characteristics, oil recovery and fatty acid composition of *Balanites aegyptiaca* Del. kernels. *Plant. Foods Hum. Nutr.* **2002**, *57*, 179–189. [CrossRef]
11. Chapagain, B.P.; Yehosha, Y.; Wiesman, Z. Desert date (*Balanitesaegyptiaca*) as an arid lands sustainable bioresource for biodiesel. *Bioresour. Technol.* **2009**, *100*, 1221–1226. [CrossRef]
12. Kubmarawa, D.; Andenyang, I.F.H.; Magomya, A.M. Amino acid profile of two non-conventional leafy vegetables, *Sesamum indicum* and *Balanites aegyptiaca*. *Afr. J. Biotechnol.* **2008**, *7*, 3502–3504.
13. Khamis, G.; Saleh, A.M.; Habeeb, T.H.; Hozzein, W.N.; Wadaan, M.A.M.; Papernbrock, J.; AbdElgawad, H. Provenance effect on bioactive phytochemicals and nutritional and health benefits of the desert date *Balanitesaegyptiaca*. *J. Food Biochem.* **2020**, *44*, e13229. [CrossRef]
14. Becker, B. The contribution of wild plants to human nutrition in the Ferlo (Northern Senegal). *Agrofor. Syst.* **1983**, *1*, 257–267. [CrossRef]
15. Li, A.N.; Li, S.; Zhang, Y.J.; Xu, X.R.; Chen, Y.M.; Li, H.B.; Li, H.B. Resources and biological activities of natural polyphenols. *Nutrients* **2014**, *6*, 6020–6047. [CrossRef] [PubMed]
16. Tsao, R. Chemistry and biochemistry of dietary polyphenols. *Nutrients* **2010**, *2*, 1231–1246. [CrossRef] [PubMed]
17. Kusch, P.; Deininger, S.; Specht, S.; Maniako, R.; Haubrich, S.; Pommerening, T.; Lin, P.K.T.; Hoerauf, A.; Kaiser, A. *In vitro* and *in vivo* antimalarial activity of seeds from *Balanitesaegyptiaca*: Compounds of the extract show growth inhibition and activity against plasmodial aminopeptidase. *J. Parsitol Res.* **2011**, 368692.
18. Hassan, D.M.; Anigo, K.M.; Umar, I.A.; Alegbejo, J.O. Evaluation of phytoconstituent of *Balanitesaegyptiaca* leaves and fruit-mesocarp extracts. *M.O.J. Bioorg. Org. Chem.* **2017**, *1*, 228–232.
19. Meda, R.N.T.; Vlase, L.; Lamien-Meda, A.; Lamien, C.E.; Muntean, D.; Tiperciuc, B.; Oniga, I.; Nacoulma, O.G. Identification and quantification of phenolic compounds from *Balanites aegyptiaca*(L.) Del (*Balanitaceae*) galls and leaves by HPLC-MS. *Nat. Prod. Res.* **2011**, *25*, 93–99. [CrossRef]
20. Farag, M.A.; Porzel, A.; Wessjohann, L.A. Unraveling the active hypoglycemic agent trigonelline in *Balanitesaegyptiaca* date fruit using metabolite fingerprinting by NMR. *J. Pharm. Biomed. Anal.* **2015**, *115*, 383–387. [CrossRef]
21. Maksoud, S.A.; Hadidi, M.N.E. The flavonoids of *Balanites aegyptiaca* (Balanitaceae) from Egypt. *Plant. Syst. Evol.* **1988**, *160*, 153–158. [CrossRef]
22. Shafik, N.H.; Shafek, R.E.; Michel, H.N.; Eskander, E.F. Phytochemical study and antihyperglycemic effect of *Balanitesaegyptiaca* kernel extract on alloxan induced diabetic male rat. *J. Chem. Pharma. Res.* **2016**, *8*, 128–136.

23. Seida, A.A.; Kinghorn, A.D.; Cordell, G.A.; Farnsworth, N.R. Isolation of bergapten and marmesin from *Balanitesaegyptiaca*. *Planta Med.* **1981**, *43*, 92–93. [CrossRef]
24. Hosny, M.; Khalifa, T.; Calis, I.; Wright, A.D.; Sticher, O. Balanitoside, a furostanal glycoside and 6-methyl-diosgenin from *Balanitesaegyptiaca*. *Phytochemistry* **1992**, *31*, 3565–3569. [CrossRef]
25. Hardman, R.; Sofowora, E.A. Isolation and characterization of yamogenin from *Balanitesaegyptiaca*. *Phytochemistry* **1970**, *9*, 645–649. [CrossRef]
26. Samuelsson, G.; Farah, M.H.; Cleason, P.; Hagos, M.; Thulin, M.; Hedberg, O.; Warfa, A.M.; Hassan, A.O.; Elmi, A.H.; Abdurahman, A.D.; et al. Inventory of plants used in traditional medicine in Somalia. I. Plants of the families of Acanthaceae-Chenopodiaceae. *J. Ethnopharmacol.* **1991**, *35*, 25–63. [CrossRef]
27. Ashaal, H.A.; Farghaly, A.A.; Abd El Aziz, M.M.; Ali, M.A. Phytochemical investigation and medicinal evaluation of fixed oil of *Balanitesaegyptiaca* fruits (Balanitaceae). *J. Ethnopharmacol.* **2010**, *127*, 495–501. [CrossRef]
28. Kamel, M.S.; Koskinen, A. Pregnane glycosides from fruits of *Balanites aegyptiaca*. *Phytochemistry* **1995**, *40*, 1773–1775. [CrossRef]
29. Liu, H.W.; Naknishi, K. The structures of balanitins, potent molluscicides isolated from *Balanites aegyptiaca*. *Tetrahedron* **1982**, *38*, 513–519. [CrossRef]
30. Pettit, G.R.; Doubek, D.L.; Herald, D.L.; Numata, A.; Takahasi, C.; Fujiki, R.; Miyamoto, T. Isolation and structure of cytostatic steroidal saponins from the African medicinal plant *Balanitesaegyptiaca*. *J. Nat. Prod.* **1991**, *54*, 1491–1502. [CrossRef]
31. Gnoula, C.; Megalizzi, V.; De Neve, N.; Sauvage, S.; Ribaucour, F.; Guissou, P.; Duez, P.; Dubois, J.; Ingrassia, L.; Lefranc, F.; et al. Balanitin-6 and −7: Diosgenyl saponins isolated from *Balanitesaegyptiaca* Del. display significant anti-tumor activity *in vitro* and *in vivo*. *Int. J. Oncol.* **2008**, *32*, 5–15. [CrossRef]
32. Beit-Yannai, E.; Ben-Shabat, S.; Goldschmidt, N.; Chapagain, B.P.; Liu, R.H.; Wiesman, Z. Antiproliferative activity of steroidal saponins from *Balanitesaegyptiaca*—An *in vtiro* study. *Phytochem. Lett.* **2011**, *4*, 43–47. [CrossRef]
33. Brimer, L.; ElSheik, S.H.; Furu, P. Preliminary investigation of the disposition of the molluscicidal saponin deltonin from *Balanites aegyptiaca* in a snail species (*Biomphalaria glabrata*) and in mice. *J. Pestic. Sci.* **2007**, *32*, 213–221. [CrossRef]
34. Farid, H.; Haslinger, E.; Kunert, O.; Wegner, C.; Hamburger, M. New steroidal glycosides from *Balanitesaegyptiaca*. *Helv. Chim. Acta* **2002**, *85*, 1019–1026. [CrossRef]
35. Kamel, M.S. A furostanolsaponin from fruits of *Balanitesaegyptiaca*. *Phytochemistry* **1998**, *48*, 755–757. [CrossRef]
36. Al-Ghannam, S.M.; Hamdy-Ahamed, H.; Zein, N.; Zahran, F. Antitumor activity of balanitoside extracted from *Balanitesaegyptiaca* fruit. *J. Appl. Pharm. Sci.* **2013**, *3*, 179–191.
37. Staerk, D.; Chapagain, B.P.; Lindin, T.; Wiesman, Z.; Jaroszewski, J.W. Structural analysis of complex saponins of *Balanitesaegyptiaca* by 800 MHz^{1}H NMR spectroscopy. *Magn. Reson. Chem.* **2006**, *44*, 923–928. [CrossRef]
38. Kamel, M.S.; Ohtani, K.; Kurokawa, T.; Assaf, M.H.; El-Shanawany, M.A.; Ali, A.A.; Kasai, R.; Ishibashi, S.; Tanaka, O. Studies on *Balanitesaegyptiaca* fruits, an antidiabetic Egyptian folk medicine. *Chem. Pharm. Bull.* **1991**, *39*, 1229–1233. [CrossRef]
39. Speroni, E.; Cervellati, R.; Innocenti, G.; Cost, S.; Guerra, M.C.; Dall Acqua, S.; Govani, P. Anti-inflammatory, anti-nociceptive and antioxidant activities of *Balanitesaegyptiaca* (L.) Delile. *J. Ethnopharmacol.* **2005**, *98*, 117–125. [CrossRef]
40. Ezzat, S.M.; Motaal, A.A.; Awdan, S.A.W.E. *In vitro* and *in vivo* antidiabetic potential of extracts and a furostanolsaponin from *Balanitesaegyptiaca*. *Pharm. Biol.* **2017**, *55*, 1931–1936. [CrossRef]
41. Motaal, A.A.; El-Askary, H.; Crockett, S.; Kunert, O.; Sakr, B.; Shaker, S.; Grigore, A.; Albulescu, R.; Bauer, R. Aldose reductase inhibition of a saponin-rich fraction and new furostanol saponin derivatives from *Balanitesaegyptiaca*. *Phytomedicine* **2015**, *22*, 829–836. [CrossRef]
42. Seida, A.A. Isolation, Identification and Structure Elucidation of Cytotoxic and Antitumor Principles from *Ailanthus integrifolia*, *Amyris pinnata* and *Balanitesaegyptiaca*. Ph.D. Thesis, University of Illinois, Chicago, IL, USA, 1979.
43. Venugopala, K.N.; Rashmi, V.; Odhav, B. Review on natural coumarin lead compounds for their pharmacological activity. *BioMed Res. Int.* **2013**, *2013*, 963248. [CrossRef]
44. Roberts, M.F.; Wink, M. *Alkaloids-Biochemistry, Ecological Functions and Medical Applications*; Plenum Press: New York, NY, USA, 1998.
45. Panda, N.; Banerjee, S.; Mandal, N.B.; Sahu, N.P. Pregnane glycosides. *Nat. Prod. Commun.* **2006**, *1*, 665–695. [CrossRef]
46. Challinor, V.L.; De Voss, J.J. Open-chain steroidal glycosides, a diverse class of plant saponins. *Nat. Prod. Rep.* **2013**, *30*, 429–454. [CrossRef] [PubMed]
47. El Aziz, M.M.A.; Ashour, A.S.; Melad, A.S.G. A review on saponins from medicinal plants: Chemistry, isolation, and determination. *J. Nanomed Res.* **2019**, *7*, 282–288.
48. Murthy, H.N.; Bapat, V.A. Importance of Underutilized Fruits and Nuts. In *Bioactive Compounds in Underutilized Fruits and Nuts, Murthy, H., Bapat, V., Eds.*; Reference Series in Phytochemistry; Springer: Cham, Switzerland, 2020; pp. 3–19.
49. Hassan, L.E.A.; Dahham, S.S.; Saghir, S.A.M.; Mohammed, A.M.A.; Eltayeb, N.M.; Majis, A.M.S.A.; Majid, A.S.A. Chemotherapeutic potentials of the stem bark of *Balanitesaegyptiaca* (L.) Delile: An antiangiogenic, antitumor and antioxidant agent. *B.M.C. Complement. Altern. Med.* **2016**, *16*, 396. [CrossRef]
50. Khalil, N.S.A.; Abou-Elhamd, A.S.; Wasfy, S.I.A.; El Mileegy, I.M.H.; Hamed, M.Y.; Ageely, H.M. Antidiabetic and antioxidant impacts of desert date (*Balanitesaegyptiaca*) and parsley (*Petroselinum sativum*) aqueous extracts: Lessons from experimental rats. *J. Diabet. Res.* **2016**, 8408326.

51. Kahsay, T.; Muluget, A.; Unnithan, C.R. Antioxidant and antibacterial activities of *Balanites aegyptiaca* Delil from Northern Ethiopia. *Am. J. Pharma Tech. Res.* **2014**, *4*, 415–422.
52. Rasouli, H.; Farzaei, M.H.; Khodarahmi, R. Polyphenols and their benefits: A review. *Int. J. Food Prop.* **2017**, *20*, S1700–S1741. [CrossRef]
53. Yoshida, Y.; Niki, E. Antioxidant effects of phytosterol and its components. *J. Nutr. Sci. Vitaminol.* **2003**, *49*, 277–280. [CrossRef]
54. Cowan, M.M. Plant Products as Antimicrobial Agents. *Clin. Microbiol. Rev.* **1999**, *12*, 564–582. [CrossRef]
55. Anani, K.; Adjrah, Y.; Ameyapoh, Y.; Karou, S.D.; Agbonon, A.; de Souza, C.; Gbeassor, M. Effects of hydroethanolic extracts of *Balanites aegyptiaca* (L.) Delile (Balanitaceae) on some resistant pathogens bacteria isolated from wounds. *J. Ethnophamacol.* **2015**, *164*, 16–21. [CrossRef]
56. Chapagain, B.P.; Wiesman, Z.; Tsror (Lahkim), L. *In vitro* study of the antifungal activity of saponin-rich extracts against prevalent phytopathogenic fungi. *Ind. Crop Prod.* **2007**, *26*, 109–115. [CrossRef]
57. Runyoro, D.K.B.; Matee, M.I.N.; Ngassapa, O.D.; Joseph, C.C.; Mbwambo, Z.H. Screening of Tanzanian medicinal plants for anti-candida activity. *B.M.C. Complement. Altern. Med.* **2006**, *6*, 11. [CrossRef] [PubMed]
58. Thirupathi, K.; Krishna, D.R.; Ravi Kumar, B.; Apparao, A.V.N.; Krishna Mohan, G. Hepatoprotective effect of leaves of *Balanitesroxburghii* against carbon tetrachloride-induced hepatic damage in rats. *Curr. Trends Biotechnol. Pharm.* **2009**, *3*, 219–224.
59. Mariam, S.; Onyenibe, N.S.; Oyelola, O.B. Aqueous extract of *Balanites aegyptiaca* Del Fruit mesocarp protects against CCl4—Induced liver damage in rats. *Br. J. Pharm. Res. Int.* **2013**, *3*, 917–928. [CrossRef]
60. Jaiprakash, B.; Rajkumar, A.; Karadi, R.V.; Savadi, R.V.; Hukkeri, V.L. Hepatoprotective activity of bark of *Balanitesaegyptiaca* Linn. *J. Nat. Rem.* **2003**, *3*, 205–207.
61. Ganesan, K.; Jayachandran, M.; Xu, B. A critical review on hepatoprotective effects of bioactive food components. *Crit. Rev. Food Sci. Nutr.* **2018**, *58*, 1165–1229. [CrossRef] [PubMed]
62. Choudhari, A.S.; Mandave, P.C.; Deshpande, M.; Ranjekar, P.; Prakash, O. Phytochemicals in cancer treatment: From preclinical studies to clinical practice. *Front. Pharmacol.* **2020**, *10*, 1614. [CrossRef]
63. Grattan, B.J., Jr. Plant sterols as anticancer nutrients: Evidence for their role in breast cancer. *Nutrients* **2013**, *5*, 359–387. [CrossRef]
64. Man, S.; Gao, W.; Zhang, Y.; Huang, L.; Liu, C. Chemical study and medical application of saponins as anti-cancer agents. *Fitoterapia* **2010**, *81*, 703–714. [CrossRef]
65. Issa, N.M.; Mansour, F.K.; El-Safti, F.A.; Nooh, H.Z.; El-Sayed, I.H. Effect of *Balanites aegyptiaca* on Ehrlich Ascitic carcinoma growth and metastasis in Swiss mice. *Exp. Toxicol. Pathol.* **2015**, *67*, 435–441. [CrossRef]
66. Yuan, G.; Wahlqvist, M.L.; He, G.; Yang, M.; Li, D. Natural products and anti-inflammatory activity. *Asia Pac. J. Clin. Nutr.* **2006**, *15*, 143–152.
67. Kim, H.P.; Mani, I.; Iversen, L.; Ziboh, V.A. Effects of naturally occurring flavonoids and bioflavonoids on epidermal cyclooxygenase and lipoxygenase from guinea-pigs. *Prostaglandins Leukot. Essent. Fatty Acids* **1998**, *58*, 17–24. [CrossRef]
68. Ahmed, M.M.; Eid, M.M. Evaluation of *Balanitesaegyptiaca* oil as untraditional source of oil and its antiinflammatory activity. *J. Drug Res. Egypt* **2015**, *36*, 1–11.
69. Grover, J.K.; Yadav, S.S.; Vats, V.J. Medicinal plants of India with anti-diabetic potential. *J. Ethnopharmacol.* **2002**, *81*, 81–100. [CrossRef]
70. Gad, M.Z.; El-Sawalhi, M.M.; Ismail, M.F.; El-Tanbouly, N.D. Biochemical study of the antidiabetic action of the Egyptian plants Fenugreek and *Balanites*. *Mol. Cell. Biochem.* **2006**, *281*, 173–183. [CrossRef] [PubMed]
71. Al-Malki, A.L.; Barbour, E.K.; Abulnaja, K.O.; Moselhy, S.S. Management of hyperglycaemia by ethyl acetate extract of *Balanitesaegyptiaca* (Desert date). *Molecules* **2015**, *20*, 14425–14434. [CrossRef]
72. Gnoula, C.; Guissou, P.; Duez, P.; Frederich, M.; Dubois, J. Nematocidal compounds from the seeds of *Balanitesaegyptiaca*: Isolation and structure elucidation. *Int. J. Phamacol.* **2007**, *3*, 280–284.
73. Makena, W.; Hamman, W.O.; Buraimoh, A.A.; Dibal, N.I.; Obaje, S.G. Therapeutic effects of balanitoside in streptozotocin-induced diabetic rats. *J. Tiabah Univ. Med. Sci.* **2018**, *13*, 402–406. [CrossRef]
74. Hassanin, K.M.A.; Mahmoud, M.A.; Hassan, H.M.; Abdel-Razik, A.H.; Aziz, L.N.; Rateb, M.E. *Balanitesaegyptiaca* ameliorates insulin secretion and decreases pancreaticapoptosis in diabetic rats: Role of SAPK/JNK pathway. *Biomed. Pharmacother.* **2018**, *102*, 1084–1091. [CrossRef]
75. Anto, F.; Aryeetey, M.E.; Anyorigiya, T.; Asoala, V.; Kpikpi, J. The relative susceptibilities of juvenile and adult *Bulinusglobosus* and *Bulinustruncatus* to the molluscicidal activities in the fruit of Ghanaian *Blighiasapida, Blighiaunijugata* and *Balanitesaegyptiaca*. *Ann. Trop. Med. Parasitol.* **2005**, *99*, 211–217. [CrossRef]
76. Molla, E.; Giday, M.; Erko, B. Laboratory assessment of the molluscicidal and cercariacidal activities of *Balanitesaegyptiaca*. *Asian Pac. J. Trop. Biomed.* **2013**, *3*, 657–662. [CrossRef]
77. Dawidar, A.E.M.; Mortada, M.M.; Raghib, H.M.; Abdel-Mogib, M. Molluscicidal activity of *Balanitesaegyptiaca* against *Monachacartusiana*. *Pharma. Biol.* **2012**, *50*, 1326–1329. [CrossRef] [PubMed]
78. Shalaby, M.A.; Moghazy, F.M.; Shalaby, H.A.; Nasr, S.M. Effect of methanoic extract of *Balanitesaegyptiaca* fruits on enteral and parenteral stages of *Trichonella spiralis* in rats. *Parasitol. Res.* **2010**, *107*, 17–25. [CrossRef] [PubMed]
79. Shalaby, H.A.; El-Namaky, A.H.; Khalil, F.A.; Kandil, O.M. Efficacy of methonolic extract of *Balanitesaegyptiaca* fruits on *Toxocaravitulorum*. *Vet. Parasitol.* **2012**, *183*, 386–392. [CrossRef] [PubMed]

Review

Euphorbia antisyphilitica Zucc: A Source of Phytochemicals with Potential Applications in Industry

Romeo Rojas [1], Julio César Tafolla-Arellano [2] and Guillermo C. G. Martínez-Ávila [1,*]

1 Laboratory of Chemistry and Biochemistry, School of Agronomy, Autonomous University of Nuevo Leon, General Escobedo, Nuevo Leon 66050, Mexico; romeo.rojasmln@uanl.edu.mx
2 Basic Sciences Department, Laboratory of Biotechnology and Molecular Biology, Antonio Narro Agrarian Autonomous University, Saltillo, Coahuila 25315, Mexico; jtafare@uaaan.edu.mx
* Correspondence: guillermo.martinezavl@uanl.edu.mx; Tel.: +52-81-8329-4000 (ext. 3511)

Abstract: *Euphorbia antisyphilitica* Zucc, better known as the candelilla plant, is one of the 10 non-timber forest products of greatest economic importance in the desert and semi-desert regions of Mexico. Moreover, it is a potential source of some functional phytochemicals such as polyphenolic compounds, wax and fiber, with potential applications in food, cosmetic and pharmaceutical industries. Thus, this review aims to describe these phytochemicals and their functional properties as antimicrobial, antioxidant, reinforcing and barrier agents. In addition, a suitable valorization of the candelilla plant and its byproducts is mandatory in order to avoid negative effects on the environment. This review provides, for the first time, an overview of the alternative methodologies for improving candelilla plant production, pointing out some of the agricultural aspects of the cultivation of this plant.

Keywords: candelilla plant; antioxidant properties; antimicrobial activity; polyphenols; candelilla cultivation; candelilla wax

Citation: Rojas, R.; Tafolla-Arellano, J.C.; Martínez-Ávila, G.C.G. *Euphorbia antisyphilitica* Zucc: A Source of Phytochemicals with Potential Applications in Industry. *Plants* **2021**, *10*, 8. https://doi.org/10.3390/plants10010008

Received: 25 November 2020
Accepted: 21 December 2020
Published: 23 December 2020

1. Introduction

The genus *Euphorbia* belongs to the *Euphorbiaceae* family, and it is one of the largest genera of higher plants, with more than 2000 recognized species, such as *Euphorbia thymifolia*, *Euphorbia neriifolia*, *Euphorbia antiquorim*, *Euphorbia chamaesyce*, *Euphorbia helioscopia* and *Euphorbia antisyphilitica* [1,2]. One of the most important species of this family is *E. antisyphilitica* Zucc, better known as the candelilla plant, which is used traditionally as an herbal remedy in countries such as India and other arid and semiarid countries [1]. It naturally grows in the desert and semi-desert regions of northern Mexico, and this non-timber plant is a very important economic resource for the people living in these areas, due to the extreme climatic conditions which restrict agricultural activities [3,4]. The candelilla plant grows in clusters, with thin wax-covered stems that protect them as thick layers giving tolerance against environmental conditions (i.e., temperature variations) and biotic agents (i.e., insects) [5,6]. Therefore, this plant is mainly used to obtain wax, which can be considered as a multipurpose agent for several industries due to its unique properties and multiple applications in the formulation of food, cosmetic and pharmaceutical products. In addition, the candelilla plant has proven to be a good source of other useful phytochemicals such as fiber, wax and polyphenolic compounds (i.e., catechin and ellagic acid), as shown in Figure 1. In this sense, this review provides interesting information generated from around the world about some constituents and bioactive compounds from the candelilla plant and their functional properties.

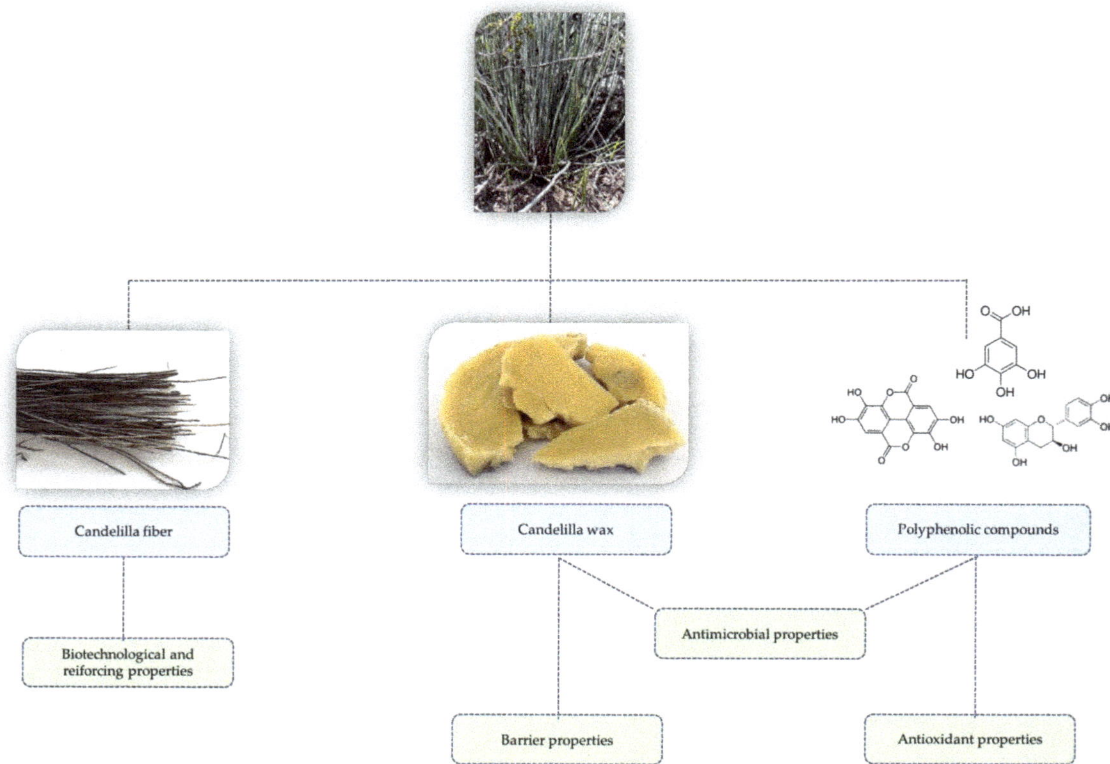

Figure 1. Phytochemicals and functional properties of E. antisyphilitica Zucc.

2. Phytochemicals Present in Candelilla Plant and Byproducts

Usually, a proximal chemical analysis is carried out to determine the chemical constituents of a plant material. In this sense, the chemical analysis of candelilla plants has been reported by Rojas-Molina et al. [5] and Ventura-Sobrevilla et al. [7], as shown in Table 1. Although the chemical composition of plants is strongly influenced by factors such as weather conditions, and season and genetic variability, it can be seen from the reported proximal analysis that lipids and ashes are the major constituents of candelilla plants, which are related to the presence of candelilla wax as mentioned above. Moreover, it has been reported that the candelilla plant possesses a large number of high-quality bioactive phytomolecules with potential applications in different industries acting as antimicrobial and antioxidant agents, and providing other technological advantages [8–13]. Figure 2 shows the structure of the main phytochemicals identified in the candelilla plant and its byproducts.

Table 2 shows the phytochemicals obtained from the candelilla plant and its byproducts, and the functional activity of these compounds. Since candelilla wax is one of the most important constituents of E. antisyphilitica, it is necessary to know its chemical composition, which has been reported as follows: n-alkanes (hentriacontane as the main component) > high molecular weight esters > alcohols and sterols > free acids (7–9%) [4,5,14–17]. Another important structural component of the candelilla plant is fiber, which can be used as a support in the production of hydrolytic enzymes and as a reinforcing agent, as explained below [11,18]. It has been established that candelilla bagasse fiber (CBF) comprises cellulose (45%), hemicellulose (16%), lignin (37%), pectin (1.8%), wax (0.5%) and water-soluble extract (8–12%), a composition that is similar to other natural fibers (i.e., sisal and jute fiber) with industrial applications [11]. In addition, the presence of some polyphenolic compounds has also been reported in the candelilla plant. Rojas-Molina et al. [5] and Ventura-Sobrevilla et al. [7] detected the presence of hydrolysable and condensed tannins, as well as the presence of catechin, ellagic and gallic acid. The first study about the extraction of ellagic acid from E. antisyphilitica was reported by Aguilera-Carbo et al. [19]; the authors reported that the candelilla plant has at least double the quantity of this phenolic compound than other plant materials such as Turnera diffusa and Jatropha dioica. Recently, the putative structure of a high molecular weight (860.7 g mol^{-1}) ellagitannin called candelitannin has been reported in candelilla residues [9]. Finally, flavonoids and other molecules (samonins and quinons) can be found in the methanolic extracts of the candelilla plant [8]; however, no further information has been provided about these components. Regardless of these reports, there is currently limited information on the chemical characterization of this plant, which provides an opportunity for innovative studies in this field due to its economic importance.

Table 1. Proximal chemical analysis carried out in candelilla plants.

	g per 100 g of Dry Material	
Component	[5]	[7]
Moisture	0.4 ± 0.0006	4.44
Total solids	99.6 ± 1.81	95.6
Lipids	15.9 ± 1.119	15.92
Crude fibers	9.0 ± 1.217	9.05
Proteins	2.3 ± 0.0571	1.42
Ashes	10.9 ± 0.315	10.89
Total sugars	0.27 ± 0.043	23.93
Reducing sugars	0.16 ± 0.021	

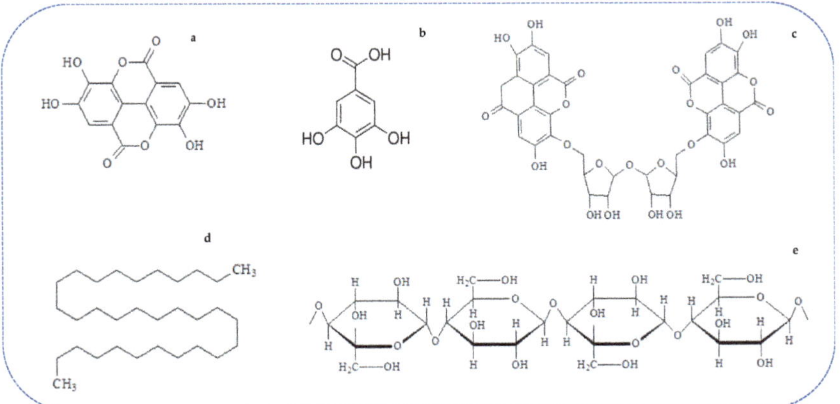

Figure 2. Main phytochemicals identified in E. antisyphilitica and byproducts: (**a**) ellagic acid; (**b**) gallic acid (candelilla whole plant); (**c**) candelitannin (candelilla byproducts); (**d**) hentriacontane (wax); (**e**) cellulose (candelilla bagasse fiber).

3. Agricultural Aspects of Cultivation

According to a very recent study published by Vargas-Pineda et al. [20], candelilla plants have a distribution area of more than 19.1 million hectares in North America under the current climatic conditions. Nevertheless, although the collection of candelilla plants is an important economic activity for communities from northern Mexico, any overharvesting should be avoided. Villa-Castorena et al. [21] conducted a comprehensive study related to the production of candelilla seedlings by cuttings. In this study, four eco-types of candelilla plants called Cuencamé, Cuatrociénegas, Tlahualilo and Viesca, four growing substrates (sandy soil, mixture of river sand and coconut fiber (1:1), mixture of river sand and peat moss (1:1) and mixture of peat moss, perlite and vermiculite (1:1:1)), and four chemical treatments (ProRoot, magic root, phenoxyiacetic acid and a treatment without chemical application) for root promotion were evaluated. According to the authors, untreated Cuatrociénegas eco-type had the highest ability to promote superior rooting of the cuttings due to its special genetic characteristics which allow this eco-type to produce more endogenous auxins for emitting more roots without the presence of other chemicals. For the other eco-types, the cuttings treated with growth media of peat moss with perlite and vermiculite and the mixture of river sand and peat most improved the percentage of rooted cuttings and the shoot growth of this plant. In this sense, the production of candelilla plants can be explored in controlled environments such as greenhouses in order to provide seedlings of good quality for the reforestation of affected areas. In addition, it has been reported that candelilla plants exhibited greater relative growth rate (0.15 $g \cdot g^{-1} \cdot d^{-1}$) and relative water content (88%), which was associated with the physiological status of this plant, allowing its consideration for suitable growth and production as green roofs in arid regions [22].

On the other hand, some stress conditions in the cultivation of candelilla plants, such as the use of lime on the cultivation soil, can improve the production of secondary metabolites (i.e., epicular wax) due to an increase in the pH of the plant tissues due to the alkaline conditions in the soil [23]. According to the authors, candelilla plants cultivated on a liming soil (10 g per pot) reach, on average, more than 54% extractable wax when compared to the control and other abiotic stress conditions (plastic cover and solar reflection). However, no additional studies related to the effect of the stress conditions on additional secondary metabolite expression of this plant were found, which enables the development of new research in this regard.

4. Functional Properties of the Phytochemicals

Due to their functional characteristics, phytochemicals of candelilla plant have the potential to be used in several processes in food, cosmetic and biotechnological industries, as they have demonstrated antimicrobial and antioxidant properties (phenolics and wax), good barrier properties (wax) and reinforcing and biotechnological properties (fiber).

4.1. Antimicrobial Properties

In the last decade, interest in the study of the phytochemical constituents of E. antisyphilitica has grown (Table 2). In addition, it has been reported that plants from northern Mexico have several secondary metabolites to which some biological activities are attributed [8,12]. Thus, the studied molecules have proven to be highly effective at acting as antifungal and antimicrobial bacterial agents [8,9,12]. In their study, Serrano-Gallardo et al. [12] evaluated non-toxic methanolic extracts of four different plants used as a traditional remedy from a semi-desert region in Mexico, including E. antisyphilitica. In their study, the authors evaluated different concentrations of the plant extracts (500, 100 and 2000 $\mu g\ mL^{-1}$) against two reference bacterial strains, Staphylococcus aureus BAA44 and Klebsiella pneumoniae 9180, and four bacteria isolated from clinical samples (S. aureus, K. pneumoniae, Pseudomonas aeruginosa and E. coli). From the obtained results, the authors found that the candelilla extracts presented antimicrobial activity against all the tested bacteria at 500 $\mu g\ mL^{-1}$ (as a minimum inhibitory concentration), which can be related to the phytochemical profile detected in this plant (saponins and quinones). This is according to the previous results reported by Vega-Menchaca et al. [8], who determined that methanolic extracts from candelilla leaves showed antimicrobial activity against clinically isolated bacteria strains such as S. aureus, E. coli O157 and Enterobacter aerogenes 9183. In comparison, in the study of Vega-Menchaca et al. [8], the minimum inhibitory concentration needed for the inhibition of S. aureus was lower (26.8 $\mu g\ mL^{-1}$) than that reported by Serrano-Gallardo et al. [12], which could be due to the presence of flavonoids in candelilla extracts evaluated by Vega-Menchaca et al. [8], as explained by the authors. However, it is important to consider other factors which can affect phytochemical composition of candelilla plants such as weather conditions and season.

Candelilla byproducts also contain other bioactive phenolic compounds that have been related to the antimicrobial activities of candelilla extracts against Erwinia amylovora, Xanthomonas axonopodis and Clavibacter michiganensis [10]. According to the authors, polyphenolic compounds from the hydro-alcoholic extracts obtained from the candelilla byproducts are responsible for the inhibitory effect against the pathogenic bacteria. This antimicrobial potential could be related to the presence of ellagitannins such as candelitannin in these plant materials which has shown effective antifungal properties against four phytopathogenic fungal strains: Alternaria alternata, Fusarium oxysporum, Colletotrichum gloeosporioides and Rhizoctonia solani [9]. The antimicrobial activities exhibited by the phenolic-rich extracts from candelilla plants may be attributed to the interaction of these compounds with the cell membrane causing several modifications to it and changes in various intracellular functions, such as interspecific permeability causing microbial death [24]. Thus, polyphenolic compounds from the candelilla plant can be an important alternative to traditional antimicrobial agents, and a complement to antibiotic therapy. On the other hand, it has been proven and hypothesized that, by itself, candelilla wax has antimicrobial effects on E. coli ATCC 10536 and some fungal strains such as Botrytis cinerea, Colletotrichum gloeosporioides and Fusarim oxysporum, when it is used in the formulation of edible coatings and films [25,26]. However, due to scarce information in the literature, the specific role of candelilla wax in this effect is still unknown.

4.2. Antioxidant Properties

Free-radical scavenging of extracts from candelilla byproducts was evaluated by Burboa et al. [10]. The authors reported that ethanolic extracts can inhibit more than 88% of the DPPH$^{\bullet}$ radicals according to the methodology used for this analysis, which was related to the presence of phenolic compounds in these extracts. This is in line with the findings reported for other species of the Euphorbiaceae family, in which polyphenolic compounds have been detected, such as tannins and flavonoids, which are well known for their antioxidant capacity [1]. Nevertheless, regardless of the high content of phenolic compounds, the evaluation and exploitation of antioxidant properties of candelilla plant extracts are still scarcely explored. In this sense, more research should be conducted with the purpose of generating new and innovative knowledge about the antioxidant properties of the polyphenolic compounds and other phytochemicals from this plant material, as they have great potential to be applied in food, pharmaceutical and cosmetic industries.

4.3. Barrier Properties of Candelilla Wax

Barrier properties of candelilla wax against water vapor and gas transfer are two of the most important features for researchers, and they are usually measured by the formulation of edible coatings and films. The moisture barrier property is the most recognized characteristic of candelilla wax, which is based on its hydrophobic nature and capacity to form a compact network in synergy with other structural compounds such as proteins and carbohydrates [6,14,27,28]. According to Kowalczyk et al. [27], candelilla wax-based films exhibit low water and oxygen permeability compared to other lipid sources, which decreases when the concentration increases (from 0.5 to 2.0%). This can be attributed to its ability to increase film surface hydrophobicity. In the same way, some authors have recorded a decrease in the weight loss of food products such as avocado, "Golden Delicious" apples, Fuji apples and Persian limes. This could be due to the morphology of the coating surface which has a homogenous particle size and less roughness, reducing the open area of the emulsified solids network and therefore avoiding loss of moisture from the fruit [6,28–30]. However, more studies must be conducted regarding the low oxygen permeability of candelilla wax-based coating and films, as it can affect some quality and sensory attributes of coated food products due to the anaerobic respiration and changes in pH [31,32].

Table 2. Phytochemicals from candelilla plant, methods for characterization, registered yields and potential functional activity.

Chemical Nature	Phytochemicals	Yield	Analysis Method for Chemical Characterization	Potential Activity	Reference
Phenolic compounds	Ellagic acid	2.18 ± 0.39 mg g^{-1}	HPLC	Anti-atherosclerotic, antimicrobial and antioxidant activity	[19]
	Hydrolysable tannins	0.56 ± 0.010	Colorimetric methods	Antioxidant activity	[5]
	Condensed tannins	0.16 ± 0.013			
	Ellagic acid	2.2 ± 0.15 mg g^{-1}	HPLC		
	Gallic acid	0.6 ± 0.03 mg g^{-1}			
	Catechin	0.2 ± 0.02 mg g^{-1}			
	Ellagic acid	6.06–7.09 mg g^{-1}		Antitumoral, antiviral, antioxidant activity	[33]
	The presence of ellagitannins and ellagic acid was hypothesized	NI	NI	Antimicrobial, antioxidant and antihemolytic activities	[10]
	Candelitannin		FT-IR and HPLC	Antifungal activity against phytopathogenic agents	[9]
Carbohydrates	Fiber		Water absorption index and critical humidity point and SEM	Great potential to be used as support in solid-state fermentation process	[18]
			FT-IR, thermogravimetry and X-ray diffraction	As a reinforcing agent for polypropylene composites	[11]
			FT-IR, thermogravimetry, X-ray diffraction and SEM	Reinforcing, structural and thermal agents	[34]
			Dynamic mechanical analysis and SEM	Improve the mechanical properties of polypropylene composites	[35]
Lipids	Candelilla wax		NI (No identified)	Excipients clinical applications	[36]
				Reduces water permeability of edible films	[15]
				In the encapsulation of fertilizers with slow-release properties	[13]
			HPLC and GC	Structural agent in food systems	[16]
					[17]
			NI	Antimicrobial and water resistance properties	[25]
		2–5 g 100 g^{-1}		Can be used to obtain petroleum	[37]
Other compounds	Saponins and quinones	NI	NI	Antimicrobial potential	[12]

SEM = scanning electron microscope; FT-IR = Fourier transform infrared spectroscopy; HPLC = high-performance liquid chromatography; GC = gas chromatography.

4.4. Other Functional Applications

Due to the candelilla wax extraction process generating large amounts of lignocellulosic waste, they can be used for different technological purposes. Regarding the biotechnological aspects, candelilla fiber has been used as a support for fungal growth to the production of ellagitannase due to its ellagitannin content [18]. In this study, the authors evaluated four agroindustrial byproducts, and, although it was concluded that candelilla stalks have great potential for use as support for solid-state fermentation (SSF), the lowest values of ellagitannase were obtained using this plant material. Thus, future studies must focus on the standardization of this biotechnological process in order to increase enzyme-activity titles. On the other hand, due to their physico-chemical characteristics, candelilla fiber has also been used as a reinforcing agent in the formulation of new composites based on polypropylene (PP) and CBF [11]. In this study, the authors found that this byproduct is stable at around 200 °C, which improves the thermal stability and tensile properties of the CBF–PP composites due to the crystallinity index of the cellulose and the disruption of the free moment of the polymeric chains in the matrix, respectively. In a more recent study, it was demonstrated that candelilla fiber contains intracuticular wax and resins, which has the benefit of being a compatibilizer between the fiber and the polypropylene [35]. This is according to a recent study by Pulido-Barragán et al. [34], who reported that CBF is a good plant material for obtaining cellulose nanocrystals, which can be used as a reinforcing, structural or thermal agent, as well as for 3D printing and as a constructive nanocellulosic paper agent, among other technical applications due its the specific physico-chemical properties. Although these studies were well conducted, limited information on the application of candelilla fiber is available, providing an opportunity for the investigation of other functional properties of this plant material.

5. Final Remarks and Perspectives

Since the candelilla plant is an endemic of the semiarid regions of northern Mexico, this country has been recognized for its potential to be the main producer of candelilla wax. However, this review provides interesting and innovative information associated with the promising applications and sustainable valorization of other phytochemicals from the candelilla plant not published elsewhere. In addition, it provides an opportunity for developing more investigations into the physicochemical characterization of polyphenolic compounds (different to ellagic acid), and the fiber present in the candelilla plant and its byproducts. Furthermore, considering its great potential to be used as a source of components with several functional applications, researchers should focus more on the evaluation, stability and application of the phytochemicals of this plant, as they could help to replace synthetic molecules used at the industrial level. In addition, the waste disposal for animal feed (after the SSF process) and the evaluation of CBF as a reinforcing agent (after the extraction of wax and phenolic compounds) could be of great interest for creating a "zero waste" process as an integral use of this plant resource.

Author Contributions: Conceptualization, writing—original draft preparation (G.C.G.M.-Á); investigation, writing—review and editing, project administration, and funding acquisition (R.R. and G.C.G.M.-Á); investigation, writing—original draft preparation, writing—review and editing (J.C.-T.-A.). All authors have read and agreed to the published version of the manuscript.

Funding: This research was funded by the National Forest Commission and the Mexican Council for Science and Technology (CONAFOR-CONACYT), through the Sectoral Fund for Forestry Research, Development and Technological Innovation with the project Diseño y construcción de equipo semiautomático para la extracción de cera de candelilla orgánica (number B-S-131466).

Acknowledgments: Authors thank the staff of the National Forest Commission for the facilities given to develop this project.

Conflicts of Interest: The authors declare no conflict of interest.

References

1. Garipelli, N.; Runja, C.; Potnuri, N.; Pigili, R.K. Anti-inflamatory and anti-oxidant activities of ethanol extract of *Euphorbia thymifolia* Linn whole plant. *Int. J. Pharm. Pharm. Sci.* **2012**, *4*, 516–519.
2. Johari, S.; Kumar, A. *Euphorbia* spp. and their use in traditional medicines: A review. *World J. Pharm. Res.* **2020**, *9*, 1477–1485. [CrossRef]
3. Martínez-Ballesté, A.; Mandujano, M. The consequences of harvesting on regeneration of a non-timber wax producing species (*Euphorbia antisyphilitica* Zucc.) of the Chihuahuan desert. *Econ. Bot.* **2013**, *67*, 121–136. [CrossRef]
4. Tinto, W.F.; Elufioye, T.O.; Roach, J. Waxes. In *Pharmacognosy Fundamentals, Applications and Strategies*; Badal, S., Delgoda, R., Eds.; Elsevier: Amsterdam, The Nederlands, 2017; pp. 443–455.
5. Rojas-Molina, R.; De León-Zapata, M.A.; Saucedo-Pompa, S.; Aguilar-Gonzalez, M.A.; Aguilar, C.N. Chemical and structural characterization of candelilla (*Euphorbia antisyphilitica* Zucc.). *J. Med. Plants Res.* **2013**, *7*, 702–705.
6. De León-Zapata, M.A.; Ventura-Sobrevilla, J.M.; Salinas-Jasso, T.A.; Flores-Gallegos, A.C.; Rodríguez-Herrera, R.; Pastrana-Castro, L.; Rua-Rodríguez, M.L.; Aguilar, C.N. Changes of the shelf life of candelilla wax/tarbush bioactive based-nanocoated apples at industrial level conditions. *Sci. Hortic.* **2018**, *231*, 43–48. [CrossRef]
7. Ventura-Sobrevilla, J.; Gutiérrez-Sánchez, G.; Bergmann, C.; Azadi, P.; Boone-Villa, D.; Rodriguez-Herrera, R.; Aguilar, C.N. Glycosylation of polyphenols in tannin-rich extracts from *Euphorbia antisyphilitica*, *Jatropha dioica*, and *Larrea tridentata*. In *Green Chemistry and Biodiversity Principles, Techniques, and Correlations*; Aguilar, C.N., Ameta, S.C., Haghi, A.K., Eds.; Apple Academic Press: New York, NY, USA, 2019; Chapter 7. [CrossRef]
8. Vega-Menchaca, M.C.; Rivas-Morales, C.; Verde-Star, J.; Oranday-Cárdenas, A.; Rubio-Morales, M.E.; Núñez-González, M.A.; Serrano-Gallardo, L.B. Antimicrobial activity of five plants from Northern Mexico on medically important bacteria. *Afr. J. Microbiol. Res.* **2013**, *7*, 5011–5017. [CrossRef]
9. Ascacio-Valdés, J.; Burboa, E.; Aguilera-Carbo, A.F.; Aparicio, M.; Pérez-Schmidt, R.; Rodríguez, R.; Aguilar, C.N. Antifungal ellagitannin isolated from *Euphorbia antisyphilitica* Zucc. *Asian Pac. J. Trop. Biomed.* **2013**, *3*, 41–46. [CrossRef]
10. Burboa, E.A.; Ascacio-Valdés, J.A.; Zugasti-Cruz, A.; Rodríguez-Herrera, R.; Aguilar, C.N. Antioxidant and antibacterial capacity of candelilla extracts residues. *Rev. Mex. Cienc. Farm.* **2014**, *45*, 51–56.
11. Morales-Cepeda, A.B.; Ponce-Medina, M.E.; Salas-Papayanopolos, H.; Lozano, T.; Zamudio, M.; Lafleur, P.G. Preparation and characterization of candelilla fiber (*Euphorbia antisyphilitica*) and its reinforcing effect in polypropylene composites. *Cellulose* **2015**, *22*, 3839–3849. [CrossRef]
12. Serrano-Gallardo, L.B.; Castillo-Maldonado, I.; Borjón-Ríos, C.G.; Rivera-Guillén, M.A.; Morán-Martínez, J.; Téllez-López, M.A.; García-Salcedo, J.J.; Pedroza-Escobar, D.; Vega-Menchaca, M.C. Antimicrobial activity and toxicity of plants from northern Mexico. *Indian J Tradit. Knowl.* **2017**, *16*, 203–207.
13. Navarro-Guajardo, N.; García-Carrillo, E.M.; Espinoza-González, C.; Téllez-Zablah, R.; Dávila-Hernández, F.; Romero-García, J.; Ledezma-Pérez, A.; Mercado-Silva, J.A.; Pérez-Torres, C.A.; Pariona, N. Candelilla wax as natural slow-release matrix for fertilizers encapsulated by spray chilling. *J. Renew. Mater.* **2018**, *6*, 226–236. [CrossRef]
14. Toro-Vazquez, J.F.; Charó-Alonso, M.A.; Pérez-Martínez, J.D.; Morales-Rueda, J.A. Candelilla wax as an organogelator for vegetable oils—An alternative to develop trans-free products for the food industry. In *Edible Oleogels, Structure and Health Implications*; Marangoni, A.G., Garti, N., Eds.; Academic Press and AOCS Press: Urbana, IL, USA, 2011; pp. 119–148. [CrossRef]
15. Ochoa, E.; Saucedo-Pompa, S.; Rojas-Molina, R.; de la Garza, H.; Charles-Rodríguez, A.V.; Aguilar, C.N. Evaluation of a candelilla wax-based edible coating to prolong the shelf-life quality and safety of apples. *Am. J. Agric. Biol. Sci.* **2011**, *6*, 92–98. [CrossRef]
16. Doan, C.D.; To, C.M.; De Vrieze, M.; Lynen, F.; Danthine, S.; Brown, A.; Dewettinck, K.; Patel, A.R. Chemical profiling of the major components in natural waxes to elucidate their role in liquid oil structuring. *Food Chem.* **2017**, *214*, 717–725. [CrossRef] [PubMed]
17. Moreau, R.A.; Harron, A.F.; Hoyt, J.L.; Powell, M.J.; Hums, M.E. Analysis of wax esters in seven commercial waxes using C30 reverse phase HPLC. *J. Liq. Chromatogr. Relat. Technol.* **2018**, *41*, 604–611. [CrossRef]
18. Buenrostro-Figueroa, J.; Ascacio-Valdés, A.; Sepúlveda, L.; De la Cruza, R.; Prado-Barragán, A.; Aguilar-González, M.A.; Rodrígueza, R.; Aguilar, C.N. Potential use of different agroindustrial by-products as supports for fungal ellagitannase production under solid-state fermentation. *Food Bioprod. Process.* **2014**, *92*, 376–382. [CrossRef]
19. Aguilera-Carbo, A.F.; Augur, C.; Prado-Barragan, L.A.; Aguilar, C.N.; Favela-Torres, E. Extraction and analysis of ellagic acid from novel complex sources. *Chem. Pap.* **2008**, *62*, 440–444. [CrossRef]
20. Vargas-Piedra, G.; Valdez-Cepeda, R.D.; López-Santos, A.; Flores-Hernández, A.; Hernández-Quiroz, N.S.; Martínez-Salvador, M. Current and Future Potential Distribution of the Xerophytic Shrub Candelilla (*Euphorbia antisyphilitica*) under Two Climate Change Scenarios. *Forests* **2020**, *11*, 530. [CrossRef]
21. Villa-Castorena, M.; Catalán-Valencia, E.A.; Inzunza-Ibarra, M.A.; González-López, M.D.; Arreola-Ávila, J.G. Production of candelilla seedlings *Euphorbia antisyphilitica* Zucc. by cuttings. *Rev. Chapingo Ser. Cienc. For. Ambiente* **2010**, *16*, 37–47. [CrossRef]
22. Gioannini, R.; Al-Ajlouni, M.; Kile, R.; VanLeeuwen, D.; Hilaire, R.S. Plant Communities Suitable for Green Roofs in Arid Regions. *Sustainability* **2018**, *10*, 1755. [CrossRef]
23. Muñoz-Ruiz, C.V.; López-Díaz, S.; Covarrubias-Villa, F.; Villa-Luna, E.; Medina-Medrano, J.R.; Barriada-Bernal, L.G. Effect of abiotic stress conditions on the wax production in candelilla (*Euphorbia antisyphilitica* zucc.). *Rev. Latinoamer. Quím.* **2016**, *44*, 26–33.

24. Besednova, N.N.; Andryukov, B.G.; Zaporozhets, T.S.; Kryzhanovsky, S.P.; Kuznetsova, T.A.; Fedyanina, L.N.; Makarenkova, I.D.; Zvyagintseva, T.N. Algae polyphenolic compounds and modern antibacterial strategies: Current achievements and immediate prospects. *Biomedicines* **2020**, *8*, 342. [CrossRef] [PubMed]
25. Chevalier, E.; Chaabani, A.; Assezat, G.; Prochazka, F.; Oulahal, N. Casein/wax blend extrusion for production of edible films as carriers of potassium sorbate—A comparative study of waxes and potassium sorbate effect. *Food Packag. Shelf Life* **2018**, *16*, 41–50. [CrossRef]
26. Ruiz-Martínez, J.; Aguirre-Joya, J.A.; Rojas, R.; Vicente, A.; Aguilar-González, M.A.; Rodríguez-Herrera, R.; Alvarez-Perez, O.B.; Torres-León, C.; Aguilar, C.N. Candelilla wax edible coating with *Flourensia cernua* bioactives to prolong the quality of tomato fruits. *Foods* **2020**, *9*, 1303. [CrossRef] [PubMed]
27. Kowalczyk, D.; Gustaw, W.; Zieba, E.; Lisiecki, S.; Stadnik, J.; Baraniak, B. Microstructure and functional properties of sorbitol-plasticized pea protein isolate emulsion films: Effect of lipid type and concentration. *Food Hydrocoll.* **2016**, *60*, 353–363. [CrossRef]
28. De León-Zapata, M.A.; Sáenz-Galindo, A.; Rojas-Molina, R.; Rodríguez-Herrera, R.; Jasso-Cantú, D.; Aguilar, C.N. Edible candelilla wax coating with fermented extract of tarbush improves the shelf life and quality of apples. *Food Packag. Shelf Life* **2015**, *3*, 70–75. [CrossRef]
29. Bosquez-Molina, E.; Guerrero-Legarreta, I.; Vernon-Carter, E.J. Moisture barrier properties and morphology of mesquite gum–candelilla wax based edible emulsion coatings. *Food Res. Int.* **2003**, *36*, 885–893. [CrossRef]
30. Aguirre-Joya, J.A.; Ventura-Sobrevilla, J.; Martínez-Vazquez, G.; Ruelas-Chacón, X.; Rojas, R.; Rodríguez-Herrera, R.; Aguilar, C.N. Effects of a natural bioactive coating on the quality and shelf life prolongation at different storage conditions of avocado (*Persea americana* Mill.) cv. Hass. *Food Res. Int.* **2017**, *14*, 102–107. [CrossRef]
31. Kowalczyk, D.; Ziezba, E.; Skrzypek, T.; Baraniak, B. Effect of carboxymethyl cellulose/candelilla wax coating containing ascorbic acid on quality of walnut (*Juglans regia* L.) kernels. *Int. J. Food Sci. Tech.* **2017**, *52*, 1425–1431. [CrossRef]
32. Kowalczyk, D.; Kordowska-Wiater, M.; Kałwa, K.; Skrzypek, T.; Sikora, M.; Łupina, K. Physiological, qualitative, and microbiological changes of minimally processed Brussels sprouts in response to coating with carboxymethyl cellulose/candelilla wax emulsion. *J. Food Process. Preserv.* **2019**, *43*, 1–8. [CrossRef]
33. Ascacio-Valdés, J.A.; Aguilera-Carbó, A.; Martínez-Hernández, J.L.; Rodríguez-Herrera, R.; Aguilar, C.N. *Euphorbia antisyphilitica* residues as a new source of ellagic acid. *Chem. Pap.* **2010**, *64*, 528–532. [CrossRef]
34. Pulido-Barragán, E.U.; Morales-Cepeda, A.B.; Castro-Guerrero, C.F.; Koschella, A.; Heinze, T. Upgrading Euphorbia Antisyphilitica fiber compost: A waste material turned into nanocrystalline cellulose. *Ind. Crops. Prod.* **2020**. [CrossRef]
35. Ponce-Medina, M.E.; Sánchez-Valdés, S.; Ángeles-San Martin, M.E.; Salas-Papayanopolos, H.; Hernández-Hernández, D.E.; Lozano-Ramírez, T.; Karami, S.; LaFleur, P.; Morales-Cepeda, A.B. Composites of polypropylene/candelilla fiber (*Euphorbia antisyphilitica*): Synergic of wax- polypropylene grafted maleic anhydride. *Cogent Eng.* **2018**, *5*, 1526861. [CrossRef]
36. Apte, S.P. Considerations for setting specifications for excipients of natural origin. *J. Excip. Food Chem.* **2015**, *6*, 101–103.
37. Kumar, A.; Roy, S. Agrotechnology, production, and demonstration of high-quality planting material for biofuels in arid and semiarid regions. In *Biofuels: Greenhouse Gas Mitigation and Global Warming*; Kumar, A., Ogita, S., Yau, Y.Y., Eds.; Springer: New Delhi, India, 2018; pp. 205–228. [CrossRef]

plants

Review

Zebrafish as a Successful Animal Model for Screening Toxicity of Medicinal Plants

Amir Modarresi Chahardehi *, **Hasni Arsad** and **Vuanghao Lim** *

Integrative Medicine Cluster, Advanced Medical and Dental Institute, Universiti Sains Malaysia, Bertam, Kepala Batas 13200, Malaysia; hasniarsad@usm.my

* Correspondence: amirmch@gmail.com (A.M.C.); vlim@usm.my (V.L.)

Received: 8 September 2020; Accepted: 7 October 2020; Published: 12 October 2020

Abstract: The zebrafish (*Danio rerio*) is used as an embryonic and larval model to perform in vitro experiments and developmental toxicity studies. Zebrafish may be used to determine the toxicity of samples in early screening assays, often in a high-throughput manner. The zebrafish embryotoxicity model is at the leading edge of toxicology research due to the short time required for analyses, transparency of embryos, short life cycle, high fertility, and genetic data similarity. Zebrafish toxicity studies range from assessing the toxicity of bioactive compounds or crude extracts from plants to determining the optimal process. Most of the studied extracts were polar, such as ethanol, methanol, and aqueous solutions, which were used to detect the toxicity and bioactivity. This review examines the latest research using zebrafish as a study model and highlights its power as a tool for detecting toxicity of medicinal plants and its effectiveness at enhancing the understanding of new drug generation. The goal of this review was to develop a link to ethnopharmacological zebrafish studies that can be used by other researchers to conduct future research.

Keywords: zebrafish; toxicity; embryotoxicity; medicinal plant; animal model

1. Introduction

Herbal plants have pharmacological and therapeutic characteristics due to the natural chemical compounds they contain [1]. Hence, they are widely utilized every day for culinary purposes and nutritional supplements to promote health [2]. This product can be toxic, so its toxicity must be measured to ensure adequate safety for human health [3]. Although many people view most medicinal plants as safe, poisoning can occur in some cases. Consumers can also be exposed to potential health risks caused by specific components or contaminants of botanical products; thus, their risk needs to be evaluated [4,5]. Plant materials and their extracts contain various toxic substances synthesized by plants as a defense against disease, insects, and other organisms [6]. Botanical toxicity studies are complicated due to expense, time, use of animals, and the complex mix of components [4,7]. The most commonly used extraction solvents, from polar to non-polar, are water, ethanol, methanol, acetone, ethyl acetate, chloroform, dichloromethane and hexane, and numerous researchers have studied the effectiveness of these solvents [8]. Various extracts may display distinctive cytotoxicity traits and a wide variety of pharmacological consequences at different concentrations [9]. Additionally, numerous factors affect compounds, such as the type and volume of extraction solvents used and varying storage environments [8]. Bioactive compounds in medicinal plants can have toxic impacts on cardiac glycosides, phorbol esters, alkaloids, cyanogenic glycosides, and lectins [10]. The statistical evaluation carried out by the National Pharmaceutical Control Bureau and Health ministry's Malaysian Adverse Drug Reaction Advisory Committee in 2013 confirmed that 11,437 cases of harmful drug reaction had been reported, and 0.2% were due to herbal remedies [1]. Thus, the evaluation of the toxicity of potential drug compounds has been enhanced in recent years [11]. Allopathy drugs and complementary and alternative medicines require a toxicology assay to identify any harmful effects that are not well-known until signs and symptoms appear after high consumption [12].

However, the main objective of toxicity research is to predict human toxicity via fast and accurate testing of many substances [13] based on model systems [14], and different routes of drug delivery systems such as using nanoparticles in the gastrointestinal tract [15] or chemical fertilizers and water retention as an example [16]. Previously, classical toxicity screening (including rodents, dogs, and rabbits) involved the compilation of data from a given laboratory for one substance at a time [17]; however, these tests are frequently expensive, time-consuming, and tedious [18]. This classic method focused on studying chemical compounds on phenotypic cell or animal results [19]. Thus, an attractive alternative approach involves applying the 3R concept of human-animal studies (i.e., reduction, replacement, and refinement), but it is not consistent with the use of rodent animal models [20]. Replacement: Zebrafish assays can be used to substitute such animal-toxicity studies using larval zebrafish, and it can be shown that larval zebrafish represent a critical system model; Reduction: zebrafish larvae can be used as a first-level model for toxicity to classify toxic drug candidates so that more stable drugs can be evaluated in mammalian models; Refinement: The model of embryonic and larval zebrafish provides a refined design to research organisms since embryos are partially fertilized and translucent in their early life [21]. It is recommended that the number of animals is limited, the test methods minimize pain and suffering of experimental animals, and approved substituted animal tests are used as much as possible [22].

The zebrafish (*Danio rerio*) is a suitable model for screening drugs for potential use to treat human diseases [23] based on phylogenetic analysis of fish and human genomes, which shows similar morphology and physiology of the nervous, cardiovascular, and digestive systems [18]. The zebrafish genome has been sequenced in full [24]. Zebrafish are a fast model for the study of genetic and de novo mutations [25]. The genes can inactivate in vivo, simulate human phenotypes, and obtain information on human diseases with genetic background through genomic editing approaches, such as CRISPR/Cas9 or artificial site-specific nucleases, zinc-finger nucleases, and transcription activator-like nucleases [25]. Zebrafish also provide a significant data file for vertebrate animals that enables researchers to anchor biochemical, genetic, and cellular hypotheses to high-performance observations at structural, functional, and behavioral levels [26]. The zebrafish embryotoxicity test, or fish embryotoxicity test (FET), is gaining popularity because it provides a total and well-defined developmental duration for a vertebrate embryo and allows the study of its early life stages [27]. As experimental models, both adults and zebrafish embryos are used [22]. For example, various compound tests (e.g., measurement of drug cardiotoxicity) have been conducted to assess drug effects in zebrafish efficiently [11]. Additionally, for the last two decades, zebrafish have been used to study angiogenesis, metastasis, anticancer drug screening, and an assessment of drug toxicity [28]. In summary, the cell structural and biochemical similarities between humans and zebrafish enable rapid forecasting of the possible impacts of chemical and other substances on human communities. Because zebrafish are becoming increasingly important as a test model, husbandry requirements to enhance the reproducibility and efficiency of this type of model in research environments are needed [29].

Toxicity studies generally begin in in vitro studies, with many different cell lines at different sample concentrations. The substance is then tested in several animal models, especially in mice and rats, before using on patients [30]. As mentioned in Table 1, there are several explanations why zebrafish as an animal model could be extremely useful for intermediate toxicity tests.

Table 1. Suggested methods for toxicological analysis [30,31]

	Cells	Fruit Fly	Zebrafish	Rodents	Humans
Benefits	Fast, simple, inexpensive	Short period of generation	Fast, simple, inexpensive	Complexity of body	High value translation
	Well-standardized	Work easily with	Complexity of body	Adequate predictiveness	Credible and logical
	Many probable readings	Low maintenance costs	Ethical examination in vivo	Administration route	Dosage details
Limitations	Modest predictivity	Genetically far from human beings	Modest flexibility	Time-consuming and costly	Time-consuming and costly
	Simplified process	Simple anatomy	Moderate foresight	Comprehensive regulation	Comprehensive regulation
	Low value translation	No efficient immune system	Moderate value translation	Ethical limits	Ethical limits

In vitro toxicity studies are inexpensive, fast, and easy, as shown in Table 1, but cultured cells are poorly associated with in vivo processes and therefore have limited translation benefit, while accurate data are obtained from laboratory rodent studies to extrapolate toxicants to humans [32]. Hence, the gold standard for predictive analysis of chemical risks to humans remains vertebrate toxicity studies; however, these studies are unsuitable for initial toxicity screening due to the reasons above and often involve significant amounts of a valuable test compound [26]. There has been a trend in the last 30 years towards the restricted use of higher animals in herbal toxicology studies in particular [22]. These animals have a series of restrictions. For example, rodents may be immune to the cardiotoxicity, especially when the endpoint is left ventricular contractile function. This may be due to rodents' ability to substitute for myocyte failure by employing alternative mechanisms [25]. The zebrafish model is especially useful for the bioassay-guided identification of bioactive secondary metabolites [33]. The National Health Institute (NIH), USA, recently promoted a zebrafish model organism to study various genetically engineered diseases [34]. The Food and Drug Administration has recognized zebrafish tests for toxicity and safety evaluations for investigative newly developing drugs [26]. Large volume screening platforms exist in the form of a multi-wave plate format for testing chemical impacts on embryo development by assessing deformities, mortality, and structural characteristics across a range of concentrations [26]. Small molecules may be introduced directly into the water in multi-well racks, where fish take them up via diffusion. The researched drugs can also be injected into the yolk sac [35]. Thus, the zebrafish is a great model for studying the premature embryonic diet on toxicants [36]. Although researchers hope that both adults and embryos will prove useful for toxicology studies, embryos are ideal for toxicity examinations because of the transparency of the egg, which makes it easy to detect developmental phases and evaluate endpoints during the toxicity test [22,37]. This enables technical and economic benefits over rodent models [26,38] to reduce the number of substances and reduce the cost of animals in the development of drugs [39]. Up to 200–300 embryos can be developed per pair of adult zebrafish, while a typical pair of rodents only produce 5–10 descendants per matching occurrence [24]. Five days after post fertilization (dpf), the heart, liver, brain, pancreas, and other organs are created [40]. The three-lobe liver of zebrafish reflects humans' biological function, including the absorption and production of lipids, vitamins, proteins, and carbohydrates [25]. One of the significant benefits of using zebrafish is that compounds can be easily supplied by adding water, similar to chemicals in the cell culture medium, which generally requires an overall amount of only 100 μL during growth [41]. Their optical clarity makes it easy to produce and recognize phenotypic properties during mutagenesis screening, and determine toxicity endpoints during toxicological analysis [32]. In addition, zebrafish embryos have been incredibly helpful in studying heart development and the functional effects of toxicants [14]. Figure 1 illustrates the temporal distinction between humans, rats, and zebrafish in early developmental life [26,41]. During development, there are still several limitations in the chemical identification of zebrafish. The fish is ectothermic and lacks heart septa, limbs, synovial joints, cancellous bones, lungs, and other organs [42]. Age is one of the most important limitations. Although chemical tests can be undertaken in adult fish, a high-performance scan using appropriate quantities of small molecules requires the animals to fit into a multi-well plate [19]. The analysis of gene expression throughout the larvae is feasible due to the optical transparency of the zebrafish tissues, which enables good penetration for light microscopy [32].

An alternative preliminary toxicity test is called the brine shrimp lethality test, which involves testing different toxicant concentrations. However, this technique does not reveal abnormalities or causes of death [43]. In contrast, using zebrafish as a screening model animal has some rational benefits, including large embryos, large numbers of embryos for testing, and simple visualization of organogenesis using fluorescence and transgenic strains [36]. Additionally, the ability to view biological processes provides the researcher with visual links to the organ system dynamics and, possibly, interorgan systems [44].

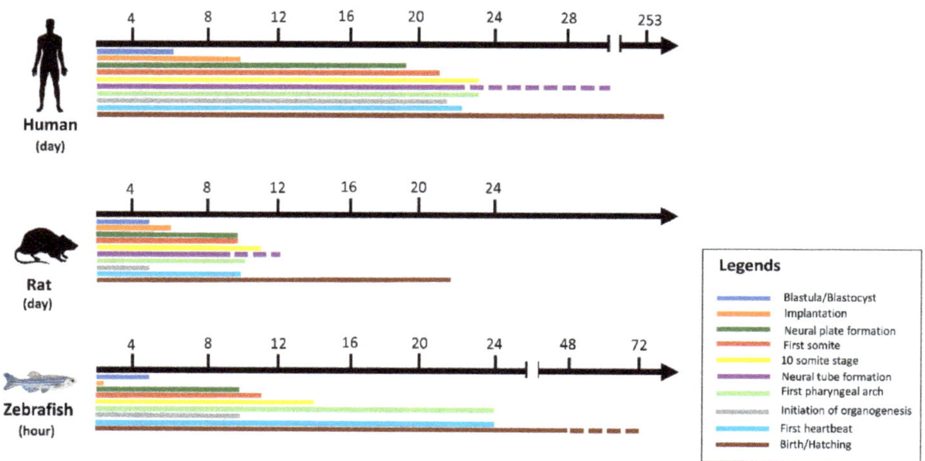

Figure 1. Comparative analysis of human, rat, and zebrafish based on the early stages of development.

Today, many herbal products claim to provide pharmaceutical health benefits but do not provide any toxicological data [1]; thus, the safety of these drugs is questionable. This review aimed to compare the results of studies of the toxicity of medicinal plants and assess the toxicity tests used to analyze some of these commercial products from plants. This paper is the first review to assess the toxic and teratogenic effects of plant extracts on the zebrafish model to the best of our knowledge.

2. Zebrafish as a Suitable Alternative Animal Model for Toxicity Tests

One of the main goals of medicinal plant toxicology research is to identify many bioactive compounds with specific toxicity within a short period [13]. For many years, higher species have been used as models to measure medicinal products based on their toxicity [11]. For instance, recent studies of acute toxicity have mainly used mice as the animal model. Nevertheless, it is difficult to achieve a thorough and immediate toxicity check with this organism because of its strong state of breeding, high expenses, complicated procedures, and ethical restrictions [45]. Instead of using rodents, fish are the best candidate for this purpose. Fish traditionally have been used in toxicity testing of individual substances and effluents [46], and today zebrafish are commonly utilized to test for developmental toxicity, general toxicity and to carry out medication screening as a credible vertebrate model [11,26]. Hence, the first large-scale studies for bioactive molecules using zebrafish embryos were published, using the merits of zebrafish as phenotypic testing models to evaluate the effects of biomolecules and explore different bioactive compounds [47]. The zebrafish is widely used in numerous subfields of biology, and without a doubt, it is one of the leading species in various research areas, including developmental biology, ecotoxicology, and genetics [48]. Zebrafish embryos can also quickly consume tiny molecular compounds, thus providing a valuable model for drug testing and evaluating teratogenic effects [49,50] of exposure to toxic compounds [38]. Zebrafish embryos and larvae are outstanding models for testing the toxicity of substances, especially if those substances are present in low quantities [51]. Zebrafish larvae are very useful in imaging studies, and they likely will prove useful for non-imaging endpoints. Scientists are discovering new innovative paths for evaluating biochemical processes [19]. Within 5–6 days dpf, zebrafish development reflects the full developmental period of a vertebrate embryo before it becomes self-sustainable and, therefore, this substitute organism is not currently recognized as a genuine in vivo form by European law [52].

Teratology is the study of unusual growth, and a teratogen is any substance that triggers the production of a congenital anomaly or enhances the occurrence of a specific hereditary deficiency [53]. Screening for teratogenicity includes introducing zebrafish embryos to the required concentration of the compound of interest, and it has become a popular model for analyzing the teratogenic effects of medicines [35]. Teratological and embryo-toxic effects are easy to detect due to the transparency of zebrafish embryos during their growth outside of their parent [54]. Teratogenic effects include tail malformation, pericardial edema, malformation of the notochord, scoliosis, yolk edema, and growth delays [22]. Therefore, it is essential to evaluate the embryotoxic and teratogenic toxicity of medicinal plants on embryo growth. Embryo malformations may be caused by activation of the Caspase-3 enzyme, which is the leading cause of apoptosis [55], or by other factors such as reactive oxygen species-induced oxidative stress [6].

The effectiveness of the zebrafish model system was evident in the 1960s and 1970s, as numerous studies used zebrafish as bioassays of chemicals affecting normal functioning and reproductive success. During the 1980s and 1990s, many researchers verified that zebrafish are a convenient and reliable option for comprehensive toxicological screening and early life and lifetime exposure tests [56]. In 2000, the first chemical screening procedure was reported that allowed researchers to test a very slight concentration of compounds on live zebrafish in 96-well plates [19]. To date, there are ten types of tests used to study toxicity in zebrafish: (1) the zebrafish embryo toxicity test (FET); carcinogenicity; (2) developmental toxicity and teratogenicity assessments; (3) reproductive toxicity assessment; (4) behavioral toxicity assessment; (5) endocrine disorders; (6) acute toxicity; (7) neurotoxicity; (8) optical (ocular toxicity); (9) cardiotoxicity; and (10) vascular toxicity [31]. Among the available tests, FET is the most useful for assessing chemical and substance toxicity in zebrafish [48]. Numerous toxicity studies of specific chemicals have shown a significant correlation between results for zebrafish embryos and acute toxicity of fish [57]. The FET test aims to determine the acute toxicity of the embryonic phases and establish variables for zebrafish [52]. It is based on mortality and teratogenesis of the zebrafish embryos [46]. FET is a valuable option for replacing the use of adult animals to evaluate toxicity and enhance existing toxicity assays [48]. Advantages of this test include:

1. A wide range of chemicals may be relevant;
2. Embryos are short-lived, susceptible, cost-effective, and have low variability;
3. The regulatory and scientific communities consider them to be standardized;
4. The tolerance of various species or organisms can be compared [58].

The possibility of tracking many toxicity endpoints, including alteration of molecular processes and malformations, is one of FET's main advantages with *Danio rerio* [59]. With the FET test, acute toxicity is assessed based on positive results, and embryo formation is examined every day for any unusual developmental phenotypes, especially morphological defects, including (i) coagulation of fertilized eggs; (ii) lack of somite formation; (iii) lack of detachment of the tail-bud from the yolk sac; and (iv) lack of a heartbeat [52]. Nagel (2002) outlined these toxicity indicators as measured using an inverse light microscope based on "yes" or "no" responses to the presence of the four factors [60]. On the other hand, Hermsen et al. [61] introduced a new model of embryotoxicity (ZET) using the general morphology score (GMS). A separate scoring list was established for teratogenic impact, which helped track slow progress, developmental delay, and teratogenicity. However, the ZET, along with GMS, operates as an essential and useful test method for screening the chemicals' embryotoxic properties in the compound groups studied [27]. Furthermore, the MolDarT was created with zebrafish eggs/larvae to develop a molecular mechanism test method [61]. It is based in theory on the DarT (*Danio rerio* teratogenicity test), established by Nagel et al. [60], which reveals and tracks the developmental impact and deformities of freshly fertilized zebrafish eggs within 48 h [62].

Experiments to test toxicity include acute, sub-chronic, and chronic studies of compounds' toxicity to specific organ pathways and hypothesis-driven research [31]. Zebrafish were found to be a useful tool for the comprehension not only of neurotoxicants' structural and chemical effects but also for the assessment of behavioral dysfunction correlated with such toxicity [38]. Several behavioral endpoints are used in neurotoxicity studies to determine the therapeutic influences of medications and their neurotoxins on neuron development in zebrafish [63]. There is an incomparable system to identify endocrine activity lacking significant morphological abnormalities [13]. For the reasons described above, zebrafish toxicity screening provides many significant experimental advantages, including embryo and larvae visibility, high-efficiency short test time, low number of necessary compounds, ease of handling, and direct delivery compounds [54]. The zebrafish embryo develops quickly outside the mother's body and enters maturity in a couple of months, among other benefits [38]. Additionally, more than 70% of the disease-associated genes are similar to those found in human diseases [50]. Furthermore, the physical, biochemical, genetic, and molecular makeup of zebrafish, including organs and tissues, has been demonstrated to be identical to their mammalian relatives. Metabolites, signaling mechanisms, and neurological and cognitive structures are similar to those in mammals [54].

A compound's toxicity is measured using two indicators: the median effective concentration (EC_{50}) and the median lethal concentration (LC_{50}). The concentration-response curve is generated using 24 h of data to obtain the EC_{50} (teratogenic effects) and LC_{50} (embryotoxic effects). The ratio of LC_{50} to EC_{50} is the corresponding therapeutic index (TI). In pharmaceutical administration, a TI value < 1 is optimal [50,64]. The strength of the association between zebrafish embryo LC_{50} values and rodent LD_{50} values for 60 different compounds was demonstrated in one particular laboratory study by Ali and colleagues [42]. Test protocols for the use of zebrafish embryos (FET) in the Organization for Economic Cooperation and Development (OECD) test guideline 236 were tested and applied [65]. As determined by OECD guidelines, the value of LC_{50} is calculated based on lethality, coagulation, lack of somite formation, heartbeat failure, and a lack of detachment of the tail. In contrast, teratogenic effects are used to compute the EC_{50} values [22]. Parng et al. (2002) showed that the toxicity of the drug tested (ethanol) was similar for zebrafish and mammals with log LC_{50} values of 4.0 and 3.8 mg/mL, respectively [66]. Throughout the screening steps, zebrafish bioassays have been used to classify the active fractions based on endpoints developed and tested in other species, which led to identification of the actual effect of extracts on different life stages in zebrafish (Table 2).

Table 2. Zebrafish bioassays and measured endpoints.

Life Stage	Measured Endpoints	References
Egg	Survival, hatching, teratology	[1,60,67]
Larvae	Survival, abnormalities, enzyme activities and behavior	[67–70]
Adults	Survival, behavior, enzyme activities and genotoxicity	[71,72]

Additionally, indicators such as delayed growth, restricted movement, irregular head-trunk angle, scoliosis/flexure, and yolk sac edema have been analyzed in developing zebrafish [73]. However, recent findings indicate that herbal products' metabolism using zebrafish can represent actual outcomes of mammalian methods, while single in vitro methods cannot do so [39]. Table 3 lists some toxicity screening studies of medicinal plants that used zebrafish as the model organism.

Table 3. Existing studies that screened toxicity of medicinal plants using the zebrafish model.

Plant/Family	Common Name	Extraction/Part of Plant	Concentrations Range (μg/mL)	LC_{50} (μg/mL)	Characteristics of Effect	Reference
Acorus calamus/Acoraceae	Sweet flag	Ethanol/rhizome	n.a.	n.a.	Fish treated with this extract returned to normal condition and revealed a decrease level in superoxide dismutase and decreased mitochondrial volume and cell viability.	[74]
Achyranthes aspera/Amaranthaceae	Chaff-flower, prickly chaff flower, devil's horsewhip, burweed	Ethanol/leaf	n.a.	n.a.	Fish treated with this extract returned to normal condition and revealed decrease in superoxide dismutase level and decreased mitochondrial and cell viability.	[74]
Achyranthes bidentate/Amaranthaceae	Ox knee, niu xi (Chinese)	Aqueous/root	0–30.01 μg/mL	No mortality observed	The extract did not exhibit noticeable lethal or severe side effects on zebrafish larvae/embryos.	[75]
Allium carinatum/Amaryllidaceae	Keeled garlic, witch's garlic	70% aqueous methanol/whole plant	1–60 μg/mL	55.8 μg/mL	The extract reduced developmental toxicity and neutropenia. No teratogenic effects were seen.	[76]
Allium flavum/Amaryllidaceae	The small yellow onion, yellow-flowered garlic	70% aqueous methanol/whole plant	1–60 μg/mL	50.3 μg/mL	The extract reduced developmental toxicity and neutropenia. No teratogenic effects were seen.	[76]
Andrographis paniculata/Acanthaceae	Green chiretta, creat	Aqueous/leaf	0–10,000 μg/mL	525.5 μg/mL (48 hpf) 525.6 μg/mL (96 hpf)	Morphological defects were observed at 96 hpf after exposure to teratogen concentration.	[2]
Annona squamosa/Annonaceae	Sugar apple, sweetsop	Ethanol/young leaf	100–1000 μg/mL	n.a.	Late hatching process and morphological deformations occurred at higher concentrations up to 800 μg.	[77]
Avicennia marina/Acanthaceae	Grey mangrove or white mangrove	Methanol/bark, leaf, stem, flower and fruit	50–100 μg/mL	n.a.	n.a.	[78]
Azadirachta indica/Meliaceae	Neem	Ethanol (cold)/fruit	1000–5000 μg/mL	n.a.	Not toxic to adult zebrafish and did not alter the locomotor system.	[79]
Bougainvillea galbra/Nyctaginaceae	The lesser bougainvillea, paperflower	Aqueous/bracts	1–300 μg/mL	85.51 μg/mL	Yolk sac edema was observed at different concentrations. Hypopigmentation in the embryo was caused by the purple bract extract at 30 μg/mL. Overall, this extract generally showed moderate embryo toxicity.	[80]

Table 3. *Cont.*

Plant/Family	Common Name	Extraction/Part of Plant	Concentrations Range (μg/mL)	LC_{50} (μg/mL)	Characteristics of Effect	Reference
Bixa Orellana/Bixaceae	Achiote	Hot water/leaf	50–10,000 μg/mL	n.a.	Plant extract affected the hatchability of embryos and was linked to delayed development. Also, a coagulated embryo, and tail malformation were observed as teratogenic effects.	[81]
Capsicum chinense/Solanaceae	Habanero type pepper	Ethanol/fruit	0.39–100 μg/mL	39.7 ± 2.1 μg/mL	The embryo demonstrated late development represented by an absence of pigmentation of the tail, and irregular formation of the somites demonstrated the noticeable tail curve with a wide end.	[82]
Carthamus tinctorius/Asteraceae	Safflower (English), Kashefeh (Persian)	Aqueous/flower	40–1000 μg/mL	345,600 μg/L	Hatching inhibition, depressed heart rate, abnormal spontaneous movement, pericardial edema, yolk sac edema, unusual head-trunk tilt, suppression of melanin release, enlarged yolk, and short body length were identified.	[83]
Clinacanthus nutans/Acanthaceae	Sabah snake grass	Hexane/leaf	15.63–500 μg/mL	75.49 μg/mL	Morphological disorders (less pigmentation, tail bending, edema, abnormal yolk sac).	[84]
Cinnamon zeylanicum/Lauraceae	True cinnamon tree, Ceylon cinnamon	Aqueous/bark	0–10,000 μg/mL	985.8 μg/mL (48 hpf) 50.58 μg/mL (96 hpf)	This plant showed minimum embryotoxicity on zebrafish. In 72 hpf embryos, higher concentration of extract provoked yolk sac and pericardial edema.	[2]
Croton tiglium/Euphorbiaceae	Purging croton, Jamaal gota (Hindi)	Aqueous/seed	4000–24,000 μg/mL	11,880 μg/mL (24 h) 8160 μg/mL (48 h)	Signs of stress, enhanced respiratory rate, jerky movements, loss of equilibrium, circular swimming before losing balance.	[10]
Curcuma longa/Zingiberaceae	Turmeric	Methanol/rhizome	7.8–125 μg/mL	92.42 μg/mL (24 hpf) 79.20 μg/mL (48 hpf) 68.32 μg/mL (72 hpf) 56.68 μg/mL (96 hpf) 55.90 μg/mL (120 hpf)	Physical anatomy deformities as teratogenic effects; tail bending, yolk sac edema extension, and a curved trunk were observed at high concentration among hatched embryos.	[1]
Curcuma longa/Zingiberaceae	Turmeric	Methanol/rhizome	No concentration ranges were mentioned (10 μg/mL used)	n.a.	No visible symptoms of toxicity were observed: no change in heart rate, body impairment, absence or delay in reaction to tactile stimuli, or mortality.	[85]

Table 3. *Cont.*

Plant/Family	Common Name	Extraction/Part of Plant	Concentrations Range (µg/mL)	LC_{50} (µg/mL)	Characteristics of Effect	Reference
Curcuma xanthorrhiza/Zingiberaceae	Javanese ginger, Temulawak	Ethanol/rhizome	100–500 µg/mL	180.52 µg/mL at 48 h 79.55 µg/mL at 96 h	Major malformations of the pericardial edema of embryos at a concentration of 100 µg/mL, yolk sac edema and tail malformation also were observed.	[43]
	Javanese ginger, Temulawak	Aqueous/rhizome	0–10,000 µg/mL	748.6 µg/mL (48 hpf) 703.7 µg/mL (96 hpf)	Decreased survival rate, organ deformity, unusual heartbeat, and delayed hatching rates.	[2]
Cynodon dactylon/Poaceae	Scutch grass	Hexane, chloroform, acetone, methanol/n.a.	10–100 µg/mL	32.6 µg/mL	The period of systole and diastole was decreased by the methanol extract.	[86]
Derris elliptica/Fabaceae	Tuba root (Indonesia), opay (Philippines)	Aqueous/leaf	50 (0.05%), 500 (0.5%) µL/mL	n.a.	The concentration of 0.5% led to the early death of embryos: unformed head, unformed tail, coagulation, and death. A delay in growth and restricted movement compared to the control group were present at 0.05%.	[53]
Dieffenbachia amoena/Araceae	Spotted dumbcane	Aqueous/leaf	0–10,000 µg/mL	1190 µg/mL	Growth retardation, no heartbeat was noticed at 500 µg/mL and higher concentrations (cardiotoxicity).	[87]
Diospyros discolor/Ebanaceae	Velvet persimmon, velvet apple, mabolo tree	Aqueous/leaf	0.05–10%	1%	Tail malformation, delayed growth, head abnormality, yolk malformations, and abdominal swelling were observed.	[88]
Eclipta prostrata/Asteraceae	False daisy, yerba de tago, Karisalankanni and bhringraj	Aqueous/leaf	0.01–40%	1%	At the concentration of 0.01% v/v and 1% v/v, did not observe any mortality and abnormality	[89]
Enydra fluctuans/Asteraceae	Marsh herb, water cress	Ethanol/leaf	12,500–400,000 µg/mL	24 h: 204,132 µg/L 48 h: 170,513 µg/L 72 h: 139,478 µg/L 96 h: 92,956 µg/L	The death rate was found as the exposure period was extended from 24 to 96 h, the median lethal concentration decreased. A negative relationship between exposure time and LC_{50} was found.	[90]
Eugenia polyantha/Myrthaceae	Bay leaf	Aqueous/leaf	0–10,000 µg/mL	921.2 µg/mL (48 hpf) 60.39 µg/mL (96 hpf)	Decreased survival rate, organ deformity, unusual heartbeat, and delayed hatching rates.	[2]
Euphorbia pekinensis Rupr/Euphorbiaceae	The Peking spurge, daji (Chinese)	Aqueous/commercial raw herb	100–250 µg/mL	250–300 µg/mL (previous study)	n.a.	[91]

Table 3. *Cont.*

Plant/Family	Common Name	Extraction/Part of Plant	Concentrations Range (μg/mL)	LC_{50} (μg/mL)	Characteristics of Effect	Reference
Euodia rutaecarpa (Tetradium ruticarpum)/Rutaceae	Goshuyu (Japanese), Bee tree	Evodiamine (Commercial product)	0.05–1.6 μg/mL	0.354 μg/mL (LC_{10})	Evodiamine (bioactive compound) had a 10% lethality at 354 ng/mL and caused heart defect, changes in heartbeat and blood flow, and pericardial malformation. Also, it could cause cardiovascular side effects involving oxidative stress.	[92]
Excoecaria agallocha/Euphorbiaceae	Back mangrove	Methanol/bark, leaf and stem	50–100 μg/mL	n.a.	Higher concentrations induced intense impacts on embryo melanogenesis. Those concentrations also reduced the eye melanin material of the embryo.	[78]
Ficus glomerate/Moraceae	Cluster fig tree, Indian fig tree, goolar fig	Hot aqueous/leaf	125–2000 μg/mL	239.88 μg/mL	Lower hatchability, body size, heartbeat rate, and morphological growth defects of the embryo were exhibited.	[73]
Garcinia hanburyi/Clusiaceae	Siam gamboge, Hanbury's Garcinia	Commercial product/dried yellow resin	0.5–1.0 μM	1.76 μM	Gambogic acid triggered a lack of fin developmental in the zebrafish embryo.	[93]
Garcinia mangostana (Xanthone crude extract)/Clusiaceae	Mangosteen	Mixture of acetone and water (80:20)	7.81–250 μg/mL	15.63 μg/mL	Among surviving zebrafish, no deformities were found.	[28]
Geissospermum reticulatum/Apocynaceae	Challua caspi (Kichwa language)	Ethanol/bark	0.1 μg/mL	n.a.	This plant was nontoxic and caused no deformations to zebrafish, even at high concentrations.	[94]
Himatanthus drasticus/Apocynaceae	Janaguba milk (latex)	Commercial product	500–1500 μg/mL	1188.54 μg/mL	This plant can be considered to have low toxicity. No teratogenic effect was observed.	[95]
Hylocereus polyrhizus/Cactaceae	Pitaya	Ethanol; water solution (70:30, v/v)/peel & pulp	100–1000 μg/mL and 20 μL)	>1000 μg/mL (96 h)	The results indicated the pulp and peel of this plant were non-toxic.	[96]
Leonurus japonicus/Lamiaceae	Oriental/Chinese motherwort	Essential oil/the aerial parts	6.25–100 μg/mL	1.67 ± 0.23 μg/mL	Motherwort essential oil was toxic to embryos. Morphological abnormalities: partial or complete absence of eye development, yolk sac oedema, curved spine, tail deformities, scattered haemorrhages in the oedematous yolk sac, incomplete heart development, and pericardial oedema.	[97]

Table 3. *Cont.*

Plant/Family	Common Name	Extraction/Part of Plant	Concentrations Range (μg/mL)	LC_{50} (μg/mL)	Characteristics of Effect	Reference
Ligusticum chuanxiong/Apiaceae	Szechuan lovage	A protein-containing polysaccharide/rhizome	0–800,000 μg/L	965 μg/mL	No significant morphological anomalies were found.	[98]
Maerua subcordata (Gilg) DeWolf/Capparaceae		Methanol/fruit, leaf, root tuber, and seed	150,000–1,500,000 μg/L	209,000 μg/L	≤5% sub-lethal abnormalities (signs of malformation of the heart).	[4]
Marsdenia tenacissima (Xiaoaiping extract)/Asclepiadaceae	Murva (Hindi), Tong-Guan-Teng (Chinese)	Commercial product	0–3200 μg/mL	2660 μg/mL (24 hpf) 2310 μg/mL (48 hpf) 1920 μg/mL (72 hpf) 1910 μg/mL (96 hpf) 1790 μg/mL (120 hpf)	Severe malformation such as impairment of the swim bladder, retaining the yolk, pericardial oedema, and tail curvature. Histopathological study: Xiaoaiping induced lesion of the liver, muscle, and heart.	[99]
Millettia pachycarpa/Fabaceae	Bokol-bih, Holsoi, Bokoa-bih (Assamese or Hindi)	Aqueous/root	1–7.5 μg/mL	4.276 μg/mL	There were a few developmental anomalies, such as yolk sac oedema, curvature of the spinal cord, pericardial oedema, swim bladder swelling, reduced heart rate, and late hatching. This plant extract is also linked to reactive oxygen species and apoptosis, which induced embryonic mortality and toxicity (identified in trunk, brain, and tail).	[6]
Momordica charantia/Cucurbitaceae	Bitter melon, bitter gourd, bitter apple	Methanol/fruit, seed	1–400 μg/mL	50 μg/mL (only seed extract)	Although it affected the heartbeat, the fruit extract is considered to be harmless and no mortality was observed. The seed extract demonstrated a slight level of developmental failure and had significant cardiac toxicity.	[100]
		Hot and cold aqueous/leaf	15.625–1000 μg/mL	144.54 μg/mL: Hot aqueous Chinese extract 199.53 μg/mL: Hot aqueous Indian extract 251.19 μg/mL: Cold aqueous Chinese extract	For all samples the heart beat differed at higher concentrations. Also, mild toxicity was observed, although consumption of this plant is assumed to be safe.	[12]

Table 3. *Cont.*

Plant/Family	Common Name	Extraction/Part of Plant	Concentrations Range (μg/mL)	LC_{50} (μg/mL)	Characteristics of Effect	Reference
Moringa oleifera/Moringaceae	Drumstick tree, Moringa	Aqueous/leaf	0.05–3%	3.0% at 24 h	Teratogenic effects such as bent back, tip of tail bent, oedema in the yolk sac, and scoliosis. Delayed development and morphological deformities of the embryo were observed.	[101]
		Hot water/leaf and bark	0.3–6 μg/mL	1.5 μg/mL at 36 h (leaves); 3 μg/mL at 36 h (bark)	Even low concentration was embryo-toxic and had teratogenic impacts on the developing embryos.	[102]
		Essential oil/seed	0.1–1000 μg/mL	21.24 ± 0.44 μg/mL	High concentrations (≥ 50 μg/mL) caused 100% mortality of embryos. The process of angiogenic blood vessel formation in zebrafish embryos was substantially disturbed by the seed oil.	[103]
		Aqueous/seed	12.5–200 μg/mL	190 μg/mL (48 h) 133 μg/mL (72 h) 49 μg/mL (96 h)	No toxic endpoint was seen after 24 h, hatching was prolonged, and the larval period at 96 h compared with the control was decreased.	[104]
Moringa peregrina/Moringaceae	Miracle tree	Essential oil/seed	0.1–1000 μg/mL	25.11 ± 0.547 μg/mL	High concentrations (≥ 50 μg/mL) caused 100% mortality of embryos. The process of angiogenic blood vessel formation in zebrafish embryos was substantially disturbed by the seed oil.	[103]
Olea europaea/Oleaceae	Olive	Raw and polar fraction (raw oil: commercial)/fruit	1–100%	Raw oil: 11.98% (EC_{50}) Polar fraction: 61.87 (EC_{50})	Abnormalities observed included non-hatching, pericardial oedema, and blood accumulation, and more were expressed at a concentration of 11.98%. Raw oil induced 50% of developmental abnormalities compared with the polar fraction. After 96 h, strong developmental retardations such as developmental delay and absence/prolongation of pigmentation formation were observed.	[105]
Onosma dichroantha/Boraginaceae	n.a.	Cyclohexane, ethyl acetate, methanol/root	0.02–50 μg/mL	n.a.	n.a.	[106]
Orthosiphon stamineus/Lamiaceae	Java tea	Aqueous/whole plant	0–10,000 μg/mL	1685 μg/mL (48 hpf) 1685 μg/mL (96 hpf)	n.a.	[2]

Table 3. *Cont.*

Plant/Family	Common Name	Extraction/Part of Plant	Concentrations Range (μg/mL)	LC_{50} (μg/mL)	Characteristics of Effect	Reference
Palicourea deflexa/Rubiaceae	n.a.	Methanol (fraction)/leaf	1–100 μg/mL	72.18 μg/mL	At high concentration (100 μg/mL), some possible anomalies, such as no development of somites, decreased pigmentation, and formation of oedema were observed. Lethality only appeared after 96 h.	[52]
Palmaria palmata/Archaeplastida	Dulse, dillisk (dilsk), red dulse, sea lettuce flakes (creathnach)	Crude protein (hydrolysate)/leaf	1–10,000 μg/mL	n.a.	At high concentration (10 mg/mL), larvae started developing spinal curvature and a deformed yolk sac. At 5 mg/mL, larvae appeared to have a swollen yolk sac and developed a curved spine.	[51]
Passiflora caerulea/Passifloraceae	Bluecrown Passionflower	Aqueous/leaf	40–120 μg/mL	80 μg/mL (40–120 μg/mL)	Mortality increased based on concentration and delay in hatching. No apparent signs of abnormal growth or morphology were detected in embryos at 96 hpf.	[107]
Peucedanum alsaticum/Apiaceae	n.a.	n-heptane, ethyl acetate, methanol, water/fruit	50–400 μg/mL	n.a.	n.a.	[33]
Phyllanthus niruri/Phyllanthaceae	Gale of the wind, stonebreaker	Aqueous/leaf	0.05–10%	-	Tail malformation and coagulation were detected as the most remarkable toxic effect from the extract; dose-dependent impacts on the heartbeat and hatchability of the embryo also were detected.	[108]
Piper betle/Piperaceae	Betel, Ikmo, Daun sirih (Malay), Paan (Hindi)	Hot water/leaf	50–10,000 μg/mL	-	The plant extract affected the hatchability of embryos and was linked to delayed development. Also, a coagulated embryo and tail malformation were observed as a teratogenic effect.	[81]
Polygonum multiflorum/Polygonaceae	Tuber fleeceflower	n-hexane, ethyl acetate, water/root	0–175,000 μg/L	TIs: 1430 (ethyl acetate), 630 μg/L	40 and 105 mg/L concentrations (all extracts) induced yolk sac oedema (heart oedema), hemovascular defects, abnormal trunk, and necrosis. The water extract induced synthesis of melanin in zebrafish through stimulation of tyrosinase.	[109]

Table 3. *Cont.*

Plant/Family	Common Name	Extraction/Part of Plant	Concentrations Range (μg/mL)	LC_{50} (μg/mL)	Characteristics of Effect	Reference
		Water, ethanol, methanol, acetone/dried roots	0–10,000 μg/mL	129.4 ± 2.7 μg/mL (water) 58.2 ± 1.9 μg/mL (30% ethanol) 39.8 ± 0.8 μg/mL (50% ethanol) 27.9 ± 2.3 μg/mL (70% ethanol) 25.5 ± 2.0 μg/mL (95% ethanol) 48.8 ± 2.8 μg/mL (methanol) 23.6 ± 2.2 μg/mL (acetone)	Teratogenic effects included coagulated embryos, lack of development of the somite, absence of tail detachment and heartbeat, and malformation of the notochord	[110]
Psoralea corylifolia/Fabaceae	Babchi	Commercial product (Psoralen)	0–6.70 μg/mL	3.40 μg/mL (LC_{50}) 2.52 μg/mL (LC_{10}) 1.98 μg/mL (LC_1)	Psoralen therapy caused hatching rate and body size to decrease and the abnormality rate of zebrafish to increase significantly. Yolk retention, pericardial oedema, swim-bladder malformation, and curved body form were observed.	[111]
Punica granatum/Lythraceae	pomegranate	Ethanol/peel	100–250 μg/mL	196,037 ± 9.2 μg/mL	No issues with the reproductive organs, heart, and androgen hormones were detected.	[112]
Rhizosphora apiculata/Rhizosphoraceae	Mangrove	Methanol/bark, leaf, stem and root	50–100 μg/mL	n.a.	Higher concentrations of the extract from bark, leaf, and stem did not cause mortality.	[78]
Salvia miltiorrhiza/Lamiaceae	Red sage	Commercial product/Tanshinone IIA (diterpene quinone)/root	0.44–7.06 μg/mL	5.44 μg/mL (72 hpf) 3.77 μg/mL (96 hpf)	Tan-IIA showed potential cardiotoxicity and growth inhibition in zebrafish embryos. Scoliosis, tail deformity, and pericardium oedema were the primary signs of teratogenicity.	[45]
Sida acuta/Malvaceae	Wireweed	Hexane, chloroform, acetone, methanol/n.a.	10–100 μg/mL	20.9 μg/mL	The methanolic extract led to heartbeat rate reductions (more significant than nebivolol as a positive control). This study reported no teratogenic effects.	[86]

Table 3. *Cont.*

Plant/Family	Common Name	Extraction/Part of Plant	Concentrations Range (µg/mL)	LC_{50} (µg/mL)	Characteristics of Effect	Reference
Solidago canadensis/Asteraceae	Canada goldenrod, Canadian goldenrod	Ethanol/leaf	0–500µg/mL	0.42 ± 0.03 µg/mL (24 h) 0.33 ± 0.04 µg/mL (48 h) 0.32 ± 0.03 µg/mL (72 h)	This paper only mentioned that low toxicity on zebrafish was observed.	[113]
Sonneratia alba/Lythraceae	Mangrove tree	Methanol/bark and leaf	50–100 µg/mL	-	-	[78]
Spilanthes acmella/Asteraceae	Paracress, toothache plant, Sichuan buttons, buzz buttons, tingflowers, electric daisy Kradhuawean (Thai)	Aqueous/leaf	0.01–20%	n.a.	No mortality was observed in the highest concentration test.	[89]
Spondias mombin/Anacardiaceae	Yellow mombin, hog plum	Hydroethanolic/leaf	1–9 g/kg (LD_{50}); 25,000–75,000 µg/L (LC_{50})	4.515 g/kg (LD_{50}: 48 h: immersion) 49.86 µg/mL (LC_{50}: 48 h: oral)	There were no recorded teratogenic effects.	[114]
Streblus asper/Moraceae	Siamese rough bush, khoi, serut, thoothbrush tree	Methanol/bark	50–100 µg/mL	2,000,000 µg/mL	Slight oedema of heart muscles at 100 µg/mL. No other severe malformations were observed.	[115]
Sutherlandia frutescens/Fabaceae	Cancer bush, balloon pea, sutherlandia	Aqueous and 80% ethanol	5–50 µg/mL	30.00 µg/mL	Both extracts showed bleeding and pericardial cyst development at high concentration, but the aqueous extract was less toxic to larvae.	[8]
		Aqueous, ethanol/aerial part	5–300 µg/mL	297.57 µg/mL (aqueous) 40.54 µg/mL (ethanol)	The high concentration of both extracts inhibited hatching rate and mortality.	[116]
Terminalia chebula/Combretaceae	Myrobalan	Ethanol/fruit, leaf, root	n.a.	n.a.	At 50 µg/L, fish treated with this extract came back to normal condition and showed a related decrease in superoxide dismutase and decreased mitochondria and cell viability.	[74]
Tetrapterys (Melanolepis) multiglandulosa/Malpighiaceae	Pakalkal (Tagalog)	Dichloromethane, hexane, ethyl acetate, methanol and water/leaf	n.a.	200 µg/mL (methanol extract)	A significant ecdysteroid was the most toxic in the zebrafish.	[117]

Table 3. *Cont.*

Plant/Family	Common Name	Extraction/Part of Plant	Concentrations Range (μg/mL)	LC_{50} (μg/mL)	Characteristics of Effect	Reference
Tinospora cordifolia/Menispermaceae	Heart-leaved moonseed, gaduchi, guduchi, giloy, Makabuhay (Filipino)	Aqueous/leaf and bark	0.01–10%	1% (leaves), 10% (barks)	Abnormalities of head and tail, delayed development, reduced mobility, stunted tail, and scoliosis/flexure were observed.	[118]
Thuja orientalis/Cupressacae	Oriental Arbor-vitae	Ethanol/leaf	150–2400 μg/mL	702.9 μg/mL	At 2.4 mg/mL embryos were coagulated; at 1.2 mg/mL they did not hatch and oedema was observed; 0.6 mg/mL resulted in skeletal deformities; at 0.3 mg/mL slight oedema was observed; at 0.15 mg/mL normal growth was observed.	[50]
Trapa natans/Lythraceae	Water chestnut, water caltrop	Acetone, methanol and ethyl acetate/leaf	100,000–1,000,000 μg/mL	50 μg/mL (methanol) 40 μg/mL (acetone) 30 μg/mL (ethyl acetate)	Methanol and acetone extracts did not show toxicity to zebrafish.	[119]
Tripterygium wilfordii/Celasteraceae	Thunder god vine, léi gōng téng (Mandarin)	Purchased commercially each compound	1/10, 1/5, 1/2 LC_{50}	Taxol: 0.10, 0.21, 0.53 μg/mL; Gambogic acid 0.03, 0.05, 0.13 μg/mL; Triptolide: 0.04, 0.07, 0.18; Auranofin: 0.84, 1.68, 4.21 μg/mL; Mycophenolic acid: 0.44, 0.95, 2.22; Curcumin: 0.43, 0.87, 2.17 μg/mL; Thalidomide: 0.71, 1.43, 3.56 μg/mL	From those 7 compounds, gambogic acid followed by taxol showed high anticancer activity	[120]
Zanthoxylum sp./Rutaceae	n.a.	Commercial product; zanthoxylum	0–1000 μg/mL	81.18 μg/mL	No teratogenic effects were reported.	[121]

Days post-fertilization: dpf; Hours post-fertilization: hpf; n.a.: not applicable.

3. The Most Relevant Behavioral Effects and Basic Methodology Based on Toxicological Assessment Observed in Zebrafish Larvae

Previously, zebrafish were used for toxicity evaluations of agrochemicals, but recently they have been used for toxicity assessments for therapeutic compounds [11]. Of note, the zebrafish response is also a sensitive predictor of irregular toxicity changes [34]. In particular, specific tasks have been created or changed to test the behavior of zebrafish compared to rodent models. Quite basic swimming steps and the ability to capture/consume can often be helpful [38]. The creation of behavioral tests that can employ repeated tests in a short window (i.e., within 24 h) may prove to be the perfect compromise, and some researchers have begun designing tasks that use multiple tests to measure their acquisition during a single session [38]. If one particular mode of action is to be detected, various experimental parameters can contribute to incoherent actions (hypo-or hyperactivity) responses of zebrafish when determining toxicity [122]. As a consequence, several behavioral evaluation methods have been developed to address different behavioral endpoints, including spontaneous tail coiling, photomotor response (PMR), locomotor response (LMR), and the alternating light/dark-induced locomotor response (LMR-L/D) [122], which we will not discuss further in this review.

Zebrafish are easy to control in toxicology screening tests using different platforms (exposure format) for detecting parameters. The type of exposure format is classified into ten categories, six of which are useful. Zebrafish body length at 5 dpf is about 4 mm, so a multiple-well plate can be used to incubate zebrafish larvae [17]. However, embryos cover a wide range of sizes based on the performing test, so 6-well to 96-well transparent plates can be used. Analytical experiments that use microplates, such as cell-based evaluation, can be conducted and applied to high-performance primary drug screening [123]. Furthermore, zebrafish larvae fit onto 96 or 386-well microplates [24]. Because the zebrafish embryo is tiny, can be handled easily [31], and requires only a small number of compounds per test, they can be screened in a 96-well microplate format [123]. However, OECD TG236 suggests using a 24-well plate (with a medium of 2 mL per well), as it overcomes the problem of high concentration developing in conjunction with evaporation [17]. Nevertheless, TG236 specifically checks for lethality and only considers developmental defects indirectly [17]. Table 4 lists several studies that analyzed embryotoxicity and apoptotic induction.

Table 4. Description of toxicological aspects of various studies.

Exposure Format	Number	Exposure Stage	Exposure Time	Parameters Evaluated (Endpoint)	Compound(s)	Reference
96-well plates	1/well	24 hpf	96 h	Embryotoxicity	Armatamide, rubecenamide, lemairamin, rubemamine, and zanthosine	[121]
	1/well (10/group)	6 hpf	96 h	Embryotoxicity and teratogenic effects	Palmitic acid, phytol, hexadecanoic acid, 1-monopalmitin, stigmast-5-ene, pentadecanoic acid, heptadecanoic acid, 1-linolenoylglycerol, and stigmasterol	[84]
	30/group (90 embryos)	n.a.	96 h	Teratogenic effects and embryotoxicity	Alkaloids, flavonoids, phenolics, and saponins	[73]
	1/well	24 hpf	72 h	Acute toxicity assay	Polysaccharides, peptides, protein, lipids, terpenoids, saponins, phenolics, and sterols	[12]
	12/group	2 hpf	96 h	Embryotoxicity and teratogenic effects	α/β-thujone, thujone	[50]
	n.a.	1 hpf (embryo)	72 h	Cytotoxicity and embryonic toxicity test (survival rate)	Xanthone	[28]
	4/group (per well)	12 hpf	48 h	Embryotoxicity (mortality, hatchability, heartbeat rate, and malformation), teratogenic activity	Ascorbic acid, flavonoids, phenolics, and carotenoids	[102]
	1/well	24 hpf	96 h	Embryotoxicity and teratogenic effects	Capsaicin, ethyl palmitate, ethyl linoleate, dihydrocapsaicin, and docosenamide	[84]
	36/group	2.5–3.7 hpf	72 h	Developmental toxicity and behavioural safety	Roots: glycoside, siaraside (pregnane glycoside); Stem bark: α-amyrin, acetate, lupeol, and β-sitosterol	[115]
	12/group	4 cell embryo stage	144 h	Embryotoxicity (heart rate and hatching time)	Spilanthal, and flavonoid	[89]
24-well plates	12 fertilized eggs/group	6 hpf	120 h	Embryotoxicity and teratogenic effects (hatching rate, heartbeat, otolith, blood circulation, tail detachment, and motility)	Catechin, epicatechin, and naringenin	[1]
	20/concentration	6 hpf	96 h	Embryotoxicity		[76]
	10/well	6 hpf	96 h	Embryotoxicity	Methanol extract: phenolic compound (p-heydroxybenzoic acid), phenolic acid, flavonoid, and flavonoid glycoside	[119]
	10/group (30 per tested sample)	64-cell (egg)	96 h	Embryotoxicity	Phenolic compounds: tyrosol, catechol. Hydroxytyrosol, gallic acid, and ascorbic acid.	[105]

Table 4. *Cont.*

Exposure Format	Number	Exposure Stage	Exposure Time	Parameters Evaluated (Endpoint)	Compound(s)	Reference
	1/well	4 hpf	96 h	Embryotoxicity	Pyrrolizidine alkaloids such as senecionine or senecivernine; phenolic compounds, terpenoids, and iridoids, glucosinolates (glucolepidin and glucobrassicin), alkaloids or amines (stachydrine and trigonelline)	[4]
	4/group	12 hpf	48 h	Embryotoxicity, teratogenic effects	n.a.	[108]
	30/group	4 hpf	96 h	Embryotoxicity	Phenols, and polyphenols	[107]
	10/well	16 cell stage	72 h	Embryotoxicity	n.a.	[82]
	6/well	24 hpf	96 h	Embryotoxicity	Sutherlandin, and sutherlandioside	[116]
	6/group (per well) (n =18)	24 hpf	216 h	Developmental toxicity and cardiotoxicity	No pure compound isolated	[8]
	15/well	2 hpf	96 h	Embryotoxicity, teratogenic effects	Alkaloids (stachydrine)	[97]
	4/well	-	48 h	Embryotoxicity, teratogenic effects	n.a.	[88]
	n.a.	6 hpf	96 h	Acute toxicity in vivo (zebrafish), *in silico* (computational chemistry)	Brevifolin, tannin, saponin, quinone, steroid/triterpenoid, and flavonoid	[112]
	15/well	4 hpf	96 h	Embryotoxicity, teratogenic effects	Psoralen	[111]
	40/group	16-cell stage	96 h	Zebrafish embryo acute toxicity (ZFET)	Emodin, rhein, physcion (dihydroxyanthraquinone), chrysophanol, aloe-emodin, rhaponticin, polygonumnolide, and 2,3,5,4′-tetrahydroxystilbene-2-O-β-D-glucopyranosede, and resveratrol-2-O-β-D-glucopyranoside	[110]
	30/group (10/well)	9 hpf	55 h	Anti-melanogenic and embryotoxicity	Lupeol (triterpenoid), hydrolysable, and 3,3′-di-*O*-methyl ellagic acid (bark of *S. alba*)	[78]
	1/well (20 embryos per concentration)	16-cell stage	168 h	Embryotoxicity	n.a.	[104]
	20 eggs/group (1 embryo per well	1 hpf (embryo)	96 h	Embryotoxicity, teratogenic effects (body axis, head, tail, blood circulation, eyes, heart, pigmentation, somite, and yolk sac)	Curcumin, desmethoxycurcumin, and bisdemethoxycurcumin	[43]

Table 4. *Cont.*

Exposure Format	Number	Exposure Stage	Exposure Time	Parameters Evaluated (Endpoint)	Compound(s)	Reference
	1/well	32-cell stage	96 h	Embryotoxicity	harman-3-carboxylic acid, βc alkaloids: β-carboline alkaloid	[52]
	30 embryos/well	2 hpf	96 h	Toxicity on cell and zebrafish (coagulation of eggs, tail detachment, presence of heartbeat, and hatching rate)	Alkanes, alkenes, amino acid, amines, and amides	[2]
	20/well	6 hpf	72 h	–	3-O-propionyl-5, 10, 14-O-triacetyl-8-O-(20 -methyl- butanoyl)-cyclomyrsinol, and mirsinance type diterpene	[91]
	20/well	2 hpf	70	Embryotoxicity, teratogenic effects	Gambogic acid	[93]
	5/group (per well)	6 hpf	120 h	Developmental toxicity and apoptosis induction	Rotenone, and saponin	[6]
	10/well	24 hpf	96 h	Embryotoxicity	n.a.	[86]
	6/well	7 dpf	24 h	Determination of maximum tolerated concentration (MTC)	ar-turmerone, α, β-turmerone, and α-atlantone	[85]
12-well plates	4/group (per vial)	1 hpf (embryo)	48 h	Embryotoxicity, teratogenic effects	Beta-momorcharin	[101]
	20/group	6 hpf	96 h	Embryotoxicity, teratogenic effects	Quinochalcones, and flavonoids	[83]
	4/well	–	48 h	Embryotoxicity, teratogenic effects	Flavonoids, terpenoids, cardiac glycosides, saponins, and tannins	[81]
	4/group (per well)	12 hpf	48 h	Embryotoxicity, teratogenic effects	Alkaloids, di-terpenoid lactones, glycosides, steroids, sesquiterpenoid, phenolics, and aliphatic	[118]
	3/well	n.a.	48 h	Embryotoxicity and teratogenicity	Rotenone	[53]
	10/well	48 hpf	48 h	Embryotoxicity	Peptides	[51]
	12/group	egg	120 h	Developmental toxicity (mortality, hatching rate, oedema, and malformations)	Alkaloids, flavonoids, terpene derivatives, glycosides, novel diazepine, and squalene	[77]
Glass beaker	24/group	n.a.	96 h	Acute toxicity test	Phyllocactins, betanins, 2′-O-apiosyl-malonyl-betanin, 6′-O-malonyl-2-descarboxybetanin, flavonoids, isorhamnetin triglycosides, quercetin-3-O-hexoside, and carbohydrates.	[96]

Table 4. *Cont.*

Exposure Format	Number	Exposure Stage	Exposure Time	Parameters Evaluated (Endpoint)	Compound(s)	Reference
	15/group	0 hpf	48 h	Acute toxicity test, histopathology	Fatty acids, lupeol, and terpenes	[114]
Glass Petri dish	6/group	adult	96 h	Fish acute toxicity	Alkaloids, phenol, and flavonoids with some chemical groups of alkanes, aromatics, hydroxyls, carbonyls, aldehyde, amines, nitriles, amides and ethers	[78]
	30/Petri dish	6 hpf	72 h	Embryotoxicity	*Moringa oleifera* seed oil: 2,2-Bipydidine,3,3-diol, myristoleic acid, palmitoleic acid, stearic acid, oleic acid, arachidic acid, methyl-9-octadecanoate, adrenic acid, erucic acid, behenic acid, decamethylcyclopentasiloxane, α-tocopherol, γ-tocopherol, β-sitosterol, camphenone, hexahydrofarnesylacetone, linoleic acid, gadoleic/eicosenoic acid *Moringa peregrina* seed oil: palmitic acid, stearic acid, oleic acid, arachidic acid, gadoleic/eicosenoic acid, tricosane, erucic acid, behenic acid, α-tocopherol or γ-tocopherol, α-sitosterol, δ-tocopherol, 2-allyl-5-t-butylhydroquinone, isamoxole, hexahydrofarnesylacetone, linoleic acid, and β-sitosterol	[103]
	n.a.	8-cell stage	5 dpf	Embryotoxicity, teratogenic effects	1,2-cyclopentanedione, 2,3-dihydro-3,5-dihydroxy-6-methyl-4h-pyran-4-one, 1,3;2,5-dimethylene-l-rhamnitol, elemol, (-)-selina-4.alpha.,11-diol, and beta-eudesmol	[100]
	20/Petri dish	1–4 cell stage	72 h	Embryotoxicity	Shikonin, β,β-dimethylacrylalkannin, and β,β-dimethylacryl shikonin	[106]
Rectangular glass tank	12/group	n.a.	48 h	Acute toxicity	Fatty acids, esters, alcohols, and the sterol sitosterol	[114]
REKO glass (artificial egg environment)	20 fertilized eggs	4 hpf	48 h	Embryotoxicity, teratogenic effects	Phenolic compounds	[94]
	10/glass tank	n.a.	48 h	Evaluation of genotoxicity and acute toxicity	Phenolic compounds, alkaloids, saponins, terpenoids, carbohydrates, and tannins	[10]
Vial	4/group (per well)	n.a.	48–72 h	Cytotoxicity, toxicity, teratogenicity (hatchability and heartbeat rate)	n.a.	[87]
Vessel	10/group	adult	72 h	Fish acute toxicity	n.a.	[113]

Apoptotic functions are identical in zebrafish and humans [66], and can be easily detected using fluorescent labeling methods [123]. The zebrafish model was used to discover the toxicity of 69 plants in this review, including 88 crude plant extracts, eight polyherbal formulates/commercial products, and two phytocompounds. In terms of exposure format, out of 58 studies described in Table 4, 25 used 24-well plates, and 10 used 96-well plates.

4. Using Zebrafish Embryos to Detect Developmental Toxicity

It seems that the number of zebrafish screening tests to evaluate the toxicity of compounds will keep increasing each year [19]. The use of fish embryos was proposed as an alternative screening method to assess fish's acute toxicity [21,57]. In transparent species, the effects of compounds on different organs, such as the brain, heart, cartilage, liver, intestine, and kidney, were identified without arduous screening [66]. This specific trait also makes it feasible to quickly measure toxicity endpoints of multiple substances [6,124], highlighting the efficacy of toxicity models using zebrafish embryos [66]. For example, after 2005, regular sewage surveillance monitoring of fish embryotoxicity has been made compulsory, substituting conventional fish tests no longer approved for standard whole effluent assessment [62].

From the egg phase, zebrafish embryos can survive by ingesting yolk and can visibly be tested for malformation for a few days in a single well of a microplate [32]. When the embryo is not clumped, has a cardiac beat, has fully shaped body sections, and the tail is detached from the yolk, the standard endpoints of embryogenesis have been said to have occurred [28]. The main characteristics of embryogenesis and the genetic basis of growth in zebrafish have been widely researched [125]. Another significant indicator for assessing toxicity is the hatching rate, and zebrafish embryos start to hatch at about 48 h post-fertilization (hpf) under ordinary conditions [8]. Delayed growth of zebrafish embryos can result in a low hatchability and can thus be among the critical aspects of the sub-lethal effects of plant extracts [53].

The essential cell structure and biochemical similarities between humans and animals enable researchers to use the zebrafish model to forecast the possible impacts of chemicals and other human communities [10]. Furthermore, it is easier to study the biochemical mechanisms of significant zebrafish organs, whereas histological analysis required in the mouse model is more difficult [66]. Testing parameters commonly used for zebrafish include embryo survival rate, lethality, behavior, and organ deformity; researchers have found that zebrafish exhibit a strong dose reaction to toxicity, making it a useful animal model for toxicity screening [123]. Previously, we mentioned FET; most studies use OECD guidelines to measure toxicity of crude extract/fractions, and those guidelines for FET set levels as follows: dangerous (10 mg/L $< LC_{50} <$ 100 mg/L), toxic (1 mg/L $< LC_{50} <$ 10 mg/L), and carcinogenic ($LC_{50} <$ 1 mg/L) [126].

Zebrafish developmental toxicity testing is a type of FET (FET can be used for any fish species), and it is primarily aimed at supplementing developmental toxicity screening in mammals [17]. When detecting developmental toxicity in zebrafish embryos, endpoint parameters are categorized in two ways compared to a standard (negative control). First, lethality is assessed based on coagulation, malformation of somites, absence of heartbeat, and lack of tail detachment. Second, teratogenicity is evaluated based on abnormal eye development, lack of spontaneous movement, unusual heart rate, lack of pigmentation, and edema. The standard shows normal development [60,127].

5. Defects in Zebrafish Organs Found in Developmental Toxicity Studies

In developmental toxicity studies, the mortality of zebrafish embryos occurs before 24 hpf or just after hatching, and cardiac deformities are sometimes detected, such as pericardial edema, abnormal heart form due to edema or aplasia, and irregular heartbeat [4]. The motility of the zebrafish embryo can also be affected by toxins, as exhibited by coagulation, heart failure, and non-development of the yolk embryo's tail [112]. Other factors, such as reactive oxygen species-induced oxidative stress, are also assumed to be linked to abnormal development during embryogenesis [128]. A teratogenicity test is carried out to assess developmental toxicity (i.e., if the compound has teratogenic effects on the embryos). After treatment, the development stages of embryos are observed under an inverted microscope to look for malformations based on the numerical system designed from 5 to 0.5, where 1 indicates severe malformation, 4 indicates slight malformation and 5 is totally normal [50]. The teratogenic score is calculated as follows:

$$\text{Teratogenicity \%} = \left[\frac{\text{Malformation score at each developmental stage}}{\textit{Total}\text{ score}}\right] \times 100 \quad (1)$$

In such analyses, it is essential to specify the observational setup and endpoint evaluation (i.e., which particular behavioral change(s) will be analyzed). Below we discuss the defects to vital organs of zebrafish that can occur during the toxicity assay.

5.1. Heart (Cardiotoxicity)

The zebrafish is an excellent model organism for studying organogenesis and the cardiovascular system [86,129]; thus, it was developed quickly as a cardiovascular research model organism [35]. The first active organ that develops in zebrafish is the heart [73,130,131]. The heartbeat is a primary and relevant sub-lethal endpoint in the embryonic fish toxicology assay [73,130]. It is regularly assessed as an indicator of toxicity in zebrafish embryos [73] and estimates fish metabolic function as a biological factor [6,132]. In zebrafish, the normal embryonic heartbeat of 120–180 beats per minute is close to that of the human heartbeat. At the embryo pharyngula level, the heartbeat pulse can be tracked because the tail is visibly pigmented [102,118]. Ismail et al. reported that the most vulnerable stage for external stimulation (e.g., chemicals, toxicants, and physical pressure) is the early growth stage of embryos [2]. In toxic environments, pericardial edema can be caused by the general response. Pericardial edema in zebrafish embryos can be caused by many different toxicants [133], and changes in heartbeat rate may be a common reaction to toxicant exposure [130]. The toxicity mechanism of a compound is due to activation of Na^+ and K^+ inhibitors, and these Na^+ channels likely trigger the heart rate. An elevated heart rate can overwhelm the myocardium and cause operational and biological damage to the heart of zebrafish [134]. For example, studies of the effects of medicinal plants such as *Andrographis paniculata*, *Cinnamon zeylanicum*, *Curcuma xanthorrhiza*, *Eugenia polyantha*, *Orthosiphon stamineus* [2], *Tinospora cordifolia* [118], *Carthamus tinctorius* [83], and *Euodia retaecara* [92] on zebrafish embryos reported a decreased heartbeat. In contrast, embryos treated with curcumin showed a heartbeat increase [1].

The heart is anteroventral in zebrafish and lies between the operculum and pectoral girdle in the thoracic cavity [35,135]. It is divided into the atrium, sinus venous, ventricle, and bulbous arteriosus, it starts beating at about 22 to 26 hpf [17,136], and produces a full set of ion channels and metabolic regulation [123]. In later stages, the normal function of the heart plays a leading role in the growth of the embryo. An abnormally formed cardiovascular system can result in the unusual overall growth of the animal, severe malformations, and a malfunctioning body [8]. Additionally, when the heart is swollen due to the active compounds in the pericardial sac, cardiac cells become irritated [43]. The following method is used to measure a normalized cardiac rate [120]:

$$\text{\% Normalized cardiac rate} = \frac{\textit{Heart rate (sample test)}}{\text{Heart rate (vehicle)}} \times 100 \quad (2)$$

Zebrafish embryos are also being used in cardiotoxicity (reduction of heartbeat) studies to model various human diseases [11]. The morbidity and mortality of people with cancer are associated with cardiotoxicity [35]. Most common drugs have cardiotoxic effects and affect the heart in zebrafish model systems [86]. A very accurate zebrafish cardiotoxicity assay assessing potential drug toxicity to the human cardiovascular system has been documented [11]. Thus, the effects of cardiotoxicity, such as minimal or no blood flow, may be explained by the fact that a large percentage of apoptotic heart cells lead to underdevelopment of the heart and pericardium, which in turn can cause an unusual heartbeat and a delay in body development (growth retardation) [137,138]. Gao et al. reported that the expression of two heart markers (*amhc* and *vmhc*) reflected the seriousness of heart failure in zebrafish [133].

At day 3 of the formation of zebrafish embryos, the primary and sprout vessels become active, and the blood vessel structure is quite like that in the human body [91]. A single endothelial cell layer surrounded by supporting cells forms the vessels that constitute the circulatory system. Endothelial precursor cells result from vasculogenesis and angiogenesis within the lateral mesoderm of mammal embryos. Vasculogenesis explains the de novo establishment of blood vessels by angioblast coalescence. In comparison, angiogenesis includes the creation of new vessels by earlier developed vessels and is typically described by the sprouting of an endothelial cell [139]. Kinna et al. reported the potential role of zebrafish *crim1* in regulating vascular and somatic development [140].

5.2. Gills

Gills are essential to fish because they are the leading site for gas exchange [141–143], and they engage in osmoregulation [90], ionic control, acid-base balance, and excretion of nitrogen waste in fish [144]. Several studies on the histological organization and physiology of gills in many fishes have been carried out [145]. Enhanced green fluorescent protein (EFGP) is a successful marker for toxic chemicals in transgenic zebrafish. One of the benefits of this zebrafish transgenic model system is that spatio-temporal patterns of EGFP expression can be assessed during initial stages after exposure to a particular toxicant [146]. A study by Seok and colleagues found that the most effective and earliest expression of EGFP was seen in zebrafish gills, indicating that the gill is the most sensitive tissue and is impacted at relatively low metal concentrations [146]. The studies conducted by Rajini et al. [147] showed that visible damage such as inflammatory cell infiltration in gills, minimum aggregation in primary lamellas, secondary lamellae fusion, diffuse epithelial hyperplasia, and multifocal cell mucosal hyperplasia occurred in zebrafish exposed to sublethal concentrations of a combination of pesticides [147]. Generally, in zebrafish, gills exposed to toxicants from 48 to 72 h show fusion, and clubbing of distal lamellar regions [148]. However, for herbal toxicity, this organ has not fully been considered.

5.3. Tail

One of the endpoints of embryonic growth in zebrafish is when the tail is detached entirely from the yolk [28], which can be affected by toxicants. The tail emerges during the larval stages from the primordium in the ventral fin fold, which coincides with a gap in the melanophore line immediately before the back tip of the notochord [149]. Tail kink and tail bending are malformations caused by toxicity. However, tail malformations and spinal axis disabilities can also be affected by predation and a drastic decrease in the food supply [6].

5.4. Yolk Sac

Like human embryos, zebrafish embryos possess a protruding yolk sac, but in zebrafish, it serves as a nutritional reservoir for the embryo [36,43]. Yolk sac edema is a common pathology in toxicity tests of zebrafish [32,150], and it is a sign of reduced nutrient adsorption by the embryo [43]. It may be affected by overhydration of osmoregulation and toxin accumulation in the yolk sac [150]. Malformation in other organ systems that help allocate nutrients may be the trigger for this deficiency. This deficit could also contribute to nutritional absorption, undernourished embryos, and embryo mortality [43]. After maternal exposure, toxicants can be placed in the yolk sac as well [36]. As they are deposited in the yolk, the embryo becomes directly exposed to these toxicants, which allows for definitive exposure timing. Hence, a finite quantity of maternally stored embryo yolk is used in the zebrafish embryo for early feeding, and it provides an understanding of embryo nutrition processes and disruptions that occur due to toxicity exposure [36].

In zebrafish developmental toxicology research, there are numerous classes of yolk phenotypes. Yolk sac edema is quite like yolk edema. Yolk retention refers to decreased mobility/use or malabsorption of the yolk. If the area of the yolk sac is significantly greater in a test fish than in the control fish, it suggests that yolk absorption has been affected. On the other hand, the rapid use of yolk leads to a smaller yolk area and indicates enhanced movement or use of the yolk [36].

5.5. Hatchability

A zebrafish embryo is assumed to have been hatched once it is entirely out of the chorion. Certain drugs and extracts can affect hatchability. Lack of hatchability may imply a growth lag or slow development [102]. For example, zebrafish embryos exposed to *Millettia pachycarpa* extract exhibited a considerable dose-dependent reduction in embryo hatching [6]. Zebrafish embryos start to hatch at 48 hpf in ordinary conditions [8], but percent hatchability after 72 h post-treatment exposure (hpte) is calculated using the following formula [73]:

$$\% \text{ Hatchability } = \frac{\text{No. of hatched embryos}}{\text{Initial no. of embryos}} \times 100 \tag{3}$$

According to the literature, this model to determine the toxicity of crude extracts or bioactive compounds, and toxicological effects can be compared.

6. Chemical Studies of Embryotoxicity

Changes in fish behavior are sensitive indicators of unintended chemical pollution [151]. The zebrafish model is vital in evaluating the effects on embryonic/larval activity of developmental exposure to harmful compounds [38]. Comprehension and monitoring of these considerations and future revision/harmonization of procedures will reduce the uncertainty of the outcomes for chemical hazard assessments [122].

6.1. Salinity

Several early life researches have been carried on the effects of salinity on zebrafish [152]. When larval zebrafish are subjected to mild environmental stressors, severe salinity shifts lead to a quick locomotive reaction [153]. Lee and colleagues showed that larvae (5 dpf) treated with NaCl exhibit dramatically increased locomotive activity and concentration-dependent reactions. A linear relationship between whole-body cortisol and sodium chloride in the concentration range of 5.84 ppt was shown in zebrafish larvae [154].

6.2. Phytochemical Compounds

Yumnamcha et al. [6] reported the presence of saponins, alkaloids, phenolic compounds, and triterpenoids in the aqueous extract of *Millettia pachycarpa*. Potent hemolytic activity was found to lead to acute cell membrane death in zebrafish. The toxic function of saponin in the zebrafish embryo is uncertain. However, prior studies have shown that it can be caused by physicochemical characteristics (surfactant) and/or membranolytic impacts on the chorion, a semipermeable membrane that covers the embryo until it hatches [6,155]. Moreover, according to Liu et al. [156] and Yumnamcha et al. [10], some bioactive compounds such as saponin mixtures and alkaloids can cause damage to DNA. *Artemisia capillaris* was evaluated for its toxicity based on embryotoxicity, and isofraxidine 7-O-(6′-O-p-coumaroyl)-β-glucopyranoside was isolated from this plant. The toxicity, death, and cardiac rates of handled zebrafish embryos are used to diagnose safe and effective concentrations [157].

In contrast to flavone, artificial flavonoids such as kaempherol, 7-hydroxyflavone, 6-methoxyflavon, and 7-methoxyflanone induced considerable toxicity in zebrafish larvae [158]. Additionally, the lack of flavonoids can potentially trigger oxidative stress to cellular molecules, which leads to lower survival rates of zebrafish embryos [12]. Bugel et al. [63] reported that 15 of 24 flavonoids affected at least one or more behavioral and developmental endpoints in zebrafish. However, they focused on two endpoints in their experiment: abnormal spastic behaviors (72 hpf) and alternations in the larval photomotor responses assay (120 hpf). Li et al. [99] exposed zebrafish to Xiaoping at high doses, and many larvae opted to swim at a medium to a low rate, which led to decreased locomotion ability. Changes in larval behavior have been confirmed to be associated with damage to the central nervous system. The function of neural population numbers in animals with a broader central nervous system is being measured using fluorescence and new devices developed over the last decade [159]. In one study, zebrafish embryos subjected to different caffeine dosages demonstrated a decrease in susceptibility to touch-induced movement. Such abnormalities were linked to changes in muscle fibers and axon projections of both primary and secondary motor neurons through immunohistochemistry [160]. Furthermore, toxicological testing may also impact other environmental variables (temperature, water, pH, total hardness or dissolved oxygen) [161].

7. Conclusions

The data collected in this study suggest that zebrafish embryotoxicity tests are able to evaluate drug toxicity and that the zebrafish model offers a suitable replacement for laboratory animals such as rats, mice, and rabbits. As a toxicology model, zebrafish can expose developmental toxicity mechanisms because they are close to mammals. Zebrafish embryos and larvae showed significantly higher susceptibility to toxins than did adult zebrafish. In this review, most of the extracts were polar, such as ethanol, methanol and aqueous extracts, which were used to detect the toxicity and bioactivity. However, the use of the zebrafish model will provide insight into the mechanisms of toxicity of medicinal plants and will help identify and discover new medications for the treatment of human diseases. The zebrafish model is planned as a replacement for models based on higher vertebrate animals to study medicinal plants' toxicity.

Author Contributions: Project administration, V.L. and H.A.; writing-original draft preparation, A.M.C.; writing-review and editing, A.M.C., V.L.; funding acquisition, V.L. All authors have read and agreed to the published version of the manuscript.

Funding: We would like to thank Ministry of Higher Education, Malaysia for funding support from Fundamental Research Grant Scheme, FRGS (203.CIPPT.6711684).

Conflicts of Interest: The authors declare no conflict of interest.

References

1. Alafiatayo, A.A.; Lai, K.S.; Syahida, A.; Mahmood, M.; Shaharuddin, N.A. Phytochemical evaluation, embryotoxicity, and teratogenic effects of *Curcuma longa* extract on zebrafish *(Danio rerio)*. *Evid. Based Complement. Altern. Med.* **2019**, *2019*. [CrossRef]
2. Ismail, H.F.; Hashim, Z.; Soon, W.T.; Ab Rahman, N.S.; Zainudin, A.N.; Abdul Majid, F.A. Comparative study of herbal plants on the phenolic and flavonoid content, antioxidant activities and toxicity on cells and zebrafish embryo. *J. Tradit. Complement. Med.* **2017**, *7*, 452–465. [CrossRef] [PubMed]
3. Falcao, M.A.P.; de Souza, L.S.; Dolabella, S.S.; Guimaraes, A.G.; Walker, C.I.B. Zebrafish as an alternative method for determining the embryo toxicity of plant products: A systematic review. *Environ. Sci. Pollut. Res. Int.* **2018**, *25*, 35015–35026. [CrossRef]
4. Gebrelibanos Hiben, M.; Kamelia, L.; de Haan, L.; Spenkelink, B.; Wesseling, S.; Vervoort, J.; Rietjens, I.M.C.M. Hazard assessment of *Maerua subcordata* (Gilg) DeWolf. for selected endpoints using a battery of *in vitro* tests. *J. Ethnopharmacol.* **2019**, *241*, 111978.
5. Fürst, R.; Zündorf, I. Evidence-based phytotherapy in Europe: Where do we stand? *Planta Med.* **2015**, *81*, 962–967. [CrossRef]
6. Yumnamcha, T.; Roy, D.; Devi, M.D.; Nongthoma, U. Evaluation of developmental toxicity and apoptotic induction of the aqueous extract of *Millettia pachycarpa* using zebrafish as model organism. *Toxicol. Environ. Chem.* **2015**, *97*, 1363–1381. [CrossRef]

7. Little, J.G.; Marsman, D.S.; Baker, T.R.; Mahony, C. In silico approach to safety of botanical dietary supplement ingredients utilizing constituent-level characterization. *Food Chem. Toxicol.* **2017**, *107*, 418–429. [CrossRef] [PubMed]
8. Chen, L.; Xu, M.; Gong, Z.; Zonyane, S.; Xu, S.; Makunga, N.P. Comparative cardio and developmental toxicity induced by the popular medicinal extract of *Sutherlandia frutescens* (L.) R.Br. detected using a zebrafish Tuebingen embryo model. *BMC Complement. Altern. Med.* **2018**, *18*, 273.
9. Mclaughlin, J.L.; Chang, C.J.; Smith, D.L. Bench Top Bioassays for the Discovery of Bioactive Natural Products: An Update. In *Studies in Natural Products Chemistry*; Atta-Ur-Rahaman, Ed.; Elsevier: Amsterdam, The Netherlands, 1991; Volume 9, pp. 383–409.
10. Yumnamcha, T.; Nongthomba, U.; Devi, M.D. Phytochemical screening and evaluation of genotoxicity and acute toxicity of aqueous extract of *Croton tiglium* L. *Int. J. Sci. Res. Publ.* **2014**, *4*, 1–5.
11. Caballero, M.V.; Candiracci, M. Zebrafish as Toxicological model for screening and recapitulate human diseases. *J. Unexplor. Med. Data* **2018**, *3*, 4. [CrossRef]
12. Thiagarajan, S.K.; Krishnan, K.R.; Ei, T.; Shafie, N.H.; Arapoc, D.J.; Bahari, H. Evaluation of the effect of aqueous *Momordica charantia* Linn. extract on zebrafish embryo model through acute toxicity assay assessment. *Evid. Based Complement. Alternat. Med.* **2019**, 9152757. [CrossRef]
13. Bambino, K.; Chu, J. Zebrafish in toxicology and environmental health. *Curr. Top. Dev. Biol.* **2017**, *124*, 331–367.
14. Heideman, W.; Antkiewicz, D.S.; Carney, S.A.; Peterson, R.E. Zebrafish and cardiac toxicology. *Cardiovasc. Toxicol.* **2005**, *5*, 203–214. [CrossRef]
15. Ayub, A.D.; Hock, I.C.; Mat Yusuf, S.N.A.; Abd Kadir, E.; Ngalim, S.H.; Lim, V. Biocompatible disulphide cross-linked sodium alginate derivative nanoparticles for oral colon-targeted drug delivery. *Artif. Cells Nanomed. Biotechnol.* **2019**, *47*, 353–369. [CrossRef]
16. Abuchenari, A.; Hardani, K.; Abazari, S.; Naghdi, F.; Ahmady Keleshteri, M.; Jamavari, A.; Modarresi Chahardehi, A. Clay-reinforced nanocomposites for the slow release of chemical fertilizers and water retention. *J. Compos. Comp.* **2020**, *2*, 85–91. [CrossRef]
17. Nishimura, Y.; Inoue, A.; Sasagawa, S.; Koiwa, J.; Kawaguchi, K.; Kawase, R.; Maruyama, T.; Kim, S.; Tanaka, T. Using zebrafish in systems toxicology for developmental toxicity testing. *Congenit. Anom. Kyoto* **2016**, *56*, 18–27.
18. Han, H.S.; Jang, G.H.; Jun, I.; Seo, H.; Park, J.; Glyn-Jones, S.; Seok, H.K.; Lee, K.H.; Montovani, D.; Kim, Y.C.; et al. Transgenic zebrafish model for quantification and visualization of tissue toxicity caused by alloying elements in newly developed biodegradable metal. *Sci. Rep.* **2018**, *8*, 13818. [CrossRef]
19. Rennekamp, A.J.; Peterson, R.T. 15 years of zebrafish chemical screening. *Curr. Opin. Chem. Biol.* **2015**, *24*, 58–70. [CrossRef]
20. Van der Laan, J.W.; Chapin, R.E.; Haenen, B.; Jacobs, A.C.; Piersma, A. Testing strategies for embryo-fetal toxicity of human pharmaceuticals. Animal models vs. *in vitro* approaches: A workshop report. *Regul. Toxicol. Pharmacol.* **2012**, *63*, 115–123. [CrossRef]
21. Cassar, S.; Adatto, I.; Freeman, J.L.; Gamse, J.T.; Iturria, I.; Lawrence, C.; Muriana, A.; Peterson, R.T.; Van Cruchten, S.; Zon, L.I. Use of zebrafish in drug discovery toxicology. *Chem. Res. Toxicol.* **2020**, *33*, 95–118. [CrossRef]
22. Jayasinghe, C.D.; Jayawardena, U.A. Toxicity Assessment of Herbal Medicine Using Zebrafish Embryos: A Systematic Review. *Evid. Based Complement. Altern. Med.* **2019**, 7272808. [CrossRef] [PubMed]
23. Mizgirev, I.V.; Revskoy, S. A new zebrafish model for experimental leukemia therapy. *Cancer Biol. Ther.* **2010**, *9*, 895–902. [CrossRef] [PubMed]
24. Pham, D.H.; Roo, B.D.; Nguyen, X.B.; Vervaele, M.; Kecskes, A.; Ny, A.; Copmans, D.; Vriens, H.; Locquet, J.P.; Hoet, P.; et al. Use of zebrafish larvae as a multi-endpoint platform to characterize the toxicity profile of *Silica Nanoparticles*. *Sci. Rep.* **2016**, *6*, 37145. [CrossRef] [PubMed]
25. Caballero, M.V.; Candiracci, M. Zebrafish as screening model for detecting toxicity and drugs efficacy. *J. Unexplor. Med. Data* **2018**, *3*, 4. [CrossRef]
26. Garcia, G.R.; Noyes, P.D.; Tanguay, R.L. Advancements in zebrafish applications for 21st century toxicology. *Pharmacol. Ther.* **2016**, *161*, 11–21. [CrossRef] [PubMed]

27. Hermsen, S.A.B.; van den Brandhof, E.J.; van der Ven, L.T.M.; Piersma, A.H. Relative embryotoxicity of two classes of chemicals in a modified zebrafish embryotoxicity test and comparison with their in vivo potencies. *Toxicol. In Vitro* **2011**, *25*, 745–753. [CrossRef]
28. Fazry, S.; Mohd Noordin, M.A.; Sanusi, S.; Mat Noor, M.; Aizat, W.M.; Mat Lazim, A.; Dyari, H.R.E.; Jamar, N.H.; Remali, J.; Othman, B.A.; et al. Cytotoxicity and toxicity evaluation of Xanthone crude extract on hypoxic human hepatocellular carcinoma and zebrafish (*Danio rerio*) embryos. *Toxics* **2018**, *6*, 60. [CrossRef]
29. De Oliveira, R. Zebrafish Early Life-stages and Adults as A Tool for Ecotoxicity Assessment. Master's Thesis, Universidade de Aveiro, Aveiro, Portugal, 2009.
30. Rizzo, L.Y.; Golombek, S.K.; Mertens, M.E.; Pan, Y.; Laaf, D.; Broda, J.; Jayapaul, J.; Mockel, D.; Subr, V.; Hennink, W.E.; et al. In vivo nanotoxicity testing using the zebrafish embryo assay. *J. Mater. Chem. B.* **2013**, *1*, 3918–3925. [CrossRef]
31. D'Amora, M.; Giordani, S. The Utility of Zebrafish as a Model for Screening Developmental Neurotoxicity. *Front. Neurosci.* **2018**, *12*, 976. [CrossRef]
32. Hill, A.J.; Teraoka, H.; Heideman, W.; Peterson, R.E. Zebrafish as a model vertebrate for investigating chemical toxicity. *Toxicol. Sci.* **2005**, *86*, 6–19. [CrossRef]
33. Koziol, E.; Deniz, F.S.S.; Orhan, I.E.; Marcourt, L.; Budzynska, B.; Wolfender, J.L.; Crawford, A.D.; Skalicka-Wozniak, K. High-performance counter-current chromatography isolation and initial neuroactivity characterization of furanocoumarin derivatives from *Peucedanum alsaticum* L (Apiaceae). *Phytomed* **2019**, *54*, 259–264. [CrossRef] [PubMed]
34. Chakraborty, C.; Sharma, A.R.; Sharma, G.; Lee, S.S. Zebrafish: A complete animal model to enumerate the nanoparticle toxicity. *J. Nanobiotechnol.* **2016**, *14*, 65. [CrossRef] [PubMed]
35. Zakaria, Z.Z.; Benslimane, F.M.; Nasrallah, G.K.; Shurbaji, S.; Younes, N.N.; Mraiche, F.; Da'as, S.I.; Yalcin, H.C. Using zebrafish for investigating the molecular mechanisms of drug-induced cardiotoxicity. *BioMed Res. Int.* **2018**, *2018*, 1642684. [CrossRef] [PubMed]
36. Sant, K.E.; Timme-Laragy, A.R. Zebrafish as a model for toxicological perturbation of yolk and nutrition in the early embryo. *Curr. Environ. Health Rep.* **2018**, *5*, 125–133. [CrossRef] [PubMed]
37. Lammer, E.; Wendler, K.; Rawling, J.M.; Belanger, S.E.; Braunbeck, T. Is the fish embryo toxicity test (FET) with the zebrafish (*Danio rerio*) a potential alternative for the fish acute toxicity test? *Comp. Biochem. Physiol. C. Toxicol. Pharmacol.* **2009**, *149*, 196–209. [CrossRef] [PubMed]
38. Bailey, J.; Oliveri, A.; Levin, E.D. Zebrafish model systems for developmental neurobehavioral toxicology. *Birth Defects Res. Part C Embryo Today Rev.* **2013**, *99*, 14–23. [CrossRef]
39. Wei, Y.; Li, C.; Wang, C.; Peng, Y.; Shu, L.; Jia, X.; Ma, W.; Wang, B. Metabolism of tanshinone IIA, cryptotanshinone and tanshinone I from *Radix Salvia miltiorrhiza* in zebrafish. *Molecules* **2012**, *17*, 8617–8632. [CrossRef]
40. Kimmel, C.B.; Ballard, W.W.; Kimmel, S.R.; Ulimann, B.; Schilling, T.F. Stages of embryonic development of the zebrafish. *Dev. Dyn.* **1995**, *203*, 253–310. [CrossRef]
41. Carlson, B.M. *Human Embryology and Developmental biology E-book*; Elsevier Health Sciences: St. Louis, MI, USA, 2018.
42. Ali, S.; van Mil, H.G.; Richardson, M.K. Large-scale assessment of the zebrafish embryo as a possible predictive model in toxicity testing. *PLoS ONE* **2011**, *6*, e21076. [CrossRef]
43. Syahbirin, G.; Mumuh, N.; Mohamad, K. Curcuminoid and toxicity levels of ethanol extract of Javanese ginger (*Curcuma xanthorriza*) on brine shrimp (*Artemia salina*) larvae and zebrafish (*Danio rerio*) embryos. *Asian J. Pharm. Clin. Res.* **2017**, *10*, 169–173. [CrossRef]
44. Littleton, R.M.; Hove, J.R. Zebrafish: A nontraditional model of traditional medicine. *J. Ethnopharmacol.* **2013**, *145*, 677–685.
45. Wang, T.; Wang, C.; Wu, Q.; Zheng, K.; Chen, J.; Lan, Y.; Qin, Y.; Mei, W.; Wang, B. Evaluation of tanshinone IIA developmental toxicity in zebrafish embryos. *Molecules* **2017**, *22*, 660. [CrossRef] [PubMed]
46. Braunbeck, T.; Boettcher, M.; Hollert, H.; Kosmehl, T.; Leist, E.; Rudolf, M.; Seitz, N. Towards an alternative for the acute fish LC_{50} test in chemical assessment: The fish embryo toxicity test goes multi-species – an update. *ALTEX* **2005**, *22*, 87–102. [PubMed]
47. Yang, L.; Ho, N.Y.; Alshut, R.; Legradi, J.; Weiss, C.; Reischl, M.; Mikut, R.; Liebel, U.; Muller, F.; Strahle, U. Zebrafish embryos as models for embryotoxic and teratological effects of chemicals. *Reprod. Toxicol.* **2009**, *28*, 245–253. [CrossRef]

48. Di Paolo, C.; Seiler, T.B.; Keiter, S.; Hu, M.; Muz, M.; Brack, W.; Hollert, H. The value of zebrafish as an integrative model in effect-directed analysis—A review. *Environ. Sci. Eur.* **2015**, *27*, 8. [CrossRef]
49. Stern, H.M.; Zon, L.I. Cancer genetics and drug discovery in the zebrafish. *Nat. Rev. Cancer* **2003**, *3*, 533–539. [CrossRef]
50. R, E.B.; Jesubatham, P.D.; Berlin Grace, V.M.; Viswanathan, S.; Srividya, S. Non-toxic and non teratogenic extract of *Thuja orientalis* L. inhibited angiogenesis in zebra fish and suppressed the growth of human lung cancer cell line. *Biomed. Pharmacother.* **2018**, *106*, 699–706. [CrossRef]
51. Fitzgerald, C.; Gallagher, E.; O'Connor, P.; Prieto, J.; Mora-soler, L.; Grealy, M.; Hayes, M. Development of a seaweed derived platelet activating factor acetylhydrolase (PAF-AH) inhibitory hydrolysate, synthesis of inhibitory peptides and assessment of their toxicity using the zebrafish larvae assay. *Peptides* **2013**, *50*, 119–124. [CrossRef]
52. Bertelli, P.R.; Biegelmeyer, R.; Rico, E.P.; Klein-Junior, L.C.; Toson, N.S.B.; Minetto, L.; Bordignon, S.A.L.; Gasper, A.L.; Moura, S.; de Oliveira, D.L.; et al. Toxicological profile and acetylcholinesterase inhibitory potential of Palicourea deflexa, a source of beta-carboline alkaloids. *Comp. Biochem. Physiol. C Toxicol. Pharmacol.* **2017**, *201*, 44–50. [CrossRef]
53. Tolentino, J.; Undan, J.R. Embryo-toxicity and teratogenicity of *Derris elliptica* leaf extract on zebraish *(Danio rerio)* embryos. *Int. J. Pure Appl. Biosci.* **2016**, *4*, 16–20. [CrossRef]
54. He, J.H.; Gao, J.M.; Huang, C.J.; Li, C.Q. Zebrafish models for assessing developmental and reproductive toxicity. *Neurotoxicol. Teratol.* **2014**, *42*, 35–42. [CrossRef] [PubMed]
55. Deng, J.; Yu, L.; Liu, C.; Yu, K.; Shi, X.; Yeung, L.W.Y.; Lam, P.K.S.; Wu, R.S.S.; Zhou, B. Hexabromocyclododecane-induced developmental toxicity and apoptosis in zebrafish embryos. *Aquat. Toxicol.* **2009**, *93*, 29–36. [CrossRef]
56. Carwan, M.J., III; Heiden, T.K.; Tomasiewicz, H.C. The utility of zebrafish as a model for toxicological research. In *Biochemistry and Molecular Biology of Fishes*; Mommsen, T.P., Moon, T.W., Eds.; Chapter 1; Elsevier: New York, NY, USA, 2005; Volume 6, pp. 3–41.
57. Knobel, M.; Busser, F.J.M.; Rico-Rico, A.; Kramer, N.I.; Hermens, J.L.M.; Hafner, C.; Tanneberger, K.; Schirmer, K.; Scholz, S. Predicting adult fish acute lethality with the zebrafish embryo: Relevance of test duration, endpoints, compound properties, and exposure concentration analysis. *Environ. Sci. Technol.* **2012**, *46*, 9690–9700. [CrossRef] [PubMed]
58. Fairbrother, A.; Lewis, M.A.; Menzer, R.E. Methods in environmental toxicology. In *Principles and Methods of Toxicology*, 4th ed.; Hayes, A.W., Ed.; Chapter 42; CRC Press, Taylor & Francis Group: London, UK, 2001; pp. 1759–1801.
59. Massei, R.; Hollert, H.; Krauss, M.; von Tumpling, W.; Weidauer, C.; Haglund, P.; Kuster, E.; Gallampois, C.; Tysklind, M.; Brack, W. Toxicity and neurotoxicity profiling of contaminated sediments from Gulf of Bothnia (Sweden): A multi-endpoint assay with Zebrafish embryos. *Environ. Sci. Eur.* **2019**, *31*, 8. [CrossRef]
60. Nagel, R. *DarT*: The embryo test with the zebrafish *Danio rerio*–a general model in ecotoxicology and toxicology. *ALTEX* **2002**, *19*, 38–48. [PubMed]
61. Liedtke, A.; Muncke, J.; Rufenacht, K.; Eggen, R.I.L. Molecular multi-effect screening of environmental pollutants using the MolDarT. *Environ. Toxicol.* **2008**, *23*, 59–67. [CrossRef] [PubMed]
62. Njiwa, J.R.K.; Suter, M.J.-F.; Eggen, R.I. Zebrafish Embryo Toxicity Assay, Combining Molecular and Integrative Endpoints at Various Developmental Stages. In *Handbook of Hydrocarbon and Lipid Microbiology*; Timmis, K.N., Ed.; Springer: Berlin/Heidelberg, Germany, 2010; pp. 4481–4489.
63. Bugel, S.M.; Bonventre, J.A.; Tanguay, R.L. Comparative developmenal toxicity of flavonoids using an integrative zebrafish system. *Toxicol. Sci.* **2016**, *154*, 55–68. [CrossRef] [PubMed]
64. Haldi, M.; HArden, M.; D'Amico, L.; DeLise, A.; Seng, W.L. Developmental toxicity assessment in zebrafish. In *Zebrafish: Methods for Assessing Drug Safety and Toxicity*; McGrath, P., Ed.; Phylonix: Cambridge, MA, USA, 2011.
65. Busquet, F.; Strecker, R.; Rawlings, J.M.; Belnger, S.E.; Braunbeck, T.; Carr, G.J.; Cenijn, P.; Fochtman, P.; Gourmelon, A.; Hubler, N.; et al. OECD validation study to assess intra and inter-laboratory reproducibility of the zebrafish embryo toxicity test for acute aquatic toxicity testing. *Regul. Toxicol. Pharmacol.* **2014**, *69*, 496–511. [CrossRef]
66. Parng, C.; Seng, W.L.; Semino, C.; McGrath, P. Zebrafish: A preclinical model for drug screening. *Assay Drug Dev. Technol.* **2002**, *1*, 41–48. [CrossRef]

67. OECD. *Fish Embryo Toxicity Test*; Organization for Economic Cooperation and Development: Paris, France, 2006.
68. Kalueff, A.V.; Gebhardt, M.; Stewart, A.M.; Cachat, J.M.; Brimmer, M.; Chawia, J.S.; Craddock, C.; Kyzar, E.; Roth, A.; Landsman, S.; et al. Towards a comprehensive catalog of zebrafish behavior 1.0 and beyond. *Zebrafish* **2013**, *10*, 70–86. [CrossRef]
69. Chueh, T.C.; Hsu, L.S.; Kao, C.M.; Hsu, T.W.; Liao, H.Y.; Wang, K.Y.; Chen, S.C. Transcriptome analysis of zebrafish embryos exposed to deltamethrin. *Environ. Toxicol.* **2017**, *32*, 1548–1557. [CrossRef] [PubMed]
70. Xiao, C.; Han, Y.; Liu, Y.; Zhang, J.; Hu, C. Relationship between fluoroquinolone structure and neurotoxicity revealed by zebrafish neurobehavior. *Chem. Res. Toxicol.* **2018**, *31*, 238–250. [CrossRef] [PubMed]
71. Fan, C.Y.; Cowden, J.; Simmons, S.O.; Padilla, S.; Ramabhadran, R. Gene expression changes in developing zebrafish as potential markers for rapid developmental neurotoxicity screening. *Neurotoxicol. Teratol.* **2009**, *32*, 91–98. [CrossRef] [PubMed]
72. OECD. *Fish Acute Toxicity Test*; Organization for Economic Cooperation and Development: Medmenham, UK, 1992.
73. Shaikh, A.; Kohale, K.; Ibrahim, M.; Khan, M. Teratogenic effects of aqueous extract of *Ficus glomerata* leaf during embryonic development in zebrafish (*Danio rerio*). *J. Appl. Pharm. Sci.* **2019**, *9*, 107–111.
74. Paramanik, V.; Basumatary, M.J.; Nagesh, G.; Kushwaha, J.K.; Kurrey, K. Ethanolic extracts of candidate Indian Traditional Medicines *Acorus calamus, Terminalia chebula* and *Achyranthes aspera* are neuroprotective in Zebrafish. *IBRO Rep.* **2019**, *6*, S247. [CrossRef]
75. Yan, C.; Zhang, S.; Wang, C.; Zhang, Q. A fructooligosaccharide from *Achyranthes bidentata* inhibits osteoporosis by stimulating bone formation. *Carbohydr. Polym.* **2019**, *210*, 110–118. [CrossRef]
76. Aleksandar, P.; Dragana, M.C.; Nebojsa, J.; Biljana, N.; Natasa, S.; Branka, V.; Jelena, K.V. Wild edible onions—*Allium flavum* and *Allium carinatum*—Successfully prevent adverse effects of chemotherapeutic drug doxorubicin. *Biomed. Pharmacother.* **2019**, *109*, 2482–2491. [CrossRef]
77. Ponrasu, T.; Ganeshkumar, M.; Suguna, L. Developmental toxicity evaluation of ethanolic extract of Annona squamosa in zebrafish *(Danio rerio)* embryo. *J. Pharm. Res.* **2012**, *5*, 277–279.
78. Grisola, M.A.C.; Fuentes, R.G. Phenotype-based screening of selected mangrove methanolic crude extracts with anti-melanogenic activity using zebrafish *(Danio rerio)* as a model. *ScienceAsia* **2017**, *43*, 163–168. [CrossRef]
79. Batista, F.L.A.; Lima, L.M.G.; Abrante, I.A.; de Araujo, J.I.F.; Batista, F.L.A.; Abrante, I.A.; Lima, M.C.L.; do Prado, B.S.; Moura, L.F.W.G.; Guedes, M.I.F.; et al. Antinociceptive activity of ethanolic extract of *Azadirachta indica A. Juss (Neem, Meliaceae)* fruit through opioid, glutamatergic and acid-sensitive ion pathways in adult zebrafish *(Danio rerio)*. *Biomed. Pharmacother.* **2018**, *108*, 408–416.
80. Teh, L.E.; Kue, C.S.; Ng, C.H.; Lau, B.F. Toxicity effect of *Bougainvillea glabra* (paper flower) water extracts on zebrafish embryo. *INNOSC Ther. Pharmacol. Sci.* **2019**, *2*, 23–26. [CrossRef]
81. De Vera, J.S.; De Castro, M.E.G.; Dulay, R.M.R. Phytochemical constituents and teratogenic effect of lyophilized extracts of *Bixa orellana* L. (Achuete) and *Piper betle* L. (Ikmo) leaves in *Danio rerio* embryos. *Der Pharma Chem.* **2016**, *8*, 432–437.
82. Vargas-Mendez, L.Y.; Rosado-Solano, D.N.; Sanabria-Florez, P.L.; Puerto-Galvis, C.E.; Kouznetsov, V. *In vitro* antioxidant and anticholinesterase activities and in vivo toxicological assessment (Zebrafish embryo model) of ethanolic extracts of *Capsicum chinense* Jacq. *J. Med. Plants Res.* **2016**, *10*, 59–66.
83. Xia, Q.; Ma, Z.; Luo, J.; Wang, Y.; Li, T.; Feng, Y.; Ni, Y.; Zou, Q.; Lin, R. Assay for the developmental toxicity of safflower *(Carthamus tinctorius L.)* to zebrafish embryos/larvae. *J. Tradit. Chin. Med. Sci.* **2017**, *4*, 71–81.
84. Murugesu, S.; Khatib, A.; Ahmad, Q.U.; Ibrahim, Z.; Uzir, B.F.; Benchoula, K.; Yusoff, N.I.N.; Perumal, V.; Aljami, M.F.; Salamah, S.; et al. Toxicity study on *Clinacanthus nutans* leaf hexane fraction using *Danio rerio* embryos. *Toxicol. Rep.* **2019**, *6*, 1148–1154. [CrossRef] [PubMed]
85. Orellana-Paucar, A.M.; Serruys, A.S.K.; Afrikanova, T.; Maes, J.; De Borggraeve, W.; Alen, J.; Leon-Tamariz, F.; Wilches-Arizabala, I.M.; Crawford, A.M.; de Witte, P.A.M.; et al. Anticonvulsant activity of bisabolene sesquiterpenoids of *Curcuma longa*. *Epilepsy Behav.* **2012**, *24*, 14–22. [CrossRef] [PubMed]
86. Kannan, R.R.; Vincent, S.G. Cynodon dactylon and Sida acuta extracts impact on the function of the cardiovascular system in zebrafish embryos. *J. Biomed. Res.* **2012**, *26*, 90–97. [CrossRef]
87. Memita, A.Y.B.; Macalinao, S.D.; Reyes, L.E.G.; Damian, E.E.; Dulay, R.M.R. Toxicity and teratogenicity of *Dieffenbachia amoena* leaf extract. *Int. J. Biol. Pharm. Allied Sci.* **2018**, *7*, 1591–1600.

88. Flaviano, J.F.; Kenniker, M.A.A.; Ganda, F.J.C.; Sacedor, J.; Lee, M.Y.S.; Cabuhat, K.S.P. Assessment of the toxic and teratogenic effects of Kamagong (*Diospyros discolor* Willd) leaves extracts to the developing embryo of zebrafish (*Danio rerio*). *Egypt. Acad. J. Biol. Sci. B. Zool.* **2018**, *10*, 53–60. [CrossRef]
89. Ponpornpisit, A.; Pirarat, N.; Suthikrai, W.; Binwihok, A. Toxicity test of Kameng (*Ecliptra prostrate* Linn.) and Kradhuawean (*Spilanthes acmella* (Linn.) Murr.) to early life stage of zebrafish (*Danio rerio*). *Thai. J. Vet. Med.* **2011**, *41*, 523–527.
90. Xavier, J.; Kripasana, K. Acute Toxicity of Leaf Extracts of *Enydra fluctuans* Lour in Zebrafish (*Danio rerio* Hamilton). *Sci. Cairo* **2020**, *2020*, 3965376. [CrossRef] [PubMed]
91. Zhang, W.; Liu, B.; Feng, Y.; Liu, J.; Ma, Z.; Zheng, J.; Xia, Q.; Ni, Y.; Li, F.; Lin, R. Anti-angiogenic activity of water extract from Euphorbia pekinensis Rupr. *J. Ethnopharmacol.* **2017**, *206*, 337–346. [CrossRef] [PubMed]
92. Yang, W.; Ma, L.; Li, S.; Cui, K.; Lei, L.; Ye, Z. Evaluation of the cardiotoxicity of evodiamine in vitro and in vivo. *Molecules* **2017**, *22*, 943. [CrossRef] [PubMed]
93. Jiang, L.L.; Li, K.; Lin, Q.H.; Ren, J.; He, Z.H.; Li, H.; Shen, N.; Wei, P.; Feng, F.; He, M.F. Gambogic acid causes fin developmental defect in zebrafish embryo partially via retinoic acid signaling. *Reprod. Toxicol.* **2016**, *63*, 161–168. [CrossRef]
94. Sajkowska-Kozielewicz, J.J.; Kozielewicz, P.; Barnes, N.M.; Wawer, I.; Paradowska, K. Antioxidant, cytotoxic, and antiproliferative activities and total polyphenol contents of the extracts of *Geissospermum reticulatum* bark. *Oxid. Med. Cell. Longev.* **2016**, *2016*, 2573580. [CrossRef]
95. Dos Santos, I.V.F.; de Souza, G.C.; Santana, G.R.; Duarte, J.L.; Fernandes, C.P.; Keia, H.; Velazquez-Moyado, J.A.; Navarrete, A.; Ferreira, I.M.; Carvalho, H.O. Histopathology in zebrafish (*Danio rerio*) to evaluate the toxicity of medicine: An anti-inflammatory phytomedicine with Janaguba milk (*Himatanthus drasticus* Plumel). *Histopathol. Update* **2018**, 39–64.
96. Lira, S.M.; Dionisio, A.P.; Holanda, M.O.; Marques, C.G.; da Silva, G.S.; Correa, L.C.; Santos, G.B.M.; de Abreu, F.A.P.; Magalhaes, F.E.A.; Reboucas, E.D.; et al. Metabolic profile of pitaya (Hylocereus polyrhizus (F.A.C. Weber) Britton & Rose) by UPLC-QTOF-MSE and assessment of its toxicity and anxiolytic-like effect in adult zebrafish. *Food Res. Intl.* **2020**, *127*, 108701.
97. He, Y.L.; Shi, J.Y.; Peng, C.; Hu, L.J.; Liu, J.; Zhou, Q.M.; Guo, L.; Xiong, L. Angiogenic effect of motherwort (*Leonurus japonicus*) alkaloids and toxicity of motherwort essential oil on zebrafish embryos. *Fitoterapia* **2018**, *128*, 36–42. [CrossRef]
98. Wang, W.; Feng, S.; Xiong, Z. Protective effect of polysaccharide from Ligusticum chuanxiong hort against H2O2-induced toxicity in zebrafish embryo. *Carbohydr. Polym.* **2019**, *221*, 73–83. [CrossRef]
99. Li, J.; Zhang, Y.; Liu, K.; He, Q.; Sun, C.; Han, J.; Han, L.; Tian, Q. Xiaoaiping induces developmental toxicity in zebrafish embryos through activation of ER stress, apoptosis and the Wnt pathway. *Front Pharmacol.* **2018**, *9*, 1250. [CrossRef]
100. Khan, M.F.; Abutaha, N.; Nasr, F.A.; Alqahtani, A.S.; Noman, O.M.; Wadaan, M.A.M. Bitter gourd (*Momordica charantia*) possess developmental toxicity as revealed by screening the seeds and fruit extracts in zebrafish embryos. *BMC Complement. Altern. Med.* **2019**, *19*, 184. [CrossRef] [PubMed]
101. Santos, M.M.R.; Agpaoa, A.R.; Sayson, A.D.; De Castro, M.E.G.; Dulay, R.M.R. Toxic and teratogenic effects of water leaf extract of *Momordica charantia* in zebrafish (*Danio rerio*) embryos. *Der Pharma Chem.* **2017**, *9*, 119–122.
102. David, C.R.S.; Angeles, A.; Angoluan, R.C.; Santos, J.P.E.; David, E.S.; Dulay, R.M.R. *Moringa oleifera* (Malunggay) water extracts exhibit embryo-toxic and teratogenic activity in zebrafish (*Danio rerio*) embryo model. *Der Pharm. Lett.* **2016**, *8*, 163–168.
103. Elsayed, E.A.; Farooq, M.; Sharaf-Eldin, M.A.; El-Enshasy, H.A.; Wadaan, M. Evaluation of developmental toxicity and anti-angiogenic potential of essential oils from Moringa *oleifera* and *Moringa peregrina* seeds in zebrafish *(Danio rerio)* model. *S. Afr. J. Bot.* **2019**, *129*, 229–237. [CrossRef]
104. De Santana Silva, L.L.; Alves, R.N.; de Paulo, D.V.; da Silva, J.D.F. Ecotoxicity of water-soluble lectin from *Moringa oleifera* seeds to zebrafish (*Danio rerio*) embryos and larvae. *Chemosphere* **2017**, *185*, 178–182. [CrossRef] [PubMed]
105. Babic, S.; Malev, O.; Pflieger, M.; Lebedev, A.T.; Mazur, D.M.; Kuzic, A.; Coz-Rakovac, R.; Trebse, P. Toxicity evaluation of olive oil mill wastewater and its polar fraction using multiple whole-organism bioassays. *Sci. Total Environ.* **2019**, *686*, 903–914. [CrossRef]

106. Safavi, F.; Farimani, M.M.; Golalipour, M.; Leung, P.-C.; Lau, K.-M.; Kwok, H.-F.; Wong, C.-W.; Bayat, H.; Lau, C.B.-S. Investigations on the wound healing properties of *Onosma dichroantha* Boiss root extracts. *S. Afr. J. Bot.* **2019**, *125*, 344–352. [CrossRef]
107. Santhoshkumar, J.; Sowmya, B.; Kumar, S.V.; Rajeshkumar, S. Toxicology evaluation and antidermatophytic activity of silver nanoparticles synthesized using leaf extract of *Passiflora caerulea*. *S. Afr. J. Chem. Eng.* **2019**, *29*, 17–23. [CrossRef]
108. Lamban, I.Q.; Balbeuna, E.S.; Lee, M.Y.S.; Sacdalan, M.; Carpio JR, A.P.; Cabuhat, K.S.P. Toxic and teratogenic effects of Sampa-Sampalukan (*Phyllanthus niruri*) leaves extract using *Danio rerio* embryo assay. *Egypt. Acad. J. Biol. Sci.* **2019**, *11*, 109–115. [CrossRef]
109. Thanh, D.T.H.; Thanh, N.L.; Thang, N.D. Toxicological and melanin synthesis effects of *Polygonum multiflorum* root extracts on zebrafish embryos and human melanocytes. *Biomed. Res. Ther.* **2016**, *3*, 808–818. [CrossRef]
110. Yang, J.B.; Li, W.F.; Liu, Y.; Wang, Q.; Cheng, X.L.; Wei, F.; Wang, A.G.; Jin, H.T.; Ma, S.C. Acute toxicity screening of different extractions, components and constituents of *Polygonum multiflorum* Thunb. on zebrafish (Danio rerio) embryos in vivo. *Biomed. Pharmacother.* **2018**, *99*, 205–213. [PubMed]
111. Xia, Q.; Wei, L.; Zhang, Y.; Kong, H.; Shi, Y.; Wang, X.; Chen, X.; Han, L.; Liu, K. Psoralen induces developmental toxicity in zebrafish embryos/larvae through oxidative stress, apoptosis, and energy metabolism disorder. *Front. Pharmacol.* **2018**, *9*, 1457. [CrossRef] [PubMed]
112. Wibowo, I.; Permadi, K.; Hartati, R.; Damayanti, S. Ethanolic extract of pomegranate (*Punica granatum* L) peel: Acute toxicity tests on zebrafish *(Danio rerio)* embryos and its toxicity prediction by *in silico*. *J. Appl. Pharm. Sci.* **2018**, *8*, 082–086.
113. Huang, Y.; Bai, Y.; Wang, Y.; Kong, H. *Solidago canadensis L.* extracts to control algal (Microcystis) blooms in ponds. *Ecol. Eng.* **2014**, *70*, 263–267. [CrossRef]
114. Dos Santos Sampaio, T.I.; de Melo, N.C.; de Freitas Paiva, B.T.; da Silva Aleluia, G.A.; da Silva Neto, F.L.P.; da Silva, H.R.; Keita, H.; Cruz, R.A.S.; Sanchez-Ortiz, B.L.; Pineda-Pena, E.A.; et al. Leaves of *Spondias mombin* L. a traditional anxiolytic and antidepressant: Pharmacological evaluation on zebrafish *(Danio rerio)*. *J. Ethnopharmacol.* **2018**, *224*, 563–578.
115. Kumar, P.B.S.; Kar, B.; Dolai, N.; Haldar, P.K. Study on developmental toxicity and behavioral safety of *Streblus asper* Lour. bark on zebrafish embryos. *Indian J. Nat. Prod. Res.* **2013**, *4*, 255–259.
116. Zonyane, S.; Chen, L.; Xu, M.J.; Gong, Z.N.; Xu, S.; Makung, N.P. Geographic-based metabolomic variation and toxicity analysis of *Sutherlandia* frutescens L. R.Br.—An emerging medicinal crop in South Africa. *Ind. Crops Prod.* **2019**, *133*, 414–423.
117. Russo, H.M.; Queiroz, E.F.; Marcourt, L.; Crawford, A.D.; Bolzani, V.; Wolfender, J.L. Chemical profile and toxicity study of Tetrapterys multiglandulosa secondary metabolites using zebrafish toxicity model. In *Brazilian Conference on Natural Products and Annual Meeting on Micromolecular Evolution, Systematics and Ecology*; GALOÁ: Campinas, Brazil, 2018.
118. Romagosa, C.M.R.; David, E.S.; Dulay, R.M.R. Embryo-toxic and teratogenic effects of *Tinospora cordifolia* leaves and bark extracts in Zebrafish *(Danio rerio)* embryos. *Asian J. Plant Sci. Res.* **2016**, *6*, 37–41.
119. Aleksic, I.; Ristivojevic, P.; Pavic, A.; Radojevic, I.; Comic, L.R.; Vasiljevic, B.; Opsenica, D.; Milojkovic-Opsenica, D.; Senerovic, L. Anti-quorum sensing activity, toxicity in zebrafish (*Danio rerio*) embryos and phytochemical characterization of *Trapa natans* leaf extracts. *J. Ethnopharmacol.* **2018**, *222*, 148–158. [CrossRef]
120. Gao, X.P.; Feng, F.; Zhang, X.Q.; Liu, X.X.; Wang, Y.B.; She, J.X.; He, Z.H.; He, M.F. Toxicity assessment of 7 anticancer compounds in zebrafish. *Int. J. Toxicol.* **2014**, *33*, 98–105. [CrossRef]
121. Puerto Galvis, C.E.; Kouznetsov, V.V. Synthesis of zanthoxylamide protoalkaloids and their *in silico* ADME-Tox screening and in vivo toxicity assessment in zebrafish embryos. *Eur. J. Pharm. Sci.* **2019**, *127*, 291–299. [CrossRef] [PubMed]
122. Ogungbemi, A.; Leuthold, D.; Scholz, S.; Kuster, E. Hypo- or hyperactivity of zebrafish embryos provoked by neuroactive substances: A review on how experimental parameters impact the predictability of behavior changes. *Environ. Sci. Eur.* **2019**, *31*, 88. [CrossRef]
123. Ma, C.; Parng, C.; Wen, L.S.; Zhang, C.; Willett, C.; McGrath, P. Zebrafish-an in vivo model for drug screening. *Innov. Pharm. Technol.* **2003**, *3*, 38–45.
124. Truong, L.; Harper, S.L.; Tanguay, R.L. Evaluation of embryotoxicity using the zebrafish model. *Methods Mol. Biol.* **2017**, *691*, 271–279.

125. Driever, W.; Stemple, D.; Schier, A.; Solnica-Krezel, L. Zebrafish: Genetic tools for studying vertebrate development. *Trends Genet.* **1994**, *10*, 152–159. [CrossRef]
126. OECD. *Guidelines for Testing of Chemical-Fish Embryo Acute Toxicity (FET) Test*; OECD Publishing: Paris, France, 2013.
127. Calienni, M.N.; Feas, D.A.; Igartua, D.E.; Chiaramoni, N.S.; Alonso, S.V.; Prieto, M.J. Nanotoxicological and teratogenic effects: A linkage between dendrimer surface charge and zebrafish developmental stages. *Toxicol. Appl. Pharmacol.* **2017**, *337*, 1–11. [CrossRef]
128. Yamashita, M. Apoptosis in zebrafish development. *Comp. Biochem. Physiol. B Biochem. Mol. Biol.* **2003**, *136*, 731–742. [CrossRef]
129. Haffter, P.; Granato, M.; Brand, M.; Mullins, M.C.; Hammerschmidt, M.; Kane, D.A.; Odenthal, J.; van Eeden, F.J.; Jiang, Y.J.; Heisenberg, C.P.; et al. The identification of genes with unique and essential functions in the development of the zebrafish, *Danio rerio*. *Development* **1996**, *123*, 1–36.
130. Li, H.; Yu, S.; Cao, F.; Wang, C.; Zheng, M.; Li, X.; Qiu, L. Developmental toxicity and potential mechanisms of pyraoxystrobin to zebrafish *(Danio rerio)*. *Ecotoxicol. Environ. Saf.* **2018**, *151*, 1–9. [CrossRef]
131. Bakkers, J. Zebrafish as a model to study cardiac development and human cardiac disease. *Cardiovasc. Res.* **2011**, *91*, 279–288. [CrossRef]
132. Thorarensen, H.; Galiugher, P.E.; Farrel, A.P. The limitations of heart rate as a predictor of metabolic rate in fish. *J. Fish Biol.* **1996**, *49*, 226–236. [CrossRef]
133. Gao, X.Y.; Li, K.; Jiang, L.L.; He, M.F.; Pu, C.H.; Kang, D.; Xie, J. Developmental toxicity of auranofin in zebrafish embryos. *J. Appl. Toxicol.* **2017**, *37*, 602–610. [CrossRef] [PubMed]
134. Ye, Q.; Liu, H.; Fang, C.; Liu, Y.; Liu, X.; Liu, J. Cardiotoxicity evaluation and comparison of diterpene alkaloids on zebrafish. *Drug Chem. Toxicol.* **2019**, 1–8. [CrossRef] [PubMed]
135. Farrell, A.P.; Pieperhoff, S. Design and physiology of the heart/cardiac anatomy in fishes. *Encycl. Fish Physiol.* **2011**, *2*, 998–1005.
136. Hu, N.; Sedmera, D.; Yost, H.J.; Clark, E.B. Structure and function of the developing zebrafish heart. *Anat. Rec.* **2000**, *260*, 148–157. [CrossRef]
137. Yamauchi, H.; Hotta, Y.; Konishi, M.; Miyake, A.; Kawahara, A.; Itoh, N. Fgf21 is essential for haematopoiesis in zebrafish. *EMBO Rep.* **2006**, *7*, 649–654. [CrossRef]
138. Wang, S.; Liu, X.; He, Q.; Chen, X. Toxic effects of celastrol on embryonic development of zebrafish *(Danio rerio)*. *Drug Chem. Toxicol.* **2011**, *34*, 61–65. [CrossRef]
139. Lawson, N.D.; Weinstein, B.M. In vivo imaging of embryonic vascular development using transgenic zebrafish. *Dev. Biol.* **2002**, *248*, 307–318. [CrossRef]
140. Kinna, G.; Kolle, G.; Carter, A.; Lieschke, G.J.; Perkins, A.; Little, M.H. Knockdown of zebrafish crim1 results in a bent tail phenotype with defects in somite and vascular development. *Mech. Dev.* **2006**, *123*, 277–287. [CrossRef]
141. Cinar, K.; Aksoy, A.; Emre, Y.; Asti, R.N. The histology and histochemical aspects of gills of the flower fish, *Pseudophoxinus antalyae*. *Vet. Res. Commun.* **2009**, *33*, 453. [CrossRef]
142. Mauceri, A.; Fossi, M.C.; Leonzio, C.; Ancora, S.; Minniti, F.; Maisano, M.; Ferrando, S.; Fasulo, S. Stress factors in the gills of *Liza aurata* (Perciformes, Mugilidae) living in polluted environments. *Ital. J. Zool.* **2005**, *72*, 285–292. [CrossRef]
143. Hughes, G. Introduction to the study of gills. *Semin. Ser.-Soc. Exp. Biol* **1982**, *16*, 1–24.ill.
144. Evans, D.H. The fish gill: Site of action and model for toxic effects of environmental pollutants. *Environ. Health Perspect.* **1987**, *71*, 47–58. [CrossRef]
145. Srivastava, N.; Kumari, U.; Rai, A.K.; Mittal, S.; Mittal, A.K. Alterations in the gill filaments and secondary lamellae of *Cirrhinus mrigala* exposed to "nuvan," an organophosphorus insecticide. *J. Histol.* **2014**, 190139. [CrossRef]
146. Seok, S.-H.; Baek, M.-W.; Lee, H.-Y.; Kim, D.-J.; Na, Y.-R.; Noh, K.-J.; Park, S.-H.; Lee, H.-K.; Lee, B.-H.; Ryu, D.-Y.; et al. Quantitative GFP fluorescence as an indicator of arsenite developmental toxicity in mosaic heat shock protein 70 transgenic zebrafish. *Toxicol. Appl. Pharmacol.* **2007**, *225*, 154–161. [CrossRef] [PubMed]
147. Rajini, A.; Revathy, K.; Selvam, G. Histopathological changes in tissues of *Danio rerio* exposed to sub lethal concentration of combination pesticide. *Indian J. Sci. Technol.* **2015**, *8*, 1–12. [CrossRef]
148. Vutukuru, S.S. Assessment of the Histological Changes in the Gill Architecture of Zebrafish *(Danio rerio)* Exposed to Nicotine and Cotinine Extracts from Cigarette Tar. *EC Pharmacol. Toxicol.* **2019**, *7*, 888–895.

149. Ertzer, R.; Müller, F.; Hadzhiev, Y.; Rathnam, S.; Fischer, N.; Rastegar, S.; Strähle, U. Cooperation of sonic hedgehog enhancers in midline expression. *Dev. Biol.* **2007**, *301*, 578–589. [CrossRef]
150. Park, H.; Lee, J.-Y.; Park, S.; Song, G.; Lim, W. Developmental toxicity and angiogenic defects of etoxazole exposed zebrafish *(Danio rerio)* larvae. *Aquat. Toxicol.* **2019**, *217*, 105324. [CrossRef]
151. Huang, Y.; Zhang, J.; Han, X.; Huang, T. The use of zebrafish *(Danio rerio)* behavioral responses in identifying sublethal exposures to deltamethrin. *Intl. J. Environ. Res. Public Health* **2014**, *11*, 3650–3660. [CrossRef]
152. Dabrowski, K.; Miller, M. Contested Paradigm in Raising Zebrafish *(Danio rerio)*. *Zebrafish* **2018**, *15*, 295–309. [CrossRef]
153. Lee, H.B.; Schwab, T.L.; Sigafoos, A.N.; Gauerke, J.L.; Krug, R.G., 2nd; Serres, M.R.; Jacobs, D.C.; Cotter, R.P.; Das, B.; Peterson, M.O.; et al. Novel zebrafish behavioral assay to identify modifiers of the rapid, nongenomic stress response. *Genes Brain Behav.* **2019**, *18*, e12549. [CrossRef] [PubMed]
154. De Marco, R.J.; Groneberg, A.H.; Yeh, C.-M.; Treviñ, M.; Ryu, S. The behavior of larval zebrafish reveals stressor-mediated anorexia during early vertebrate development. *Front. Behav. Neurosci.* **2014**, *8*, 367. [CrossRef] [PubMed]
155. Price, K.R.; Johnson, I.T.; Fenwick, G.R. The chemistry and biological significance of saponins in foods and feedingstuffs. *Crit. Rev. Food Sci. Nutr.* **1987**, *26*, 27–135. [CrossRef] [PubMed]
156. Liu, W.; Di iorgio, C.; Lamidi, M.; Elias, R.; Ollivier, E.; De Meo, M.P. Genotoxic and clastogenic activity of saponins extracted from Nauclea bark as assessed by the micronucleus and the comet assays in Chinese hamster ovary cells. *J. Ethnopharmacol.* **2011**, *137*, 176–183. [CrossRef]
157. Yim, S.-H.; Tabassum, N.; Kim, W.H.; Cho, H.; Lee, J.H.; Batkhuu, G.J.; Kim, H.J.; Oh, W.K.; Jung, D.W.; Williams, D.R. Isolation and characterization of isofraxidin 7-O-(6′-Op-Coumaroyl)-β-glucopyranoside from *Artemisia capillaris* Thunberg: A novel, nontoxic hyperpigmentation agent that is effective in vivo. *Evid. Based Complement. Altern. Med.* **2017**, 1401279. [CrossRef]
158. Chen, Y.H.; Yang, Z.S.; Wen, C.C.; Chang, Y.S.; Wang, B.C.; Hsiao, C.A.; Shih, T.L. Evaluation of the structure-activity relationship of flavonoids as antioxidants and toxicants of zebrafish larvae. *Food Chem.* **2012**, *134*, 717–724. [CrossRef]
159. Yoon, Y.; Park, J.; Taniguchi, A.; Kohsaka, H.; Nakae, K.; Nonaka, S.; Ishii, S.; Nose, A. System level analysis of motor-related neural activities in larval *Drosophila*. *J. Neurogenet.* **2019**, *33*, 179–189. [CrossRef]
160. Chen, Y.H.; Huang, Y.H.; Wen, C.C.; Wang, Y.H.; Chen, W.L.; Chen, L.C.; Tsay, H.J. Movement disorder and neuromuscular change in zebrafish embryos after exposure to caffeine. *Neurotoxicol. Teratol.* **2008**, *30*, 440–447. [CrossRef]
161. Mendis, J.C.; Tennakoon, T.K.; Jayasinghe, C.D. Zebrafish embryo toxicity of a binary mixture of pyrethroid insecticides: D-tetramethrin and cyphenothrin. *J. Toxicol.* **2018**, *2018*, 4182694. [CrossRef]

MDPI
St. Alban-Anlage 66
4052 Basel
Switzerland
Tel. +41 61 683 77 34
Fax +41 61 302 89 18
www.mdpi.com

Plants Editorial Office
E-mail: plants@mdpi.com
www.mdpi.com/journal/plants